DYNAMICS OF FOREST INSECT POPULATIONS

Patterns, Causes, Implications

POPULATION ECOLOGY: Theory and Application
Series Editor: **Alan A. Berryman**
Washington State University
Pullman, Washington

POPULATION SYSTEMS: A General Introduction
Alan A. Berryman

FOREST INSECTS: Principles and Practice of Population Management
Alan A. Berryman

DYNAMICS OF FOREST INSECT POPULATIONS: Patterns, Causes, Implications
Edited by Alan A. Berryman

DYNAMICS OF FOREST INSECT POPULATIONS

Patterns, Causes, Implications

Edited by
Alan A. Berryman
Washington State University
Pullman, Washington

Plenum Press • New York and London

Library of Congress Cataloging in Publication Data

Dynamics of forest insect populations.

(Population ecology)
Includes bibliographies and indexes.
1. Forest insects—Control. 2. Forest insects—Ecology. 3. Forest insects—Host plants. 4. Insect populations. 5. Forest ecology. I. Berryman, A. A. (Alan Andrew), 1937- . II. Series.
SB761.D96 1988 634.9′67 88-12631
ISBN 0-306-42745-1

A Division of Plenum Publishing Corporation
233 Spring Street, New York, NY 10013

Printed in the United States of America

In Memory of

George C. Varley

First Chairman of IUFRO
Working Party S2.07-06
"Population Dynamics of Forest Insects"

CONTRIBUTORS

ALF BAKKE Norwegian Forest Research Institute, N-1432 Ås-NLH, Norway

WERNER BALTENSWEILER Department of Plant Sciences, Swiss Federal Institute of Technology, CH-8092 Zurich, Switzerland

YURI N. BARANCHIKOV V.N. Sukachev Institute of Forestry and Wood, Siberian Branch, U.S.S.R. Academy of Sciences, Krasnoyarsk, 660036 U.S.S.R.

DAVID A. BARBOUR British Forestry Commission, Forest Research Station, Wrecclesham, Farnham, Surrey GU10 4LH, England

BRODER BEJER Department of Forest Entomology and Ecology, Royal Veterinary and Agricultural University, Copenhagen, Denmark, DK 1870

C. WAYNE BERISFORD Department of Entomology, University of Georgia, Athens, Georgia 30602

ALAN A. BERRYMAN Department of Entomology, Washington State University, Pullman, Washington 99164–6432

JOHN H. BORDEN Center for Pest Management, Department of Biological Sciences, Simon Fraser University, Burnaby, British Columbia, V5A 1S6 Canada

ERIK CHRISTIANSEN Norwegian Forest Research Institute, N-1432 Ås-NLH, Norway

ROBERT N. COULSON Department of Entomology, Texas A&M University, College Station, Texas 77843

GEORGE T. FERRELL U.S. Forest Service, Pacific Southwest Forest and Range Experiment Station, Berkeley, California 94701

ANDREAS FISCHLIN Department of Plant Sciences, Project Centre IDA/Celtia, Swiss Federal Institute of Technology, CH-8092 Zurich, Switzerland

RICHARD O. FLAMM Department of Entomology, Texas A&M University, College Station, Texas 77843

IMOGEN M. GATE British Forestry Commission, Forest Research Station, Wrecclesham, Farnham, Surrey GU10 4LH, England

CLAUDE GERI Institut National de la Recherche Agronomique, INRA-CRF-Station de Zoologie Forestière, Ardon 45160, Olivet, France

JEAN-CLAUDE GRÉGOIRE Biologie Animale-C.P. 160, Université Libre de Bruxelles, B-1050 Brussels, Belgium

FRED PAUL HAIN Departments of Entomology and Forestry, North Carolina State University, Raleigh, North Carolina 27695–7626

SINIKKA HANHIMÄKI Department of Biology, University of Turku, SF-20500 Turku 50, Finland

ERKKI HAUKIOJA Department of Biology, University of Turku, SF-20500 Turku 50, Finland

ALEXANDER S. ISAEV V. N. Sukachev Institute of Forestry and Wood, Siberian Branch, U.S.S.R. Academy of Science, Krasnoyarsk, 660036 U.S.S.R.

NEIL A. C. KIDD Department of Zoology, University College, Cardiff CF1 1XL, Wales

FUJIO KOBAYASHI Forestry and Forest Products Research Institute, Tsukuba, Ibaraki, 305 Japan

SIMON R. LEATHER Forestry Commission, Northern Research Station, Roslin, Midlothian EH26 9SJ, Scotland

GARRELL E. LONG Department of Entomology, Washington State University, Pullman, Washington 99164-6432

JOHN L. MADDEN Faculty of Agricultural Science, University of Tasmania, Hobart, Tasmania 7000, Australia

VERA S. MALUTINA V. N. Sukachev Institute of Forestry and Wood, Siberian Branch, U.S.S.R. Academy of Sciences, Krasnoyarsk, 660036 U.S.S.R.

RICHARD R. MASON U.S. Forest Service, Forestry and Range Sciences Laboratory, La Grande, Oregon 97850

WILLIAM J. MATTSON U.S.D.A. Forest Service, North Central Forest Experiment Station, Pesticide Research Center, Michigan State University, East Lansing, Michigan 48824

MARK S. McCLURE The Connecticut Agricultural Experiment Station, Valley Laboratory, Windsor, Connecticut 06095

MICHAEL E. MONTGOMERY U.S.D.A. Forest Service, Northeastern Forest Experiment Station, Center for Biological Control of Northeastern Forest Insects and Diseases, Hamden, Connecticut 06514

F. DAVID MORGAN Department of Entomology, Waite Agricultural Research Institute, University of Adelaide, Glen Osmond, South Australia 5064, Australia

K. S. S. NAIR Division of Entomology, Kerala Forest Research Institute, Peechi 680 653, Kerala, India

SEPPO NEUVONEN Department of Biology, University of Turku, SF-20500 Turku 50, Finland

PEKKA NIEMELÄ Department of Biology, University of Turku, SF-20500 Turku 50, Finland

THOMAS L. PAYNE Department of Entomology, Virginia Polytechnic Institute and State University, Blacksburg, Virginia 24061

KENNETH F. RAFFA Department of Entomology, University of Wisconsin, Madison, Wisconsin 53706

ALAIN ROQUES Institut National de la Recherche Agronomique, INRA-CRF-Station de Zoologie Forestière, Ardon 45160, Olivet, France

GARY A. SIMMONS Department of Entomology, Michigan State University, East Lansing, Michigan 48824

GARY S. TAYLOR Department of Entomology, Waite Agricultural Research Institute, University of Adelaide, Glen Osmond, South Australia 5064, Australia

DAVID WAINHOUSE British Forestry Commission, Forest Research Station, Wrecclesham, Farnham, Surrey GU10 4LH, England

WILLIAM E. WALLNER U.S.D.A. Forest Service, Northeastern Forest Experiment Station, Center for Biological Control of Northeastern Forest Insects and Diseases, Hamden, Connecticut 06514

ALLAN D. WATT Natural Environment Research Council, Institute of Terrestrial Ecology, Edinburgh Research Station, Penicuik, Midlothian EH26 OQB, Scotland

BOYD E. WICKMAN U.S. Forest Service, Forestry and Range Sciences Laboratory, La Grande, Oregon 97850

JOHN A. WITTER School of Natural Resources, University of Michigan, Ann Arbor, Michigan 48109

PREFACE

> Insects multiply. Destruction reigns. There is dismay, followed by outcry, and demands to Authority. Authority remembers its experts or appoints some: they ought to know. The experts advise a Cure. The Cure can be almost anything: holy water from Mecca, a Government Commission, a culture of bacteria, poison, prayers denunciatory or tactful, a new god, a trap, a Pied Piper. The Cures have only one thing in common: with a little patience *they always work*. They have never been known entirely to fail. Likewise they have never been known to prevent the next outbreak. For the cycle of abundance and scarcity has a rhythm of its own, and the Cures are applied just when the plague of insects is going to abate through its own loss of momentum.
> —Abridged, with insects in place of voles, from C. Elton, 1924, *Voles, Mice and Lemmings*, with permission of Oxford University Press

This book is an enquiry into the "natural rhythms" of insect abundance in forested ecosystems and into the forces that give rise to these rhythms. Forests form unique environments for such studies because one can find them growing under relatively natural (primeval) conditions as well as under the domination of human actions. Also, the slow growth and turnover rates of forested ecosystems enable us to investigate insect population dynamics in a plant environment that remains relatively constant or changes only slowly, this in contrast to agricultural systems, where change is often drastic and frequent. These characteristics enhance our ability to investigate the stability of rapidly changing insect populations, as well as the stabilizing and destabilizing forces, something that is difficult or impossible to do in most agricultural settings. Finally, again because of slow turnover, the genetics of tree species occupying most forests have not been significantly altered by human selection. This provides us with an opportunity to study the interactions between trees and insects that have evolved over eons of coadaptation.

This book, being composed of a number of selected studies on individual forest insects, is of an empirical nature. We attempt to present the facts, both biological and ecological, on which our interpretations are based. Yet the aim of the book is to provide empirical support for a general unified theory of population dynamics, a theory that grew out of the "Great Debate" between Nicholson and Andrewartha during the 1950s. Although this theory has been consolidated in recent years from somewhat different perspectives, its underlying construct rests on the same critical question that haunted the "Great Debate": Are natural populations stabilized by density-induced negative feedback processes? Even today the literature bears testament to those who resist this notion. Does the "natural rhythm" emerge from some intrinsic properties of the system or is it merely an expression of chaotic external forces? The reader of this book is challenged to draw personal conclusions.

Out of the general unified theory of population dynamics has emerged a theory of outbreaks that focuses on the instabilities that arise in natural systems and the forces that tend to destabilize them. Critical questions can also be posed about the roots of this theory: Are natural populations destabilized by density-induced positive feedback processes and/or by the action of delayed negative feedback? In other words, is the apparent chaos that pest outbreaks seem to generate also an intrinsic property of the natural system—and, therefore, amenable to human manipulation?

This book should also be read with the view of understanding the human impact on the global ecosystem. Forests are just a reflection of other ecosystems, some of which are much more heavily affected by humans, some less so. Relevant questions are: Have human activities led to more or less stable interactions between forests and insects? What kinds of activities tend to destabilize these ecosystems? How can we mitigate the interaction between man and forests and so create more stable ecosystems? If this book can help answer these questions, our time will have been well spent.

This book was conceived and written while I was chairman of a working party entitled "Population Dynamics of Forest Insects" within the International Union of Forestry Research Organizations (IUFRO), and many of the contributors have been active participants in this group. Working party S2.07-06 was formed by G. C. Varley, F. Schwerdtfeger, and A. D. Voûte at a meeting in Arnheim in 1955. George Varley, to whom we dedicate this volume, became the first chairman of S2.07-06 the following year at the IUFRO World Congress in Oxford. When George retired in 1976 at the Oslo World Congress, Werner Baltensweiler assumed the chair and I became his co-chairman. My tenure as co-organizer of "Population Dynamics of Forest Insects" ended 10 years later at the World Congress in Ljubljana (Yugoslavia) and I view this book as the culmination of this enjoyable and productive experience.

Finally, this book represents the work and thoughts of numerous individuals from many countries. Each chapter tells a story in itself that is both interesting and provocative. Yet out of this diversity emerge common themes, generalities, and conclusions. The reader is invited to join us in the search for those commonalities that affect us all as we play our part in the dynamic patterns of nature.

Alan A. Berryman

Pullman, Washington

CONTENTS

CHAPTER 1

THE LARCH CONE FLY IN THE FRENCH ALPS

Alain Roques

CHAPTER 2

THE LARCH GALL MIDGE IN SEED ORCHARDS OF SOUTH SIBERIA

Alexander S. Isaev, Yuri N. Baranchikov, and Vera S. Malutina

CHAPTER 3

THE ARMORED SCALES OF HEMLOCK

Mark S. McClure

CHAPTER 4

THE BEECH SCALE

David Wainhouse and Imogen M. Gate

CHAPTER 5

THE BALSAM WOOLLY ADELGID IN NORTH AMERICA

Fred Paul Hain

CHAPTER 6

THE LARGE PINE APHID ON SCOTS PINE IN BRITAIN

Neil A. C. Kidd

CHAPTER 7

THE WHITE LACE LERP IN SOUTHEASTERN AUSTRALIA

F. David Morgan and Gary S. Taylor

CHAPTER 8

THE NANTUCKET PINE TIP MOTH

C. Wayne Berisford

CHAPTER 9

THE AUTUMNAL MOTH IN FENNOSCANDIA

Erkki Haukioja, Seppo Neuvonen, Sinikka Hanhimäki, and Pekka Niemalä

CHAPTER 10

THE DOUGLAS-FIR TUSSOCK MOTH IN THE INTERIOR PACIFIC NORTHWEST

Richard R. Mason and Boyd E. Wickman

CHAPTER 11

THE NUN MOTH IN EUROPEAN SPRUCE FORESTS

Broder Bejer

CHAPTER 12

THE LARCH CASEBEARER IN THE INTERMOUNTAIN NORTHWEST

Garrell E. Long

CHAPTER 13

THE PINE BEAUTY IN SCOTTISH LODGEPOLE PINE PLANTATIONS

Allan D. Watt and Simon R. Leather

CHAPTER 14

THE TEAK DEFOLIATOR IN KERALA, INDIA

K. S. S. Nair

CHAPTER 15
THE PINE LOOPER IN BRITAIN AND EUROPE
David A. Barbour

CHAPTER 16
THE SPRUCE BUDWORM IN EASTERN NORTH AMERICA
William J. Mattson, Gary A. Simmons, and John A. Witter

CHAPTER 17
THE LARCH BUDMOTH IN THE ALPS
Werner Baltensweiler and Andreas Fischlin

CHAPTER 18
THE GYPSY MOTH: A WESTWARD MIGRANT
Michael E. Montgomery and William E. Wallner

CHAPTER 19

THE PINE SAWFLY IN CENTRAL FRANCE

Claude Geri

CHAPTER 20

SIREX IN AUSTRALASIA

John L. Madden

CHAPTER 21

THE JAPANESE PINE SAWYER

Fujio Kobayashi

CHAPTER 22

THE GREATER EUROPEAN SPRUCE BEETLE

Jean-Claude Grégoire

CHAPTER 23

THE SPRUCE BARK BEETLE OF EURASIA

Erik Christiansen and Alf Bakke

CHAPTER 24

THE MOUNTAIN PINE BEETLE IN WESTERN NORTH AMERICA

Kenneth F. Raffa

CHAPTER 25

THE SOUTHERN PINE BEETLE

Richard O. Flamm, Robert N. Coulson, and Thomas L. Payne

CHAPTER 26

THE FIR ENGRAVER BEETLE IN WESTERN STATES

Alan A. Berryman and George T. Ferrell

CHAPTER 27
THE STRIPED AMBROSIA BEETLE
John H. Borden

CHAPTER 1

THE LARCH CONE FLY IN THE FRENCH ALPS

Alain Roques

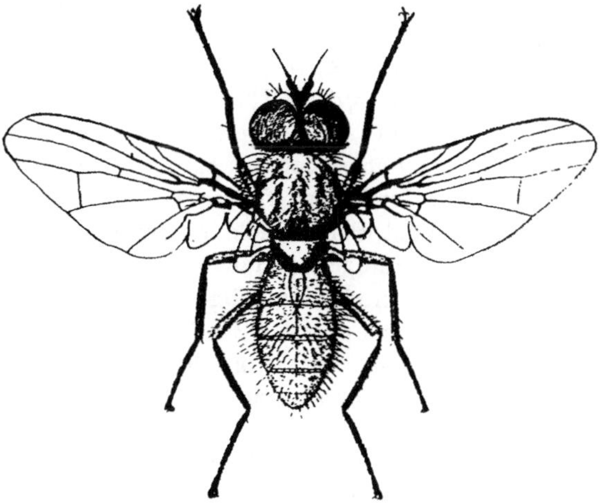

Alain Roques • Institut National de la Recherche Agronomique, INRA-CRF-Station de Zoologie Forestière, Ardon 45160, Olivet, France.

1. INTRODUCTION

Among the pest species that damage larch cones in Eurasia, the larch cone fly, *Lasiomma melania* Ackland (Diptera: Anthomyiidae), is undoubtedly one of the most serious. The genus *Lasiomma* contains at least three closely related species that attack larch cones in the same areas: *L. melania, L. laricicola* Karl, and *L. infrequens* Ackland. These three species, previously known under the name of *Lasiomma* (= *Chortophila* = *Phorbia* = *Hylemya*) *laricicola* Karl, were separated in 1965 on the basis of male characters[1] and later using female characters.[34] Thus, much cone damage attributed to the single species, *L. laricicola,* in Europe and the Soviet Union during the first half of the twentieth century, was probably also attributable to the two other species. For example, *L. melania* is generally the dominant species in the French Alps, even though it was only discovered in 1975 in this region.[34] Today, the presence of *L. melania* is well established in Western Europe,[1,17,34,35] in the Soviet Union from the European part to the Far Eastern regions,[8,12,31,48,49] and in China.[9] The larch cone fly is restricted to *Larix* cones but damages various species and subspecies of this genus according to region, including European larch (*L. decidua*), Siberian larch (*L. sibirica*), Dahurian larch (*L. gmelini*), and Japanese larch (*L. leptolepis* = *L. kaempferi*). Thus, its geographic range probably covers the major part of the Eurasian natural distribution area of the genus *Larix,* as well as the western European regions in which these conifers have been introduced.

Considerable damage from larch cone fly larvae is observed wherever accurate investigations have been carried out. For example, on average, 40–60% of the Siberian larch seed crop is destroyed in the European part of the Soviet Union; this percentage may be as high as 85–95% when cone production is light.[49] Our studies in the French Alps assess the damage to European larch cones at 30–75%, varying with years.[37] As the average seed consumption per larva varies from 30 to 40 per cone, the percentage destruction increases significantly when several larvae are present in the same cone—from 20 to 92%.[37]

Lasiomma melania appears to be one of the major factors limiting the natural regeneration of larch in many areas. The extensive establishment of larch seed orchards, for the production of genetically improved seeds, increases its economic impact, particularly since the pest can colonize artificial stands distant from natural larch forests.[47]

In addition to its economic importance, the larch cone fly is an interesting model of a highly host-specific pest. Cones can be regarded as dynamic plant microunits, with a limited life span and highly variable spatial and temporal distribution. Cone production is characterized by very irregular fluctuations from year to year. Thus, the qualitative and quantitative composition of the pest complex within the ecological niche depends on the adaptability of the various species to these fluctuations. Some cone insects, such as many *Dioryctria* (Lepidoptera: Phycitidae), can feed on other tree structures, such as buds and shoots, when cones are scarce.[33,35] By contrast, species such as *L. melania,* which attack only cones, are strictly dependent on cone production and have evolved adaptative strategies, such as extended diapause, to compensate for cone crop variation. Their population dynamics, as we shall see, is a graded response to this driving variable, the cone crop.

2. BIOLOGY AND LIFE HISTORY

One of the most typical features of the life cycle of *L. melania* is the close synchrony between the various development stages of the insect and cone growth (Fig. 1). Adult emergence always coincides, at whatever altitude, with the onset of ovuliferous scale growth (as they become externally visible). In the Alps, this period varies from early May, at 1200 m altitude, to late June at 2200 m. Egg-laying occurs 12–15 days later following a feeding phase in which the ovaries mature.

Host recognition seems to be determined by several different signals.[36,38] At a distance, insects distinguish cone-bearing trees from the reflectance contrast between cônes and larch foliage, Then, olfactory stimuli emitted by attractive cones induce females to land. Egg laying occurs after an orientation motion that is probably related to tactile stimuli. The whole process ensures the selection of a particular cone type, exhibiting distinctly developed scales that are still half-covered by bracts. The characteristic ivory white eggs are wedged in at the junction between scales and bracts, generally in the mid-section of the cone.

Lasiomma melania has three larval instars (Fig. 1), but only two are free-living.[37] The first instar remains within the eggshell, a behavior that is also observed in several other species of the same genus.[50,52] The ovolarval stage lasts 8–13 days.

The second instar larva, after emerging from the egg, immediately enters the cone and then follows a complex course. Penetration into the scale is indicated by a circular hole from which liquids mixed with resin exude. The larva bores a linear gallery into the

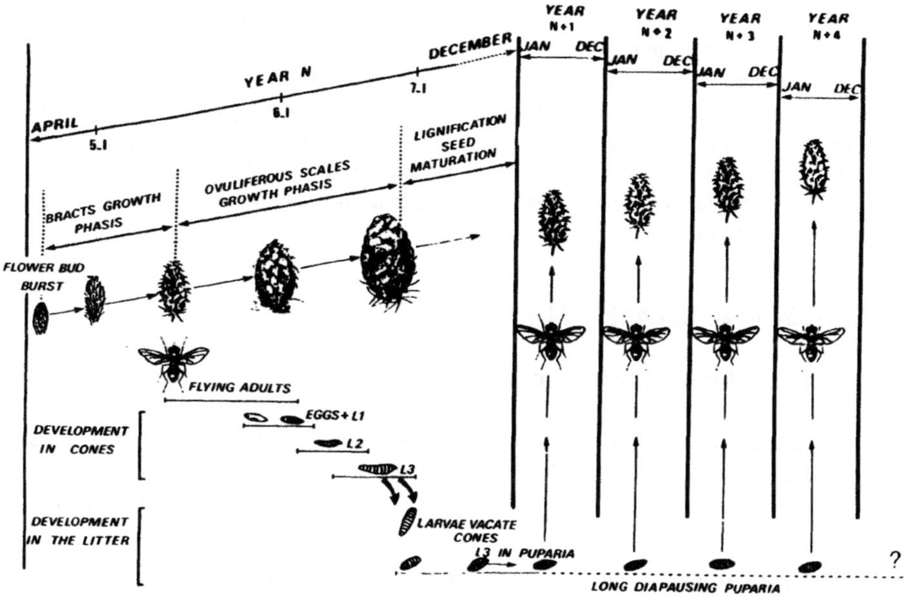

FIGURE 1. Biological cycle of *Lasiomma melania* and phenological relationships with larch cone development in the French Alps (1200-m altitude).

inner side of the scale and quickly reaches and devours the two basal seeds. It then penetrates deeper into the underlying scale and bores a gallery down toward the cone base, feeding on seeds it encounters on its way. The initially straight gallery then spirals around the cone axis. The second instar lasts 6–12 days, depending on altitude.

Molting to the third instar occurs after 1½ spiral whorls. If the egg has been laid in the mid- or basal part of the cone, as is usual, the larva reaches this stage in the cone base. The third instar larva then spirals up toward the apex of the cone, generally without entering the axis. Spirals become irregular and confused when several larvae are present simultaneously. If egg laying has taken place near the apex of the cone, the larva follows the opposite course. The development period within the cone terminates 18–25 days after molting to the third instar. At this time, the larva bores an emergence hole of very irregular outline and escapes to the litter on the forest floor. The exit period varies from late June at low elevations to mid-September at high elevations. Some observations indicate that emergence can take place 15 days later, under very cold weather conditions.[37,49]

The emergence period seems to coincide with a sudden decrease in cone moisture content that occurs after the initiation of lignification. Thus, the total duration of larval development within the cone varies broadly with climatic conditions and altitude. We estimate that development ranges from 38–50 days at an altitude of 1200 m to 32–43 days at an altitude of 2200 m in the French Alps.[37] This relative acceleration of development at high altitudes can be attributed to the summer weather conditions, which favor late initiated development. By contrast, larvae developing in larch stands above 2000-m altitude only become mature in late August or mid-September, so that their development can be stopped by early drops in temperature.

Once it is on the ground, the larva enters the litter and builds a puparium at a depth of a few centimeters. Obligatory winter diapause occurs while the larva is in its puparium[29]; it must be exposed to a total low-temperature summation of about −100°C day-degrees before pupation.[49] Adult emergence is observed the following spring, but maggots frequently remain in prolonged diapause for up to 4 years (Fig. 1). Most imagos, however, seem to emerge within 2 years of the onset of development.[37] The physiological determinants of diapause length are not known except that photoperiod and temperature do not seem to be involved.[29]

3. ECOLOGICAL RELATIONSHIPS

3.1. Relationships with the Cone

Penetration of the cone by *L. melania* larvae induces alterations in cone development, the extent of which depends on the growth stage of the cone at the time of insect attack. These changes feed back to affect larval development. Two main types of reaction can be distinguished[37]:

1. A local resinous secretion begins at the entrance hole made by the second instar larva and spreads into the galleries as the maggot penetrates deeper into the cone. In sufficient quantities, these resinous secretions can cause the death of the delicate second instar larvae.

2. Acceleration of the lignification process may occur from the middle of the second instar, when boring in the cone base causes early drying. This process continues during the third instar and produces a lignified and dessicated cone about 1 month earlier than normal. In these cases, vegetative growth is limited and the final cone is only 79–90% of the standard size.

Sudden decreases in moisture content, related to this accelerated hardening, can hinder and even block larval exit; the fully dessicated larva then dies inside the cone. Lignification and drying frequently occur when several maggots simultaneously attack the cone or when the cone has been previously damaged by other insects.

3.2. Relationships with Other Organisms

3.2.1. Interspecific Competition

The strict specificity of many cone insects, together with the often limited availability of cones relative to other tree structures, e.g., foliage, suggest that special attention be paid to interspecific competition between cone insects. A total of 26 insect species are encountered in larch cones within the distribution of *L. melania.*[1,9,17,31,34,35,37,39,42,44,45,47–49,52,53] Knowledge of the precise geographical range of these insects is still fragmentary, but it appears that the number of species seldom exceeds 10 in any given area.[35,45,47–49]

Three insect groups can be distinguished according to the degree of competition with the larch cone fly. Several tortricid (Lepidoptera) larvae are the earliest insects that can damage newly opening flower buds. These insects, whose species differ according to region, feed essentially as defoliators but can also develop on female flowers or cones. Their attack rate is generally less than 10%, but the larch bud moth *Zeiraphera diniana* Guenee can attack up to 75% of the cone crop when populations reach the peak of their cyclical outbreaks.[35] Damage by these species is largely superficial, but *Choristoneura diversana* Hübner may injure cone scales and the axis, causing growth arrest and drying of the young cones.[49] Larvae of the other species feed on the external parts of bracts and scales and cover the cone with silky threads but never penetrate. Consequently, they do not usually disturb cone growth and do not compete directly with *L. melania.* External resin flow caused by the feeding of these insects can hinder egg laying by the female larch cone fly. The main effect of these insects is the reduction of cone availability to the cone maggot, which attacks later.

The second group consists of other *Lasiomma* species, mainly *L. laricicola.* The distribution of this species is more extensive and completely overlaps that of *L. melania.* Thus, the two species are frequently present in larch stands from Great Britain to the Far Eastern Soviet Union.[1,9,31,34,35,49] *L. laricicola* lays its eggs during bract growth, coinciding with pollination, 2 or 3 weeks earlier than *L. melania.* The hatching larvae enter the cone and spiral down around the axis toward the cone base. After this, the third instar larvae penetrates the axis, hollowing it out almost completely. The larvae finally exit the cones and complete their development in the litter. Because of their early attack, cone growth ceases prematurely, and the final size is less than that of a cone damaged by *L. melania.*[37] Colonization of cones previously attacked by *L. laricicola* seems to be unfavorable for the larch cone fly because hatching occurs when its competitor is in the second or third instar and has destroyed the cone base. Resin flow and early hardening restrict the normal development of *L. melania.* Another species, *L. jurtschenkoï* Elberg, is apparently limited to the Far Eastern Soviet Union[49] and appears to play a similar role in this area. Egg laying occurs, on the average, 1–2 weeks after larch flowering in the Amur region. Various Lonchaeidae (*Diptera*) may also compete with the larch cone fly.[45,49]

All the other species attack the cone later than *L. melania,* after most of the cone has been consumed, and when the larch cone fly is at the end of the second larval instar. These insects compete only weakly, even though some of them destroy the same cone structures. *L. infrequens,* whose distribution is similar to that of *L. melania,* invades the cone during the growth of ovuliferous scales, when bracts are almost entirely hidden.[35,37] This delay of 8–12 days, relative to the larch cone fly, is maintained during the rest of the developmental period. Damage by *L. infrequens* becomes noticeable after 3 weeks and only disturbs larch cone maggots during the last 10 days of its existence in the cone. Two other *Lasiomma* species, *L. baicalensis* Elberg and *L. luteoforceps* Fan and Fang, have been recorded with L. *melania* in the Lake Baïkal area[31] and in northeastern China,[9] respectively, but no accurate information exists on their biology.

Various species of Cecidomyiidae and Tortricidae, which attack the cone toward the end of scale growth, 2–4 weeks after *L. melania* lays its eggs, have limited impact on the cone fly. These insects develop when maggots are about to leave the cone.[35] Finally, some seed-eating chalcids (Torymidae and Eurytomidae) and pyralid larvae (Pyralidae) use the cone after the onset of lignification, as *L. melania* larvae exit to the litter.[31]

3.2.2. Parasites and Predators

The recent separation of the three species, originally known under the name of *L. laricicola,* has delayed identification of their respective parasite complexes. However, this has not been a serious problem because all species seem to be preyed on by similar natural enemies. In the French Alps, it may be safely assumed that most of the insects attacking the various *Lasiomma* species are also able to feed on *L. melania*[12,32,39,42,48,49] (Fig. 2). In fact, all parasites of larch cone insects are polyphagous and prey on other insects, including some from different orders that develop outside the cone.[48,49] This seems to be typical of cone insect parasites, allowing them to compensate for annual fluctuations of their hosts.[33,35,46] Prolonged diapause is also a common feature among parasites. Diversity of the parasite complex appears to be reduced in any given location, with the number of species rarely exceeding seven. Nevertheless, the different parasite species are distributed on all the developmental stages of the host in the French Alps (Fig. 2).

Eggs are parasitized by the chalcidid *Trichogramma embryophagum* Hartig (Hymenoptera: Trichogrammatidae), whose oviposition coincides with that of *L. melania.* The polyembryonic larvae (up to 6 larvae per egg) develop over 4 weeks; the second-generation adults emerge in July–August, according to altitude. These adults have to find another host in order to complete their life cycle. Although this insect is polyphagous,[49] it does not seem to attack the other two *Lasiomma* species. This could be explained by differences in the disposition of the eggs on the cone, *L. melania* eggs generally being externally visible, differing in this respect from other species of *Lasiomma.*[37]

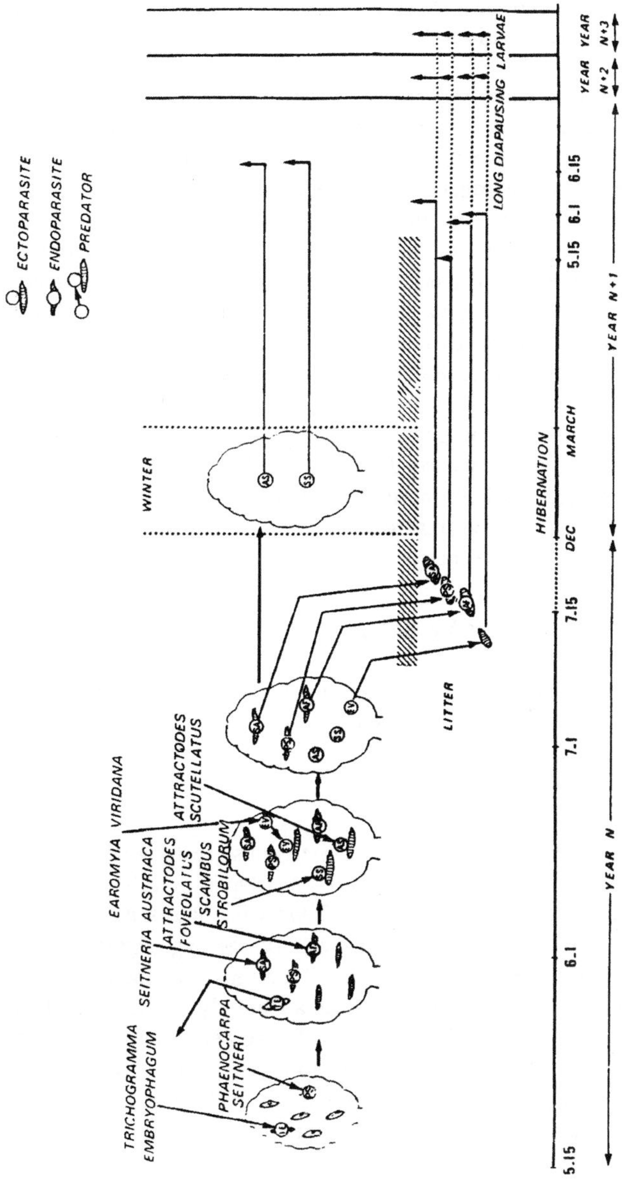

FIGURE 2. Attack periods and biological cycles of parasites and predators of *L. melania* in the French Alps (1200-m altitude).

The following parasites, however, prey on all three species. The hymenopterous parasite, *Phaenocarpa seitneri* Fahringer (Braconidae), is probably an ovolarval parasite whose flight period coincides with that of *L. melania*. Larval development takes place inside host larvae that continue to grow until the formation of puparia in the litter. Parasite adults emerge from the puparium the following spring but can show prolonged diapause for 2 or 3 years.

Second-instar larvae are parasitized by two other hymenopterous endoparasites, *Attractodes foveolatus* Gravenhorst (Ichneumonidae) and *Seitneria austriaca* Tavares (Cynipidae). Their larval development is similar to that of *Phaenocarpa.*

Third-instar larvae are attacked by a predatory dipterous larva, *Earomyia viridana* Meigen (Lonchaeidae), which has mistakenly been considered a pest. The adult fly lays one or more eggs on the inner side of the cone scales. The young larvae then search out cone maggots and devour their internal tissues. Third instars leave the cone and pupate in the litter. Prolonged diapause may also occur during this stage. Third-instar *Lasiomma* larvae are also preyed on by two ectoparasitic ichneumonids, *Attractodes scutellatus* Hellen and *Scambus strobilorum* Ratzeburg, whose larvae hibernate in the cone after they have consumed the paralyzed host.

No specific parasite of the puparium stage has yet been discovered. Parasitism by each species does not generally exceed 5%,[48] with the notable exception of the predator, *E. viridana,* which can result in mortality as high as 35% in some years. Unidentified pathogens and rodents as well as unfavorable weather appear to decimate overwintering or diapausing puparia in some years.[49]

4. NUMERICAL POPULATION BEHAVIOR

In discussing the dynamics of *L. melania* populations, consideration must be given to density variation within cones, trees, and stands.

4.1. Variations within Cones

The number of eggs laid per attacked cone varies greatly with cone production (Fig. 3). These data were obtained in 20 neighboring stands in 1981, after a late frost had destroyed a large proportion of the female flowers in some stands. When cone yield is high or very high, most attacked cones have only one egg deposited on them, and the maximum is three. By contrast, an average of four eggs per cone is characteristic of light cone crops, with a maximum of 15. Females apparently distribute all their eggs on the available cones; consequently, the number of larvae per damaged cone varies inversely with crop size (Fig. 3). Various mortality factors, however, reduce this variation. Thus, the number of damaged cones is only a rough index of the total number of larvae because it represents the functional as well as the numerical responses of the larch cone fly to changes in cone abundance.[20]

4.2. Variations within Trees

Larch cones are not homogeneously distributed in the crown, particularly if the crop is medium or light. Mid- and upper parts of the crown generally bear most of the cones,

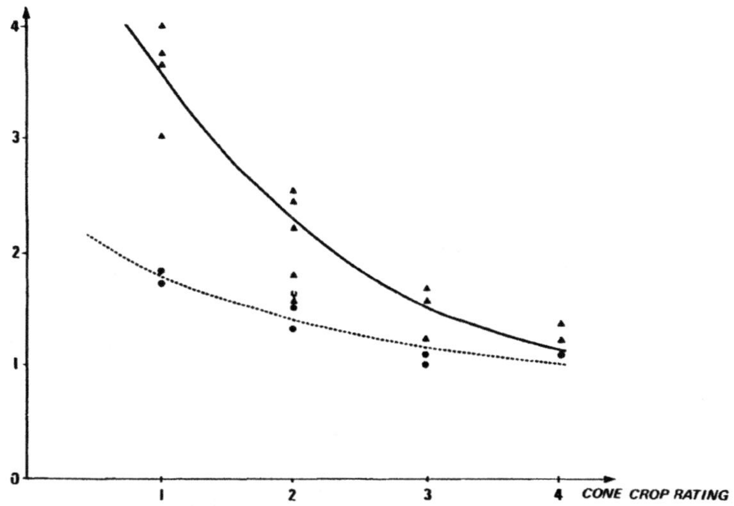

FIGURE 3. Variations in the average number of eggs per damaged cone ($y1$, ▲) and in the number of third instar larvae per damaged cone ($y2$, ●) in relationship to cone crop size rating (x) of plots sampled the same year (200 damaged cones collected per plot). Qualitative cone crop ratings: 0, no cone; 1, light crop (1–500 cones per tree); 2, medium crop (500–1000 cones per tree); 3, heavy crop (1000–2000 cones per tree); 4, very heavy crop (more than 2000 cones per tree). Fitted curves: $y1 = 7.92\ (x + 1)^{-0.82} - 1 (r = 0.89; p < 0.001)$; $y2 = 3.63\ (x + 1)^{-0.38} - 1. (r = 0.93; p < 0.001)$.

and sunny parts produce better than do shaded ones. Climatic conditions more favorable for the initiation and development of female flowers also affect cone production.[23] Figure 4 compares the mean distribution of damaged cones and cone abundance in four strata of the crown. The data are the means of systematic samples obtained from two groups of 10 neighboring trees presenting low and high cone production, respectively. Cone damage is significantly higher in the middle and upper crown and follows the same distribution pattern as the cone crop, irregardless of cone abundance. Larch cone fly attacks are directly proportional to the total number of cones on a branch, as has been already established for some other cone pests.[25] Cone position on the branch does not appear to affect the probability of attack. Thus, representative samples of the entire cone population can be obtained from any crown level, if it is referred to the total cone production from that level. A simple random sample of 10% of the middle or lower crown cones, which are more accessible, makes it possible to estimate larch cone fly damage and the larval population of a given tree.

4.3. Variations within Stands

The effect of various environmental factors on the distribution of damaged cones has been analyzed by principal component analysis of standardized initial values followed by regression analysis of the number of damaged cones against the principal components. Only the first two components, explaining 58.6% of the total variation, appear significant.

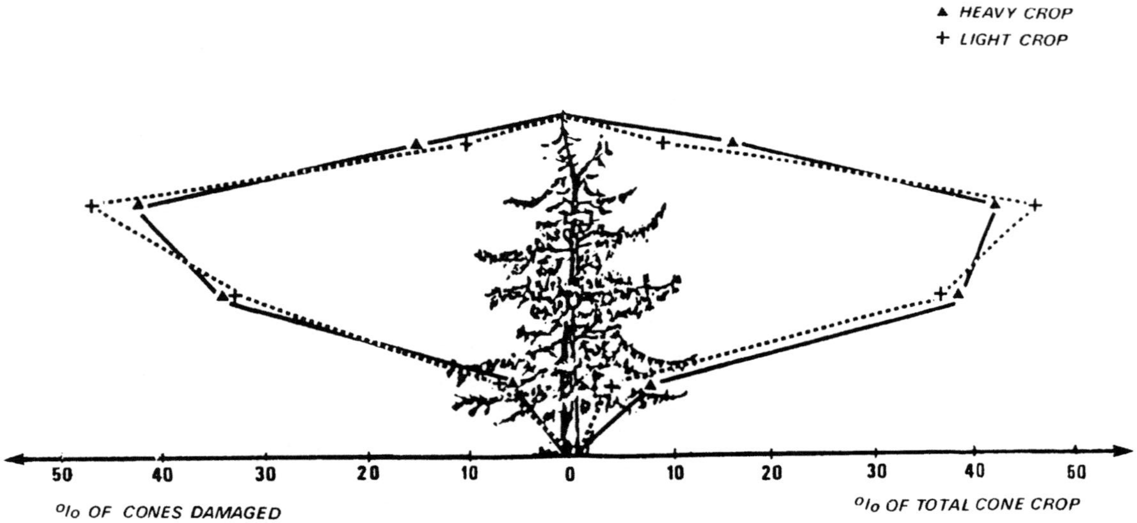

FIGURE 4. Relationship between vertical distribution of cones damaged by *L. melania* and cone distribution in the same tree for two extreme cases of cone crop; heavy crop, 2000 cones per tree; light crop, 400 cones per tree.

By plotting correlation circles before and after varimax rotation (Fig. 5), these components can be interpreted as follows:

1. The first component includes the biological characteristics of the habitat, i.e., the number of available cones per tree, the number of cones previously damaged by *L. laricicola,* and the number of cones damaged by *L. melania.*
2. The second component includes stand factors in which altitude and tree height have opposite effects.

These results suggest that the number of damaged cones per tree varies in relationship to the joint effect of cone crop size and previous damage by *L. laricicola.* This insect appears to have a positive influence on *L. melania* oviposition. Perhaps kairomonal attraction exists between the two species or between previously attacked cones and larch cone fly responses. There also seems to be a less significant positive relationship between the number of damaged cones and tree height, which is also negatively correlated with altitude.

Previous cone colonization by the larch bud moth, *Zeiraphera diniana,* does not seem to have any effect on larch cone fly distribution. Regression analysis on the 1979 data, when the budmoth population was at its peak and damaged 40–75% of the cones, indicates no significant correlation between damage by the two insects.

4.4. Action of Biotic Mortality Factors

Typical life tables for three successive generations of the cone fly during large cone crop fluctuations are shown in Table I. Numbers of diapausing larvae and males are

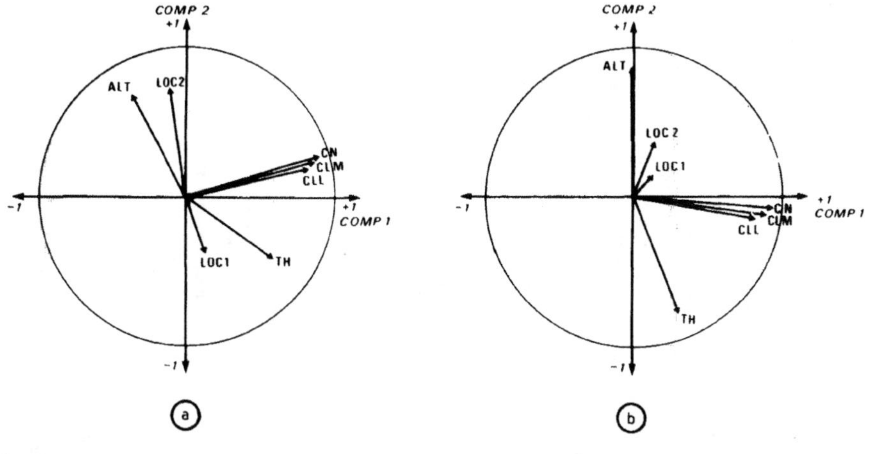

FIGURE 5. Principal component analysis of environmental variables influencing larch cone fly distribution among trees. Correlation circles in the first two components space. (a) Before varimax rotation. (b) After varimax rotation. ALT, altitude; C, number of cones per tree; TH, tree height; CLL, number of cones damaged by *L. laricicola* per tree; CLM, number of cones damaged by *L. melania* per tree; LOC 1, location variable (LOC 1 = 1 if the sampled tree is isolated; 0 in other cases); LOC 2, location variable (LOC 2 = 1 if the sampled tree is inside the stand, 0 in other cases).

TABLE I
Typical Life Tables for Three Successive Generations of *L. melania* in the Same Larch Stand[a] during a Large Cone Crop Fluctuation[b,c]

		1978				1979				1980			
x	DxF	L*x*	D*x*	100D*x*/L*x*	S*x*	L*x*	D*x*	100D*x*/L*x*	S*x*	L*x*	D*x*	100D*x*/L*x*	S*x*
Egg and first larval instar		53.2				30.6				50.1			
	Infertility		10.1	19.0			3.5	11.4			8.0	15.9	
	Trichogramma		4.5	8.5			1.0	3.3			3.5	7.0	
	Unknown embryophagum		1.6	3.0			0.7	2.3			0.8	1.6	
	Total		16.2	30.5	69.5		5.2	17.0	83.0		12.3	24.5	75.5
Second larval instar		37.0				25.4				37.8			
	Resinosis		4.5	12.2			3.2	12.6			4.0	10.6	
	Unknown		5.2	14.0			1.3	5.1			2.1	5.6	
	Total		9.7	26.2	73.8		4.5	17.7	82.3		6.1	16.2	83.8
Third larval instar (in cone)		27.3				20.9				31.7			
	Earomyia viridana		2.8	10.3			1.9	9.1			4.4	13.9	
	Scambus strobilorum		0.7	2.6			1.3	6.2			0.6	1.9	
	Attractodes scutellatus		0.1	0.4			0.4	1.9			0.1	0.3	
	Cone drying		1.5	5.5			1.0	4.8			3.5	11.0	
	Unknown		3.6	13.1			1.0	4.8			1.7	5.4	
	Total		8.7	31.9	68.1		5.6	26.8	73.2		10.3	32.5	67.5
Puparium[d]		18.6				15.3				21.4			
	Seitneria austriaca		0.9	4.8			0.3	2.0			1.8	8.4	
	Attractodes foveolatus		0.6	3.2			1.0	6.5			0.5	2.3	
	Phaenocarpa seitnari		4.0	21.5			2.0	13.1			4.6	21.5	
	Diapausing parasites		0.6	3.2			2.6	16.9			0.8	3.7	
	Unknown		0.9	4.8			0.4	2.6			0.7	3.3	
	Total		7.0	37.5	62.5		6.3	41.1	58.9		8.4	39.2	60.8
Diapausing larvae		11.6	2.9			9.0	4.3			13.0	3.0		
Male ratio		8.7	4.5			4.7	2.3			10.0	5.0		
N + 1 emerging Females		4.2				2.4				5.0			

[a] Altitude: 1200 m.
[b] Average cone number per tree: 1978; 1800 cones; 1979, 2500 cones; 1980, 750 cones; 500 cones studied per age interval.
[c] *x*, age interval; DxF, mortality factors within *x*; L*x*, number alive per lot of 100 cones; D*x*, number dying within *x*; 100D*x*/L*x*, D*x* as a percentage of L*x*; S*x*, survival rate within *x*.
[d] Estimation based on lots stocked in laboratory.

considered mortality factors because variations in diapause or sex ratio can affect population fluctuations. Changes in female fecundity within generations were not observed. It should also be noted that pupal mortality was probably underestimated because of the procedure used to rear this stage, i.e., 500 cones were brought into the laboratory before the exit of the third instar larvae and placed on wet sand, and the puparia were stored in a photoeclector at outdoor temperature. Under these conditions, neither parasitism nor predation after larval exit to the litter, nor the various mortality factors acting overwinter (e.g., virus, rodents), was estimated. Nevertheless, some interesting inferences were obtained from these partial life tables. Calculating *k* factors[51] show that, despite variations in the survival of different stages, total mortality (*K*) is similar for different initial densities (Table II). This compensating effect is mainly due to diapause (*k*5) (see Section 4.5). It is also significant that total larval mortality is higher in years of low cone production and dense occupation. This mortality seems to occur mainly in the egg stage (*k*1) and well before third larval instar mortality in the (*k*3). Fluctuations in *Trichogramma* parasitism and egg abortion are the major components of *k*1 variation. Second instar mortality, mainly due to resinosis, seems to have little relationship to population density. By contrast, *k*3 and *k*4 seem to include the activity of at least two density-dependent factors. Cone drying after early lignification is aggravated when cone production falls because the mean number of insects per cone increases. In addition, the qualitative and quantitative composition of the parasitic complex is also modified (Fig. 6), depending on the concentration of host eggs, which is itself correlated with cone production (see Fig. 3). Parasites with efficient host-finding abilities, mainly ectoparasitic ichneumonids, are in the majority when cone production is heavy and cone fly eggs are dispersed. They are replaced by endoparasitic braconids (and *Trichogramma*), which exhibit a more limited range of action, during light cone crops, when the number of eggs per cone increases. However, total parasitism although higher than previously believed,[12,48] remains limited and relatively constant with time.

TABLE II
Variations in *k* Factors According to Values of Table I [a]

Age interval	*k* factor	Year 1978	Year 1979	Year 1980
Egg				
First larval instar	*k*1	0.158	0.081	0.122
Second larval instar	*k*2	0.132	0.085	0.076
Third larval instar (in cone)	*k*3	0.167	0.136	0.171
Puparium	*k*4	0.205	0.230	0.217
Diapausing larvae	*k*5	0.125	0.282	0.269
Sex ratio	*k*6	0.316	0.292	0.301
Total larval mortality	(*k*1−*k*4)	0.661	0.531	0.586
Total mortality	*K* (*k*1−*k*6)	1.103	1.106	1.156
Mean number of eggs per cone		0.11	0.06	0.10
Mean number of cones per tree		1800	2500	750

[a] $ki = \log(N_{i+1}/N_i)$, where N_i is the number of larvae in the i th stage.

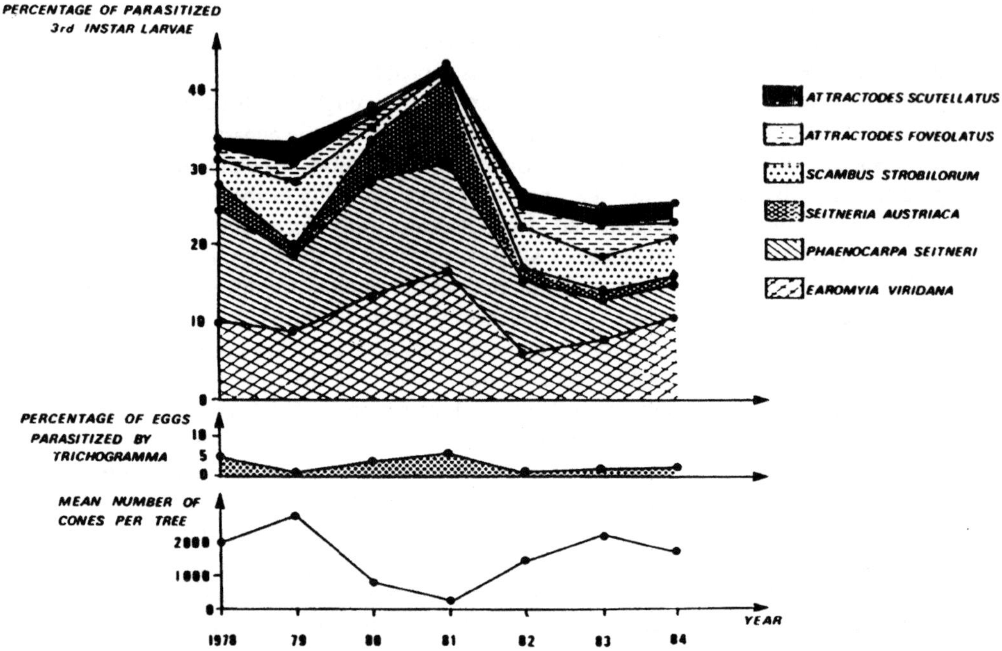

FIGURE 6. Evolution of respective attack rates of parasites and predators in relationship to cone crop fluctuation during six consecutive years in the same larch stand in the French Alps (1200-m altitude).

4.5. Variation in Prolonged Diapause

The mean percentage of larvae entering prolonged diapause in a given year is highly variable. Adult emergence during the first year following egg hatch varies from 52 to 95% of the total population (Fig. 7). Consequently, prolonged diapause varies from 5 to about 50%. No significant relationship can be established between prolonged diapause and the size of the current cone crop. However, the incidence of prolonged diapause is inversely correlated with the rate of change in cone yield from one year to the next and, less significantly, with the size of the cone crop the following year. This finding confirms the hypothesis that prolonged diapause permits survival of the population during periods of low food supply.

Prolonged diapause can also be influenced by the action of certain weather factors during larval feeding. For example, total rainfall during June and July is inversely correlated with the percentage entering prolonged diapause in that year. Prolonged diapause also increases with total solar insolation.

Termination of prolonged diapause has been measured only under artificial conditions. It is therefore not surprising that no relationship with cone crop size or weather factors could be established. Under these breeding conditions, most imagos emerge during the 2 years following puparia formation, although some individuals emerged 3 or 4 years later (Fig. 7).

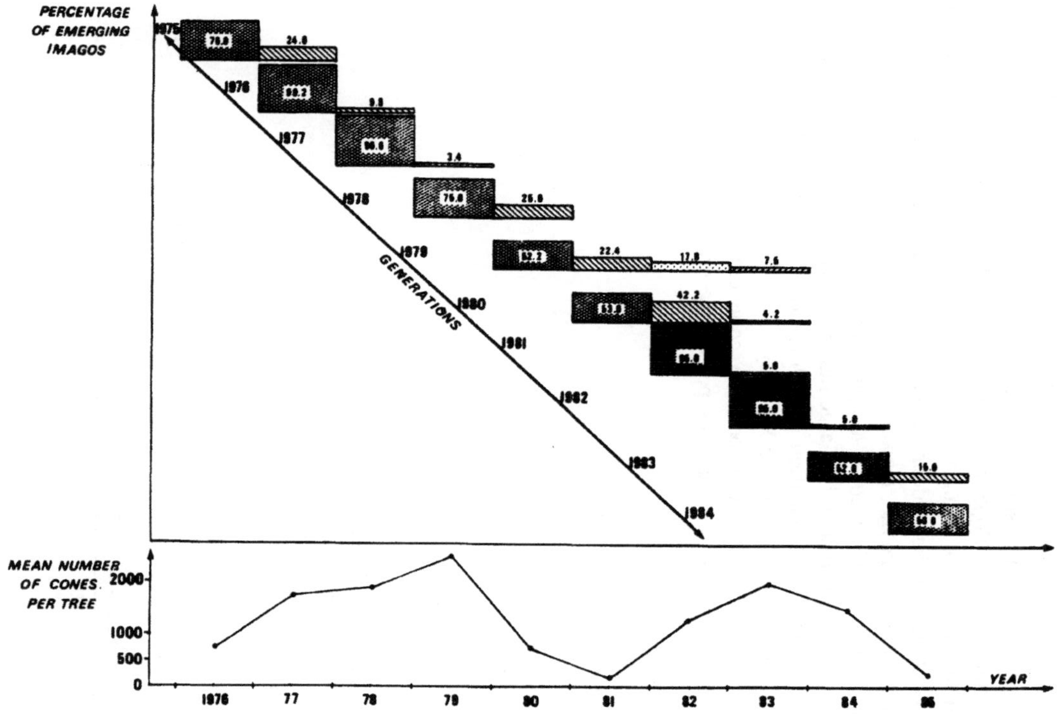

FIGURE 7. Relationship between adult emergence and cone crop fluctuations for 10 successive generations in the same larch stand of the French Alps (1200-m altitude).

4.6. Annual Fluctuations in Larch Cone Fly Populations

Fluctuations in the number of third instar larvae per tree during 10 consecutive years in three larch stands of similar age (about 30 years), but located at different altitudes, are shown in Fig. 8. Also provided for comparison are the annual changes in cone yield per tree.

Larval density per tree varies considerably from year to year, but the trends of all three populations appear quite similar, except that the overall abundance of the high-altitude population is lower. Populations clearly seem to respond, in all stands, to the irregular fluctuations of cone abundance. For example, populations increase regularly when cone production rises, but sudden decreases in cone abundance lead to an immediate reduction in the number of larvae. However, if the cone crop increases again during the next year, populations quickly recover to their previous levels. Population development seems different when late frosts kill a large number of cones of an initially heavy crop, particularly at high altitude. Population growth is slower than normal, if the production increases again the following year (see Fig. 8c).

If we examine the relationship between cone crop size and the percentage of damaged cones (Fig. 9), we see that this percentage drops to a low level when cone

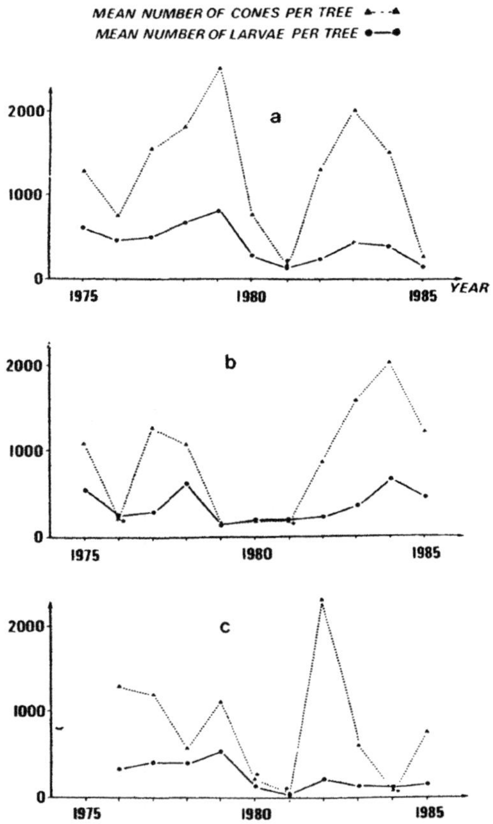

FIGURE 8. Comparative dynamics of larch cone fly populations and cone crops in three larch stands located at different altitudes. (a) 1200-m altitude; (b) 1560-m altitude. (c) 2200-m altitude. Late frosts(*).

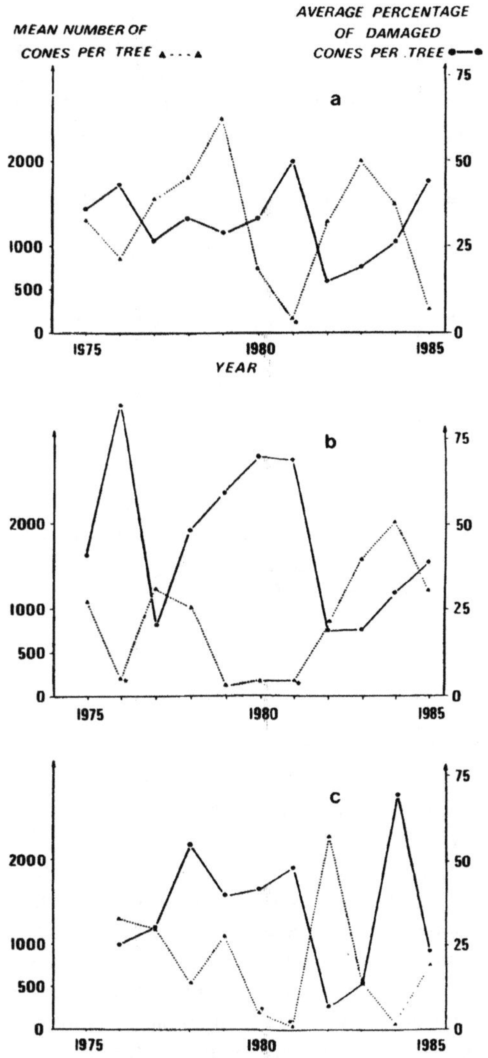

FIGURE 9. Comparative dynamics of percentage of damaged cones and cone crops in the same larch stands as in Fig. 8. Late frosts (*).

abundance rises and increases again when cone abundance stabilizes or declines. In years during which cone abundance is drastically reduced, say by late frosts, almost all the cones are infested.

5. HYPOTHESIS FOR THE CAUSE OF POPULATION FLUCTUATION

5.1. Cone Crop Variation: The Decisive Factor

The close correspondence between cone crop size and the reproductive success of *L. melania* suggests that the major factor governing the dynamics of cone fly populations is the relative availability of cones in a given year, i.e., the number of cones available per female. A similar hypothesis has been proposed for some cone insect populations in North America.[20,22,24] To test this hypothesis, I examined the relationship between the per-capita (specific) rate of change of the larval population and an index of relative cone availability (Fig. 10). This analysis indicates that the availability of cones explains 74–92% of the variation in larval population change from year to year. Thus, the rate of change of the larval population can be defined by the following equation:

$$\ln(N_t/N_{t-1}) = a + b \ln(C_t/N_{t-1}) \tag{1}$$

where N_t is the density of larvae per tree in year t, and C_t is the density of cones per tree the same year.

Equation (1) can also be expressed in logistic form:

$$N_t = N_{t-1}\, \lambda^{1-\beta\ln(N_{t-1}/C_t)} \tag{2}$$

where the finite rate of increase is $\lambda = e^a$ and $\beta = b/a$.

It is apparent that the cone fly population will be in equilibrium when its specific rate of increase is zero ($y = 0$ in Fig. 10). Hence, the cone fly carrying capacity, N^*, can be calculated directly from Eq. (1):

$$0 = a + b \ln(C_t/N^*)$$
$$N^* = \alpha C_t \tag{3}$$

where $\alpha = e^{a/b}$. In the three test stands, α varies from 0.33 to 0.49, with a mean of 0.39. This means that the cone fly population will remain static, or in equilibrium, when $1/\alpha =$ 2–3 cones available in the current year for each larva present in the previous generation. By contrast, the population will grow or decline depending on whether more or fewer cones are available per larva, respectively.

This analysis leads to the conclusion that cone fly populations are regulated by competitive interactions for a limited food supply, the cone crop. It also suggests that consistently high cone production provides little selective advantage to trees, since it allows pest populations to multiply and consume much of the reproductive output.[11] Very large crops followed by small ones would seem to be a viable strategy for minimizing seed predation. These large variations are nevertheless difficult to forecast because cone production in larch varies irregularly. European larch produces a very heavy cone crop about once every 3–10 years.[40] Apart from intrinsic feedback control, numerous exogenous factors could also be involved,[22] e.g., tree genetics, site quality, cone position in the tree crown, size of past and present crops, weather conditions during the period of primordia initiation and differentiation, adequacy of pollination, and biotic factors, particularly insects. Large cone crops are often preceded by, and are believed to cause, small crops during the following year, presumably because they preferentially mobilize and use most nutrients and photosynthates. Even during years of very low cone production, some cones are generally available, often on trees at stand edges. Cone insects are therefore largely restricted to edge trees in poor cone years but distribute themselves over the entire stand as cone production increases.

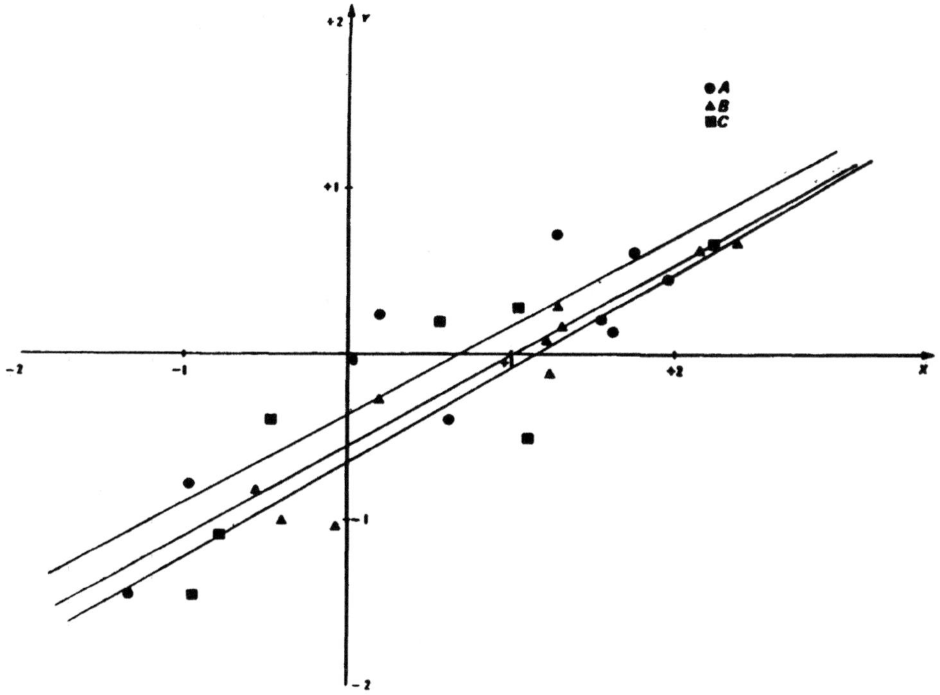

FIGURE 10. Relationship between the specific rate of change of cone fly populations, $y = \ln(N_t) - \ln(N_{t-1})$, where N_t is the density of fly larvae per tree in year t, and the relative availability of cones for ovipositing females. $x = \ln(C_t) - \ln(N_{t-1})$, where C_t is the density of cones per tree in year t. (Larch stands are the same as in Fig. 8.) Fitted curves: A, $y = 0.50x - 0.671$ ($r = 0.96$; $p < 0.001$); B, $y = 0.50x - 0.354$ ($r = 0.92$; $p < 0.001$); C, $y = 0.54x - 0.554$ ($r = 0.92$; $p < 0.001$).

5.2. Prolonged Diapause: An Important Adaptive Factor

We have seen that the percentage of larvae entering prolonged diapause is inversely correlated with the rate of change in cone yield and with the number of cones available the following year (see Section 4.5). Therefore, prolonged diapause can be considered an adaptive mechanism that enables larch cone flies to compensate for drastic fluctuations in cone production. The addition of a variable related to prolonged diapause, the percentage of nondiapausing insects in the previous generation (which equals the proportion of flies of this generation emerging during the current year), was tested in the previous model (Table III). It increases the percentage of total variance explained by the model. The gain is limited, however, and presents reduced significance because cone availability and the percentage of nondiapausing insects are highly correlated ($r = 0.884$). It does not seem necessary to introduce this latter variable in the population dynamics model. It would perhaps be more interesting to test the introduction of another variable, the total number of

TABLE III
Stepwise Multiple Regression of the Rate of Change of Larval Population $\ln(N_t/N_{t-1})$ against Different Variables[a]

Variable	Step entered	R^2	Regression statistics				$F^{(1)}$ to enter	p
			a	b_1	b_2	b_3		
x_1: Cone availability $\ln(C_t/N_{t-1})$	1	0.918	−0.67	0.60	—	—	89.16	<0.001
x_2: Annual percentage of flies emerging the current year from $(t-1)$ generation $\ln(1 - DIA_{t-1})$	2	0.964	−0.01	0.35	1.22	—	8.94	<0.05
x_3: Percentage of larval survival from parasitism	3	0.966	−0.02	0.34	1.30	−0.15	0.35	NS
x_4: $\ln(C_{t-1}/N_{t-1})$	4	0.996	−0.01	0.14	1.32	0.33	0.0	NS

[a]Values collected during 10 consecutive years in the same stand ($N = 10$).

flies terminating their prolonged diapause and so emerging in the current year, after at least 2 years in the puparium. However, this must await further experimentation.

Unfortunately, the factors that induce prolonged diapause in larch cone fly populations are not clearly understood. Rainfall and solar insolation during the larval feeding period may be involved in the onset or intensity of prolonged diapause; inverse correlations between temperature during the feeding stages and prolonged diapause have been observed for some other cone insects.[6,7,15,18] However, temperature has a marginal effect on the diapause of *L. melania*.[29] Cone production is also known to be correlated with weather variables, but the relationship is complex and involves climatic patterns over the 2 years prior to the year of production.[15] We can speculate that as a result of coevolution, coinciding weather factors have opposite effects and determine both the incidence of prolonged diapause and future cone crop size. This could explain the inverse relationship observed between these two phenomena.

The possibility that either water content or chemical composition of the cones during larval development, or both, could affect prolonged diapause has been noted for some other species.[6,7,15] Larval consumption of chemicals associated with seed (cone) production could cue the insects to the size of seed (cone) crop to be expected the following year,[27] but the incidence of prolonged diapause does not appear to be dependent on the larval density of *L. melania* in the cone during the feeding stages.

The factors involved in terminating prolonged diapause in cone insects are also poorly studied. Emerging insects have been positively correlated with the flowering intensity of the host tree.[3–5] Thus, factors that affect cone production, especially weather factors, could also be involved in terminating diapause.[6,27]

5.3. Minor Factors

The addition of "percentage of larvae surviving from parasitism and predation" only increased the multiple correlation by a marginal, and nonsignificant, amount (Table III). Thus, this factor probably has only a minor influence on the population fluctuation of *L. melania*.[48]

Competition between cone inhabitants occurs mainly during years of light cone crops, but *L. melania* appears to be a superior competitor because of its attack phenology and feeding preferences. Only *L. laricicola* seems to be a real competitor, but its effects are complicated by its positive influence on subsequent cone colonization by *L. melania*, possibly due to kairomones. By contrast, the simultaneous presence of larvae of both species accelerates cone hardening and increases the mortality of the younger one. Interspecific competition also seems to play a minor role in the population fluctuation of *L. melania*.

Weather factors, as we have seen, indirectly affect cone fly populations through their coincident effect on cone production and prolonged diapause induction and termination. In addition, late frosts may play an important role in desynchronizing fluctuations of cone crops and larch cone fly populations. Massive destruction of cones is often induced by these frosts, while insect emergence does not seem to be affected. When the initial cone production is high, most diapausing insects emerge normally and cannot compensate for cone reduction caused by the late frosts. These climatic accidents, which are particularly prevalent at high altitude (Fig. 8c), could be one cause for the lower populations observed in this zone.

Climatic conditions can have a direct effect on population fluctuations by acting as a mortality factor during development. High temperatures in the cones during larval development can induce significant mortality to some cone insects.[30,41] Although we have not evaluated these effects on *L. melania,* they are probably small because most larvae are already in their puparia during the heat of summer. Only larvae living at high altitude (above 1850 m) are still within the cones during August and sometimes have to endure abnormally high temperatures. Very low temperatures in September could also induce some mortality to the maggot population living above 2200-m altitude before it migrates toward the litter (mid-September).

5.4. Conclusion

The number of larch cone fly larvae present in the cones appears to be regulated primarily by annual fluctuations of an external factor, the cone crop. However, variations in the total number of larvae, including those diapausing from previous years, is affected by an adaptive internal factor, the ability to remain in prolonged diapause, which is somehow synchronized with cone crop fluctuations. It has been hypothesized that fluctuations in seed production evolve in response to seed predation.[19] Prolonged diapause, then, is a mechanism that cone and seed insects evolve to counter this defensive adaptation of the host.[27] This adaptation enables populations to survive at higher densities than the food supply permits, particularly when cone production is low. Nevertheless, large annual variations in cone yield effectively prevent the continuous increase of the pest population. The critical question that remains is how emergence from diapause is synchronized with the size of the current year's cone crop.

6. MANAGEMENT IMPLICATIONS

The economic significance of seed for natural regeneration and genetic improvement often requires control of cone insects in selected seed-production stands and seed orchards. These artificial stands permit the use of different management practices than those normally applied in natural forests. More intensive pest control can be employed because of higher product value per unit area and because of physical characteristics that permit the use of mechanical equipment and constant survey.[16] By contrast, orchard locations selected to produce large cone crops (e.g., sunny exposure, tree spacing) generally encourage pest population development.[35] Certain management practices (fertilizers, tree pruning) designed to create regular cone yields also lead to higher pest population levels. Under these circumstances, the development of a control strategy requires (1) a seed crop inventory, (2) the prediction and evaluation of damage caused by the pest as related to crop size, and (3) an available methodology for reducing the pest population. No accurate prognosis methods exist in most cases, and orchardists usually rely on routine preventive insecticides sprays. Annual repetition of such treatments could cause adverse side effects, such as phytoxicity, pest resistance to insecticides, and outbreaks of other pests.[2]

6.1. Damage Prognosis

Any prognosis of larch cone fly damage must consider whether the newly emerging population will cause economically significant damage to the forthcoming seed crop. Because larval feeding in the cone results in total seed consumption, expected seed losses can be correctly evaluated by estimating the percentage of damaged cones.

Long-term prognosis is practically impossible, since the main limiting factor, cone crop size, fluctuates unpredictably. However, it is possible to predict cone damage during the winter preceding cone harvest.[49] This method uses data on the relative colonization of last year's cones by larvae of *L. melania* and winter forecasts of cone harvest for the current year. The method indicates, for example, that when the percentage of damaged cones does not exceed 25% in the past year, there is no need for pest control if the expected cone crop is very large. This is only a rough qualitative prognosis of damage expectations. A more accurate prediction can be obtained from the relationship between the annual specific rate of population change and the relative availability of cones (Fig. 10). The use of this population model, however, requires numerous samples to estimate precisely both cone crop and damage on a per-tree basis. To reduce labor costs, we can approximate these values by a relative cone crop size rating at the stand level (see Fig. 3 for crop-rating system). This substitution generally leads to underestimation of real damage, a problem that can be overcome by determining the potential damage range rather than a single expected damage value (Table IV).

A reasonable prognosis of damage expectations can usually be obtained. The exceptions are essentially due to emergence of diapausing insects that are not considered by this method and to late frosts modifying expected yields. In practice, it is possible to remove this latter difficulty. Early cone-crop prediction is performed in winter by counting female spikes on 2–3-year-old shoots of 10–15 larches. This approach provides the first estimate of potential damage in relationship to past year cone colonization. A second cone crop projection must then be made after flowering in order to take into account the impact of late frosts and to furnish a final damage prognosis. Afterward, the decision of whether to implement protective measures can be taken by comparing expected financial losses and treatment costs. This method obviously needs continuous and accurate recording of real cone damage, which must be performed before larval exit toward the litter. Random samplings of 100 cones per lot, collected on 10 trees, is sufficient.

Other prognosis methods, such as the use of attractive visuals traps, are still in the experimental stage.[36,38] These methods could provide a more accurate pest prognosis, including the emergence of diapausing insects. It would not be necessary to record cone crop size and real damage continuously but only to associate the number of trapped female flies with the total population present in the plot, hence its damage potential. However, this approach is only feasible shortly before damage occurs (15 days before egg laying).[38]

6.2. Preventive and Protective Measures

Noninsecticidal control of cone insects has received little attention, even though seed orchards offer possibilities for employing simple mechanical and cultural methods. Burning the litter, hence the fly puparia, can theoretically limit larch cone fly populations.

TABLE IV

Comparison between Observed Percentages of Cones Damaged by *L. melania* in 1985 and Provisional Estimates by the Prognosis Method in Larch Stands of Various Altitudes in the French Alps

Stand	Altitude (m)	Observed damage (1984)	Relative cone crop size rating (1985/1984)	Provisional damage (1985)[a]		Observed damage (1985)	Observations
				Max.	Min.		
1	1100	28.0	3/3	41.0	32.1	40.8	
2	1300	24.1	3/3	35.2	26.1	29.2	
3	1750	18.5	3/3[b]	27.0	20.0	58.4	Late frosts in 1985
4	1900	40.0	2/1	31.0	22.5	23.1	
5	2000	72.0	3/1	21.8	15.2	20.8	
6	2100	8.3	1/1	12.2	9.5	12.0	
7	2200	42.5	3/1	12.8	9.0	18.2	Massive emergence of flies diapausing since 1983

[a]Maximal percentage of damaged cones: $PDC_{t-\max} = PDC_{t-1} \times [1.54 - 0.33 \times (CR_t/CR_{t-1})]^2$. Minimal percentage of damaged cones: $PDC_{t-\min} = PDC_{t-1} \times [1.33 - 0.29 \times (CR_t/CR_{t-1})]^2$, where PDC_{t-1} is the observed damage in the year $t-1$, and CR_t is the cone crop size rating in the year t.
[b]After late frosts.

This technique is likely to succeed only if reinfestation of the orchard from outside is negligible.[16] Orchards on sites remote from trees of the same species will be less susceptible to heavy invasions than those surrounded by naturally growing stands, but *L. melania* also seems capable of long-range migration.[48]

Other biological methods could be useful as well. For example, hibernating puparia could be destroyed by massive infusions of microbial preparations into the litter during autumn or winter.[49] In spite of the general ineffectiveness of entomophages, the release of *Trichogramma* also appears feasible for controlling seed orchard pests. Attractive traps, using visual stimuli, cone kairomones, or fly pheromones, offer still unemployed possibilities.

Insecticidal control remains the only practical method when the need for control is immediate. Contact insecticides (organochlorine preparations) have been used in aerial sprays against the larch cone fly in the U.S.S.R.[47,49] They are effective only during the flight period, as larvae are protected by cone tissues. This period coincides with the end of flowering, a time at which chemical treatments should be avoided because they often have detrimental effects on pollen germination.[2] Systemic insecticides, which are translocated to internal cone tissue from an external application, offer some advantages. The timing of application is less restricted but must be synchronized so that cone tissues become poisonous to the larvae before they are irreparably damaged (i.e., the end of the second instar). The optimum time appears to be the onset of visible scale emergence. The concentrations must be considerably higher than for ordinary use (at least 0.5%). but excessive concentrations (more than 1%) should be avoided because of toxicity to trees and other nontarget organisms.[2] Phosphamidon has been used successfully in the U.S.S.R.,[48] but some chemicals, such as Dimethoate, are likely to control larch cone flies.[2,14,26] The insecticides are applied by aerial or ground sprays. Stem injection (Dicrotophos, Oxydemetonmethyl) can also be used.[10] Special attention should finally be paid to soil-injected insecticides, which would destroy hibernating as well as long-diapausing puparia.

Insecticides are only one of the tools that can be used to control larch cone fly populations, even if they are the most practical one at the present time. Their use should be part of an integrated control strategy, based primarily on damage prognosis that avoids routine treatment. A priority for future research is to develop and test noninsecticidal methods on a large scale.

ACKNOWLEDGMENTS. Grateful acknowledgments are given to Dr. K. Van Achterberg, Leiden, Holland; Dr. J. F. Aubert, Paris, France, M. M. Martinez, Versailles, France; and M. Voegele and B. Pintureau, Antibes, France, for identifying parasite specimens.

REFERENCES

1. Ackland, D. M., 1965, Two new British species of *Anthomyiidae* (*Diptera*) with taxonomic notes on related pests of conifers, *Entomol. Mon. Mag.* **100:**136–144.
2. Annila, E., 1973, Chemical control of spruce cone insects in seed orchards. *Comm. Inst. For. Fenn.* **78:**1–25.
3. Annila, E., 1981, Fluctuations in cone and seed insect populations in Norway spruce, *Comm. Inst. For. Fenn.* **101:**1–32 (in Norwegian).

4. Annila, E., 1982, Diapause and population fluctuations in *Megastigmus specularis* Walley and *Megastigmus spermotrophus* Wachtl. (*Hymenoptera, Torymidae*), *Ann. Entomol. Fenn.* **48:**33–36.
5. Annila, E., 1984, Population fluctuation of some cone and seed insects in Norway Spruce, in: *Proceedings of the Cone and Seed Insects Working Party Conference, International Union of Forestry Research Organizations, July 31–August 6, 1983* (H. O. Yates, ed.), pp. 57–84, Southeastern Forest Experiment Station, Athens, U.S.A.
6. Bakke, A., 1983, Studies on the spruce cone insects *Laspeyresia strobilella* (L.) (*Lepidoptera, Tortricidae*), *Kaltenbachiola strobi* (Winn.) (*Diptera, Itonidae*) and their parasites (*Hymenoptera*) in Norway, *Medd. Norsk. Skogforsoksv.* **19:**1–151.
7. Bakke, A., 1971, Distribution of prolonged diapausing larvae in populations of *Laspeyresia strobilella* L. (Lep. *Tortricidae*) from spruce cones, *Nor. Entomol. Tidsskr.* **18:**89–93.
8. Elin, A. V., 1973, Dangerous pests on larch seeds and cones, *Lesn. Khoz.* **2:**71–73 (in Russian).
9. Fan, Z., Ge, F., Zhang, R. and Fang, S.,1982, On Chinese species of genus *Lasiomma* especially those injurious to the coniferous cones (*Diptera: Anthomyiidae*), *J. NE For. Inst.* **1:**1–12 (in Chinese).
10. Fogal, W. H., and Lopushanski, S. M.,1984, Stem injections of insecticides for control of white spruce seed and cone insects, in: *Proceedings of the Cone and Seed Insects Working Party Conference, International Union of Forestry Research Organizations, July 31–August 6, 1983,* (H. O. Yates, III, ed.), pp. 157–117, Southeastern Forest Experiment Station, Athens, U.S.A.
11. Forcella, F., 1980, Cone predation by pinyon cone beetle (*Conophthorus edulis: Scolytidae*): Dependence on frequency and magnitude of cone production, *Am. Nat.* **116:**594–598.
12. Golutvina, L. S., 1973, Entomophages of pests of the reproductive organs of Siberian larch in Irkustsk province, *Lesn. Khoz.* **155:**14–18 (in Russian).
13. Hedlin, A. F., 1964, A six-year plot study on Douglas-fir cone insect population fluctuations, *For. Sci.* **10:**124–128.
14. Hedlin, A. F., 1973, Spruce cone insects in British Columbia and their control, *Can. Entomol.* **105:**113–122.
15. Hedlin, A. F., Miller, G. E., and Ruth, D. A., 1982, Induction of prolonged diapause in *Barbara colfaxiana* (*Lepidoptera: Olethreutidae*): Correlations with cone crops and weather, *Can. Entomol.* **114:**465–471.
16. Hedlin, A. F., Yates, H. O., Tovar, D. C., Ebel, B. H., Koerbez,T. W., and Merkel, E. P., 1981, Cone and seed insects of North American conifers, *Can. For. Serv., U.S.D. A , Secr. Agric. Rec. Hidraul. Mexico.*
17. Hennig, W., 1967–1976, *Anthomyiidae,* in: *Die fliegen der palearktichen Region* (E. Lindner, ed.), pp. 1–974, Stuttgart.
18. Hussey, N. W., 1956, The life history of *Megastigmus spermotrophus* Wachtl. (*Hymenoptera, Chalcidoidea*) and its principal parasites, with descriptions of the developmental stages, *Trans. R. Entomol. Soc.* (Lond.) **106:**133–151.
19. Janzen, D. H., 1971, Seed predation by animals, *Annu. Rev. Ecol. Syst.* **2:**465–492.
20. Mattson, W. J., 1971, Relationship between cone crop size and cone damage by insects in red pine seed-production areas. *Can. Entomol.* **103:**617–621.
21. Mattson, W. J., 1976, Distribution of the cone insect, *Dioryctria disclusa,* in red pine, *U.S.D.A. For. Serv. Res. Pap. NC* 136, 7 pp.
22. Mattson, W. J., 1978, The role of insects in the dynamics of cone production of red pine, *Oecology* **33:**327–349.
23. Mattson, W. J., 1979, Red pine cones: Distribution within trees and methods for sampling, *Can. J. For.* **9:**257–262.
24. Mattson, W. J., 1980, Cone resources and the ecology of the red pine cone beetle, *Conophthorus resinosae* Hopkins (*Coleoptera, Scolytidae*), *Ann. Entomol. Soc. Am.* **73:**390–396.
25. Mattson, W. J., Tabashnik, B. E., and Miller, J. R., 1984, Developing a conceptual model of cone-finding behavior by the red pine cone beetle, *Conophthorus resinosae* Hopkins (*Coleoptera: Scolytidae*), in: *Proceedings of the Cone and Seed Insects Working Party Conference, International Union of Forestry Research Organizations, July 31–August 6, 1983, Athens* (H. O. Yates III, ed.), pp. 65–76, Southeastern Forest Experiment Station, Athens, U.S.A.
26. Miller, G. E., and Hedlin, A. F.,1984, Douglas-fir cone moth and cone gall midge: Relation of damage and prolonged diapause to seed cone abundance in British Columbia, in: *Proceedings of the Cone and Seed Insects Working Party Conference, International Union of Forestry Research Organizations, July 31–August 6, 1983, Athens* (H. O. Yates III, ed.), pp. 91–99, Southeastern Forest Experiment Station, Athens, U.S.A.

27. Miller, G. E., and Hutcheson, D. W., 1981, Aerial spraying for control of the spiral spruce cone borer *Hylemia anthracina (Diptera, Anthomyiidae)*. *J. Entomol. Soc. Br. Columbia* **789:**3–6.
28. Nebeker, T. E., 1974, Population dynamics of the Douglas-fir cone moth, *Barbara colfaxiana* (Kft.) (*Lepidoptera, Oleuthreutidae*), doctoral dissertation, Oregon State University, 84 pp.
29. Nesin, A. P., 1984, To the knowledge of diapause of some pests of the cones and seeds of the conifers, *Rev. Entomol. U.R.S.S.* **63**(2)**:**225–230 (in Russian).
30. Odera, J. A., 1971, The effect of solar radiations on cone insects of Eastern white pine (*Pinus strobus*) in the Fredericton area, New Brunswick, *Can. Entomol.* **103:**605–608.
31. Popova, A. A., and Elberg, K. Y. U., 1970, Species of the genus *Lasiomma* (*Diptera, Anthomyiidae*) which damage larch seeds and cones in the baykal area, *Entomol. Rev.* **64**(3)**:**555–562.
32. Rimsky-Korzakov, M. I., 1935, Parasites of the larch fly (*Chlorophila laricicola* Karl.) *Entomol. Obozr.* **25:**221–222 (in Russian).
33. Roques, A., 1977, Etude du complexe entomologique lié aux cônes de pin sylvestre en forêt de Fontainebleau, *Bull. Ecol.* **8:**415–434.
34. Roques, A., 1983, *Les insectes ravageurs des cônes et graines de conifères en France*, Institut National de la Recherche Agronomique, Versailles, 150 pp.
35. Roques, A., 1984, Etudes sur le rôle des stimulii visuels dans la reconnaissance des cônes de mélèze par les adultes de *Lasiomma* (*Diptera, Anthomyiidae*), in: *Proceedings of the Cone and Seed Insects Working Party Conference, International Union of Forestry Research Organizations, July 31–August 6, 1983* (H. O. Yates III, ed.), pp. 46–56, Southeastern Forest Experiment Station, Athens, U.S.A.
36. Roques, A., 1986, Réponses des adultes de *Lasiomma melania*, ravageur des cônes de *Larix decidua*, à des pièges colorés de différents types, *Entomol. Exp. Appl.*, **40:**177–187.
37. Roques, A., Martinez, M., and Delplanque, A., 1983, Les diptères *Anthomyiidae* du genre *Lasiomma* Stein., ravageurs des cônes et graines de Mélèze d'Europe (*Larix decidua* Mill.) en France. I. Taxonomie et répartition, *Z. Angew. Entomol.* **95:**429–439.
38. Roques, A., Raimbault, J. P., and Delplanque, A., 1984, Les Diptères *Anthomyiidae* du genre *Lasiomma* Stein. ravageurs des cônes et graines mélèze d'Europe (*Larix decidua* Mill.) en France. II. Cycles biologiques et dégâts, *Z. Angew. Entomol.* **98:**350–367.
39. Rozhkov, A. A., 1966, Pests of Siberian larch, *Akad. Nauk SSSR.* (in Russian).
40. Rudolf, P. O., 1974, *Larix* Mill. Larch, in: *Seeds of Woody Plants in the United States*, U.S.D.A. Forest Service Handbook, 450:478–485.
41. Schaefer, C. H., 1963, Factors affecting the distribution of the monterey pine cone beetle (*Conophthorus radiatae* Hopkins) in Central California, *Hilgardia* **34:**79–103.
42. Seitner, M., 1929, *Chortophila laricicola* Karl. Die Lärchenzapfen-Samenfliege und ihre Feinde, *Zentralbl. Gesamte Forstwes.* **55:**153–167.
43. Šindelář, J., 1981, Heritability of some reproduction characteristics of European larch (*Larix decidua*), *Comm. Inst. For. Čechoslov.* **12:**103–118.
44. Skrzypczyńska, M., 1973, Uszkodzenia szyszed modrzewiowych przez smietkę modrzewiowa *Hylemyia laricicola* Karl. w., *Pol. Pismo Entomol.* **2:**283–394.
45. Skrzypczyńska, M., 1984, Insects of cones and seeds of the European Larch, *Larix decidua* Mill., and the Polish larch, *L. polonica* Rac. in Poland, in: *Proceedings of the Cone and Seed Insects Working Party Conferennce, International Union of Forestry Research Organizations, July 31–August 6, 1983, Athens*, (H. O. Yates III, ed.), pp. 6–14, Southeastern Forest Experiment Station, Athens, U.S.A.
46. Stadnitskii, G. W., 1972, Characteristics of cones of Norway Spruce (*Picea abies* (L.) Karst. as a habitat and the complex of insects infesting them, *Entomol. Rev.* **50:**50–55.
47. Stadnitskii, G. W., and Grebenshchikova, V. P., 1977, The formations of an insect fauna in the reproductive organs of siberian larch in artificial plantations, *Entomol. Rev.* **56**(4)**:**28–32.
48. Stadnitskii, G. W., Grebenshchikova, P., Naumov, F. V., and Bortnik, A. M., 1974, Management and monitoring of pests of the reproductive organs of coniferous species and control of such pests in the seed sectors and plantations of the european part of the U.S.S.R. Scientific Research Institute of Forestry, Leningrad (in Russian).
49. Stadnitskii, G. W., Iurchenko, G. I., Smetanin, A. N., Grebenshchikova, V. P., and Pribylova, M. V., 1978, *Les ravageurs des cônes et graines de conifères*, Lesnaia Promyshlennost, Moscow, 168 pp. (in Russian).
50. Tripp, H. A., 1954, The instars of a Maggot (*Pegohylemyia*) inhabiting white spruce cones, *Can. Entomol.* **86**(4)**:**184–189.

51. G. C. Varley, and Gradwell, G. R., 1960, Key factors in population studies, *J. Anim. Ecol.* **29**:251–273.
52. Yamada, F., Kobayashi, K., Yamazaki, S., and Ozawa, T., 1972, Studies on the larch cone Maggot, *Hylemyia laricicola* Karl. (*Diptera. Muscidae*), *Bull. Gov. For. Exp. Sta. Tokyo* **247**:15–68.
53. Zonova, Y. E. V., 1935, The larch fly (*Chortophila laricicola* Karl.) as a larch seed pest, *Entomol. Obozr.* **25**(3–4):206–220 (in Russian).

CHAPTER 2

THE LARCH GALL MIDGE IN SEED ORCHARDS OF SOUTH SIBERIA

Alexander S. Isaev, Yuri N. Baranchikov, and Vera S. Malutina

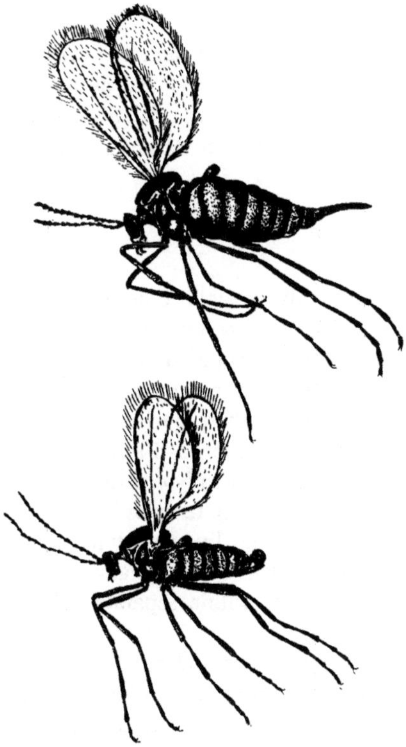

Alexander S. Isaev, Yuri N. Baranchikov and Vera S. Malutina • V. N. Sukachev Institute of Forestry and Wood, Siberian Branch, U.S.S.R. Academy of Sciences, Krasnoyarsk, 660036 U.S.S.R.

1. INTRODUCTION

The development of new practices in agriculture and forestry frequently alters the habitats of herbivorous insects, turning normally nondamaging species into economically significant pests. Seed orchard plantations in Siberia provide an example wherein cultural practices have given rise to problems of protecting larch seed from pests, in particular, the larch gall midge, *Dasyneura laricis* F. Loew (*Diptera: Cecidomyiidae*).

Damage to larch seed by the larch gall midge was first observed in Germany in 1875[13,20] and the taxonomic status of Siberian populations was determined in 1955.[18] The species is widely distributed in central Europe,[35] central and eastern Siberia (Tomsk, Irkutsk, Chita Regions, Krasnoyarsk Territory, and the Buryat ASSR), and in the Mongolian People's Republic.[15,43]

In nature, the species is found on European larch (*Larix decidua*),[19] Siberian larch (*L. sibirica*), and Daurian larch (*Larix gmelini*), but in the grafted plantations of middle Siberia, it also infests many other larches (e.g., *L. sukaczevii, L. cajanderi, L. ochotensis, L. lubarskii, L. czekanowskii, L. polonica,* and *L. laricina*).[1]

2. BIOLOGY AND BEHAVIOR

The larch gall midge is a highly specialized herbivore with a 1-year life cycle.[18,35] The flight period of adult midges is synchronized with the initiation of larch needle growth, usually in April in European populations[19] and in late May in Siberia. The duration and intensity of flight are determined by weather, with midges emerging over a 2–3-day period when temperatures rise sharply to 12°C or more. At lower average daily temperatures (6–12°), emergence may take 12–14 days, but 60% of the insects usually emerge within 1 day (Fig. 1). Midges generally emerge from galls during the first half of the day.[15] There are usually twice as many females as males. The life-span of the male midge averages 1–3 days, while females live somewhat longer. Fecundity is relatively constant, at 60–85 eggs per female. This small variation in fecundity seems to be due to the rather constant feeding conditions within separate galls and is not affected by the density of the population (Table I).

Females oviposit between the needle and bract scales on the buds of spur shoots (short lateral shoots on old growth that bear needle clusters). Larvae hatch in 6–9 days and crawl between the needles to the vegetative cone of the next year's bud, which is situated at the center of the needle cluster. Larvae enter the clusters during initiation, and the growth of the bract scales and substances secreted by the insect cause dramatic changes in bud morphogenesis, e.g., an increase in the number of foliar primordia and a 6–12-fold increase in the growth rate of scales. As new scales initiate and grow, they shift the previously formed scales to the periphery, forming an artichokelike gall at the center of the needle cluster. Galls become visible in early June, then grow for 1 or 2 months, at the end of which they are 5–12 by 4–8 mm in size with dry weights 10–15 times that of normal buds (Fig. 2). The base of the galled bud increases 2.5 times, and the scales, closing over it, form a larval chamber, in which the insect develops.

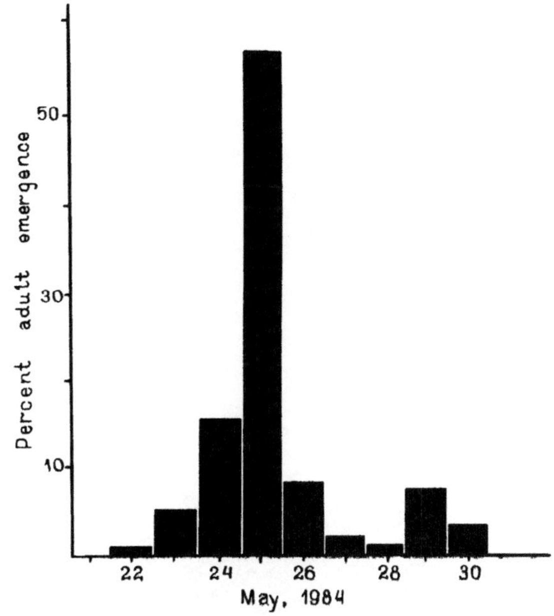

FIGURE 1. Typical distribution of adult gall midges emerging from a single Siberian larch tree. (From Baranchikov and Malutina.[3])

All larval stages feed at the bud apex by sucking food substances from plant tissues. The first instar is 0.4–0.5 mm long and grows very little. In the second instar, however, growth increases sharply, and the developing fat body gives it a bright yellow color. By late August, the larva is in the fourth instar and 3–3.2 mm long. It then leaves the expanded gall base and weaves a thick white cocoon between the external gall scales. Here it overwinters and pupates by late May of the next year. After 5–7 days, the adult midge emerges, leaving the empty pupa, which can persist intact for several days, at the top of the gall.

3. EFFECTS ON THE HOST TREE

Significant economic damage caused by the larch gall midge was first observed during the late 1960s, when larch seed orchards were founded in southern Siberia. Many

TABLE I
Fecundity of Gall Midge Females Emerging from Trees with Different Crown Infestation Densities

Number of trees in group	Buds infested (%)	Females examined	Mean fecundity (eggs/female)
5	<1	26	77.3 ± 2,2
5	10–30	25	78.3 ± 1.9
5	50–70	26	78.3 ± 2.5

FIGURE 2. Healthy buds and galls formed by the gall midge on a 2-year-old twig of Siberian larch. Also shown is a longitudinal axial section across a mature gall showing the gall base (shaded).

FIGURE 3. A heavy infestation by the gall midge on Siberian larch.

of these seed orchards were repeatedly and intensely infested, causing the gall-bearing buds to die. In persistent and heavy infestations, trees may lose up to 90% of their buds (Fig. 3). Such trees produce no flower buds, so commercial seed production is impossible. However, despite sharp declines in functional foliage, even badly infested trees do not die because new twigs are formed annually on gall-bearing branches.

Larches heavily infested by gall midges differ considerably from healthy trees. This is especially true in early spring, when the initiation of needle growth is delayed in infested trees and they also have fewer vegetative buds. The branches of midge-infested larches are panicle-shaped because gall formation leads to the shrinkage of long shoots and to an increase in branching.

The apical growth of young infested larches is greater than uninfested trees during the first years of attack. This is because the terminal shoots increase in length,[39] and the needles of galled buds are much longer than those of ordinary buds on the same tree. On heavily infested trees, the weight of needle clusters may increase by 40% and the number of needles in the cluster by 7–11%.[31] Under persistent heavy infestation, however, height increment and annual ring width decrease.[39]

Gall midges affect the metabolism of their host tree and induce a number of biochemical changes in tree tissues. Galls contain three to four times more monosaccharides and seven times more starch than do the tissues of undamaged short shoots,[40] and the concentration of free amino acids increases by 29%.[31] The needles of damaged trees accumulate both mono- and disaccharides during the feeding period of the first and second instars, but the concentration of oligosaccharides varies very little.[31]

Permanent damage to the tree is accompanied by accumulation of volatile oils in their needles and twigs. For example, the concentration of α-pinene increases, Δ^3-carene decreases, and β-phellandrene gradually accumulates in the vegetative organs and galls. As the galls age, the quantity of limonene and Δ^3-carene decreases, and the pinene fraction increases.[38]

Total phosphorus and nitrogen is lower in the bark and sapwood of infested trees, but starch increases, particularly in mid-summer.[40] Changes also occur in the quantity and quality of indolic substances in the plant tissues. Metabolites secreted by the gall midge larva cause increases in the activity and diversity of these indolic substances as well as cytokinines, but the activity of indolylacetic acid oxidase remains unchanged.[23,24,41]

4. FACTORS AFFECTING POPULATION DYNAMICS

4.1. Dispersal and Oviposition

The notion of mass migration by gall midges, often cited in the literature,[28] was not confirmed by our research. We found the distribution of female midges emerging from a point source to be highly aggregated (Fig. 4). In this experiment, a bunch of twigs containing about 3000 galls was suspended at the center of the margin of a dense larch stand. At the time of the experiment, there were no midges in this stand, but 8–10-year-old galls showed that the edge trees were susceptible to attack. Dispersal distances of females was estimated by counting the galls formed in autumn at different distances from the source. Nearly 80% of the new galls were located within 2 m of the source and no galls were more than 9 m away (Fig. 4).

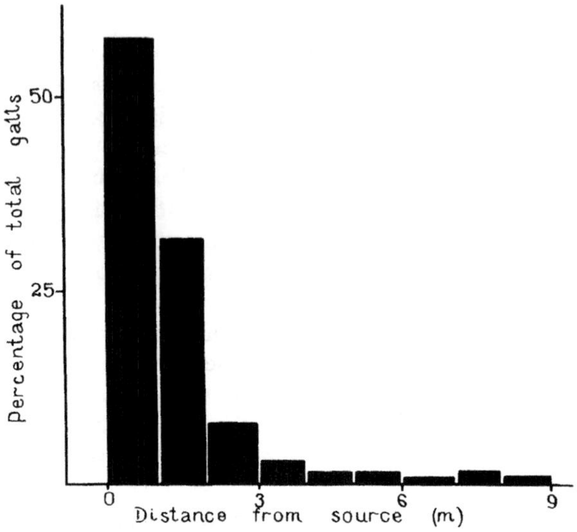

FIGURE 4. Distribution of newly formed galls after migration of adult gall midges from a point source; a bunch of galled twigs placed at the margin of a susceptible larch stand. (From Baranchikov and Malutina.[3])

Observing the behavior of this insect in the laboratory and in nature showed that males are much more active than females. The latter seldom use their wings and are mainly seen walking over the branches. Thus, the same tree is often used as a food supply for successive generations, females often ovipositing in buds adjacent to their birthplaces. Only a small proportion of the hatching females disperse and, if taken by the wind, these individuals may infest distant susceptible trees. However, because of the large number of factors influencing gall formation and growth,[14] one may surmise that successful invasion of a new host is a rather random and unpredictable process. This conclusion is supported by the distribution of eggs that can be described by a Poisson series.[29] Thus, female midges are apparently unable to recognize previously infested buds, so that the number of reinvaded buds increases with increasing density of the midge population. When 50% of the buds are infested, each contains an average of two eggs, but up to 14 eggs may be found in a single bud. Because of the limited flight capacity of females and their inability to determine previously infested buds, the distribution of midges within the forest is highly clumped.

In larch stands inhabited by this insect, 1–20% of trees remain free from galls for

TABLE II
Effect of Distance from the Source of Adult Midge Emergence on the Extent of Oviposition in Spurshoots of Resistant and Susceptible Larches

Locality of trees	Tree category	Number analyzed		Infestation (eggs/spurshoot)	p
		Twigs	Buds		
Crowns overlapping	Resistant	10	30	6.0 ± 0.6	>0.1
	Susceptible	10	30	6.4 ± 0.5	
Crowns 4 m distant	Resistant	9	105	0.1 ± 0.06	<0.001
	Susceptible	10	172	1.1 ± 0.2	

their entire lives. Our studies showed that gall midges oviposit at the same rate in the buds of both infested and gall-free trees, indicating that host selection behavior is not the determining process and that resistant trees are not repellent (Table II). Note that resistant trees 4 m distant from susceptible ones had significantly lower oviposition, again demonstrating the limited dispersal of this insect.

4.2. Tree Resistance

Although they are infested with eggs at about the same rate, resistant larch trees do not form galls, either because the larvae cannot reach the vegetative cone of the buds or, if they can, because they cannot initiate gall formation.

4.2.1. Morphological Resistance

The resistance of most gall-free trees is related to morphological characteristics of spur shoot buds during infestation.[21,22] At the time of egg hatching, 8–10 bract scales are produced on the next-year-bud of resistant trees. These densely formed domelike scales close the bud apex and prevent the newly hatched larva from reaching the vegetative cone. By contrast, susceptible larches form only 4–8 scales by this time. These do not completely cover the apex and fail to stop the larva from reaching the meristem. Thus, almost all larvae die on the closed scales of resistant trees and most succeed on susceptible ones. Gall midges, however, seem to be capable of adapting to this morphological resistance. Adult emergence can vary from generation to generation, and races can evolve in which the timing of larval hatching coincides with the optimal period of bud morphogenesis. Thus, gall midge populations may arise that are capable of reaching the bud apex before the scales close, thereby infesting trees previously free from galls. For this reason, morphological resistance is not recommended for seed orchard plantations. One may obtain temporary resistance, but midge races will presumably evolve to infest such plantations.

4.2.2. Functional Resistance

The activity of plant growth hormones is known to be higher in tissues in which the physiological–biochemical processes are most intense. Active vegetative growth is directly related to the activity of auxins and gibberelins in the growing tissues.[11,17] Thus, hormonal activity at any given moment may be estimated by measuring growth rates. It has also been shown that growth hormone activity maxima within the bud coincides with maximum needle growth.[23,24] Thus, the dynamics of growth-promoting activities in buds, in general, and in meristematic tissues of the apex, in particular, may be estimated from the dynamics of needle growth.

We measured the rate of needle growth in 30 larches (about 40 years old) with different levels of gall infestation. The dynamics of relative needle growth differed between trees, with the majority having a unimodal growth-rate curve (Fig. 5). We identified four main types of needle growth on the basis of their maximum relative growth rates, and the timing of this maximum with respect to the time that larvae reach the vegetative bud apexes:

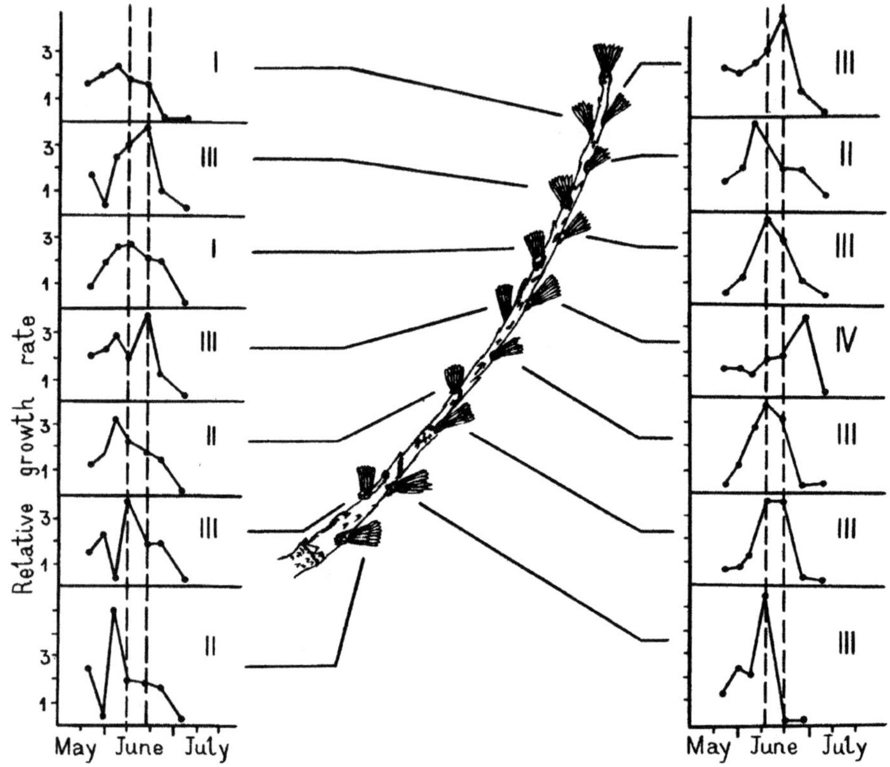

FIGURE 5. Polymorphism among buds on one twig of a Siberian larch in terms of relative needle growth rates (growth on that day/total needle length). The dashed lines indicate the period when midge larvae reach the vegetative cones.

Type I: Slow growth, in which maximum needle growth rates are always less than 3% of the final needle length

Type II: Fast, early growth, in which maximum relative growth rates exceed 3% but occur before larvae contact the bud apex

Type III: Fast, average growth, in which maximum relative growth rates coincide with the time that larvae reach the apex

Type IV: Fast, late growth, in which maximum relative growth rates occur long after the larvae reach the apex

Such variability in needle growth can be found on a single twig (Fig. 5); therefore, the general pattern of needle development on a given tree depends on the proportion of the four types of buds in the crown.

We examined the relative needle growth rates over time on trees with different infestation intensities (Table III). It turned out that the most heavily infested trees were those in which the maximum relative needle growth rates occurred at the time when larvae were contacting the bud apexes (type III buds predominate). Resistant larch trees bore only type I buds that have the slowest rates of relative needle growth.

Although bud polymorphism results in short shoots suitable for gall midge develop-

TABLE III
Relative Growth Rates (% of the Final Length of the Needles per Day) of Larch Needles during the Growing Season on Trees of Different Bud Types and Infestation Levels.

Colonization by galls (%)	Number of trees	Date of relative growth rates					Bud type prevailing[a]
		May 18	May 25	June 1	June 6	June 20	
0	2	0.5	2.9	2.0	2.4	2.2	I
1–10	3	1.2	3.8	3.4	1.5	0.8	I, II
11–30	5	0.7	2.5	3.7	2.7	1.0	II
31–50	4	0.9	3.0	2.8	3.8	0.4	II, III
51–70	8	0.7	2.4	2.5	4.1	0.8	III
>70	8	0.6	2.1	3.8	3.1	0.8	III

[a]See the description of the bud types in the text.

ment on most trees, the more resistant trees bear a high proportion of type I and II buds. A possible explanation for this lack of success on trees with slow relative growth rates or early-peaking growth is that the lower activity of growth-promoting substances required for gall formation is found in these buds.

The success of midge infestations of larch short shoots presumably depends on the levels of growth-promoting substances in the buds; this is determined by the activity of growth hormones in the meristematic tissue of growing apexes. Larvae contacting the bud apex can presumably promote the growth of vegetative tissue if hormonal activities are sufficiently high at the moment of contact. This is the critical moment of bud development, when its morphogenesis may be modified by the insect. Bud polymorphism in growth activity decreases the probability that larvae will contact apexes during this critical period. This variability impedes the development of infestations within individual trees as well as within stands and determines the degree of resistance in the host population.

4.3. Parasites and Predators

In southern Siberia, the gall midge is parasitized by two species of Hymenoptera in the superfamily Chalcidoidea: *Tetrastihus* sp. and *Torymus* sp.[8,29] These chalcids are ectoparasites on the gall midge larvae and seldom infect more than 30% of the population. Their role in the regulation of gall midge populations is probably not significant.[18]

Galls are often destroyed by birds during winter, particularly pine finches and bullfinches. These birds feed on vegetative tissue at the gall base rather than on the insect themselves, but many cocoons are inadvertently destroyed. Some years birds may destroy up to 80% of the hibernating midges.[36] Birds also peck midges from branches during the flight period, when spiders also take their toll.

4.4. Intra- and Interspecific Competition

More eggs are laid on spur shoots when midge populations are large so that more larvae reach the vegetative cone of each bud. By the autumn, however, there is usually only one larva present in each gall. Galls with two to three larvae are occasionally found but comprise no more than 1% of the total. Mortality of the extra larvae usually occurs during the second and third instars, when 10–60% die within their galls, depending on the density of bud infestation. Intraspecific competition is therefore an important factor in the regulation of high-density populations. The precise cause of larval mortality, whether cannibalism or starvation, has not been determined.

Another species of gall midge, *Dasyneura* sp., is also found on Siberian and Daurian larch.[37] Adults of this insect emerge in early June and the larvae, after reaching the bud apex of spur shoots, initiate a small gall 2–3 mm high. This species is not as numerous as *D. laricis,* but we find no larch tree to be resistant to it. On trees with buds of type III prevailing, *D. laricis* completely eliminates its competitor because of its earlier emergence and gall initiation.

4.5. Microorganisms

The epiphytic microflora associated with the larch gall midge is rather variable and depends on the properties of the habitat and the developmental phase of the insect. Some microorganisms occur only in gall tissue (e.g., *Mycobacterium phlei, Bacillus subcuticularis, Arthrobacter terregens*), some are found only on the larva (e.g., *Bacillus foliacens, A. globiformis*), while others occur everywhere, on buds, galls, needles, pupae, and larvae.

Some authorities have suggested that gall formation is induced by the symbiotic action of larval secretions and microflora.[25] In support of this possibility, epiphytic microflora have been shown to provide an additional source of indolic compounds, possible participants in gall induction.[12]

4.6. Interaction with Other Insect Herbivores

Infestations by gall midges increase the nutritional quality of larch needles for other needle-eating insects. For example, gypsy moth caterpillars fed on such needles grow twice as fast, at the same consumption and digestion rates, as those fed normal needles.[5] The higher metabolic cost of using normal larch needles seems to be due to the cost of detoxifying Δ^3-carene, a monoterpene prevailing in the volatile oil of normal larch needles.[4] Concentrations of this allelochemic are much lower in the needles of infested trees.[38]

Thus, permanent intense infestations by the gall midge can increase the nutritional quality of larch needles and result in the possible enhancement of needle-eating insect populations. We observed this situation in the south of Krasnoyarsk Territory, where local outbreaks of the Siberian larch casebearer (*Coleophora sibiricella*) were recorded in a number of permanent gall midge plots. It is interesting that heavy defoliation by casebearer caterpillars reduced gall midge infestation, by causing a delay in the onset of vegetative growth in the spring following defoliation,[33] so that gall midge larvae died on the bud apexes.

5. CAUSES OF POPULATION FLUCTUATIONS

5.1. Outbreak Zones

The distribution and population dynamics of the larch gall midge in the forests of southern Siberia are determined by temperature accumulations during the vegetative period. For larvae to complete development, a sum of 1600°C day-degrees above zero is required.[32] In the mountainous areas of Krasnoyarsk Territory, three regions can be distinguished where the population dynamics are clearly different. In the first region, at 200–500 m above sea level, the accumulated day-degrees vary between 1600 and 1900°C. Forests in this region are known for repeated outbreaks because the heat summation is always sufficient for the insect to complete its full cycle. In this region, therefore, gall midge populations fluctuate asynchronously in response to local variations in the phenology of midge-susceptible larches.

In regions with altitudes of 500–700 m and day-degree summations of 1500–1650°C, large gall midge populations are noted only in warm years. Characteristic population fluctuations arise simultaneously in all stands within this zone, and higher amplitude fluctuations occur in years in which temperature summations exceed 1600°C.

In the third region, at 700–900 m above sea level, gall midge populations are rarely found and the species has never been observed in larch forests above 900 m.

5.2. Host-Plant Imposed Fluctuations

The dynamics of the larch gall midge infestations depends largely on the relative abundance of buds of types II, III, and IV in the crown. Because the ratio of buds of different types remains relatively constant on new twigs formed each year, the response of a particular tree to similar changes in its environment remains much the same during its entire life.

Hormonal levels in larch tissue are influenced by temperature conditions,[27] as is the development of the gall midge. However, the temperature thresholds needed to initiate vital activity in larch trees do not seem to coincide with those for gall midge development. Thus, environmental changes periodically upset the stability of the midge–plant interaction by altering the resistance–susceptibility profile of the forest.

Spring temperatures cause considerable fluctuations in gall midge populations on trees with type II and IV buds. As a rule, these trees have relatively low infestation rates, usually less than 50% of the buds being galled. Unusually warm early springs, as well as prolonged cold ones, shift the flight period of the gall midge so that, in the first case, larvae reach the bud apexes during the critical period of type II buds, and in the second case, of type IV buds. For example, the critical periods for three larches with prevailing buds of type II and III are shown in Fig. 6. If the larvae reach the apexes during the period BF, all buds on larches 2 and 3 are susceptible but less than 50% of the buds of larch 1 are susceptible. However, a warm spring would shift larval hatching to the left, so that all three trees become susceptible.

Although warm spring temperatures enable the gall midge population to increase, a return to normal temperatures in the next year will cause the infestation to decrease to its normal level. By contrast, the abnormally cold spring of 1974 caused the gall midge

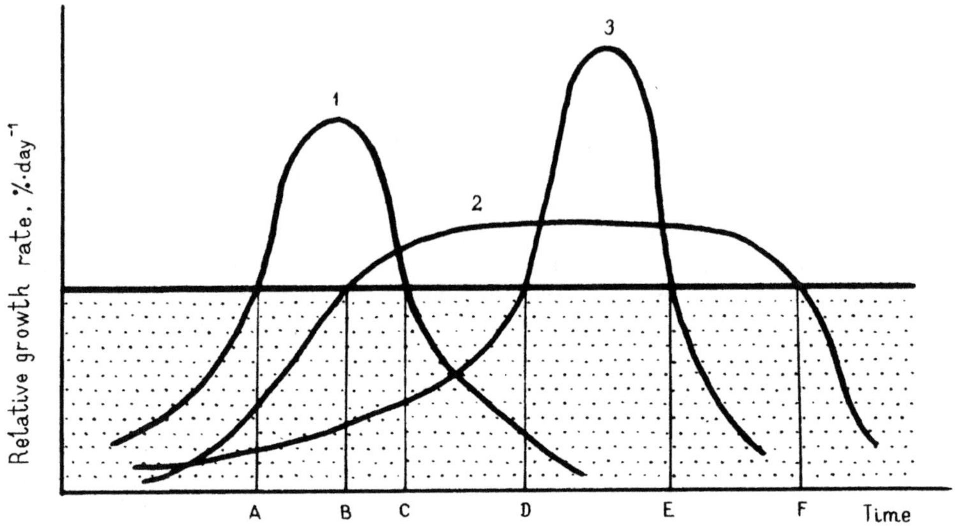

FIGURE 6. Hypothetical relative growth rates of buds on three larch trees. Tree 1 dominated by type II buds and trees 2 and 3 by type III buds, but the critical period for most buds on tree 2 is longer than that of tree 3. In the shaded area, the growth-promoting ability of the bud is too low to sustain gall development.

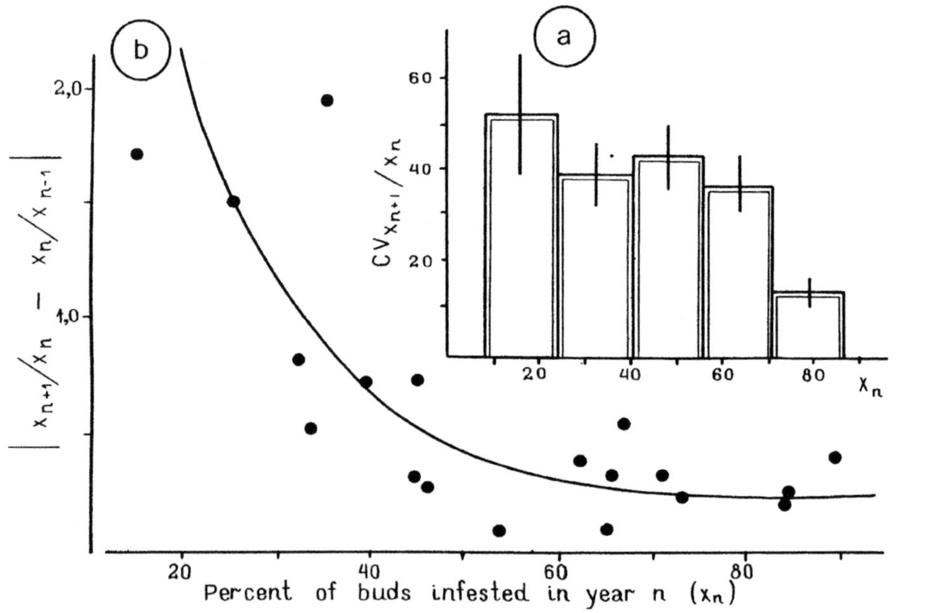

FIGURE 7. (a) Coefficient of variation of the recruitment rate (x_{n+1}/x_n) of larch gall midge demes inhabiting trees with different intensities of gall colonization (x_n) in the year n. Vertical lines indicate standard deviations of the coefficient of variation. (From Baranchikov and Malutina.[3]) (b) Differences in the absolute replacement rates of larch gall midge demes between 2 years ($n-1$ to $n+1$) on separate trees infested at different intensities in the year n.

population to disappear completely in some of our plots. Warm springs cause infestations to fall sharply on trees with type IV buds.

Larches with type III buds are almost always infested, and the intensity of infestation depends mainly on the duration of the critical period of bud morphogenesis. So on tree 3 (Fig. 6), gall formation is only possible if larvae reach the bud apexes in the period DE, while on tree 2 the critical period BF is more than twice as long. If larvae reach the apexes in the period that includes the interval DE, both trees become infested, while if they reach them in the periods BD or EF then only tree 2 can be infested. Thus, the infestation of larches with long critical periods (tree 2) are not so dependent on fluctuating spring temperatures but remain relatively constant from year to year.

These circumstances explain the variability observed in the rates of increase of gall midge populations on larches colonized at different intensities (Fig. 7a). The least variation is found in densely colonized larches in which type III buds prevail. The insect population on these trees is rigidly restricted by the annual recovery of a constant number of type III buds. Therefore, the replacement rate of the midge population on these trees is close to unity, and annual fluctuations are insignificant.

Variability in the average replacement rate becomes larger as the infestation rate decreases (Fig. 7a). In these cases, asynchrony between gall midge emergence and bud development causes changes in the number of susceptible buds each year and, consequently, the replacement rate of the gall midge population can vary considerably from year to year (Fig. 7b).

5.3. Adaptation to Individual Hosts

The ratio of buds of different types within a larch crown determines its relative susceptibility to galling, and is also the key factor affecting gall midge mortality. Thus, the major mortality factor on trees with type III buds is intraspecific competition during the second and third larval instars. On larches with type II and IV buds, however, most of the first instar larvae die on the bract scales before reaching the bud apexes or on the apexes after being unable to initiate gall development. In addition, first and second instars also die in abortive galls that cease growing for unknown reasons. Abortive galls occur much more frequently on trees with type II and IV buds.

Differences between larches also affect the ratio of the two chalcid parasites. The mortality of first and second instar gall midge larvae in abortive galls causes the death of one of the parasites (*Tetrastychus* sp.), which attacks the early instar midge larvae. The second chalcid (*Torymus* sp.) oviposits on later instars and does not suffer such high mortality rates on trees with type II and IV buds.

Because of the variability between individual host trees and the limited dispersal of female midges, we might expect gall midge demes to adapt to the individual growth characteristics of their hosts. The driving force for this adaptive process is the timing of bud development, which inflicts high mortality on larvae that are poorly synchronized with bud growth. This will tend to select for midge genotypes that reach the bud apexes synchronously with the critical period of development on that particular tree. This shift in larval development can be attained by changes in the emergence and activity thresholds of ovipositing females.

Adaptations of this type may be surprisingly rapid. For example, gall midges infest five adjacent larches of different species in a plantation at the Institute of Forest and Wood, Krasnoyarsk Territory, all of which differ in the time of bud burst and maximum needle growth rates. The initial infestation of these trees came from local gall midge populations inhabiting *L. sibirica* in the vicinity. In fewer than 17 generations, the gall midge colonized the crowns of *L. gmelinii, L. ochotensis, L. sukaczovii,* and *L. decidua,* and the dynamics of adult emergence was observed to differ between species. Adaptations to specific individual hosts have been observed in scale insects[9,10,26,42] (see Chapters 3 and 4), and the observed shifts in *D. laricis* emergence may indicate that these midges also form isolated demes adapted to individual host phenology.

5.4. Summary

The major factor affecting the populational dynamics of larch gall midges is polymorphism in larch bud development with respect to its synchrony with midge emergence and oviposition. This insect has little impact on the survival of its host plant, does not disperse far from its birthplace, and is not strongly affected by natural enemies. Thus, *D. laricis* belongs to that group of forest insects that exhibit prodromal[16] or gradient[6,7] outbreaks. Sustained gradient outbreaks occur on larch trees or stands bearing predominantly type III buds (those that attain maximum needle growth at the time midge larvae are hatching). The infestation of trees with slow needle growth (type II) or late flushing buds (type IV), however, depends on spring temperatures so that gall midge populations may fluctuate unpredictably, exhibiting pulse gradient outbreaks.

6. MANAGEMENT ALTERNATIVES

Management of gall midge populations is economically feasible only in seed orchards, where the protection of buds guarantees future seed yields. Unfortunately, most seed-tree plantations in southern Siberia are located in larch forests of the foothills and valleys, i.e., in the zone of sustained gradient outbreaks and consistently high densities of gall midges.

A number of insecticides have been used to combat the pest in these seed orchards.[30,34,41] The cryptic habits of the insect, however, reduces the efficacy of summer spray operations. During this period, only systemic insecticides can provide satisfactory control.[30] Treatments during the first week after adult emergence are generally most effective.[2]

Chemical treatment of seed orchards is expensive and has to be repeated at least once every 3 years to achieve acceptable levels of bud protection. In the long run, the most promising way to protect orchards from gall midge infestations is to place them on high-elevation sites, which are free from midges, or to maintain their genetic diversity. The apparant ability of the midge to adapt to host phenology should warn against planting monocultures of genetically similar larches.

ACKNOWLEDGMENTS. We thank all our colleagues who have worked on the larch gall midge project in Siberia, V. Uskin and S. Pusir for translating the manuscript from the Russian, and T. Baranchikova and E. Skorobogatko for drawing the figures. Our deep appreciation is extended to Alan Berryman for his invitation to participate in this volume and for his critical editorial comments on an earlier draft. We also greatly appreciate the review of the manuscript by Seri Rudolph.

REFERENCES

1. Avrov, F. D., and Nikolsky, V. I., 1984, Larch infestations by the bud gall midge on draft plantations in Siberia, Lesovedenie **4:**42–52 (in Russian).
2. Baranchikov, Y. N., and Malutina, V. S., 1986, Modification of hormonal balance as a prospective way to control larch gall midge populations, in: *Protection of Steppe Forests from Pests and Diseases* (A.I. Vorontson, ed.), pp. 98–101, Gosleshas, Volgograd (in Russian).
3. Baranchikov, Y. N., and Malutina, V. S., 1986, *Factors of Population Dynamics of the Larch Gall Midge,* Institute of Forest and Wood Press, Krasnoyarsk (in Russian).
4. Baranchikov, Y. N., Safonova, L. V., and Ryzhkova, T. S., 1986, Relation of the volatile oil components of larch needles and the nutritional and growth indices of needle-eating insects, in: *Extract Substances of Woody Plants* (G. N. Chernyaeva and V. A. Raldugin, eds.), pp. 147–149, Institute of Organic Chemistry Press, Novosibirsk (in Russian).
5. Baranchikov, Y. N., Safonova, L. V., Ryzhkova, T. S., and Matryonina, R. M., 1985, Variation of the food quality of needles on larches infested by the gall midge, in: *Optimal Habitats of Herbivores* (Y. N. Baranchikov, ed.), pp. 21–23, Institute of Forest and Wood Press, Krasnoyarsk (in Russian).
6. Berryman, A. A., 1986, *Forest Insects: Principles and Practices of Population Management,* Plenum, New York.
7. Berryman, A. A., 1987, The theory and classification of outbreaks, in: *Insect Outbreaks* (P. Barbosa and J. C. Schultz, eds.), pp. 3–24, Academic, San Diego, California.
8. Boldarue, V. O., 1969, Larch gall midge *Dasyneura laricis* F. LW. (*Diptera, Cecidomyiidae*) in the forests of Buryatia, *Trudi Buryat. Inst. Estestven. Nauk.* **7:**20–28 (in Russian).
9. Edmunds, G. F., Jr., and Alstad, D. N., 1978, Coevolution in insect herbivores and conifers, *Science* **199:**941–945.
10. Edmunds, G. F., Jr., and Alstad, D. N., 1981, Responses of black pineleaf scales to host plant variability, in: *Insect Life History Patterns* (R. F. Denno and H. Dingle, eds.), pp. 29–38, Springer-Verlag, New York.
11. Fyedorova, A. I., 1982, *Phytohormones and Tree Growth,* Nauka, Novosibirsk (in Russian).
12. Gustelyeva, L. A., and Matryonina, R. M., 1983, The role of epiphytic microflora of the larch gall midge in pathological spreading of vegetative growth, in: *Microbial Associations in Forest Biogeocenosis* (A. B. Gukasyan, ed.), pp. 18–27, Institute of Forest and Wood Press, Krasnoyarsk (in Russian).
13. Henshel, S., 1875, Die Lärchenknospen—Gallmücke (*Cecidomyia* Kellneri), *Zentrbl. Gesamte Forstw.* **1:**183–185.
14. Isaev, A. S., Nikolsky, V. I., and Matryonina, R. M., 1983, Biology and ecology of larch gall midge in the forests of Middle Siberia, *Forest Entomology* in: (O. A. Kataev, ed.), pp. 108–122, Nauka, Leningrad (in Russian).
15. Isaev, A. S., Khlebopros, R. G., Nedorezov, L. B., Knodakov, Y. P., and Kiselyev, V. V., 1984, *Population Dynamics of Forest Insects,* Nauka Novosibirsk (in Russian).
16. Isaev, A. S., Baranchikov, Y. N., and Malutina, V. S., 1985, Ecological basis of monitoring the larch gall midge in the seed orchards of Southern Siberia, in: *Monitoring System in Forest Protection* (A. S. Isaev, ed.), pp. 89–90, Institute of Forest and Wood Press, Krasnoyarsk (in Russian).
17. Ivonis, I., Yu, Shulyakovskaya, T. A., and Anisimovene, N. A., 1984, *Auxins and Gibberellins of Coniferous Trees,* Nauka, Leningrad (in Russian).
18. Kolomietz, N. G., 1955, Larch gall midge *Dasyneura laricis* F. Lw. (*Diptera, Cecidomyiidae*) in the forests of Siberia, *Zool. Zhurn.* **342:**347–350 (in Russian).
19. Lanfer, K. F., 1933, Die Lärcheenknospen-Gallmücke *Dasyneura laricis* F. Lw., *Mitt. Deutsch. Dendrol. Les.* **45:**247–257.
20. Löw, F., 1878, Mitteilungen über Gallmücken, *Verh. Zool. Bot. Gesl.* (*Wein*) **24:** 387–406.
21. Malutina, V. S., 1986, Morphological barrier of larch resistance as a mortality factor of larch gall midge, in: *Herbivorous Animals in Land Biocenosis* (B. D. Abaturov and A. A. Tishkov, Eds.), pp. 91–94, Nauka, Moscow (in Russian).
22. Malutina, V. S., and Baranchikov, Y. N., 1984, Phenology of bud morphogenesis as a factor of larch

resistance to infestation by gall midge, in: *Study of the Structural Elements of Siberian Forests* (Y. N. Baranchikov, ed.), pp. 31–35, Institute of Forest and Wood Press, Krasnoyarsk (in Russian).

23. Matryonina, R. M., 1978, Variation of auxin activity in larch damaged by Siberian gall midge, *Regularities of Distribution and Population Dynamics of Forest Insects,* in: (A. S. Isaev, Ed.), pp. 60–70, Institute of Forest and Wood Press, Krasnoyarsk (in Russian).
24. Matryonina, R. M., 1979, Indole compounds in buds of Siberian larch during infestation by gall midge, in: *Response of Conifers to the Impact of Harmful Agents* (A. S. Isaev and G. I. Girs, eds.), pp. 59–75, Institute of Forest and Wood Press, Krasnoyarsk (in Russian).
25. Matryonina, R. M., and Gustelyeva, L. A., 1984, Epiphytic microorganisms of larch gall midge as auxin producers, *Izvest. Sib. Otdel. AN ASSR Ser. Biol. nauk.* **18**(3):80–85 (in Russian).
26. McClure, M. S., 1985, Reproduction and adaptation of exotic hemlock scale (*Homoptera: Diaspididae*) on their new and native hosts, *Environ. Entomol.* **12:**1811–1815.
27. Minina, E. G., and Tretyakova, I. N., 1983, *Geotropism and Sex in Conifers,* Nauka, Novosibirsk (in Russian).
28. Nikolsky, V. I., 1980, Larch gall midge in the forests of Middle Siberia, in: *Role of Insects in Taiga Ecosystems* (A. S. Isaev, ed.), pp. 97–99, Institute of Forest and Wood Press, Krasnoyarsk (in Russian).
29. Nikolsky, V. I., 1982, Development of gall midge on larches in Siberia, *Zool. Zhurn.* **61**(1):27–35 (in Russian).
30. Nikolsky, V. I., and Kubenin, V. V., 1981, Protection of seed orchards from larch gall midge, *Lesnoe Khozyaistvo* **6:**51–53 (in Russian).
31. Nikolsky, V. I., and Matveeva, I. B., 1978, Habitat preference and population dynamics of larch gall midge in the forests of the south part of Krasnoyarsk territory, in: *Regularities of Distribution and Population Dynamics of Forest Insects* (A. S. Isaev ed.), pp. 5–19, Institute of Forest and Wood Press, Krasnoyarsk (in Russian).
32. Nikolsky, V. I., Matryonina, R. M., and Ryzkhova, T. S., 1981, Endoparasitic interrelations between the larch bud gall midge and the Siberian larch, in: *The Role of Insect–Plant Relationship in the Population Dynamics of Forest Pests* (A. S. Isaev ed.), pp. 74–76, Institute of Forest and Wood Press, Krasnoyarsk (in Russian).
33. Pleshanov, A. S., 1982, *Insects, Defoliating Larch Forests in East Siberia,* Nauka, Novosibirsk (in Russian).
34. Postner, M., 1962, Zur Lebensweise und Bekämpfung der Lärchengallmü *Dasyneura laricis* F. Lw. (*Cecidomyiidae, Diptera*), *Anz. Schädlingsk.* **35:**169–174.
35. Postner, M., 1982, *Cecidomyiidae* (= *Honididae*), Gallmücken, in: *Die Forstschädlinge Europas,* Vol. IV Hautflüger und Zweifluger (Schwenke, ed.), pp. 291–356, Paul Parly, Hamburg.
36. Prokophyev, S. M., and Nikolsky, V. I., 1981, Birds predating on larch gall midge in the forests of the Krasnoyarsk territory, in: *Protection of Forest Resources in the Middle Siberia* (E. N. Savin, ed.), pp. 143–149, Institute of Forest and Wood Press, Krasnoyarsk (in Russian).
37. Rozhkov, A. S., 1966, Diptera, in: *The Pests of Siberian Larch* (A. S. Rozhkov, ed.), pp. 219–224, Moscow (in Russian).
38. Ryzhkova, T. S., 1979, Distribution of volatile oils in different organs of Siberian larch during infestation by larch gall midge. in: *Response of Conifers to the impact of Harmful Agents* (A. S. Isaev and G. I. Girs, eds.), pp. 75–83, Institute of Forest and Wood Press, Krasnoyarsk (in Russian).
39. Verzhutckyi, B. N., 1972, Morphological changes of trees damaged by the larch gall midge, in: *Anatomical, Histological and Biochemical Changes in Larches after Pests Invasions* (A. S. Rozhkov, ed.), pp. 75–83, SIFIBR, Irkutsk (in Russian).
40. Verzhutckyi, B. N., and Dokiyichuk, T. A., 1971, Larch resistance to damage by gall midge, in: *Problems of Protecting Taiga Forests* (A. S. Isaev, ed.), pp. 26–29, Institute of Forest and Wood Press, Krasnoyarsk (in Russian).
41. Wachtendorf, W., 1958, Über den Einflus einiger Insecticide auf in Gallen lebende Insektenformen, *Fw. Zentralbl.* **77:**279–286.
42. Wainhouse, D., and Howell, R. S., 1983, Intraspecific variation in beech scale populations and in susceptibility of their host *Fagus sylvatica, Ecol. Entomol.* **8:**351–359.
43. Yanovski, V. M., 1980, Main forest pests of Mongolian People's Republic, in: *Forests of MPR* (*Economy*) (E. N. Savin, ed.), pp. 116–137, Nauka, Moscow (in Russian).

CHAPTER 3

THE ARMORED SCALES OF HEMLOCK

Mark S. McClure

Mark S. McClure • The Connecticut Agricultural Experiment Station, Valley Laboratory, Windsor, Connecticut 06095.

1. INTRODUCTION

Two armored scales, *Fiorinia externa* Ferris and *Nuculaspis* (formerly *Tsugaspidiotus*[4]) *tsugae* (Marlatt) (Homoptera: Diaspididae), native to Japan,[44,45] were accidentally introduced into the vicinity of New York City at the turn of the nineteenth century.[8,42,48] They have since spread into several northeastern states and have become destructive pests of Eastern hemlock, *Tsuga canadensis* Carriere, their primary host. In addition to Eastern hemlock, these two scales attack many other native and exotic species of conifers.[23] A survey that I conducted at two arboretums in Connecticut demonstrated that *F. externa* can mature and reproduce on *Abies* (9 species), *Cedrus* (3 species), *Picea* (12 species), *Pinus* (12 species), *Pseudotsuga* (1 species), *Taxus* (2 species), and *Tsuga* (5 species), while *N. tsugae* matured and reproduced on *Abies* (2 species), *Cedrus* (1 species), *Picea* (3 species), and *Tsuga* (5 species). The lower number of conifers exploited by *N. tsugae* probably reflects the much lower abundance of this scale in hemlock forests surrounding the two arboretums, rather than a more restricted host range.[23] Both scales were generally more abundant on *Abies* and *Picea* than on *Tsuga,*[23] which suggests that the pests may become even more important as they spread farther northward into the natural ranges of these other conifers.

Eastern hemlock is distributed widely throughout the northeastern United States and southeastern Canada, where, in association with numerous species of conifers and hardwoods, it is a major component of four forest types.[10] *Tsuga canadensis* is an important timber species which is used primarily for construction lumber and pulp. Eastern hemlock is seldom planted in monocultures because of its superior ability to regenerate even under conditions of very low light. Yet pure natural stands of *T. canadensis* occur throughout its range. In addition to its importance as a forest species, Eastern hemlock is also among the most widely used evergreens in landscape and ornamental plantings.

Both scales feed on the needles of their hosts by removing cell fluids from the mesophyll through piercing and sucking mouth parts. Although the youngest needles of the lower crown are highly preferred as feeding sites, scales can generally be found on needles of every age (hemlock retains foliage for 4 or 5 years) and throughout the crown.[23] Feeding causes foliage to discolor and drop prematurely and branches to die. Scale densities that commonly exceed 10 individuals per needle on eastern hemlock have resulted in the death of many ornamental and forest trees within 10 years. The size and age of the tree has little to do with its susceptibility to attack. Some trees have endured very high population densities of these scales for several years, although in a much weakened condition.

In Japan, *F. externa* and *N. tsugae* occur throughout the natural distributions of their primary native hosts, *Tsuga diversifolia* Masters and *Tsuga sieboldii* Carriere, on the island of Honshu.[15,37] Northern Japanese hemlock, *T. diversifolia,* is an important element of the climax evergreen forest of northern and central Honshu, where it grows naturally between 800 and 2200 m[14] (Fig. 1). *T. sieboldii,* southern Japanese hemlock, is an important component of the climax forest of the intermediate zone between the evergreen and deciduous broad-leaf forests of central and southern Honshu, where it occurs naturally at 300–1500 m[14] (Fig. 1). The natural ranges of these two hemlocks overlap in

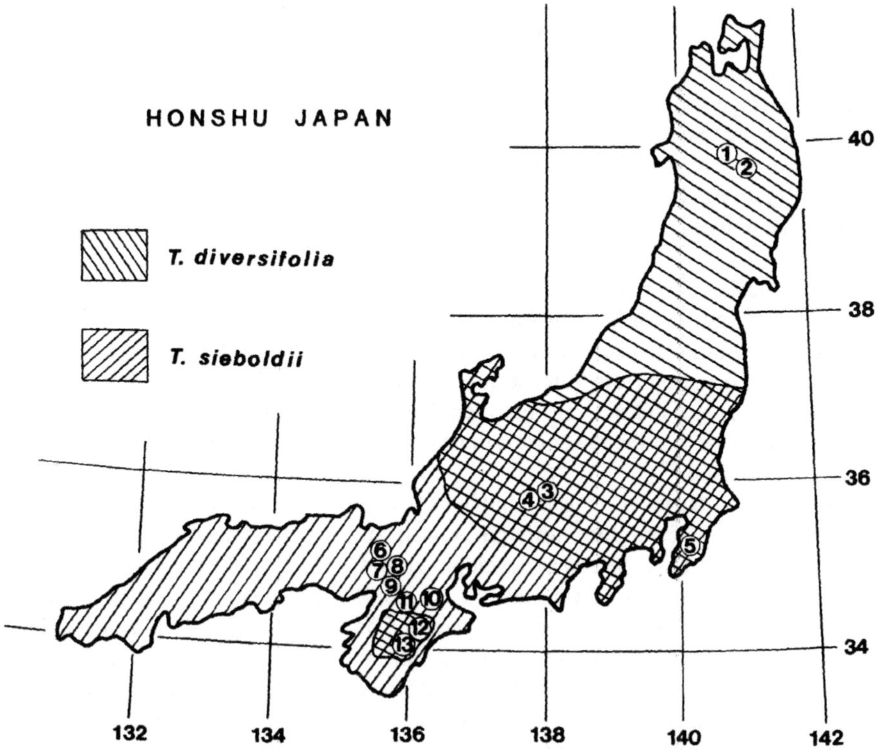

FIGURE 1. Natural distribution of *Tsuga diversifolia* and *Tsuga sieboldii* and the location of 13 sites where natural and cultivated trees were sampled for hemlock scales in Japan. (From McClure.[37])

central Honshu and at midelevations (ca. 1200 m) in the Kii mountains of southern Honshu[36] (Fig. 1). In addition both hemlock species have been planted at elevations lower than their natural ranges throughout Honshu, particularly in experimental forests, botanical gardens, and parks and around shrines and temples. In 1984, I found *F. externa* and *N. tsugae* in 14 of 15 natural and cultivated hemlock stands at 13 different locations in Honshu (Fig. 1), but scale densities were always very low ($\leqslant$10 scales per 100 needles). There was also no apparent injury from hemlock scales in any of these stands. As we will see, natural enemies were effective in maintaining populations of hemlock scales at innocuous densities in Japan.

2. BIOLOGY AND LIFE HISTORY

2.1. Life Stages and Voltinism

The stages of development and life cycles of *F. externa* and *N. tsugae* have been described in detail,[9,12,16–18,24,45–47] so only a brief outline is given here. As is characteristic of all armored scales,[2] males of both species have five postembryonic stages of development and females have three. The first and second postembryonic stages are similar in both sexes and are called first- and second-instar nymphs. The third and fourth instars in the male are nonfeeding prepupal and pupal stages, while the fifth and final stage is the winged adult. The third and final developmental stage of the female is a sessile adult. Following insemination, females of both species deposit their complement of eggs within a cover that is firmly fixed to the hemlock needle; shortly thereafter they die alongside their egg mass. The shriveled dead body of the female beneath her cover signifies that oviposition has been completed, providing a reliable method for determining fecundity.

Both species of hemlock scales were bivoltine at each of the 13 study sites in Honshu, which ranged in latitude from 34° to 40°N (Fig. 1) and in elevation from 80 to 1695 m. The summer generation was present from May through July, and the overwintering generation occurred from August through the following April. Time and duration of developmental stages did not vary between *Tsuga* species and did not vary by more than 10 days between sites at the lower elevations of southern Honshu and the higher elevations of northern Honshu.[37]

As it does in Japan, *N. tsugae* completes two generations each year in Connecticut: a summer generation from June through August and an overwintering generation from September through May. Seasonal development in Connecticut (latitude 41°N) was 7–10 days later than that in Kyoto (latitude 35°N). Unlike the situation in Japan, *F. externa* completes only a single generation each year in Connecticut; most of the few individuals that initiate a second generation in autumn die the subsequent winter.[24] The inability of *F. externa* to complete a second generation probably reflects the colder climate in Connecticut relative to its natural habitats in Japan.[15] As we shall see, the univoltine nature of *F. externa* in the northeastern United States undermines the effectiveness of its natural enemies. Voltinism of *F. externa* is less distinct in Maryland (latitude 39°N), where developmental stages overlap substantially.[5]

2.2. Foliar Nitrogen and the Survival and Fecundity of Hemlock Scales

The availability of nitrogen to feeding nymphs is an important factor in determining the suitability of hemlock and other hosts for *F. externa* and *N. tsugae*. For example, fertilization of hemlock trees with ammonium nitrate (17.5 g nitrogen) significantly

TABLE I
Effect of Fertilization on Nitrogen Concentration of Youngest Hemlock Needles and Subsequent Effects on the Survival and Fecundity of *Fiorinia externa*[a]

Treatment	Nitrogen in needles (% dry wt.)	Survival of nymphs (%)	Eggs per female (number)
Fertilized	5.6 ± 0.4	81.5 ± 4.6	13.3 ± 2.1
Unfertilized	4.3 ± 0.4	68.5 ± 7.9	9.3 ± 1.9

[a]Differences in fitness parameters between fertilized and unfertilized trees were all significant at $p < 0.005$ by ANOVA.

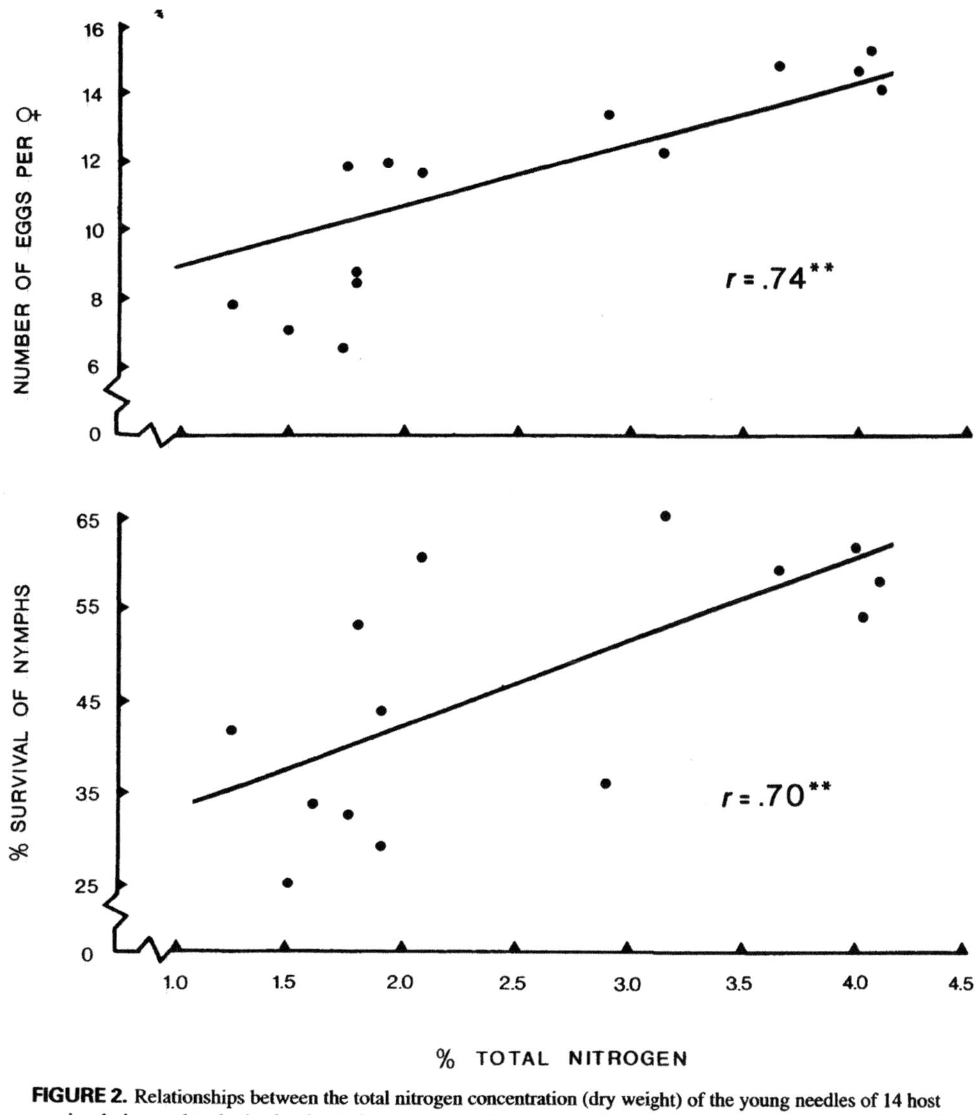

FIGURE 2. Relationships between the total nitrogen concentration (dry weight) of the young needles of 14 host species during peak colonization by scale nymphs and the survival of nymphs to maturity and the fecundity of adult females. By linear regression analysis. (**$p<0.01$.) (From McClure.[29])

enhanced the success of scales residing on those trees.[29] Nymphs incurred 16% less mortality, and adults produced 30% more offspring on hemlocks whose foliar nitrogen concentration had been elevated by only 1% (Table I). Similarly, survival and fecundity of hemlock scales were strongly positively associated with concentrations of nitrogen in the young needles of individual forest hemlocks[29] and in the foliage of 14 other coniferous host species from five genera at an arboretum (Fig. 2). On individual hemlocks and on host species whose needles contained higher amounts of nitrogen, nymphs incurred up to 40% less mortality, and adults produced twice as many eggs than did scales on trees having the lower foliar nitrogen concentrations.[29] Clearly, nitrogen has an important role in the relative suitability of trees as hosts for hemlock scales.

2.3. Host Phenology and Scale Colonization

In each of 10 hemlock forests sampled in Connecticut, foliar nitrogen and water were highest immediately following needle flush in May and then quickly became less concen-

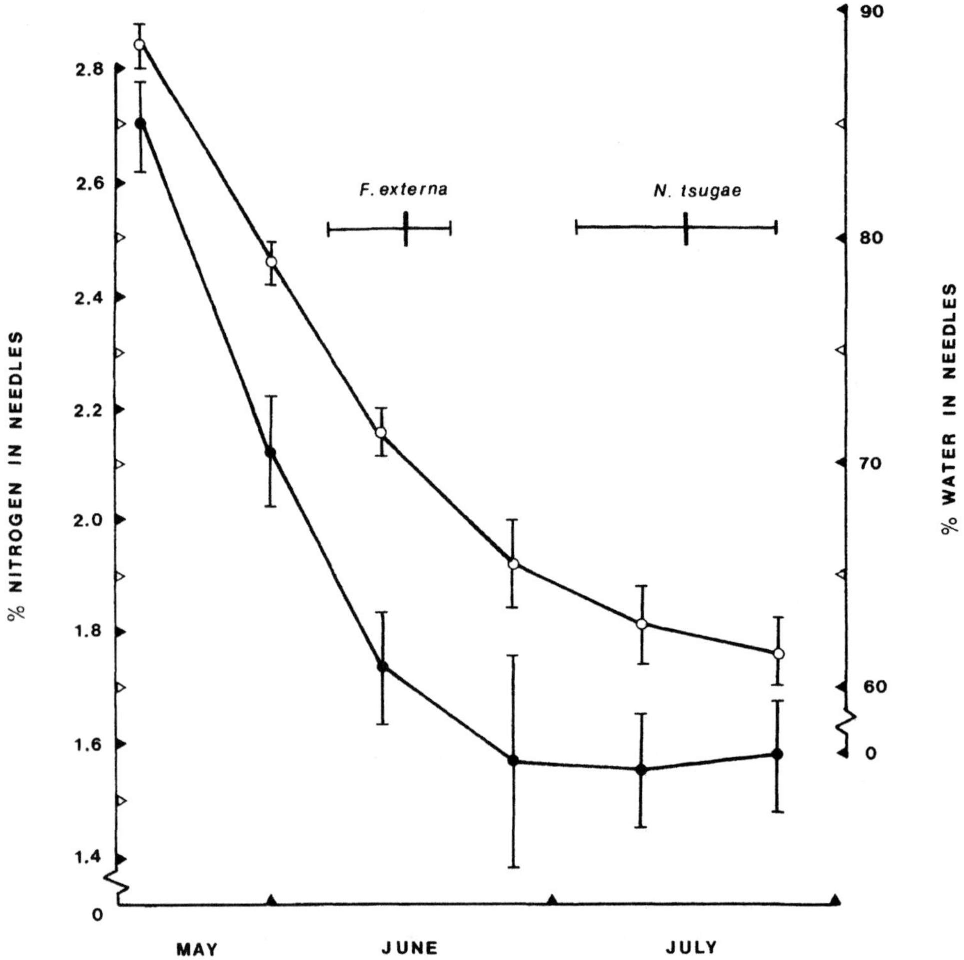

FIGURE 3. Mean (±SD) percentage nitrogen (●) and water content (○) in hemlock needles and colonization times of *Fiorinia externa* and *Nuculaspis tsugae* in 10 cohabited hemlock forests of southwestern Connecticut. (From McClure.[30])

trated during June and July[29] (Fig. 3) as in many other species.[19] The opposite trend existed for foliar terpenoids, a group of defensive chemicals, which declined as the season progressed.[35] Thus, it is advantageous for herbivores to commence feeding as soon as the first flush begins.

Throughout Japan and the United States, first-instar nymphs of *F. externa* hatch and colonize young hemlock needles 2–4 weeks earlier than those of *N. tsugae* (Fig. 3). During this period, the concentration of nitrogen and water in young foliage decreased 20%, and concentrations of volatile leaf oils doubled. Because the survival, development, and fecundity of hemlock scales are strongly associated with concentrations of nitrogen, water, and terpenes in host foliage, *F. externa*, by virtue of its earlier colonization time, has an obvious nutritional advantage over its competitor, *N. tsugae*. This plays an important role in the outcome of the interaction between these scales.

3. FACTORS AFFECTING SCALE ABUNDANCE

3.1. Density-Dependent Feedback

There is substantial evidence that dense hemlock scale populations significantly limit the success of individuals of subsequent generations. On heavily infested hemlocks, nymphal mortality was four times greater, several more days were required to complete development, and up to 30% fewer eggs were produced per female than on sparsely infested trees.[26,27] Further indication that density adversely affects hemlock scale reproductive rates is seen when scale populations rapidly resurge following pesticide spraying.[20] Scales that survive on sprayed trees have significantly higher fecundity than do those on controls, probably because of improved host quality following reduced herbivore pressure[20] (Table II).

Studies conducted in Connecticut forests showed that scale feeding significantly reduced the quantity and quality of food available to nymphs later that season as well as during the following year.[26,29] Concentrations of foliar nitrogen in infested 14-year-old hemlocks in the field plot were reduced by an average of 18% relative to uninfested trees after 7 weeks of feeding.[29] In a mature forest, the biomass of the youngest needles and

TABLE II
Density of Hemlock Scales before and after Application of Dimethoate Insecticide, and Fecundity and Parasitism of Scales That Survived

	No. of scales per 100 needles			Survival scales	
	1 week before treatment	3 weeks after treatment	17 weeks after treatment	No. of eggs per female	Mortality from parasitoids (%)
Insecticide sprayed	288.0 ± 47.2	623.7 ± 28.8	1285.7 ± 65.0	14.5 ± 0.6	0.6 ± 0.2
Unsprayed control	284.4 ± 51.0	1056.6 ± 65.7	1325.2 ± 70.2	12.8 ± 0.4	35.1 ± 2.4
p	NS	<0.01	NS	<0.025	<0.005

their nitrogen concentration were significantly negatively correlated with scale densities on the youngest needles during the previous year (Fig. 4). Scales that colonized previously lightly infested trees whose needles were high in nitrogen suffered significantly less mortality and were significantly more fecund than were those that colonized previously heavily infested trees whose needles were low in nitrogen.[29]

3.2. Interspecific Competition and Competitive Exclusion

The success of *F. externa* and *N. tsugae* on their numerous coniferous hosts is clearly related to the quality and availability of food, which are functions of host phenology and herbivory. Experiments conducted in the greenhouse and in hemlock forests in Connecticut established that these two scales compete for food and space and that *F. externa* is the superior competitor.[29,33] Comparison of mortality data from solitary and coexisting populations demonstrated that *F. externa* had a greater adverse effect on the survival of its competitor than *N. tsugae* had on itself, while *N. tsugae* had a less significant effect on *F.*

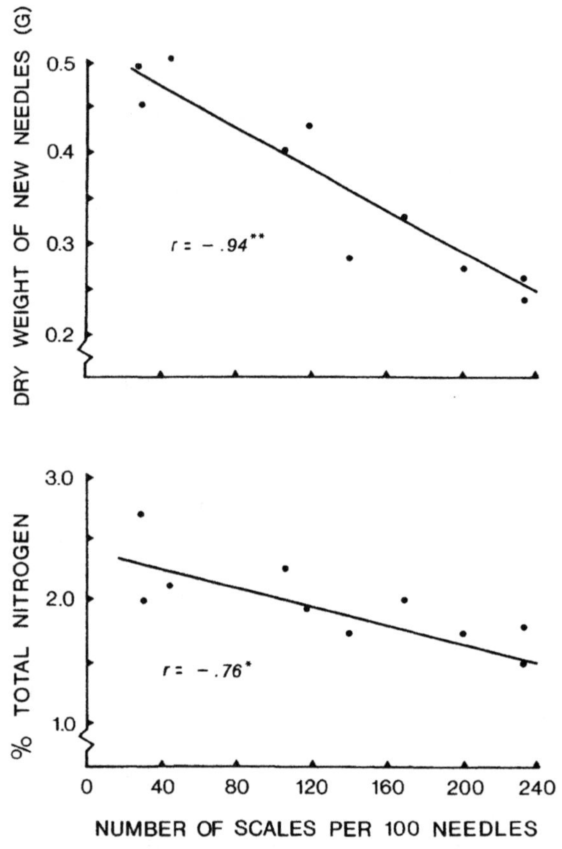

FIGURE 4. Relationships between the density of scales on 10 mature forest hemlocks and the mean biomass and nitrogen concentration of the young needles of those same trees during the following year. By linear regression analysis. (*$p<0.05$; **$p<0.01$.) (From McClure.[29])

TABLE III
Effects of Cohabitation on Selection of Colonization Sites by Hemlock Scales and on Subsequent Survivorship

Species	Infestation	Colonization (%) Youngest needles	Colonization (%) 1-year-old needles	Mortality (%) Youngest needles	Mortality (%) 1-year-old needles
Fiorinia externa	Pure	90.5 ± 15.6	9.5 ± 7.0	16.2 ± 2.2	18.3 ± 1.2
	Mixed	74.5 ± 19.7	25.5 ± 8.7	10.6 ± 3.0	13.3 ± 4.3
Nuculaspis tsugae	Pure	74.2 ± 13.0	25.8 ± 6.8	23.8 ± 3.3	30.6 ± 3.8
	Mixed	42.0 ± 7.0	58.0 ± 9.1	37.2 ± 4.1	46.4 ± 3.4

externa survival than *F. externa* had on itself.[29] Therefore, intraspecific competition has a greater impact on populations of *F. externa,* whereas interspecific competition is more significant in *N. tsugae* populations. This superior competitive ability of *F. externa* results from the nutritional advantage gained by early host colonization when foliar nitrogen and water concentrations are highest, i.e., 2–4 weeks earlier than its competitor. Early feeding by *F. externa* not only reduced the amount of foliar nitrogen by the time that *N. tsugae* colonized the needles but also forced *N. tsugae* to colonize the less nutritious older growth where scale success was significantly reduced[30] (Table III).

Because of its competitive superiority, *F. externa* was quickly able to exclude *N. tsugae* from mixed species infestations in a field plot and in 20 forests in southwestern Connecticut (Fig. 5). In the 12 forests in which *F. externa* outnumbered its competitor in the initial census, *N. tsugae* was eliminated after only 3 years (Fig. 5a). However, even in the eight forests in which densities of *N. tsugae* were initially sevenfold higher than those of *F. externa, N. tsugae* was eliminated after only 4 years[30] (Fig. 5b). Similar results were obtained in the field plot study, wherein the relative abundance of *N. tsugae* on cohabited trees was reduced from 66.9% of the total scales present to less than 1% in 4 years.[30] The host-finding behavior of a parasitoid common to both scales also contributed to the rapid decline of *N. tsugae* populations (see Section 3.4.2).

3.3. Interactions with Natural Enemies

By far the most important natural enemy of *F. externa* and *N. tsugae* in Japan is the aphelinid parasitoid, *Aspidiotiphagus citrinus* Craw, which consistently killed 91–94% of both scales at each of the natural and cultivated hemlock stands I had sampled in Japan (Table IV). This species comprised 88.4% of the total number of parasitoids collected (N=1953); the remainder consisted of the encyrtid, *Arrhenophagus albitibiae* Gerault.

The impact of these parasitoids on the growth of endemic hemlock scale populations in Japan was determined from life-table data gathered in 15 hemlock stands throughout Honshu. Life tables were constructed for summer and autumn generations following standard methods.[13,32,38,39] The potential growth of scale populations from generation to

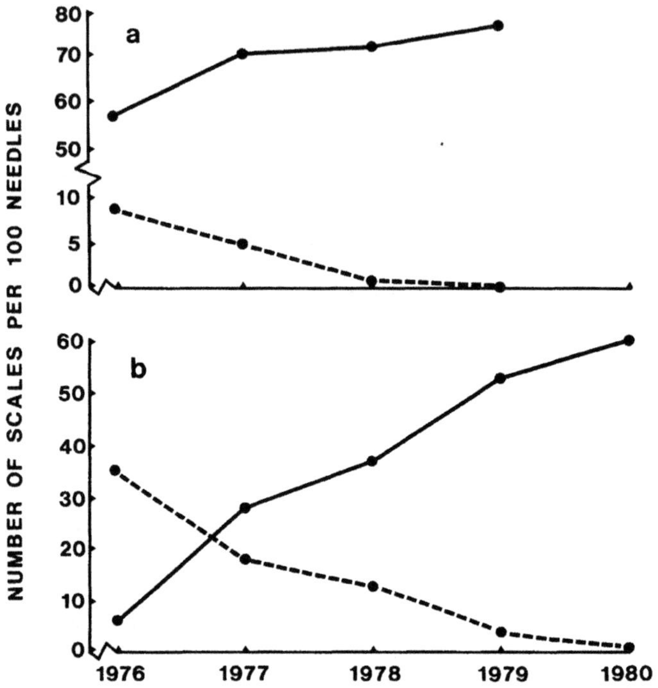

FIGURE 5. Population densities of *Fiorinia externa* (——) and *Nuculaspis tsugae* (- - -) in 12 cohabited hemlock forests in Connecticut in which *F. externa* was initially predominant (a) and in 8 forests in which *N. tsugae* was initially predominant (b).

generation was then estimated by calculating the trend index[39] (see Table IV). The effect of parasitoids on the population trend was estimated by comparing the actual values (incorporating all life-table data) with hypothetical values that incorporated all data except mortality caused by parasitoids (Table IV).

When parasitism was included in the trend calculations, *F. externa* exhibited a relatively stable population ($I \approx 1$) on cultivated *T. diversifolia* and a declining growth rate ($I < 1$) on cultivated *T. sieboldii* and on natural *T. sieboldii* (Table IV). Populations of *N. tsugae* declined sharply in natural stands of both hemlock species when parasitism was considered, but in cultivated stands populations showed an increasing trend ($I > 1$). By contrast, when population trend values were calculated without mortality due to parasitoids, both scales showed a high potential rate of increase on both natural and cultivated trees[37] (Table IV).

An experiment conducted in Japan confirmed the prediction that hemlock scale populations have the potential to increase rapidly in the absence of parasitoids. Scale population growth was monitored on individual hemlocks enclosed in nylon mesh cages to protect the scales from parasitoids and on other uncaged trees of the same hosts (Table V). This experiment demonstrated that populations were relatively stable in the presence of *A. citrinus*, but increased 8–25-fold when protected from the parasitoid. Clearly parasitoids,

TABLE IV
Percentage of Parasitism of Hemlock Scales in Japan and Population Trend calculated with (+P) and without (−P) Mortality Due to Parasitoids[a]

			Fiorinia externa			*Nuculaspis tsugae*		
			Parasitism	Population trend[b]		Parasitism	Population trend[b]	
Tsuga sp.	Habitat	No. of stands	(%)	$+P$	$-P$	(%)	$+P$	$-P$
T. diversifolia	Natural	5	—	—	—	93.7 ± 1.4	0.4 ± 0.2	6.7 ± 1.1
	Cultivated	1	91.2 ± 3.6	1.1 ± 0.1	12.8 ± 1.4	94.2 ± 4.4	1.5 ± 0.1	25.2 ± 1.3
T. sieboldii	Natural	3	91.1 ± 2.2	0.1 ± 0.0	1.4 ± 0.1	92.3 ± 1.9	0.7 ± 0.3	8.5 ± 1.3
	Cultivated	6	93.6 ± 1.2	0.6 ± 0.2	7.9 ± 0.5	91.4 ± 1.1	1.7 ± 0.6	20.3 ± 0.8

[a]From McClure.[37]
[b]Population trend index $I = N_t/N_{t-1}$.

TABLE V
Effects of Parasitism on Population Growth of *Fiorinia externa* and *Nuculaspis tsugae* on *Tsuga* Species in Japan[a]

		Fiorinia externa		*Nuculaspis tsugae*	
Tsuga sp.	Parasitoids	Parasitism (%)	Population trend[b]	Parasitism (%)	Population trend[b]
T. diversifolia	Present	91.2 ± 3.6	1.1	94.2 ± 4.2	1.2
	Excluded	0.0 ± 0.0	13.3	0.0 ± 0.0	25.2
T. sieboldii	Present	89.8 ± 2.6	1.0	93.4 ± 5.0	1.1
	Excluded	0.0 ± 0.0	8.9	0.0 ± 0.0	21.0

[a]From McClure.[37]
[b]Population trend index: $I = N_t/N_{t-1}$.

in particular *A. citrinus,* play an important role in the regulation of scale populations in Japan.

Several species of predators and parasitoids attack populations of *F. externa* and *N. tsugae* in the northeastern United States. Two mirid bugs, *Atractotomus magnicornis* nr. *buenoi* Knight and *Phytocorus* sp., the conipoterygid, *Conwentzia pineticola* Enderlein, the coccinellid, *Chilocorus stigma* (Say), and the aphelinid, *Aphytis* nr. *aonidiae* (Mercet), are common inhabitants of scale-infested hemlocks in Connecticut,[28] but they exerted no significant regulatory pressure on scale populations.[32] The most important natural enemy of exotic hemlock scales is *A. citrinus,* the same aphelinid parasitoid that has a major effect on scale populations in Japan. Studies in Connecticut have shown that *A. citrinus* responds to the density of its hosts both within and among hemlock trees[21] and that it can kill up to 96% of the scales during some generations.[24,32] However, for reasons to be discussed, parasitism is inconsistent from generation to generation and introduced scale populations remain unregulated.

Despite the inability of *A. citrinus* to consistently kill a high proportion of scales in Connecticut this parasitoid does retard the buildup of scale populations. For example, hemlock scales resurge rapidly in infested hemlock forests that are sprayed with pesticides (see Table II). This is partly due to reduced intraspecific competition, but the virtual elimination of natural enemies by the pesticide spray contributed as well.[20]

4. HYPOTHESES FOR THE CAUSES OF OBSERVED POPULATION PATTERNS

4.1. Host Vigor, Stress, and Scale Fitness

A substantial body of literature supports the idea that population outbreaks of herbivorous insects are stimulated by stress-induced increases in food quality (e.g., nitrogen, water, terpenes).[1,19,29,43,49,50] My studies in Connecticut have shown that hemlock scales respond to changes in food quality associated with host phenology,[29] herbivory,[26] and host stress from edaphic conditions.[22] Studies in Japan also support the hypothesis that hemlocks under stress are more suitable for *F. externa* and *N. tsugae* than are healthy trees. The fitness of these scales varied significantly within and beyond the natural distributions of *T. diversifolia* and *T. sieboldii* and along an elevational gradient. Density, survivorship, and fecundity of both scales were significantly higher on cultivated hemlocks growing outside the natural range of *Tsuga* than on natural montane trees (Table VI). Within the natural range of the hosts, scales were significantly more successful on hemlocks growing at the edges of the natural distribution than in the center.[36]

Additional evidence supporting the stress hypothesis is seen in the fecundity of hemlock scales on native and exotic *Tsuga* species in Japan and the United States (Table VII). In Connecticut (latitude 41°N), both scales were significantly more fecund on exotic *T. sieboldii*, whose natural latitudinal range in Japan is approximately 35–37°N,[14] than on native *T. canadensis*, whose natural range is approximately 35–47°N.[10,34] In Kyoto, Japan (latitude 35°N), the opposite pattern of scale fecundity was observed. Therefore, in both countries the scales were more successful on the more latitudinally displaced *Tsuga* species, which was presumably under greater climatic stress.

4.2. Importance of Parasitoid–Host Synchrony

Parasitism by *A. citrinus*, and to a much lesser extent by *A. albitibiae*, exceeded 90% at each of the natural and cultivated sites in Japan and was the major factor regulating populations of *F. externa* and *N. tsugae*. In the northeastern United States, however, introduced populations of hemlock scales are unregulated and injure and kill their new host, *T. canadensis*, even though *A. citrinus* has been established for several decades. I hypothesize that these differences are due to a lack of synchrony in the life cycles of parasitoid and hosts in the United States.

Throughout Japan and the northeastern United States, *A. citrinus* is bivoltine, completing two generations annually on both hemlock scales.[24,37] Peak emergence of adult parasitoids occurs in June and August in both countries and coincides with the peak abundance of susceptible first-instar nymphs of the bivoltine *N. tsugae* (Fig. 6). Similarly, peak emergence of adult parasitoids coincides with the peak of susceptible second-instar

TABLE VI
Fitness Parameters of Hemlock Scales on Natural and Cultivated *Tsuga* species in Japan[a,b]

		Fiorinia externa			*Nuculaspis tsugae*		
Tsuga sp.	Habitat	Scales per 1000 needles	Survival (%)	Fecundity	Scales per 1000 needles	Survival (%)	Fecundity
T. diversifolia	Natural	0.0 ± 0.0	—	—	5.7 ± 0.3	33.6 ± 2.3	40.5 ± 0.9
	Cultivated	38.6 ± 2.0	86.8 ± 2.8	29.4 ± 2.0	21.8 ± 3.6	74.9 ± 3.2	66.8 ± 2.6
T. sieboldii	Natural	1.5 ± 1.0	24.2 ± 1.6	11.4 ± 0.6	5.5 ± 1.0	39.1 ± 2.4	43.7 ± 0.4
	Cultivated	21.9 ± 1.0	78.6 ± 1.7	20.1 ± 0.4	13.5 ± 0.8	69.5 ± 1.1	48.3 ± 1.2

[a]From McClure.[37]
[b]Differences in fitness parameters between natural and cultivated habitats were all significant at $p < 0.0005$ by ANOVA.

TABLE VII
Number of Eggs Produced by Hemlock Scales on Cultivated *Tsuga* Species in the United States and Japan

Scale species	*Tsuga* sp.	Connecticut, United States	Kyoto, Japan
Fiorinia externa	*T. canadensis*	16.9 ± 0.7	25.7 ± 1.6
	T. sieboldii	29.8 ± 1.1	22.8 ± 0.8
p		<0.0005	<0.05
Nuclaspis tsugae	*T. canadensis*	49.8 ± 1.0	65.8 ± 2.7
	T. sieboldii	73.0 ± 1.0	56.5 ± 1.0
p		<0.0005	<0.005

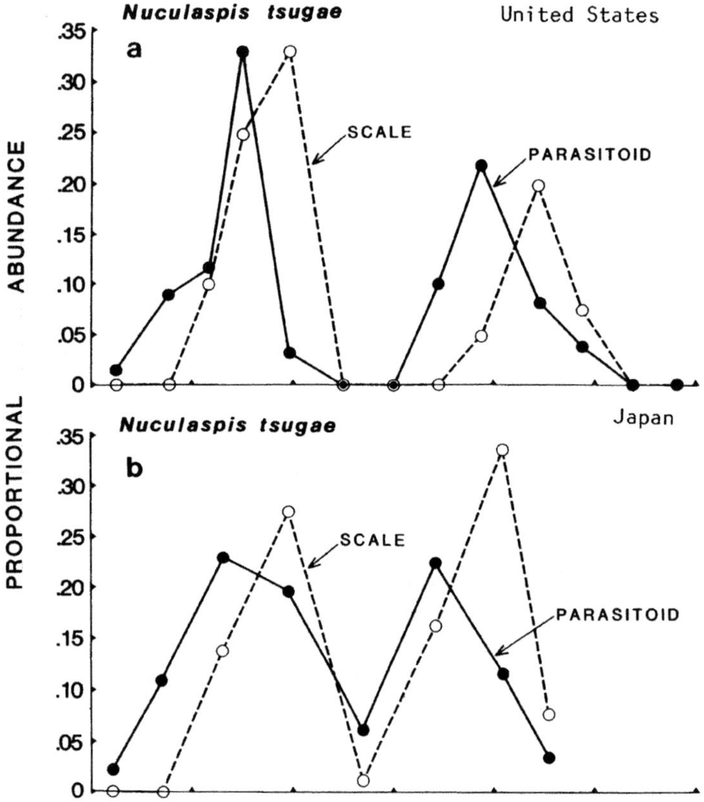

FIGURE 6. Seasonal abundance of adult parasitoids and susceptible host stages of *Nuculaspis tsugae* and *Fiorinia externa* on hemlock in the United States (a) and in Japan (b). (From McClure.[37])

nymphs of *F. externa* in Japan (Fig. 6b). In America, however, peak emergence of the bivoltine *A. citrinus* occurs nearly 8 weeks after peak abundance of the univoltine *F. externa* (Fig. 6a). Thus, because *F. externa* is unable to complete a second generation in the northeastern United States, probably because of the colder climate,[15,37] most of the adult parasitoids emerging from summer generations fail to find suitable hosts, and few adult parasitoids are present the following spring to exploit the abundant hosts present at the time.[24] This interaction between *A. citrinus* and its hemlock scale hosts is much more complex in forests cohabited by *F. externa* and *N. tsugae* because *A. citrinus* is able to shift from one host to the other when *F. externa* becomes rare in autumn[32] (Fig. 7). Although *N. tsugae* mortality in mixed populations was due to interspecific competition,[32] at least 70% was due to the shift in parasitism from *F. externa* to *N. tsugae* (see the shaded portion of Fig. 7). This high rate of parasitism offset the numerical advantage of the two generations of *N. tsugae* and accelerated its exclusion from cohabited forests.[32]

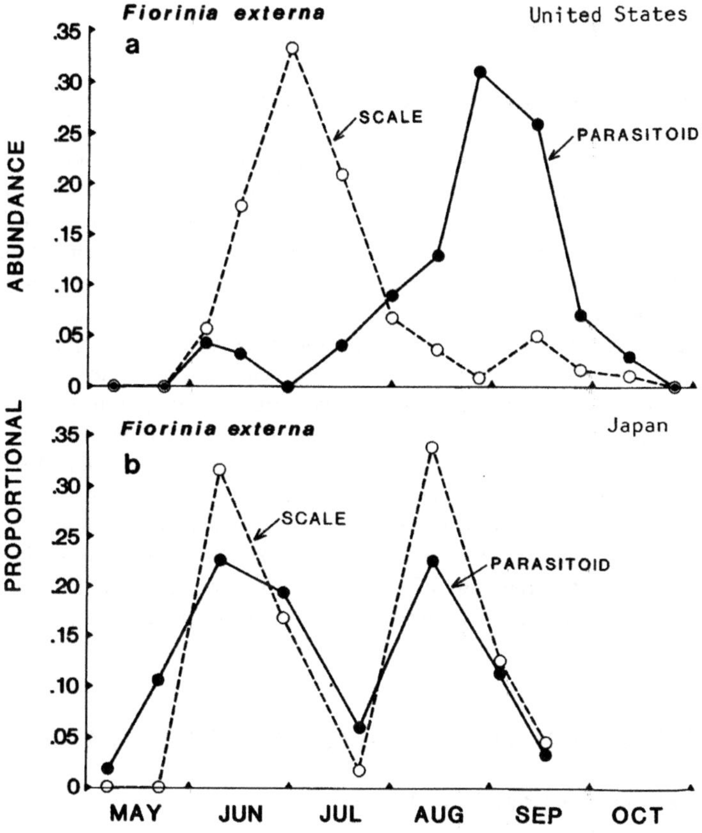

FIGURE 6. *(continued)*

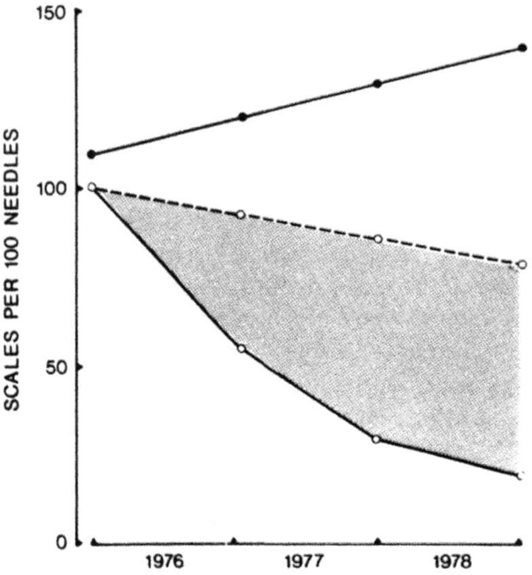

FIGURE 7. Population densities of *Fiorinia externa* (●) and *Nuculaspis tsugae* (○) during a 3-year period in a cohabited hemlock forest in Connecticut. (- - -) Hypothetical population trend for *N. tsugae* when parasitism by parasitoids emerging from *F. externa* was removed. (Shaded region) Reduction in density resulting from the parasitoid–host shift.

4.3. Interactive Structure of the Scale–Hemlock System

The armored scales on hemlock are a highly interactive community that can be illustrated nicely by a network of feedback loops (Fig. 8). Extraneous factors that can affect the quality (e.g., nitrogen, water, terpenes) and quantity (biomass) of hemlock foliage include silvicultural practices (e.g., cultivation, fertilization) and climatic stress. Interactions between herbivore and host produce delayed negative feedback (loops −ab and −cd) with hemlock inducing cycles of abundance because of the lag in host response. A stabilizing influence on the system is the rapid negative feedback (loops −*e* and −*f*) attributable to intraspecific competition among scales of the same generation. Interspecific competition, however, is a disruptive influence that produces a positive feedback (loop +gh) leading to the competitive exclusion of *N. tsugae,* because −g > −h. Interactions between the scales and their principal parasitoid (loops −ij and −lk) produce delayed negative feedback due to density-dependent numerical responses of *A. citrinus.* This creates cycles of abundance if the delay in the parasitoid response is long. In the United States, the interaction of −j is broken or weakened because parasitoids die before they can attack the next generation of *F. externa.* However, the interaction (+i) (−l) = (−il) causes an additional negative impact of *F. externa* on *N. tsugae,* which accentuates exclusion.

5. IMPLICATIONS FOR DEVELOPING MANAGEMENT STRATEGIES

5.1. Maintenance of Host Plant Vigor

Studies on the population of dynamics of *F. externa* and *N. tsugae* in Japan show that scales are maintained at low densities (≪10 scales per 100 needles) and seldom if ever cause noticeable injury to hemlock. However, when parasitism was factored out of life-

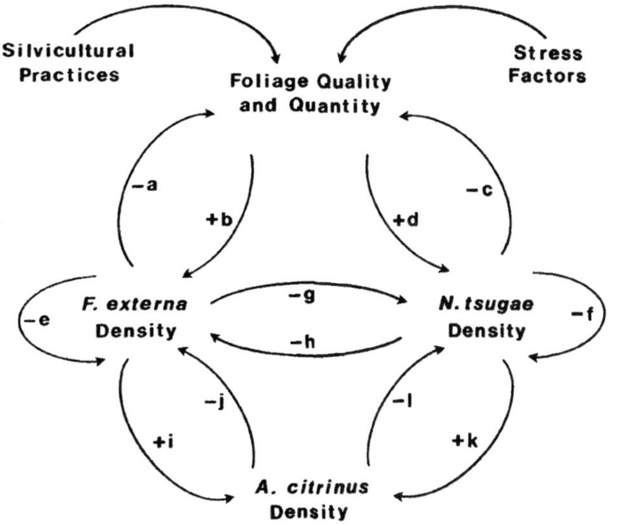

FIGURE 8. Feedback diagram illustrating the highly interactive community involving the armored scales of hemlock. See Section 4.3 for a description and discussion of feedback loops.

table data, scale populations were found to have the potential to increase rapidly to injurious levels on cultivated hemlocks growing outside the natural range. Further studies are needed to verify that hemlocks growing in cultivation or outside their natural range in Japan are stressed and are nutritionally more suitable for herbivores than are naturally growing trees. Nevertheless, the significantly higher rates of increase of both scales on cultivated hemlocks (see Table VI) emphasizes the need to maintain host vigor, as well as the danger of planting tree species outside their natural ranges.

A lengthy discussion of the various silvicultural practices that can be used to maintain the vigor of trees and thereby aid in the management of insect pest populations is beyond the scope of this chapter. However, one practice for improving the vigor of plants—fertilization—warrants some consideration in view of my own findings. Sucking insects are generally favored by increases in soluble nitrogen component of their food.[7,11,41] Fertilization of young hemlocks with ammonium nitrate (17.5 g nitrogen) prior to flush significantly increases the total nitrogen concentration of new needles and enhances the success of hemlock scales feeding on them[29] (see Table I). Fertilization should therefore be used with discretion.

5.2. Establishment and Preservation of Natural Enemies

Many of our most serious insect pests are species that have been accidentally introduced from another area of the world. These pests often arrive without natural enemies that may have important regulatory affects in the homeland. The invading species often increases rapidly to injurious population densities on its new host and causes widespread destruction to previously healthy stands.[33] Obviously, the rapid spread of these introduced pest populations is not merely a function of host stress but also involves major alterations in the plant–herbivore community.

A popular approach for controlling injurious populations of exotic species is the introduction and establishment of natural enemies from the homeland of the pest.[3,6,40,51] Examples of successful biological control, however, are relatively scarce despite great efforts over many years. Considering the time and resources expended during the twentieth century for the biological control of introduced pests, it is surprising that so little effort has been applied to the study of pest population dynamics in their native communities. The failure of many biological control programs is perhaps attributable to a lack of understanding of how the various components of the endemic community interact.

The armored scales of hemlock are a good example. Clearly, *A. citrinus* plays a major role in the regulation of native populations of both scales in Japan. Even though *A. citrinus* is already well established in the northeastern United States, *F. externa* populations increase to such high densities that they frequently injure and kill their new host. It would be prudent to bear in mind that establishment of parasitoids is not the only condition necessary for the biological control of introduced pests. This is because conditions in the new environment may have quite different effects on the population dynamics of the pest and parasitoid. Studies that compare the dynamics of introduced and native populations can (1) provide valuable insight into factors that regulate herbivore populations, (2) identify factors that can be manipulated to control damaging populations of introduced species, and (3) provide a sound basis for evaluating the performance of natural enemies that have been introduced for biological control.[37]

5.3. Management of Hemlock Scale Populations

Populations of *F. externa* and *N. tsugae* in the natural montane habitats of Japan are maintained at innocuous densities by the combined influence of host vigor (resistance) and natural enemies. Even on cultivated hemlocks growing outside their natural ranges, populations are always held in check by parasitoids. Managing native populations of hemlock scales may simply involve adherence to silvicultural practices that maintain plant vigor and preserve the natural enemy community.

The management of introduced populations of hemlock scales in the northeastern United States is a much more formidable task. Even vigorous *T. canadensis* forests show little resistance to the buildup of injurious scale populations, and introduced parasitoids are poorly synchronized with one of the scales, *F. externa*. Pesticides can provide effective control of *F. externa* and *N. tsugae* on cultivated trees where complete coverage of the foliage can be obtained.[20,25] However, it is virtually impossible to obtain complete coverage, while pesticide spraying often results in the elimination of natural enemies and the rapid resurgence of scale populations.[20]

Various techniques have been used to enhance the effectiveness of pesticides, while reducing their deleterious impact on natural enemies. Soil injection of systemic insecticides, such as acephate and dimethoate, control *F. externa* and *N. tsugae* with minimal adverse effects on parasites and predators.[31] Alternatively, foliar sprays can be applied with precise timing to maximize scale mortality with minimal impact on beneficial natural enemies.[28]

The best solution to the problem of introduced hemlock scales in the United States may be the identification or development, through genetic manipulation, of univoltine strains of *A. citrinus* or other natural enemies. Until more effective and persistent natural enemies are found, we must strive to maintain vigorous hemlock forests and to use pesticides occasionally as prudently as possible to minimize their deleterious impact on existing parasitoids and predators.

ACKNOWLEDGMENTS. I would like to thank David Wainhouse and David Morgan for their helpful reviews of this chapter and the Ecological Society of America for permission to reprint Figs. 1–4 and 6 and Tables IV–VI.

REFERENCES

1. Baltensweiler, W., Benz, G., Bovey, P., and Delucchi, V., 1977, Dynamics of larch budmoth populations, *Annu. Rev. Entomol.* **22**:79–100.
2. Beardsley, J. W., and Gonzalez, R. H., 1975, The biology and ecology of armored scales, *Annu. Rev. Entomol.* **20**:47–73.
3. Clausen, C. P. (ed.), 1978, Introduced parasites and predators of arthropod pests and weeds: A world review, U.S. Department of Agriculture Forest Service, Agriculture Handbook #480.
4. Danzig, E. M., 1977, An ecological and geographical review of the scale insects (Homoptera: Coccoidea) of south of the far east, *Zool. Inst. Tr.* **70**:37–60.
5. Davidson, J. A., and McComb, C. W., 1958, Notes on the biology and control of *Fiorinia externa* Ferris, *J. Econ. Entomol.* **51**:405–406.
6. Debach, P., 1974, *Biological Control by Natural Enemies,* Cambridge University Press, London.
7. Feeny, P., 1970, Seasonal changes in oak leaf tannins and nutrients as a cause of spring feeding by winter moth caterpillars, *Ecology* **51**:565–581.
8. Ferris, G. F., 1936, Contributions to our knowledge of the Coccoidea (Homoptera), *Microentomology* **1**:1–16.
9. Ferris, G. F., 1942, *Atlas of the Scale Insects of North America.* Vol. IV, Stanford University Press, Stanford, California.
10. Fowells, H. A., 1965, Silvics of forest trees of the United States, U.S. Department of Agriculture Forest Service, Agriculture Handbook #271.
11. Friend, W. G., 1958, Nutritional requirements of phytophagous insects, *Annu. Rev. Entomol.* **3**:57–74.
12. Garrett, W. T., 1965, The biology and control of the *Fiorinia* hemlock scale, *Fiorinia externa* (Homoptera: Diaspididae), M.S. thesis, University of Maryland, College Park, Maryland.
13. Harcourt, D. G., 1963, Major mortality factors in the population dynamics of the diamond back moth, *Plutella maculipennis* (Curt.) (Lepidoptera: Plutellidae), *Mem. Entomol. Soc. Can.* **32**:55–66.
14. Hayashi, Y., 1960, *Taxonomical and Phytogeographical Study of Japanese Conifers,* Norin Shuppan, Tokyo.
15. Kawai, S., 1980, *Scale Insects of Japan in Colors,* Tokyo University Press, Tokyo, Japan.
16. Kuwana, I., 1933, The Diaspine Coccidae of Japan. VII. Ministry of Agriculture and Forestry, *Jpn. Sci. Bull.* **3**:1–42.
17. MacGillivary, A. D., 1921, *The Coccidae,* Scarab Co., Urbana, Illinois.
18. Marlatt, C. L., 1911, A newly imported scale pest on Japanese hemlock (Rhynch.), *Entomol. News* **22**:385–387.
19. Mattson, W. J., Jr., 1980, Herbivory in relation to plant nitrogen content, *Annu. Rev. Ecol. Syst.* **11**:119–161.
20. McClure, M. S., 1977, Resurgence of the scale, *Fiorinia externa* (Homoptera: Diaspididae), on hemlock following insecticide application, *Environ. Entomol.* **6**:480–484.
21. McClure, M. S., 1977, Parasitism of the scale insect, *Fiorinia externa* (Homoptera: Diaspididae), by *Aspidiotiphagus citrinus* (Hymenoptera: Eulophidae) in a hemlock forest: Density-dependence, *Environ. Entomol.* **6**:551–555.
22. McClure, M. S., 1977, Dispersal of the scale *Fiorinia externa* (Homoptera: Diaspididae) and effects of edaphic factors on its establishment on hemlock, *Environ. Entomol.* **6**:539–544.
23. McClure, M. S., and Fergione, M. B., 1977, *Fiorinia externa* and *Tsugaspidiotus tsugae* (Homoptera:

Diaspididae): Distribution, abundance, and new hosts of two destructive scale insects of eastern hemlock in Connecticut, *Environ. Entomol.* **6**:807–811.

24. McClure, M. S., 1978, Seasonal development of *Fiorinia externa, Tsugaspidiotus tsugae* (Homoptera: Diaspididae), and their parasite, *Aspidiotiphagus citrinus* (Hymenoptera: Aphelinide): Importance of parasite–host synchronism to the population dynamics of two scale pests of hemlock, *Environ. Entomol.* **7**:863–870.
25. McClure, M. S., 1978, Effectiveness of foliar systemic insecticides for controlling scales on forest and ornamental hemlock, *Insecticide Acaricide Tests* **3**:146–147.
26. McClure, M. S., 1979, Self-regulation in populations of the elongate hemlock scale, *Fiorinia externa* (Homoptera: Diaspididae), *Oecologia* **39**:25–36.
27. McClure, M. S., 1979, Self-regulation in hemlock scale populations: Role of food quantity and quality, *Entomol. Soc. Am. Misc. Publ.* **11**(3):33–40.
28. McClure, M. S., 1979, Spatial and seasonal distribution of disseminating stages of *Fiorinia externa* Ferris (Homoptera: Diaspididae) and natural enemies in a hemlock forest, *Environ. Entomol.* **8**:869–873.
29. McClure, M. S., 1980, Foliar nitrogen: A basis for host suitability for elongate hemlock scale, *Fiorinia externa* (Homoptera: Diaspididae), *Ecology* **61**:72–79.
30. McClure, M. S., 1980, Competition between exotic species: Scale insects on hemlock, *Ecology* **61**:1391–1401.
31. McClure, M. S., 1980, Efficacy of soil-injecting a systemic insecticide for controlling two armored scales on hemlock, *Insecticide Acaricide Tests* **5**:204.
32. McClure, M. S., 1981, Effects of voltinism, interspecific competition and parasitism on the population dynamics of the hemlock scales, *Fiorinia externa* and *Tsugaspidiotus tsugae* (Homoptera: Diaspididae), *Ecol. Entomol.* **6**:47–54.
33. McClure, M. S., 1983, Competition between herbivores and increased resource heterogeneity, in: *Variable Plants and Herbivores in Natural and Managed Systems* (R. F. Denno and M. S. McClure, eds.), pp. 125–153, Academic, New York.
34. McClure, M. S., 1983, Reproduction and adaptation of exotic hemlock scales (Homoptera: Diaspididae) on their new and native hosts, *Environ. Entomol.* **12**:811–815.
35. McClure, M. S., and Hare, J. D., 1984, Foliar terpenoids in *Tsuga* species and the fecundity of scale insects, *Oecologia* **63**:185–193.
36. McClure, M. S., 1985, Patterns of abundance, survivorship, and fecundity of *Nuculaspis tsugae* (Homoptera: Diaspididae) on *Tsuga* species in Japan in relation to elevation, *Environ. Entomol.* **14**:413–415.
37. McClure, M. S., 1986, Population dynamics of Japanese hemlock scales: A comparison of endemic and exotic communities, *Ecology* **67**:1411–1421.
38. Morris, R. F., 1963, The dynamics of epidemic spruce budworm populations, *Mem. Entomol. Soc. Can.* **31**:1–332.
39. Morris, R. F., and Miller, C. A., 1954, The development of life tables for the spruce budworm, *Can. J. Zool.* **32**:283–301.
40. Ridgeway, R. L., and Vinson, S. B. (eds.), 1977, *Biological Control by Augmentation of Natural Enemies*, Plenum, New York.
41. Rodriguez, J. G., 1960, Nutrition of the host and reaction to pests, in: *Biological and Chemical Control of Plants and Animal Pests* (L. P. Reitz, ed.), pp. 149–167, Publ. 61, American Association of Advanced Sciences, Washington, D. C.
42. Sasscer, E. R., 1912, The genus *Fiorinia* in the United States, *U.S. Dep. Agric. Bur. Entomol. Tech. Ser.* **16**:75–82.
43. Scriber, J. M., and Slansky, F., Jr., 1981, The nutritional ecology of immature insects, *Annu. Rev. Entomol.* **26**:183–211.
44. Takagi, S., 1963, *Fiorinia externa* on *Tsuga sieboldii* in Japan, *Insecta Matsum.* **26**:115–117.
45. Takahashi, R., and Takagi, S., 1957, A new genus of Diaspididae from Japan (Coccoidea, Homoptera), *Kontyu* **25**:102–105.
46. Tippins, H. H., 1970, The second instar males of three species of *Fiorinia* (Homoptera: Diaspididae), *J. Georgia Entomol. Soc.* **5**:94–99.
47. Wallner, W. E., Jr., 1965, Biology and control of *Fiorinia* hemlock scale, *Fiorinia externa* Ferris, Ph.D thesis, Cornell University, Ithaca, New York.

48. Weiss, H. B., 1914, Notes on three imported insects occurring in New Jersey, *J. Econ. Entomol.* **7**:250–251.
49. White, T. C. R., 1978, The importance of a relative shortage of food in animal ecology, *Oecologia* **33**:71–86.
50. White, T. C. R., 1984, The abundance of invertebrate herbivores in relation to the availability of nitrogen in stressed food plants, *Oecologia* **63**:90–105.
51. Wilson, C. L., and Graham, C. L., (eds.), 1983, *Exotic Plant Pests and North American Agriculture*. Academic, New York.

CHAPTER 4

THE BEECH SCALE

David Wainhouse and Imogen M. Gate

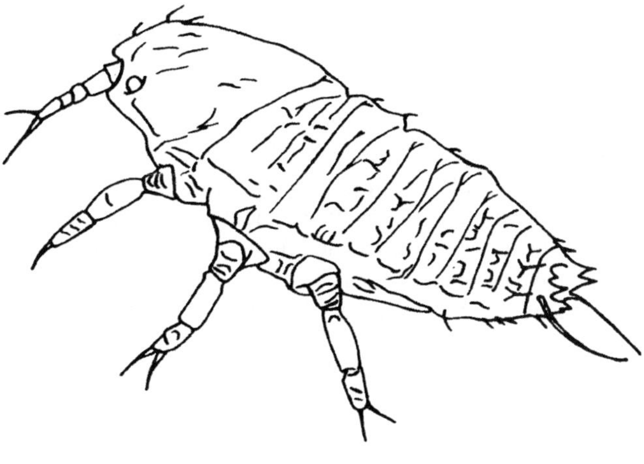

David Wainhouse and Imogen M. Gate • British Forestry Commission, Forest Research Station, Wrecclesham, Farnham, Surrey GU10 4LH, England

1. INTRODUCTION

Scale insects are among our most important pests, but relatively few have been studied in detail. These insects are often small and inconspicuous, and identification is usually a highly specialized activity. The beech scale *Cryptococcus fagisuga* Lindinger is to some extent an exception because the white wax secreted by this insect makes it highly conspicuous and the absence of any similar coccid on beech largely eliminates the chore of identification. Nevertheless, during the nineteenth century, this organism was described as a fungus (*Psilonia nivea*), the curly wax threads being taken for fungal hyphae.[8,20]

The beech scale is presumed to have originated in Europe, on its native host *Fagus sylvatica*, but the international movement of plant material has resulted in its establishment in the New World. The scale is believed to have been carried to the botanic gardens of Halifax, Nova Scotia around 1890, on imported nursery stock of the European copper beech.[28] After a period of gradual spread in the early 1900s, there followed a fairly rapid expansion of the infestation into the indigenous American beech (*Fagus grandifolia*) forests.

Although heavy scale populations can cause serious damage to trees, resulting in increment loss,[41] emphasis has been placed on the association between the scale and a fungus, *Nectria* spp., as being jointly responsible for a complex syndrome called beech bark disease. The death of bark, resulting largely from infection by *Nectria*,[20] is often extensive enough to kill the tree. In considering the impact of the disease, we have made no attempt to separate the effects of insect and fungus. Although the death of trees is only indirectly attributable to *Cryptococcus*, it is the important first stage in a disease syndrome that is ultimately dependent on the effects of both organisms.

Trees have apparently been dying from the beech bark disease in Europe for well over a century.[8,27] Early fears that "beech all over England are doomed"[73] have fortunately never materialised, but considerable destruction of beech forests has occurred within Europe as a whole.

The European beech (*F. sylvatica*) grows throughout much of Europe, and *Cryptococcus* is endemic over most of its range, extending to the southern limit of its host in Sicily.[16] For reasons that are not entirely clear, however, outbreaks of the disease are more restricted; e.g., outbreaks in France that started during the mid-1960s were apparently occurring there for the first time despite a long record of the disease in Europe.[54]

1.1. Impact of Beech Bark Disease in Europe

Beech is the second most important hardwood in Britain, occupying some 74,000 ha, about one half of which is grown on the North and South Downs of Southern England, areas important for conservation and recreation.[2,51] Commercial thinnings are usually used for pulp, and timber from good-quality final crop trees can be used for furniture making.

Timber losses caused by the beech bark disease in Britain have not been quantified. Certainly, thousands of trees have died and, in a few cases, whole forest compartments have been clear-felled. The most significant losses occurred during the 1960s and 1970s, partly because the postwar planting boom resulted in peak numbers of susceptible trees 20–30 years later. In continental Europe, timber of high quality is used to produce veneer and, in general, beechwood is used for an increasing number of industrial purposes.[52]

Within continental Europe, losses have been most noticeable in mature stands, partly because of their greater commercial value. In Normandy, where many of the forests derive from plantations, the disease has caused considerable disruption of beech silviculture.[34] During the early 1970s, average losses reached 25% of the growth increment in affected forests, and more severe local damage occurred later. In the forest of Lyon, for example, twice the annual growth increment was reported lost during the mid-1970s.[34,53,56] Damage occurred in stands aged 30–40 years and upward, reaching a maximum at 90–100 years. It is important to note, however, that older trees stressed by drought may develop symptoms similar to those of later stages of the disease in the absence of heavy infestations of *Cryptococcus*. Thus, the severe drought that occurred throughout much of Europe during 1975 and 1976 may have indirectly influenced beech mortality.[56]

The disease has been known in Germany since the late 1800s.[27] Current damage occurs mainly in stands >80 years old, although some younger trees are affected in northern Bavaria.[39] It has been estimated that 26% of trees moderately to heavily infested with scale ultimately die, although among dominant trees only 5% are moderately and 1% heavily attacked.[7] It seems reasonable to conclude that the impact of the disease in Germany is not too serious at present.[40]

When estimating economic losses, it is important to realize that the timber quality of surviving trees can be impaired.[74] The wood underlying necrotic bark can become stained or defective, causing problems in veneer-quality timber. Growth of the trees subsequent to damage incorporates the necrotic defect into the timber, and this must be clipped from the veneer. Volume losses from such defects can be as much as 12%.

The main impact of the disease has been on *Fagus sylvatica* in central Europe. However, the oriental beech (*F. orientalis*) is also susceptible; in Iran, for example, monocultures of this tree form about 18% of the managed forest, and it is considered the most economically important tree species there.[1] Attack by beech scale appears to be locally important.

1.2. Impact of Beech Bark Disease in North America

The American beech (*F. grandifolia*) is highly susceptible to attack by the introduced scale; the pattern of damage in North America has largely followed the spread of the scale into the indigenous beech population from its point of introduction in Nova Scotia (Fig. 1). Beech is a significant component of the northern hardwood forests, which occupy about one third of the commercial forest land in the New England and middle Atlantic states.[18] As much of the beech population still remains to be colonized (Fig. 1), the losses incurred in the New England states should be expected in other parts of eastern North America.

Following an initial period of concern during the 1930s, the disease subsequently received less attention because beech timber was considered less valuable than the other major hardwoods.[34] Since the 1960s, however, the wood industry has developed methods

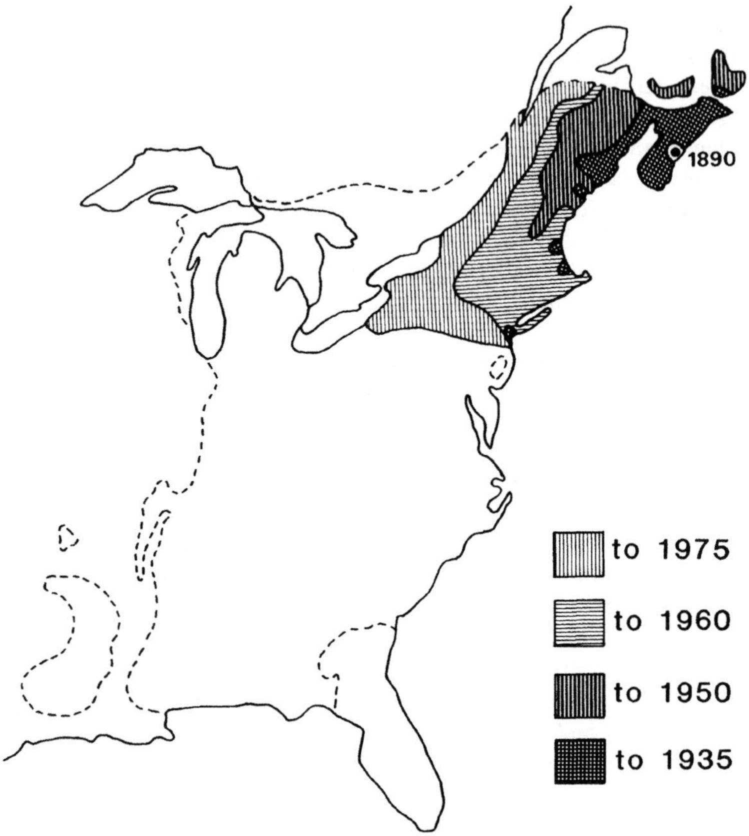

FIGURE 1. Distribution of *Cryptococcus fagisuga* in North America. Following its introduction into Halifax, Nova Scotia during the late nineteenth century, *C. fagisuga* has progressively extended its range and is now found throughout eastern North America. The distribution of its host (*Fagus grandifolia*) is delimited by a dashed line. (After Houston *et al.*[33])

for using beechwood profitably, and recent losses now threaten that supply.[62] Products currently made from beech include furniture, plywood, flooring, and containers, while beechnuts provide food for forest animals.[22]

In common with the work in Europe, most American studies show that mortality from the disease is higher among larger trees, which are often the remnants of earlier cuttings. Thus, initial losses may be high, with total mortality occasionally reaching 80 or 90%.[20,34] In Fredericton, New Brunswick, about 20% of the trees, representing 40% of the beech volume, were killed within 15 years of the discovery of the scale in 1937, and many of the surviving trees had scarred and misshapen boles.[4] These sublethal effects have degraded some 6.3 million m^3 of high-quality beech in the maritime provinces of Canada.[46]

Unlike the European beech, the American species is usually just one component of a mixed hardwood forest, and the death of trees may initiate changes in stand composition.[22,65]

2. BIOLOGY AND LIFE HISTORY

In common with many other scales, *Cryptococcus fagisuga* has a simple univoltine life cycle. The neotenous adults are entirely parthenogenetic, each laying about 50 eggs. The first instars, or crawlers, are active from July to November, during which some aerial dispersal from the host tree occurs. After settling to feed, however, they become distended and immobile and overwinter in this form. In Europe, the sessile second instars and adults occur from April to June and from May to September, respectively.

Effective dispersal is confined to the crawler stage. The dorsoventrally flattened shape of the crawlers, together with numerous protruding setae, aids airborne dispersal by reducing their terminal velocity. This is an important adaptation for passive dispersal where the pattern of wind movement determines the distance and direction of dispersal. Even so, the mean distance dispersed by crawlers within the forest is only around 10 m from an infested tree.[68] Temperature rather than windspeed has the major influence on aerial density, reflecting the increased numbers and activity of crawlers on the bark surface during warmer weather.

Crawlers are deposited inefficiently on large tree trunks because of their small size ($100 \times 300\ \mu m$) and the low windspeeds within the forest. They tend to be carried in the airstream diverted around the trunk rather than impacting onto its surface.[26] Small-diameter trees are therefore more efficient collectors of crawlers, and the density of deposition is greater on these trees. Nevertheless, larger trees are often observed to be the most heavily infested, suggesting that, at least in Europe, host plant factors have a predominant influence on settling success or on subsequent survival.

Air movements within forests can be complex, involving vertical as well as horizontal components; as a result, some crawlers are carried above the canopy into more turbulent and faster-moving air (Fig. 2). About 1% of the crawlers disperse in this way and have potentially a much greater range of dispersal compared with those below the canopy. This long-range dispersal is probably responsible for the rapid spread of beech scale at 6–16 km/year after its introduction into North America.[33,38,64,68]

2.1. Interactions with Fungi, Lichens, and Algae

Three different kinds of *Nectria* are associated with scale infestations. In North America, *Nectria coccinea* var. *faginata* is the main species, but *N. galligena* is associated with isolated infestations in West Virginia.[20,48] In Europe, *N. coccinea* is the only fungus associated with the disease.

Nectria coccinea appears to be widely distributed as a saprophyte and does not need to be carried by dispersing insects to new hosts. Little is known of its preinfection ecology, but it occurs on the surface of scale-infested bark and has been isolated from the small necrotic flecks caused by insect feeding within the bark.[44] *N. coccinea* is thus well situated to take advantage of a substrate damaged by insect feeding.

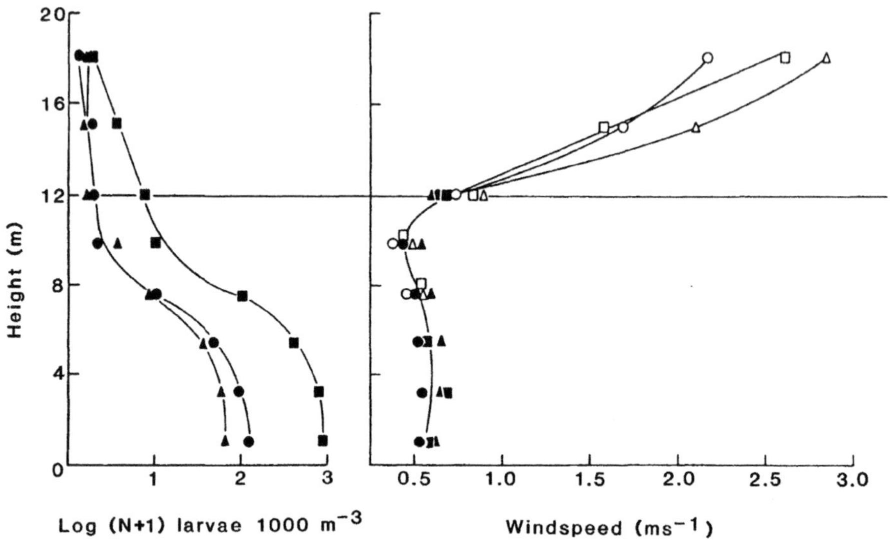

FIGURE 2. Vertical profiles of aerial density of crawlers and of windspeed on three occasions during the autumn dispersal period. Horizontal line represents the top of the canopy. (From Wainhouse.[68])

Invasion of the bark by *Nectria* is usually associated with moderate to heavy *Cryptococcus* infestations and appears to be a direct consequence of insect-induced changes in the bark. Lesions on trees representing a range of insect densities showed a significant increase in bark lesion length with increasing insect density[41] (Fig. 3). Interestingly, trees that showed all the symptoms of previously high *Cryptococcus* density, but on which current insect populations were low, developed only small lesions. Similar relationships may hold for variations in insect density within individual trees.[55] Thus, although insect feeding may temporarily impair the defense mechanisms in the bark, these can be regained once feeding ceases.

The symptoms of *Nectria* attack on beech in the Old World and the New World show distinct differences. In Europe, dark exudations from dead bark areas, called tarry spots or slime fluxes, are usually the first signs of *Nectria* infection. Although trees normally recover from a few infections, girdling kills the tree. In American beech, penetration of *Nectria* to the cambium is less frequent, and circular lesions may be restricted to the outer bark.

In North America, where *Cryptococcus* is still actively spreading through the beech population, three phases of the epidemic are recognized: (1) the invading front of scale insects; (2) the killing front of scales plus *Nectria;* and (3) the aftermath zone, containing defective surviving trees, resistant trees, and relatively low levels of active disease.[62]

Other organisms colonizing beech bark may interact with scale populations and thereby influence the course of the disease. In the Canadian maritimes, unattacked trees seem to owe their apparent resistance to a covering of crustose lichens on the bark surface, some of which, e.g., *Graphis scripta,* seem to be strongly antagonistic to scale settling and feeding.[32] Other lichens, such as the common crustose lichen *Lecanora conizaeoides*

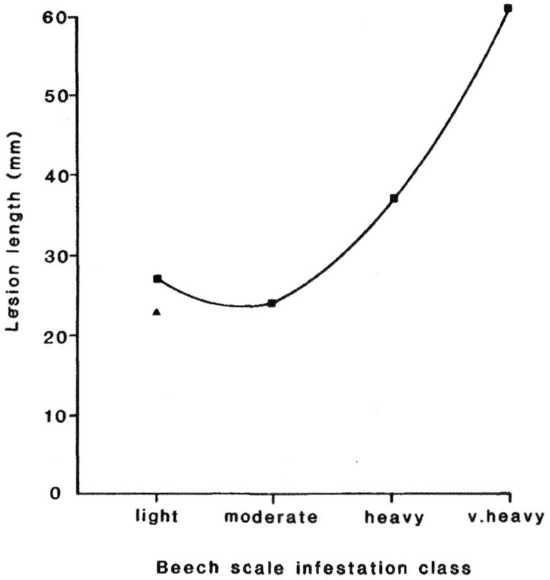

FIGURE 3. Relationship between the size of bark lesions caused by *Nectria coccinea* and *Cryptococcus fagisuga* infestation of the trees. (▲) Previously heavily infested trees now bearing light scale infestation. (From Lonsdale.[41])

in southern England, may provide protected settling sites for crawlers, effectively increasing scale infestation.[33] Algae colonizing beech bark may also protect settling crawlers.[72]

The fungus *Ascodichaena rugosa* colonizes the outer bark of some beech trees,[12] preventing the establishment of scale crawlers.[33] Cover of the bark by this slow-growing fungus is rarely sufficiently dense or extensive, however, to protect the entire trunk; consequently, the two organisms often occur together on the same tree.

2.2. Interactions with the Host Plant

Scales feed intracellularly within the bark parenchyma. Cell complexes ("galls") may be formed just below the feeding stylet,[36,42] possibly caused by plant hormonelike compounds in the saliva.[23] These areas of cell proliferation are thought to enhance scale nutrition, but local exhaustion of host-defensive responses may also occur. Scales placed onto bark from which naturally occurring colonies have been removed tend to establish themselves on the previously occupied areas rather than on the adjacent uninfested bark.[72] Eventually, however, intensive feeding under large colonies results in necrotic flecks and sometimes a more general discoloration in the cortex of infested bark, which becomes unsuitable for scale colonization. Cambial function may be so imparied by the indirect effects of scale feeding that abnormal, very narrow, xylem rings are produced.[42] This eventually results in an externally visible depression, or "dimple," in the bark, as well as considerable stem distortion on some trees. The development of dimples may be due to hypersensitive responses of individual trees, as not all attacked trees develop this symptom.

Beech is unusual among temperate hardwood species in that the bark surface usually remains smooth into old age. This results from the retention of the primary periderm and the consequent absence of a rhytidome incorporating dead cortical tissue. During intense attack by the beech scale, however, wound periderms are formed, so that necrotic areas may eventually be isolated in an outer secondary bark or pathological rhytidome.[9,10,50] When this occurs, the outer bark becomes characteristically rougher and more fissured, and it is more difficult for the scales to penetrate living bark to feed. The scales become progressively confined to the newly formed vertical fissures, where live tissues are nearer to the surface. This nonspecific defensive reaction by the bark is often effective in reducing population density of scales on attacked trees.[36,42]

A striking illustration of the influence of the host plant on *Cryptococcus* populations is shown in (Fig. 4). On this grafted tree, a dense population developed on the stock and ended abruptly at its junction with the scion. The scion was clearly resistant or perhaps lacked some factor necessary for insect development as a result of stock–scion interaction. Within the attacked bark, the activity of wound phellogens eventually produced a fissured rhytidome (Fig. 4b), resulting in the virtual elimination of the scale population.

Most species of *Fagus* appear to be susceptible to *Cryptococcus,* but there is anecdotal evidence of varietal differences in susceptibility among European beeches. For example, it has been suggested that the weeping beech (*F. sylvatica* var. *pendula*) is more resistant than the ordinary form.[15,49] Similar reports have been made about purple or copper beech (var. *purpurea*)[58] (but see Green[25]). In North America, differences in susceptibility between types of *F. grandifolia* are partly attributable to the suitability of bark for colonization.[13,14]

Studies on the distribution of attack within forests provide circumstantial evidence for variation between trees in their susceptibility to *Cryptococcus,* and a number of recent studies provide supporting experimental evidence. For example, trees of clonal origin in a seed orchard in southern England were shown to vary in resistance to *Cryptococcus*[69,70]; a survey of the distribution of natural infestation among progeny of single beech trees planted in several different sites suggested consistent genetically based differences in susceptibility.[71] In North America, inoculation experiments on trees that had remained free from natural infestation confirmed that some of them were resistant to *Cryptococcus.*[31] In a survey of two stands in Nova Scotia, about 0.7% of the current population of young beech trees growing in the aftermath of an outbreak remained free from attack by the endemic scale and fungus populations.[31] These trees occurred in groups and may have come from seed or had a common root-sprout origin.

The mechanisms of resistance have not been studied in detail, but both physical and chemical factors appear to be involved.[72] It has long been known that heavily infested trees may survive and appear to develop resistance in the bark to subsequent attack.[11] This is partly related to the nonspecific defensive reactions in the bark described previously. In addition, stone cells may limit penetration of the stylets into the deeper bark tissues[20,36,59] and are probably involved to some extent in resistance. The amount of lignin seems to increase in attacked bark, and there is an apparent relationship between resistance and the continuity and depth of the stone cell layers, although other factors are probably involved as well.[42]

The experimental inoculation of trees demonstrates an interesting relationship between the density of natural scale infestations and the survival of inoculated insects (Fig. 5) that cannot be attributed simply to increased survival on sites previously occupied by

a

b

FIGURE 4. (a) Stock of grafted beech infested with *Cryptococcus fagisuga* in 1982. The infestation ends abruptly at the junction of the stock and scion. (b) Residual infestation on the same stock in 1984. Fissures in the bark resulting from pathological rhytidome formation are visible.

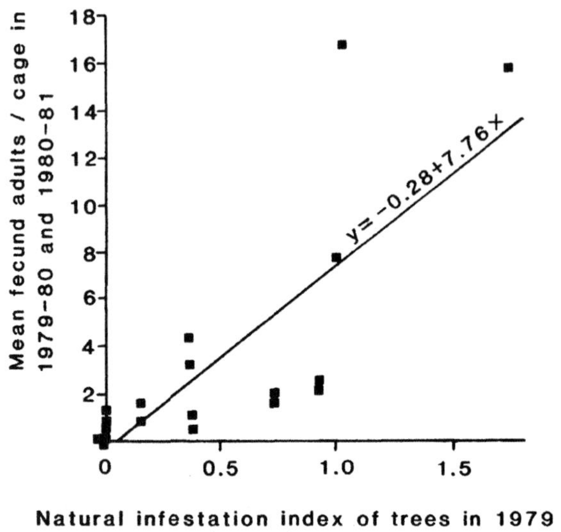

FIGURE 5. Relationship between the number of fecund adults developing in cages inoculated with crawlers and the extent of natural infestation of the tree. (From Wainhouse and Howell.[70])

indigenous colonies.[70] This means that, in European beech forests, the degree of natural infestation reflects current susceptibility, and trees remaining uninfested or only lightly infested for more than a few years within infested forests are likely to be at least temporarily resistant.

Of greater significance for population dynamics is the evidence that scale populations become adapted to specific host individuals.[70] For example, when crawlers obtained from an infested tree are transferred to other trees with similar infestation density, as well as being reinoculated onto their own original host, the proportion reaching the adult stage is usually lower on the new host compared with the original tree. One possible explanation for this phenomenon is that biochemical defenses are highly heterogeneous within a population of trees, so that survival of *Cryptococcus* populations is enhanced by adaptation to the defense characteristics of an individual tree.[19] Three features of the life cycle of *Cryptococcus* appear to favor the development of host-adapted populations: (1) limited powers of dispersal, (2) the sexual isolation resulting from parthenogenetic reproduction, and (3) the development of many successive generations on the same tree. It seems likely that the characteristic distribution of single heavily infested trees in European forests is partly due to the occurrence of isolated adapted scale populations.

2.3. Interactions with Natural Enemies

No insect parasites have been recorded from beech scale. In this respect, it is like the balsam woolly aphid (*Adelges piceae*) that occupies a similar habitat on fir trees (see Chapter 5).

Of the many predaceous insects recorded in association with *Cryptococcus*, the coccinellids (*Chilocorus* spp. and *Exochomus* spp.) and the cecidomyid (*Lestodiplosis* spp.) are among the most common.[5,66] With the exception of *Lestodiplosis*, they are all generalist predators and their abundance is usually proportional to that of their prey.[47,61] Thus, they are rarely seen on lightly infested trees. Heavy predation has occasionally been noted on single trees.[61] The same is true of *Lestodiplosis* whose larvae are predatory on all stages of the scale. This specific predator can often be fairly abundant and widely distributed within a stand but is usually confined to trees with moderate to heavy scale infestations.

3. POPULATION DYNAMICS

3.1. Population Change on Single Trees

Few investigators have considered the population dynamics of *C. fagisuga* on individual trees. The following account is based largely on unpublished work.[24] In this study, populations were followed for two complete generations on three trees growing on the same site in southern England. These trees were initially heavily, moderately, or lightly infested with *Cryptococcus*. Failure of adults to achieve their potential fecundity and mortality of crawlers were, numerically, the most important factors affecting populations. The maximum observed fecundity was 43, but the average realized fecundity per tree ranged from about 4 to 16 eggs per female, and many died without laying eggs. About 10% of adults that had not layed eggs overwintered and some of them oviposited in the following summer.

Mortality of crawlers was estimated to be about 86%, but comparatively few (<1%) were washed off the bark during rainfall, contrary to previous assumptions (e.g., see Malphettes[45]). Aerial dispersal from one of the trees was studied and accounted for a loss of 10–20% of the crawlers. Thus, most mortality occurred among larvae that remained on the host, through failure to become established. A lack of suitable settling sites for crawlers may have contributed to this high mortality, especially on heavily infested trees where large areas of bark may be unavailable, being already colonized by *Cryptococcus*. Other organisms, such as epiphytic lichens and algae, as well as host physiological factors, may influence the availability of settling sites as well. Too few generations have been studied to determine whether these numerically important mortalities are the key factors determining population change.

Following establishment of crawlers, mortality of the later sessile stages was much lower, averaging ~16% for second instars. Overall generation mortality, including loss of potential fecundity, was estimated to be 99%.

Other than a lack of settling places, there was no clear evidence of density-dependent mortality for any of the stages in the life cycle. But studies on the interaction of insect and host suggest that scale populations would ultimately decline as a result of changes in the bark induced by high populations of scale or bark death resulting from *Nectria* attack.

3.2. Population Change within the Forest

In European beech forests, where *Cryptococcus* is endemic, the scale can be found on most trees older than about 15 years, although typically only at very low densities.

Trees aged 20–40 years are considered most susceptible to attack, at least in Britain. The scale is usually more abundant on larger trees,[3,20,69,70] but in naturally regenerating forests this may be partly due to differences in age. However, size-related susceptibility has also been observed in even-aged plantations (Fig. 6), so the cause of this effect remains unclear. Heavily infested trees are usually scattered throughout infested forests and often occur next to uninfested trees.[8,37,57,63]

The degree of infestation within the forest partly depends on the stage of the infestation, which, in North America, corresponds to the invading front, the killing front, or the aftermath zone.[62] At the invading front, infestation density is initially light and scattered. There follows an increase both in population density on individual trees and in the number of trees infested, so that eventually more than 85% are attacked.[20,34] At the killing front, populations on moderate and heavily infested trees crash as the bark becomes infected with *Nectria*. Following the death of susceptible trees, the uninfested or lightly infested ones that remain are assumed to be resistant. However, sprout regeneration from the roots of dead susceptible trees may eventually restructure the forest for a new outbreak so that a cyclical pattern of population dynamics could develop.[30]

In Europe many more trees survive the initial outbreak than in North America. Nevertheless, it is possible to recognise early peak and aftermath stages, but only within even-aged beech plantations, where most trees reach a susceptible stage at about the same time.[43] In naturally regenerating forests, trees of different ages become infested, die, or recover at different times; thus, phases of the disease are not synchronized. Those that have apparently developed resistance and have currently low densities of *Cryptococcus* may possibly become susceptible again later in the rotation as the normal processes of growth increase the amount of previously unattacked and presumably susceptible tissues within the bark.

In a series of long-term observation plots, determination of population-density

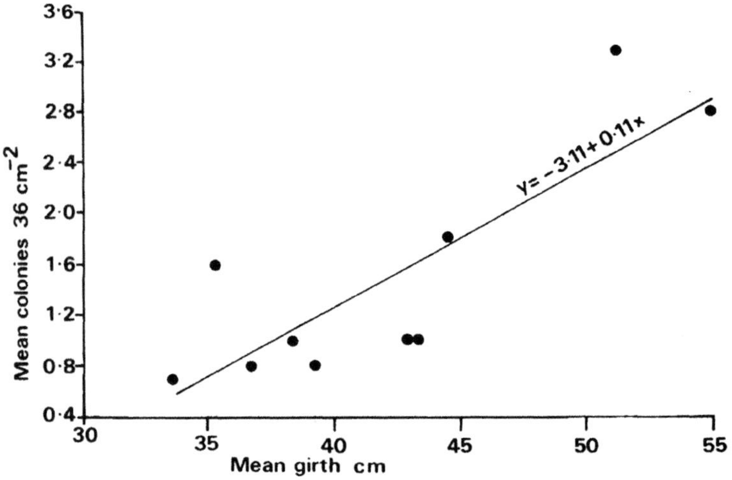

FIGURE 6. Relationship between the number of colonies of *Cryptococcus fagisuga* per 36 cm^2 and the girth of 16-year-old trees. Each point is a mean value for plots of about 30 trees. (From Wainhouse and Howell.[71])

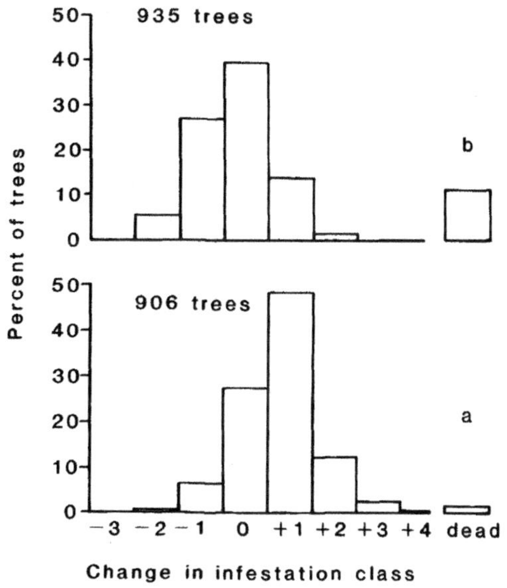

FIGURE 7. Population change in long-term observation plots. Trees in plots in an early (a) and late (b) stage of disease development occupy one of six infestation classes from uninfested to very heavily infested. The change in infestation class of trees from an initial assessment to one 7 years later is shown.

changes of *Cryptococcus* over a period of 7 years shows that many trees support more or less stable populations during this period (Fig. 7). Relative population density was estimated by visual assessment of the extent of wax cover on the tree trunks. As expected, population density increased in plots in an early stage of disease development and declined in those in which the disease was initially well advanced (Fig. 7). However, population change on most trees is restricted to a maximum of one infestation class, representing a relatively small change in population density. Few trees support rapid population growth (Fig. 7a). In the later stages of disease (Fig. 7b), about 10% of the trees died and, since there was a decrease in infestation on surviving trees, the resistance of these plots to beech-scale attack appears to have increased.

4. HYPOTHESES FOR THE CAUSES OF OBSERVED POPULATION BEHAVIOR

Dramatic reductions in population density may be caused by low winter temperatures,[17,63] and water-table variations may indirectly affect populations through the host plant.[29,60,61,63] Our studies indicate, however, that population densities on individual trees can remain remarkably constant from year to year and that this is largely determined by the host–insect relationship.

The main factors and feedback structure[6] affecting scale population growth on initially susceptible trees are shown in Fig. 8. Improved nutrition at colonial feeding sites and the adaptation of scales to individual trees form unstable positive feedback loops that drive population growth. The system is stabilized, however, by negative feedback resulting from host reactions to intensive feeding, competition for settling space, and *Nectria* infections, all of which cause deterioration of the substrate as population density rises on

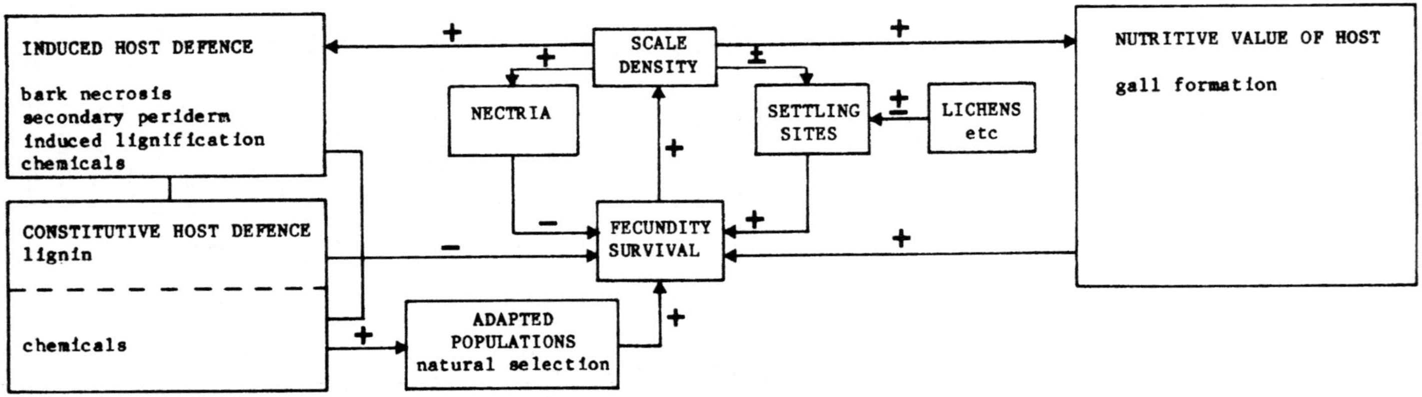

FIGURE 8. Summary of the main interactions among *Cryptococcus*, *Nectria*, and individual *Fagus sylvatica*.

individual trees. On a forestwide scale, the adaptation of populations to individual hosts probably reduces the rate of colonization of other trees and acts as a constraint on disease development in European forests. However, when stands are genetically similar, rapid population increase may occur.

A generalized sequence of scale population changes on trees, and probably in forest stands as a whole, may be derived from the principal interactions depicted in Fig. 8. During an initial establishment phase, while population density is low, the processes of nutritional facilitation and possible adaptation to host genotypes occur. A period of population growth on susceptible trees follows. Population density peaks and begins to decline as competition for settling sites becomes important and induced host-resistance factors cause increased mortality. *Nectria* infections on heavily infested trees in the killing phase of the disease contribute to the decline in scale population. The aftermath stage is characterized by low scale densities on resistant trees. Natural regeneration, by producing more susceptible trees, may result in a new cycle of scale population growth.

The development of populations of insects adapted to individual hosts was first described for the black pineleaf scale (*Nuculaspis californica*).[19] In beech bark disease, this process of adaptation enhances population growth on individual trees, increases the likelihood of *Nectria* infection and subsequent host death, and with it the particular subpopulation of scales.

We believe that adaptation reflects an underlying heterogeneity in qualitative[21] host defenses.[19] In North America it may be that the absence of a coevolved relationship between insect and host has resulted in a more uniformly susceptible tree population that has not encouraged subpopulation development in *Cryptococcus*.[67]

5. IMPLICATIONS FOR THE DEVELOPMENT OF MANAGEMENT STRATEGIES

Chemical control can play only a minor role in the management of beech bark disease. Although impractical in forestry, chemical treatment of a few valuable trees in towns or parks may be acceptable and appears to be followed by a slow recolonization rate. Scrubbing bark with detergent, although somewhat eccentric and certainly labor intensive, is nevertheless highly effective and should be considered when few trees are involved and chemicals best avoided.

In forest stands, salvage cutting of diseased trees before the timber is degraded by insects and fungi minimizes commercial losses. There is currently no evidence that sanitation removal of diseased trees from plantations reduces disease rates, since both *Nectria* and *Cryptococcus* are widely distributed in the forest. In naturally regenerating forests, however, the removal of such susceptible genotypes may be beneficial. Diseased trees would be removed during normal thinning or salvage cuts. Of the trees that remain, current work suggests that, in Europe at least, scale populations will not inevitably rise to damaging levels and that moderately attacked trees may recover without succumbing to *Nectria* attack.

In Europe, the interactions between insect and host populations (Fig. 8) that have produced genetically diverse host and insect populations should reduce the likelihood of widespread damaging outbreaks. This relationship appears not to be present in North

America but may evolve in the long run. Our studies suggest that the maintenance of genetic diversity within stands is all-important in minimizing disease losses. Although some trees appear to be entirely resistant to attack by *Cryptococcus* and could theoretically form the basis of a breeding program, the possibility of the development of adapted scale populations effectively precludes this approach.

Lichens on the bark surface may afford some protection against infestation by *Cryptococcus,* especially in parts of North America, and lichen-covered trees should be favored during silvicultural operations. In some parts of Europe, however, the common lichen *Lecanora conizaeoides* may encourage scale establishment,[33] so that such methods are not appropriate in all forests.

Recent studies of predators suggest that their effects are sporadic and usually confined to trees that are at least moderately infested with scales and that may well already be infected with *Nectria*. There seems no way of increasing the contribution of predators to natural control.

In contrast to the European situation, 80% or more of the beech trees in North American forests eventually become heavily infested with scale and succumb to *Nectria* infection. In the invading front, therefore, badly affected trees, which are often the largest ones in the stand, should be salvaged. However, this may increase the proportion of susceptible beech stems as a result of suckering from the root stocks of felled trees.

In the Adirondacks, heavy shading in the forest favors both sugar maple and beech. Beech is also favored when browsing deer severely limit the height growth of more palatable but silviculturally more desirable species.[35] Successful forest management depends on reduction of deer density by shooting, removal of the understory by herbicide application, and then selective felling to reduce the overstory basal area, with the result that the density of yellow birch and sugar maple is increased. Whether such methods can be economically applied to large areas of forest needs to be considered, as does the environmental contamination that could result from the widespread use of herbicides.

ACKNOWLEDGMENTS. We would like to thank in particular our colleagues David Lonsdale and David Mobbs for help and advice during various stages of the work. Figure 7 was produced from unpublished data of D. Lonsdale, D. Wainhouse, and E. J. Parker. We are indebted to Dr. Berryman for his many helpful and stimulating comments, which have greatly improved our manuscript. Part of this work was funded by an N.E.R.C. research studentship to I.M.G.

REFERENCES

1. Adeli, E., 1975, Insects on Oriental beech (*Fagus orientalis* ssp. *macrophylla*) in Iran and their importance for forestry practices and wood utilisation, *Second World Technical Consultation on Forest Diseases and Insects, New Delhi, April 7–12, 1975*. F.A.O. special paper No. 8.
2. Anonymous, 1984, *Great Britain. Census of Woodlands and Trees 1979–82,* Forestry Commission publications, pp. 1–61.
3. Barter, G. W., 1947, The beech scale, *Bi-mon. Prog. Rep. Forest Insect Invest.* **3**(4):1.
4. Barter, G. W., 1953, Additional observations on the beech scale, *Bi-mon. Prog. Rep. Forest Insect Invest.* **9**(3):1–2.

5. Baylac, M., 1980, Faune associée à *Cryptococcus fagi* (Baer) (*Homoptera: Coccoidea*) dans quelques hêtraies du nord de la France, *Acta Oecol. Oecol. Appl.* **1**:199–208.
6. Berryman, A. A., 1981, *Population systems. A General Introduction,* Plenum, New York.
7. Bogenschutz, H., 1983, Management of beech stands infected by *Cryptococcus fagisuga* in West Germany, in: *Proceedings, of the I.U.F.R.O. Beech Bark Disease Working Party Conference* D. Houston and D. Wainhouse, eds.), pp. 115–119, U.S.D.A. For. Ser. Gen. Tech. Rep. WO-37.
8. Boodle, L. A., and W. Dallimore, 1911, Report on investigations made regarding "Beech coccus" (*Cryptococcus fagi,* Bärensprung), *Kew Ry. Bot. Gard. Bull. Misc. Inform.* **8**:332–343.
9. Braun, H. J., 1976, Das Rindensterben der Buche, *Fagus sylvatica* L, verursacht durch die Buchenwollschildlaus *Cryptococcus fagi* Bar. I. Die Anatomie der Buchenrinde als Basis-Ursache, *Eur. J. Forest Pathol.* **6**:136–146.
10. Braun, H. J., 1977, Das Rindensterben der Buche, *Fagus sylvatica* L, verursacht durch die Buchenwollschildlaus *Cryptococcus fagi* Bar. II. Ablauf der Krankeit, *Eur. J. Forest Pathol.* **7**:76–93.
11. Brower, A. E., 1949, The beech scale and beech bark disease in Acadia national park, *J. Econ. Entomol.* **42**:226–228.
12. Butin, H., 1977, Taxonomy and morphology of *Ascodichaena rugosa* gen. et. sp. nov, *Trans. Br. Mycol. Soc.* **69**:249–254.
13. Camp, W. H., 1950, A biogeographic and paragenetic analysis of the American beech (*Fagus*), *Acad. Natl. Sci. Philos. Am. Philos. Yearbk.* 166–169.
14. Camp, W. H., 1951, A study of relative virulence of beech bark disease on the several types of forest beech present in eastern Canada and Maine, *Acad. Natl. Sci. Phil. Am. Philos. Ybk.* 180–183.
15. Carpenter, G. H., 1903, Woolly scales of the beech. *Cryptococcus fagi* Bärensp. *in* Injurious insects and other animals observed in Ireland during the year 1902, *R. Dublin Soc. Econ. Proc.* **1**:207.
16. Covassi, M., 1975, Nuovi reperti sulla corologia de *Cryptococcus fagisuga* Lndgr. in Italia e in Corsica (Homoptera, Cryptococcidae), *Redia* **56**:555–564.
17. Crosby, D., and Jones, T. W., 1950, Beech scale-*Nectria. Treepest leaflets* No. 4, 4pp. Committee on tree pest leaflets, New England section, Society of American Foresters.
18. Crosby, D., and Bjorkbom, J. C., 1958, Timely salvage can reduce losses from beech scale-*Nectria* attack, *For. Res. Notes, NE For. Exp. Sta.* No. 82, 4pp.
19. Edmunds, G. F., and Alstad, D. N., 1978, Coevolution in insect herbivores and conifers, *Science* **199**:941–945.
20. Ehrlich, J., 1934, The beech bark disease. A *Nectria* disease of *Fagus* following *Cryptococcus fagi* (Baer), *Can. J. Res.* **10**:593–692.
21. Feeny, P., 1975, Biochemical coevolution between plants and their insect herbivores, in: *Coevolution of Animals and Plants* (L. E. Gilbert and P. H. Raven, eds.), pp. 3–19, University of Texas Press, Austin.
22. Filip, S. M., 1978, *Impact of beech bark disease on uneven-aged management of a northern hardwood forest (1952 to 1976).* For Ser. Gen. Tech. Rep. NE-45. 1–7.
23. Fink, S., and Braun, H. J., 1980, Die Bedeutung anatomischer Veränderungen bei der Buchenrindennekrose, *Ann. Sci. Forest.* **37**:333–343.
24. Gate, I. M., 1988, *Population ecology of the beech scale* (*Cryptococcus fagisuga Lind.*), doctoral thesis, University of London.
25. Green, E. E., 1915, *Cryptococcus fagi* Baerensp, in: Observations on British coccidae in 1914, with descriptions of new species. *Entomologist's Mon. Mag.* **51**:180–181.
26. Gregory, P. H., 1961, *The Microbiology of the Atmosphere,* Plant Science Monographs, Leonard Hill Publ, London.
27. Hartig, R., 1880, Die Buchen-Wollaus, *Chermes fagi* Kltb, in: *Untersuchungen aus dem Forstbotanischen Institut zu Munchen,* **1**:156–162.
28. Hawboldt, L. S., 1944, History of the spread of the beech scale *Cryptococcus fagi* (Baerensprung), an insect introduced into the maritime provinces, *Acad. Natl. Bull. Nat. Hist. Soc. N. Bruns.* **1**:137–146.
29. Holstener-Jorgensen, H., and Eiselstein, L. M., 1969, Investigation into the possibility of root drowning as a cause of *Cryptococcus fagi* attacks and other diseases in beech stands, *Det forstlige Forsogsvaesen.* **32**:250.
30. Houston, D. R., 1975, Beech bark disease, the aftermath forests are structured for a new outbreak, *J. For.* **73**:660–663.

31. Houston, D. R., 1983, American beech resistance to *Cryptococcus fagisuga*, in: *Proceedings of the I.U.F.R.O. Beech Bark Disease Working Party Conference* (D. Houston and D. Wainhouse, eds.), pp. 38–42, U.S.D.A. For. Ser. Gen. Tech. Rep. WO-37.
32. Houston, D. R., 1983, Influence of lichen species on colonisation of *Fagus grandifolia* by *Cryptococcus fagisuga:* Preliminary observations from certain Nova Scotian forests, in: *Proceedings of the I.U.F.R.O. Beech Bark Disease Working Party Conference* (D. Houston and D. Wainhouse, eds.), pp. 105–108, U.S.D.A. For. Ser. Gen. Tech. Rep. WO-37.
33. Houston, D. R., Parker, E. J., and Lonsdale, D., 1979, Beech bark disease: Patterns of spread and development of the initiating agent *Cryptococcus fagisuga, Can. J. For. Res.* **9:**336–344.
34. Houston, D. R., Parker, E. J., Perrin, R., and Lang, K. J., 1979, Beech bark disease: A comparison of the disease in North America, Great Britain, France, and Germany, *Eur. J. Forest Pathol.* **9:**199–211.
35. Kelty, M. J., and Nyland, R. K., 1981, Regenerating Adirondack northern hardwoods by shelterwood cutting and control of deer density, *J. For.* **79:**22–26.
36. Kunkel, H., 1968, Untersuchungen über die Buchenwollschildlaus *Cryptococcus fagisuga* Bär. (Insecta Coccina), einen Vertreter der Rindenparenchymsauger, *Z. Angew. Entomol.* **61:**373–380.
37. Kuthe, K., and Kramer, K., 1961. Zur Bekämpfung der Buchenwollschildlaus (*Cryptococcus fagi* Bär) mit chemischen Mitteln, *Anz. Schadlingsk.* **34:**42–43.
38. Lachance, D., 1983, Status of beech bark disease in the province of Quebec, in: *Proceedings, I.U.F.R.O. beech bark disease working party conference.* (D. Houston and D. Wainhouse, Eds.), pp. 18–20, U.S.D.A. For. Ser. Gen. Tech. Rep. WO-37.
39. Lang, K. J., 1983, Present state of beech bark disease in Germany, in: *Proceedings of the I.U.F.R.O. Beech Bark Disease Working Party Conference* (D. Houston and D. Wainhouse, eds.), pp. 10–12, U.S.D.A. For. Ser. Gen. Tech. Rep. WO-37.
40. Lang, K. J., 1983, Development of beech bark disease during a seven year period on two plots in northern Bavaria, in: *Proceedings of the I.U.F.R.O. Beech Bark Disease Working Party Conference.* (D. Houston and D. Wainhouse, eds.), pp. 120–126, U.S.D.A. For. Ser. Gen. Tech. Rep. WO-37.
41. Lonsdale, D., 1980, *Nectria* infection of beech bark in relation to infestation by *Cryptococcus fagisuga* Lindinger, *Eur. J. Forest Pathol.* **10:**161–168.
42. Lonsdale, D., 1983, Wood and bark anatomy of young beech in relation to *Cryptococcus* attack, in: *Proceedings of the I.U.F.R.O. Beech Bark Disease Working Party Conference.* (D. Houston and D. Wainhouse, eds.), pp. 43–49, U.S.D.A. For. Ser. Gen. Tech. Rep. WO-37.
43. Lonsdale, D., and Parker, E. J., 1981, Forest Pathology: Beech bark disease, *Rep. Forest Res. Edin.* **1981:**34–35.
44. Lonsdale, D., and Sherriff, C., 1983, Some aspects of the ecology of *Nectria* on beech, in: *Proceedings of the I.U.F.R.O. Beech Bark Disease Working Party Conference,* (D. Houston and D. Wainhouse, eds.) pp. 59–68, U.S.D.A. For. Ser. Gen. Tech. Rep. WO-37.
45. Malphettes, C. B., 1977, *Cryptococcus fagi* (Barspr.) et dépérissement du hêtre en foret domaniale de Lyons (departements de l'Eure et de Seine-Maritime (France)), *Ann. Sci. Forest.* **34:**159–173.
46. Magasi, L. P., and Newell, W. R., 1983, The status of beech bark disease in the Maritime Provinces of Canada in 1980, in: *Proceedings of the I.U.F.R.O. Beech Bark Disease Working Party Conference,* (D. Houston and D. Wainhouse, eds.) pp. 13–17, U.S.D.A. For. Ser. Gen. Tech. Rep. WO-37.
47. Mayer, M., and Allen, D. C., 1983, *Chilocorus stigma* (Coleoptera: Coccinellidae) and other predators of beech scale in central New York, in: *Proceedings of the I.U.F.R.O. Beech Bark Disease Working Party Conference,* (D. Houston and D. Wainhouse, eds.) pp. 89–98, U.S.D.A. For. Ser. Gen. Tech. Rep. WO-37.
48. Mielke, M. E., and Houston, D. R., 1983, Beech bark disease in West Virginia: status and impact on the Monongahela national forest, in: *Proceedings of the I.U.F.R.O. Beech Bark Disease Working Party Conference,* (D. Houston and D. Wainhouse, eds.) pp. 27–30, U.S.D.A. For. Ser. Gen. Tech. Rep. WO-37.
49. Newstead, R., 1901, Monograph of the Coccidae of the British Isles, *Ann. Vol. Ray. Soc.* 1900 **1:**1, 4, 28–29, 41.
50. Ostrofsky, W. D., and Blanchard, R. O., 1983, Characteristics and development of necrophylactic periderms in mature bark of American beech, in: *Proceedings of the I.U.F.R.O. Beech Bark Disease Working Party Conference,* (D. Houston and D. Wainhouse, eds.), pp. 69–79, U.S.D.A. For. Ser. Gen. Tech. Rep. WO-37.

51. Parker, E. J., 1975, Some investigations with Beech bark disease *Nectria* in southern England, *Eur. J. Forest Pathol.* **5:**118–124.
52. Paucke, H., 1966, Stand der Problematic der Buchenrindennekrose, *Soz. Forstwirtsch.* **16:**179–181.
53. Perrin, R., 1977, Le dépérissement du hêtre, *Rev. For. Fr.* **29:**103–126.
54. Perrin, R., 1979, Contribution à la connaissance de l'étiologie de la maladie de l'ecorce du hetre. I. Etat sanitaire des hetraies francaises. Rôle de *Nectria coccinea* (Pers ex Fries) Fries, *Eur. J. Forest Pathol.* **9:**148–166.
55. Perrin, R., 1980, Contribution à la connaissance de l'étiologie de la maladie de l'écorce du hetre. II. Etude experimentale de l'association *Cryptococcus fagisuga* lind-*Nectria coccinea* (Pers ex Fries) Fries. Rôle respectif des deux organismes, *Ann. Sci. Forest.* **37:**319–331.
56. Perrin, R., 1983, Current status of beech bark disease in France, in: *Proceedings of the I.U.F.R.O. Beech Bark Disease Working Party Conference,* (D. Houston and D. Wainhouse, eds.), pp. 7–9, U.S.D.A. For. Ser. Gen. Tech. Rep. WO-37.
57. Rhumbler, L., 1914, *Die Buchenrinden-Wollaus (Cryptococcus fagi) und ihre Bekampfung,* J. Neumann, Neudamm, pp. 1–32.
58. Schenk, P. J., 1924, De Beukenwolluis, *Floralia* **45:**508.
59. Schimitschek, E., 1980, *Cryptococcus fagi* Barensp. und die Buchen-Rindennekrose, *Anz. Schadlingskde., Pflanzenschutz, Umweltschutz.* **53:**97–98.
60. Schindler, U., 1960, Einige Ergebnisse mehrjahriger Beobachtungen uber Buchensterben und Buchenwollaus (*Cryptococcus fagi* Baer.) *Der Forst-u. Holzwirt* **15:**196–198.
61. Schindler, U., 1962, Erfahrungen mit der Buchenwollschildlaus, *Der Forst-u. Holzwirt* **17:**290–294.
62. Shigo, A. L., 1972, The beech bark disease today in the Northeastern U.S., *J. For.* **70:**286–289.
63. Thomsen, M., Buchwald, N. F., and Hauberg, P., 1949, Angreb af *Cryptococcus fagi, Nectria galligena* og andre parasiter paa bog i Danmark 1939–43, *Det. forstlige Forsogsvaesen.* **18:**97–326.
64. Towers, B., 1983, Status of beech bark disease in Pennsylvania, in: *Proceedings of the I.U.F.R.O. Beech Bark Disease Working Party Conference,* (D. Houston and D. Wainhouse, eds.), pp. 24–26, U.S.D.A. For. Ser. Gen. Tech. Rep. WO-37.
65. Twery, M. J., and Patterson, W. A., 1984, Variations in beech bark disease and its effects on species composition and structure of northern harwood stands in central New England, *Can. J. For. Res.* **14:**565–574.
66. Wainhouse, D., 1976, Forest Entomology: beech scale *Cryptococcus fagisuga, Rep. Forest Res. Edin.* **1976:**36–37.
67. Wainhouse, D., 1983, Interaction between beech and beech scale, in: *Proceedings of the I.U.F.R.O. Beech Bark Disease Working Party Conference,* (D. Houston and D. Wainhouse, eds.), pp. 31–32, U.S.D.A. For. Ser. Gen. Tech. Rep. WO-37.
68. Wainhouse, D., 1980, Dispersal of first instar larvae of the felted beech scale, *Cryptococcus fagisuga, J. Appl. Ecol.* **17:**523–532.
69. Wainhouse, D., and Deeble, R., 1980, Variation in susceptibility of beech (*Fagus* spp) to beech scale (*Cryptococcus fagisuga*), *Ann. Sci. Forest.* **37:**279–289.
70. Wainhouse, D., and Howell, R. S., 1983, Intraspecific variation in beech scale populations and in susceptibility of their host *Fagus sylvatica, Ecol. Ent.* **8:**351–359.
71. Wainhouse, D., and Howell, R. S., 1983, Distribution of attack by beech scale *Cryptococcus fagisuga* in beech progeny trials, in: *Proceedings of the I.U.F.R.O. Beech Bark Disease Working Party Conference,* (D. Houston and D. Wainhouse, eds.), pp. 33–37, U.S.D.A. For. Ser. Gen. Tech. Rep. WO-37.
72. Wainhouse, D., Gate, I. M., and Lonsdale, D., 1987, Host resistance to beech scale; a variety of defences, *I.U.F.R.O. Working Party Conference. Plant resistance mechanisms to insects and pathogens* (in press), Springer-Verlag, New York.
73. Wilkes, W., 1902, The beech tree in England, *J. Roy. Hort. Soc.* **26:**598–599.
74. Wujciak, R., 1976, Buchenschleimfluss und Holzqualitat, *Forstarchiv* **47:**71–78.

CHAPTER 5

THE BALSAM WOOLLY ADELGID IN NORTH AMERICA

Fred Paul Hain

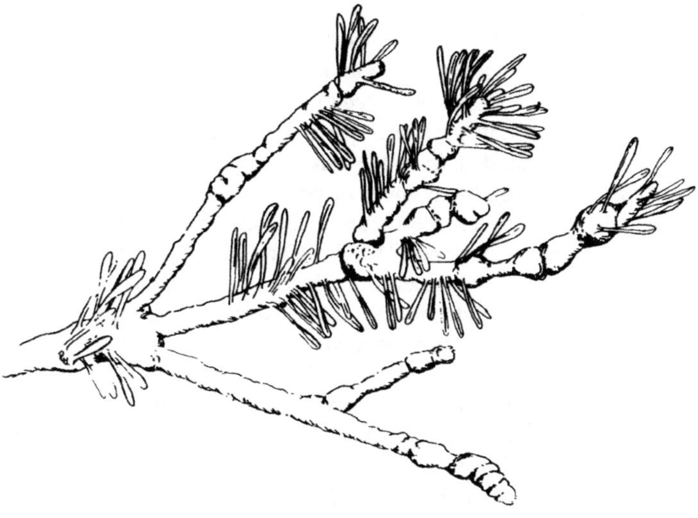

Fred Paul Hain • Departments of Entomology and Forestry, North Carolina State University, Raleigh, North Carolina 27695–7626.

1. INTRODUCTION

The balsam woolly adelgid, also known as the balsam woolly aphid, is native to the silver fir (*Abies alba*) forests of central Europe. The insect was introduced into North America around 1900. Although European firs are not seriously affected by this adelgid, North American firs frequently experience either crown dieback or tree death, or both.

There is a great deal of variety in the biology of the adelgid, the host's reaction to its attack, and, in the opinion of scientists regarding its taxonomic status. For example, Europeans attribute much of the variation in adelgid biology to differences among species and subspecies,[57,92] while in North America the variation may be caused by specific subspecies.[63,64] Europeans place the true fir infesting adelgids in the genus *Dreyfusia*.[23] North American entomologists follow a classification that places all the sub-family Adelginae into two genera, *Pineus* and *Adelges*, based on the number of abdominal spiricles.[11] This latter classification is used here.

Of the four *Adelges* species found in Europe only *A. nordmannianae* Eckstein and *A. piceae* (Ratzeburg) occur in North America.[34] *A. nordmannianae* is the more important economic pest in Europe, where it is also an introduced species, but relatively unimportant in North America. By contrast, *A. piceae* is a very important economic pest in North America but innocuous on European fir, where the insect is a native species.

Adelges piceae is separated from the other adelgids on the basis of several morphological features, including the number of pores on the dorsal sclerites of first instar nymphs, the length of the stylet, the shape of the terminal antennal segments, and the structure of the abdominal wax glands.[34,44,117] Three subspecies of *A. piceae* have been identified on the basis of morphometric analysis and revisions by European taxonomists[63,64]: *A. piceae piceae* (Ratzeburg), Pacific Northwest, British Columbia, and the southeastern United States; *A. piceae canadensis* (Merker and Eichhorn), eastern Canada and New England; and *A. piceae occidentalis* (Foottit and Mackauer), British Columbia.

The balsam woolly adelgid was accidentally introduced into Maine and Nova Scotia, probably on nursery stock. It became established on balsam fir, *A. balsamea* in Maine and Maritime Canada by 1908.[18] A separate introduction occurred in the Pacific Northwest, where the principal hosts are grand fir (*A. grandis*), subalpine fir (*A. lasiocarpa*), and Pacific silver fir (*A. amabilis*).[81] The adelgid was discovered in the southern Appalachians in 1955 on Fraser fir (*A. fraseri*) in North Carolina, Virginia, and Tennessee, and on bracted balsam fir (*A. balsamea* var. *phanerolepis*) in Virginia and West Virginia.[2] Balsam fir has the most extensive range of all the North American true firs, occurring from Newfoundland to Alberta, Labrador to Pennsylvania; however, the adelgid is only found on balsam fir in New England (as far south as New York), Newfoundland, Nova Scotia, New Brunswick and Quebec.[66,86] Adelgid-caused damage in these areas has been extensive and, at times, intense.

2. BIOLOGY AND ECOLOGICAL RELATIONSHIPS

2.1. Life History

Most fir-infesting adelgids have complex life cycles composed of several forms, sexual and parthenogenetic reproduction, and primary (usually spruce) and secondary hosts. However, *A. piceae* has lost the capacity to produce migrant forms and occurs only on true firs (the secondary host). Males and sexual reproduction are not known to occur.

The life cycle consists of the egg, three larval instars and the adult. The first instar, about 0.4 mm in length, is the only form capable of movement and is commonly called the crawler stage (Fig. 1). The crawler locates a suitable feeding site, usually within 24–48 hr after hatching, inserts its stylet into the bark of the host, and transforms, without molting, into a flattened, wax-fringed resting stage (neosistens, singular; neosistentes, plural (Fig. 1). Thereafter, the insect remains permanently attached to the feeding site (sessile). Feeding sites are usually bark lenticels and other roughened areas of the main stem, around branch and twig nodes, and the base of buds.[33] Tree-to-tree dispersal occurs when the crawlers or eggs are passively transported by wind, birds, or other animals. The

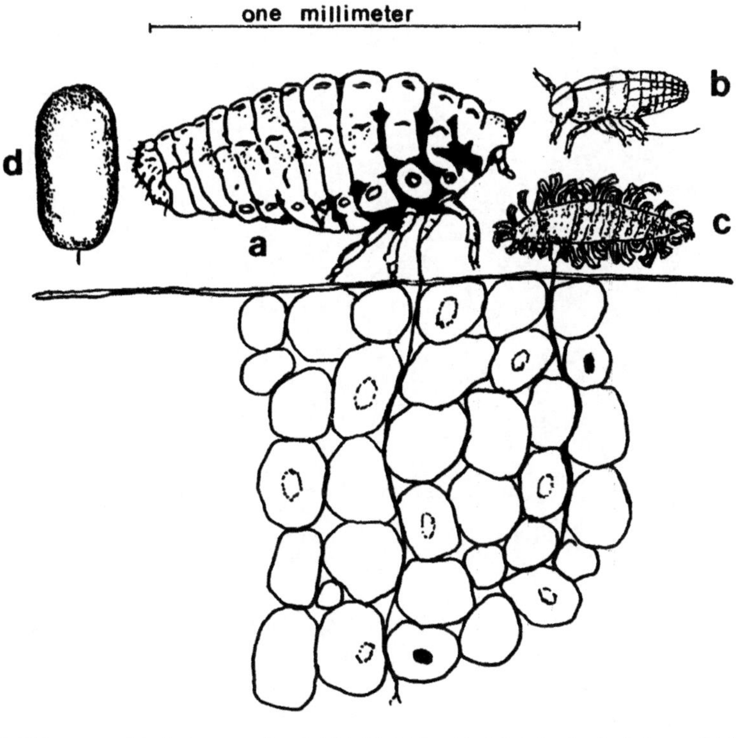

FIGURE 1. Life stages of the balsam woolly adelgid. (a) Adult with stylet piercing bark tissue. (b) Crawler. (c) Neosisten with wax fringe and stylet piercing bark tissue. (d) Egg.

second and third instars are about 0.5 and 0.65 mm in length, respectively, closely resemble the adult, and are covered with secretions of waxy threads that appear as a dense white wool mass (Fig. 2). These stages and the adult are collectively known as the sistentes (sistens, singular). The adult females are about 0.80 mm in length, dark purple to black, nearly spherical, and wingless. As many as 248 amber-colored eggs (Fig. 1) are

FIGURE 2. Typical bole infestation of the balsam woolly adelgid.

laid within the woolly mass. The dense woolly secretions provide protection for all of the life stages except the crawler.[18] In Europe and the Maritime Provinces of Canada, a rare stage has been observed and is known as the progrediens.[86] There are winged and wingless forms of the progredientes (plural). The wingless form is similar to the sistentes, while the winged form has conspicuous membranous wings and five-segmented antennae and generally lacks wax pores.

The winter is passed as a dormant first instar nymph, and this generation is referred to as hiemosistens. Development is completed in the spring, with adult numbers peaking in late May or early June.[18,66] The first instar nymphs of the second generation also undergo a dormant period (summer aestivation) ranging from 2 to 8 weeks.[6] The adults of the aestivosisten generation peak in mid-August and produce the overwintering generation. At lower altitudes or warmer climates, additional generations may be produced.[13,84] A third generation usually results from faster development, and a partial fourth may occur during years of extremely early development.

2.2. Dispersal

With the exception of the rarely observed winged form, the only dispersing stage of the adelgid is the crawler, which is passively dispersed by wind, by humans on infested trees or logs[17] or by phoresy. Insects, mostly ground beetles; mammals, mostly *Sorex* sp.; and amphibians, a mixture of frogs, newts, and salamers, were found to carry crawlers.[121] However, wind is considered the main dispersal agent.

Crawlers show no distinct pattern of movement on the host tree and usually stay near their point of origin or drop from the tree.[16] The crawlers are photokinetic; i.e., they display a light-stimulated increase in walking velocity but avoid going directly toward light and an increased rate of dropping from the tree.[16,56] Therefore, dispersal in closed-canopy stands is confined to within-tree dispersal and movement to neighboring trees. The higher temperatures and light intensities associated with stand openings increase crawler activity and dropping from the tree. These crawlers are more likely involved in long distance dispersal because they are picked up by the daytime outflow of air along the margin and are then carried aloft by the convection currents generated in the clearing.[16]

2.3. Symptoms and Host Colonization

The amount of damage sustained by an infested tree is related to the size of the attacking adelgid population, to the part(s) of the host being attacked, and to the physiological state of the tree. Adelgids generally concentrate in the outer portions of tree crowns or on the main stem and large branches. The precise location of the infestation depends on the tree species and geographical location. Stem infestations usually cause more damage and tree mortality than do crown infestations.

Crown infestations are characterized by abnormal drooping of the current shoots and gouting of the outer twigs. Gouting (Fig. 3) refers to the swelling of nodes and buds and retarded new growth caused by an increase of thick-walled xylem cells and an enlargement of phloem and bark cells.[81] The crown becomes increasingly thin, and dieback may

occur. A persistent crown infestation can kill the tree over a number of years. Trees in which crown tops have been killed assume a typical flat-topped appearance.[10]

Stem attacks are characterized by the conspicuous presence of white woolly masses that, under heavy attack, give the lower bole a whitewashed appearance (Fig. 5.2). In some cases, stem infestations may virtually disappear after a few years without killing the tree—this is especially true in Europe. However, most North American firs are usually killed within 2–6 years of a sustained infestation.

2.4. Host Response to Adelgid Attack

The anatomical and structural changes in host tissue caused by an *A. piceae* infestation include an increase in the number and size of parenchyma cells in the feeding zone and the production of *rotholz* or redwood in the xylem tissues (Fig. 4). This reaction may be the result of salivary secretions from the adelgid stylet.[19] The stimulation of the cambium to produce *rotholz* is particularly intriguing, since the physical presence of the stylet only extends to the living bark cells (see Fig. 1). The stylet is inserted intercellularly through the bark epidermis into the cortex and the parenchyma tissue. The adelgid feeds from cortical tissues, including the annual phelloderm layers, the medullary rays and the epithelium of the resin canals. In young and thin bark tissue, the stylets occasionally reach the phloem but do not enter it. At the base of buds, the stylets do enter the meristematic tissue.[18]

Because of *rotholz* formation and the increased volume of bark parenchyma, the radial increment of infested trees in the area of infestation may increase for several years. In places on the stem where adelgid feeding has been concentrated, the bark surface may bulge outward. After a few years, however, the width of the annual rings will be reduced below that of a normal uninfested tree.

Rotholz is reddish in color and superficially resembles compression wood.[18,38,117] The intensity of the red coloration varies, however, from undetectable to dark red. The xylem tissue from infested fir contains higher amounts of ray tissue,[83,115] thickened cell walls, shorter trachaeids,[55,65] a reduced number of conducting pit pores in the trachaeids,[101] and encrusted pit membranes that resemble those of normal heartwood.[102]

Water flow in the sapwood of infested trees is greatly reduced,[82] almost to the level

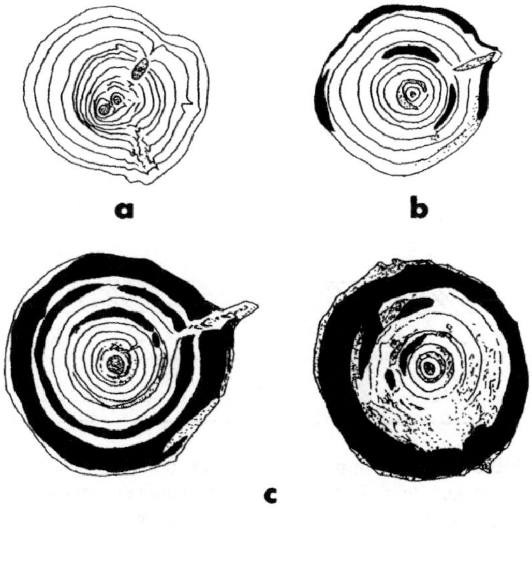

FIGURE 4. Various degrees of rotholz or redwood caused by a bole infestation of the balsam woolly adelgid. (a) None. (b) Slight. (c) Heavy.

of normal heartwood.[98] There is no distinguishable difference between normal heartwood and the wood formed as a response to adelgid attack.[98,100,101] Since the cambium and ray cells are not physically damaged, it appears that the adelgid merely enhances the normal pattern of heartwood formation (premature heartwood) and causes the production of new chemical compounds in the sapwood.[100] In contrast to uninfested trees, water transport in infested trees is limited to narrower bands of early wood and in fewer growth rings and does not ascend as high.[82] The number of cellular protoplasmic starch granules decreases as the period of adelgid feeding increases.[92,103] The tree is in a state of physiological drought. As fluid movement is impaired, photosynthesis and respiration are reduced, contributing to the death of the host.[99]

The initial feeding of the adelgid occurs near lenticels or bark fissures where the tissue protein content is naturally high. On European firs, continued feeding in these areas causes increased protein levels in adjacent tissues which leads to the subsequent spread of the adelgid onto these areas.[78] In time, the infested tissue becomes exhausted, the cells die and are no longer suitable for adelgid feeding. European firs do not form *rotholz* in response to attack,[81] which may explain the innocuous status of the pest.

The formation of *rotholz* in North American firs appears to be a continuation of the defensive system (allergy). A healthy defensive response to stem invasion involves three processes[21]: (1) wound cleansing by primary resin flow, (2) wound containment by a hypersensitive response that includes the concentration of additional resin in the cells adjacent to the wound site, and (3) wound healing by the formation of necrophylactic periderm and callus tissue.

When the periderm tissue of conifers is injured, either by biotic or abiotic means, necrophylactic periderm (secondary periderm) is formed internal to the wound.[89,90] This tissue isolates the necrotic cells of the hypersensitive response from the healthy portion of the stem and is part of the normal wound healing process. A layer of impervious tissue precedes formation of necrophylactic periderm.[22,87] In a susceptible response, the formation of the impervious layer (Fig. 5) may be inhibited or delayed,[88] perhaps as a result of

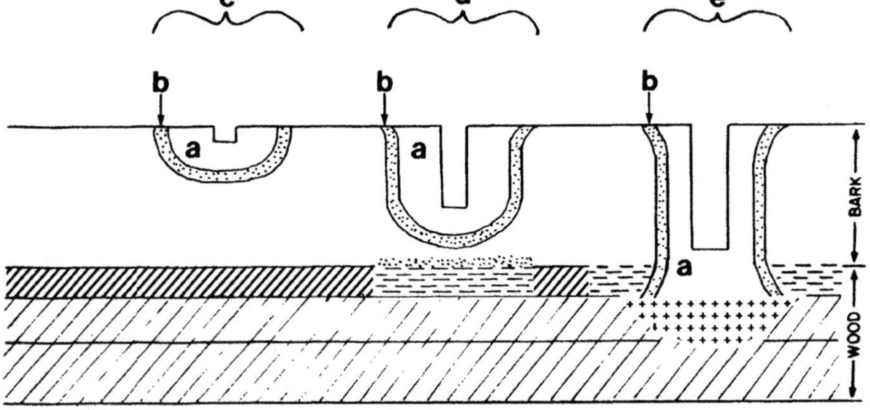

FIGURE 5. Host response to wounding. (a) Hypersensitive response. (b) Formation of impervious tissue around the wound. (c,d) Responses contained within the bark with little or no effect to underlying tissue. (e) Response to a deeper wound resulting in the formation of impervious xylem tissue. (From Mullick.[87])

stylet secretions. The hypersensitive response continues into deeper tissue layers and ultimately into the xylem, where *rotholz* is formed. By contrast, European firs may complete the formation of the impervious tissue without delay and thus avoid any permanent damage to the underlying cells.

The formation of impervious tissue internal to a wound can occur within 3 weeks in midsummer or 6 weeks in the spring and fall.[91] Water stress can greatly retard the formation of impervious tissue at artificial wound sites.[105]

A number of stylet secretions or other chemicals associated with *A. piceae* infestations have been identified. These include pectinase,[1] auxinlike compounds,[19] and juvabionelike substances.[104] However, there is no conclusive evidence that these compounds inhibit the formation of impervious tissue.[14] In fact, the juvabionelike substances may be produced by the host in response to attack. The substances are not produced in uninfested fir, nor are they found in infested fir except in the vicinity of the adelgid.[104] Other chemicals concentrated within the hypersensitive response, such as monoterpenes, may also play a role in host resistance.[12]

2.5. Effects on the Host

Living trees can recover from all levels of adelgid damage. This is particularly true of crown infestations where damage surveys have been developed for describing the extent of damage from observable symptoms.[110,118] Light damage may consist of node swelling with some stunting or distortion, while severe damage consists of prominent distortion with the terminals and some branches bare for more than 30 cm. Even trees severely damaged by a crown infestation can recover, but the likelihood of recovery decreases with age.[110]

Recovery of top-killed trees may proceed with lateral growth but no vertical development, resulting in flat-topped trees. The recovery of younger trees usually results in the reorientation of a branch to the vertical position. Recovery may be permanent with no reoccurrence of the adelgid, or temporary with the tree passing through several cycles of damage.[108] Recovery can also occur from stem infested trees; however, the likelihood decreases with tree age and adelgid density.

The impact of crown infestations on seed and cone production is significant only on severely damaged trees.[107,109] Firs with stem infestations may still produce cones,[60] but the cones are shorter and the seeds ligther, smaller, less viable, and more readily infested by seed chalcids.[61] The amount of damage to seed crops appears to be related to the cumulative damage sustained by the tree rather than to the level of infestation during the year that cones are produced.

Whatever the impact of the adelgid on seed and cone production, understory fir reproduction is usually plentiful.[72,110,120] Understory seedlings and saplings also appear to be more resistant to attack (perhaps because of smooth bark, fewer lenticels, and poor nutrient status of the tree) and sustain an adelgid population only when there is a heavily infested overstory. Once the overstory is killed, the adelgids apparently have difficulty surviving on the remaining seedlings and saplings.

Adelgid infestations may predispose the tree to infection by *Armellaria* root rot. Disease incidence and intensity can increase dramatically with increasing levels of adelgid damage,[75,76] and may be an important factor contributing to the mortality of infested trees. Wood wasps and their fungal symbionts have also been associated with adelgid-infested trees[20,74,116] and may kill heavily infested firs. A bark fungus, *Lachnellula agassizii,* closely associated with stem-infested fir that are irreversibly damaged,[62] may be a reliable predictor of the imminent death of green trees.

Infested trees are more susceptible to ice or wind damage. Those trees that have recovered from crown infestations are especially susceptible, because the new crown growth is abnormally exposed (crown breakage) or root development is retarded (uprooted).

2.6. Natural Controls

Numerous projects have involved the collection and importation of balsam woolly adelgid predators (no parasites are known) from Europe (especially Germany), Australia, Japan, India, and Pakistan for release into North America.[3,8,9,25–30,37,45–51,59,85,112–14] Six of the European predators have become established in various parts of North America.[86] These are *Laricobius erichsonii* (Rosenhauer) (Coleoptera: Derondontidae), *Pullus impexus* (Mulsant) (Coleoptera: Coccinellidae), *Aphidecta obliterata* (Linnaeus) (Coleoptera: Coccinellidae), *Aphidoletes thompsoni* Möhn (Diptera: Cecidomyiidae), *Cremifania nigrocellulata* (Czerny) (Diptera: Chamaemyiidae), and *Leucopis obscura* Haliday (Diptera: Chamaemyiidae). Numerous native North American predators have also been noted.[29,47,80] However, none of the predators, native or introduced, has had a major impact on adelgid populations. Apparently the predators are general feeders that neither concentrate on adelgid infestations nor fluctuate with adelgid population levels, or the predators feed mainly on the egg stage but do not affect subsequent life stages.[7]

Several naturally occurring fungal pathogens have been isolated from *A. piceae.*[111] *Fusarium larvarum* showed a high degree of pathogenicity and may be found on most infested trees,[58] but trees with heavily infected adelgid populations are uncommon and scattered.

Cold winter temperatures can have a dramatic effect on adelgid survival. Only individuals below the snowline survive temperatures of −30°F,[18] and substantial mortality occurs at −5°F or less.[4,66] Even in the southern Appalachians, cold winter temperatures can periodically decimate adelgid populations. For example, Mt. Mitchell, North Carolina, recorded a temperature of −35°F with very little snow cover in January 1985 and the following summer adelgid infestations decreased by 50%. The apparent lack of adelgid activity in the northeastern United States during recent years is also attributed to cold winter temperatures.

Late spring frost occurring after adelgid development begins or early frost in the fall before the crawlers have settled can kill a high proportion of the population.[31,73,110] Winter survival is also influenced by the site at which the adelgid settles on the tree, those settling on the lower bole and overwintering below the snow having the greatest survival and those settling in the crown, especially at the shoot tips, having the least.[32]

Temperature and relative humidity have a direct effect on survival, especially during the egg and crawler stages.[5] Although egg hatch is similar over a wide range of temperatures (11–20°C) and relative humidities (75–98%), the longevity of crawlers varies considerably and is generally reduced by high humidities. Optimum conditions for incubating eggs and the subsequent longevity of crawlers is 75% RH and 15°C.[5] At cooler temperatures (<7°C), egg incubation is prolonged and crawler longevity reduced. Crawlers produced under suboptimum conditions will probably have greater difficulty finding suitable feeding sites. Low humidities are also thought to increase egg mortality.[5,7]

Although the dormant overwintering nymph is capable of surviving temperatures well below freezing, the other life stages, including the overwintering nymph that has broken dormancy, are killed at temperatures near freezing.[18,66] Apparently a portion of the adelgid population has the capacity to break dormancy early, when temperatures are near freezing.[15] These individuals run the risk of high mortality if conditions become unfavorable. However, if temperatures remain above freezing, development is completed sooner compared with the case in individuals that break dormancy later, and an additional generation may be completed during the year. Although high temperatures reportedly killed adelgids during hot summers in Canada,[18] this has not been observed in North Carolina.[7]

3. FACTORS AFFECTING DISTRIBUTION AND ABUNDANCE

The host and its geographical distribution are the two most important factors determining the numerical behavior of balsam woolly adelgid populations. Although the insect is found in eastern and western North America, it has not extended its range into the interior, probably because of the harsh winter climate. In the east, fir–spruce forests extend down the crest of the Appalachians from Maine to North Carolina and Tennessee and merge with the great boreal coniferous forests of the north. Southward, however, the spruce–fir type is poorly developed through the relatively low Allegheny Mountains of Pennsylvania, West Virginia, and Virginia. In the high southern Appalachians of North Carolina and Tennessee, spruce–fir forests occur as a glacial remnant from the Pleistocene (see refs. 24,35,36,39,40,54, and 95–97 cited in ref. 93) only on the high mountain and ridge tops, surrounded on all sides by deciduous forests of the lower slopes and valleys.

The spruce–fir forests of the northern and southern parts of the Appalachians are quite similar.[93] However, many of the southern species have been isolated long enough to have developed different characteristics from their more widespread northern relatives. In the case of fir, different species and varieties have evolved, such as Fraser fir in North Carolina and Tennessee and bracted balsam fir in West Virginia.[106]

In Oregon, Washington, and British Columbia, most adelgid populations are found in the warmer climate west of the Cascade Mountains. The few infestations east of the Cascades are always at the lowest elevations, where fir will grow naturally or in low-elevation ornamental plantings.

3.1. Within the Tree

The patterns of infestation (crown or stem attack) differ with host species and geographical area. For example, Fraser fir is damaged primarily from stem infestations while bracted balsam fir sustain mainly crown infestations.[10] On Fraser fir, stem infestations usually begin near mid-bole, but this varies with height and age of the tree.[7] Infestations usually start higher on older trees with rough bark and at the base of young trees where the rough bark is located.

In the Pacific Northwest, the main hosts for the adelgid are grand fir, subalpine fir, and Pacific silver fir. These species occupy a variety of sites and are sometimes found growing together. The pattern of infestation and the effects on the host differ markedly with species and location.

Grand firs typically sustain stem infestations but, being the most resistant of the Pacific species, only 20–30% of the infested trees die.[81] Trees become susceptible at about 15 years of age; the greatest mortality occurs at 25–35 years. Older trees that become infested are usually reinfestations. Attacks usually start on the lower 5 ft of the main bole and, as nutrients are exhausted from this region in 2–5 years, the adelgid slowly moves up the stem. Hypersensitive responses may produce such copious quantities of resin that it flows down the bole. Adelgid-infested grand firs do not always produce *rotholz*,[101] although the anatomical and chemical features of the xylem are altered and permeability reduced.[99–102]

Subalpine fir is the most susceptible Pacific Northwest species.[81] Up to 90% mortality may occur within 5 years of stem infestation. Unlike grand fir, stem infestations typically start high on the tree (mid-crown) and move down. Most trees die within 1 year after the infestation reaches the base of the tree. Although resinosis may be associated with the last stages of decline, it is not as common as in grand fir.

Pacific silver fir is somewhat less susceptible than subalpine fir. Although the less lethal crown attack is more common in Pacific silver fir, stem attacks do occur and can kill up to 70% of a stand within a few years.[81] When stem attacks occur, the pattern of infestation is unpredictable, sometimes starting high on the bole and working down, or vice versa. Again, stem resinosis is less common than it is on grand fir.

Other fir species growing in natural stands, arboretums, or ornamental plantings show varying degrees of susceptibility.[81] In general, North American firs are very sensitive (Table I), with Fraser fir in the Southeast and subalpine fir in the Pacific Northwest showing the greatest damage. Oriental firs are moderately susceptible, while European firs are practically immune.

The size of the adelgid population and the degree of damage is usually the greatest on fast-growing dominant and codominant trees growing on the best sites.[7,71,77] This finding suggests that the most vigorous trees provide the most nutritious substrate for the adelgid.

The following scenario is typical of tree growth while being stem-infested by the balsam woolly adelgid[7]:

1. The initial infestation causes an increase in diameter growth of fast-growing trees because of adelgid stimulation.
2. Decline in wood and branch growth occurs as adelgid populations increase and cause changes in the nutritional quality of the bark.[78]
3. As the adelgid population peaks, adelgid survival and tree growth decline until the adelgid population eventually collapses.
4. The tree dies 1 or 2 years after most of the adelgids have disappeared.

TABLE I
Damage Severity of 18 Species of *Abies* Found Naturally Infested with Balsam Woolly Adelgid in the Pacific Northwest[a]

Tree species	Tree origin	Damage			
		Severe[b]	Moderate[c]	Slight[d]	Nil[e]
Abies lasiocarpa (Hooker) Nuttall Subalpine fir	Western North America	X	—	—	—
A. fraseri (Pursch) Poiret Fraser fir	Eastern U.S.	X	—	—	—
A. balsamea (Linnaeus) Miller Balsam fir	Northeastern North America	X	—	—	—
A. amabilis (Douglas) Forbes Pacific silver fir	Northwestern North America	X	—	—	—
A. grandis (Douglas) Lindley Grand fir	Western North America	—	X	—	—
A. lasiocarpa var. *arizonica* (Merriam) Lemmon Corkbark fir	Southwestern North America	—	X	—	—
A. magnifica var. *shastensis* Lemmon Shasta red fir	Western U.S.	—	X	—	—
A. koreana Wilson Korean fir	Korea	—	X	—	—
A. sachalinensis Masters Sakhalin fir	Northeastern Asia	—	X	—	—
A. religiosa Lindley Sacred fir	Southern Mexico	—	—	X	—
A. procera Rehder Noble fir	Western U.S.	—	—	X[f]	—
A. concolor (Gordon and Glendining) Lindley White fir	Western U.S.	—	—	X[f]	—
A. alba Miller European silver fir	Western Europe	—	—	—	X
A. cephalonica Loudon Grecian fir	Greece	—	—	—	X
A. pinsapo Boissier Spanish fir	Spain	—	—	—	X
A. sibirica Ledebour Siberian fir	Northern Asia	—	—	—	X
A. firma Siebold and Zuccarini Momi fir	Japan	—	—	—	X
A. veitchii Lindley Veitch's silver fir	Japan	—	—	—	X

[a]From Mitchell.[81]
[b]Trees often killed; gouting severe.
[c]Gouting moderate to severe; trees occasionally killed.
[d]Gouting moderate to light; trees not observed killed.
[e]Gouting not apparent; trees not killed.
[f]Moderate to light infestations.

3.2. Within Stands

Fir mortality usually begins at low elevations (near the spruce–fir transition with hardwoods), along stand edges, or in open stands of fir in association with hardwoods. The infestation then spreads toward the summit and deeper into the stand as the adelgid population increases.[7,41,72,77,94,120]

Damage is usually more severe on dry well-drained sites than on moist poorly drained sites.[94,110] In fact, sites conducive to the development of moisture stress are commonly associated with severe adelgid damage. The inhibition of water transport as a result of adelgid attack apparently causes the rapid development of severe symptoms on water-stressed sites.[110] The fact that severe damage occurs most frequently on the older more mature trees may also be related to increasing water stress. Larger populations will develop on trees in the more moist sites, perhaps because of better nutritional substrates and/or a longer period before tree death.

3.3. Geographical Areas

Infestation patterns on balsam fir vary with geographical location and climate.[66] The climate of the maritime provinces of Canada changes dramatically from a mild maritime climate to a harsher continental climate in the interior. In the maritime regions, winter temperatures are relatively mild and do not influence population trends of the adelgid. Since temperatures rarely fall below −15°F, adelgids survive in both crowns and stems, and crown infestations disperse freely, so that damage in the region is extensive.

In the continental interior, temperatures frequently drop below −15°F. Therefore, adelgids attacking the crown usually sustain high overwintering mortality. Since wind velocities are low under the canopy, dispersal is limited. Infestations are usually small and isolated with damage from stem attack accumulating slowly. As one might expect, the infestation and damage patterns as well as the severity of the winters are much less regular in the transitional climatic regions.[66]

Grand fir stands are uniformly susceptible to attack on the west side of the Cascade Mountains. However, on the east side and along the mountain tops, grand firs are usually uninfested even when intermingled with infested subalpine fir.[81]

Both host characteristics and meteorological factors are important considerations in evaluating the patterns of numerical behavior of adelgid populations within a geographical area. Furthermore, the great boreal spruce–fir forest of the north is so extensive that a uniformly distributed food supply is available. In many areas of Europe, the native fir are restricted to specific bioclimatic zones along the slopes of rugged mountain terrain. In such areas, adelgid dispersal is probably dependent on the cross-valley circulation flow that occurs as the sun warms the air.[119] During such dispersal, especially if frontal weather alters the circulation, high mortality rates can occur. Mortality during dispersal, and the coevolved resistance factors of the native hosts, may explain why *A. piceae* is not destructive in its native home. When the adelgid was introduced into the gentler hills and plateaus of northeastern North America, it was released from the cross-valley transport system and provided with a large supply of susceptible hosts.[119]

4. HYPOTHESES FOR THE CAUSES OF OBSERVED POPULATION DYNAMICS

The population behavior of the balsam woolly adelgid is affected by four major factors: (1) host species and tolerance, (2) stress factors acting on the host plant (site and weather factors), (3) meteorological conditions acting on the insect, and (4) tree size (Fig. 6).

4.1. Host Tolerance

Resistant host species tolerate adelgid attack and possess efficient mechanisms for containing the infestation and preventing additional injury. Regardless of its defensive capacity, the tree provides a suitable nutrient substrate for a period of time that is made even more suitable by the adelgid feeding. As the attack begins, protein levels in the surrounding bark tissue increase, making the tissue more nutritive. Populations continue to increase until the nutrient content of the bark is depleted, at which time the adelgid disappears.

In tolerant host species, hypersensitive responses may contain chemicals that inhibit adelgid development. Necrophylactic periderm is then produced interior to the wound, and the underlying conductive tissue is relatively unharmed.

In susceptible species, the hypersensitive responses appear to continue unabated into the conductive tissue (allergenic responses). Necrophylactic periderm either is not formed or is incompletely formed. The conducive tissue becomes impregnated with chemicals, perhaps the same that occur in tolerant species, and these impede water flow. Thus, the inability of the tree to control its defense response results in increased water stress and ultimate death. This type of effect seems analogous to the allergenic overreaction of animal defenses to invasion by foreign matter.

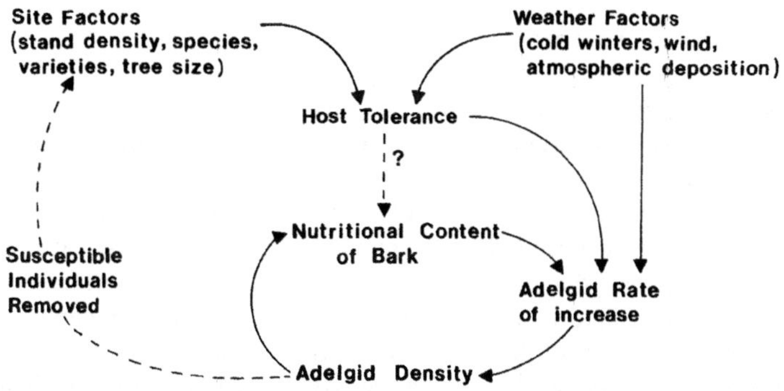

FIGURE 6. The interactions of variables that influence the development of balsam woolly adelgid populations.

4.2. Host Stress and Tree Size

Water-stressed trees or species that are more susceptible to water stress usually exhibit greater damage from adelgid attack. This may be due to impaired defensive responses resulting from their stressed condition and/or impaired capacity to translocate fluids.

Although other stress factors have not been investigated in detail, the mortality patterns in the southern Appalachians suggest that additional stress factors may be involved.[68] Six discontinuous areas of Fraser fir occur in the southern Appalachian chain: the Great Smokey Mountains, the Plott Balsam Mountains, the Black Mountains (including Mt. Mitchell), Roan Mountain, Grandfather Mountain, and Mt. Rogers. Greatest Fraser fir mortality has occurred in the more southerly areas surrounding the heavily industrialized and populated French Broad River Basin. Fir stands to the north of the Black Mountains have experienced considerably less mortality and, until recently, Mt. Rogers was considered uninfested. However, Mt. Rogers has been infested for 22 years with little tree mortality.[70] In contrast to Mt. Mitchell, Mt. Rogers is located in a remote part of the southern Appalachians. A number of hypotheses have been offered to explain the observed mortality pattern of Fraser fir.[68] First, climatological differences may have an effect on host vigor, the southerly mountains being higher and subject to greater extremes of temperature, wind velocity, and cloud interception. Second, the six zones of Fraser fir are genetically distinct and may possess wide variation in their tolerance to infestation. Fraser fir is basically a glacial remnant that survives on six mountaintop islands with little genetic interchange. It is possible that certain subpopulations of the host may not have evolved tolerance to bark-feeding insects. Currently, evolution may be selecting for tolerant varieties (those Fraser fir that survive attack). Finally, air pollution or atmospheric deposition may be predisposing Fraser fir to adelgid-caused mortality or, conversely, atmospheric deposition may be adding a final stress to heavily infested trees. Those more southerly mountaintops that have the greatest amount of Fraser fir mortality may also sustain the highest levels of atmospheric deposition because of their elevation, high frequency of cloud interception (clouds provide a concentrated source of many pollutants), and proximity to areas of dense human population and industrialization.

4.3. Meteorological Factors

Cold winter temperatures clearly have an influence on adelgid survival and on the subsequent within-tree and within-region infestation patterns. In some parts of its range, the adelgid can only survive under the snow line. In other parts, where winter temperatures are usually warmer, a periodic cold snap can greatly reduce the overwintering insects, or a late spring frost can reduce the number of first instars that have just emerged from dormancy. It appears that cold winter temperature is an important factor in preventing the spread of the adelgid into the interior of the North American continent and may have contributed to the current demise of the adelgid in the northeast.

Meteorological factors may also affect the host, which may affect adelgid establishment. For example, Pacific silver fir is susceptible to attack at 1500-ft elevation but is rarely attacked at 4000 ft when surrounded by infested subalpine fir.

5. MANAGEMENT IMPLICATIONS

On the basis of our understanding of the factors influencing the numerical behavior of *A. piceae* populations, the main implications for management are (1) minimizing tree stress, (2) encouraging tolerant species and varieties, and (3) monitoring environmental factors that influence adelgid development.

5.1. Minimizing Tree Stress

Tree stress may be minimized by thinning overstocked stands, fertilizing sites of poor nutrient quality, harvesting mature stands in high hazard areas, and planting in low hazard areas. Thinning may not reduce the adelgid population but could increase the tolerance of residual trees or retard their death. Fertilization has been shown to reduce adelgid populations by inhibiting larval settling,[42] but some forms of nitrogenous fertilizer such as urea increase survival.[43]

Harvesting schedules should be applied to high hazard stands and to those already infested.[77,110] A damage hazard-rating system, based on the site and stand characteristics associated with severe adelgid damage, has been developed for Newfoundland,[94] but it is probably applicable (with modifications) throughout the adelgid range. The main variables of the hazard-rating system are site elevation, soil moisture regime, percentage balsam fir by basal area, total basal area of balsam fir, and stand breast–height age.[94] The system has three parts, representing three elevation classes: (1) below the maximum elevations for occurrence of severe damage and tree mortality, (2) above the preceding but below the maximum elevation of any class of damage, and (3) above the preceding where the probabilty of adelgid damage is low. In some regions, class 1 may be considered those elevations where infestations are likely to originate, class 2 elevations where infestations are likely to spread into, and class 3 elevations where little damage occurs. In some areas, such as the southern Appalachians, class 3 may not exist. Within each elevation, the other parameters mentioned are used to rate the damage hazard (Table II).

Cutting infested trees can facilitate adelgid dispersal if the trees are cut when the crawler life stage predominates.[79] Even winter cutting may not be effective if the logs are not used immediately, since the adelgids are able to diapause on cut logs.[15]

5.2. Encouraging Tolerant Species and Varieties

After harvesting, resistant and immune species already growing in the aphid-infested area should be encouraged. Depending on the region, these species could include red spruce, black spruce, Douglas-fir, western hemlock, Pacific silver fir, noble fir, or Shasta red fir. Planting apparently tolerant species outside their native range may alter the degree of tolerance.[81] However, in Newfoundland two strains of Veitch's silver fir from Japan were found to be a suitable substitute for balsam fir[69] when growth and resistance factors were compared. Veitch's silver fir is totally resistant to *A. piceae* (see Table I). In some areas, such as the southern Appalachians, converting to another species would be impractical because of the low commercial value of fir in this restricted range. However, certain

TABLE II
Severe Damage-Hazard Rating for Sites and Stands in Elevation Classes 1 and 2[a]

Elevation class	Moisture regime	Percentage balsam fir by basal area	Total balsam fir basal area (m^2/ha)	Breast-height age classes (years)		
				<25	25–45	>45
1	Dry–fresh	<40	Any value	Low	Low-moderate	Moderate-high
		>40	<25	Moderate	Moderate-high	High
		>40	>25	Moderate	High	High
	Moist–wet	<40	Any value	Very low	Very low	Low-moderate
		>40	<25	Very low	Moderate	Moderate
		>40	>25	Low	Moderate-high	Moderate-high
2	Dry–fresh	<40	Any value	Nil	Very low	Low-moderate
		>40	<25	Low	Low-moderate	Moderate
		>40	>25	Low	Moderate	Moderate
	Moist–wet	<40	Any value	Nil	Nil	Very low
		>40	<25	Nil	Low	Low
		>40	>25	Nil	Low-moderate	Low-moderate

[a]From Page.[94]

varieties of Fraser fir, such as those found at Mt. Rogers, may have increased tolerance of the adelgid.

5.3. Monitoring Environmental Factors

Monitoring environmental factors that stress both the host and the insect, such as drought or winter temperature, may serve as a predictor of adelgid populations. Thus high-hazard stands (susceptibility) can also be rated for risk (i.e., probability of an infestation).

5.4. Other Controls

Attempts at artificial application of the fungal pathogen, *F. larvarum,* proved unsuccessful.[58] Treatment with combinations of cultures or with sublethal doses of chemical insecticides may show promise. Secondary metabolites of the fungus have shown toxicity toward mosquito larvae and adult blowflies.[52,53,67] Chemical control is impractical over large forested areas but can be effective on ornamental trees, Christmas tree plantations, or seed production areas. Although the adelgid population may be completely controlled, the damage to a tree that has been infested over a number of years is irreversible, and the tree may not recover. Therefore, chemical treatments should only be applied to recently infested trees; additional spraying may not be necessary for many years.

ACKNOWLEDGMENTS. I wish to thank G. D. Amman and R. G. Mitchell for reviewing an earlier draft of this chapter. The drawing of gouting of fir twigs was done by R. L. Livingston. I am also grateful to the following for permission to reprint previously published material: Plenum Publishing Corporation for Fig. 5, the *Canadian Journal of Forest Research* for Table II, and the U.S.D.A. Pacific NW Forest and Range Experiment Station for Table I.

REFERENCES

1. Adams, J. B., and McAllan, J. W., 1958, Pectinase in certain insects, *Can. J. Zool.* **36:**305–308.
2. Amman, G. D., 1962, Seasonal biology of the balsam woolly aphid on Mt. Mitchell, North Carolina, *J. Econ. Entomol.* **55:**96–98.
3. Amman, G. D., 1966, *Aphidecta obliterata* (Coleoptera: Coccinellidae), an introduced predator of the balsam woolly aphid, *Chermes piceae* (Homoptera: Chermidae), established in North Carolina, *J. Econ. Entomol.* **59:**506–511.
4. Amman, G. D., 1967, Effect of minus 29°F on overwintering populations of the balsam woolly aphid in North Carolina, *J. Econ. Entomol.* **60:**1765–1766.
5. Amman, G. D., 1968, Effects of temperature and humidity on development and hatching of eggs of *Adelges piceae, Ann. Entomol. Soc. Am.* **61:**1606–1611.
6. Amman, G. D., 1969, A method of sampling the balsam woolly aphid on Fraser fir in North Carolina, *Can. Entomol.* **101:**883–889.
7. Amman, G. D., 1970, Phenomena of *Adelges piceae* populations (Homoptera: Phylloxeridae) in North Carolina, *Ann. Entomol. Soc. Am.* **63:**1727–1734.
8. Amman, G. D., and Speers, C. F., 1964, Release of predators of the balsam woolly aphid in North Carolina, U.S. Forest Serv. Res. Note SE-32.
9. Amman, G. D., and Speers, C. F., 1971, Introduction and evaluation of predators from India and Pakistan for control of the balsam woolly aphid (Homoptera: Adelgidae) in North Carolina, *Can. Entomol.* **103:**528–533.
10. Amman, G. D., and Talerico, R. L., 1967, Symptoms of infestation by the balsam woolly aphid displayed by Fraser fir and bracted balsam fir, U.S. Forest Serv. Res. Note SE-85.
11. Annand, P. W., 1928, *A Contribution Toward a Monograph of the Adelgidae (Phylloxeridae) of North America,* Stanford University Press, Palo Alto, California.
12. Arthur, F. H., 1985, Physiological characteristics of Fraser fir in relation to mortality caused by the balsam woolly adelgid, Ph.D. thesis, North Carolina State University, Raleigh, Norh Carolina.
13. Arthur, F. H., and Hain, F. P., 1984, Seasonal history of the balsam woolly adelgid (Homoptera: Adelgidae) in natural stands and plantations of Fraser fir, *J. Econ. Entomol.* **77:**1154–1158.
14. Arthur, F. H., and Hain, F. P., 1985, Development of wound tissue in the bark of Fraser fir and its relation to injury by the balsam woolly adelgid, *J. Entomol. Sci.* **20:**129–135.
15. Atkins, M. D., 1972, Developmental variability among laboratory reared balsam woolly aphid (Hemiptera: Chermidae), *Can. Entomol.* **104:**203–208.
16. Atkins, M. D., and Hall, A. A., 1969, Effect of light and temperature on the activity of balsam woolly aphid crawlers, *Can. Entomol.* **101:**481–488.
17. Atkins, M. D., and Woods, T. A. D., 1968, Survival of the balsam woolly aphid on *Abies* logs, *Can. Entomol.* **100:**412–420.
18. Balch, R. E., 1952, Studies of the balsam woolly aphid, *Adelges piceae* (Ratz.) and its effects on balsam fir, *Abies,* (L.) Mill, Canadian Department of Agriculture Publ. 867.
19. Balch, R. E., Clarke, J., and Bonga, J. M., 1964, Hormonal action in products of tumors and compression wood (in *Abies balsamea*) by an aphid (*Adelges piceae*), *Nature (Lond.)* **202:**721–722.
20. Basham, J. T., Etheridge, D. E., Hudak, J., LaChance, D., Maqasi, L. P., and Stillwell, M. A., 1974, The deterioration of balsam fir trees killed by girdling in eastern Canada, Can. For. Serv., Sault St. Marie, Ont. Inf. Rep. O-X-214.

21. Berryman, A. A., 1972, Resistance of conifers to invasion by bark beetle–fungus associations, *BioScience* **22**:598–602.
22. Biggs, A. R., 1985, Detection of impervious tissue in the tree bark with selective histochemistry and fluorescence microscopy, *Stain Technol.* **60**:299–304.
23. Borner, C., 1908, Theories on the biology of Chermidae, *Zool. Anz.* **33**:647–663 (in German).
24. Brown, C. A., 1938, The flora of Pleistocene deposits in the Western Florida Parishes, West Feliciana Parish, and East Baton Rouge Parish, Louisiana, *Louisiana Geol. Surv., Geol. Bull.* **12**:59–94.
25. Brown, N. R., and Clark, R. C., 1956, Studies of predators of the balsam woolly aphid, *Adelges piceae* (Ratz.). I. Field identification of *Neoleucopis obscura* (Hal.), *Leucopina americana* (Mall) and *Cremefania nigrocelluata* (Cz.) (Diptera, Chamaemiidae), *Can. Entomol.* **88**:272–279.
26. Brown, N. R., and Clark, R. C., 1956, Studies of predators of the balsam woolly aphid, *Adelges piceae* (Ratz.). II. An annotated list of the predators associated with the balsam woolly aphid in eastern Canada, *Can. Entomol.* **88**:678–683.
27. Brown, N. R., and Clark, R. C., 1957, Studies of predators of the balsam woolly aphid, *Adelges piceae* (Ratz.). IV. *Neoleucopis obscura* (Hal.), an introduced predator in eastern Canada, *Can. Entomol.* **89**:533–546.
28. Brown, N. R., and Clark, R. C., 1959. Studies of predators of the balsam woolly aphid, *Adelges piceae* (Ratz.). VI. *Aphidecta obliterata* (L.) (Coleoptera: Coccinellidae), an introduced predator in eastern Canada. *Can. Entomol.* **91**:596–599.
29. Brown, N. R., and Clark, R. C., 1960. Studies of predators of the balsam woolly aphid, *Adelges piceae* (Ratz.). VII. *Syrphidae* (Diptera), *Can. Entomol.* **92**:801–811.
30. Brown, N. R., and Clark, R. C., 1962, Studies of predators of the balsam woolly aphid, *Adelges piceae* (Ratz.). X. Field identification of *Laricobius erichsonii* Rosen. (Coleoptera: Derodontidae), *Can. Entomol.* **94**:191–193.
31. Bryant, D. G., 1971, Balsam woolly aphid, *Adelges piceae* (Homoptera: Phylloxeridae), seasonal and spatial development in crowns of balsam fir, *Abies balsamea, Can. Entomol.* **103**:1411–1420.
32. Bryant, D. G., 1972, The measurement of population density of the balsam woolly aphid, *Adelges piceae* (Ratz.) (Homoptera: Phylloxeridae), a highly aggregate insect, Ph.D. dissertation, Yale University, New Haven, Connecticut.
33. Bryant, D. G., 1974, Distribution of first instar nymphs of *Adelges piceae* (Homoptera: Phylloxeridae) on branches of balsam fir, *Abies balsamea,* after colonization, *Can. Entomol.* **106**:1075–1080.
34. Bryant, D. G., 1974, A review of the taxonomy, biology and importance of the balsam woolly aphid, *Adelges piceae* (Homoptera: Phylloxeridae) on branches of balsam fir, *Abies balsamea, Can. Entomol.* **108**:1097–1111.
35. Buell, M. F., 1945, Late Pleistocene forests of southeastern North Carolina, *Torreya* **45**:117–118.
36. Buell, M. F., 1946, The age of Jerome bog, "A Carolina bay," *Science* **103**:14–15.
37. Buffam, P. E., 1962, Observations on the effectiveness and biology of the European predator *Laricobius erichsonii* Rosen. (Coleoptera: Derodontidae) in Oregon and Washington, *Can. Entomol.* **94**:461–472.
38. Busby, R. J. N., 1964, A new adelgid on *Abies grandis* causing compression wood, *Q. J. For.* **58**:160–162.
39. Cain, S. A., 1944, Pollen analysis of some buried soils, Spartanburg County, South Carolina, *Bull. Torr. Bot. Club* **71**:11–22.
40. Cain, S. A., 1944, Size frequency characteristics of *Abies fraseri* pollen as influenced by different methods of preparation, *Am. Midl. Nat.* **31**:232–236.
41. Carrow, J. R., 1973, Establishment and survival of balsam woolly aphid on second growth Amabilis fir at intermediate elevations, *Bi-Mon. Res. Notes, Can. For. Serv.* **29**:10–11.
42. Carrow, J. R., and Graham, K., 1968, Nitrogen fertilization of the host tree and population growth of the balsam woolly aphid, *Adelges piceae, Can. Entomol.* **100**:478–485.
43. Carrow, J. R., and Betts, R. E., 1973, Effects of different foliar-applied nitrogen fertilizers on balsam woolly aphid, *Can. J. For. Res.* **3**:122–139.
44. Carter, C. I., 1971, Conifer woolly aphids (Adelgidae) in Britain, British Forestry Commission Bulletin #42
45. Clark, R. C., and Brown, N. R., 1957, Studies of predators of the balsam woolly aphid, *Adelges piceae* (Ratz.). III. Field identification and some notes on the biology of *Neoleucopis pinicola* Mall. (Dipteria: Chamaemyiidae), *Can. Entomol.* **89**:404–409.

46. Clark, R. C., and Brown, N. R., 1958, Studies of predators of balsam woolly aphid. V. *Laricobius erichsonii* Rosen (Coleoptera: Derodontidae) an introduced predator in eastern Canada, *Can. Entomol.* **90:**657–672.
47. Clark, R. C., and Brown, N. R., 1959, Predator introduction for balsam woolly aphid control, *Can. Dept. Agric. Bi-Mon. Prog. Rep.* **15:**1.
48. Clark, R. C., and Brown, N. R., 1960, Studies of predators of the balsam woolly aphid, *Adelges piceae* (Ratz.). VII. *Laricobius subsidus* Lec. (Coleoptera: Derodontidae), a predator on *Pineus strobi* (Homoptera: Adelgidae), *Can. Entomol.* **92:**237–240.
49. Clark, R. C., and Brown, N. R., 1960, A laboratory method for rearing predators of the balsam woolly aphid, *Adelges piceae* (Ratz.), *Can. Entomol.* **92:**696–697.
50. Clark, R. C., and Brown, N. R., 1961, Studies of balsam woolly aphid, *Adelges piceae* (Ratz.). IX. *Pullus impexus* (Muls.) (Coleoptera: Coccinellidae), an introduced predator in eastern Canada, *Can. Entomol.* **93:**1162–1168.
51. Clark, R. C., and Brown, N. R., 1962, Studies of predators of the balsam woolly aphid, *Adelges piceae* (Ratz.). XI. *Cremifania nigrocelluata* Cz. (Diptera: Chamaemyiidae), an introduced predator in eastern Canada, *Can. Entomol.* **94:**1171–1175.
52. Claydon, N., Grove, J. F., and Pople, M., 1977, Insecticidal secondary metabolic products from the entomogenous fungus *Fusarium solani, J. Invert. Pathol.* **30:**216–233.
53. Claydon, N., Grove, J. F., and Pople, M., 1979, Insecticidal secondary metabolic products from the entomophagous fungus *Fusarium larvarum, J. Invert. Pathol.* **33:**364–367.
54. Davis, J. H., 1946, The Peat Deposits of Florida, Florida Geol. Surv., Geol. Bull. 30.
55. Doerksen, A. H., and Mitchell, R. G., 1965, Effects of the balsam woolly aphid upon wood anatomy of some western true firs, *For. Sci.* **11:**181–188.
56. Edwards, D. K., 1966, Observations on the crawlers of the balsam woolly aphid, *Adelges piceae* (Ratz.), *Can. Dept. Forestry and Rural Development, Bi-Mon. Res. Notes* **22:**4–5.
57. Eichhorn, O., 1968, Problems of the population dynamics of silver fir woolly aphids, genus Adelges (=Dreyfusia), Adelgidae, *Z. Angew. Entomol.* **61:**157–214.
58. Fedde, G. F., 1971, A parasitic fungus disease of *Adelges piceae* (Homoptera: Phylloxeridae) in North Carolina, *Ann. Entomol. Soc. Am.* **64:**749–750.
59. Fedde, G. F., 1972, Status of imported and native predators of the balsam woolly aphid on Mt. Mitchell, North Carolina, U.S. Dept. Agric. For. Serv. Res. Note SE-175.
60. Fedde, G. F., 1973, Cone production in Fraser firs infested by the balsam woolly aphid, *Adelges piceae* (Homoptera: Phylloxeridae), *J. Ga. Entomol. Soc.* **8:**127–130.
61. Fedde, G. F., 1973, Impact of the balsam woolly aphid (Homoptera: Phylloxeridae) on cones and seed produced by infested Fraser fir, *Can. Entomol.* **105:**673–680.
62. Fedde, G. F., 1974, A bark fungus for identifying Fraser fir irreversibly damaged by the balsam woolly aphid, *Adelges piceae* (Homoptera: Phylloxeridae), *J. Ga. Entomol. Soc.* **9:**64–68.
63. Foottit, R. G., and Mackauer, M., 1980, Morphometric variation between populations of the balsam woolly aphid, *Adelges piceae* (Ratzeburg) (Homoptera: Adelgidae) in North America, *Can. J. Zool.* **58:**1494–1503.
64. Foottit, R. G., and Mackauer, M., 1983, Sub-species of the balsam woolly aphid, *Adelges piceae* (Homoptera: Adelgidae), in North America, *Ann. Entomol. Soc. Am.* **76:**299–304.
65. Foulger, A. N., 1968, Effect of aphid infestation on properties of grand fir, *For. Prod. J.* **18:**43–47.
66. Greenbank, D. O., 1970, Climate and ecology of the balsam woolly aphid, *Can. Entomol.* **102:**546–548.
67. Grove, J. F., and Hosken, M., 1975, The larvicidal activity of some 12,13-epoxytrichothec-9-enes, *Biochem. Pharmacol.* **24:**959–962.
68. Hain, F. P., and Arthur, F. H., 1985, The role of atmospheric deposition in the latitudinal variation of Fraser fir mortality caused by the balsam woolly adelgid, *Adelges piceae* (Ratz.) (Hemipt; Adelgidae): A hypothesis, *Z. Angew. Entomol.* **99:**145–152.
69. Hall, J. P., and Schooley, H. O., 1981, Growth and development of *Abies veitchii* in western Newfoundland, Environ. Can. For. Serv. Inform. Rep. N-X-198.
70. Haneman, D. M., Lambert, H. L., and Johnston, D. D., 1981, Detection and evaluation of the balsam woolly aphid infestations on Mount Rogers, Virginia, 1980. USDA For. Serv. For. Insect and Disease Management Rep. 81.
71. Harris, J. W. E., 1973, Tree and site characteristics relative to progressive balsam woolly aphid damage to

Abies spp., British Columbia, 1959–1970, Dept. Environ. Can. For. Serv. Pacific For. Res. Centre, Victoria, B.C., Internal Rep. BC-46.

72. Hay, R. L., Eager, C. C., and Johnson, K. D., 1976, Status of the balsam woolly aphid in the Great Smoky Mountains National Park. USDI, Natl. Park Serv., SE Region, Res./Resource Management Rep. 20.
73. Hemmerick, G. M., and Kendall, G. R., 1972, Frost data 1941–1970, Can. Atmospheric Environ. Serv. CLI 5-72.
74. Hudak, J., 1975, Microbial deterioration of balsam fir damaged by the balsam woolly aphid in western Newfoundland. Ph.D. dissertation, State University of New York, Syracuse, College of Environmental Science and Forestry.
75. Hudak, J., and Singh, P., 1970, Incidence of Armillaria root rot in balsam fir infested by balsam woolly aphid, *Can. Dept. Agric. Plant Dis. Serv.* **50:**99–101.
76. Hudak, J., and Wells, R. E., 1974, Armillaria root rot in aphid-damaged balsam fir in Newfoundland, *For. Chron.* **50:**74–76.
77. Johnson, N. E., Mitchell, R. G., and Wright, K. H., 1963, Mortality and damage to Pacific silver fir by the balsam woolly aphid in southwestern Washington, *Am. J. For.* **61:**854–860.
78. Kloft, W., 1957, Further investigations concerning the inter-relationship between bark condition of *Abies alba* and infestation by *Adelges picea typica* and *A. nusslini schneideri*, *Z. Angew. Entomol.* **41:**438–442.
79. Lambert, H. L., and Ciesla, W. M., 1967, Impact of summer cutting on the dispersal of the balsam woolly aphid, *J. Econ. Entomol.* **60:**613–614.
80. Mitchell, R. G., 1962, Balsam woolly aphid predators native to Oregon and Washington, Agr. Exp. Sta., Oregon State Univ., Tech. Bull. 62.
81. Mitchell, R. G., 1966, Infestation characteristics of the balsam woolly aphid in the Pacific Northwest, USDA For. Serv. Res. Paper PNW-35.
82. Mitchell, R. G., 1967, Translocation of dye in grand and subalpine firs infested by the balsam woolly aphid, USDA For. Serv. Res. Note PNW-46.
83. Mitchell, R. G., 1967, Abnormal ray tissue in three true firs infested by balsam woolly aphid, *For. Sci.* **13:**327–332.
84. Mitchell, R. G., Johnson, N. E., and Rudinsky, J. A., 1961, Seasonal history of the balsam woolly aphid (*Adelges piceae*) in the Pacific Northwest, *Can. Entomol.* **93:**794–798.
85. Mitchell, R. G., and Wright, K. H., 1967, Foreign predator introductions for control of the balsam woolly aphid in the Pacific Northwest, *J. Econ. Entomol.* **60:**140–147.
86. Mitchell, R. G., Amman, G. D., and Waters, W. E., 1970, Balsam woolly aphid, USDA For. Serv. For. Pest Leaflet 118.
87. Mullick, D. B., 1975, A new tissue essential to necrophylactic periderm formation in the bark of four conifers, *Can. J. Bot.* **53:**2443–2457.
88. Mullick, D. B., 1977, The nonspecific nature of defense in bark and wood during wounding, insect and pathogen attack, in: *Recent Advances in Phytochemistry*. Vol II (F. A. Loewus and V. C. Runeckles, eds.), pp. 395–441, Plenum, New York.
89. Mullick, D. B., and Jensen, G. D., 1973, Cryofixation reveals uniqueness of reddish-purple sequent periderm and equivalence between brown first and brown sequent periderms of three conifers, *Can. J. Bot.* **51:**135–143.
90. Mullick, D. B., and Jensen, G. D., 1973, New concepts and terminology of coniferous periderms: Necrophylactic and exophylactic periderms, *Can. J. Bot.* **51:**1459–1470.
91. Mullick, D. B., and Jensen, G. D., 1976, Rates of nonsuberized impervious tissue development after wounding at different times of the year in three conifer species, *Can. J. Bot.* **54:**881–892.
92. Oechssler, von G., 1962, Sucking injuries to the tissues of native and exotic firs caused by Central European fir aphids, *Z. Angew. Entomol.* **50:**408–454 (in German; transl. Can. Dept. For. No. 57, 1966.)
93. Oosting, H. J., and Billings, W. D., 1951, A comparison of virgin spruce–fir forest in the northern and southern Appalachian system, *Ecology* **32:**84–103.
94. Page, G., 1975, The impact of balsam woolly aphid damage on balsam fir stands in Newfoundland, *Can. J. For. Res.* **5:**195–209.
95. Potzger, J. E., 1945, The pine barrens of New Jersey, a refugium during Pleistocene times, *Butler Univ. Bot. Stud.* **7:**182–196.

96. Potzger, J. E., and Otto, J. H., 1943, Post-glacial forest succession in northern New Jersey as shown by pollen records from five bogs, *Am. J. Bot.* **30**:83–87.
97. Potzger, J. E., and Tharp, B. C., 1947, Pollen profile from a Texas bog, *Ecology* **28**:274–280.
98. Puritch, G. S., 1971, Water permeability of the wood of grand fir (*Abies grandis* (Doug.) Lindl.) in relation to infestation by the balsam woolly aphid, *Adelges piceae* (Ratz.), *J. Exp. Bot.* **22**:936–945.
99. Puritch, G. S., 1973, Effect on water stress on photosynthesis, respiration and transpiration of four *Abies* species, *Can. J. For. Res.* **3**:293–298.
100. Puritch, G. S., 1977, Distribution and phenolic composition of sapwood and heartwood in *Abies grandis* and effects of the balsam woolly aphid, *Can. J. For. Res.* **7**:54–62.
101. Puritch, G. S., and Petty, J. A., 1971, Effect of balsam woolly aphid, *Adelges piceae* (Ratz.), infestation on the xylem of *Abies grandis* (Doug.) Lindl, *J. Exp. Bot.* **22**:946–952.
102. Puritch, G. S., and Johnson, R. P. C., 1971, Effects of infestation by balsam woolly aphid, *Adelges piceae* (Ratz.) on the ultrastructure of bordered-pit membranes of grand fir, *Abies grandis* (Doug.) Lindl, *J. Exp. Bot.* **22**:953–958.
103. Puritch, G. S., and Talmon-de l'Armée, M., 1971, Effect of the balsam woolly aphid, *Adelges piceae*, infestation on the food reserves of grand fir, *Abies grandis*, *Can. J. Bot.* **49**:1219–1223.
104. Puritch, G. S., and Nijholt, W. W., 1974, Occurrence of juvabione-related compounds in grand fir and Pacific silver fir infested by balsam woolly aphid, *Can. J. Bot.* **52**:585–587.
105. Puritch, G. S., and Mullick, D. B., 1975, Effect of water stress on the rate of non-suberized impervious tissue (NIT) formation following wound healing in *Abies grandis*, *J. Exp. Bot.* **26**:903–910.
106. Robinson, J. F., and Thor, E., 1969, Natural variation in *Abies* of the southern Appalachians, *For. Sci.* **15**:238–245.
107. Schooley, H. O., 1975, Cone production of balsam fir damaged by balsam woolly aphid, *For. Chron.* **51**:105–107.
108. Schooley, H. O., 1976, Effect of balsam woolly aphid on cone and seed production by balsam fir, *For. Chron.* **52**:237–239.
109. Schooley, H. O., 1976, Recovery of young balsam fir trees damaged by balsam woolly aphid, *For. Chron.* **52**:143–144.
110. Schooley, H. O., and Bryant, D. G., 1978, The balsam woolly aphid in Newfoundland, Can. For. Serv. Inf. Rep. N-X-160.
111. Smirnoff, W. A., 1970, Fungus diseases affecting *Adelges piceae* in the fir forests of the Gaspe Peninsula, Quebec, *Can. Entomol.* **102**:799–805.
112. Smith, B. C., 1958, Response to light and influences of light and temperature on locomotion of the crawlers of balsam woolly aphid, *Adelges piceae* (Ratz.) and of insect predators of this species, *Can. Entomol.* **90**:193–201.
113. Smith, B. C., 1958, Development, feeding habits, and predator-prey relations of insect predators of the balsam woolly aphid, *Adelges piceae* recently introduced into Canada, *Can. Entomol.* **90**:441–449.
114. Smith, B. C., and Coppel, H. C., 1957, Releases in North America and reviews of bionomics in Europe of insect predators of the balsam woolly aphid, *Adelges piceae* (Ratz.), *Can. Entomol.* **89**:410–420.
115. Smith, F. H., 1967, Effects of balsam woolly aphid (*Adelges piceae*) infestation on cambial activity in *Abies grandis*, *Am. J. Bot.* **54**:1215–1223.
116. Stillwell, M. A., 1966, Woodwasps (Siricidae) in conifers and the associated fungus, *Stereum chailletii*, in eastern Canada, *For. Sci.* **12**:121–128.
117. Varty, I. W., 1956, *Adelges* insects of silver firs, *G. B. For. Comm. Bull.* **26**:1–75.
118. Warren, G. L., Parrott, W. C., and Cochran, S. G., 1967, Balsam woolly aphid dispersal and damage in balsam fir stands in Newfoundland, Can. For. Serv. Inform. Rep. N-X-15.
119. Wellington, W. G., 1976, Life at the cloud line, *Nat. Hist.* **85**:100–105.
120. Witter, J. A., and Ragenovitch, I. R., 1986, Impact and ecology of Fraser fir subjected to depredations by the balsam woolly adelgid, *For. Sci.* **32**:585–594.
121. Woods, T. A. D., and Atkins, M. D., 1967, A study of the dispersal of balsam woolly aphid crawlers by small animals, *Can. Dept. For. Rural Devel. Bi-Mon. Res. Notes* **23**:44.

CHAPTER 6

THE LARGE PINE APHID ON SCOTS PINE IN BRITAIN

Neil A. C. Kidd

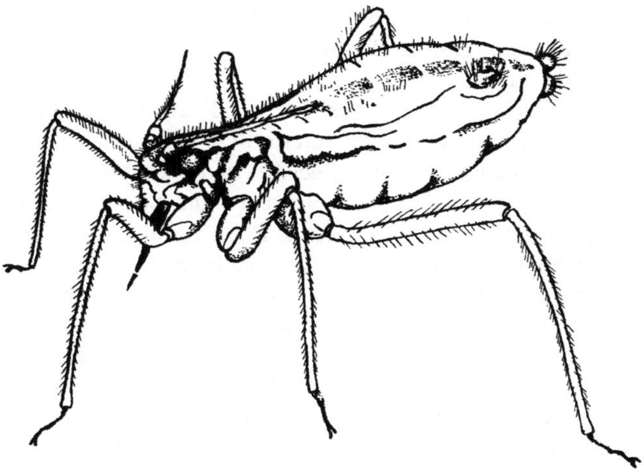

Neil A. C. Kidd • Department of Zoology, University College, Cardiff CF1 1XL, Wales.

1. INTRODUCTION

The large pine aphid, *Cinara pinea* (Mordvilko), infests a number of pines but is found most commonly on Scots pine (*Pinus sylvestris*).[16] It is widely distributed throughout the Palaearctic, coinciding with the range of its main host plant.[6] There are also numerous records from North America, where Scots pine has been introduced and grown on a commercial basis.

Cinara pinea belongs to a mainly tree-dwelling family of aphids, the Lachnidae, many of which have damaging effects on their hosts. In common with most aphid species, they usually feed by extracting phloem sap, an activity that involves the production of a copious sugary honeydew. Aphid feeding may cause defoliation,[19,20,37,38,55,68,70] reduced or disturbed growth,[1,18,22–24,30,36,49,61,65] and reduced resistance to other pathogens.[33,34] Occasionally, after high or repeated infestations, this may lead to tree death.[27] Injury may be caused by the direct nutrient drain, the influence of salivary secretions, or the growth of sooty moulds on honeydew deposits.[18,30,61,64] Anatomical changes can also be produced at some distance from the site of feeding, including uneven tapering of stems, nodal swelling, and splitting of bark, all of which may lead to a reduction in the economic value of the timber.[61]

While numerous lachnid species infest hardwood trees,[16,17,60] it is those that attack conifers that are of greatest concern to the forester. On European cedars, for example, *C. lapportei* induces defoliation of 1-year old needles and branch dieback,[19] while *C. cedri* defoliates important shade trees in parts of France and Spain.[55] In South Africa, *C. cronartii* has recently become a serious pest of *Pinus taeda* and *P. patula*.[61]

The most damaging lachnid attacks have usually been recorded on young trees or seedlings. The spruce aphid, *C. piceae,* for example, can be a serious pest of young European fir trees,[64] while in Japan, infestation of young spruce plantations may reduce tree growth by as much as 60–70%.[23,24] On Scots pine, a number of species may be considered injurious. The spotted pine aphid (*Eulachnus agilis*) has caused serious defoliation of commercially grown Christmas trees in some parts of North America,[37–39] and similar problems are now becoming apparent in parts of Britain. Both *C. pinea* and the gray pine aphid (*Shizolachnus pineti*) are known to reduce the growth rate of Scots pine,[39,49,65] and it has been suggested that *C. pini,* a stem-feeding species, may transmit a virus that induces shoot-growth disturbances.[30] All these last four species may periodically infest nursery-raised pines,[6] often reducing their sturdiness and suitability for planting. There is evidence that the survival and subsequent performance of pine seedlings in the field are related to their size at planting, so any retardation of growth by aphids at the nursery stage is likely to have serious implications.[49,65]

Despite the economic importance of conifer lachnids, there have been surprisingly few investigations of the factors determining their abundance. This neglect may be partly due to difficulties in obtaining accurate population estimates, as many species frequently change their feeding sites within the canopy in apparently unpredictable ways. Some, e.g., *C. cronartii,* may also move on a seasonal basis between canopy and roots.[66] These habits make even the collection of basic life-history information an arduous task. Fortunately, however, there are no such problems with the large pine aphid, *C. pinea*. Its simple life cycle, predictable feeding habits, and widespread distribution have helped make it the most studied, and perhaps best understood, of the tree-dwelling lachnids.

2. BIOLOGY AND LIFE HISTORY

Cinara pinea overwinters as an egg, although in certain warmer parts of its range the adult aphids may survive until spring. In Britain, egg hatch starts at the beginning of March and generally continues until the end of that month (Fig. 1). The first generation adults (fundatrices) are found until late May, after they have passed through four instars. These virgin females (virginoparae) produce a sequence of asexual (parthenogenetic) generations (usually six), the adults of which are all wingless (apterous), except for a short period during June and July, when a proportion are winged (alate) (Fig. 1). In autumn, some of the aphids develop as either winged males or wingless oviparae, the latter producing the overwintering eggs, which are deposited sparsely on the current year's needles. The aphids are thus confined to pine throughout the year.

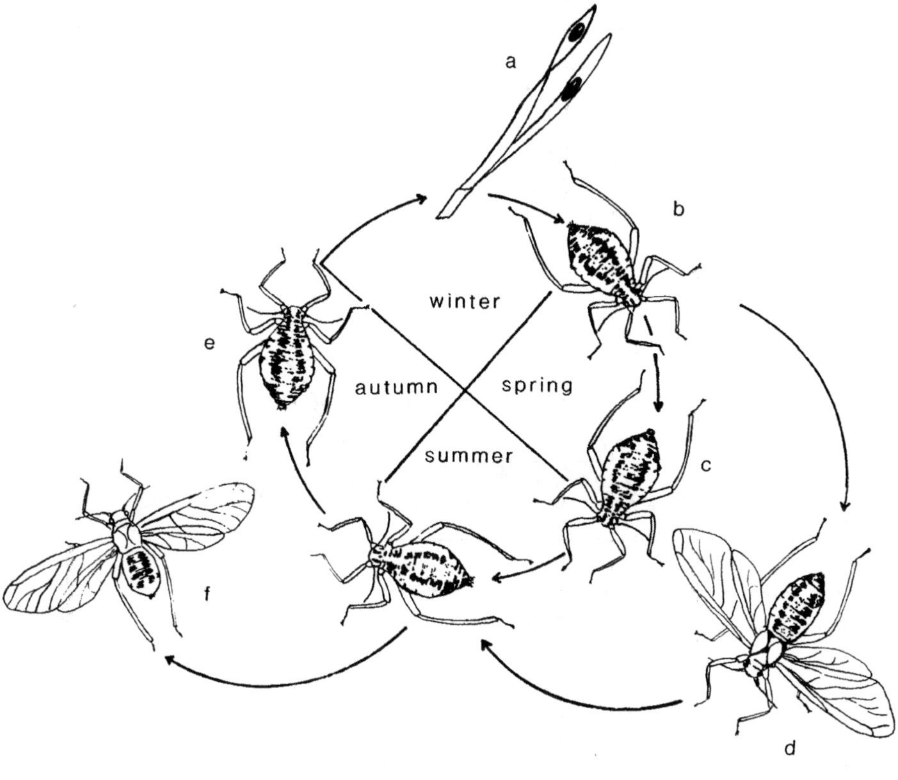

FIGURE 1. Life cycle of the large pine aphid. (a) Egg. (b) Fundatrix. (c) Aptera. (d) Alate. (e) Oviparae. (f) Male.

The life cycle of *C. pinea* is partly controlled by temperature and daylength. For example, the rate of egg development after diapause increased with temperature, while egg hatch in the laboratory increased with longer photoperiods and lower temperatures. The temperature threshold for egg development is 0°C, with a development time of 532 day-degrees to the point at which 50% of the eggs hatched. These results point to a possible daylength trigger that reactivates development after diapause, with subsequent direct thermal control. A computer-simulation model of egg development and hatch, based on this hypothesis, predicted a pattern of egg hatch close to that observed in the field.[46]

Changing photoperiod and temperatures, however, do not appear to influence the production of winged females during midsummer.[44] Laboratory populations, kept at constant temperature and long daylength, show a pattern of alate production similar to that in the field. By contrast, crowding experience during nymphal life significantly increases the frequency of alatae, but only if the tree is in the stage of active shoot growth. Later in the season, when shoots are mature, alatae are suppressed, irrespective of crowding.[44]

2.1. Distribution and Behavior

Pine aphids feed on the foliated shoots of the current and previous year[45] (Fig. 2), where they form loose aggregations, as opposed to the contact-clustered colonies characteristic of some other lachnids.[41] Certain parts of the canopy tend to be more heavily infested than others, particularly those where the shoots are of small diameter and the phloem is relatively close to the surface.[45] Whether this distribution is due to active selection of thinner shoots for easier penetration or to a response to other associated differences within the canopy (e.g., light intensity) is unknown. What is clear, however,

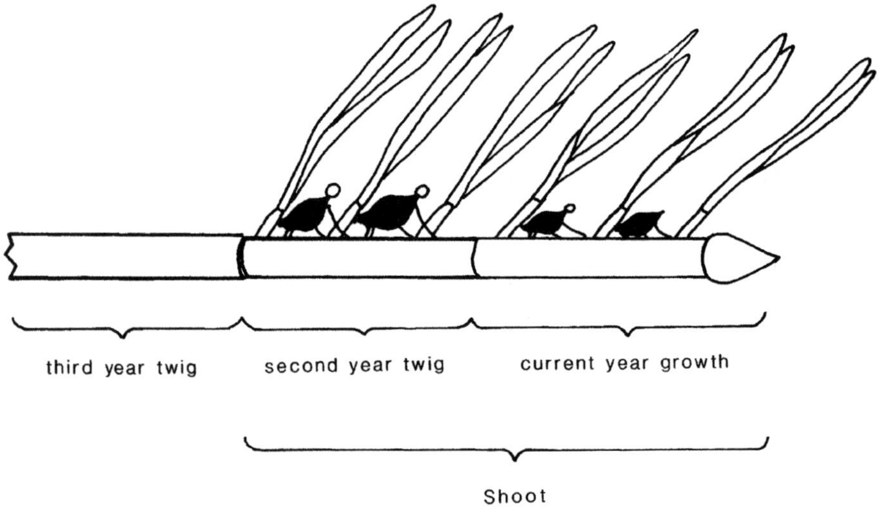

FIGURE 2. Diagrammatic representation of the feeding positions of *C. pinea* on the terminal shoots of Scotts pine. (After Kidd and Tozer.[45])

is that the aphids are just as likely to be found at any height in the canopy, and they show no discrimination between shoots bearing male cones and those bearing female cones.

Although there is no evidence that gregariousness between aphids is involved in aggregation, nymphal inertia may create localized high-density centers around the reproducing adults. Such aggregations tend to occur in those aphid species in which reproducing adults remain at one feeding site long enough to produce a cluster of sedentary nymphs. *S. pineti,* a needle-feeding species on Scots pine, is a good example.[41] Conversely, *C. pinea* is extremely active and easily disturbed, so that tight aggregations are usually observed only in midsummer, when population densities are high.[45]

Prior to shoot extension in late June and early July, the aphids feed exclusively on the previous year's growth. Then, as the buds burst and shoots begin to lengthen, they move to feed on new growth. Not all parts of the shoot are exploited to the same degree, however. The smaller instars favour the distal end of the current year's shoot, despite the fact that they have to penetrate further to reach the phloem in this region. Presumably there are nutritional benefits from feeding next to the bud, which acts as a physiological sink for translocated nutrients.[47,58] The fourth instars show no preference for any particular part of the shoot, but the adults retain a strong tendency to select the previous year's growth for feeding.[45] Why this should be is not clear. Perhaps their large size makes them especially vulnerable to bird predators such as Coal Tits and Goldcrests, which are known to take large numbers of pine aphids at certain times of the year.[25] If this is so, feeding among the darker and thicker foliage of the previous year's growth may render them as cryptic to searching birds as they are to the human observer.

On those parts of shoots with needles, the aphids invariably feed within the needle axils, with their heads toward the proximal end of the shoot (Fig. 2). The phloem is relatively near the surface within the axils and this supports the suggestion that ease of shoot penetration may be an important criterion in feeding-site selection.[45]

2.2. Natural Enemies

Cinara pinea is attacked by a large variety of parasites and predators, including the bird species already mentioned. In East Germany, as many as four species of primary parasitoids (*Pauesia* spp., Hymenoptera: Aphidiidae) have been recorded.[59] In our own study in South Wales,[43–46] only one primary parasitoid (*Pauesia picta* (Haliday)) is present but is attacked by two hyperparasitoids (*Dendrocerus* sp. and *Asaphes* sp., Pteromalidae). In the north of England, the large pine aphid is again attacked by only one primary parasitoid, in this case, *P. pini* (Haliday), which has three hyperparasitoids: (*Asaphes vulgaris* Walker, *Alloxysta* sp., Charipidae and *Dendrocerus serricornis* (Boheman), Megaspilidae.

The most common insect predators associated with *C. pinea* in South Wales are larvae and adults of the ladybird beetle, *Anatis ocellata* Linnaeus (Coccinellidae), the bugs *Acompocoris pygmaeus* (Fallen), *Phytocoris pini* Kirschbaum and *P. tiliae* (Fabricius) (Heteroptera: Miridae), and the larvae of the hoverflies *Episyrphus baltiatus* (Degeer), *Syrphus ribesii* (Linnaeus), *S. vitripennis* Meigen, and *Meliscaeva cinctella* (Zetterstedt) (Syrphidae). In the north of England, aphids are also attacked by Hemerobiid larvae (*Kimminsia* spp.) and adult Mecoptera (*Panorpa* spp.).[62] Clearly, *C. pinea* is associated with a considerable number of insect predators, many of which are generalists, feeding on a variety of prey types. Nevertheless, the same predator groups tend to recur in different localities, even though the particular species may differ.

Scots pine trees also harbor many spiders; some authorities consider them major predators of *Cinara* species.[5,21,59] Mortality from spiders is considered highest during midsummer, when warm weather induces greater activity in the aphids, increasing their chances of falling or flying into webs. However, detailed quantitative studies indicate that less than 1% of a *C. pinea* population was trapped each day by lyniphiid sheet-web-spinning species, despite the fact that these spiders were among the most common predators on the trees. This mortality was found to be too low to have any significant effect in reducing aphid population numbers.[52,53]

2.3. Relationships with Other Species

Conifer lachnids frequently form mutualistic associations with ants, particularly wood ants (*Formica* spp.),[71] which tend the aphids for their sugary honeydew. The aphids in turn may benefit from the activity of the ants, which keep the colony clean of honeydew deposits[6] and protect it from natural enemies.[10,59] In fact, some lachnid species, such as *C. kochiana* on European larch, are thought to depend on ants for their very survival, as the aphids are found only in their presence.[6]

Ants may also be found tending *C. pinea* at certain times of the year, but in this case the relationship is not obligatory. Excluding ants from the canopy of Scots pine trees appears to have no effect on *C. pinea* population densities, on honeydew production, or on the number of insect predators.[59]

A number of other insects, including adult parasitoids and syrphids, also use conifer honeydew as food.[59,73] However, perhaps the most notable insects in this category are honeybees, which in parts of Central and Eastern Europe provide a useful crop of forest honey for beekeepers. Scots pine is one of the three conifer species considered suitable for the provision of forest honey, and *C. pinea* is undoubtedly a substantial contributor to the honeydew used by bees, particularly in midsummer, when aphid numbers are high.[6] It is interesting that insects considered to be damaging by the forester should prove beneficial to the beekeeper. Truly one man's meat is another's poison!

3. POPULATION DYNAMICS

3.1. Patterns of Population Change

Although aphids live all the year round on pine, they only become abundant in midsummer, when the new shoots and needles are growing (Fig. 3). Populations usually show a single annual peak in abundance, the magnitude of which varies from year to year and from tree to tree.[43] After the peak, population numbers generally decline and remain low, until the aphids disappear from the trees during November. This pattern appears to be typical throughout the aphid's geographical range, having been reported in populations

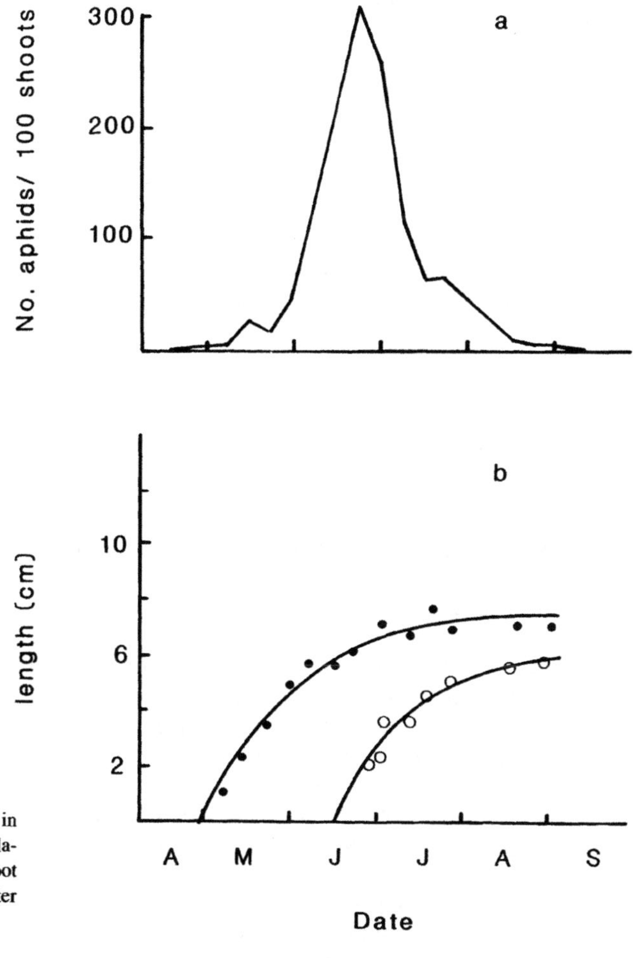

FIGURE 3. (a) Seasonal changes in numbers of *Cinara pinea* in relationship to (b) the pattern of shoot (●) and needle (○) growth (After Kidd and Tozer.[44])

from North America,[39] East Germany,[59] and Britain.[43,62] Unfortunately, no conclusion can be drawn regarding the comparability of population densities in these localities, as different measurements were used by the various investigators.

I have used as a sample the smallest, most convenient host-plant unit, the shoot, comprising that part of the terminal growth bearing needles developed during the current and previous year (Fig. 2). Knowing the average number of shoots per branch and the number of branches per tree, it becomes possible to estimate the total number of *C. pinea* on each sample tree. In my studies the highest recorded populations peaked at an average of three aphids per shoot, or approximately 45,000 aphids to a 12-m tree. Populations were also monitored continuously over a number of years to detect any long-term numerical trends.[43] Peak densities showed an apparently cyclical pattern over a 9-year period, varying by as much as fourfold (Fig. 4).

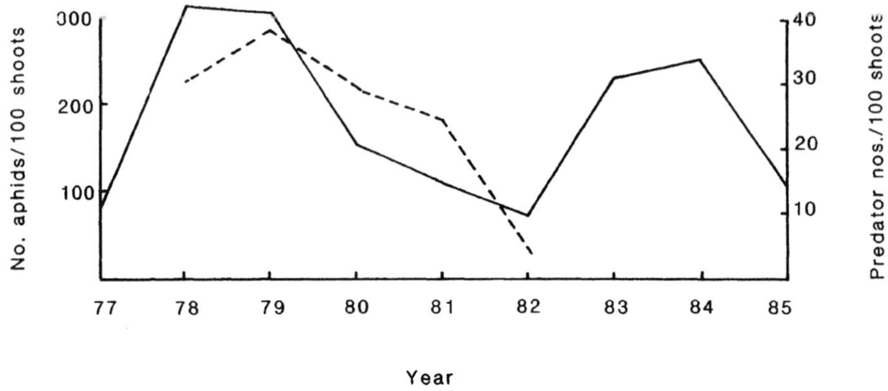

FIGURE 4. Variations in annual peak density of *Cinara pinea* (——) and its insect predators (- - -) on a single pine tree. Predators not recorded after 1982.

3.2. Factors Influencing Population Numbers

3.2.1. Crowding

The sizes (live weight) of adult apterae vary seasonally, being largest in spring (>12 mg) and smallest in August (<5 mg) following peak population densities. These size changes have implications for population growth because small adults produce fewer offspring and take longer to develop to reproductive maturity. Small apterae also have a shorter life span than do large ones.[43]

As in other aphids,[40,54] these size variations are strongly influenced by crowding during nymphal life. Aphids reared under crowded conditions in sleeve cages tend to become smaller adults than those reared at sparse densities.[43] Crowding increases the frequency, speed, and duration of walking bouts, hence the frequency of disturbing encounters between aphids.[40,43] This in turn reduces the amount of time spent feeding, adversely affecting nymphal growth rates. There is no evidence, however, that development times are influenced by crowding.[43]

3.2.2. Tree Quality

Seasonal changes in the host trees also influence adult size.[43] Adults reared from mature foliage in the laboratory are considerably smaller than those reared on actively growing shoots. On trees in the field, successive generations show a decline in weight independent of rearing density after the new growth has ceased.[43] However, this trend is reversed in late summer, when weights again begin to increase. These size changes reflect an underlying alteration in the pattern of aphid growth and development. Although aphids take longer to develop in midsummer, this is not enough to compensate for a simultaneous rapid decline in growth rates.

There is no evidence that physical changes in pine shoots can account for these observations, as aphids can attain large sizes in the spring while feeding on the previous year's growth. Presumably, chemical changes contribute to the variations in aphid growth and development, but it is not clear which aspects of tree chemistry are important. Aphid performance has most frequently been interpreted in relationship to the nutritional quality of the phloem sap amino-nitrogen (amines, amides, amino acids, proteins),[50,51] but other substances (e.g., allelochemics) may be equally important.[43] We are currently attempting to identify those chemicals (e.g., amino acid, phenol, terpenes) that affect *C. pinea* growth rates.

3.2.3. Mortality

Overwintering eggs are subject to unknown mortality (disappearance and collapse) and mould infection.[46] The most likely sources of this unknown mortality seem to be predators and weather.[15,28,35,57] Egg mortality, however, was estimated to be relatively constant (~50%), with no evidence of density dependence.[46] It seems unlikely that egg mortality helps regulate pine aphid population numbers, nor does it appear to act as a density-disturbing factor.

Apart from natural enemies and old age, most aphid mortality comes from falling out of the canopy. Although the proportion of nymphs falling from day to day varies considerably, there is no evidence that it is correlated with windspeed or rainfall.[43] However, there is a strong negative correlation between the daily proportion falling and prevailing nymphal growth rates.[43] When growth rates were high, fewer fell off.

When confined in sleeve cages, aphids obviously cannot fall off, but they can register the effects of their surroundings by dying *in situ*. The mortality of caged nymphs was higher in the later generations as well as in high-density cages (Fig. 5a), suggesting that nymphal loss from the trees may be density dependent. This conclusion was supported by laboratory experiments in which the proportion of nymphs walking off small saplings was found to increase with aphid denisty (Fig. 5b). Plant quality therefore also appears to affect the degree of restlessness of the aphids. When conditions for nymphal growth are suboptimal, they move about more on the trees, increasing their chances of emigration either by walking down the stems of small trees or by falling out of the canopy of large trees. Restlessness is further promoted at high densities as a result of the disturbing effects of interactions between aphids.

3.2.4. Behavior of Flying Adults

Alatae are suppressed during the first spring generation, but in the second and third, their production is density dependent. Irrespective of crowding experience, 91% of all alatae take off with 24 hr of attaining adulthood, unless cloudy conditions or low temperatures prevail. Only when prevented from flying for 3 days are alatae destined to remain on the tree, with 56% dying before reproduction. For those that survive to reproduce, there is a reproductive delay of 7 days, during which time they increase their body weight by as much as 2.5 times. Heavy alatae are more fecund than light ones, although reproduction is highly variable, with each aphid producing on average one nymph per day.

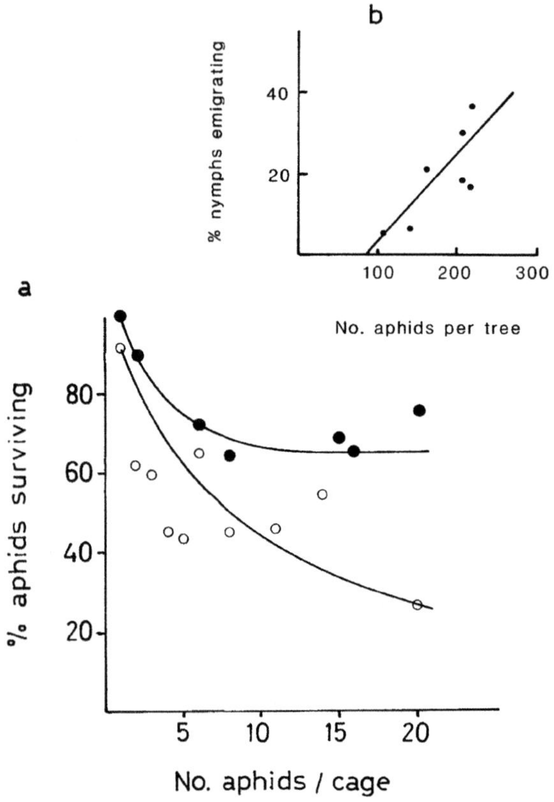

FIGURE 5. (a) Survival of nymphs in generations 2 and 3 (●) and 4 and 5 (○) caged on field trees at different densities. (b) Weekly emigration of nymphs from a laboratory sapling, as percentage of the total number born during the time period, in relationship to total densities prevailing.

3.3. Modeling Pine Aphid Populations

Although there are well-established methods for examining the population dynamics of insects with discrete generations,[63,67] such methods are not generally applicable to aphids, since they often have partially overlapping or fully overlapping generations. The difficulty lies in separating the factors influencing recruitment to, and loss from, the population. While methods have been devised to model aphid populations in the field,[12,13,31,32,56,69] they are only really applicable to populations with fully overlapping generations and stable age distributions.[8]

The use of computer-simulation models enables us to separate all the factors influencing recruitment and loss.[2,9,26] These methods can be applied to populations with either nonoverlapping, overlapping, or partially overlapping generations.[42] They also have the advantage over the traditional life-table-based techniques in that they require less time for an understanding to be gained of the processes influencing population change.

We constructed a simulation model similar to others used in aphid population dynamics.[43,46] The components of the model and their relationships are illustrated in Fig. 6. The initial input to the model is the number of autumn eggs on the tree in 1 year and the final output autumn eggs in the next year. Predation and parasitism were excluded at this stage,

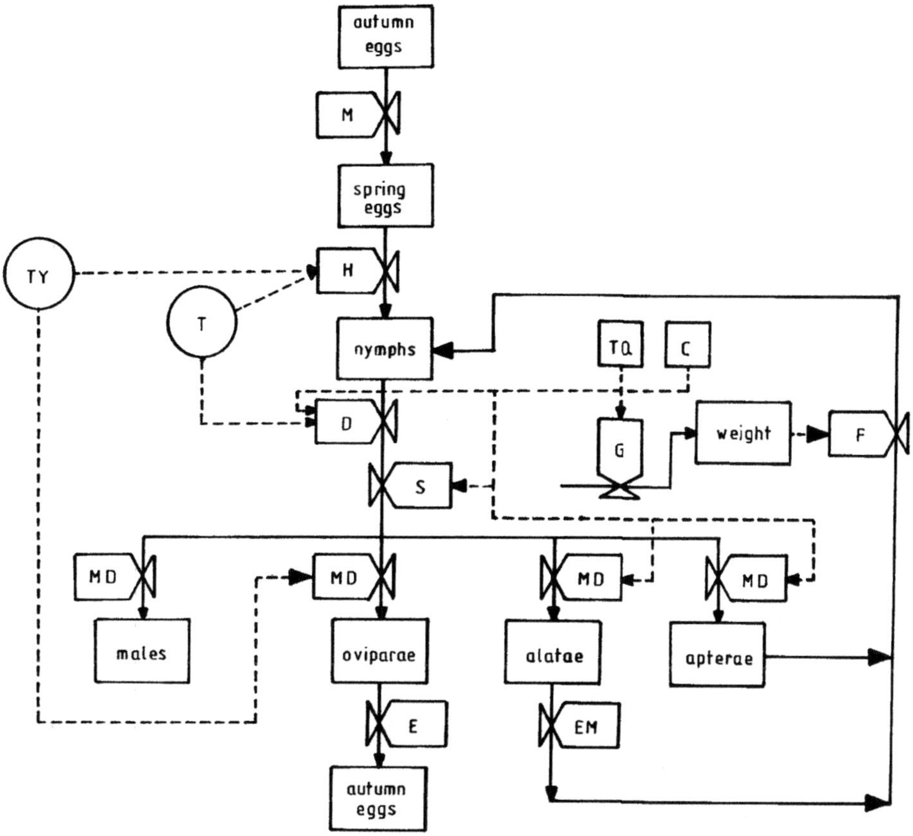

FIGURE 6. Relational diagram of the pine aphid model. (Rectangles) state variables; (squares) subsidiary variables; (circles) driving variables; (valve symbols) factors affecting rates of flow; (→) flow of energy or matter; (----➔) effect of one variable on another. M, egg mortality rate; H, hatching rate; D, nymphal development rate; S, nymphal survival rate; G, nymphal growth rate; R, reproductive rate; MD, adult morph determination; EM, alate emigration rate; E, oviposition rate; T, temperature; TY, time of year; C, crowding during development; TQ, tree favorability to aphids. (After Barlow and Dixon.[2])

their effects being deduced from the difference between the behavior of the model and the field populations. Nevertheless, our confidence in the model was verified by the fact that predictions agreed closely with population changes on saplings from which natural enemies had been excluded.[43]

The agreement between the model predictions and population data from trees in the field was also good (Fig. 7). Discrepancies were only observed after the midsummer peak, the difference reflecting the impact of natural enemies which become abundant at this time. This finding demonstrates that parasitism and predation play an insignificant part in determining population growth in early summer, but this is not the case in late summer and autumn, when natural enemies are likely to have an important influence on the number of eggs laid.

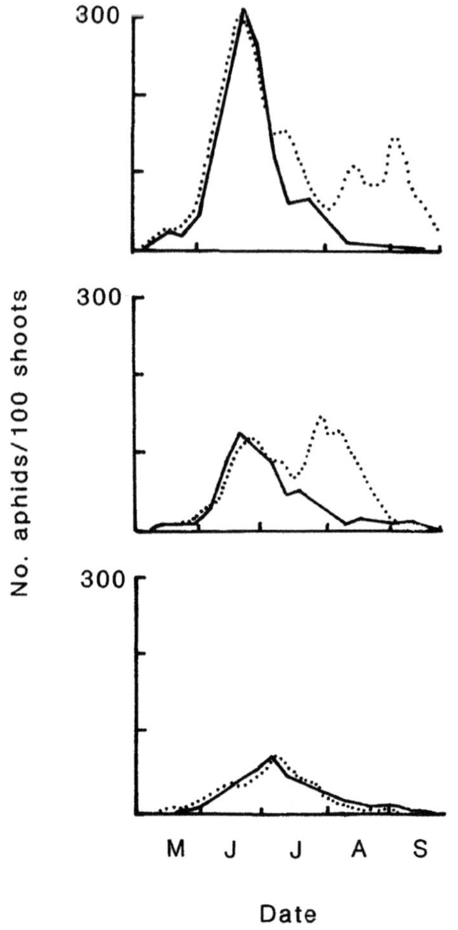

FIGURE 7. Aphid population dynamics during one season on three field trees (——), together with densities predicted by the simulation model in the absence of predation (· · · ·). (After Kidd.[43])

Sensitivity analysis was used to assess the relative contributions of each factor to population change. Spring egg numbers, for example, were found to be an important determinant of the peak numbers achieved in any year. For this reason, predation the previous year is likely to be an important source of variability in peak densities. With predation removed, population numbers the following year can increase by as much as 10-fold, limited by tree-induced effects. However, the constraints set by the tree are not constant from year to year and these may impose a severe restriction on the peak number attained, even in the absence of predation. Thus, populations will tend to track variations in host plant quality between trees and between years, with natural enemies lowering the average number of aphids involved in the tracking response.

Intrinsic density-dependent processes, such as the effects of crowding on the size and fecundity of females, appear to be of secondary importance in determining aphid numbers. When predation and tree influences on nymphal growth, development, and mortality were removed from the model, numbers were stabilized at about 30 aphids per shoot,

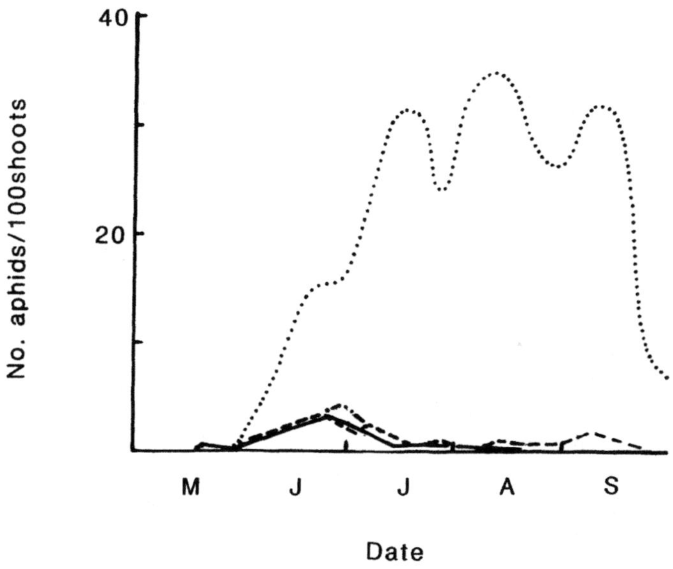

FIGURE 8. Seasonal changes in aphid numbers on one tree (——) together with those predicted by the model in the absence of either predation alone (- - -), predation and alate production (-···-), or predation and tree-induced changes in size and mortality (· · · ·).

approximately 10 times the maximum densities seen in the field (Fig. 8). Clearly, populations are constrained at densities well below those which trigger the full effects of such density-dependent mechanisms.

Similarly, density-dependent emigration by flight has little effect in reducing population numbers. Alatae are not produced for long enough or in adequate quantities at the prevailing densities to have any regulative significance (Fig. 8).

The variables driving *C. pinea* populations are therefore (1) seasonal and between-tree differences in the host pines, and (2) the effects of natural enemies, either of which is theoretically capable of inducing the observed cyclical population pattern (see Fig. 4). Although cyclical fluctuations in herbivore numbers can be created by induced plant defenses,[3,29] there is no evidence that aphid feeding can induce chemical defenses that affect their performance in subsequent years. Another explanation of cyclical behavior can be provided by examining insect predator densities. These tended to be highest in years in which aphids were particularly abundant, but there is also evidence for a delay in the predator response (Fig. 4). This underlying delayed density dependence could generate aphid population cycles.[13,72] Clearly, more detailed research is needed to test these alternative hypotheses.

4. IMPLICATIONS FOR MANAGEMENT

Besides improving our general understanding of the population dynamics of the large pine aphid, the simulation model is a useful tool for predicting and, it is hoped, for preventing harmful infestations. Management recommendations based on simulation are also likely to be of relevance to other lachnid pest problems, at least in the absence of more detailed studies on those species.

The particular growth stages of Scots pine most at risk from *C. pinea* are 2-year-old nursery trees and young plantation saplings. Our own experience with nursery-raised stock suggests that these trees are especially favorable for infestation and tend to foster high population growth rates. Unfortunately, it is in the nursery environment that a varied complement of natural enemies is likely to be absent, creating the combination of conditions that the model predicts would result in population outbreaks. The longer trees are kept under such conditions (e.g., over a period of 2 or 3 years), the greater the severity of attack. In Britain, pine trees are normally planted out 2 years of age, which may not be soon enough to prevent significant growth reductions.

Short of expensive insecticide treatment, the only feasible recommendation at this stage would be to encourage general predators, by siting nursery plants close to mature forests, or even arable crops. This approach worked effectively in 1982, when a high infestation of *S. pineti* occurred on a plot of experimental pine saplings.[49] Immigration of large numbers of ladybirds and hoverflies from surrounding farmland reduced aphid numbers dramatically.

After saplings are planted in the forest, the feasibility of controlling aphids by chemical means or by manipulation of natural enemies is reduced. Certainly regular or even periodic use of insecticides would be considered uneconomical on a crop that may take many decades to reach marketable size. Growth reductions caused by aphids such as *C. pinea* are likely in this case to be considered an acceptable loss, particularly where there are no noticeable symptoms of damage.

Only where pine trees are grown for maximum commercial returns, for example, in the Christmas tree industry, could the use of insecticides prove realistic.[37–39] Here, high populations of aphids may be especially undesirable if the cosmetic value of the tree is reduced by honeydew and sooty mould deposits. In this case, the estimation of egg densities on the needles (a fairly simple task) could provide reasonable predictions of summer population peaks, providing a sound basis for the decision of whether to spray. Spraying should be conducted in the spring after egg hatch, but before the rapid increase in numbers in June, to prevent the possibility of honeydew contamination as well as to avoid killing bees. By using a nonpersistent contact insecticide, the grower would also benefit from the activity of natural enemies in removing any residual survivors or summer immigrants. Consequently, with *C. pinea* oviposition in the autumn unlikely, protection from aphid problems the following year would also be ensured. Only in the third year would egg densities again need to be monitored, after the possibility of significant immigration of alatae the previous summer.

The sensitivity of pine aphid populations to variations in tree chemistry offers perhaps the best long-term solution to their effective control. While seasonal variations are important in driving population changes, it is between-tree chemical differences that define the resistance or susceptibility of a tree to aphid attack. Research aimed at the detection and identification of the responsible chemicals is currently under way. Assuming that these tree differences have a significant genetic component, the resulting information should be of potential use in both provenance selection and tree breeding.

Site characteristics can also influence host plant susceptibility to infestation.[4] Tree phenologies or growth characteristics that are altered, for example, by prevailing temperatures or soil conditions, may prove beneficial to herbivorous insects.[7] By feeding directly on the translocation system, aphids are in an excellent position to respond quickly to any resulting chemical changes. This may have been the reason for the *C. pinea* outbreaks on Scots pines in South Carolina.[39] These populations showed an unusual resurgence in numbers during October and November, which, assuming a holocyclical life cycle, would have resulted in very high egg numbers. This, as predicted by the model, is one major prerequisite for outbreak population levels the following year. Care must therefore be exercised, when planting nonindigenous provenances or species, that site conditions do not alter the trees in the aphid's favour.

Clearly, the modeling approach to pine aphid population dynamics has been of considerable benefit, not only in providing a deeper understanding of the factors determining their abundance, but also in directing research towards areas of practical benefit to the forester.

ACKNOWLEDGMENTS. It is a pleasure to thank Clive Carter, Gavin Lewis, and Mark Jervis for criticizing the manuscript and for providing much useful discussion. Debbie Tozer and the late Ronald Middleton contributed in large measure to the collection of field data on *C. pinea* and its natural enemies, and to them I am truly grateful. Thanks are also due to Simon Leather and Dr. M. J. P. Shaw for information on the pest status of lachnids in Scotland and South Africa, respectively. My own work was generously funded by the Natural Environment Research Council.

REFERENCES

1. Baker, W. L., 1972, *Eastern Forest Insects,* Misc. Publ. U.S. Dept. Agric. Forest Service. 1175.
2. Barlow, N. D., and Dixon, A. F. G., 1980, *Simulation of Lime Aphid Population Dynamics,* Pudoc, Wageningen.
3. Benz, G., 1974, Negative Rückkoppelung durch Raum-und Nahrungskonkurrenz sowie zyklische Veranderung der Nahrungsgrundlage als Regelprinzip in der Populations-dynamik des Grauen Larchenwicklers, Zeiraphera diniana (Guenee), *Z. Angew Entomol.* **76:**196–228.
4. Berryman, A. A., 1985, Site characteristics and population dynamics: A theoretical perspective, in: *Site Characteristics and Population Dynamics of Lepidopteran and Hymenopteran Forest Pests* (D. Bevan and J. T. Stoakley, eds.), pp. 1–7, For. Comm. Res. Dev. Paper 135.
5. Bradley, G. A., and Hincks, J. D., 1968, Ants, aphids and Jack pine in Manitoba, *Can Entomol.* **100:**40–50.
6. Carter, C. I., and Maslen, N. R., 1982, *Conifer Lachnids,* For. Comm. Bull. 58.
7. Carter, C. I., and Nichols, J. F. A., 1985, Some resistance features of trees that influence the establishment and development of aphid colonies, *Z. Angew. Entomol.* **99:**64–67.
8. Carter, N., Aikman, D. P., and Dixon, A. F. G., 1978, An appraisal of Hughes' time-specific life table analysis for determining aphid reproductive and mortality rates, *J. Anim. Ecol.* **47:**677–688.
9. Carter, N., Dixon, A. F. G., and Rabbinge, R., 1982, *Cereal Aphid Populations: Biology, Simulation and Prediction,* Pudoc, Wageningen.
10. Czechowski, W., 1975, Bionomics of *Formica* (Coptoformica) *pressilabris* Nyl. (Hymenoptera, Formicidae), *Ann. Zool.* **33:**103–125.
11. Dixon, A. F. G., 1970, Quality and availability of food for a sycamore aphid population, in: *Animal Populations in Relation to Their Food Resources* (A. Watson, ed.), pp. 271–287, Blackwell, Oxford.
12. Dixon, A. F. G., 1970, Stabilisation of aphid populations by an aphid induced plant factor, *Nature (Lond.)* **227:**1368–1369.

13. Dixon, A. F. G., 1971, The role of intra-specific mechanisms and predation in regulating the numbers of the lime aphid, *Eucallipterus tiliae* L. *Oecologia (Berl.)* **8:**179–193.
14. Dixon, A. F. G., 1979, Sycamore aphid numbers: The role of weather, host and aphid, in: *Population Dynamics* (R. M. Anderson, B. D. Turner, and L. R. Taylor, eds.), pp. 105–121, Blackwell, Oxford.
15. Dunn, J. A., and Wright, D. W., 1955, Overwintering egg populations of the pea aphid in East Anglia, *Bull. Entomol. Res.* **46:**389–392.
16. Eastop, V. F., 1972, A taxonomic review of the genus *Cinara* Curtis occurring in Britain (Hemiptera: Aphididae), *Bull. Br. Mus. Nat. Hist. (Entomol.)* **27:**104–186.
17. Eastop, V. F., 1976, A review of *Cinara* subgenus *Cinarella* (Hemiptera: Aphididae), *Bull. Br. Mus. Nat. Hist. (Entomol.)* **35**(1)**:**1–23.
18. Eckloff, W., 1972, Studies on the ecology and economic significance of *Cinara piceicola* (Hom., Lachnidae), *Z. Angew. Entomol.* **70:**134–157.
19. Emmonot, P., Gayraud, Y., Leclant, F., and Remaudière, G., 1967, Sur la presence en France de *Cedrobium lapportei* Remaudière, puceron nuisible au cedre. Extrait du Proces-verbal Séance du June 28, 1967, Academie d'Agriculture de France.
20. Fabre, J. P., 1967, Note on the presence of *Cinara cedri* (an aphid that damages cedar) in France, *C. R. Seances Acad. Agric. France* **62:**771–775 (in French).
21. Fox, R. C., and Griffiths, K. H., 1976, Predation of pine cinaran aphids by spiders, *J. Georgia Entomol. Soc.* **11**(3)**:**241–243.
22. Fox, R. C., and Griffiths, K. H., 1977, Pine seedling growth loss caused by cinaran aphids in S. Carolina, *J. Georgia Entomol. Soc.* **12**(1)**:**29–34.
23. Furuta, K., and Takai, M., 1983, Population dynamics of *Cinara bogdanowi ezoana* Inouye (Hom. Lachnidae) in plantations of *Picea glenhii* Masters and *P. jezoensis* Carriere, *Z. Angew. Entomol.* **95:**238–249.
24. Furuta, K., Takai, M., and Funatsu, T., 1983, Effects of an infestation of *Cinara bogdanowi ezoana* (Inouye) (Hemiptera, Lachnidae) on the growth of *Picea glehnii* Mast, *J. Jpn. For. Soc.* **65:**166–171.
25. Gibb, J. A., 1960, Populations of tits and goldcrests and their food supply in pine plantations, *Ibis* **102:**163–208.
26. Gilbert, N., Gutierrez, A. P., Frazer, B. D., and Jones, R. E., 1976, *Ecological Relationships,* Freeman, Reading.
27. Gunkel, W., 1963, *Cupressobium juniperum* Mordv. (Homoptera: Lachnidae) ein Schadling an *Thuja occidentalis* L., *Z. Angew Zool.* **50:**1–48.
28. Hand, S. C., 1983, The effect of temperature and humidity on the duration of development and hatching success of the eggs of the aphid, *Sitobion avenae, Entomol. Exp. Appl.* **33:**220–222.
29. Haukioja, E., 1980, On the role of plant defences in the fluctuation of herbivore populations, *Oikos* **35:**202–213.
30. Holopainen, J., 1984, Occurrence of *Cinara pini* (Homoptera, Lachnidae) in Scots pine (*Pinus sylvestris*) seedlings with disturbed growth, *Ann. Entomol. Fenn.* **50:**108–110.
31. Hughes, R. D., 1962, A method for estimating the effect of mortality on aphid populations, *J. Anim. Ecol.* **31:**389–396.
32. Hughes, R. D., 1963, Population dynamics of the cabbage aphid, *Brevicoryne brassicae* L., *J. Anim. Ecol.* **32:**393–424.
33. Inouye, M., 1958, Studies on the silvicultural control of conifer aphids in Hokkaido, Japan, *Proc. Xth Int. Cong. Entomol. Montreal.* **4:**163–170.
34. Inouye, M., and Yamaguchi, H., 1955, Aphid problems of the Todo-fir plantations in Hokkaido. 1. The relationship of ants and other organisms to certain aphid pests on Todo-fir plantations at Hononai Iburi Province in Hokkaido, *Spec. Rep. For. Exp. Sta. Hokkaido* **3:**28–41.
35. James, B. D., and Luff, M. L., 1982, Cold-hardiness and development of eggs of *Rhopalosiphum insertum, Ecol. Entomol.* **7:**277–282.
36. Johnson, N. E., 1958, Reduced growth associated with infestations of Douglas fir seedlings by *Cinara* species (Homoptera: Aphididae), *Can. Entomol.* **97:**113–119.
37. Kearby, W. H., 1967, Helicopter spraying for aphids in Scotch pine plantations, *J. Econ. Entomol.* **60:**1453.
38. Kearby, W. H., and Bliss, M., 1968, Field evaluation of five insecticides for the control of *Eulachnus agilis,* an aphid of conifers, *J. Econ. Entomol.* **61:**1124–1125.

39. Kearby, W. H., and Bliss, M., 1969, Field evaluation of three granular systemic insecticides for control of the aphids, *Eulachnus agilis* and *Cinara pinea* on Scotch pine, *J. Econ. Entomol.* **62:**60–62.
40. Kidd, N. A. C., 1977, The influence of population density on the flight behaviour of the lime aphid, *Eucallipterus tiliae, Entomol. Exp. Appl.* **22:**251–261.
41. Kidd, N. A. C., 1982, Predator avoidance as a result of aggregation in the grey pine aphid, *Schizolachnus pineti, J. Anim. Ecol.* **51:**397–412.
42. Kidd, N. A. C., 1984, A BASIC programme for use in teaching population dynamics, *J. Biol. Educ.* **18**(3)**:**227–228.
43. Kidd, N. A. C., 1985, The role of the host plant in the population dynamics of the large pine aphid, *Cinara pinea, Oikos* **44:**114–122.
44. Kidd, N. A. C., and Tozer, D. J., 1984, Host plant and crowding effects in the induction of alatae in the large pine aphid, *Cinara pinea, Entomol. Exp. Appl.* **35:**37–42.
45. Kidd, N. A. C., and Tozer, D. L., 1985, The distribution of the large pine aphid, *Cinara pinea* (Mordv.) within the canopy of Scots pine, *Pinus sylvestris* L., *Z. Angew. Entomol.* **99:**341–350.
46. Kidd, N. A. C., and Tozer, D. J., 1985, Distribution, survival and hatching of overwintering eggs in the large pine aphid, *Cinara pinea* (Mordv.) (Hom., Lachnidae), *Z. Angew. Entomol.* **100:**17–23.
47. Kozlowski, T. T., 1971, *Growth and Development of Trees,* Academic, New York.
48. Leather, S. R., 1980, Egg survival in the bird cherry-oat, *Rhopalosiphum padi, Entomol. Exp. Appl.* **27:**96–97.
49. Lewis, G. B., 1987, Regulating interactions between pine aphid colonies (*Schizolachnus pineti*) and host plant growth, Ph.D. thesis, University of Wales, Cardiff.
50. Mattson, W. J., 1980, Herbivory in relation to plant nitrogen content, *Annu. Rev. Ecol. Syst.* **11:**119–161.
51. McNeill, S., and Southwood, T. R. E., 1978, The role of nitrogen in the development of insect/plant relationships, in: *Biochemical Aspects of Plant and Animal Coevolution* (J. B. Harborne, ed.), pp. 77–98, Academic, London.
52. Middleton, R. J., 1984, The distribution and feeding ecology of web-spinning spiders living in the canopy of Scots pine, (*Pinus sylvestris* L.), Ph.D. thesis, University of Wales, Cardiff.
53. Middleton, R. J., and Kidd, N. A. C., 1986, Aphid–spider interactions within the canopy of Scots pine, *Pinus sylvestris, Acta Xth Congr. Int. Arachnol. Jaca, Espana.* **1:**274.
54. Murdie, G., 1969, Some causes of size variation in the pea aphid, *Acyrthosiphon pisum* Harris, *Trans. R. Entomol. Soc. Lond.* **121:**423–442.
55. Noratorio, A., Cadierno, D., and Mijares, A., 1978, Presencia en hoyo de Manzanares (Madrid) de un pulgon que ataca a los cedros *Cinara cedri* Mimeur, *An. Inst. Natl. Invest. Agrar.* **8**(5)**:**59–64.
56. Perrin, R. M., 1976, The population dynamics of the stinging nettle aphid, *Microlophium carnosum* (Buckt.), *Ecol. Entomol.* **1:**31–40.
57. Peterson, A., 1920, Some studies on the influence of environmental factors on the hatching of the eggs of *Aphis avenae* Fabricius and *Aphis pomi* De Geer, *Ann. Entomol. Soc. Am.* **13:**391–401.
58. Sacher, J. A., 1954, Structure and seasonal activity of the shoot apices of *Pinus lambertiana* and *Pinus ponderosa, Am. J. Bot.* **41:**749–759.
59. Scheurer, S., 1971, Der Einfluss der ameisen und der naturlichen feinde auf einige an *Pinus sylvestris* L. lebende Cinarinen in der Dubener-Heide (DDR), *Pol. Pismo Entomol.* **41:**197–229.
60. Shaposhnikov, G. K., 1967, Aphidinea, in: *Keys to the Insects of the European U.S.S.R.*, Vol. 1 (G. V. BeiBienko, ed.), pp. 616–799, Israel Program for Scientific Translation Ltd., Jerusalem.
61. Shaw, M. J. P., 1983, Some effects of infestation by the black pine aphid, *Cinara cronartii* (T and P) (Hemiptera, Aphididae), in: *Proceedings of the Fourth Entomology Congress of the Entomology Society of Africa, Johannesburg, September 1983.*
62. Sommerville, A. H. C., 1972, Studies on the distribution and dynamics of the Lachnidae of Scots pine. Ph.D. thesis, University of Leeds, Leeds, England.
63. Southwood, T. R. E., 1979, *Ecological Methods,* Methuen, London.
64. Stary, P., 1976, *Cinara piceae* (Panz.) (Hom., Lachnidae), a pest of young European fir trees (*Abies alba* Mill.) and its natural enemy complex in Czechoslovakia, *Stud. Entomol. For.* **2**(10)**:**171–180.
65. Thompson, S., 1977, The effect of an attack by the aphid *Schizolachnus pineti* Fabricius on the growth of young Scots pine trees. *Scott. For.* **31:**161–164.
66. Van Rensburg, N. J., 1979, *Cinara cronartii* on the roots of pine trees (Homoptera: Aphididae), *J. Entomol. Soc. S. Afr.* **42:**151–152.

67. Varley, G. C., and Gradwell, G. R., 1970, Recent advances in insect population dynamics, *Annu. Rev. Entomol.* **15**:1–24.
68. Varty, I. W., 1953, *Cinaropsis pilicornis*, a rare aphid attacking spruce transplants, *Scott. For.* **7**:86–87.
69. Voegtlin, D. J., and Dahlsten, D. L., 1982, Observations on the biology of *Cinara ponderosae* (Williams) (Homoptera: Aphididae) in the westside forests of the Sierra Nevada, *Hilgardia* **50**(5):1–19.
70. Wang, E. L. H., and Hughes, I. W., 1976, Juniper aphid on cedar, *F.A.O. Plant Protection Bull.* **24**:27–28.
71. Way, M. J., 1963, Mutualism between ants and honeydew producing Homoptera, *Annu. Rev. Entomol.* **8**:307–344.
72. Wratten, S. D., 1973, The effectiveness of the coccinellid beetle *Adalia bipunctata* (L.) as a predator of the lime aphid, *Eucallipterus tiliae* L., *J. Anim. Ecol.* **42**:785–802.
73. Zoebelein, G., 1956, Der Honigtau als Nahrung der Insekten, *Z. Angew. Entomol.* **38**:369–416.

CHAPTER 7

THE WHITE LACE LERP IN SOUTHEASTERN AUSTRALIA

F. David Morgan and Gary S. Taylor

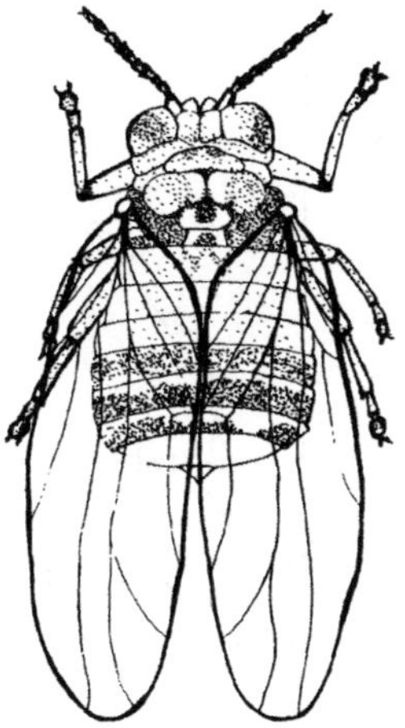

F. David Morgan and Gary S. Taylor • Department of Entomology, Waite Agricultural Research Institute, University of Adelaide, Glen Osmond, South Australia 5064, Australia.

1. INTRODUCTION

The white lace lerp, *Cardiaspina albitextura* Taylor (Psylloidea: Spondyliaspididae), causes severe defoliation, branch dieback, and sometimes death of its food plants, *Eucalyptus blakelyi* Maiden and *E. camaldulensis camaldulensis* Dehnhardt in southeastern Australia.[2,22] Its distribution covers the entire range of *E. blakelyi* in the montane and riverine zones of southern New South Wales, northeastern Victoria, and the Australian Capital Territory around the city of Canberra. Infestations in river red gum (*E. camaldulensis*) forests are generally restricted to southern areas (south of the dotted line in Fig. 1), where the subspecies *E. camaldulensis camaldulensis* dominates. The northern subspecies, *E. camaldulensis obtusa,* is not attacked by *C. albitextura* but supports outbreaks of at least two undescribed species of the spondyliaspidid genus *Creiis*. These psyllids cause necrosis of the leaves similar to, but not as deeply colored as, those caused by the nymphs of *C. albitextura* and its close relatives.[22] Severe prolonged outbreaks of

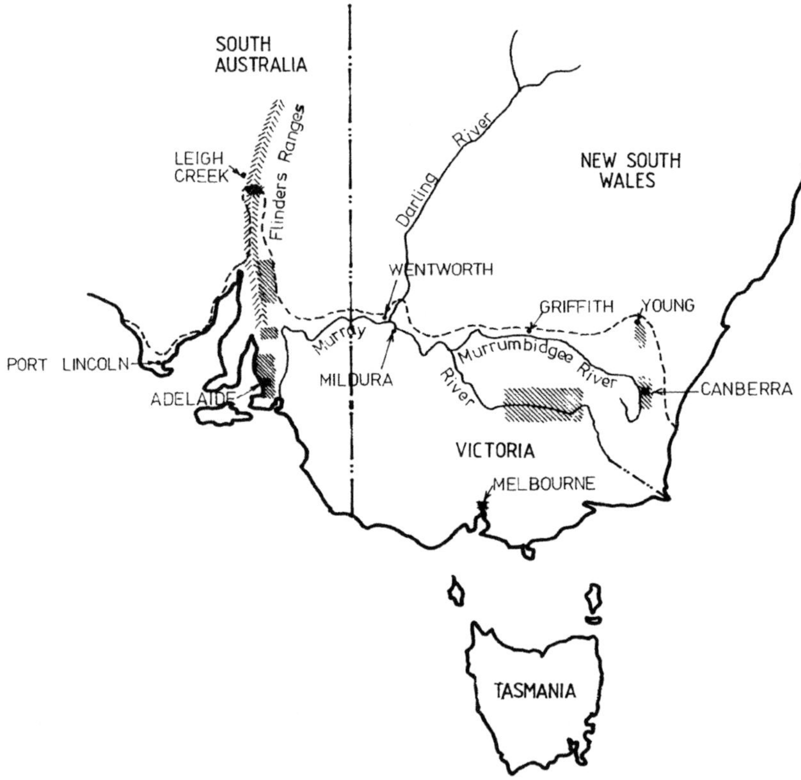

FIGURE 1. Distribution of the major outbreak areas (hatched) of *Cardiaspina albitextura* in southeastern Australia with approximate northern boundary of *Eucalyptus camaldulensis camaldulensis* (dotted line).

C. albitextura are mainly confined to certain zones within its range (Fig. 1, hatched areas), but shorter localized outbreaks may severely damage trees outside these areas.

2. BIOLOGY AND ECOLOGY

The white lace lerp has a bivoltine life cycle in the cooler parts of its range, is trivoltine over most of its distribution, and occasionally has up to five generations in years with warm winters.[2,22] The species is bisexual and few, if any, eggs are produced by unmated females. The eggs are ovoid, smooth, glabrous and reddish brown in color. They have basal stalks plugged into the surfaces of mature leaves of the food plants and from which the eggs derive moisture.[2] Eggs do not develop if they are laid into leaf surfaces, which die shortly thereafter. Patterns of oviposition vary somewhat with leaf shape—the narrower the leaf the more basal the oviposition sites. On broad leaves, oviposition may occur on any part of the surface but most eggs are laid near the base of the lamina on either leaf surface (Fig. 2). Females respond to tactile stimuli, which results in clustering of eggs about prominences or the edges of holes in the leaf lamina. Up to 2000 eggs may be deposited on a single leaf.

Fecundity ranges up to 290, with a mean of about 45 eggs per female. Fecundity is influenced by temperature which also determines adult longevity. At moderate temperatures (18–25°C), female longevity averages 14 days, during which about 60 eggs are laid in South Australia. At higher temperatures (25–29°C), females live about 10 days, with a fecundity of 45, while at very high temperatures (>33°C) longevity is about 4 days and fecundity 38. Adult longevity at very low temperatures (<10°C) often exceeds 60 days and even longer in the case of overwintering adults.[19] Near Adelaide, South Australia, overwintering females will oviposit when daily temperatures exceed 18°C for 4 hr or more.[22] Indeed, additional generations can be produced during warm winters.

Incubation of eggs is also influenced by temperatures, varying from 7 weeks at 15°C to 1 week at 30°C. Upon eclosion, first-instar nymphs search for settling sites, usually on the same leaf on which they were born. When suitable feeding sites cannot be found, young nymphs move to other leaves. The period from hatching to settling varies from about 1 hr to 48 hr, with nymphs feeding intermittently before settling. Immediately upon settling they begin to construct their lerps, or secreted covers (Fig. 2). The first stadium lasts from 5 (30°C) to 60 days (10°C); so thus, in cold winters only two or three stadia may be completed in 4 months. When grown at the same temperatures, the five nymphal stadia do not differ significantly,[22] so that generation time can be predicted precisely if oviposition time and temperature are known.

Adults, which are active and can jump more than 1 foot, begin feeding shortly after molting from fifth-instar nymphs. Males signal their presence by repeated short bursts of stridulation. Females, responding with more subdued stridulation, are found by males, which then court them until they are ready to copulate. The first complement of eggs is ready to be laid shortly after females mate. They select oviposition sites on mature leaves, usually with full turgor and low in soluble nitrogen and phenolics.[2,3,9,22] Development from egg to adult takes about 1 month in warm temperatures (30–35°C) and can be extended for 7 or more months in cool regions (<15°C).

There is some disagreement on the dispersal capabilities of the species.[2,22] Life-table

a

b

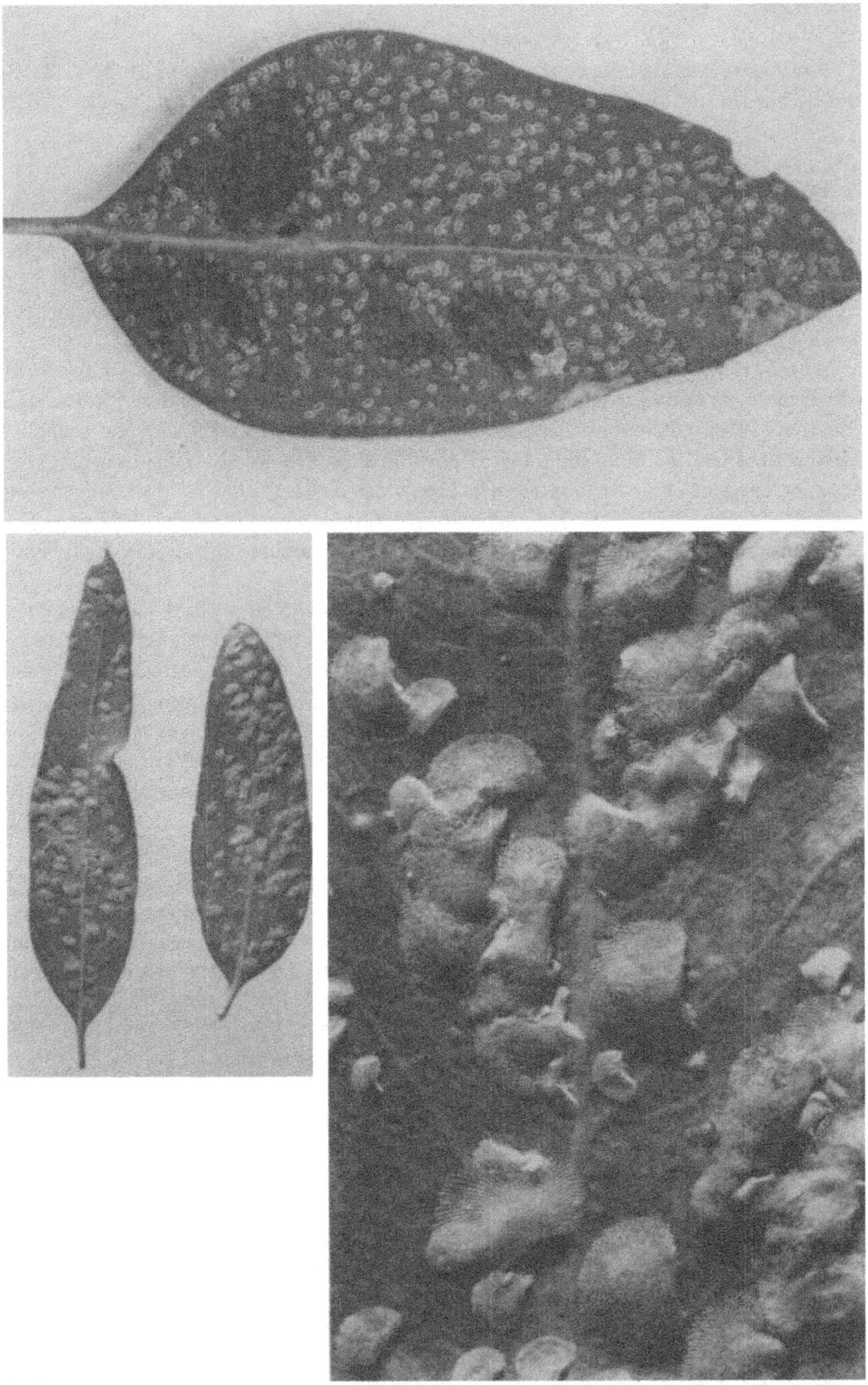

FIGURE 2. Leaf infestations by *Cardiaspina albitextura*. (a) Hatched eggs (dark patches) and lerps of first instar nymphs on mature leaf of *Eucalyptus camaldulensis camaldulensis*. (b) Infested leaves of *Eucalyptus camaldulensis calmaldulensis* with a range of nymphal lerps.

studies, however, record crepuscular migrations and considerable losses of adults from sample plots every generation.[22] Emigrating adults may spread the infestation in a wave-like process typical of eruptive outbreaks.[1]

2.1. Interactions with the Host Plant

Oviposition sites and settling sites of first instar nymphs are usually found on the same mature leaves. Low psyllid densities have little initial effect on the leaves, except that the area directly under the lerp and around the feeding site develops a pale yellow coloration, changing to pink, then reddish-purple, as the nymph completes feeding. Necrotic leaves subsequently die and take on a fawn-brown color. Necroses can be prevented by the addition of nitrogen fertilizer, which maintains a phenol-to-nitrogen ratio similar to that in the undamaged leaf.[22] As damage continues over each generation, soluble nitrogen generally declines and phenolics increase (Fig. 3). These chemical changes do not appear to affect the development of the insects that produced them but, in the next generation, first instars attempting to settle on such leaves die. Nymphs can therefore tolerate gradual changes in the chemistry of their feeding sites, but sudden changes, such as those due to fertilization of infested trees, result in the cessation of feeding, starvation of young nymphs, and an inability of older nymphs to produce normal secretions and covering lerps.[22]

High phenolic concentrations in leaf tissues apparently have deleterious effects on the survival and digestion of psyllid nymphs[22] as well as on other eucalyptus defoliators.[11,25] Quercetin is the phenol that is most effective in influencing utilization of nutrients by a lepidopteran defoliator (*Uraba lugens* Walker); this phenol has also been shown to vary greatly with time in eucalypt foliage.[12–17]

When more than 60% of its surface is damaged by psyllid feeding, the leaf usually falls by the following spring. Severe defoliation may also result in the death of affected twigs and, as damage increases toward the end of an outbreak, which may involve nine generations of the insect, branches may die. Some trees may also die, depending on the severity of defoliation and the vigor of the plant, but this is rare in most outbreaks.

Trees respond to psyllid attack by producing increasing amounts of phenolics at the apparent expense of nitrogenous compounds and sugars (Fig. 3). These chemical changes tend to persist,[20] and transfer experiments show that defended leaves are unsuitable for the establishment of first-instar nymphs. Other species, such as *Glycaspis brimblecombei* Moore seem to be more tolerant of the defensive chemicals and outbreaks of this species are observed following recovery from *C. albitextura* defoliation. *Glycaspis brimblecombei* typically begins feeding on immature foliage, which has about twice the soluble nitrogen and 1.3 times the phenolics that is optimal for the first instars of *C. albitextura*.[22]

3. CAUSES OF POPULATION FLUCTUATIONS

The temporal abundance of this lace lerp has been intensively studied over several decades in two zones of eastern Australia.[2,9,22,23] Fluctuations in numbers are related to weather, natural enemies, water stress in the food plant, and changes in the quality or chemical components of its food.

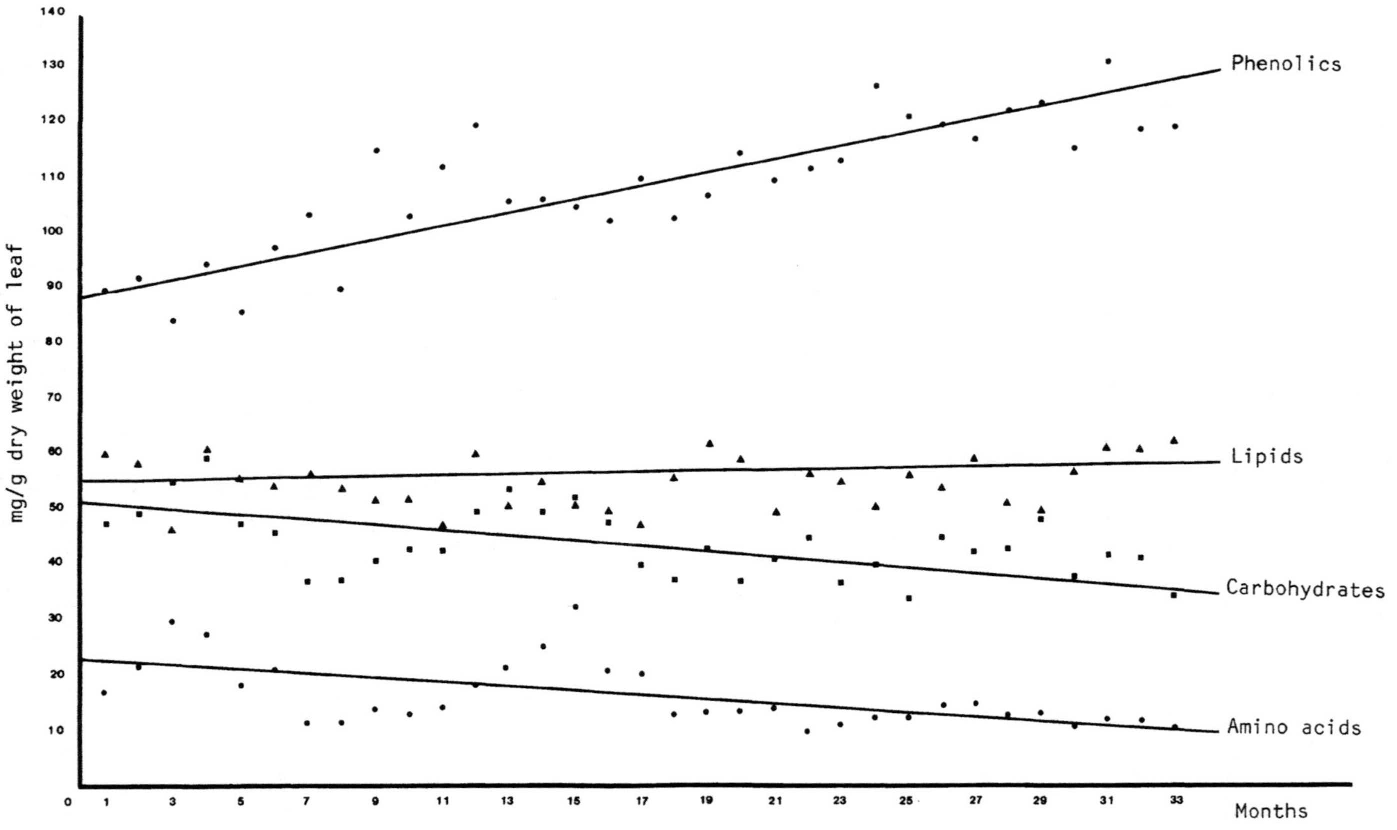

FIGURE 3. Variation in the concentrations of phenolics, lipids, carbohydrates and amino acids in the mature leaves of *Eucalyptus camaldulensis camaldulensis* at Adelaide from 1979 to 1981 during eight generations of *Cardiaspina albitextura*.

3.1. Mortality Factors

Egg mortality is generally low and sporadic and is largely due to insect predation, leaf death, and desiccation. Losses of first-instar nymphs is highly variable, with most occurring during the period between eclosion and the 2 days following settlement. Losses average 30% and range from 5% to 50%. Parasitism and predation are both rare on this instar, and few losses can be attributed to the inability of individuals to feed. Mortality usually occurs after feeding has been initiated, with individuals either dying while their mouthparts are still inserted in the leaf or after their mouthparts have been withdrawn. The inference is that individuals have ingested toxins or antifeedants. Similar symptoms are observed when first instars are transfered to mature leaves recently recovered from severe attack, or to young leaves.

Second-instar losses are usually quite low, except when populations have been dense for at least 2 years. Such populations are subjected to increasing bird predation, particularly by pardalotes and silvereyes. Many second instars also die from a lethal plant factor, presumably associated with secondary plant compounds, in the same way as first instars. In contrast to the early instars, late nymphs are strongly affected by parasitoids, such as *Psyllaephagus* spp., and by predators such as birds, syrphid flies, coccinellid beetles, and lacewings.

While total losses rarely exceed 85% in either outbreak or low density populations, the suppression of incipient outbreaks is invariably due to nymphal mortality.[22] Frequently, these rapid declines in population density occur in places where general predators have increased in abundance on an alternative host. As this alternative food declines, they switch their attention to *C. albitextura* populations. Several predator groups may be involved in suppressing incipient outbreaks but birds seem to be the most effective. The efficiency of the parasitoids, *Psyllaephagus* spp., is reduced by hyperparasitism,[6,26] while syrphid predators seem to have little influence on psyllid abundance.[5]

By contrast, bird predation is considered a major determinant of the abundance of this lerp insect.[7,22] Life-table studies indicate that 77–83% of the nymphs are removed by birds, while 84–99% of the adults are lost to predation or emigration. Furthermore, when bell miners, which feed on psyllid honeydew and defend their territories against other birds, are removed from an area, the psyllid population is quickly eliminated by other species of birds.[21] During outbreaks, this feeding pressure may not necessarily result in population collapse because of continuous psyllid immigration into areas where birds are concentrated.

3.2. Outbreak Dynamics

Outbreaks of *C. albitextura* are typified by phases of latency, outbreak initiation, population growth and spread, and finally decline (Fig. 4a–c). Outbreaks occur at rather irregular intervals at any one place (Fig. 5) and seem to be initiated by the synchronous occurrence of favorable weather, food quality and quantity, ineffective natural enemies, and sufficient numbers of psyllids to exploit this favorable environment.

The white lace lerp often remains at very low densities for many years over very large regions of eucalypt forest. For example, no outbreaks have been recorded in the

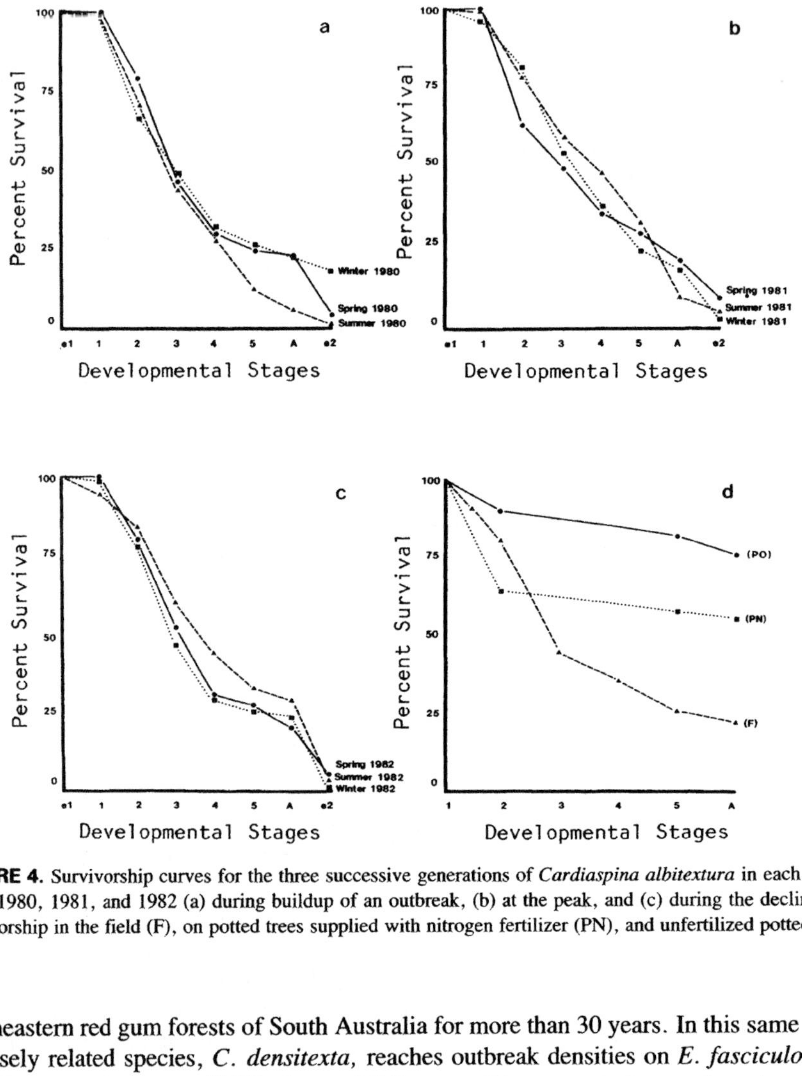

FIGURE 4. Survivorship curves for the three successive generations of *Cardiaspina albitextura* in each of the years 1980, 1981, and 1982 (a) during buildup of an outbreak, (b) at the peak, and (c) during the decline; (d) survivorship in the field (F), on potted trees supplied with nitrogen fertilizer (PN), and unfertilized potted trees (PO).

southeastern red gum forests of South Australia for more than 30 years. In this same area, a closely related species, *C. densitexta,* reaches outbreak densities on *E. fasciculosa* F. Mueller every 4–6 years.[22] Weather and foliage quality appear to be as adequate as they are in most other parts of the range of *C. albitextura*. Predation, particularly by certain birds, seems to be the main stabilizing factor in this area. These forests support moderate to high densities of various psylloid genera, including *Lasiopsylla, Glycaspis, Creiis, Schedotrioza, Spondyliaspis,* and *Cardiaspina,* that support a guild of general avian predators that readily move from one prey species to another as their relative abundances vary in time and space. *Glycaspis* spp. reach outbreak numbers from time to time in areas of this region but are quickly suppressed by predators.

In many areas, populations of *C. albitextura* occasionally increase rapidly from low densities to outbreak levels. They typically arise from initial aggregations of adults, which cluster their eggs on the lower foliage of the northern aspect of a few trees. Some of these

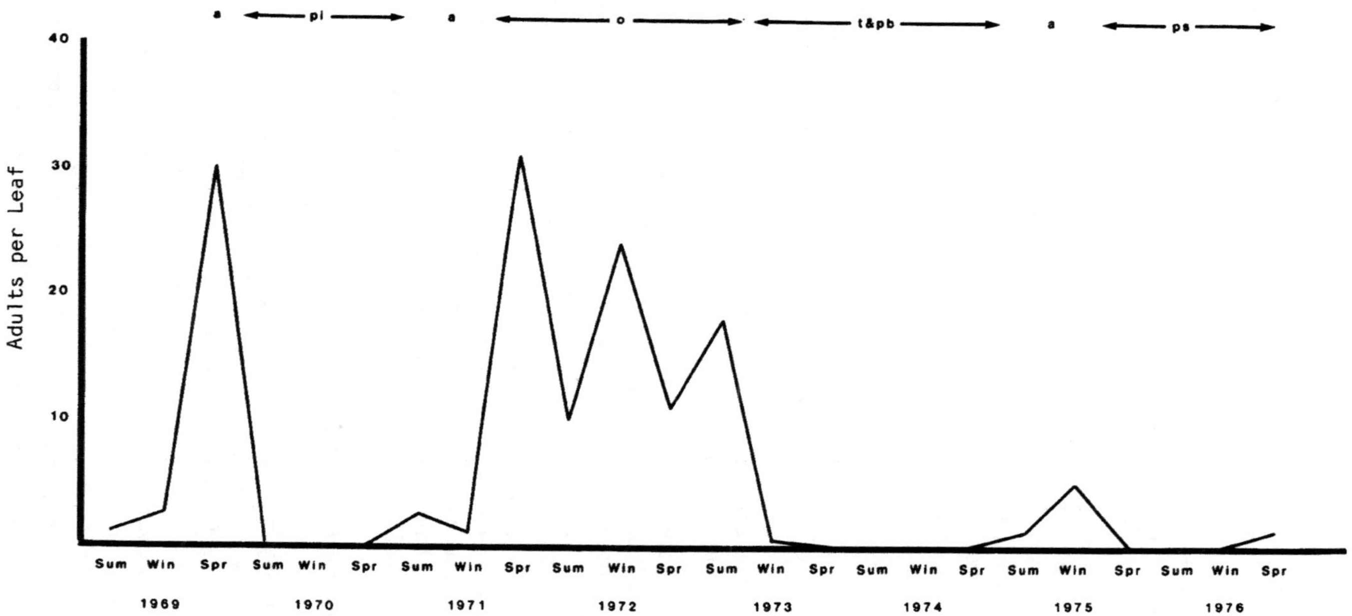

FIGURE 5. Abundance of *Cardiaspina albitextura* for 24 successive generations on *Eucalyptus camaldulensis* in South Australia. a, incipient, or short-term, outbreak; o, sustained outbreak; pb, bird predation; pi, insect parasitism and predation; ps, bird and insect predation; t, leaf toxins (phenolics).

populations escape predation and grow rapidly into an outbreak epicenter, which then increases in size over successive generations. This expansion into adjacent stands often follows valley bottoms and roadsides.[22]

Population growth and spread can be spectacular and, although local populations may be eliminated by chance bird predation, the limited number of birds cannot keep up with the rapidly expanding front. Their functional responses have been saturated, and the psyllid population escapes from their constraining influence.[18] Sometimes the flow of the outbreak is interrupted for several generations at certain locations by intense predation, but this rarely has any effect on the overall trend of the outbreak as adult psyllids continuously move into the area.

The dynamic nature of outbreak spread often involves recolonization of previously defoliated sites. Successful recolonization, however, is usually delayed by three or more generations after peak defoliation, presumably because it takes time for elevated levels of defensive chemicals, particularly phenolics, to return to their previous levels.[22]

There is general agreement on the factors that trigger outbreaks and that cause their collapse.[2–5,9,22] Outbreak initiation is invariably associated with the following factors:

1. Aggregation of adults into small loci containing a range of host plant sizes.
2. An abundance of fully turgid, mature, undamaged leaves with low to moderate concentrations of soluble nitrogen (amino acids), phenolics, antifeedants, and toxins.
3. Mortality factors during the first two or three generations, which do not reduce the population by more than 96%.

The decline of outbreaks is invariably associated with a drastic reduction in the number of mature, undamaged leaves on the host plant. These heavily infested trees may not become suitable for reestablishment of first instar *C. albitextura* nymphs for many months because of high concentrations of phenolics, and possibly other defensive chemicals. We should note, however, that other psyllids, which normally use such leaves, can exhibit outbreaks following those of *C. albitextura.*[22]

The decline in leaf quality is accompanied by heavy losses of adults, mainly through emigration because of an absence of feeding and oviposition sites, but also because of increased predation by birds. Thus, the collapse of the outbreak may be quite sudden, even though minor loci of high populations densities may remain in parts of the outbreak zone.[2,22]

We should also mention another theory, which states that water stress in trees is a progenitor of *C. albitextura* population explosions.[24,27,28] In fact, drought periods seem to occur quite frequently in the year before the start of an outbreak cycle.[9,22] However, the theory assumes that drought acts on the population dynamics of the insect through long-term changes in leaf nitrogen availability. Studies on water stress and nitrogen fertilization of potted trees do not support this assumption, however.[22] Rather, the more recent studies on *Eucalyptus camaldulensis camaldulensis* as the food plant, indicate that insects do not respond to water stress or fertilization in ways predicted by the stress–nitrogen hypothesis.[10,11,22] Thus, insect defoliators may respond in different ways to a specific change in the quality of their food. This explains why different species of the same genus (e.g., *Cardiaspina*) do not undergo outbreaks at the same time on the same plants, even though both are present in the area. It also explains why different species of the Spondyliaspididae exhibit successive outbreaks on the same plants[8,22] and why outbreaks of one species develop synchronously in widely separated zones despite exposure to differing weather patterns. In areas in which weather and food conditions are often suitable, such as in the red gum forests of southeastern Australia, it seems likely that factors other than food quality are more important in permitting outbreaks to develop and to run their course, e.g., the reduction of bird populations by human actions.

4. MANAGEMENT IMPLICATIONS

Realistically, there must always be a reasonable benefit–cost ratio before attempts are made to control natural outbreaks of pest species. A dispassionate view of the red gum–white lace lerp problem would suggest that the best option is usually to leave it alone, for few trees are killed and the forest has low commercial value. If individual trees or sites are particularly valuable, silvicultural and chemical practices can be employed.[22] It is a fact that most outbreaks of pests in Australian forests have occurred in natural or human-induced monocultures. Examples from the early 1980s are *Cardispina albitextura* outbreaks on red gums in three areas of South Australia—two in New South Wales and one in the Australian Capital Territory near Canberra. Outbreaks of *Creiis* spp. have occurred on *E. cladocalyx* F. Mueller and *E. microcarpa* Maiden in the Flinders Ranges of South Australia, as well as on *E. camaldulensis obtusa* and *E. largiflorens* F. Mueller along the Darling River Valley. *Glycaspis brimblecombei* has been in outbreak numbers on red gum in the hills and plains around Adelaide since 1985, following recovery of the trees from a *C. albitextura* outbreak the previous year. Not all outbreaks have been by Psylloidea, however; *Teara contraria* Walker (Notodontidae), the processionary caterpillar, having completely defoliated extensive forests of *Acacia pendula* A. Cunningham on the plains of southern New South Wales during the same period.

The large areas of forest concerned make any attempt to convert these monocultures into mixed stands difficult to contemplate. Until eucalyptus stands are more seriously threatened by psyllid outbreaks or the value of these forests increases considerably, control of psyllid outbreaks will be a low-priority item in Australian forestry.

REFERENCES

1. Berryman, A. A., 1986, *Forest Insects: Principles and Practice of Population Management*, Plenum, New York.
2. Clark, L. R., 1962, The general biology of *Cardiaspina albitextura* (Psyllidae), *Aust. J. Zool.* **10:**537–586.
3. Clark, L. R., 1963, Factors affecting the attractiveness of foliage for oviposition by *Cardiaspina albitextura* (Psyllidae), *Aust. J. Zool.* **11:**20–34.
4. Clark, L. R., 1963, The influence of population density on the number of eggs laid by females of *Cardiaspina albitextura* (Psyllidae), *Aust. J. Zool.* **11:**190–201.
5. Clark, L. R., 1963, The influence of predation by *Syrphus* sp. on the numbers of *Cardiaspina albitextura* (Psyllidae), *Aust. J. Zool.* **11:**470–487.
6. Clark, L. R., 1964, The intensity of parasite attack in relation to the abundance of *Cardiaspina albitextura* (Psyllidae), *Aust. J. Zool.* **12:**150–173.
7. Clark, L. R., 1964, Predation by birds in relation to the population density of *Cardiaspina albitextura* (Psyllidae), *Aust. J. Zool.* **12:**349–461.

8. Clark, L. R., and Dallwitz, M. J., 1974, On the relative abundance of some Australian Psyllidae that co-exist on *Eucalyptus blakelyi, Aust. J. Zool.* **22:**387–415.
9. Clark, L. R., and Dallwitz, M. J., 1975, Life system of *Cardiaspina albitextura* Taylor, *Aust. J. Zool.* **23:**523–561.
10. Cobbinah, J. R., Morgan, F. D., and Douglas, T. J., 1982, Response of the gum leaf skeletoniser *Uraba lugens* WLK. to sugars, amino acids, lipids, sterols, salts, vitamins, and certain extracts of eucalypt leaves, *Aust. J. Entomol. Soc.* **21:**225–236.
11. Farr, J. D., 1985, The performance of *Uraba lugens* WLK. (Lepidoptera, Nolidae) in relation to nitrogen and phenolics in its food, Ph.D. thesis, University of Adelaide.
12. Hillis, W. E., 1966, Variation in polyphenol composition within the species of *Eucalyptus* L'Herit, *Phytochemistry* **5:**541–556.
13. Hillis, W. E., 1966. Polyphenols in the leaves of *Eucalyptus* L'Herit: A chemotaxonomic survey. I. An introduction to the series Globulares, *Phytochemistry* **5:**1075–1090.
14. Hillis, W. E., 1967, A chemotaxonomic survey. II. The sections Renantheroideae and Renantherae, *Phytochemistry* **6:**259–274.
15. Hillis, W. E., 1967, A chemotaxonomic survey. III. The series Transversae, Exsertae, Subexsertae, Microcaspae, Semiunicolores, Viminales, Argyrophyllae and Pamiculatae of the Macrantherae, *Phytochemistry* **6:**275–286.
16. Hillis, W. E., 1967, A chemotaxonomic survey. IV. The sections Porantheroideae and Terminales, *Phytochemistry* **6:**373–382.
17. Hillis, W. E., 1967, A chemotaxonomic survey. V. The series Cornutae and Subcornutae of the section Macrantherae and the section Platyantherae, *Phytochemistry* **6:**845–856.
18. Holling, C. S., 1965, The functional response of predators to prey density and its role in mimicry and population regulation, *Mem. Entomol. Soc. Can.* **45:**1–60.
19. Hodkinson, I. D., and White, I. M., 1979, Homoptera: Psylloidea. Handbooks for the identification of British Insects, *R. Entomol. Soc. (Lond.)* **11**(5)**:**1–100.
20. Kramer, P. J., and Kozlowski, T. T., 1960, *Physiology of Trees,* McGraw-Hill, New York.
21. Loyn, R. H., Runnalls, R. G., and Forward, G. Y., 1983, Territorial bell miners and other birds affecting populations of insect prey, *Science* **221:**1411–1413.
22. Morgan, F. D., 1984, *Psylloidea of South Australia,* Govt. Printer, Adelaide.
23. Morgan, F. D., and Bungey, R. S., 1981, Dynamics of population outbreaks of Psyllidae (Lerp Insects) on eucalypts, in: *Eucalypt Dieback in Forest and Woodland* (K. M. Old, G. A. Kile, and C. P. Ohmart, eds.), pp. 127–129, Proceedings of the CSIRO Division of Forest Research Canberra, 1980, CSIRO, Canberra.
24. McClure, M. S., 1980, Foliar nitrogen: A basis for host suitability for elongate hemlock scale, *Forina externa* (Homoptera, Diaspididae), *Ecology* **61:**72–79.
25. Reese, J. C., 1978, Chronic effects of plant allelochemics on insect nutritional physiology, *Entomol. Exptl. Appl.* **24:**625–631.
26. Reik, E. F., 1963, A new species of *Echthroplexis,* on encyrtid hyperparasite of lerp forming psyllids on eucalypts (Hymenoptera: Chalcidoidea), *Proc. Linn. Soc. N.S.W.* **87:**189–190.
27. White, T. C. R., 1966, Food and outbreaks of phytophagous insects and special reference to *Cardiaspina densitexta* Taylor (Psyllidae: Homoptera) on *Eucalyptus fasciculosa* F.U.M. (Myrtaceae) in South Australia, Ph.D. thesis, University of Adelaide.
28. White, T. C. R., 1978, The importance of a relative shortage of food in ecology, *Oecologia (Berl.)* **33:**71–86.

CHAPTER 8

THE NANTUCKET PINE TIP MOTH

C. Wayne Berisford

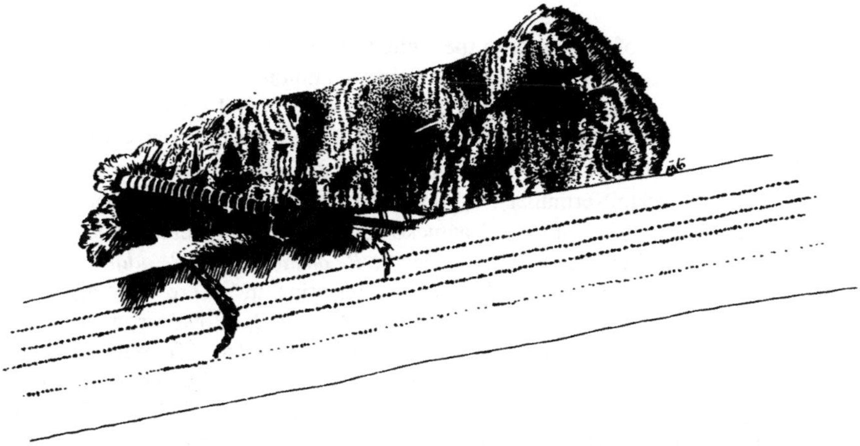

C. Wayne Berisford • Department of Entomology, University of Georgia, Athens, Georgia 30602.

1. INTRODUCTION

The Nantucket pine tip moth, *Rhyacionia frustrana* (Comstock), is one of the most common forest insects in the eastern United States. It was first discovered and studied on Nantucket Island, Massachusetts,[63] where it was causing severe damage to pines and was described as the "frustrating *Retinia*" (= *Rhyacionia*).[23] The taxonomic status of *R. frustrana* is generally clear except for its relationship with *R. bushnelli* Miller, thought by some to be a subspecies.[59] The primary difference is that *R. frustrana* overwinters in infested shoots and *R. bushnelli* overwinters in the duff and soil beneath the trees.

The range of the moth includes most of the eastern half of the United States plus a small part of southern California and New Mexico where it was accidentally introduced. It also occurs in southern Mexico and Central America.[59]

Most hard pines are susceptible to attack, but there are considerable differences in their relative susceptibilities. Among the southern pines, shortleaf (*Pinus echinata*), loblolly (*P. taeda*), and Virginia pines (*P. virginiana*) are highly susceptible, while slash pine (*P. elliottii*) is highly resistant and longleaf pine (*P. palustris*) is virtually immune to attack.

The natural range of the insect in the United States overlaps with at least four other common species of *Rhyacionia* (Fig. 1). Its most common associate, the pitch pine tip moth, *R. rigidana* (Fernald), often shares the same hosts,[2,56] and at least some of their life stages overlap.[5,20] To a lesser degree, the range of *R. frustrana* overlaps with that of the subtropical pine tip moth, *R. subtropica* Miller, and the introduced European pine shoot moth, *R. buoliana* (Schiffermüller). A few other lesser known and apparently rarer species are also sympatric with the Nantucket pine tip moth, including *R. buscskana* Heinrich, *R. aktita* Miller, and to a lesser degree *R. bushnelli* and *R. adana* Heinrich.[59]

Damage by *R. frustrana* may be serious, resulting in severe stunting and stem deformation, and in extreme cases, death of the host[80] (Fig. 2). However, damage is often transitory or negligible in many forest stands. The reasons for different stands having radically different tip moth populations and/or damage are not fully understood, but they are at least partly related to tree species, site quality, natural enemies, and competing vegetation. The latter is often a function of the intensity of site preparation prior to planting.

Tip moth damage is most severe on seedlings and saplings, usually under 5 years of age, and may be manifested in several ways: Growth loss may be permanent, or it may be temporary and quickly regained after moth populations decrease as stands approach crown closure. Loss of tree form is common where heavy attacks result in forks, crooks, or multiple trunks. The excessive branching and competition among lateral limbs for dominance following heavy attacks can result in significant increases in compression wood and loss of wood quality.[41] Trees grown for special purposes, such as Christmas trees, seed orchard trees, and progeny tests, are often regarded as high risks for tip moth attacks and damage. Attacks can also reduce cone crops by injuring conlets and killing potential cone-bearing shoots.[79] Simulated tip moth attacks reduced root growth of shortleaf pine seedlings by more than 50%.[24] Form loss is particularly important on ornamentals and Christ-

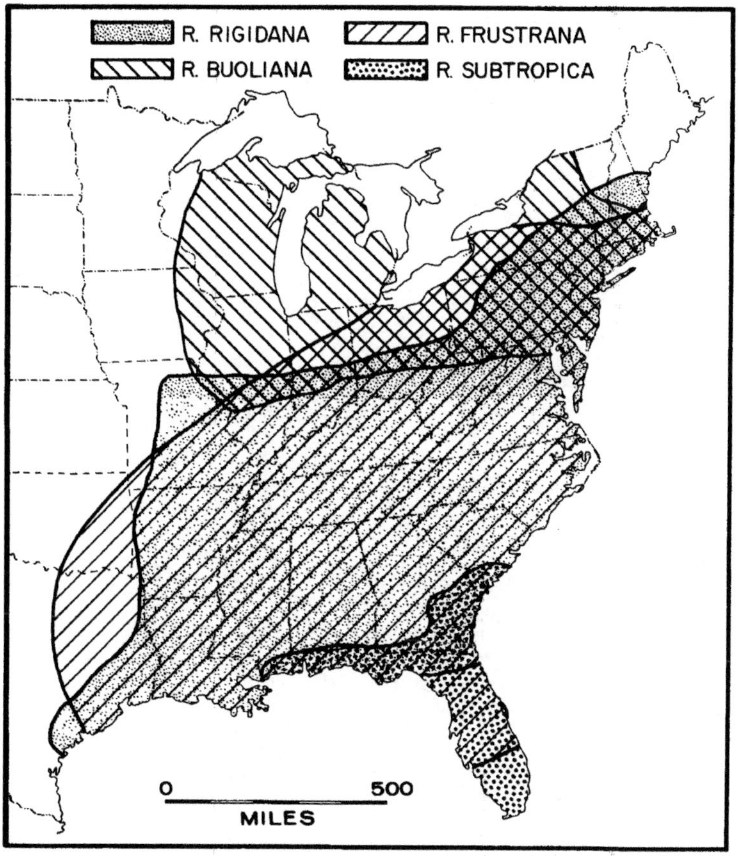

FIGURE 1. Approximate eastern United States ranges of *Rhyacionia frustrana, R. rigidana, R. subtropica,* and *R. buoliana.* (From Berisford *et al.*[13])

mas trees, which may become virtually worthless if tip moth attacks are not controlled.

Studies of the impact of *R. frustrana,* particularly on tree height growth, have produced contradictory results, with some indicating permanent losses, some showing temporary losses that are regained, and others indicating no effect.[73] For example, significant growth loss occurred in areas with three or more annual tip moth generations while those with less than three generations showed little effect.[15] Other studies, however, found significant growth losses in areas in which only two annual generations occur.[50] Unfortunately, most studies on growth impact are based on inadequate data on moth population densities, although reliable methods for estimating populations are currently available.[29,33] Also, such factors as the species of tip moths involved, the numbers of annual generations, the efficacy of chemical control, and the long-term effects of damage have usually not been adequately determined. An exception is a 20-year study in Arkansas, where it was found that stands in which tip moths were controlled yielded an average

FIGURE 2. Loblolly pine, severely deformed and stunted by heavy and repeated tip moth attacks. (USDA Forest Service photograph.)

$2.5\ m^3$ of wood per acre more than stands where tip moths were not controlled.[66] Finally, most impact studies have concentrated on height growth losses and have ignored volume increment. The catchup phenomenon, whereby attacked trees that show early height growth losses ultimately become as tall as protected trees, may be an artifact. Height growth of unattacked trees may be slowed by competition and physiological growth limits, allowing previously attacked trees to catch up. Although the height differential is lost, the diameter differentials and therefore much of the volume are not regained.[17]

At times, a relationship has been observed between heavy tip moth attack and infection of trees by the pitch canker fungus.[19] Tip moth feeding apparently creates infection courts for the fungus; adults may mechanically carry the spores to uninfected trees.

2. BIOLOGY AND LIFE HISTORY

Within the natural range of *R. frustrana,* the life cycle is synchronized with its primary hosts to produce a new generation of egg-laying adults at each time that a new growth flush occurs. This synchrony provides larvae with soft and nutritious meristematic tissue in which to feed during their early development.

The moth has two to five generations annually, depending on the climate. Generations are usually discernible, but considerable overlap may occur even in areas with as few as three annual generations.[31]

Nantucket pine tip moth pupae overwinter inside the shoots of host trees that have been killed by larval feeding (Fig. 3a). Development continues above a threshold of 9.5°C,[40] and adults emerge in late December or January in the southern part of its range. Adults are normally crepuscular, but late afternoon flights may precede evening temperatures that fall below the flight temperature threshold of 10°C. This early flight behavior is apparently triggered by daytime temperatures just above the threshold.[71] Males emerge before females and large numbers are often present when the first females appear. Mating is mediated by sex pheromones emitted from saclike glands located beneath the eighth abdominal tergite of females.[3] Large numbers of males may circle or hover near branches having one or more calling females.[8] Females call for only a short time each day, usually beginning at dusk and ending less than 2 hr later.[4]

In some regions, the pitch pine tip moth may also emerge coincidentally with *R. frustrana.*[5,20] Although the sex pheromones of the two species are mutually inhibitory,[6,9,12,14] the moths minimize interference by calling at different times. *R. rigidana* begins calling at about 1 hr after dark, approximately the time that *R. frustrana* ceases. All the pheromones identified so far from *Rhyacionia* spp. are straight-chain 12-carbon acetates or alcohols,[7] the pheromones of *R. frustrana* being (E)-9-dodecenyl acetate and (E)-9,11-dodecadienyl acetate (in a ratio of 96 : 4),[44] of *R. rigidana* being (E,E)-8,10-dodecadienyl acetate,[43] and of the subtropical pine tip moth, *R. subtropica,* and the European pine shoot moth, *R. buoliana,* being (E)-9-dodecenyl acetate. *R. frustrana* responds weakly to the pheromones of other species.[13,61,64]

Mating occurs shortly after emergence, and eggs are deposited on shoots and needles. *Rhyacionia frustrana* and *R. rigidana* may be active at the same time, and eggs of both species are sometimes found on the same shoots at the same time. However, eggs of

a

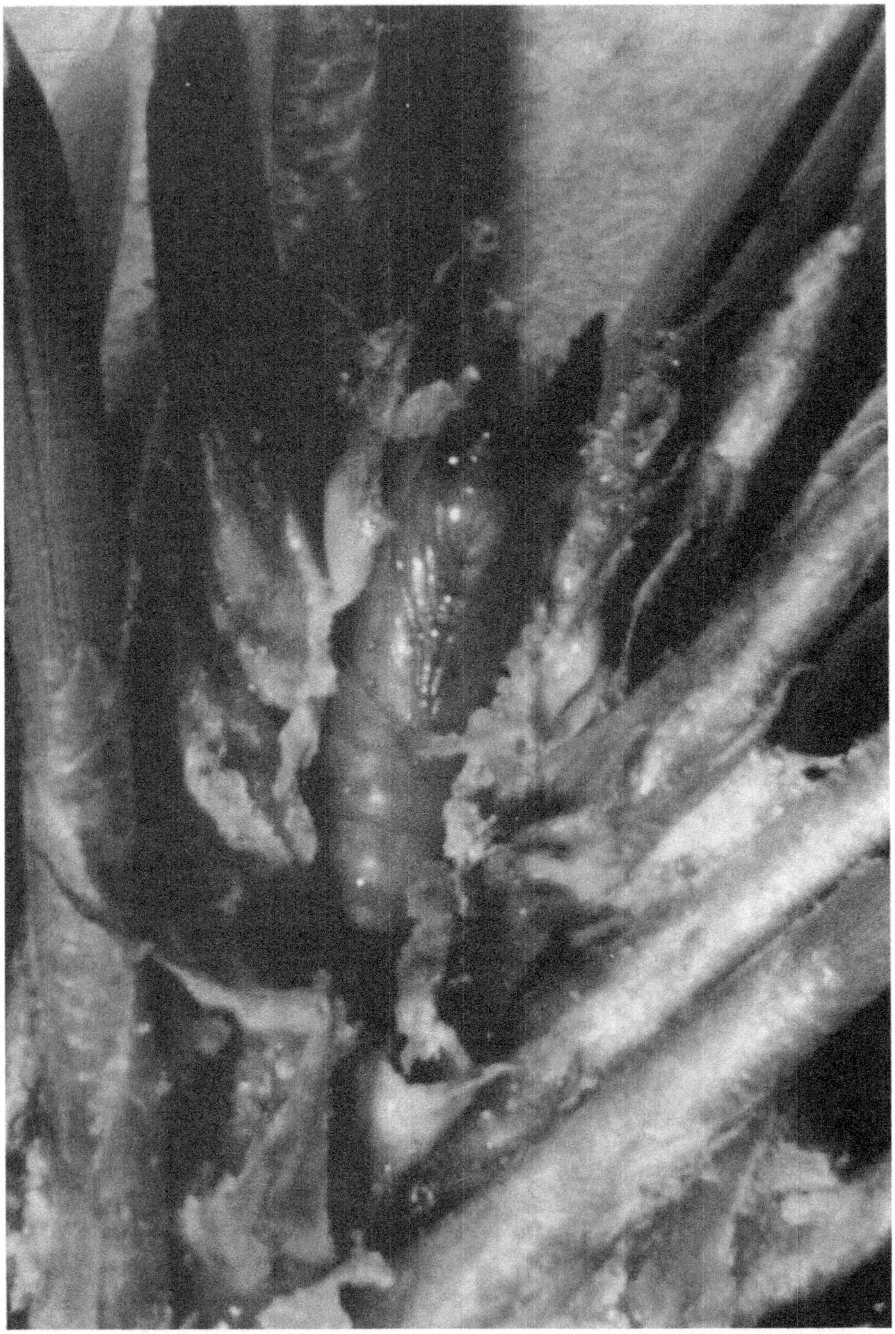

FIGURE 3. (a) Tip moth pupa inside a shoot killed by larval feeding. (USDA Forest Service photograph.) (b) Evidence of early attack showing a bead of resin at the base of a needle where a first-instar larva bored in. (USDA Forest Service photograph.) (c) Characteristic "tent" formed by silk which becomes covered with resin as larvae attack buds and shoots. (USDA Forest Service photograph.) (d) Late-instar tip moth larvae inside a dead shoot. (USDA Forest Service photograph.)

b

FIGURE 3. *(continued)*

C

FIGURE 3. *(continued)*

d

FIGURE 3. (*continued*)

R. frustrana are usually laid singly while pitch pine tip moth eggs are deposited in clusters.[30] Nantucket pine tip moth females can discriminate between susceptible and resistant host species and lay eggs preferentially on loblolly pine, even though slash pine may be available.[48] Egg development begins at about 9.5°C and, depending on prevailing temperatures, hatching occurs within 14 days. Egg hatch follows a sigmoid pattern initially but becomes progressively linear as temperatures increase in later generations.[31] First-instar larvae are also active above 9.5°C, when they frequently mine inside needles, shoots, or buds.[30] The first visible evidence of attack is small drops of resin exuding from needle bases, where first instars have bored in (Fig. 3b). Second instars feed at needle or bud axils, where they construct a characteristic tent of silk covered with resin, which becomes larger as the larva grows[80] (Fig. 3c). Subsequent instars feed inside buds and shoots (Fig. 3d). After five instars, fully grown larvae pupate inside the dead shoots (Fig. 3a). Multiple infestations are common with up to 15 individuals per shoot, depending on population density and shoot diameter.

3. NATURAL ENEMIES

Rhyacionia frustrana has a large complement of parasites and predators.[25,26,28,37,70,77] Most data have come from mass rearings of tip moth-infested shoots, but a few studies have attempted to determine the precise roles of the natural enemies.

3.1. Predators

Fourteen species of insects and seven spiders have been observed to prey on the Nantucket pine tip moth in the field and laboratory.[26] No estimates of impact were made, but spiders were the most abundant predators during the summer, and two species of clerid beetles (*Phyllobaenus* spp.) were apparently most important during the spring (first) generation.

3.2. Parasites

Although the list of parasites associated with *R. frustrana* is rather extensive, including 26 Hymenoptera and one Diptera species, only a few are sufficiently abundant to have any real impact on tip moth populations. Parasites from individual Nantucket pine tip moth larvae and pupae have been reared to help determine their specific parasitic roles, particularly those species reported to be cleptoparasites and hyperparasites.[28] One of the more abundant species found in rearings (*Eurytoma pini* Bugbee: Eurytomidae) is primarily cleptoparasitic (i.e., attacking only previously parasitized larvae), although it may also be a primary parasite or a hyperparasite.

Virtually all studies on larval and pupal parasites have shown that the ichneumonid wasp, *Campoplex frustranae* Cushman, and the tachinid fly, *Lixophaga mediocris* Aldrich, were the most abundant. They constitute 23–45% of all parasites, respectively, when averaged over four generations in Georgia.[28] The next most abundant parasites were

Eurytoma pini (11%) and *Perilampus fulvicornis* Ashmead (6%). The role of *P. flavicornis* is unclear, as it appears to function both as a primary parasite and as a hyperparasite.

Egg parasites (*Trichogramma* spp.) appear to be important in regulating tip moth populations, with up to 64.5% of the eggs being parasitized.[75] Parasitism is usually heavier in summer (47%) than in the spring generation, when only 12% of the eggs were observed to be parasitized.[30] Low egg parasitism in the first generation could be due to poor survival of overwintering *Trichogramma* wasps or to the presence of many alternate hosts in the spring. Egg mortality was also found to be density-dependent (Fig. 4), and most of this mortality was due to *Trichogramma* parasitism.[30] By contrast, no evidence of density-dependent responses has been observed among larval and pupal parasites.[27]

Analysis of life tables for six consecutive tip moth generations (Table I) showed that most tip moth mortality occurred during the egg, late larval, and pupal stages (Fig. 5) and that mortality of early instars (1 to 4) was low and constant.[30,57] However, late larval mortality was relatively constant from generation to generation and from year to year, while egg and pupal mortality varied in direct proportion to total mortality (Fig. 5). This finding suggests that key mortality factors[68] act during the egg and pupal stages. Egg mortality was mostly due to *Trichogramma*, while considerable pupal mortality was caused by the parasite *Campoplex frustranae* (Table I).

Campoplex frustranae has twice been successfully introduced into epidemic tip moth populations, with subsequent substantial decreases in shoot damage. An introduction of *C. frustranae* from Virginia into ponderosa pine (*P. ponderosa*) stands infested with *R. bushnelli* in 1924 reduced moth numbers by up to 80%.[67] Introductions from Georgia and

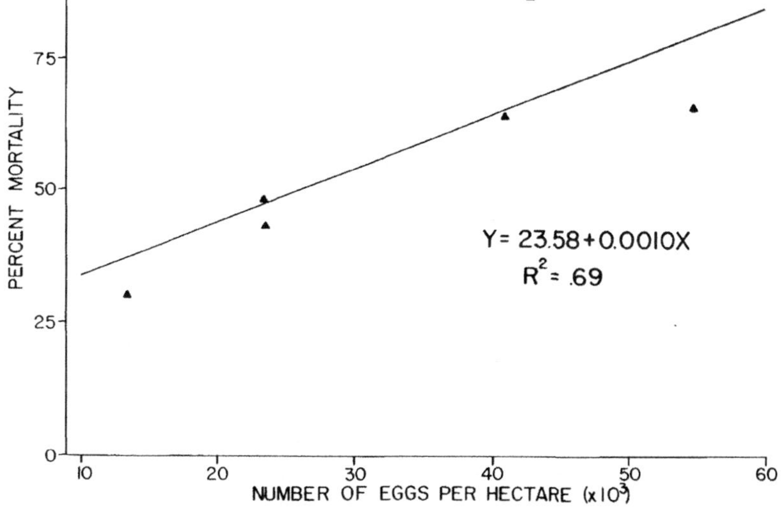

FIGURE 4. Percentage mortality of Nantucket pine tip moth eggs related to egg density. (From Gargiullo and Berisford.[30])

TABLE I
Life Table for the Second *Rhyacionia frustrana* Generation in 1979, Oglethorpe County, Georgia[a,b]

Developmental stage (x)	Number per ha (l_x)	Duration (days)	Mortality factor (d_xF)	No. dying (d_x)	Percentage mortality (100 g_x)
Eggs	23,425	8.2	*Trichogramma* sp.	11,236	48.0
Larvae					
Instar 1	12,189	6.7	Resin	881	7.2
			Other	845	7.0
			Total	1,726	14.2
Instar 2	10,463	6.2	Resin	121	1.2
			Other	1,257	12.0
			Total	1,378	13.2
Instar 3	9,085	5.0	Resin	64	0.7
			Other	912	10.0
			Total	976	10.7
Instar 4	8,109	5.1	Unknown	881	10.9
Instar 5	7,228	12.8	Parasites	1,153	15.9
			Other	1,024	14.2
			Total	2,177	30.1
Pupae	5,051	11.0	Parasites	569	11.3
			Other	1,942	38.4
			Total	2,511	49.7
Moths	2,540		Sex (M/F = 50:50)	0	0
Females ×2	2,540				
Generation				20,885	89.2

[a]From Gargiullo and Berisford.[30]
[b]Mean number of eggs per female: 26.2 ± 13.4 (95% CI); CI: confidence interval; expected eggs: 33,274 ± 14,419 (±2 SE); actual eggs: 18,051 ± 6,445 (±2 SE); trend index: 77%.

Arkansas into *R. frustrana* populations in California enabled infested ornamental Monterry pines (*P. radiata*) to resume growth and regain some form.[62] This parasite therefore appears to play an important role in the population dynamics of tip moths.

4. HOST INTERACTIONS

The abundance of *R. frustrana* is also affected by the availability of preferred hosts, which are in susceptible age classes (2–5 years). Even-aged plantations are capable of producing large populations within 1 year of establishment. Colonization of pine plantations is often rapid, with initial infestations being random but becoming clumped in later tip moth generations.[22] Peak infestations occur 3–5 years after establishment (Fig. 6), when population levels are positively correlated with the number of available shoots.[49] Although the number of available shoots increases with time and tree height, the proportion of shoots infested declines (Fig. 6). This finding suggests that other factors, such as host resistance or parasitism, may be affecting the success of the moth as trees become

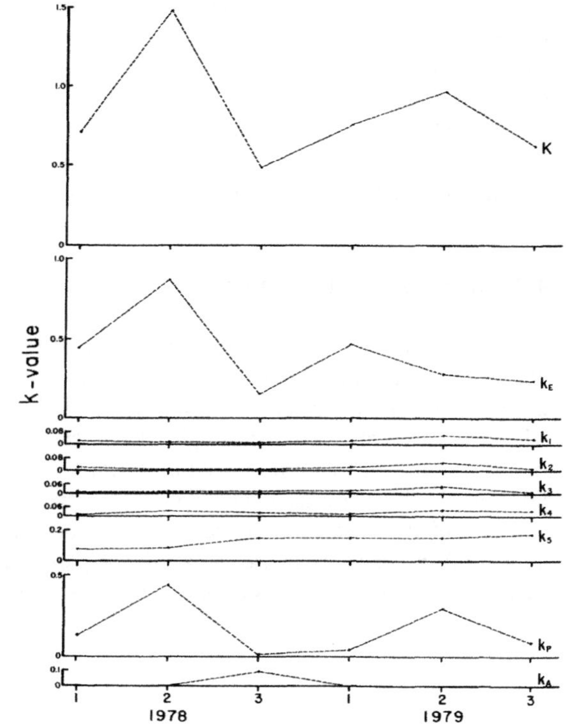

FIGURE 5. Key factor analysis (57) of six consecutive Nantucket pine tip moth generations showing stage specific mortality. K, total mortality; k_E, mortality of eggs; k_i, $i=1,2,3,4,5$ (mortality of larval instars); k_P, mortality of pupae; k_A, mortality of adults. (From Gargiullo and Berisford.[30])

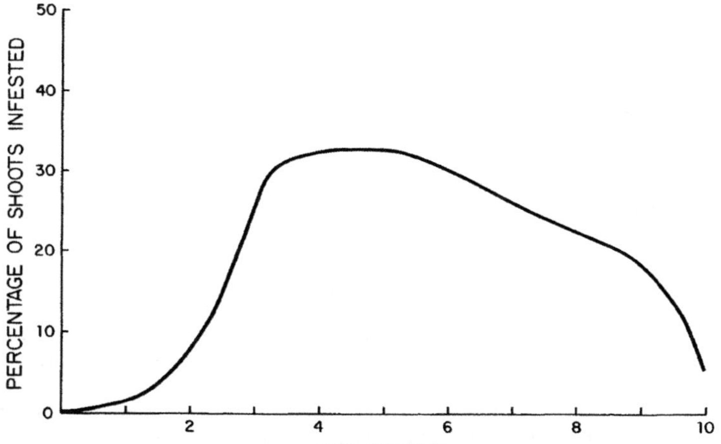

FIGURE 6. Typical pattern of Nantucket pine tip moth invasion, rapid buildup, and decline relative to the age of host stands.

older. As trees become larger and approach crown closure, pitch pine tip moths occupy higher percentages of the infested shoots; e.g., *R. frustrana* occupies 99% of infested shoots in loblolly pines 3–5 years old but *R. rigidana* occupies 75% of infested shoots in 9–11-year old trees.[1]

The Nantucket pine tip moth does not occupy all available shoots and rarely exceeds 75% infestation. Thus, other factors, such as host resistance and natural enemies, usually maintain populations below their carrying capacities.[30]

4.1. Stand and Site Influences

The highest tip moth numbers and damage tend to occur in even-aged stands with wide spacing and little competing vegetation.[15] Thus, plantations in old fields generally have higher infestations than do natural stands with the same average spacing, particularly when they are growing under a light pine overstory (Fig. 7). Other studies have shown increased tip moth numbers concomitant with decreases in competing vegetation,[42,47,53,69,72] and all data indicate that practices that reduce the diversity and/or structure of the vegetation increase tip moth populations as well. These practices may, among other things, decrease the availability of alternate hosts and other food sources, such as pollen and nectar, for natural enemies. For example, it has been shown that tip

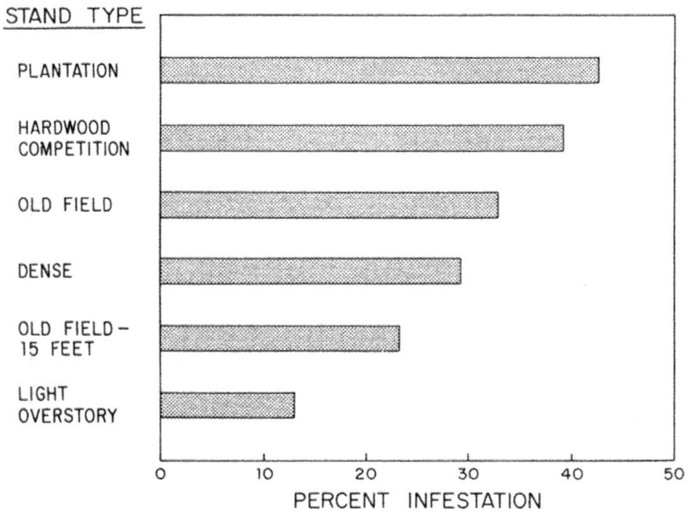

FIGURE 7. Mean percentage of loblolly pine shoots infested by tip moths for six consecutive generations in Virginia. Trees averaged 6 ft tall at the initiation of the study. Plantation: planted old fields with 7 × 8-ft spacing; hardwood competition: plantations in cutover areas with dense stump sprouts; old field: naturally seeded old fields with average spacing of ~8 × 8 ft; dense: direct seeded areas with average spacing of 3 × 3 ft or less; old fields to 15 ft: naturally seeded old fields with 6-ft trees growing among taller (~15 ft) trees with ~8 × 8 ft spacing; light overstory: pines with ~8 × 8-ft spacing growing under a light pine overstory resulting from a shelterwood-type harvest.

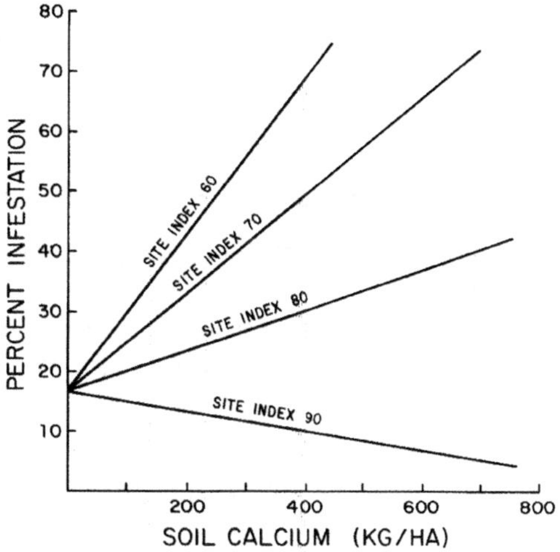

FIGURE 8. Percentage tip moth infestation relative to site index and soil calcium content in loblolly pine plantations with minimal site preparation in the South Carolina Piedmont Plateau. (From Hood.[47])

moth parasitism by Hymenoptera is lower in loblolly pine plantations, where competing vegetation has been mowed or treated with herbicides.[35]

Poor sites also tend to have higher tip moth populations in the Southern Piedmont Plateau, with site index being negatively correlated with tip moth infestations.[47] However, high tip-moth populations were found associated with higher site indices in Texas.[72] Specific soil characteristics, such as texture, depth of "A" horizon, calcium content, pH, and percentage and/or depth of clay, were correlated with infestation levels, but site index and soil calcium accounted for much of the variation (Fig. 8).

Although there are no data on the effects of site quality on resistance to tip moth attack, it seems likely that root establishment is faster on better sites, thereby increasing the general health of the seedlings. Resistance is probably due to high resin flow[38,45,46,55,74] or to induced hypersensitive reactions, as observed in bark beetles[16,36,58,60] and other shoot-infesting insects.[52,54] Certain strains of susceptible pine species have been found to be resistant to shoot moths, apparently due to different relative quantities of terpenes in the shoots.[21] Rapid growth on good sites also reduces the period during which trees are highly susceptible, since infestation decreases with tree height (see Fig. 6). Although the relationship between site quality and tip moth infestation is clear, the precise factors involved have not been elucidated. Is the link to poor site conditions due to effects on host resistance, parasites, or lower densities of competing vegetation, or is it due to slower growth rates?

High tip moth populations and severe damage occasionally occur in isolated stands in areas in which otherwise moderate populations prevail. This phenomenon may be related to localized soil conditions and to hosts such as loblolly pine being planted on sites more suitable for slash pine growth.

Since the life cycle of *R. frustrana* is closely tied to the phenology of its primary hosts, population densities may be affected by factors that determine the timing of growth flushes. Drought or flooding could eliminate or delay flushing and force tip moth larvae to feed on tissues that have hardened off, thereby reducing survival. It has also been found that different provenances of loblolly pine differ significantly in susceptibility to tip moths and that much of the resistance is due to asynchrony between tip moth activity and growth flushes.[39]

5. HYPOTHESES FOR POPULATION BEHAVIOR

5.1. Invasion and Early Population Growth

The rapid invasion[22] and subsequent exponential growth of tip moth populations during the first 2–3 years of colonization are apparently influenced by several factors (Fig. 6):

1. Sparcity of competing vegetation in old fields or intensively prepared sites allows ovipositing females unimpeded access to shoots and also reduces parasitism.[35]
2. Parasitism is generally lower in younger stands[49] and may be virtually absent during establishment.
3. Resistance of young trees is initially low but increases with age. For instance, slash pine is highly susceptible during its first year after planting but is almost completely immune by age 3.[48,76] Normally susceptible species, such as loblolly pine, may take longer to manifest any substantial resistance.

5.2. Population Stagnation and Decline

Nantucket pine tip moth populations usually level off at 3–5 years, gradually decline, and then rapidly diminish as crowns close (Fig. 6). As stands become older, competing vegetation may mechanically disrupt access to pine shoots by ovipositing females,[69] and it also provides alternate hosts and nectar for parasitic insects.[35] The more complex environment created by vertical stratification of older stands may also favor natural enemies. As stands age, therefore, natural enemy populations increase in density and diversity and, since some parasites respond in a density-dependent manner,[30] they may be capable of regulating the tip moth population.

In addition, older trees seem to be more resistant to Nantucket pine tip moth attack.[49] This phenomenon is probably related to the following internal and external factors:

1. Higher resin flow from the needles of older trees may repel the invasion of shoot moth larvae (Fig. 3b); resin can result in the mortality of many invading (first-instar) larvae (Table I).
2. The percentage of shaded shoots increases in older stands as the crowns close. Shaded loblolly pines have low tip moth infestations but attacks increase rapidly if the shading is removed.[15,35] This is apparently a physical phenomenon unrelated to the physiological status of the trees.
3. As crown closure approaches, competition for available shoots between the Nantucket pine tip moth and the pitch pine tip moth increases. The pitch pine tip moth is the more successful invader at this time and occupies a greater proportion of

available shoots, so that progressively fewer susceptible shoots are available to the Nantucket pine tip moth.[1]

6. MANAGEMENT STRATEGIES

Since tip moths rarely kill trees and their impact on volume and form is highly variable, management practices have generally aimed at maintaining stand vigor through appropriate silvicultural treatments. This approach is based on the assumption that vigorous trees are more resistant to, or more tolerant of, tip moth attacks. However, as foresters seek to maximize yields from a shrinking land base, management practices will intensify. Some workers believe that there will be a concomitant increase in the pest status of the Nantucket pine tip moth.[65] Practices in the future may include hazard-rating to make pre-establishment predictions of potential damage. Once high hazard sites are identified, managers may elect to establish less preferred tree species or to modify the site by such practices as fertilization or subsoiling.

Another management option is to establish stands that are unfavorable for the tip moth, i.e., increasing species and/or structural diversity of the stand. This option might include site preparation that maintains some competing vegetation, natural regeneration or direct seeding in lieu of planting, close spacing, and reducing the size of regeneration blocks.

Chemical control in forest stands may be selected as a last resort if substantial mortality or severe permanent form and/or growth loss is likely to occur. This is usually only economically feasible in cases in which damage is especially severe and the unit value of trees high, e.g., seed orchards and Christmas tree plantations.

Systemic insecticides applied to the soil can be effective.[18,78] However, insufficient soil moisture prevents the pesticide from being absorbed by the roots; whereas too much moisture, especially in sandy soils, leaches and dilutes the pesticide. Foliar applications of pesticides can provide good control, but precise timing is required to kill larvae before they enter the buds and shoots where they are protected from the toxicants. Spray-timing models have been developed based on the biology, development rate, and vulnerable life stages of the insect,[11,32,34] and pheromone-baited traps are used to detect the initiation of moth flight.

Some preliminary studies indicate that it may be feasible to control tip moths with biological insecticides, such as viruses[51] or bacteria. These methods are desirable, since they generally do not adversely affect natural enemies or trigger outbreaks of secondary pests, such as scale insects, mealybugs, or aphids. Synthetic pheromone components have also shown some promise for direct control by disrupting adult communication and mating, but the results have not been consistent enough for commercial adoption.[10]

In the future, breeding-resistant strains of the more susceptible species appear to be a strong possibility, e.g., by selecting trees with high resin flow,[38,45,46,55,74] strong induced responses to attack,[52,54] oviposition deterrents,[48] or asynchronous growth flushes.[39] Unfortunately, many progeny tests are routinely protected from tip moth attack by pesticides, and there is no opportunity to identify and select for resistance or to study resistance mechanisms. In fact, the use of pesticides in progeny tests could lead to the selection of tree strains that are highly susceptible to the moth.

As the value of pine stands increases, management practices will become increasingly sophisticated and intensive. Some practices, such as the control of competing vegetation, may increase tree growth but will also increase tip moth populations. A balance will ultimately have to be struck among management options to maximize both yield and quality of wood products. As management becomes more intense, it is hoped that emphasis will shift from a reactive "firefighting" approach to tip moth control to a proactive integration of tip moth management options into an overall stand management–pest management system.

ACKNOWLEDGMENTS. I am grateful to Stephen C. Cade, Weyerhaeuser Company, Hot Springs, Arkansas; to Paul M. Gross, University of Maryland, College Park, Maryland; Roy L. Hedden, Clemson University, Clemson, South Carolina; to David L. Kulhavy, Stephen F. Austin University, Nacogdoches, Texas; and to Harry O. Yates III, Southeastern Forest Experiment Station, Athens, Georgia, for access to unpublished data and for reviews of the manuscript. I would also like to acknowledge the *Journal of Chemical Ecology*, the Entomological Society of America, University of Georgia, and the U.S.D.A. Forest Service for permission to reproduce illustrations.

REFERENCES

1. Allen, D., and Hedden, R. L., Personal communication, Department of Forestry, Clemson University, Clemson, South Carolina.
2. Baer, R. G., and Berisford, C. W., 1975, Species composition of pine tip moth, *Rhyacionia* spp., infestations in northeast Georgia, *J. Ga. Entomol. Soc.* **10**:64–67.
3. Baer, R. G., and Berisford, C. W., 1976, Bioassay, histology, and morphology of pheromone-producing glands of *Rhyacionia frustrana, R. rigidana* and *R. subtropica, Ann. Entomol. Soc. Am.* **2**:307–310.
4. Berisford, C. W., 1974, Species isolation mechanisms in *Rhyacionia frustrana* and *R. rigidana, Ann. Entomol. Soc. Am.* **67**:292–294.
5. Berisford, C. W., 1974. Comparisons of emergence periods and generations of pine tip moths, *Rhyacionia frustrana* and *R. rigidana, Ann. Entomol. Soc. Am.* **67**:666–668.
6. Berisford, C. W., 1977, Inhibition of male attraction of the Nantucket pine tip moth by a synthetic pheromone for the pine tip moth, *Environ. Entomol.* **6**:245–246.
7. Berisford, C. W., 1982, Pheromones of *Rhyacionia* spp. identification, function and utility, *J. Ga. Entomol. Soc.* **17**:23–30 (second suppl.).
8. Berisford, C. W., and Brady, U. E., 1972, Attraction of Nantucket pine tip moth males to the female sex pheromone, *J. Econ. Entomol.* **65**:430–435.
9. Berisford, C. W., and Brady, U. E., 1973, Specificity and inhibition of attraction of *Rhyacionia frustrana* and *R. rigidana* to their female sex pheromones, *Nature (Lond.)* **241**:68–69.
10. Berisford, C. W., Debarr, G. L., and Payne, T. L., 1985, Utilization of pheromones in forest pest management, in: *Insects and Diseases of Southern Forests* (R. A. Goyer and J. P. Jones, eds.), pp. 92–96, *Proceedings of the Thirty-fourth Annual Forestry Symposium,* Louisiana Agric. Expt. Sta.
11. Berisford, C. W., Gargiullo, P. M., and Canalos, C. G., 1984, Optimum timing for insecticidal control of the Nantucket pine tip moth (Lepidoptera: Tortricidae), *J. Econ. Entomol.* **77**:174–177.
12. Berisford, C. W., Godbee, J. F., and Baer, R. G., 1974, Inhibition of Nantucket pine tip moth (Lepidoptera: Olethreutidae) attraction by small proportion of pitch pine tip moth pheromone, *Can. Entomol.* **106**:1109–1110.
13. Berisford, C. W., Harman, D. M., Freeman, B. L., Wilkinson, R. C., and McGraw, J. R., 1979, Sex

pheromone cross-attraction among four species of pine tip moths, *Rhyacionia* species, *J. Chem. Ecol.* **5**:205–210.
14. Berisford, C. W., and Hedden, R. L., 1978, Suppression of male *Rhyacionia frustrana* response to live females by the sex pheromone of *R. buoliana* and *R. subtropica, Environ. Entomol.* **7**:532–533.
15. Berisford, C. W., and Kulman, H. M., 1967, Infestation rate and damage by the Nantucket pine tip moth in six loblolly pine stand categories, *For. Sci.* **13**:428–438.
16. Berryman, A. A., 1972, Resistance of conifers to invasion by bark beetle–fungus associations, *Bioscience* **22**:598–602.
17. Cade, S. C., and Hedden, R. L., 1987, Growth impact of pine tip moth on loblolly pine plantations in the Ouachita mountains of Arkansas, *South J. Appl. For.* **11**:128–133.
18. Cade, S. C., and Heikkenen, H. J., 1965, Control of pine tip moths on loblolly pine with systemic insecticides, *Ga. For. Res. Pap. No. 32*.
19. Cade, S. C., Runion, G. B., and Bruck, R. I., 1986, Pine tip moth/pitch canker relationships in a coastal North Carolina (USA) loblolly pine plantation, *Proc. 18th IUFRO World Cong. Ljuhljana Div.* **2**:7–21.
20. Canalos, C. G., and Berisford, C. W., 1981, Seasonal activity of two sympatric *Rhyacionia* species determined by pheromone traps, *J. Ga. Entomol. Soc.* **16**:219–222.
21. Charles, P. J., Delplanque, A., Marpeau, A., Bernard-Dagan, C., and Arbez, M., 1982, Susceptibility of European black pine (*Pinus nigra*) to the European pine shoot moth (*Rhyacionia buoliana*): variations of susceptibility at the provenance and individual level of the pine and effect of terpine composition, pp. 206–212, in: *Proceedings on Resistance to Diseases and Pests in Forest Trees Workshop, 1980, Pudoe Wageningen, Netherlands.*
22. Clarke, S. R., 1982, Dispersion of the Nantucket pine tip moth, *Rhyacionia frustrana* (Comstock), in newly established loblolly pine plantation, Masters thesis, University of Arkansas, Fayetteville, Arkansas.
23. Comstock, J. H., 1880, Report (1879) of the entomologist, U.S. Dep. Agric. Bur. Entomol., pp. 236–237.
24. Coyne, J. F., 1968, Simulated Nantucket pine tip moth (*Rhyacionia frustrana*) attacks reduce root development of shortleaf pine seedlings, *J. Econ. Entomol.* **61**:319–320.
25. Eikenbary, R. D., and Fox, R. C., 1965, The parasites of the Nantucket pine tip moth in South Carolina, S.C. Agric. Expt. Sta. Tech. Bull. 1017.
26. Eikenbary, R. D., and Fox, R. C., 1968, Arthropod predators of the Nantucket pine tip moth *Rhyacionia frustrana, Ann. Entomol. Soc. Am.* **61**:1218–1221.
27. Eikenbary, R. D., and Fox, R. C., 1968, Responses of Nantucket pine tip moth parasites to tree level, orientation and hosts per pine tip, *Ann. Entomol. Soc. Am.* **61**:1380–1384.
28. Freeman, B. L., and Berisford, C. W., 1979, Abundance and parasitic habits of some parasitoids of the Nantucket pine tip moth (Lepidoptera: Tortricidae), *Can. Entomol.* **111**:509–514.
29. Gargiullo, P. M., and Berisford, C. W., 1981, Sampling for pine tip moths—A procedural guide, Univ. Ga. Agric. Expt. Sta. Bull. #272.
30. Gargiullo, P. M., and Berisford, C. W., 1983, Life tables for the Nantucket pine tip moth, *Rhyacionia frustrana* (Comstock), and the pitch pine tip moth, *Rhyacionia rigidana* (Fernald) (Lepidoptera: Tortricidae), *Environ. Entomol.* **12**:1391–1402.
31. Gargiullo, P. M., Berisford, C. W., Canalos, C. G., Richmond, J. A., and Cade, S. C., 1984, Mathematical descriptions of *Rhyacionia frustrana* (Lepidoptera: Tortricidae) cumulative catches in pheromone traps, cumulative eggs hatching, and their use in timing of chemical control, *Environ. Entomol.* **13**:1681–1685.
32. Gargiullo, P. M., Berisford, C. W., and Godbee, J. F., 1985, Prediction of optimal timing for chemical control of the Nantucket pine tip moth, *Rhyacionia frustrana* (Comstock) (Lepidoptera: Tortricidae) in the southeastern coastal plain, *J. Econ. Entomol.* **78**:148–154.
33. Gargiullo, P. M., Berisford, C. W., and Pienaar, L. V., 1983, Two-stage cluster sampling for pine tip moths, *Environ. Entomol.* **12**:81–90.
34. Gargiullo, P. M., Berisford, C. W., and Richmond, J. A., 1983, How to time insecticide sprays against the Nantucket pine tip moth, Ga. For. Comm. Ga. For. Res. Pap. #44.
35. Gross, P. M. Personal communication, Department of Entomology, University of Maryland, College Park, Maryland.
36. Hain, F. P., Cook, S. P., Matson, P. A., and Wilson, K. G., 1985, Factors contributing to southern pine beetle host resistance, in: *Proceedings of the Integrated Pest Management Research Symposium* (S. J. Branham and R. C. Thatcher, eds.), pp. 154–160, USDA For. Serv. Gen. Tech. Rep. SO-56.

37. Harman, D. M., and Kulman, H. M., (eds.), 1973, A world survey of the parasites and predators of the genus *Rhyacionia*, parts I–IV, University of Maryland Nat. Res. Inst. Contrib. #527.
38. Harris, P., 1960, Production of pine resin and its effect on survival of *Rhyacionia buoliana* (Schiff.) (Lepidoptera: Olethreutidae), *Can. J. Zool.* **38:**121–130.
39. Haugen, D. A., and Hedden, R. L. Personal Communication, Department of Forestry, Clemson University, Clemson, South Carolina, and S. C. Cade, Weyerhauser Company, Hot Springs, Arkansas.
40. Haugen, D. A., and Stephen, F. M., 1983, Developmental rates of Nantucket pine tip moth (*Rhyacionia frustrana* (Comstock) (Lepidoptera: Tortricidae)) life stages in relation to temperature, *Environ. Entomol.* **13:**56–60.
41. Hedden, R. L., and Clason, T., 1980, Nantucket pine tip moth impact on loblolly pine wood and product quality, Forestry Res. Rep. 1979, L.S.U. Agric. Exp. Sta. N. La. Hill Farm Exp. Sta., Homer, La.
42. Hertel, G. D., and Benjamin, D. M., 1977, Intensity of site preparation influences on pine webworm and tip moth infestations of pine seedlings in North-central Florida, *Environ. Entomol.* **6:**118–122.
43. Hill, A. S., Berisford, C. W., Brady, U. E., and Roelofs, W. L., 1976, Sex pheromone of the pitch pine tip moth, *Rhyacionia rigidana, Environ. Entomol.* **5:**959–960.
44. Hill, A. S., Berisford, C. W., Brady, U. E., and Roelofs, W. L., 1981, Nantucket pine tip moth, *Rhyacionia frustrana:* Identification of two sex pheromone components, *J. Chem. Ecol.* **7:**517–528.
45. Holst, M., 1963, Breeding resistance in pines to *Rhyacionia* moths. World consultation on forest genetics and tree improvement, Food and Agriculture Organization, Stockholm, Sweden.
46. Holst, M., and Heimburger, C., 1955, The breeding of hard pine types resistant to European pine shoot moth [*Rhyacionia buoliana* (Schiff.)], *For. Chron.* **31:**162–169.
47. Hood, W. M., 1986, Hazard rating forest sites for pine tip moth, *Rhyacionia* spp. in Northwestern South Carolina, Ph.D. dissertation, University of Georgia, Athens, Georgia.
48. Hood, W. M., Berisford, C. W., and Hedden, R. L., 1985, Oviposition preferences of the Nantucket pine tip moth (Lepidoptera: Tortricidae) on loblolly and slash pine, *J. Entomol. Sci.* **20:**204–206.
49. Lashomb, J. H., Steinhauer, A. L., and Diveby, G., 1980, Comparison of parasitism and infestation of Nantucket pine tip moth in different aged stands of loblolly pine, *Environ. Entomol.* **9:**397–402.
50. Lashomb, J. H., Steinhauer, A. L., and Douglass, L., 1978, Impact studies of Nantucket pine tip moth populations on loblolly pine, *Environ. Entomol.* **7:**910–912.
51. McLeod, P. J., Wallis, G. W., Yearian, W. C., Stephen, F. M., and Young, S. Y., 1983, Evaluation of codling moth granulosis virus for Nantucket pine tip moth suppression, *J. Ga. Entomol. Soc.* **18:**424–427.
52. Merkel, E. P., Squillace, A. E., and Bengtson, G. W., 1965, Evidence of inherent resistance to *Dioryctria* infestation pp. 96–99, in: *Proceedings of the Eighth Southern Forest Tree Improvement Conference 1965.*
53. Miller, F. D., and Stephen, F. M., 1983, Effects of competing vegetation on Nantucket pine tip moth *(Lepidoptera: Tortricidae)* populations in loblolly pine plantations in Arkansas, *Environ. Entomol.* **12:**101–105.
54. Miller, J. M., 1950, Resistance of pine hybrids to the pine reproduction weevil, USDA For. Serv. Res. Note No. 68.
55. Miller, W. E., 1967, The European pine shoot moth—ecology and control in the lake states, *For. Sci. Monog.* **14:**1–72.
56. Miller, W. E., and Wilson, L. F., 1964, Composition and diagnosis of pine tip moth infestations in the Southeast, *J. Econ. Entomol.* **57:**722–726.
57. Morris, R. F., 1959, Single-factor analysis in population dynamics, *Ecology* **40:**580–588.
58. Paine, T. D., Stephen, F. M., and Cates, R. G., 1985, Induced defenses against *Dendroctonus frontalis* and associated fungi: Variation in loblolly pine resistance, in: *Proceedings of the Integrated Pest Management Research Symposium* (S. J. Branham and R. C. Thatcher, eds.), pp. 169–176, USDA For. Serv. Gen. Tech. Rep. SO-56.
59. Powell, J. A., and Miller, W. E., 1978, Nearctic pine tip moths of the genus *Rhyacionia:* Biosystematic review (Lepidoptera: Tortricidae, Olethreutidae), USDA Agric. Handbook No. 514.
60. Raffa, K. F., and Berryman, A. A., 1983, The role of host plant resistance in the colonization behavior and ecology of bark beetles (Coleoptera: Scolytidae), *Ecol. Monog.* **53:**27–49.
61. Roelofs, W. L., Hill, A. S., Berisford, C. W., and Godbee, J. F., 1979, Sex pheromone of the subtropical pine tip moth, *Rhyacionia subtropica, Environ. Entomol.* **8:**894–895.
62. Scriven, G. T., and Luck, R. F., 1978, Natural enemy promises control of Nantucket pine tip moth, *Calif. Agric.* **32:**19–20.

63. Scudder, S. H., 1883, The pine moth of Nantucket—*Retinia frustrana*, Pub. Mass. Soc. Promotion of Agric.
64. Smith, R. C., Daterman, G. E., Daves, G. D., McMurtrey, K. D., and Roelofs, W. L., 1974, Sex pheromone of the European pine shoot moth: Chemical identification and field tests, *J. Insect Physiol.* **20**:661–668.
65. Stephen, F. M., 1983, Nantucket pine tip moth in forest stands, in: *History, Status, and Future Needs for Entomology Research in Southern Forests. Proceedings of the Tenth Anniversary of the East Texas Forest and Entomology Seminar, Kurth Lake, Texas* (T. L. Payne and R. N. Coulson, eds.), pp. 53–58.
66. Stephen, F. M., Wallis, G. W., Colvin, R. J., Young, J. F., and Warren, L. O., 1982, Pine tree growth and yield: Influence of species, plant spacings, vegetation, and pine tip moth control, *Ark. Farm Res.* **31**:10.
67. Wadley, F. M., 1932, Minutes of the 434th regular meeting of the Entomological Society of Washington, Jan. 7, 1932, *Proc. Wash. Entomol. Soc.* **34**:26–28.
68. Varley, G. C., and Gradwell, G. R., 1960, Key factors in population studies, *J. Anim. Ecol.* **31**:917–943.
69. Warren, L. O., 1963, Nantucket pine tip moth infestations, severity of attack as influenced by vegetative competition in pine stands, *Ark. Farm Res.* **12**:2.
70. Warren, L. O., 1985, Primary hymenopterous parasites of Nantucket pine tip moth, *Rhyacionia frustrana* (Comstock), *J. Entomol. Sci.* **20**:383–389.
71. Webb, J. W., and Berisford, C. W., 1978, Temperature modification of flight and response to pheromones in *Rhyacionia frustrana, Environ. Entomol.* **7**:278–280.
72. White, M. N., Kulhavy, D. L., and Conner, R. N., 1984, Nantucket pine tip moth (Lepidoptera: Tortricidae) infestation rates related to site and stand characteristics in Nacogdoches county Texas, *Environ. Entomol.* **13**:1598–1601.
73. Williston, H. L., and Barras, S. J., 1977, Impact of tip moth injury on growth and yield on 16-year old loblolly and shortleaf pine, USDA For. Serv. Res. Note SO-221.
74. Yates, H. O. III, 1962, Influence of tip moth larvae on oleoresin crystallization of southern pine, Southeast. For. Exp. Stn., Asheville, NC Res. Note No. 174.
75. Yates, H. O. III, 1966, *Rhyacionia* egg parasitism by *Trichogramma minutum* Riley, *J. Econ. Entomol.* **59**:967–968.
76. Yates, H. O. III, 1966, Susceptibility of loblolly and slash pines to *Rhyacionia* spp. oviposition, injury and damage, *J. Econ. Entomol.* **59**:1461–1464.
77. Yates, H. O. III, 1967, Key to nearctic parasites of the genus *Rhyacionia* with species annotations, USDA For. Serv. Publ. USDA U.S. For. Serv. Asheville, NC.
78. Yates, H. O. III, 1970, Control of pine tip moths *Rhyacionia* spp. (Lepidoptera: Olethreutidae) on seed orchards with phorate, *J. Ga. Entomol. Soc.* **5**:100–104.
79. Yates, H. O. III, and Ebel, B. H., 1972, Shortleaf pine conelet loss caused by the Nantucket pine tip moth. *Rhyacionia frustrana* (Lepidoptera: Olethreutidae), *Ann. Entomol. Soc. Am.* **65**:100–104.
80. Yates, H. O. III, Overgaard, N. A., and Koerber, T. W., 1981, Nantucket pine tip moth, USDA For. Ser., For. Insect and Disease Leaflet, No. 70.

CHAPTER 9

THE AUTUMNAL MOTH IN FENNOSCANDIA

Erkki Haukioja, Seppo Neuvonen, Sinikka Hanhimäki, and Pekka Niemelä

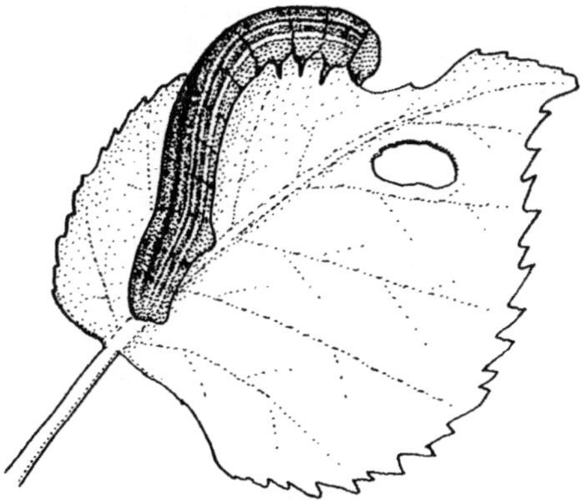

Erkki Haukioja, Seppo Neuvonen, Sinikka Hanhimäki, and Pekka Niemelä • Department of Biology, University of Turku, SF-20500 Turku 50, Finland.

1. INTRODUCTION

Mountain birch, *Betula pubescens* Ehrhart ssp. *tortuosa* (Ledebour) Nyman, forests form the tree line in northwestern Europe and are periodically defoliated by the autumnal moth, *Epirrita* =*Oporinia autumnata* (Borkhausen). This chapter analyzes the population dynamics of *Epirrita*, but many of our conclusions may apply to other defoliators, such as *Operophtera brumata* (Linnaeus), *O. fagata* (Scharfenberg), *O. bruceata* (Hulst), and *Erannis defoliaria* (Clerck), which often exhibit synchronous fluctuations in the same or adjacent areas.[38,62]

Epirrita outbreaks are restricted to northern and mountainous regions of northwestern Europe (Fig. 1), but the range of both moth and birch cover much larger areas.[62] During some outbreaks, *Epirrita* defoliation causes extensive birch mortality at the tree line[35,36,49]; as a result, vast areas of forest have been changed to treeless tundra (e.g., 500 km^2 in the borough of Utsjoki, northern Finland, in 1964–1965.[35,36] It is more common for defoliated trees to recover after defoliation, but still drastic changes may occur in the ground cover as grasses and herbs increase and flower more frequently.[35,39,43,62] These changes may cause long-term effects in the functioning of the ecosystem (both terrestrial and aquatic) and in human activities (e.g., grouse hunting and reindeer husbandry, the traditional way of land use).

Mountain birch usually grows as a shrub or a low polycormic tree and is not commercially exploited, except as firewood when growing near permanent settlements.

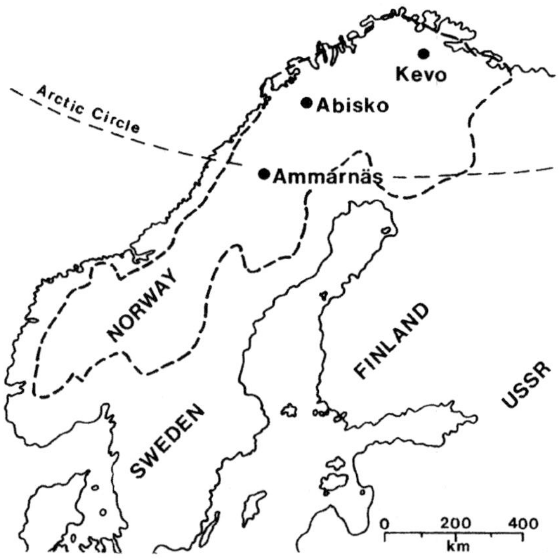

FIGURE 1. Area of *Epirrita autumnata* outbreaks in Scandinavia and Finland. The main study sites mentioned in the text are shown. (Based on Tenow.[62])

At low altitudes, scrublike birches precede pine in the natural succession,[30,59] while birches serve as windbreaks on the Atlantic coast.

2. BIOLOGY AND ECOLOGY

2.1. Phenology

Epirrita has one generation a year, and the duration of its life stages depends on latitude (Fig. 2). The greatest difference between northern and southern populations is during the pupal period which may be twice as long in the south. *Epirrita* overwinters in the egg stage, and larvae hatch in early spring at the time of host bud burst. Synchrony of these events is important for the insect because larvae hatching too early are in danger of starving. Bud break in mountain birch is quite variable especially from one year to another but also among trees. For example, the standard deviation of leaf flushing in a birch population ranged from 1.0 to 3.4 days in different years. Although larvae can endure fasting for long periods at low temperatures, they only remain active for a few days or hours at temperatures typical during leaf flush (Fig. 3). By contrast, larvae eclosing too late may encounter poor-quality foliage.[22] We often observed higher mortality rates when larval hatching was artificially delayed, but during prolonged bud-break periods caused by cold spells even long delays in larval hatching did not necessarily increase mortality. We experimentally delayed larval eclosion on several birch provenances grown in a common garden and found that delaying hatching caused lower survival, especially in larvae reared

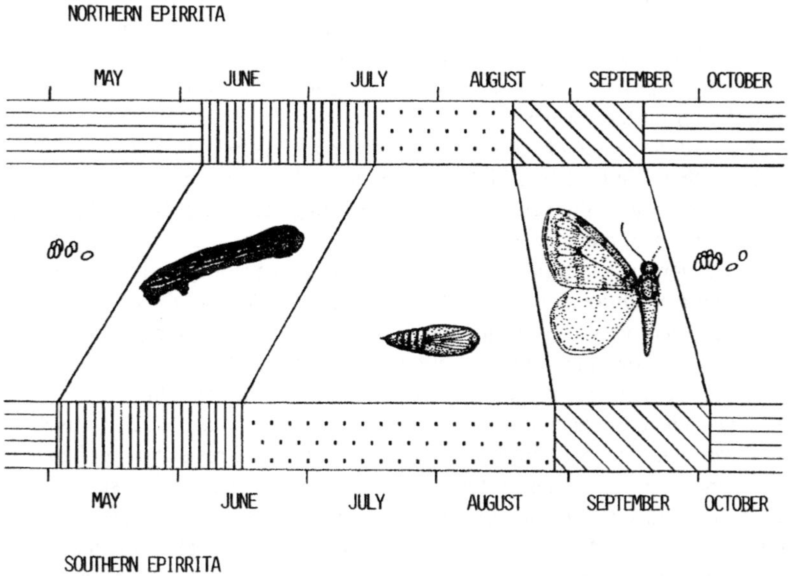

FIGURE 2. Timing of life cycle stages of *Epirrita autumnata* in northern (top) and southern (bottom) Finland.

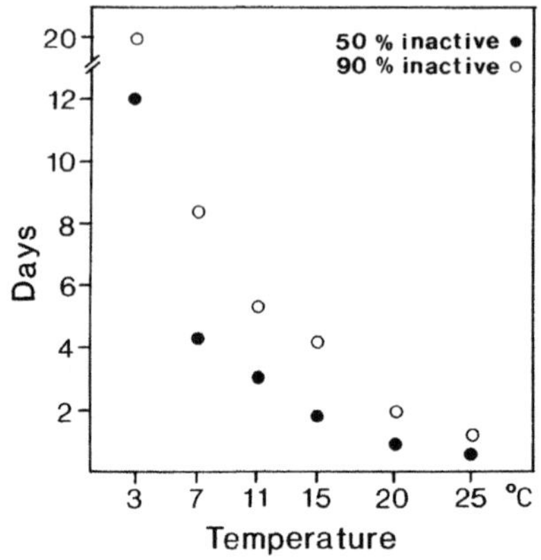

FIGURE 3. Duration of activity in starving 1st instar *Epirrita* larvae in relationship to temperature. Most or all inactivated larvae did not recover.

on southern birch provenances,[14] probably due to a shorter period of high leaf quality in spring.[16]

2.2. Interactions with Host Plants

Although birch is the main host in Fennoscandia, autumnal moth larvae are highly polyphagous.[56] For example, *Epirrita* defoliates larch forests in the Alps[55] and will complete its life cycle on several tree and shrub species (Fig. 4). In fact, its performance was equal or better on bird cherry (*Prunus padus* Linnaeus) and willow (*Salix phylicifolia* Linnaeus), even though all the larvae originated from mountain birch trees. It is also possible that autumnal moths become adapted to certain hosts. For example, dwarf birch (*Betula nana* Linnaeus) often supports dense populations but, in two experiments, the progeny of moths grown on mountain birch failed to survive on dwarf birth, while their siblings did well on the former species.

We have studied interactions between *Epirrita* and its main host, the mountain birch, at Kevo in northern Finland (see Fig. 1) since 1975. We have been particularly interested in temporal variation in the resistance of birch foliage to caterpillar feeding, especially seasonal, annual, and damage-induced variations.

Annual differences in foliage quality are difficult to quantify by bioassays because of variable climatic conditions during the larval period. In addition, the ability of animals to process foliage may vary from year to year. For example, the average weight of female pupae in enclosures on unmanipulated birches varied from 45 to 75 mg in different years. This indicates a potential fourfold year to year variation in fecundity.[18]

Damage-induced resistance is triggered by loss of leaves, which is itself strongly correlated with the density of feeding larvae. Thus, induced resistance probably acts in a density-dependent fashion. The effects of induced resistance on the stability of insect populations depend on how rapidly induction and relaxation take place. For practical purposes, we classify induced resistance into two types: (1) a rapidly induced and rapidly relaxing response (RIR) experienced by the insect generation that triggers the response, so that RIR tends to stabilize insect population densities; and (2) a delayed induced resistance (DIR) whose initiation and/or relaxation time is long enough to affect successive insect

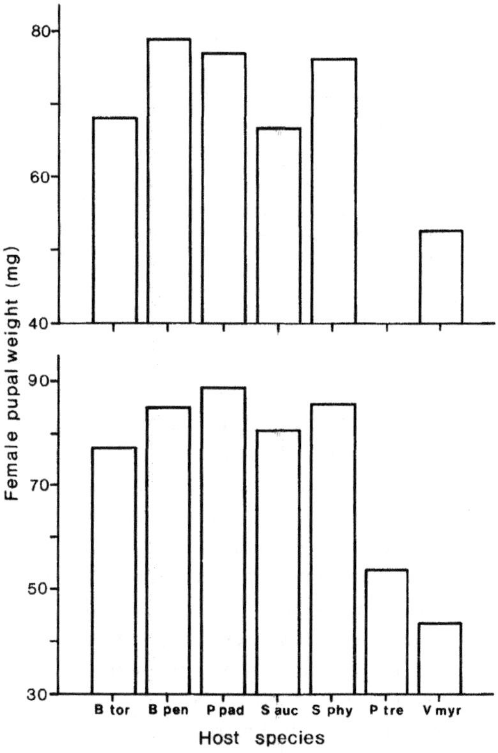

FIGURE 4. Weight of *Epirrita* female pupae reared on different host plants. (Top) 1978. (Bottom) 1979. B tor, mountain birch; B pen, silver birch; P pad, bird cherry; S auc, mountain ash; S phy, willow; P tre, aspen; V myr, bilberry.

generation(s). Thus, DIR may introduce time lags into the feedback processes acting on the population. In addition, DIR offers a mechanism that produces annual variations in foliage quality. For example, previous defoliation of mountain birch may modify the suitability of foliage for years in such a way that this phenotypic change exceeds the average differences between host species (Fig. 5); e.g., undefoliated birches differed more from previously defoliated birches than from mountain ash.

Both RIR and DIR are fairly easy to demonstrate, but quantifying their effects on insect performance is difficult. This is particularly true of the RIR because rearing larvae on control trees may stimulate them to exhibit an RIR.[45]

To study the RIR, we simulated *Epirrita* damage by tearing the lamina of birch leaves and, after 2 days, fed detached damaged leaves and their adjacent undamaged neighbors to larvae. Both previously damaged and adjacent intact leaves were less suitable for *Epirrita* development than control leaves.[16,21]

Foliage damage was the major stimulus inducing higher resistance in birch foliage in subsequent years (DIR). In addition, a nutrient subsidy added to the soil, either in commercial fertilizer or in insect frass, enhanced the response.[25] These two factors together depressed egg production by 70–78%.[25] In addition, male fertility was adversely affected,[18] and insect-caused defoliations were, for unknown reasons, more effective in inducing DIR than manual damage.[17] Furthermore, our treatment consisted of only a single partial defoliation, but cumulative defoliation may cause much greater deterioration of food quality.[19]

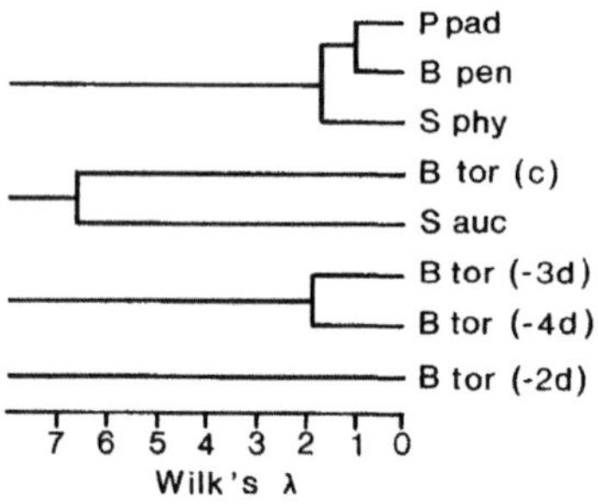

FIGURE 5. Dendrogram based on a cluster analysis of *Epirrita* larval performance (pupal weight and duration of larval period) when reared on different host plants. P pad, bird cherry; B pen, silver birch; S phy, willow; S auc, mountain ash; B tor (c), control mountain birch; (not defoliated the previous year); B tor (−2d), mountain birch defoliated 2 years ago; B tor (−3d), mountain birch defoliated 3 years ago; B tor (−4d), mountain birch defoliated 4 years ago.

The nutritional suitability of birches growing in common gardens increased with the distance from the original native site of the birch provenance.[23] *Epirrita* success was twice as high on trees from distant localities than on local birches (Fig. 6). The reason for this difference is not clear but may be due to the inability of introduced trees to produce defensive chemicals, perhaps because of stress caused by unnatural growing conditions. Egg production on local trees that had been defoliated in the previous year was more than 90% less than on introduced trees (Fig. 6). This is our best estimate of the magnitude of the effect of phenotypic variations in birch foliage.[19]

The exact relationship between DIR and previous insect density (or damage caused by them) is not known. However, even 15% leaf-area loss reduced the egg production of *Epirrita* to one-half that on control trees.[19] Two years later, leaves from defoliated trees were still poor as food for *Epirrita* larvae.[15]

Comparison of *Epirrita* performance on birches from outbreak and nonoutbreak areas showed that the RIR was significant in birch provenances from both areas.[16] However, a single provenance from southern Finland, clearly outside the outbreak range, did not exhibit the DIR associated with birches inside the outbreak range. Moths originating within and outside the outbreak area were equally susceptible to the RIR.[16]

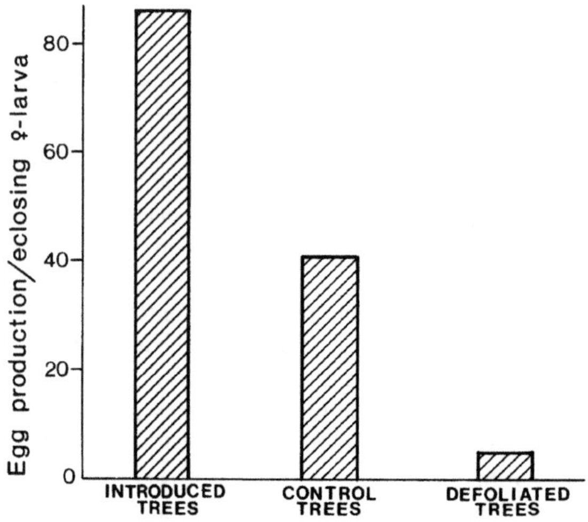

FIGURE 6. Egg production indices, (i.e., pupal weight-dependent fecundity * larval survivorship), of *Epirrita* reared on birch with different treatments. Introduced trees were grown about 1000 km from their native site. Control trees were unmanipulated local trees. Defoliated trees were local ones from which about 45% of the foliage was removed during the previous season.

2.3. Interactions with Natural Enemies

2.3.1. Parasitoids

Itamoplex armator (Fabricius) (=*Trachysphyrus* = *Cryptus albatorius* auct., non Müller) (Ichneumonidae), *Apanteles solitarius* (Ratzeburg) (Braconidae) and *Eulophus larvarum* (Linnaeus) (Eulophidae) are the most important parasites of *Epirrita* in Abisko, Sweden; the first two species contributed significantly to the crash of the 1955 outbreak.[60,61] Although a 3.3-fold increase in the number of the pupal parasitoid, *I. armator,* may occur from a peak to a post peak year,[60] data on the population growth rates of parasitoids are generally scanty.

Trapping studies during autumnal moth outbreaks provide some information on the numerical responses of ichneumonid parasitoids.[32,50,51] The species showing the most dramatic responses were the pupal parasitoids, *Itamoplex armator* and *Pimpla* (=*Coccycomimus*) *sodalis* Ruthe. The rapid crash of the parasitoids was also remarkable with no ichneumonids being caught a year after the *Epirrita* outbreak collapsed. It is also significant that during 1956, 1959, and 1961, years with low *Epirrita* density, *I. armator* and *P. sodalis* were not found among the 1798 ichneumonids collected by sweep nets in the Kevojoki area.[32] Consequently, Jussila and Nuorteva concluded that "Polyphagous parasites capable of attacking *O. autumnata* are evidently absent from the subarctic area studied."[32]

The only published estimates of the impact of parasitism on *Epirrita* seem to be those for a postpeak situation.[60] In this case more than 90% of the larvae were parasitized by *A. solitarius* and *E. larvarum*. Pupal parasitism by *I. armator* was only studied on an atypical (vegetation-free) spot and yielded the best available estimate of 51% parasitism.[60]

2.3.2. Birds

Bird densities were about 80% higher in heath birch stands at Kevo during the first year of an *Epirrita* outbreak (1964) than the mean of the preceding 10-year period (194 versus 107 pairs/km^2).[58] Densities of the two most abundant species, the brambling (*Fringilla montifringilla* Linnaeus) and the willow warbler [*Phylloscopus trochilus* (Linnaeus)], were twice as high as the 10-year average.

Bramblings also exhibited a strong numerical response to an *Epirrita* outbreak in a subalpine birch forest near Ammarnäs.[9] The density was about 150 pairs/km^2 during the outbreak in contrast to densities around 50 and 75 pairs/km^2 in other years. In addition, increases in brambling densities during *Epirrita* peaks were also observed at Budal in central Norway.[27,28]

The response of the willow warbler, the dominant passerine species in these communities, is more difficult to evaluate. At Ammarnäs, the density of willow warblers declined over the 20-year study period but in 1974 (the *Epirrita* peak year) it was 180 pairs/km^2, or about 20% higher than the expected 150 pairs/km^2 (see Fig. 5 in reference 9). At Budal, annual densities of willow warblers correlated negatively with those of the brambling.[28]

Although the brambling, and perhaps the willow warbler, respond numerically to *Epirrita* peaks, most other bird species only seem to respond functionally. Functional responses have been demonstrated with the foliage gleaners, the willow warbler, reed bunting [*Emberiza schoeniclus* (Linnaeus)],[44] and redpoll [*Carduelis flammea* (Linnaeus].[8] The redwing (*Turdus iliacus* Linnaeus), a ground-foraging thrush that normally eats earthworms, may also feed heavily on *Epirrita* larvae during the outbreak peak.[4] Since redwings forage exclusively on ground, *Epirrita* larvae are not included in the diet until they drop to the ground to pupate.

2.3.3. Small Mammals

Microtines exhibit dramatic density fluctuations with 3–5-year periodicity in the subarctic birch forest zone.[34,40,41] In northern Fennoscandia, there is an apparent synchrony in the population fluctuations of several species of small vertebrate herbivores, their predators, and alternate prey.[3,31,57] Alternate prey species include many insectivorous birds and shrews,[12,13,26,33] which are natural enemies of *Epirrita*.

Most *Epirrita* population crashes seem to coincide with microtine troughs in northern Fennoscandia, in the subarctic birch forest zone, and microtine peaks occurred during several *Epirrita* increases. Thus, high densities of small mammals do not seem necessary to cause *Epirrita* population crashes, nor are they sufficient to prevent outbreaks.

2.4. Effects of Winter Temperatures

Healthy patches of birches are often observed along river canyons in otherwise heavily defoliated birch forests.[35,62–64] This phenomenon was explained by the freezing and killing of overwintering *Epirrita* eggs by temperature inversions, during which minimum temperatures may be 20°C lower in low-lying depressions.[35] *Epirrita* eggs are killed by temperatures below −37°C.[46] Consequently, exposure of eggs to low winter temperatures is a factor that probably explains certain topographical damage patterns. In areas in which minimum temperatures regularly fall below −40°C and where *Epirrita* damage still occurs,[53] eggs must be more cold tolerant or must overwinter in sheltered sites.

3. PATTERNS OF NUMERICAL POPULATION BEHAVIOR

An extensive record of geometrid (*Epirrita autumnata* and *Operophtera* spp.) defoliation of Fennoscandian birch forests has been compiled from 1862 to 1968.[62] During this period, outbreak episodes occurred at regular 9- or 10-year intervals (Fig. 7). Both synchronous outbreak periods, with more or less simultaneous defoliations over the entire Scandinavian mountain chain, and moving outbreaks, with local defoliation spreading as a wave over the area, can be recognized.[62] Defoliations tended to be more synchronous in the northern than in the southern part of Fennoscandia (see Fig. 12 in reference 62).

Epirrita densities have been measured in mountain birch forests near Ammarnäs, Swedish Lapland, since 1967.[1,9] These data support the view that autumnal moth populations cycle with a periodicity of 9–10 years at a single locality. There was a peak in the mid-1960s, a pronounced density peak of 300–500 larvae/1000 dwarf shoots in 1974–

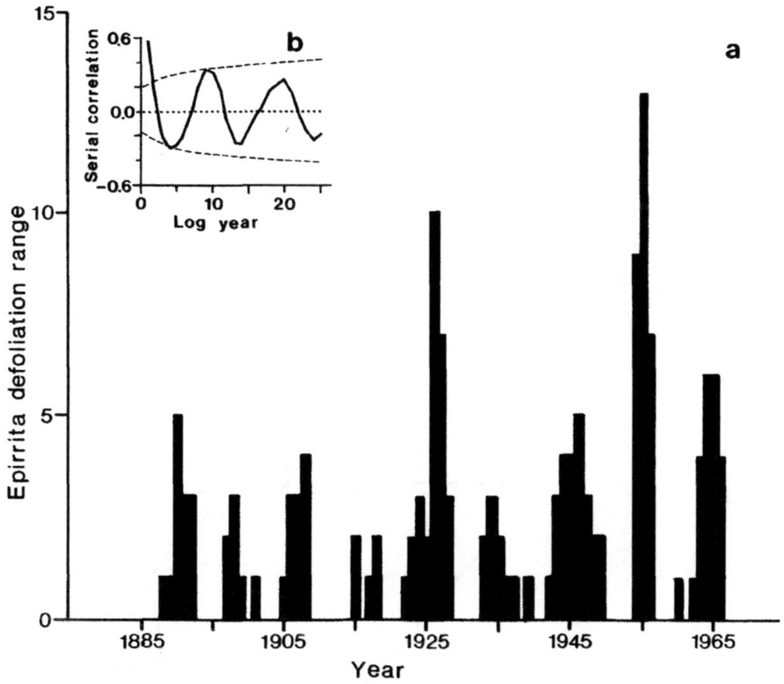

FIGURE 7. *Epirrita autumnata* damage reported in Fennoscandia north of the 64° parallel. (a) The extent of defoliation is expressed as incidence in 0.2° latitudinal belts. (b) Serial correlation analysis shows cycle periodicity to be 9 years (dashed lines give the 95% confidence level). (Based on Tenow.[62])

1975, and a new peak in 1984–1985. *Epirrita* larval densities during the troughs were at least three orders of magnitude less than during the peaks.

Site factors seem to influence the amplitude of autumnal moth cycles (Fig. 8), as well as those of other insects.[6] Peak larval densities were roughly twice as high in a birch stand growing on less fertile heath soil than in meadow birch forests (Fig. 8a). Although the per-capita rate of increase phase plane is incomplete at low densities, it demonstrates the characteristic orbit of a cyclic population (Fig. 8b); also shown is a hypothetical cycle simulated by a linear second-order density-dependent model.[6] The simulation may not have been totally successful because the linear regression method used to extrapolate r_m attained its maximum value close to 1.5. Consequently, the low phase may be shorter than in real cycles; in field populations, a measurable increase was found only 4 years after the crash in the Swedish Lapland.[9]

Using data on mean birch leaf sizes[20] and densities of birch leaves at Fennoscandian IBP sites,[37] we estimated peak larval densities to be about 1.5–2.5 million/hectare (ha) at Ammarnäs in 1974–1975. Even higher densities of about 10 million larvae/ha have been observed at Abisko in northern Sweden.[61] However, during the next maximum (in 1965) peak densities in the same area were much lower (about 0.25 million larvae/ha). Peak densities of *Epirrita* may therefore differ by more than an order of magnitude from one outbreak to the next.

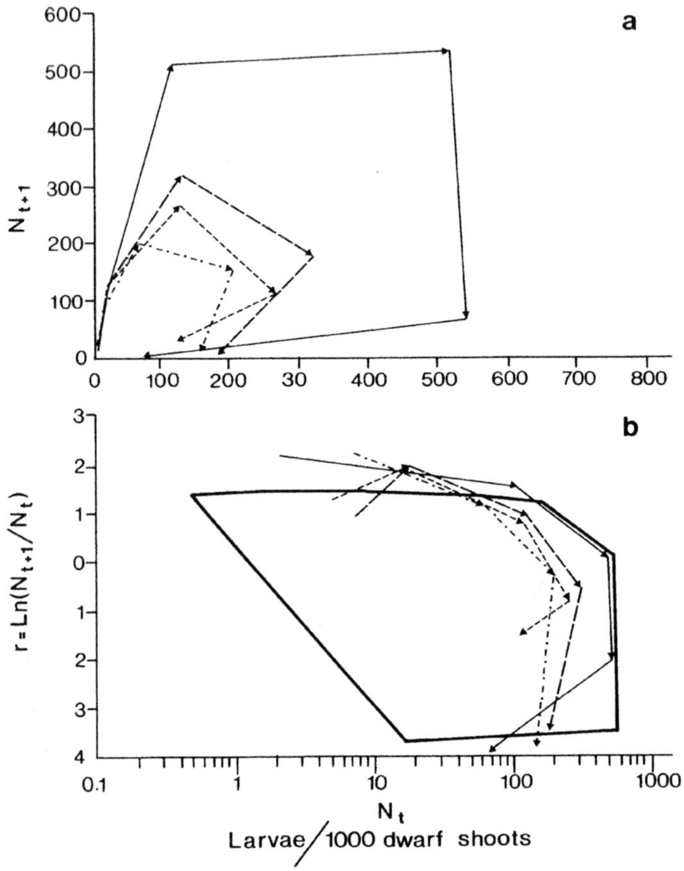

FIGURE 8. (a) Population trends of *Epirrita autumnata* at Ammarnäs, northern Sweden, during the mid-1970s. (Based on Andersson and Jonasson.[1]) The outer line refers to a heath forest site and the inner lines to meadow birch forest sites. (b) Per-capita rate of increase (r_t) phase planes for (a). (——) Full cycle at the heath forest site based on simulations with the model $r_t = r_m (1 - N_{t-1}/K)$. (Based on Berryman.[6]) Parameters were estimated by linear regressions on data in (1), plotting r_t against N_{t-1}; the x intercept gave an estimate of K (178) and the y intercept of r_m (1.554).

4. HYPOTHESES FOR THE CAUSES OF POPULATION CYCLES

Generations of *Epirrita* are nonoverlapping. Consequently, population growth occurs in discrete steps, and the appropriate models are difference equations of the general form

$$r_t = F(N_t, N_{t-1}, N_{t-2}, \ldots N_{t-k})\, G(X_1, X_2, \ldots, X_j) \tag{1}$$

where r_t is the specific per-capita rate of increase of the population [$r_t = \ln (N_{t+1}/N_t)$], F is a density-dependent feedback effect expressed as a function of density in current and previous generations, and G is a density-independent function expressed in terms of environmental variables X_j.

Our aim is not to suggest any specific form for the functions, but to use the equation as a conceptual framework for visualizing the relative importance and interplay of different factors during *Epirrita* cycles. We hypothesize that the cause of cyclic population dynamics in *Epirrita* is the high potential growth rate of the moth population together with delayed density-dependent feedback factors.[5]

4.1. Potential Growth Rates of Epirrita Populations: $G(X_1, X_2, \ldots, X_i)$

Epirrita populations have high potential growth rates with females laying, on average, 120 (max 250) eggs, which means a potential 60-fold (max 125) increase per generation, i.e., maximum values of r_m of 4–5.

This maximum rate of increase is reduced, however, by density-independent factors affecting fecundity, fertility, and mortality. Our field experiments enabled us to obtain estimates of female fecundity, as a function of pupal weight, and larval survival in field cages that excluded most natural enemies. We found that the maximum per-capita rates of increase varied from 2.9 to 3.7. These values are somewhat greater than the highest r_t values (~2.0) recorded at Ammarnäs during the early 1970s.[1]

Autumnal moths originating from outside outbreak areas generally attain larger sizes and higher fecundity than do northern populations.[16] This is the opposite of what would be expected if fecundity was a major cause of high-amplitude cycles. Accordingly, the potential rate of increase does not explain why northern but not southern *Epirrita* populations fluctuate so dramatically.

Annual climatic variations are much greater in the subarctic than in the temperate regions.[29] Climate obviously affects *Epirrita* population dynamics, but we cannot see how these effects might give rise to regular population cycles. On the contrary, weather usually causes stochastic variations that should mask the basic cyclic pattern. For example, the height of the peaks and the interval between them often varies (Fig. 7), and large-scale defoliation episodes often coincide with cooler than average summer temperatures.[47] Historical records of defoliations, however, depend on both the amount of foliage consumed and on its visibility. This leads to two complicating factors. First, *Epirrita* may consume a disproportionately large fraction of birch leaves in cold seasons, if larvae grow and develop more rapidly than leaves at low temperatures.[54,60] If this is true, the same number of larvae will cause more conspicuous damage during cold summers. Second, recovery of birch trees from defoliation is poor after cool summers.[24,35,54] Consequently, the signs of damage remain visible for a longer period and are easier to perceive during and after cool summers. If there is a real correlation between the intensity of *Epirrita* outbreaks and low summer temperatures, it can be explained, for instance, by assuming that birch may not be as resistant in cold seasons. This does not presume that cold weather favors *Epirrita,* but that birch is more sensitive to low temperatures. Further, it is possible that *Epirrita* is less susceptible to low temperatures than its predators and parasitoids. In preliminary experiments, we have cooled birches by shading the canopy and insulating the ground beneath the trees and have not found larvae to grow better on these trees relative to control trees.

A further factor that could reduce the regularity of *Epirrita* cycles is the age of the trees. Mature forests may be less resistant[7,48] and are more likely to experience heavy damage. In such cases, defoliation and tree-killing would not occur during successive peaks at the same site.

4.2. Rapid Density-Dependent Factors: $F(N_t)$

Practically all biotic factors may function in a density-dependent way. It is not necessary that they respond over the whole range of population densities, but just that density dependence starts to affect population growth after some density threshold.

Food availability determines the absolute upper limit for *Epirrita* numbers. The limit is achieved during the highest, but not necessarily during all, peaks.[60,62] The threshold density at which food shortage begins to act is relatively high and also depends on the density of other defoliators and on foliage biomass. The latter is a function of the number of buds and the rate of leaf expansion.

Food quality may also act in a density-dependent manner. Induced resistance (RIR) in birch foliage is triggered by foliage damage and will operate in a density-dependent way, at least in an on–off manner. At low densities, for example, *Epirrita* larvae may move to other (ungrazed) parts of the tree, or even to other trees, minimizing the negative effects of the RIR.

Larval density can directly modify *Epirrita* performance. For example, larvae reared in crowded conditions on high-quality foliage produced smaller pupae than those reared alone. But on poor-quality foliage there was no difference between the performance of crowded and solitary larvae; solitary larvae also performed badly.[14]

Generalist vertebrate enemies of *Epirrita* might respond in a density-dependent way, although their effects are probably small or negligible in the subarctic birch ecosystem. Threshold densities for numerical and functional responses of birds are presumably too high or too mild, respectively, to prevent defoliation. In northern Fennoscandia, *Epirrita* is vulnerable to small mammal predation during the pupal stage (in the ground) only, which lasts 1–1.5 months[11] (Fig. 2). Generalist invertebrate predators and parasites may not be important in most of the birch forest zone which exhibit pronounced *Epirrita* fluctuations.[24] This may not apply to spiders, and predation by ants sometimes creates green "islands" around ant mounds.[42]

Although rapid density-dependent agents (RIR, generalist natural enemies) seem to be unable to prevent *Epirrita* outbreaks in the north, they may help explain the geographical distribution of outbreaks as well as local topographic patterns of defoliation. It has been shown that the RIR operates in birches originating from both within and outside the outbreak zones.[16] Thus, hereditary differences in the RIR between birch provenances do not explain differences in the geographical distribution of cycles. A possible, but untested, explanation is that the lack of *Epirrita* cycles in southern Finland is caused by higher densities of generalist natural enemies such as parasitoids, predatory insects, birds, and shrews.

4.3. Delayed Density-Dependent Factors: $F(N_{t-1}, N_{t-2} \ldots N_{t-k})$

Very high densities of *Epirrita* combined with poor regrowth of birch foliage can kill dwarf shoots and, ultimately, the trees themselves. The year following defoliation, reduction of foliage may lead to food shortage.[60] Shoot mortality is not associated with all *Epirrita* peaks.

As we have seen earlier, partial defoliation causes a dramatic reduction in egg production the year following defoliation (Fig. 6). We calculated the per-capita growth rate to be $r = 1.1$ in previously defoliated trees, compared with 3.7 in stressed trees and 2.9 in undamaged local trees. These values may still underestimate the effects of DIR because all larvae were reared in groups and crowding reduces the reproductive potential of *Epirrita* on the best but not on poor-quality foliage,[14] yielding relatively high success on poor diets and/or low success on good diets.

DIR in birch provenances in the outbreak zone tends to introduce time lags into the negative feedback loops that regulate *Epirrita* populations, while delayed resistance has not been demonstrated in birches from the nonoutbreak zone. Because time lags automatically create the conditions for cyclic dynamics (overshooting and undershooting the equilibrium point), we hypothesize that DIR in northern birches is the main factor responsible for the outbreak cycles. It remains unclear as to what extent the DIR is a consequence and to what a cause of *Epirrita* cycles.

The numerical responses of specialist parasitoids can also create a feedback time delay. The r values for the pupal parasitoid *Itamoplex armator* (calculated from refs. 52 and 60) were 1.2 and 1.5, i.e., lower than that of *Epirrita* at Ammarnäs ($r = 2$). This finding indicates that under favorable growth conditions *Epirrita autumnata* populations can escape their parasitoids.

Although diseases contribute to the decline of *Epirrita* populations,[60,62] no specific information is available on their role in population cycles. However, models of insect–virus interactions may produce population cycles of the correct periodicity (e.g., 9–10 years), mainly because virus remains in the soil in infective stages for many years.[2,10] It is difficult to evaluate the real effects of diseases because of possible synergistic interactions between diseases and food quality. For instance, gypsy moth larvae reared on previously defoliated trees were more susceptible to a nuclear polyhedrosis virus than were larvae reared on undefoliated trees.[65]

Density-dependent factors with time lags of 1 or more years seem to be the main reason for the 9–10-year interval between successive peaks of *Epirrita* density. The DIR functions the year after defoliation and has a relaxation time of at least 2 years. It also has an effect on *Epirrita* r values of sufficient magnitude to induce significant population changes. The killing power of specialist parasitoids is also greatest the year after the *Epirrita* peak. We therefore hypothesize that the time lag introduced by delayed induced host defense and specialist parasitoids is the main factor causing regular population cycles of *E. autumnata* in northern Fennoscandia.

Time-delay models are sensitive to variation in the maximum rate of increase r_m.[6] To get more realistic models we should know whether r_m is maintained following the population crash or whether it is only reached temporarily; e.g., if, under some conditions, constitutive or rapid induced resistance becomes inefficient.

5. MANAGEMENT IMPLICATONS

Because *Epirrita* cycles are an inherent property of the plant–herbivore–natural enemy system, we doubt whether any effective preventive measures can be applied. In addition, *Epirrita* damage occurs on such large and economically marginal areas that any intensive management strategies are not feasible. Limited areas can be protected from damage by insecticides but we assume that repeated treatments will be needed. Artificial induction of DIR, over wide areas, might be an effective management strategy if biologically possible and economically feasible. Other alternatives might be to maintain populations of specialist parasitoids and disease vectors or to encourage generalist natural enemies.

ACKNOWLEDGMENTS. We thank all persons who have worked in the *Epirrita*-birch project and G. Anderson, A. A. Berryman, R. R. Mason, Olle Tenow, and Boyd Wickman for constructive criticism on earlier versions of the manuscript. We also thank G. Anderson and O. Tenow for allowing us to use unpublished information, particularly data used in describing population patterns (Section 9.3). The research was made possible by grants from the Academy of Finland.

REFERENCES

1. Andersson, G., and Jonasson, J. A., 1980, Relative abundance of insects on mountain birch at Ammarnäs, Swedish Lapland, *Entomol. Tidskr.* **1019:**61–69 (in Swedish with English summary).
2. Anderson, R. M., and May, R. M., 1980, Infectious diseases and population cycles of forest insects, *Science* **210:**658–661.
3. Angelstam, P., Lindström, E., and Widen, P., 1984, Role of predation in short-term population fluctuations of some birds and mammals in Fennoscandia, *Oecologia (Berl.)* **62:**199–208.
4. Arheimer, O., 1978, Choise of food and rate of feeding young in the Redwing Turdus iliacus in a subalpine birch forest at Ammarnäs in Swedish Lapland, *Anser Suppl.* **3:**31–46 (in Swedish with English summary).
5. Berryman, A. A., 1978, Population cycles of the Douglas-fir tussock moth (Lepidoptera: Lymantriidae): The time-delay hypothesis, *Can. Entomol.* **110:**513–518.
6. Berryman, A. A., 1980, Site characteristics and population dynamics: A theoretical perspective, in: *Site Characteristics and Population Dynamics of Lepidopteran and Hymenopteran Forest Pests* (D. Bevan and J. T. Stoakley, eds.), Forestry Commission Research and Development Paper 135.
7. Blais, J. R., 1958, The vulnerability of balsam fir to spruce budworm attack in northwestern Ontario, with special reference to the physiological age of the tree, *For. Chron.* **34:**405–422.
8. Enemar, A., and Nyström, B., 1981, Om gråsiskans Craduelis flammea bestondsväxlingar, föda ock häcking i fjällbjörkskog, södra Lappland, *Vår Fågelvärd* **40:**409–426.
9. Enemar, A., Nilsson, L., and Sjöstrand, B., 1984, The composition and dynamics of the passerine bird community in a subalpine birch forest, Swedish Lapland, A 20-year study, *Ann. Zool. Fenn.* **21:**321–338.
10. Entwistle, P. F., Adams, P. H. W., and Evans, H. F., 1978, Epizootiology of a nuclear polyhedrosis virus in European spruce sawfly (*Gilpinia hercyniae*): The rate of passage of infective virus through the gut of birds during cage tests, *J. Invert. Pathol.* **31:**307–312.
11. Hanski, I., and Otronen, M., 1985, Food quality induced variance in larval performance: comparison between rare and common pine-feeding sawflies (Diprionidae), *Oikos* **44:**165–174.
12. Hansson, L., 1969, Spring populations of small mammals in central Swedish Lapland in 1968–1968, *Oikos* **20:**431–450.
13. Hansson, L., 1984, Predation as the factor causing extended low densities in microtine cycles, *Oikos* **43:**255–256.
14. Haukioja, E., 1980, On the role of plant defences in the fluctuation of herbivore populations, *Oikos* **35:**202–289.
15. Haukioja, E., 1982, Inducible defences of white birch to a geometrid defoliator, *Epirrita autumnata*, in: *Proceedings of the Fifth International Symposium on Insect–Plant Relationships* (J. H. Visser and A. K. Minks, eds.), pp. 199–203, Pudoc, Wageningen.
16. Haukioja, E., and Hanhimäki, S., 1985, Rapid wound-induced resistance in white birch (*Betula pubescens*) foliage to the geometrid *Epirrita autumnata:* A comparison of trees and moths within and outside the outbreak range of the moth, *Oecologia (Berl.)* **65:**223–228.

17. Haukioja, E., and Neuvonen, S., 1985, Induced long-term resistance of birch foliage against defoliators: Defensive or incidental?, *Ecology* **66**:1303–1308.
18. Haukioja, E., and Neuvonen, S., 1985, The relationship between male size and reproductive potential in *Epirrita autumnata* (Lep., Geometridae), *Ecol. Entomol.* **10**:267–270.
19. Haukioja, E., and Neuvonen, S., 1986, Insect population dynamics and induction of plant resistance: The testing of hypotheses, in: *Insect Outbreaks* (P. Barbosa and J. Schultz, eds.), Academic, Orlando, Florida.
20. Haukioja, E., and Niemelä, P., 1974, Growth and energy requirements of the larvae of *Dineura virididorsata* (Retz.) (Hym., Tenthredinidae) and *Oporinia autumnata* (Bkh.) (Lep., Geometridae) feeding on birch, *Ann. Zool. Fenn.* **11**:207–211.
21. Haukioja, E., and Niemelä, P., 1977, Retarded growth of a geometrid larva after mechanical damage to leaves of its host tree, *Ann. Zool. Fenn.* **14**:48–52.
22. Haukioja, E., and Niemelä, P., Iso-Iivari, L., Ojala, H., and Aro, E-M., 1978, Birch leaves as a resource for herbivores. I. Variation in the suitability of leaves, *Rep. Kevo Subarctic Res. Sta.* **14**:5–12.
23. Haukioja, E., Niemelä, P., and Kapiainen, K., 1983, Herbivory and tree-line birches, in: *Tree-Line Ecology. Proceedings of the Northern Quebec Tree-Line Conference* (P. Morisset and S. Payette, eds.), pp. 153–158, Centre d'études nordicues, Université Laval, Quebec.
24. Haukioja, E., Niemelä, P., and Siren, S., 1985, Foliage phenols and nitrogen in relation to growth, insect damage, and ability to recover after defoliation, in mountain birch *Betula pubescens* ssp *tortuosa*, *Oecologia (Berl.)* **65**:214–222.
25. Haukioja, E., Suomela, J., and Neuvonen, S., 1985, Long-term inducible resistance in birch foliage: Triggering cues and efficacy on a defoliator, *Oecologia (Berl.)* **65**: 363–369.
26. Henttonen, H., 1985, Predation causing extended low densities in microtine cycles: Further evidence from shrew dynamics, *Oikos* **45**:156–157.
27. Hogstad, O., 1969, Breeding bird populations in two subalpine habitats in the middle of Norway during the years 1966–68, *Nytt Mag. Zool.* **17**:81–91.
28. Hogstad, O., 1975, Interspecific relation between Willow Warbler (*Phylloscopus trochilus*) and Brambling (*Fringilla montifringilla*) in subalpine forests, *Norw. J. Zool.* **23**:223–234.
29. Hustich, I., 1950, Yields of cereals in Finland and the recent climatic fluctuation, *Fennia* **73**:1–32.
30. Hustich, I., 1966, On the forest-tundra at the northern tree-lines, *Rep. Kevo Subarctic Res. Sta.* **3**:7–47.
31. Hörnfeldt, B., 1978, Synchronous population fluctuations in voles, small game, owls and Tularemia in northern Sweden. *Oecologia (Berl.)* **32**:141–152.
32. Jussila, R., and Nuorteva, P., 1968, The ichneumonid fauna in relation to an outbreak of *Oporinia autumnata* (Bkh.) (Lep., Geometridae) on subarctic birches, *Ann. Zool. Fenn*, **5**:273–275.
33. Kaikusalo, A., and Hanski, I., 1985, Population dynamics of *Sorex araneus* and *S. caecutiens* in Finnish Lapland, *Acta Zool. Fenn.* **173**:283–285.
34. Kalela, O., 1962, On the fluctuations in the numbers of arctic and boreal small rodents as a problem of production biology, *Ann. Acad. Sci. Fenn. Ser. A4* **66**:1–38.
35. Kallio, P., and Lehtonen, J., 1973, Birch forest damage caused by *Oporinia autumnata* (Bkh.) in 1965–66 in Utsjoki, N Finland. *Rep. Kevo Subarctic Res. Sta.*, **10**:55–69.
36. Kallio, P., and Lehtonen, J., 1975, On the ecocatastrophe of birch forests caused by *Oporinia autumnata* (Bkh.) and the problem of reforestation, in: *Fennoscandian Tundra Ecosystems 2*, (F. E. Wielgolaski, ed.) pp. 175–180.
37. Kjelvik, S., and Kärenlampi, L., 1975, Plant biomass and primary production of Fennoscandian subarctic and subalpine forests and of alpine willow and heath ecosystems in: *Fennoscandian Tundra Ecosystems 1*, (F. E. Wielgolaski, ed.) pp. 111–120.
38. Koponen, S., 1983, Phytophagous insects of birch foliage in northernmost woodlands of Europe and eastern North America, in: *Tree-Line Ecology. Proceedings of the Northern Quebec Tree-Line Conference* (P. Morisset and S. Payette, eds.), pp. 165–176, Centre d'études nordicues, Université Laval, Quebec.
39. Kärenlampi, L., and Kauhanen, H., 1972, A direct gradient analysis of the vegetation of the surroundings of the Kevo Subarctic Station, *Rep. Kevo Subarctic Res. Sta.* **9**:82–98.
40. Lahti, S., Tast, J., and Uotila, H., 1976, Fluctuations in small rodent populations in the Kilpisjärvi area in 1950–75, *Luonnon Tutkija* **80**:97–107 (in Finnish with English summary).
41. Laine, K., and Henttonen, H., 1983, The role of plant production in microtine cycles in northern Fennoscandia, *Oikos* **40**:407–418.

42. Laine, K., and Niemelä, P., 1980. The influence of ants on the survival of mountain birches during an *Oporinia autumnata* (Lep., Geometridae) outbreak, *Oecologia (Berl.)* **47**:39–41.
43. Lehtonen, J., and Yli-Rekola, M., 1979, Field and ground layer vegetation in birch forests after *Oporinia* damage, *Rep. Kevo Subarctic Res. Sta.* **15**:27–32.
44. Lennerstedt, I., 1973, Night rest during nestling period in four passerine species under subarctic summer conditions, *Ornis Scand.* **4**:17–23.
45. Neuvonen, S., and Haukioja, E., 1985, How to study induced plant resistance? *Oecologia (Berl.)* **66**:456–457.
46. Niemelä, P., 1979, Topographical delimitation of *Oporinia*-damages: Experimental evidence of the effect of winter temperature. *Rep. Kevo Subarctic Res. Sta.* **15**:33–36.
47. Niemelä, P., 1980, Dependence of *Oporinia autumnata* (Lep., Geometridae) outbreaks on summer temperature, *Rep. Kevo Subarctic Res. Sta.* **16**:27–30.
48. Niemelä, P., Tuomi, J., and Haukioja, E., 1980, Age-specific resistance in trees: Defoliation of tamaracks (*Larix laricina*) by larch bud moth (*Zeiraphera improbana*) (Lep., Tortricidae). *Rep. Kevo Subarctic Res. Sta.* **16**:49–57.
49. Nuorteva, P., 1963, The influence of *Oporinia autumnata* (Bkh.) (Lep., Geometridae) on the timber-line in subarctic conditions, *Ann. Ent. Fenn.* **29**:270–277.
50. Nuorteva, P., 1971, Decline of the parasite population of the geometrid moth *Oporinia autumnata* during the second year after a calamity on birches, *Ann. Ent. Fenn.* **37**:96.
51. Nuorteva, P., and Jussila, R., 1967, Seasonal and zonal distribution of Ichneumonidae (Hym.) on a subarctic fell during a calamity of the geometrid moth *Oporinia autumnata* (Bkh.) on birches, *Ann. Ent. Fenn.* **33**:155–163.
52. Nuorteva, P., and Jussila, R., 1969, Incidence of ichneumonids on a subarctic fell after a calamity of the geometrid moth *Oporinia autumnata* (Bkh.) on birches, *Ann. Ent. Fenn.* **35**:153–160.
53. Pulliainen, E., 1976, Birch forest damage caused by *Oporinia autumnata* Bkh. (Lep., Geometridae) in 1965–66 in eastern Itäkaira, northeastern Lapland, *Ann. Ent. Fenn.* **42**:166–170.
54. Sandberg, G., 1963, Växtvärlden i Abisko nationalpark, in: *Natur i Lappland* (K. Curry-Lindahl, ed.), pp. 885–909, Almqvist & Wiksell, Uppsala.
55. Schwenke, W., 1978, *Die Forstschädlinge Europas. Bd. 3: Schmetterlinge,* Paul Parey, Hamburg und Berlin.
56. Seppänen, E., 1970, Suurperhostoukkien ravintokasvit. (The food-plants of the larvae of the Macrolepidoptera of Finland), in: *Animalia Fennica 14.* Werner Söderström, Porvoo-Helsinki.
57. Siivonen, L., 1950, Some observations on the short-term fluctuations in numbers of mammals and birds in the sphere of the northernmost Atlantic. *Papers Game Res.* **4**:1–31.
58. Silvola, T., 1967, Changes in the bird populations in Utsjoki, Finnish Lapland in 1964–1966, caused by the mass-occurrence of the caterpillar *Oporinia autumnata, Ornis Fenn.* **44**:65–67 (in Finnish with English summary).
59. Siren, G., 1958, Hajapoimintoja Lapin metsien tulevaisuudesta, *Terra* **70**:57–73.
60. Tenow, O., 1956, Fjällbjörkmätarens härjningar i abiskodalen sommaren 1955, *Sveriges Natur* **6**:165–173, 184–187.
61. Tenow, O., 1963, Leaf-eating insects on the mountain birch at Abisko (Swedish Lapland). With notes on bionomics and parasites, *Zool. Bidr. Uppsala* **35**: 545–567.
62. Tenow, O., 1972, The outbreaks of *Oporinia autumnata* Bkh. and *Operophthera* spp. (Lep., Geometridae), in the Scandinavian mountain chain and northern Finland 1862–1968. *Zool. Bidrag Uppsala.* **2(suppl.)**:1–107.
63. Tenow, O., 1975, Topographical dependence of an outbreak of *Oporinia autumnata* Bkh. (Lep., Geometridae) in a mountain birch forest in northern Sweden, *Zoon* **3**:85–110.
64. Tenow, O. 1983, Topoclimatic limitations to the outbreaks of *Epirrita* (= *Oporinia*) *autumnata* (Bkh.) (Lepidoptera: Geometridae) near the forest limit of the mountain birch in Fennoscandia, in: *Tree-Line Ecology. Proceedings of the Northern Quebec Tree-Line Conference.* (P. Morisset and S. Payette, eds.), pp. 159–164, Centre d'études nordicues, Université Laval, Quebec.
65. Wallner, W. E., 1983, Gypsy moth host interactions: A concept of room and board, in: *Proc. Forest Defoliator—Host Interactions: A Comparison between Gypsy Moth and Spruce Budworms* pp. 5–8, USDA Forest Service Northeastern Station, Gen. Tech. Report NE-85.

CHAPTER 10

THE DOUGLAS-FIR TUSSOCK MOTH IN THE INTERIOR PACIFIC NORTHWEST

Richard R. Mason and Boyd E. Wickman

Richard R. Mason and Boyd E. Wickman • U.S. Forest Service, Forestry and Range Sciences Laboratory, La Grande, Oregon 97850.

1. INTRODUCTION

The Douglas-fir tussock moth, *Orgyia pseudotsugata* (McDunnough), is a common defoliator of fir in the interior forests of western North America. It is one of four western species of *Orgyia*, but it is the only member of the group that occasionally reaches outbreak numbers while feeding exclusively on conifers.[24] Because of the explosive and destructive nature of its outbreaks, the Douglas-fir tussock moth has achieved much deserved notoriety among forest managers. For this reason, considerable effort has been made in recent years to understand the dynamics of tussock moth populations and to develop methods for minimizing their impact.[11]

The Douglas-fir tussock moth was first recorded in 1900 as *Orgyia* (*Notolophus*) *oslari* from an adult male collected at Poncha Springs, Colorado.[5] It was next mentioned in 1906, when larvae were seen defoliating white fir (*Abies concolor*) near Yosemite National Park in California.[12,22] The species was officially described in 1921 as *Hemerocampa pseudotsugata* McDunnough.[38]

The earliest observation of a severe infestation was in 1916 when the first of a series of outbreaks that have occurred periodically in southern British Columbia was recorded.[2,26,81] The first details of a large infestation in the United States are from an extensive outbreak that occurred in 1927–1930 at widely separated points in the Northwest.[33] A variety of major and minor outbreaks have continued to be recorded throughout the interior forests of the western United States, the most recent being a widespread infestation in the Pacific Northwest during the early 1970's.[25]

The Douglas-fir tussock moth is found throughout the range of its major hosts: Douglas-fir (*Pseudotsuga menziesii* var. *glauca*), grand fir (*Abies grandis*), and white fir in western North America (Fig. 1). Although the species has been collected west of the Cascade range, it is not abundant there and is also rare east of the continental divide.[20] Severe outbreaks have occurred several times in the last half-century in British Columbia, Idaho, Montana, Washington, Oregon, California, Nevada, Arizona, and New Mexico (Fig. 1). Millions of board feet of timber have been killed and thousands of trees have suffered top kill and growth reduction (Fig. 2). Outbreaks have also created major problems in salvage, forest regeneration, and fire prevention.[24,32]

Attempts were made to study the broader economic impacts caused by tussock moth infestations after the 1972–1974 outbreak in the Pacific Northwest. Despite the obvious short-term impacts caused by tree mortality, growth loss, and costs of pest management, no precise negative impacts could be quantified for water quality or quantity, recreation, wildlife, or fire prevention and protection.[69] By contrast, some of the short- and long-term biological effects, such as increased big game forage, streamflow, and growth recovery of defoliated stands, were found to be beneficial.[34] Obviously, additional research and analyses of the socioeconomic consequences of outbreaks are needed to elucidate the true impact of this forest pest.

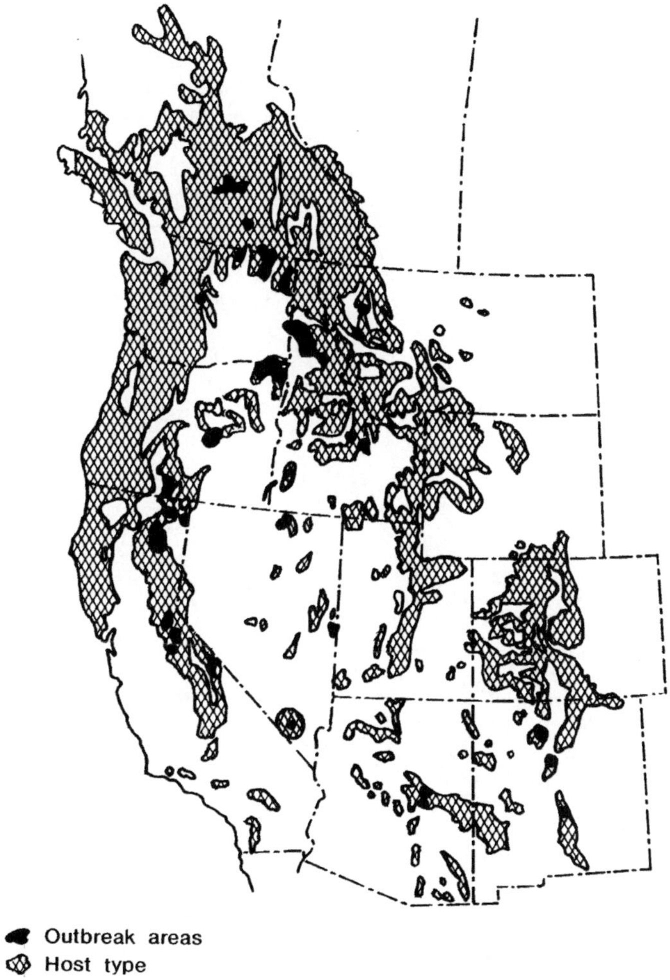

FIGURE 1. Approximate geographic range of the Douglas-fir tussock moth (after 7).

1. BIOLOGY AND LIFE HISTORY

2.1 Life Cycle

The Douglas-fir tussock moth is a univoltine insect that overwinters in egg masses containing 150–250 eggs, usually on the underside of small branches.[102] Eggs hatch in the late spring (Fig. 3A), about the time that buds break and new shoots begin to expand. These events are closely synchronized, giving new larvae the food they need for growth and development.[94] Shoot elongation progresses rapidly in June and July, resulting in a

FIGURE 2. Clumped tree mortality in a grand fir stand after severe defoliation by the Douglas-fir tussock moth.

FIGURE 3. Life stages of the Douglas-fir tussock moth: A, newly hatched egg mass; B, first instar; C, mature larva; D, female adult; and E, male adult.

B

FIGURE 3. *(continued)*

C

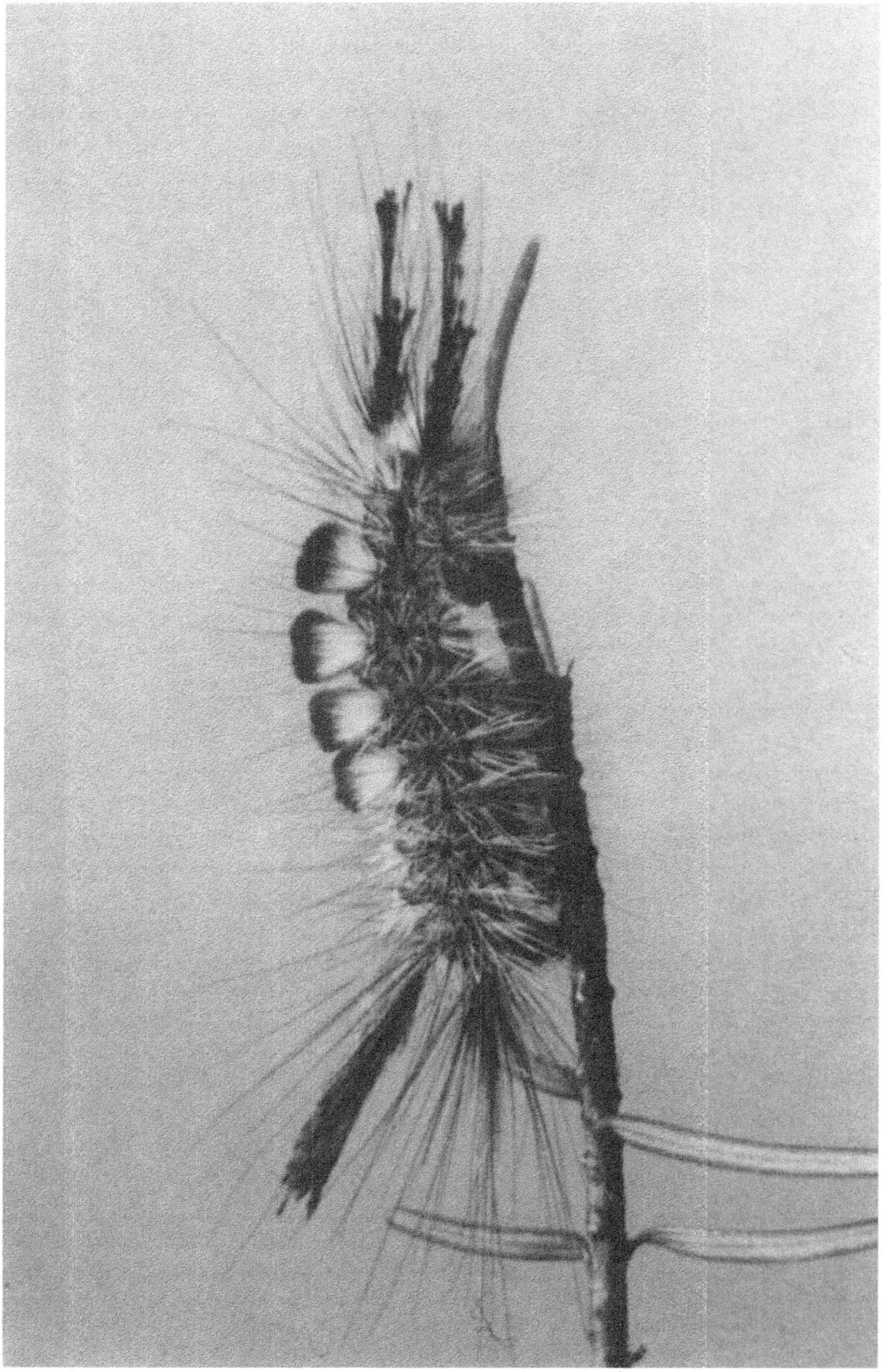

FIGURE 3. *(continued)*

D

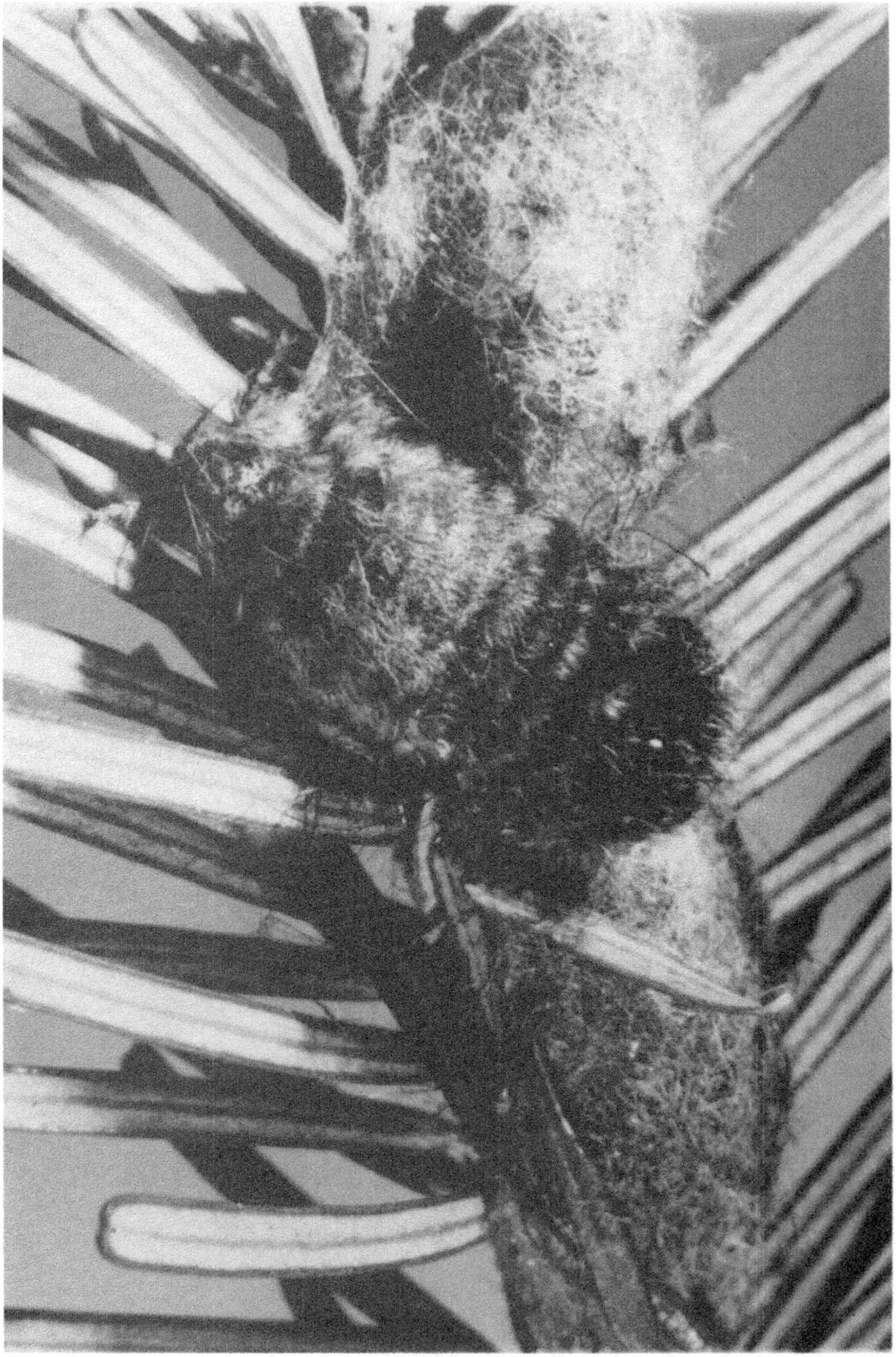

FIGURE 3. *(continued)*

E

FIGURE 3. *(continued)*

supply of new foliage for the larvae to feed upon. Neonate larvae normally leave their egg masses at the time shoot elongation is about 50% completed; by the third instar, shoot growth has usually terminated.[94,97] Small larvae produce silk strands on which they can be dispersed by wind throughout the stand. First and second instars (Fig. 3B) feed exclusively on the underside of soft, new needles. Such feeding kills the needles, and the foilage turns red-brown by midsummer, although it may remain on the tree for the rest of the season. Faded foliage in the tops of trees is the first conspicuous evidence of an infestation. After the new foliage has been destroyed, later instars feed on old foliage; during an outbreak, large portions or even the entire crown may be defoliated. The five to six larval instars feed for about 60 days (Fig. 3C) and then spin cocoons in foliated portions of the crown, sometimes in crevices of the bark, or on dead branches and twigs in the lower crown. The moths emerge in late summer, about 2 weeks after pupating. The wingless adult females (Fig. 3D) remain on the cocoon, where they mate with the winged male (Fig. 3E). The cycle is complete when the female deposits a new egg mass on or near her cocoon.

2.2. Primary Hosts and Food Quality

Tussock moth larvae can feed on many plants, but only three species—Douglas-fir, grand fir, and white fir—are primary hosts. Early records reported the insect feeding on subalpine fir (*Abies lasiocarpa*), but this tree is now considered only a minor host.[4]

The primary host species varies with geographical location. In British Columbia, it is Douglas-fir, but white fir is the principal species fed on in southern Oregon and California. In other Western states, the primary host depends on the species composition of local stands. Both Douglas-fir and grand fir are severely defoliated during outbreaks in the Blue Mountains of Oregon and Washington, but often under different stand conditions. For instance, if Douglas-fir happens to be the main host species in one area, grand fir is damaged only when Douglas-fir comprises most of the stand, but in another area the reverse may be true.[95]

The foliage of less susceptible host species is also eaten during outbreaks, particularly when they are surrounded by primary hosts that have been completely defoliated. In overcrowded populations, understory trees of all species are often fed on by larvae that drop from the host overstory.[32,95] New foliage of ponderosa pine (*Pinus ponderosa*) is sometimes eaten before the less palatable old foliage of white fir.[49] Larvae may also feed on the needles of other species of pines, western hemlock (*Tsuga heterophylla*), western larch (*Larix occidentalis*), and Engelmann spruce (*Picea engelmannii*), after the principal hosts have been stripped.[4,22] Plantings of ornamental spruces and firs may also be defoliated in montane cities and towns, even when they are miles from the nearest infested forest. The observed variability of food habits might be explained by any number of environmental, genetic, and coevolutionary factors that influence host selection.[78]

The feeding habits of larvae are also affected by the age of foliage and its position in the crown. First and second instars require new foliage to survive. As foliage is depleted, the later instars shift to the older needles.[49] Preference for new needles may be in response to higher nitrogen content, tenderness, or both.[6] Early in the season, small larvae tend to concentrate in the upper crown, where the proportion of new foliage is larger. The superior quality of food in the top of the tree is borne out by the fact that females produce significantly more eggs when fed foliage exclusively from the upper crown.[6,46] Early depletion of new growth causes considerable nutritional stress on the remaining population, which is forced to consume old needles. This stress causes a reduction in the size of larvae and pupae, a decline in female sex ratio, lower fecundity, and longer development, resulting in increased exposure to natural enemies.[6,42,56]

2.3. Major Causes of Mortality

All life stages suffer considerable natural attrition during their development. Mortality from egg to adult must be at least 99% for a population with average fecundity to remain stable from one generation to the next. Even in increasing populations, mortality is usually greater than 90% within a generation. The effectiveness of natural mortality in reducing numbers during each generation determines to a large extent the numerical behavior of tussock moth populations. When tussock moth numbers are low and food is plentiful, insect parasites and vertebrate and invertebrate predators are the primary causes of mortality. The significance of these natural enemies has frequently been demonstrated in field-rearing studies, where high survival is observed in cages which protect larvae from predators and parasites.[46,52,57] In overcrowded populations, however, dispersal loss, starvation, infertility, and disease become increasingly important.

A comparison of the main mortality factors for three population modes is given in Fig. 4. The disappearance of larvae, a sizable portion of which is caused by predation, exceeds by several times the other causes of mortality. Predation may therefore be a critical factor in regulating nonoutbreak populations. Larvae exposed on the foliage are vulnerable prey, especially when small, for a variety of arboreal spiders and predaceous insects, particularly ants.[19,52] Insectivorous birds are also common predators on all life stages.[18,86,87] At least 30 species of Diptera and Hymenoptera are known to parasitize tussock moth developmental stages,[84] but parasitization of either eggs or pupae usually contributes more to generation mortality than does larval parasitism (Fig. 4).

2.4. Site and Stand Relationships

The Douglas-fir tussock moth survives and multiplies under a variety of site and stand conditions over its geographical range, but outbreaks are much more common at the warm-dry edge of the distribution of its major host species (Fig. 1). The most severe infestations tend to develop in forest–grassland ecotones that may be only marginal for the growth of fir.[72,81] These sites frequently occur on upper slopes and ridgetops and are characterized by shallow soils.[46,79,103] They were often occupied, in former times, by old-growth ponderosa pine, which was slowly replaced by shade-tolerant fir as a result of past harvesting practices, livestock grazing, and the exclusion of fire.[96,104] Exceptions occasionally occur to the above generalizations. In the Southwest, for example, outbreaks are sometimes confined to deep canyons when these are the only sites where the host tree species can grow.

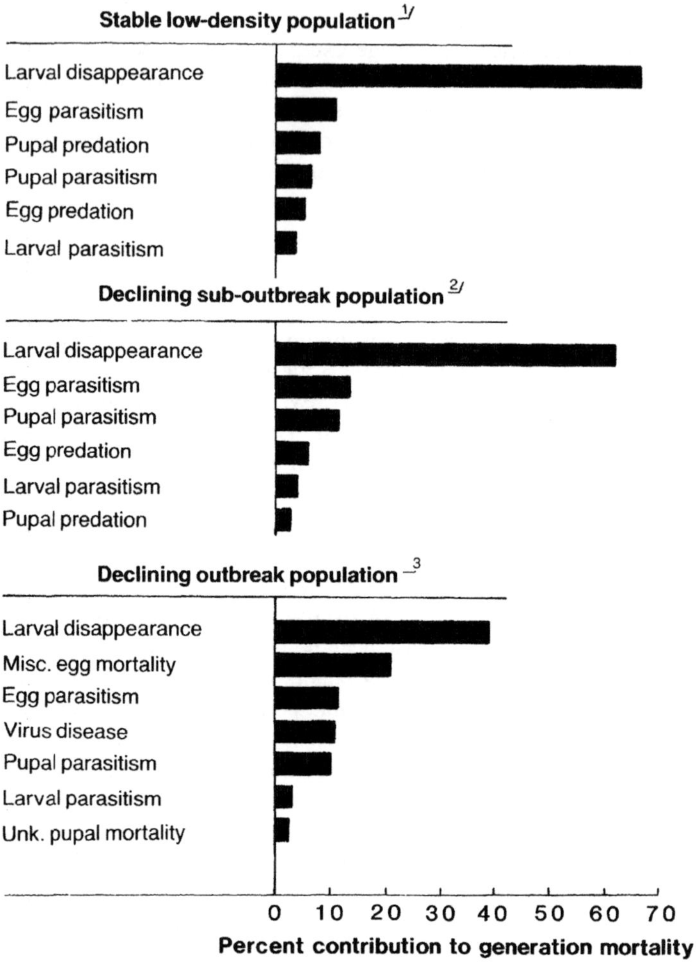

FIGURE 4. The contribution of different mortality factors to total generation mortality for three population modes of the Douglas-fir tussock moth. Calculations are from an analysis of k-values based on the killing power of each factor as a percentage of total mortality in the generation (90) ([1]after 54, [2]after 57, [3]after 42).

2.5. Impact of Outbreaks

From a purely economic perspective, tussock moth outbreaks can have serious short- and long-term effects on timber production. From the ecological viewpoint, however, the tussock moth contributes to structuring forest communities and to the stability of forest systems through its effects on nutrient cycling, tree growth and survival, and forest succession.[34]

Data on tree mortality, top-kill, and stand development after outbreaks show a consistent relationship between the degree of defoliation and the amount of damage to

Douglas-fir and grand fir.[95] As a general rule, 90% of the trees that die have been defoliated at least 90% (Fig. 5). Trees that lose 50–75% of their foliage rarely die from the effects of defoliation alone, but in some outbreaks many trees are attacked and killed by bark beetles.[105] Indirect mortality from bark beetles usually occurs within 3 years. More than one-half the total tree mortality is concentrated in patches (Fig. 2) that make up a relatively small portion of the outbreak area (usually 10–14%). For example, measurements made 5 years after an outbreak in the Modoc National Forest in California indicate that concentrated tree mortality occurred primarily on 37 of the 450 acres included in the outbreak while the remaining area only had scattered mortality.[96] Mortality in the Blue Mountain outbreak in northeastern Oregon and Washington (1972–1974) was also concentrated in patches, amounting to about 14% of the outbreak area.[98] Top-kill is common in trees defoliated more than 50%. Pole size or larger trees with dieback of half or more of the crown become more susceptible to bark beetle attacks and heart rot.[1] Top-kill damage during the Blue Mountains outbreak was 35% in heavily defoliated, 12% in moderately defoliated, and 5% in lightly defoliated stands.[95]

Both radial and height growth of trees are sharply reduced during and immediately

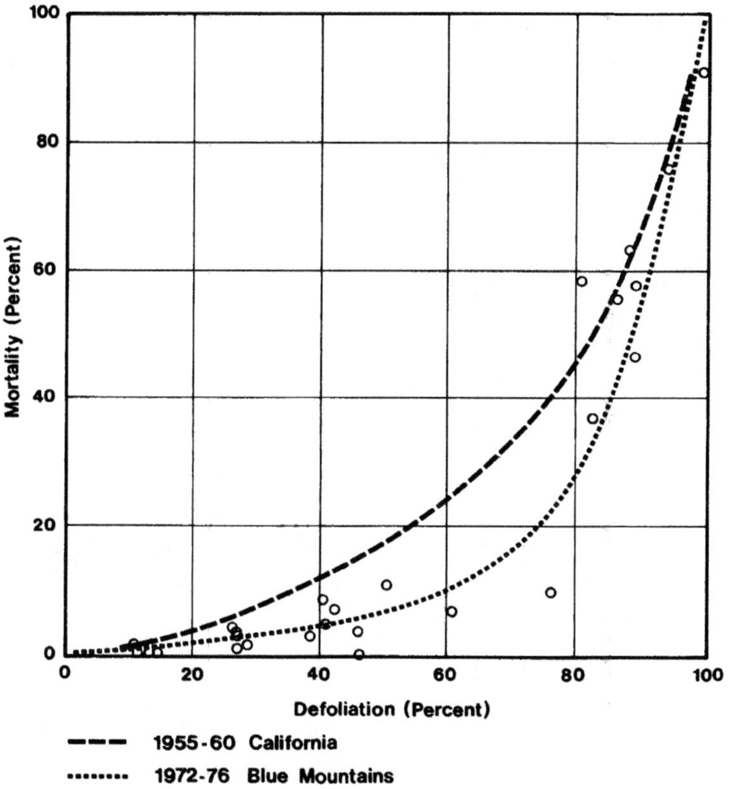

FIGURE 5. Tree mortality related to degree of defoliation (after 98).

after an outbreak. Growth reduction is proportional to the amount of defoliation and most pronounced in trees defoliated 50% or more.[100] However, recovery from defoliation is often dramatic and growth usually returns to preoutbreak rates within 5 years. Furthermore, after 10 years, growth rates may surpass preoutbreak rates[96] (Fig. 6).

The remarkable recovery of individual trees is recorded in their annual rings. During the 36 years after one outbreak, defoliated white fir grew significantly faster than both pine in the outbreak area and nondefoliated white fir growing nearby (Fig. 6); the pine also had a spurt of enhanced growth for 20 years.[99] Enhanced growth appears to be caused by increased nutrient recycling as a result of defoliation, and the thinning effect of tree mortality. Fallen needles and insect frass (a mixture of excrement and partly digested needles) have been estimated to increase available nitrogen and other nutrients by nearly 10 times that returned with normal litterfall.[34] As the more severely defoliated trees die, the remaining trees also benefit from increased nutrients, moisture, and sunlight. Only rarely do consecutive outbreaks cause damage in the same stands, and the acreage of any one outbreak is small in proportion to the total area of host type. These observations suggest that the tussock moth plays a beneficial role, over the long term, as a regulator of forest primary production and is an essential part of the forest ecosystem.[34,58,89]

3. PATTERNS OF NUMERICAL BEHAVIOR

Within the range of its host types, tussock moth abundance is extremely variable over time. Population densities often fluctuate between extremes that differ by three to four orders of magnitude (1000–10,000-fold) in susceptible stands. Peak numbers that cause conspicuous defoliation are easily identified as outbreaks, but lesser peaks may go un-

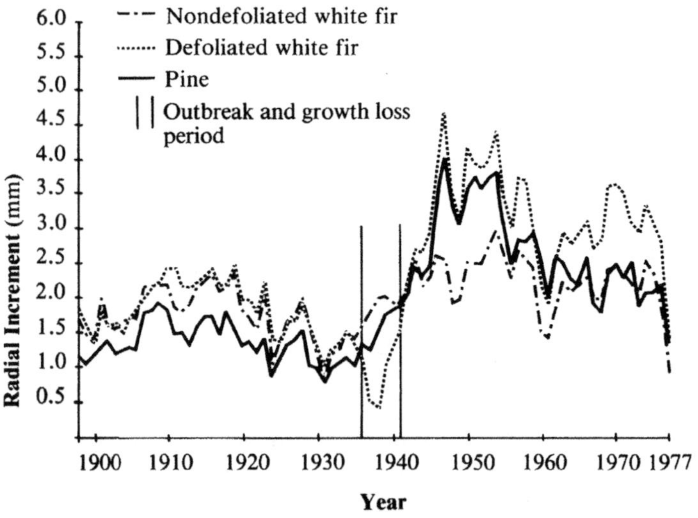

FIGURE 6. Long-term radial growth of white fir after a tussock moth outbreak (after 99).

noticed by the casual observer. In most stands, populations persist at low fluctuating numbers that seldom or never increase to outbreak densities.

Many years ago, entomologists observed that tussock moth outbreaks often recurred over large areas at the same time although not necessarily in the same stands.[4,81] Surveys from interior British Columbia clearly demonstrate the periodicity of defoliation over the past 70 years (Fig. 7). Recurrent patterns of defoliation have also been recorded in other regions, particularly in Idaho and northern California.[22,57,88] A spectral analysis of all historical data in the West showed that a major peak in tussock moth activity has occurred every 8–9 years.[15]

The consistent pattern of short, intense outbreaks followed by longer periods of low density suggests that tussock moth numbers change with some degree of regularity; in other words, numbers fluctuate around a mean population density in a systematic rather than a random manner. Although mathematical purists may have a more strict definition, such regular fluctuations are commonly referred to as cycles.[23,65] A problem with such usage is that the recurrent outbreaks of some forest insects may resemble population cycles when they are in fact random.[70] However, it has been recommended that all fluctuations believed not to be random should be called cycles until enough data are available to disprove this assumption.[68] Because tussock moth populations appear to fit this definition, we will hereafter refer to their fluctuations as cycles.

3.1. Measures of Abundance and Spatial Distribution

With the development of field-sampling techniques, tussock moth abundance is now commonly expressed in terms of the available food supply or so-called population intensity.[66] This expression usually takes the form of numbers of eggs, larvae, or pupae per m^2 (1550 in.^{2}) of mid-crown branch area, and this unit has become the standard measure for temporal and spatial comparison of population density.[40,45] Because of the relative ease with which small larvae can be sampled in the field, their density is often used for tracking populations over time when only one insect stage is sampled annually.[50] The number of

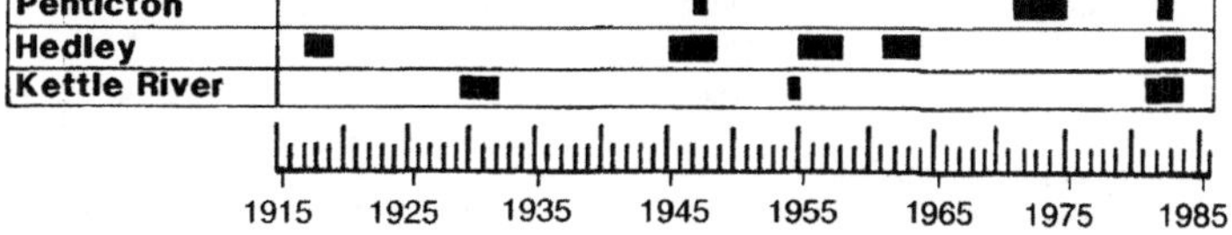

FIGURE 7. Pattern of infestation of the Douglas-fir tussock moth at twelve locations in southern British Columbia, 1916–1984 (modified from 26).

male moths caught in pheromone traps is also a useful index for monitoring population trends, although its statistical properties and precise relation to insect density are not firmly established.[21,74]

Despite their wide distribution, tussock moths usually persist in the forest at very low densities. The species is seldom noticed until its numbers increase to suboutbreak proportions (Table I). For this reason, sudden defoliation of trees gives outbreaks the appearance of having developed spontaneously when, in fact, numbers have been building up continuously over several years.

The term *outbreak* is commonly used, in its economic sense, to describe conditions that exist when trees become conspicously defoliated. The density at which defoliation is first noticed varies considerably according to larval survival, condition of the tree, and perceptiveness of the observer, but on average it occurs at about 30 small larvae per m^2 of branch area. This density has been widely used to define outbreak areas and especially for planning pest-management activities.[39,44,73,85]

Tussock moth populations are distributed nonrandomly in the forest, as determined by the interaction of the species with its local environment. This distribution is illustrated statistically by the relationship between the mean and variance over a wide range of densities from many years of field sampling (Fig. 8). The linear relationship between log variance and log mean shows that variance is related to the mean by a simple power law in which the regression coefficient b reflects the aggregative behavior of individuals in the samples.[82] This relationship is a useful basis for developing sampling programs and predicting the range of population densities that are likely to exist in the forest at any particular time.

3.2. Temporal Patterns

The best way to gain insight into the dynamic behavior of tussock moth populations is to monitor their density in the same stands over an extended period. Such a long-term commitment is difficult, especially when tussock moth populations are at very low densities and not attracting public attention. Fortunately, we have been able to collect data on

TABLE I
Range of Densities in Natural Populations of Small Tussock Moth Larvae[a]

Population status	Abundance of larvae (% of branches with larvae)	Value of N_t (larvae/m^2)[b]
Very low	Rare, 2%	$0 < N_t \leq 0.3$
Low	Uncommon, 2–25%	$0.3 < N_t \leq 3$
Suboutbreak	Common, 25–50%	$3 < N_t \leq 30$
Outbreak	Abundant, 50%	$N_t > 30$

[a]Modified from Mason.[43]
[b]Statistically reliable estimates of N_t in sparse populations can be determined from the relationship $N = -17.75\, d^{-0.6}\ln(1 - p)$, where d is the average age of larvae in days and p is the proportion of lower branch samples with larvae.[48]

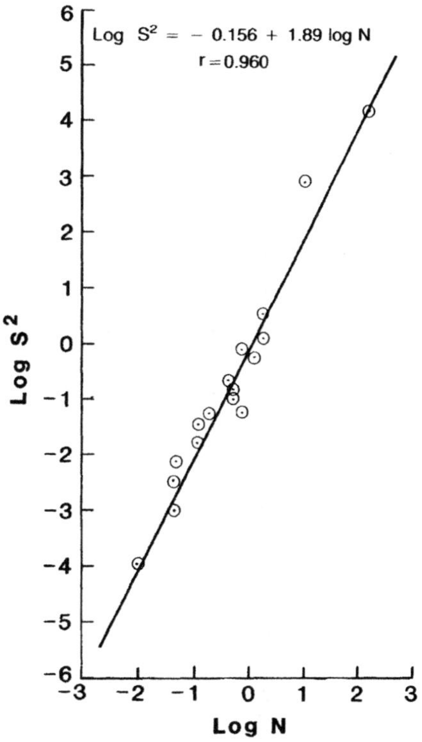

FIGURE 8. Relationship between mean density (N) and interplot variance (S^2) for samples of small tussock moth larvae.

three independent tussock moth populations from widely separated locations over a period of 15 years (Fig. 9). These data illustrate the mean densities and patterns of fluctuation that are probably common to many natural populations of the Douglas-fir tussock moth. Regular fluctuations in density occurred at each site, although not necessarily in phase with one another nor at the same amplitude and frequency. The populations appear to represent three types of dynamic behavior: (A) oscillations of similar amplitude resulting in recurrent periodic outbreaks; (B) oscillations of variable amplitude that produce only occasional outbreaks; and (C) oscillations of relatively constant and low amplitude that do not attain outbreak densities. Population A has an additional history of recurrent defoliation every 8–10 years since 1953.[57] Although the 1971 peak caused visible defoliation, the population never reached outbreak densities during 1978–1979. Population B underwent a severe outbreak in 1946–1947, but no further defoliation was observed until the widespread outbreak of 1971–1973.[101] The peak in 1983 failed to reach an outbreak density before numbers declined again. Population C has no known outbreak history and appears to be oscillating with narrower amplitude and shorter frequency than the other two populations. In this area, individual larvae are rare and are unlikely to be noticed without intensive sampling of the foliage.[54]

The consistent trends in the population data (Fig. 9) along with the historical outbreak records leaves little doubt about the regular cyclical behavior of tussock moth

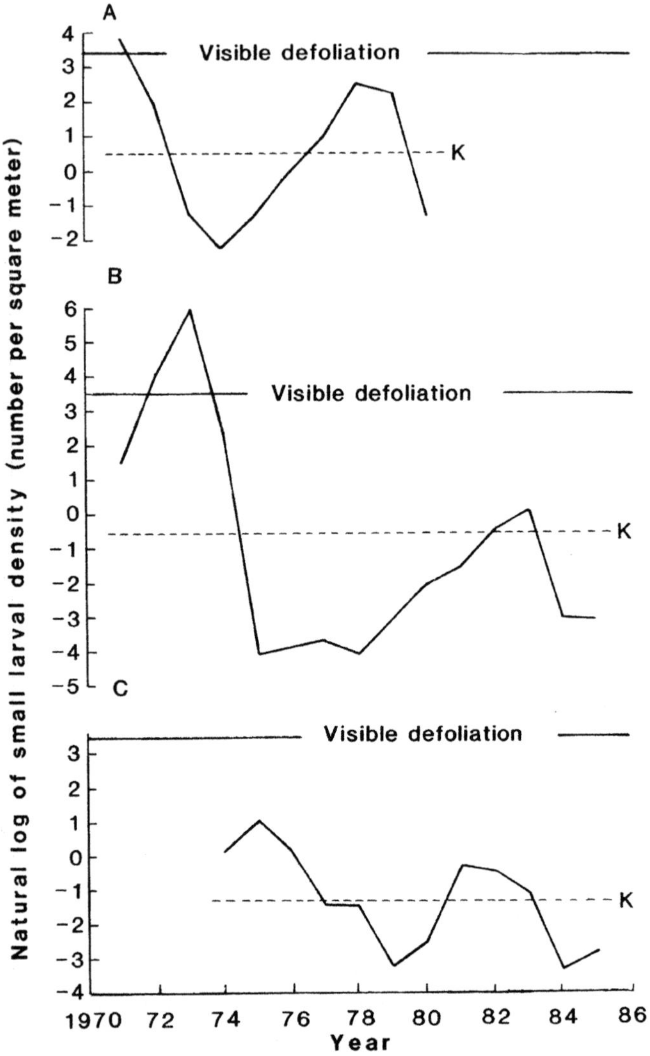

FIGURE 9. Trends in the density of small larvae for three independent populations of the Douglas-fir tussock moth: A, Sierra Nevada in northern California (approximately 8.0 km^2); B, Blue Mountains in northeastern Oregon (approximately 40.0 km^2); and C, Cascades in south-central Oregon (approximately 2.5 km^2). K is the geometric mean for each population and the approximate equilibrium density.

populations. By contrast, cyclical peaks do not always reach outbreak densities, even in populations with a history of severe outbreaks. Considering the vast areas of forest occupied by the species, outbreaks may actually be somewhat of an anomaly, occurring only during the most favorable conditions for insect reproduction, development, and survival.

3.3. Characteristics of Outbreaks

Outbreak densities rarely persist in the same stand for more than two generations; the number of larvae may rise to extremely high densities, however, and cause severe tree damage. For model simulation, the typical outbreak episode is decomposed into a sequence of 1-year phases that can usually be identified by the density of small larvae in that generation (Fig. 10). The critical year in this sequence is the first phase, when numbers increase rapidly from an innocuous suboutbreak status to outbreak densities. As we saw earlier, many suboutbreak populations never reach outbreak densities, but decline and continue in their normal cycle (Fig. 9C). Phase 1 is difficult to forecast without survival estimates,[53] but, once it develops, subsequent phases and trends are fairly predictable. Population densities usually peak in the second year of the outbreak and may remain high for another year before declining.

The duration of outbreaks and their rates of decline depend, to a large extent, on the intensity of negative feedback processes such as starvation and increased mortality from natural enemies and disease.[47,51] The generalized outbreak phases are not always in perfect synchrony, however, and variations usually occur over a large outbreak area.

4. HYPOTHESES FOR THE CAUSES OF OBSERVED POPULATION BEHAVIOR

Numerous theories have been proposed to explain the apparent cycles observed in some insect populations. Classic examples are processes intrinsic to the system, such as predator–prey interactions, quality changes and self-regulatory behavior in the insect population itself, cycling of nutrients or production of defensive chemicals, and delayed density-dependent responses.[9,14,27,29,92] All these explanations involve some form of negative feedback directly between the density of the insect and the condition of its food, natural enemies, or physiological state. Although weather has an obvious effect on insect abundance and may influence fluctuations, it is not generally accepted as a cause of cycles.[35] Until a lot more detailed data are accumulated, we cannot completely eliminate any of the above explanations as playing some role in the cycling of tussock moth numbers. On the basis of what is known, however, certain hypotheses appear to be more likely than others.

4.1. Evidence of Delayed Density Dependence

The growth of any population will ultimately be limited by a superabundance of individuals and its resultant effects. Severe defoliation obviously causes a shortage of food, resulting in increased dispersal and larval starvation, higher susceptibility to disease and natural enemies, and reduced fecundity.[42,47,51,56] Tussock moth survival rates during outbreaks are inversely related to the degree of defoliation, so that outbreaks collapse faster in heavily defoliated than in lightly defoliated stands (Fig. 11). Although virus-caused diseases are also commonly associated with declining outbreaks, they are usually

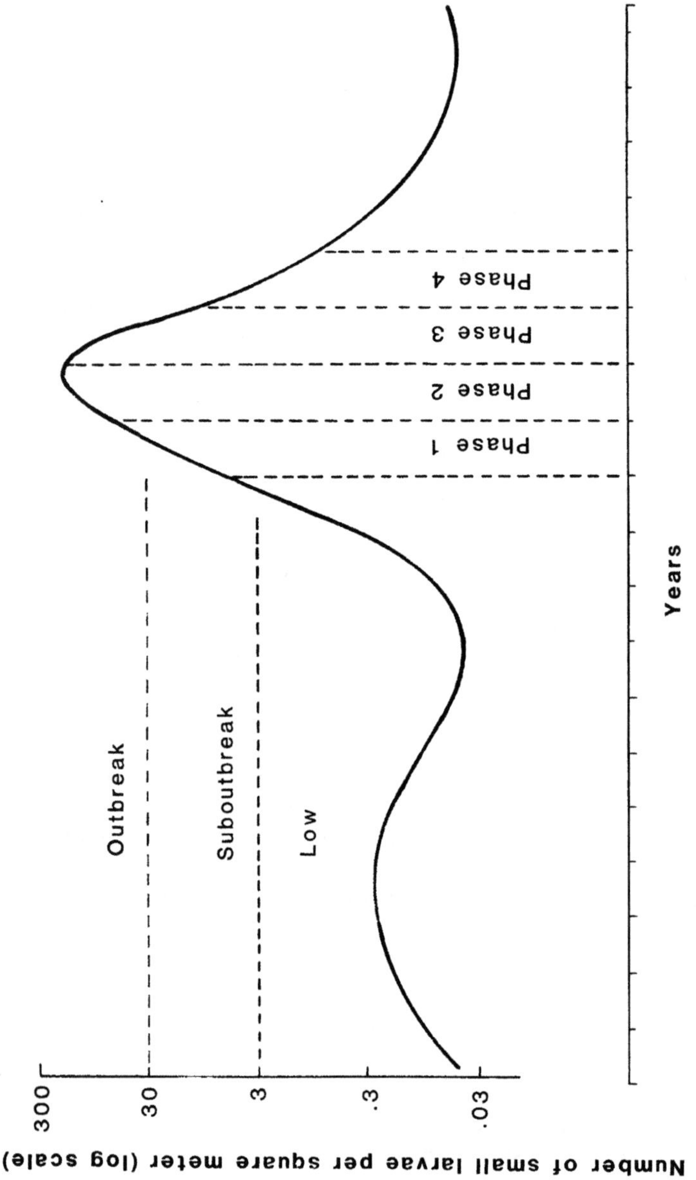

FIGURE 10. Sequence of phases in a typical outbreak of the Douglas-fir tussock moth.

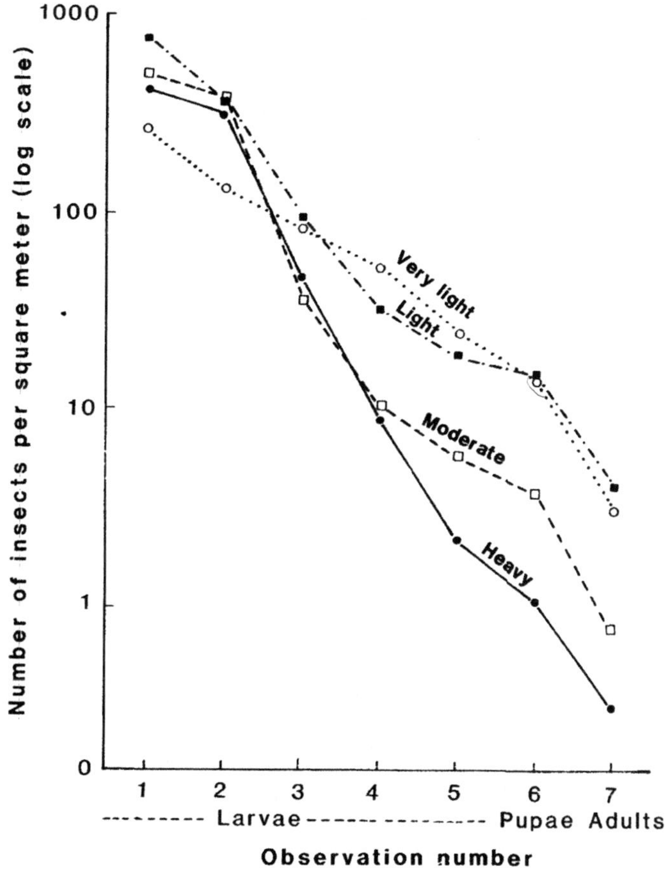

FIGURE 11. Survivorship of tussock moth from small larvae to adults under four conditions of previous stand defoliation (after 42).

triggered suddenly in stressed populations rather than building up gradually with increasing population density.[83]

Density-dependent feedback is obviously not restricted to high densities because most of the densities shown in Fig. 9 were far below levels which seriously affect the food supply. Density dependence can be demonstrated for each of these populations by comparing the per-capita rate of population change with population size.

A least-squares analysis shows an inverse linear relationship between rate of increase and population density in the two previous generations (Table II). The strongest relationship, as measured by the slope of the regression, is between r_t and the logarithm of N_{t-1} for each population (Fig. 12). Population change is thus a decreasing function of the density in the previous generation. Not only are the populations ultimately limited in size by the destructiveness of their own abundance, but they are also under some form of density-dependent regulation, even at lower numbers. Many of the factors that determine

TABLE II
Statistics from Least-Squares Analysis of the Relationship between the Rate of Population Increase (r_t)[a] and Larval Density ($\ln N_{t-\tau}$)[b] for Populations A, B, and C in Fig. 9

Time lag τ	Population	N	Intercept	Slope	Correlation coefficient	Standard error of estimate
0	A	9	−0.411	−0.564	−0.601	1.682
	B	10	−0.734	−0.312	−0.325	2.984
	C	11	−0.938	−0.428	−0.457	1.267
1	A	8	−0.329	−0.852	−0.900	0.963
	B	7	−0.782	−0.804	−0.713	2.512
	C	10	−0.980	−0.652	−0.628	1.125
2	A	7	−0.077	−0.407	−0.489	1.725
	B	7	−2.890	−0.768	−0.786	1.972
	C	9	−0.788	−0.339	−0.328	1.448
3	A	6	0.165	0.060	0.074	2.147
	B	6	−2.601	−0.595	−0.526	2.854
	C	8	−0.027	0.106	0.103	1.892

[a] $r_t = \ln(N_{t+1}/N_t)$, where N_t is the density of small larvae in year t.
[b] τ: the time lag varies from 0 to 3, so that r_t is analyzed segmentally as a function of N_t, N_{t-1}, N_{t-2}, N_{t-3}.

population growth rate, e.g., predation, parasitism, disease, dispersal, and changes in reproductive ability, are affected individually and collectively, although often in subtle ways, by the size of the population.[28,54]

Density-dependent negative feedback, as roughly described by the well-known logistic equation and its variations, is necessary for populations to reach stable equilibria.[31] If the response of the feedback mechanism to changes in population density is not instantaneous, oscillations or cycles are produced.[30,59] These occur when populations overshoot or undershoot their equilibrium density because of delays in the feedback response. How the apparent cycles observed in natural populations of the Douglas-fir tussock moth could be caused by such time lags is easy to see. Quantitative responses by many of the natural enemies, as well as changes in insect quality and fecundity, may take at least one generation to respond. In fact a simple time-delayed version of the discrete logistic equation produces cycles with the same frequency and period as natural tussock moth populations.[9] The numerical patterns that have emerged from our new data also seem to support this hypothesis (Figs. 9 and 12).

4.2. Importance of the Intrinsic Rate of Increase and Density-Independent Factors

Although lags in the response of negative feedback factors may be responsible for generating cycles, they do not completely explain other dynamic features, such as the maximum and minimum values (amplitude) that a population reaches during a cycle, or the period of time between cycles (periodicity). These features are especially important in

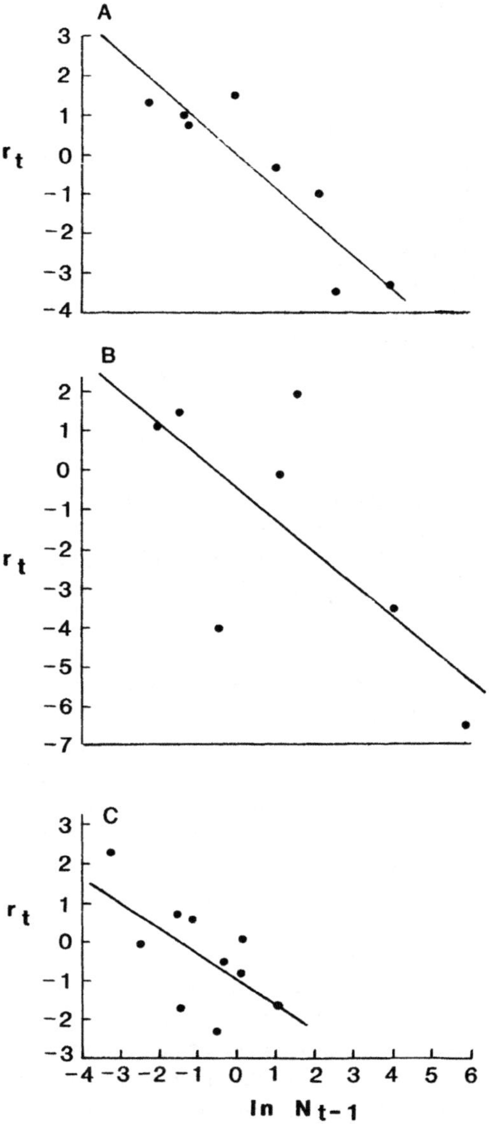

FIGURE 12. Relationship between the rate of increase (r_t) and log density of small larvae in previous generation ($\ln N_{t-1}$) (for populations A, B, and C in Fig. 9).

pest management because amplitude determines whether larvae reach outbreak densities, and periodicity measures the frequency with which outbreaks might occur. Amplitude and periodicity in logistic models are determined by the maximum rate at which populations increase free of intraspecific competition.[59] This is the intrinsic rate of increase of the population, generally designated r_m to separate it from r_t, the actual rate of increase.[3] It varies according to the birth and death rates in a given environment and can be used to describe the performance of a species under different conditions. The r_m for each population in Fig. 12 can be estimated by the maximum rate of increase predicted at very low densities. This rate is relatively high for populations A and B and low for population C.

A wide range of numerical behavior occurs in populations when different values of r_m are substituted in one of the logistic equations.[9,31,60,61] In general, the amplitude and periodicity of cycles increase directly with r_m until population behavior becomes erratic and unpredictable at large values of r_m. Under optimal conditions for survival, such species as the Douglas-fir tussock moth that have high potential fecundity are apt to have relatively large r_m values and can occasionally be expected to reach outbreak numbers. Such conditions probably exist on sites that have a history of periodic outbreaks. In less favorable environments, r_m will be smaller, so that oscillations have a lower amplitude and are closer to the stable equilibrium density (e.g., Fig. 9C). The real lesson from this theory is that the dynamic behavior of tussock moth populations is probably an intrinsic property of the species and site, generated by delayed density-dependent regulation and a potentially high reproductive ability.

This hypothesis is an oversimplification of tussock moth populations in nature. Density-dependent factors are never perfect, nor are population growth rates constant.[62,80] If they were, many populations would approach and remain at a stable equilibrium density. In the real world, random factors independent of density are constantly helping drive populations away from that equilibrium. Over the long term, however, population fluctuations of the tussock moth are intrinsically cyclical despite the fact that they are continually being disturbed by random factors.[59]

Most researchers agree that cycles of abundance in animal populations result from a combination of density-dependent and -independent factors.[23,70,71] Independent phenomena such as weather affect the growth rate of the population—either directly or indirectly, through natural enemies and the host tree—and ultimately the amplitude of the oscillations. Cycles are already set in motion by delayed density-dependent mechanisms intrinsic to each population, so that random disturbances simply modify or perpetuate these oscillations. By contrast, weather that simultaneously affects many locations may act to synchronize the cycles of isolated populations that are otherwise oscillating independently.[64]

Outbreaks of tussock moths have sometimes followed above-average temperatures and below-normal precipitation, suggesting that warm, dry conditions favor an increase in r_m and the maximum density reached in a cycle.[15,91,93] To trigger an outbreak, optimum weather obviously must coincide with a natural upswing in the population cycle. The ways in which weather affects reproduction and mortality, and ultimately the rate of population growth, however, is a circuitous and complex process that will most likely remain a subject of much speculation.[8,36]

5. IMPLICATIONS FOR DEVELOPING MANAGEMENT STRATEGIES

Our conclusion that intrinsic cycles underlie the basic numerical behavior of tussock moth populations provides valuable clues for dealing with the practical aspects of pest control. Most important is the implication that the amplitude and periodicity of cycles are basic properties of each population and are only temporarily affected by independent factors such as weather. These conclusions are based on considerable empirical evidence that makes predictability easier than if changes in abundance were viewed simply as random fluctuations.

5.1. Monitoring and Forecasting

Continuous surveillance of tussock moth abundance should have high priority in managed forests in which the insect has been a problem in the past. Annual censuses are important for determining population trends in terms of the natural cycle. Monitoring can be relatively easy by using standard pheromone trapping techniques or by sampling larvae.[21,45] When data are accumulated over a period of years, a pattern of population behavior can be established that is useful for predicting the status and probable trend of the population. Although trends are often similar over whole forests or even geographical regions, data for individual stands may have to be scrutinized to detect possible out-of-phase populations. The probability that such aberrant populations will occur can be estimated for any mean population density from the expected sample variance (Fig. 8).

Viewing populations as being cyclical has several advantages. The most obvious is that anxiety should be low when overall densities are at the low ebb of an oscillation. Even during the most favorable environmental conditions, at least 4 years are required for populations to build up to outbreak densities. Populations at intermediate densities present different concerns depending on their past history. Those with a history of cycles with wide amplitude should be monitored more intensively than populations with a history of more constrained behavior. Optimum conditions of temperature and moisture may accentuate the cycle when they coincide with the phase of population increase. The high rates of growth associated with preoutbreak populations depends on a relatively high rate of larval survival that is itself a good predictor of trend.[50] Simple methods are available for estimating larval survival if an early prediction of trend is desired.[53]

5.2. Outbreak Prognosis Models

When the peak density of a tussock moth population cycle exceeds the visible defoliation level, the population enters an outbreak episode that evolves through a sequence of predictable phases (Fig. 10). The expected dynamics of tussock moth populations during an outbreak episode have been described in a computer model that simulates the probable numerical behavior of populations under a variety of environmental conditions.[17] Larval densities generated by the outbreak model are converted to estimates of tree defoliation, which are then translated into growth loss, tree mortality, and top kill. These outputs from the outbreak model can also be linked to a stand-prognosis model to determine the ultimate long-term effects of the outbreak on stand growth and development.[16,63]

The computer model is a synthesis of all existing knowledge concerning the behavior of Douglas-fir tussock moth outbreaks and, thus, is a valuable aid in making rational pest-management decisions.[13] As the upward trend of a population approaches its cyclic peak, the probability of an outbreak increases and the possible needs for control must be evaluated. By simulating population growth and tree defoliation over a range of conditions, managers can anticipate short-term impact and be prepared to take appropriate action, should it be needed. The model will not forecast where or when outbreaks will occur, but simulations can be used in long-range forest planning to determine the likely economic consequence of periodic outbreaks and to evaluate alternative strategies for dealing with the pest over the long run.[10]

5.3. Outbreak Suppression and Prevention

Direct suppression has been, and will probably remain, the preferred means of dealing with tussock moth outbreaks. Several chemical and microbial insecticides effective in reducing population numbers have been approved for field use.[37,67,76] The objective of direct control is to reduce peak densities before significant defoliation takes place. Because the increase of a population to outbreak status is not predictable, direct treatment is usually not practical until after the outbreak has developed. Precise timing of the treatment to an early phase of the outbreak is critical for maximizing benefits; it depends strongly on up-to-date surveys and monitoring of population density. A promising suppression technique is early application of formulations of the same nuclear polyhedrosis virus that eventually occurs naturally in most outbreak populations.[76,77] Large quantities of the virus produced in laboratory-reared insects have been stockpiled. Outbreaks may be aborted prematurely by early treatment of the population with this biological agent.[75]

Preventive treatments are usually silvicultural practices that either reduce or eliminate the threat of outbreaks. Because cycles of abundance are fundamentally intrinsic to the tussock moth system, changing these basic oscillations is unlikely. By contrast, the amplitude of oscillations is variable and is determined by the population's intrinsic rate of increase r_m. Because r_m is primarily a product of the birth and death rate in each population, any silvicultural system designed to reduce the probability of tussock moth outbreaks can theoretically be judged by its effect on these two factors.

Natural enemies are the major causes of mortality in each generation and possibly the principal regulators of density in low populations.[54] This regulatory effect suggests that, at the very least, susceptible stands should be managed to conserve the integrity of these important natural controls. The physical environment of a site is less manageable than its biotic components, however, so some sites will always be high risk for outbreaks as long as they are occupied by susceptible host type. Examples are those sites in the warm, dry zone of the distribution of fir, where the most severe tussock moth outbreaks have occurred.[72,79,96,104] Although the reasons are unclear, these sites obviously favor tussock moth reproduction and survival, which enables the species to maintain a relatively high r_m. Wherever fir is growing under these conditions, outbreaks will occasionally be generated by forces that are largely beyond our control. Such practices as fire exclusion and selective cutting of pine that favor the establishment of fir only exacerbate the situation by perpetuating stands with high risk of future infestation.[55] Conversion of these sites to seral nonhost species is probably the only effective prescription for reducing the risk of tussock moth outbreaks.

REFERENCES

1. Aho, P. E., Wickman, B. E., and Roe, L., 1979, Decay in tops killed by Douglas-fir tussock moth in the Blue Mountains, *USDA For. Serv. Res. Pap. PNW*-250, Pac. Northwest For. Range Exp. Stn., Portland, Oregon.
2. Anderson, W. B., 1919, Notes on the tussock moth, *Hemerocampa vetusta gulosa* Hy. Edw., in British Columbia, *Agric. Gaz. Can.* **6**:139.
3. Andrewartha, H. G., and Birch, L. C., 1954, *The Distribution and Abundance of Animals*, The University of Chicago Press, Chicago.
4. Balch, R. E., 1932, The fir tussock moth (*Hemerocampa pseudotsugata* McD.), *J. Econ. Entomol.* **25**:1143–1148.
5. Barnes, H. F., 1900, New species and varieties of North American Lepidoptera, *Can. Entomol.* **32**:42–48.
6. Beckwith, R. C., 1976, Influence of host foliage on the Douglas-fir tussock moth, *Environ. Entomol.* **5**:73–77.
7. Beckwith, R. C., 1978, Biology of the insect: Introduction, in: *The Douglas-Fir Tussock Moth: A Synthesis* (M. Brookes, R. W. Stark, and R. W. Campbell, eds.), pp. 25–30, U.S. Dep. Agric. Tech. Bull. 1585.
8. Beckwith, R. C., 1983, The effect of temperature and food deprivation on survival of first-instar Douglas-fir tussock moths (*Orgyia pseudotsugata*) (Lepidoptera: Lymantriidae), *Can. Entomol.* **115**:663–666.
9. Berryman, A. A., 1978, Population cycles of the Douglas-fir tussock moth (Lepidoptera: Lymantriidae): The time-delay hypothesis, *Can. Entomol.* **110**:513–518.
10. Bousfield, W. E., Brickell, J. E., Cleeves, J. C., Dezellem, R. L., Gregg, T. F., Hofacker, T. H., Johnson, R. R., Lewis, J. W., Lightner, G. M., Quinn, T. P., and Wiitala, M. R., 1984, Economics of Douglas-fir tussock moth control: An analysis using the combined stand prognosis/Douglas-fir tussock moth outbreak model and the FORPLAN linear programing model on the Clearwater and Malheur National Forests. U.S. Dept. Agric., For. Serv. For. Pest Management, Washington, D.C.
11. Brookes, M. H., Stark, R. W., and Campbell, R. W., (eds.), 1978, *The Douglas-fir tussock moth: A synthesis*, U.S. Dep. Agric. Tech. Bull. 1585.
12. Burke, H. E., 1906, Notes on the natural history of *Notolophus oslari*, Minutes of the 209th regular meeting of the Entomological Society of Washington, *Proc. Entomol. Soc. Wash.* **8**:142–143.
13. Campbell, R. W., and Stark, R. W., 1980, The Douglas-fir tussock moth management system, *U.S. Dep. Agric. Handb.* 568.
14. Chitty, D., 1971, The natural selection of self-regulatory behavior in animal populations, in: *Natural Regulation of Animal Populations* (I. A. McLaren, ed.), pp. 136–170, Atherton, New York.
15. Clendenen, G.. Gallucci, V. F., and Gara, R. I., 1978, On the spectral analysis of cyclical tussock moth epidemics and corresponding climatic indices, with a critical discussion of the underlying hypotheses, in: *Time Series and Ecological Processes* (H. H. Shugart, Jr., ed.), pp. 279–293, Proceedings of a Conference sponsored by SIAM Institute for Mathematics and Society, Philadelphia.
16. Colbert, J. J., and Campbell, R. W., 1978, The integrated model, in: *The Douglas-fir tussock moth: A synthesis* (M. H. Brookes, R. W. Stark, and R. W. Campbell, eds.), pp. 216–230, U.S. Dep. Agric. Tech. Bull. 1585.
17. Colbert, J. J., Overton, W. S., and White, C., 1978, Documentation of the Douglas-fir tussock moth outbreak population model, *USDA For. Serv. Gen. Tech. Rep. PNW*-89.
18. Dahlsten, D. L., and Copper, W. A., 1979, The use of nesting boxes to study the biology of the mountain chickadee (*Parus gambeli*) and its impact on selected forest insects, in: *The Role of Insectivorus Birds in Forest Ecosystems* (J. G. Dickson, R. N. Conner, R. R. Fleet, J. C. Kroll, and J. A. Jackson, eds.), pp. 217–260, Academic, New York.
19. Dahlsten, D. L., Luck, R. F., Schlinger, E. I., Wenz, J. M., and Copper, W. A., 1977, Parasitoids and predators of the Douglas-fir tussock moth, *Orgyia pseudotsugata* (Lepidoptera: Lymantriidae), in low to moderate populations in central California, *Can. Entomol.* **109**:727–746.
20. Daterman, G. E., Livingston, R. L., and Robbins, R. G., 1977, How to identify tussock moths caught in pheromone traps, *U.S. Dep. Agric. Handb.* 517.
21. Daterman, G. E., Livingston, R. L., Wenz, J. M., and Sower, L. L., 1979, How to use pheromone traps to determine outbreak potential, *U.S. Dep. Agric. Handb.* 546.

22. Eaton, C. B., and Struble, G. R., 1957, The Douglas-fir tussock moth in California (Lepidoptera: Liparidae), *Pan Pac. Entomol.* **33:**105–108.
23. Finerty, J. P., 1980, *The Population Ecology of Cycles in Small Mammals,* Yale University Press, New Haven, Connecticut.
24. Furniss, R. L., and Carolin, V. M., 1977, Western forest insects, U.S. Dept. Agric. Misc. Publ. 1339.
25. Graham, D. A., Mounts, J., and Almos, D., 1975, Cooperative Douglas-fir tussock moth control project, USDA Forest Service, Pacific Northwest Region, Portland, Oregon.
26. Harris, J. W. E., Dawson, A. F., and Brown, R. G., 1985, The Douglas-fir tussock moth in British Columbia 1916–1984, *Can. For. Serv., Pac. For. Res. Cent. Rep.* BC-X-268. Victoria.
27. Hassell, M. P., 1978, *The Dynamics of Arthropod Predator–Prey Systems,* Princeton University Press, Princeton, New Jersey.
28. Hassell, M. P., 1985, Insect natural enemies as regulating factors, *J. Anim. Ecol.* **54:**323–334.
29. Haukioja, E., 1980, On the role of plant defenses in the fluctuation of herbivore populations. *Oikos* **35:**202–213.
30. Hutchinson, G. E., 1948, Circular causal systems in ecology. *Ann NY Acad. Sci.* **50:**221–246.
31. Hutchinson, G. E., 1978, *An Introduction to Population Ecology,* Yale University Press, New Haven.
32. Johnson, P. C., and Ross, D. A., 1967, Douglas-fir tussock moth, *Hemerocampa (Orgyia) pseudotsugata* McDunnough, in: *Important Forest Insects and Diseases of Mutual concern to Canada, the United States, and Mexico* (A. G. Davidson and R. M. Prentice, eds.), pp. 105–107, Can. Dept. For. Rural Dev., Ottawa.
33. Keen, F. P., 1952, Insect enemies of western forests, USDA For. Serv. Misc. Publ. 283.
34. Klock, G. O., and Wickman, B. E., 1978, Ecosystem effects, Pages 90–95 in M. H. Brookes, R. W. Stark, and R. W. Campbell (Eds.). *The Douglas-fir tussock moth: A Synthesis* U.S. Dep. Agric. Tech. Bull. 1585.
35. Klomp, H., 1962, The influence of climate and weather on the mean density level, the fluctuations and the regulation of animal populations, *Arch. Neerl. Zool.* **15:**68–109.
36. Lessard, E. D., 1974, Climatic, host tree, and site factors affecting the population dynamics of the Douglas-fir tussock moth, *Orgyia pseudotsugata* McDunnough, M. S. thesis. University of Washington, Seattle.
37. Linnane, J. P., and Stelzer, M. J., 1982, Protecting ornamental and shade trees, *U.S. Dept. Agric. Handb.* 604.
38. McDunnough, J. C., 1921, New British Columbia tussock moth, *Hemerocampa pseudotsugata* McD., *Can. Entomol.* **53:**53–56.
39. Mason, R. R., 1969, Sequential sampling of Douglas-fir tussock moth populations, *USDA For. Serv. Res. Note PNW*-102, Pac. Northwest For. and Range Exp. Stn.
40. Mason, R. R., 1970, Development of sampling methods for the Douglas-fir tussock moth, *Hemerocampa pseudotsugata* (Lepidoptera: Lymantriidae), *Can. Entomol.* **102:**836–845.
41. Mason, R. R., 1974, Population change in an outbreak of the Douglas-fir tussock moth, *Orgyia pseudotsugata* (Lepidoptera: Lymantriidae) in central Arizona, *Can. Entomol.* **106:**1171–1174.
42. Mason, R. R., 1976, Life tables for a declining population of the Douglas-fir tussock moth in northeastern Oregon, *Ann. Entomol. Soc. Am.* **69:**948–958.
43. Mason, R. R., 1977, Advances in understanding population dynamics of the Douglas-fir tussock moth, *Bull. Entomol. Soc. Am.* **23:**168–171.
44. Mason, R. R., 1978, Synchronous patterns in an outbreak of the Douglas-fir tussock moth, *Environ. Entomol.* **7:**672–675.
45. Mason, R. R., 1979, How to sample larvae of the Douglas-fir tussock moth, *U.S. Dept. Agric. Handb.* 547.
46. Mason, R. R., 1981, Host foliage in the susceptibility of forest sites in central California to outbreaks of the Douglas-fir tussock moth, *Orgyia pseudotsugata* (Lepidoptera: Lymantriidae), *Can. Entomol.* **113:**325–332.
47. Mason, R. R., 1981, A numerical analysis of the causes of population collapse in a severe outbreak of the Douglas-fir tussock moth, *Ann. Entomol. Soc. Am.* **74:**51–57.
48. Mason, R. R., 1987, Frequency sampling to predict densities in sparse populations of the Douglas-fir tussock moth, *For. Sci.* **33:**145–156.

49. Mason, R. R., and Baxter, J. W., 1970, Food preference in a natural population of the Douglas-fir tussock moth, *J. Econ. Entomol.* **63:**1257–1259.
50. Mason, R. R., and Overton, W. S., 1983, Predicting size and change in nonoutbreak populations of the Douglas-fir tussock moth (Lepidoptera: Lymantriidae), *Environ. Entomol.* **12:**799–803.
51. Mason, R. R., and Thompson, C. G., 1971, Collapse of an outbreak of the Douglas-fir tussock moth, *Hemerocampa pseudotsugata* (Lepidoptera: Lymantriidae), *USDA For. Serv. Res. Note PNW*-139. Pac. Northwest For. and Range Exp. Stn., Portland, Oregon.
52. Mason, R. R., and Torgersen, T. R., 1983, Mortality of larvae in stocked cohorts of the Douglas-fir tussock moth, *Orgyia pseudotsugata* (Lepidoptera: Lymantriidae), *Can. Entomol.* **115:**1119–1127.
53. Mason, R. R., and Torgersen, T. R., 1983, How to predict population trends. *U.S. Dep. Agric. Handb.* 610.
54. Mason, R. R., and Torgersen, T. R., 1987, Dynamics of a non-outbreak population of the Douglas-fir tussock moth (Lepidoptera: Lymantriidae) in southern Oregon, *Environ. Entomol.* **16:**1217–1227.
55. Mason, R. R., and Wickman, B. E., 1984, The Douglas-fir tussock moth: Ecological relationships, impact, and hazard reduction, *Silvicultural Management Strategies for Pests of the Interior Douglas-Fir and Grand Fir Forest Types:* Proc. Symp., Feb. 14–16, 1983, Spokane, Washington (D. M. Baumgartner and R. Mitchell, (eds.), pp. 95–102, Washington State University, Pullman.
56. Mason, R. R., Beckwith, R. C., and Paul, H. G., 1977, Fecundity reduction during collapse of a Douglas-fir tussock moth outbreak in northeast Oregon, *Environ. Entomol.* **6:**623–626.
57. Mason, R. R., Torgersen, T. R.. Wickman, B. E., and Paul, H. G., 1983, Natural regulation of a Douglas-fir tussock (Lepidoptera: Lymantriidae) population in the Sierra Nevada, *Environ. Entomol.* **12:**587–594.
58. Mattson, W. J., and Addy, N. D., 1975, Phytophagous insects as regulators of forest primary production, *Science* **190:**515–522.
59. May, R. M., 1973. *Stability and Complexity in Model Ecosystems,* Princeton University Press, Princeton, New Jersey.
60. May, R. M., 1974, Biological populations with nonoverlaping generations: Stable points, stable cycles, and chaos, *Science* **186:** 645–647.
61. May, R. M. 1975. Biological populations obeying difference equations: Stable points, stable cycles, and chaos, *J. Theor. Biol.* **51:**511–524.
62. Milne, A., 1957, Theories of natural control of insect populations, *Cold Spring Herb. Symp. Quant. Biol.* **22:**253–267.
63. Monserud, R. A., and Crookston, N. L., 1982, A user's guide to the combined stand prognosis and Douglas-fir tussock moth outbreak model, *USDA For. Serv. Gen. Tech. Rep. INT*-127.
64. Moran, P. A. P., 1953, The statistical analysis of the Canadian lynx cycle. II. Synchronization and meterology, *Aust. J. Zool.* **1:**291–298.
65. Moran, P. A. P., 1954, The logic of the mathematical theory of animal populations. *J. Wildl. Mgmt.* **18:**60–66.
66. Morris, R. F., 1955, The development of sampling techniques for forest insect defoliators, with particular reference to the spruce budworm, *Can. J. Zool.* **33:**225–294.
67. Neisess, J., Markin, G. P., and Schaefer, R., 1976, Field evaluations of acephate and Dimilin against the Douglas-fir tussock moth, *J. Econ. Entomol.* **69:**783–786.
68. Pielou, E. C., 1974, *Population and Community Ecology: Principles and Methods,* Gordon and Breach, New York.
69. Schreuder, G. F., 1978, The socioeconomic model, in: *The Douglas-Fir Tussock Moth: A Synthesis* (M. H. Brookes, R. W. Stark, and R. W. Campbell, eds.), pp. 215–216, U. S. Dept. Agric. Tech. Bull. 1585.
70. Schwerdtfeger, F., 1958, Is the density of animal populations regulated by mechanisms or by chance?, *Proc. Tenth Internatl. Congr. Entomol.* **4:**115–122.
71. Schwerdtfeger, F., 1968, Eine integrierte Theorie zur Abundanzdynamik tierischer Populationen, *Oecologia* **1:**265–295.
72. Shepherd, R. F., 1977, A classification of western Canadian defoliating forest insects by outbreak spread characteristics and habitat restriction, in: *Insect Ecology Papers Presented in the A. C. Hodson Ecology Lectures* (H. M. Kulman and H. C. Chiang, eds.), pp. 80–88, Agric. Exp. Stn. Tech. Bull. 310, University of Minnesota.

73. Shepherd, R. F., 1985, Pest management of Douglas-fir tussock moth: Estimating larval density by sequential sampling, *Can. Entomol.* **117**:1111–1115.
74. Shepherd, R. F., Gray, T. G., Chorney, R. J., and Daterman, G. E., 1985, Pest management of Douglas-fir tussock moth, *Orgyia pseudotsugata* (Lepidoptera: Lymantriidae): Monitoring edemic populations with pheromone traps to detect incipient outbreaks, *Can. Entomol.* **117**:839–848.
75. Shepherd, R. F., Otvos, I. S., Chorney, R. J., and Cunningham, J. C., 1984, Pest management of Douglas-fir tussock moth (Lepidoptera: Lymantriidae): Prevention of an outbreak through early treatment with a nuclear polyhedrosis virus by ground and aerial applications, *Can. Entomol.* **116**: 1533–1542.
76. Stelzer, M. J., Neisess, J., and Thompson, C. G., 1975, Aerial applications of a nucleopolyhedrosis virus and *Bacillus thuringiensis* against the Douglas-fir tussock moth, *J. Econ. Entomol.* **68**:269–272.
77. Stelzer, M., Neisess, J., Cunningham, J. C., and McPhee, J. R., 1977, Field evaluation of baculovirus stocks against Douglas-fir tussock moth in British Columbia, *J. Econ. Entomol.* **70**:243–246.
78. Stock, M. W., and Robertson, J. C., 1979, Differential response of Douglas-fir tussock moth, *Orgyia pseudotsugata* (Lepidoptera: Lymantriidae), populations and sibling groups to acephate and carbaryl: Toxicological and genetic analysis, *Can. Entomol.* **111**:1231–1239.
79. Stoszek, K. J., Mika, P. G., Moore, J. A., and Osborne, H. L., 1981, Relationships of Douglas-fir tussock moth defoliation to site and stand characteristics in northern Idaho, *For. Sci.* **27**:431–442.
80. Strong, D. R., 1984, Density-vague ecology and liberal population regulation in insects, in: *A New Ecology, Novel Approaches to Interactive Systems* (P. W. Price, C. N. Slobodchikoff, and W. S. Gaud, eds.), pp. 313–327. Wiley, New York.
81. Sugden, B. A., 1957, A brief history of outbreaks of Douglas-fir tussockmoth, *Hemerocampa pseudotsugata* McD., in British Columbia, *Proc. Entomol. Soc. BC* **54**:37–39.
82. Taylor, L. R., Woiwod, I. P., and Perry, J. N., 1978, The density-dependence of spatial behaviour and the rarity of randomness, *J. Anim. Ecol.* **47**:383–406.
83. Thompson, C. G., 1978, Nuclear polyhedrosis epizootiology, in: *The Douglas-Fir Tussock Moth: A Synthesis* (M. H. Brookes, R. W. Stark, and R. W. Campbell, eds.), pp. 136–140, U.S. Dept. Agric. Tech. Bull. 1585.
84. Torgersen, T. R., 1977, Identification of parasites of the Douglas-fir tussock moth, based on adults, cocoons, and puparia, *USDA For. Serv. Res. Pap. PNW*-215.
85. Torgersen, T. R., 1978, Biological evaluations, in: *The Douglas-Fir Tussock Moth: A Synthesis* (M. H. Brookes, R. W. Stark, and R. W. Campbell, eds.), pp. 105–108, U.S. Dept. Agric. Tech. Bull. 1585.
86. Torgersen, T. R., Mason, R. R., and Paul, H. G., 1983, Predation on pupae of Douglas-fir tussock moth, *Orgyia pseudotsugata* (McDunnough (Lepidoptera: Lymantriidae), *Environ. Entomol.* **12**:1678–1682.
87. Torgersen, T. R., Thomas, J. W., Mason, R. R., and Van Horn, D., 1984, Avian predators of Douglas-fir tussock moth, *Orgyia pseudotsugata* (McD.), in southwestern Oregon, *Environ. Entomol.* **13**:1018–1022.
88. Tunnock, S., 1973, The Douglas-fir tussock moth in the northern region—A cartographic history of the outbreaks from 1928–1973, *USDA For. Serv. Div. State and Priv. For. Rep.* 73–27. Missoula, Montana.
89. Turnbull, A. L., 1969, The ecological role of pest populations, in: *Proceedings Tall Timbers Conference on Ecological Animal Control by Habitat Management*, pp. 219–232, Tallahassee, Fla., Feb. 27–28, 1969.
90. Varley, G. C., and Gradwell, G. R., 1960, Key factors in population studies, *J. Anim. Ecol.* **29**:399–401.
91. Watt, K. E. F., 1968, A computer approach to analysis of data on weather, population fluctuations and disease, in: *Biometerology. Proceedings of the 28th Annual Biology Colloquium*, 1967 (W. P. Lowry, ed.). pp. 145–159, Oregon State University Press.
92. Wellington, W. G., 1960, Qualitative changes in natural populations during changes in abundance, *Can. J. Zool.* **38**:289–314.
93. Wickman, B. E., 1963, Mortality and growth reduction of white fir following defoliation by the Douglas-fir tussock moth, *USDA For. Serv. Res. Pap. PSW*-7. Pac. Southwest For. and Range Exp. Stn., Berkeley, California.
94. Wickman, B. E., 1976, Phenology of white fir and Douglas-fir tussock moth egg hatch and larval development in California, *Environ. Entomol.* **5**:316–322.
95. Wickman, B. E., 1978, Tree injury, in: *The Douglas-Fir Tussock Moth: A Synthesis* (M. H. Brookes, R. W. Stark, and R. W. Campbell, eds.), pp. 66–75, U.S. Dept. Agric. Tech. Bull. 1585.
96. Wickman, B. E., 1978, A case study of a Douglas-fir tussock moth outbreak and stand conditions 10 years later, *USDA For. Serv. Res. Pap. PNW*-244. Pac. Northwest For. and Range Exp. Stn., Portland, Oregon.

97. Wickman, B. E., 1978, Phenology of tussock moth, in: *Forest Ecology. McGraw-Hill Encyclopedia of Science and Technology,* 1977 Yearbook, pp. 183–184, McGraw-Hill, New York.
98. Wickman, B. E., 1978, Tree mortality and top-kill related to defoliation by the Douglas-fir tussock moth in the Blue Mountains outbreak, *USDA For. Serv. Res. Pap. PNW*-233. Pac. Northwest For. and Range Exp. Stn., Portland, Oregon.
99. Wickman, B. E., 1980, Increased growth of white fir after a Douglas-fir tussock moth outbreak, *J. For.* **78:**31–33.
100. Wickman, B. E., Henshaw, D. L., and Gollob, S. K., 1980, Radial growth in grand fir and Douglas-fir related to defoliation by the Douglas-fir tussock moth in the Blue Mountains outbreak, *USDA For. Serv. Res. Pap. PNW*-269. Pac. Northwest For. and Range Exp. Stn., Portland, Oregon.
101. Wickman, B. E., Mason, R. R.. and Thompson, C. G., 1973, Major outbreaks of the Douglas-fir tussock moth in Oregon and California, *USDA For. Serv. Gen. Tech. Rep. PNW*-5, Pac. Northwest For. and Range Exp. Stn., Portland, Oregon.
102. Wickman, B. E., Mason, R. R., and Trostle, G. C., 1981, Douglas-fir tussock moth, *USDA For. Serv. For. Ins. and Dis. Leafl.* 86.
103. Williams, C. B., Jr., Wenz, J. M., Dahlsten, D. L., and Norick, N. X., 1979, Relation of forest site and stand characteristics to Douglas-fir tussock moth (Lep. Lymantriidae) outbreaks in California, in: *Dispersal of Forest Insects: Evaluation, Theory and Management Implications* (V. Delucchi and W. Baltensweiler, eds.), pp. 297–307, *Proc. IUFRO Conf.,* Sept. 1978, Zurich and Zuoz, Switzerland.
104. Williams, J. T., Martin, R. E., and Pickford, S. G., 1980, Silvicultural and fire management implications from a timber type evaluation of tussock moth outbreak areas, in: *Proceedings of the Sixth Conference on Fire and Forest Meterology, Society of American Forestry, Bethesda, Maryland,* pp. 191–196.
105. Wright, L. C., Berryman, A. A., and Wickman, B. E., 1984, Abundance of the fir engraver, *Scolytus ventralis,* and the Douglas-fir beetle, *Dendroctonus pseudotsugae,* following tree defoliation by the Douglas-fir tussock moth, *Orgyia pseudotsugata, Can. Entomol.* **116:**293–305.

CHAPTER 11

THE NUN MOTH IN EUROPEAN SPRUCE FORESTS

Broder Bejer

Broder Bejer • Department of Forest Entomology and Ecology, Royal Veterinary and Agricultural University, Copenhagen, Denmark, DK 1870.

1. INTRODUCTION

The Nun Moth, *Lymantria monacha* (Linnaeus) (Lepidoptera: Lymantridae), owes its name to its similarity to another moth, the monk (*Panthea coenobita*), which also has a black and white monk-cloak coloring. It is the classic insect pest of the spruce forests of Europe. Why this is so can be understood by the following description of an outbreak in (then) East Prussia (now Poland–Russia):

> On the 27th of July 1853 the forests were invaded by clouds of moths carried by the south winds from Poland. The swarms spread out and it looked as if the forests had been hit by a snowstorm. The outbreak reached its height in 1855 and the larvae then no more discriminated between conifers and decidious trees. They ate everything and of all ages, even in the afforestations, so at last no green leaf was to be seen. The branches of both pine and spruce were so heavy with the weight of larvae that they drooped towards the ground and the falling of excrement sounded like persistent rain. In June and July 1855 great swarms of parasitic flies and wasps began to show up. They stung the Nun Moth larvae which then became sick and fell from the trees. Lying on the forest floor in a layer more than one foot thick they filled the forests with an abominable stench. When the Nun stopped *Ips typographus* started.

The total calamity of nun moth and the following bark beetle outbreak damaged or destroyed about 403,000 km^2 of forest (about 155,000 sq mi) in Russia and East Prussia between 1853 and 1863.

The nun moth is a tree-killing species and is therefore important to forest owners,

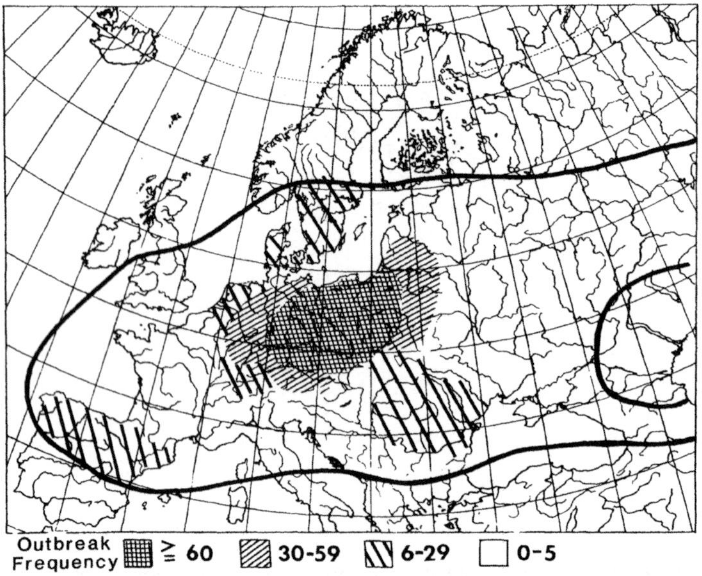

FIGURE 1. Nun moth distribution in Europe and the Frequency of outbreaks, 1885–1985. Distribution after Wilke[31] supplemented from Carter.[9] Frequency from Wellenstein[34], supplemented. No data from U.S.S.R.

even when outbreaks only affect limited areas. In more recent decades, it has given rise to large scale control operations. It defoliates many species of conifers and deciduous trees, but spruces are most severely affected and will not survive defoliation exceeding 70%. Enormous economic losses have resulted from the outbreaks because it has not been possible to harvest, transport, or sell the large quantities of timber before it was rendered worthless by fungi and woodboring insects. In the immense outbreak of 1853–1863 about 147,000,000 m^3 of timber was killed, and this led to the conversion of large areas of destroyed forest to agricultural uses. The forest was permanently lost.

The nun moth distribution covers most of Europe and within a band between N. latitudes 43–57 with isolated populations in Eastern Asia as well.[9] It evades the warmer and colder regions and its main importance as a forest pest lies in Central and Eastern Europe and, more recently, in Spain (Fig. 1).

2. BIOLOGY AND LIFE HISTORY

2.1. Life History

Nun moth adults fly from the end of July to the beginning of September to begin their 1-year life cycle. The distinctive black and white moths (Fig. 2) are most active in the hours around midnight and the males considerably more so than the females. The latter usually sit on the stems to await the male and mating. Afterward the females lay eggs mainly low on the boles of conifers, but if female density is high, egg laying will extend into the upper bole and branches. About 75% of the eggs are deposited within 2 days in clusters containing an average of 40 eggs. Preferred oviposition sites are bark crevices or under lichens on the thick-barked part of the stem. Females lay an average of 120–180 eggs, although the number may vary from about 70 to 300. After depositing most of its eggs, the female may fly more actively.

The embryo completes development 2–6 weeks after egg laying (depending on temperature) and then enters diapause for about 10 weeks. Diapause lasts into mid-winter. From then on further activity depends on temperature. At above 5.4°C the embryo metabolizes its gut yolk and then hatches. Hatching usually occurs in the beginning of May with considerable variation due to temperature and exposure.

After hatching, young larvae rest for hours or days on their egg cluster while eating pieces of the eggshells. They then start moving upward on the stem and spread out in search of newly opened buds. However, such open buds or young shoots are often absent this early in spring and this presents a risk for the larvae. Larvae are unable to use old foliage until the second or third stage (instar), and even then prefer new foliage.

When food is not available, first- and second-instar larvae spin down on silk threads and may then be transported by the wind over considerable distances. Although older larvae are not wind-dispersed, they are able to consume old needles as well as the foliage of other tree species.

There are 5–7 larval instars, the female larvae generally having more developmental stages. Pupation takes place in July, usually on the stem, but in dense populations most pupae are found in the crown. Adults generally emerge on warm, dry days in mid-July to early August. The flight period lasts about 3–5 weeks, while a single individual is

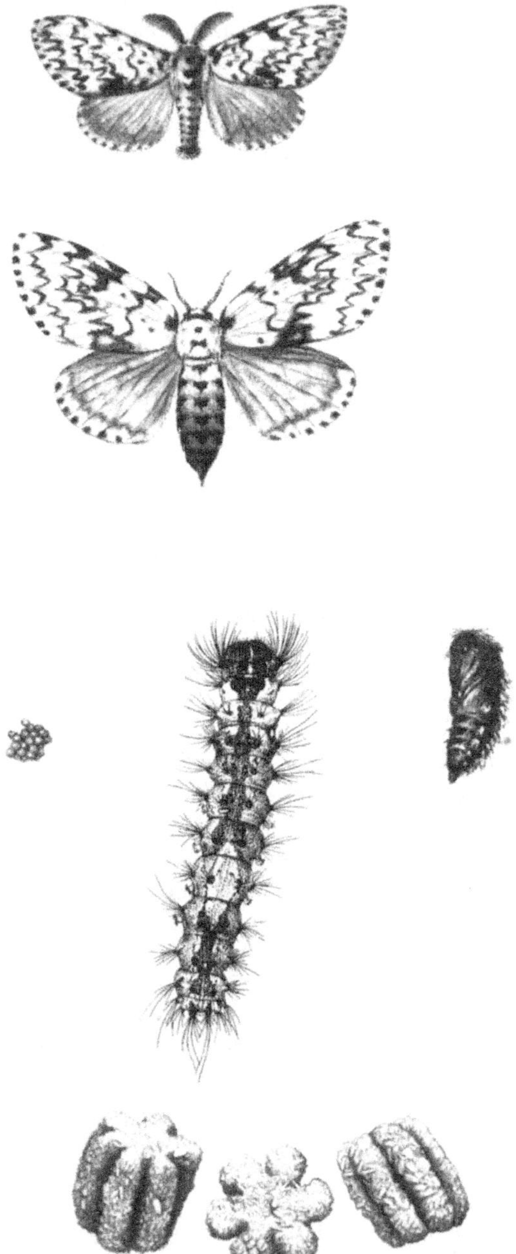

FIGURE 2. The nun moth. Male, female, eggs, larva, pupa, and frass droppings (From Boas.[5,6])

believed to live about 10 days, depending on temperature and constitution. Recent experience with tagged males, however, has shown that males may live up to 24 days and fly up to 3500 m.[25] Unmated females live somewhat longer, while the more active unmated males have a shorter life-span.

2.2. Effects of Host Foliage

Observations in the field show that nun moths can feed and thrive on the foliage of many trees, both coniferous and deciduous. Laboratory investigations of the developmental rates on various tree species have given somewhat contradictory results. One study[21] found development to be fastest on larch and beech, intermediate on oak and spruce, and slowest on pine, while another [15] registered slow development on spruce as compared to beech (Fig. 3). Mortality was lowest when larvae were raised on oak, spruce, and larch; intermediate on beech, and highest on pine. Fecundity was highest on oak and larch, intermediate on beech and spruce, and lowest on pine.[21]

The host spectrum of the nun moth in the field does not fit too well with the result of food preference tests and laboratory rearing. Even though larch, a preferred species, was conspicuous on outbreak localities in northwestern Germany in 1977–1980,[1] large populations always arise in mature stands of spruce and pine. Laboratory tests rank these species as intermediate and low in food value, respectively.

2.3. Adaption to Host Phenology

Newly hatched larvae are dependent on finding new foliage quickly. Even in cool weather, larva can only tolerate starvation for about 10 days. Considerable research has concentrated on the relationships between bud burst of the main hosts, Norway spruce and Scots pine, and hatching of nun moth larvae.

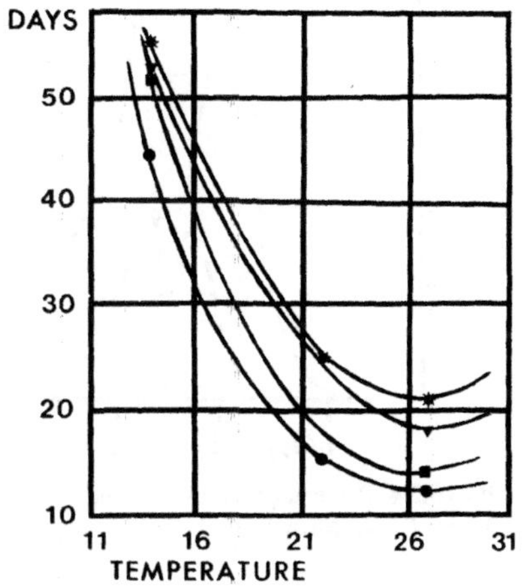

FIGURE 3. Duration of nun moth development on, in order, spruce (top), pine, apple, beech (bottom). (After Schwerdtfeger,[24] from Maercks.[15])

There is considerable variation in the time of spruce bud burst, and the coincidence between larval hatching and host foliage flush is not perfect.[29] In normal years, most larvae hatch in the first week of May; flower buds of early spruce may open on May 10th, and their vegetative buds on May 17th (Fig. 4). Male flowering, when it occurs, is a valuable food source and an important determinant of population success. On very late trees there might be a delay of about 3 weeks. The consequence is that early flushers are most threatened. Larvae that hatch on these trees will remain and, if there are flower buds, they will survive in great numbers. These trees are likely to be severely defoliated.

On late trees, however, larvae tend to descend to the forest floor by spinning silk threads. They then seek out the nearest tree and, in 24 hours, can climb to the top of a high spruce. During this process, however, larvae are exposed to many mortality factors, and great losses are often incurred. This may be more extreme in pine stands, which flush later than spruce. In moderately dense larval populations, the early spruces are more heavily defoliated and endangered than the late trees. In dense populations, however, late-flushing trees are in no way safe because larvae will leave defoliated early trees and move as older larvae to late-flushing ones, which they still have time to defoliate.

2.4. Consequences for the Hosts

It is generally acknowledged that spruce cannot survive 70% defoliation and is often killed by 50% foliage loss. Pine, which flushes its foliage later and has less foliage to catch migrating small larvae, often survives in stands in which spruce is completely destroyed by nun moth feeding. Outbreaks also do not usually last as long in pine stands. For all tree species, the risk of death is related to foliage retention; the fewer years foliage the species normally carries, the better it will survive defoliation.

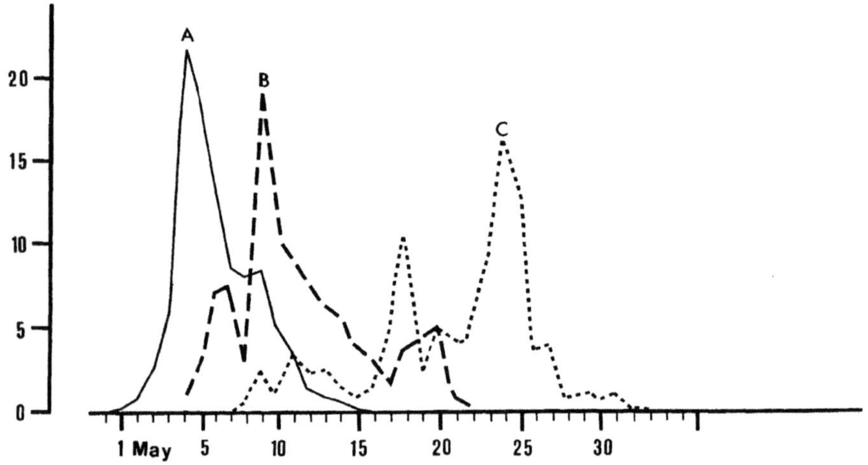

FIGURE 4. Percentage larval hatching (A), flower bud opening (B), and foliage bud opening (C) on spruce. (Adapted from Mors.[17])

2.5. Interactions with Predators, Parasites, and Pathogens

In general, birds are the most important predators of nun moths. Many species are involved; tits are known to feed on the eggs, while woodpeckers and other larger birds will take the larvae. Observations of woodpeckers showed that 94% of the food presented to nestlings were nun moth larvae.[28] During a 75-hr period, about 2200 larvae were brought to the nest, which was equivalent to the population on about six trees.

Traditional nun moth outbreak localities have a low density of breeding birds. In such localities, it is concluded that birds have little influence on nun moth density.

In areas in which outbreaks rarely occur, birds may be crucial in maintaining low nun moth density; for example, nun moths are scarce in deciduous forest where leaves of several tree species are known to be good food. One explanation advanced is that many of these trees—but not oak—have stems with little cover against bird predation on eggs.

Invertebrate predators generally seem to be of little importance in regulating nun moth populations. Ants are a possible exception, as green islands may be left around ant hills in otherwise defoliated forests.[22]

Ichneumon parasites (Hymenoptera; Ichneumonidae) are found attacking nun moth larvae but do not seem effective, as parasitism seldom exceeds 1%.[20] Parasitic flies (Diptera; Tachinidae) are much more significant; one species, *Parasetigena segregata* Rond., parasitizes up to 15–20% of the nun moth caterpillars. However, the rate of increase of this species is rather slow, having but one generation per year; it also experiences considerable losses due to predation by birds and to the death of hosts from starvation and disease.

From our earliest experiences it is known that outbreaks of the nun moth may end with the larvae climbing to the tree tops and dying there; this has been called top disease, or *Wipfelkrankheit* (Fig. 5). The disease, which is caused by a nuclear polyhedrosis virus, seems to play a major role in ending some outbreaks, especially when starvation is also occurring. The virus has been used in control operations and has, in some cases, been aided by the application of so-called stressors; e.g., copper sulfate.[33] Although the nuclear polyhedrosis virus is the most important pathogen, a cytoplasmic polyhedrosis virus, two unidentified bacterioses, and a fungus are also recorded as nun moth pathogens.[34]

2.6. Effects of Temperature

Newly hatched nun moth larvae, which can be obtained all winter, have been used extensively in autecological research. Temperature-dependent activity curves have been constructed (Fig. 6) as well as classic temperature-development rate curves.[39]

Egg production by laboratory-reared nun moths shows the important effects of temperature on fecundity. Optimum temperatures are in the high range, which rarely occur in the field, and there is a steep gradient at temperatures normal for field conditions (Fig. 7). Experience from laboratory rearings also show a similar optimum for female survival.

In general, all laboratory experiments indicate a marked direct response of nun moth to higher temperatures.

FIGURE 5. Virus disease (*Wipfelkrankheit*) on spruce top with accumulating dead and live larvae. (From Boas.[6])

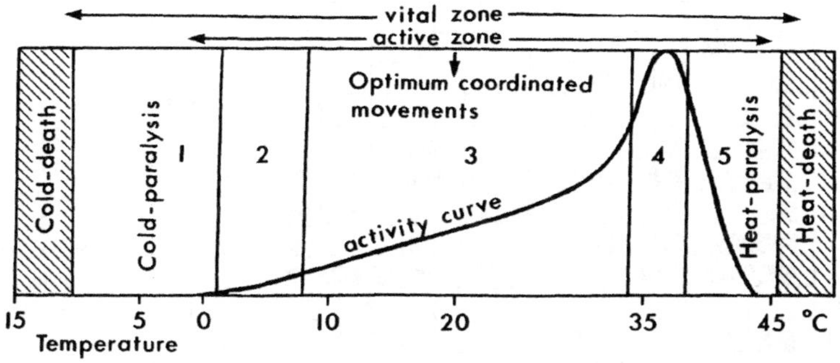

FIGURE 6. Temperature and activity levels of newly hatched nun moth larvae. (Adapted from Schwerdtfeger,[24] from Arnim.[2])

3. PATTERNS OF NUMERICAL POPULATION BEHAVIOR

3.1. Outbreak Zones

Most nun moth outbreaks have occurred in regions of Europe now known as West Russia, but are also frequent in Siberia, Poland, East and West Germany, Czechoslovakia, and Austria (see Fig. 1). Outside this region, outbreaks are less frequent and less intense. The outbreak areas can be described as semicontinental or semiarid (e.g., Spain), with cool to cold winters and warm summers. The outbreak zone generally is bounded by the July isotherm of 16°C and September isotherm of 10.5°C. In Middle Europe, this corresponds to an elevation above sea level of about 800m. Within this temperature zone, outbreaks are more frequent on low precipitation sites (400–700 mm rainfall per year).

There is a strong tendency for outbreaks to occur simultaneously or with a very short delay at different localities which may be quite widely separated (Figs. 8 and 9). Population counts and monitoring with pheromone traps also indicate parallel population trends over wide geographical areas. Migrant moths are normally males, while females fly little until they have deposited most of their eggs. This finding suggests that migration is not a major factor in synchronizing outbreaks.

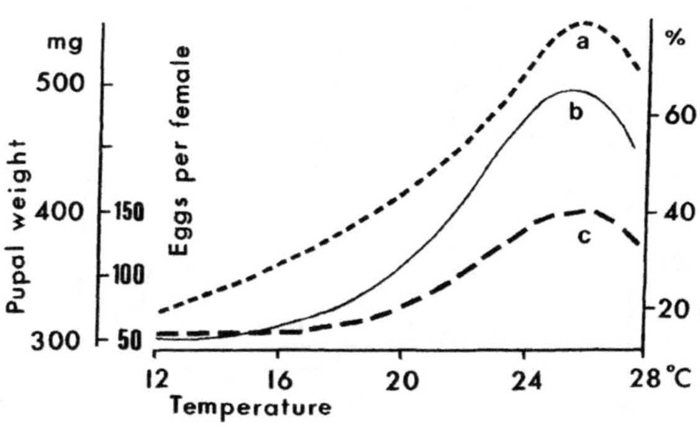

FIGURE 7. Effect of temperature on (a) eggs per female, (b) pupal weight, and (c) percentage females in laboratory rearings of the nun moth, from above. (Adapted from Zwölfer.[38])

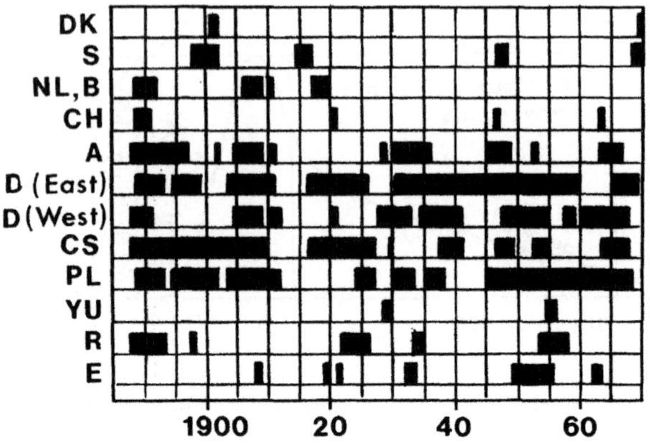

FIGURE 8. Nun moth outbreaks in Europe, 1880–1970. Outbreaks are given per state and not for single localities, thereby hiding some of the simultaneousness. DK, Denmark; S, Sweden; NL, Netherlands; B, Belgium; CH, Switzerland; A, Austria; D, East and West Germany; CS, Czechoslovakia; PL, Poland; YU, Yugoslavia; R, Romania; E, Spain. (Adapted mainly from Wellenstein.[34])

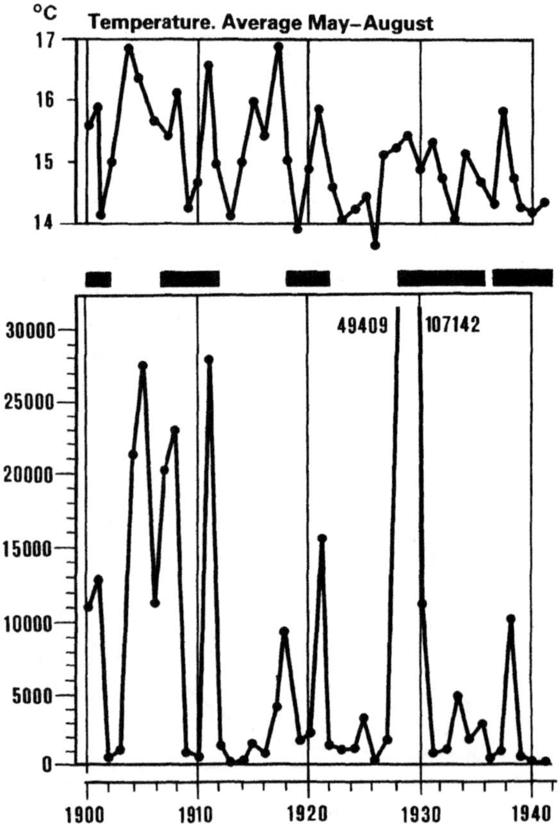

FIGURE 9. Local (endemic) nun moth fluctuation in relationship to summer temperature and outbreaks (bars) elsewhere in Germany. Figures are a sampling of months in a 12,000-ha spruce forest. (Adapted from Wellenstein.[32])

3.2. Effects of Site and Stand Conditions

The preference of the nun moth for conifers has already been mentioned. Scots pine is the major natural species in the high-risk climatic zones, but extensive spruce plantations have been regenerated there beginning in the nineteenth century. These areas are outside the natural range of spruce, which normally occurs in regions of higher precipitation.

It has been proposed that outbreaks are mainly associated with large evenaged monocultures. This is probably inaccurate and is based on the fact that heavy damage only occurs in such situations. Outbreaks are often observed even in small stands of a few hectares, but in these situations damage is of less economic significance.

The preference of the nun moth for old stands generally results in greater damage to old even-aged stands than to uneven-aged ones. In this respect, silviculture based on clear-cutting aggravates the problem.

When the actual locality of outbreaks is analyzed inside a relatively small geographical area with the same type of forest conditions (e.g., Denmark), it becomes clear that meagre sandy soils are the most susceptible sites, while stands on good, fertile, and moist soils are almost completely invulnerable.[3] This observation agrees with the general tendency for outbreaks to occur in the drier regions of Europe where spruce has been planted on sandy sites too xeric for the species.[1]

There are also differences in outbreak intensity that are related to wind exposure, generally being less severe in parts of the forest exposed to the wind. Consequently, outbreak localities are normally flat basins somewhat remote from forest edges.

3.3. Outbreak Patterns

Outbreaks in spruce forests normally last 7 years.[32] During the first 2 years, the presence of the moth is usually overlooked. During year 3, some defoliation and very strong male flight is noticed. During the fourth year, complete defoliation occurs locally, and moth flights increase. Year 5 is the first year of intense defoliation; starvation of old larvae occurs in epicenters, the first signs of virus disease are noted, tachinids move out from the epicenters, and there is some mass emigration of moths. In year 6, the second year of heavy defoliation, the affected areas enlarge, but the intensity of defoliation lessens, virus disease kills many larvae locally, and moth flight is small and restricted to the edge of the outbreak areas. In year 7, many larvae may hatch locally, but they cause little defoliation and soon disappear, often without signs of disease.

In pine forests, often only 4 years are involved instead of the 7 in spruce (3 years of upward trend and 1 year of collapse).

The number of larvae involved in the peak populations is well known from control operations. A mature spruce can bear as many as 30,000 young larvae, when only a few thousand mature larvae would be sufficient to defoliate the tree completely. In the outbreak centers, the numbers of young larvae per hectare may reach 20 million.[3]

During nonoutbreak periods, population density is often estimated at about one pair of adults per hectare. If one assumes a fecundity of 150 per female during the years of outbreak buildup, which is high, and equal numbers of males and females, it will take about 4 years to reach peak levels (75^4). If one assumes 33% mortality, it will take 5 years (50^5).[3]

Pheromone trap catches indicate that the latent population is actually considerably more than one pair per hectare, and is probably quite variable.[12]

There is typically a local expansion of the population during an outbreak (Fig. 10). The direction of expansion usually being with the wind. In centers of the outbreak area, there are fewer pupae per trunk, more dead pupae, a lower proportion of females, and a lower pupal weight (fecundity), probably due to competition for food.

The sex ratio of the moth population changes during the course of the outbreak. During the latent period, the proportion of females in the egg mass is about 50%, and at female emergence from pupae, about 25% (due to lower survival of female larvae). During the early stages of the outbreak, the sex proportion may rise to 80% females, and the average number of eggs laid per female may surpass 200. During peak population levels, the proportion of females has been found to decrease; e.g., in 5 years from 56 to 51 to 46 to 26 to 20, with average fecundity declining simultaneously from 218 to 176 to 149 to 127 to 133.[16]

Egg mortality also increases somewhat (from 4 to 22% in one case), as does mortality in other stages. When total mortality was calculated from intensive population sampling, the result was a total generation mortality rising from 95% (the population is then still expanding) to 99.99% in the fifth year.[31]

During the outbreak, there may be an increasing trend in predation, but this response is not density dependent or is only weakly so. Density-dependent parasitism often occurs but does not generally reach a high level until the year before collapse.[34] Before parasitism can have a decisive effect, however, the moth population invariably collapses from disease due to a nuclear polyhedrosis virus. This is the spectacular *Wipfelkrankheit* (see Fig. 5). The virus is presumed to be always latent in the nun moth population but not to become epidemic until two conditions are fulfilled: (1) the moth population is dense enough to facilitate the spread of virus, and (2) the larvae are in a stressed condition, with many starving from lack of food. Starvation is by itself a major source of density-mortality in the nun moth outbreak epicenters.

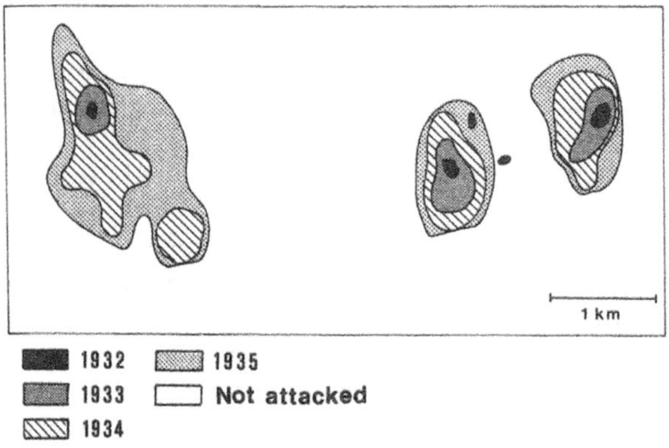

FIGURE 10. Typical expansion of nun moth outbreaks in pure pine forest during the period 1932–1935. (Adapted from Wellenstein.[32])

4. HYPOTHESES FOR THE CAUSES OF OUTBREAKS

4.1. The Occurrence of Outbreaks in Time

There is a tendency for outbreaks to occur at the same time over large though discontinuate geographical areas. This holds both within a particular country and over major parts of Europe, and is most clear at the fringe of the outbreak zone where outbreaks are rare (see Fig. 1). Here outbreaks only occur after long time intervals, e.g., 70 years in Denmark. When these outer-zone outbreaks occur, there always seem to be outbreaks in the inner zone as well, while the opposite is not always the case.

It has been suggested that outbreaks might synchronize with the sunspot fluctuation. Analysis of data for a part of Germany for the period 1810–1970 has, however, disproved this hypothesis.[13]

4.2. Synchrony and Migration

Nun moths are able to fly great distances. Male flights of $\geq$ 80 km have been recorded, and females carrying many viable eggs are known to have flown at least 40 km.[32] However, long-distance flight of gravid females is not the norm, as shown by the local expansion of outbreaks (Fig. 10). Also, the migration hypothesis fails to explain the simultaneous emergence of five outbreaks on small separated Danish localities (1978–1983), when pheromone catches revealed nun moth populations over the whole country and from year to year.[11] This consistent low-density population has also been detected with pheromone traps in other localities.

Consequently, the evidence suggests that outbreaks rise synchronously from local nun moth populations existing far from each other. These synchronous outbreaks are probably initiated by a weather phenomenon that separate populations experience in common, as also is indicated by Figure 5.

4.3. The Role of Weather in Triggering Outbreaks

Because the buildup of defoliation levels takes about 4 years, we should look for factors responsible for initiating outbreaks at this time. An analysis of Danish data for six outbreak periods provides several clues to important weather patterns (Table I):

Summers (June to August): This season is significantly warmer and drier than normal. Temperatures are generally above normal for 2–3 years before the outbreak, and precipitation is below normal 1–4 years prior to the outbreak.
Springs (April to May): This season is significantly cooler than normal.

The actual differences from normal are not very great, and the significance probably lies more in the repetition of several years with certain combinations of temperature and precipitation. These findings are consistent with older German experience.[31]

TABLE I
Deviations from Normal Temperature and Rainfall in the 4 Years before 6 Nun Moth Outbreaks in Denmark[a]

Period	Total	Temperature (>norm)	Rainfall (<norm)
	Months in period[b]		
June, July, August			
All six 4-year intervals	72	46[d,f]	48[d]
In year 1	18	10[e]	11
2	18	14[c]	13
3	18	13[c,e]	12
4	18	9	12
April, May	48	16[c,f]	21[e]
	Years in period		
May to July (period of spruce growth)	24	13	15
June to August (period of moth activity)	24	19[d]	15

[a]From Bejer.[3]
[b]χ^2 test expecting equal numbers above and below the norm.
[c]$p < 0.05$.
[d]$p < 0.01$.
[e]Excludes 1 year identical with norm.
[f]Excludes 2 years identical with norm.

High temperature has a positive effect on nun moth fecundity but the actual increase in fecundity is quite small (Fig. 7). Low rainfall may also affect the insect directly by reducing mortality of larvae and adults. However, the actual reduction is not very great. Consequently, the direct influence of weather on the insect does not seem to be the major factor precipitating outbreaks, and we must search for indirect effects.

Flowering of spruce is induced by warm summers. Repeated warm summers prior to nun moth outbreaks tend to cause increased flowering, thereby creating a food resource that coincides with the critical early larval period.

While warm dry weather in the years preceding outbreaks is rather easily explained as favorable for the insect, the benefit of repeatedly cold springs is less obvious. However, these should be examined in conjunction with the following warmer, drier summer. The result of this combination is the suppression of larval hatching until the weather becomes warm. This again secures the important coincidence of larval hatching and food.

Weather may also have an effect on the nutritive value of the food supply. We have noted a tendency for outbreaks to occur on meager and drought-sensitive soils. Furthermore, there is evidence that the fertilization of such soils, or feeding larvae with needles from fertilized trees, will cause increased larval mortality. It may be possible, therefore, for weather to influence the nutritional or defensive value of trees and needles through lower precipitation during the preoutbreak years. The mechanisms involved have been discussed,[27] and several possibilities have been mentioned; e.g., changes in nutritional value, changes in osmotic pressure, and changes in oleoresin pressure. However, none of these postulated effects has been scientifically proved, thus, we can only conclude that drought acts in much the same way as nutritional deficiencies to diminish the defensive capability of needles.

The quality of the needles as food may vary in a similar way as spruce needles do for the spruce needle miner (*Epinotia tedella*). In this species, the larva is heavier and therefore more fecund and is more successful in overcoming internal parasites when fed on drought-stressed trees.[19]

4.4. The Role of Silviculture in Outbreak Causation

It has been suggested that the nun moth could be a species adapted to deciduous trees that has only recently become a pest of coniferous forests.[35] Original species descriptions are too imprecise to prove this, and the fact that nun moths can be raised on various leaves of deciduous trees does not prove this theory.[34]

Another hypothesis is that the nun moth is a pest created by monoculture. It seems to be true that the incidence of nun moth outbreaks during the past 200 years has remained stable in pine forests but has doubled in spruce forests.[34] This has been taken as evidence that spruce monocultures are the cause. The problem is better addressed by considering several factors: Two trends have occurred over this time in Central and Northern Europe. First, the area of conifer plantations has been greatly increased, partly at the expense of deciduous forests and partly by afforesting of heathland. Spruce has played a major role in the changing forest composition. It is reasonable that outbreak incidence is related, in part, to the enlargement of the susceptible forest area.

Norway spruce grows naturally in higher, cooler, and moister locations than do those to which it has been transferred by forestry. The transference of a species out of its native range, and in many cases onto poorer sites, has also contributed to the increased incidence of outbreaks in spruce forests. Outbreak epicenters are usually old, even-aged spruce stands of fairly large area, but they can also be small stands (e.g., 5–10 ha). The tendency of forestry to move toward monocultures and even-aged stands therefore increases the incidence of moth outbreaks. But the most important factor has probably been the establishment of such large areas of susceptible forest as to make outbreaks a serious affair.

It is true that the area of pine forests has also increased in recent times, but this has occurred on sites more similar to its natural range. Nun moth outbreaks have involved tremendous areas; e.g., about 3 million ha in Poland from 1978 to 1984.[26] But these outbreaks have not lasted as long as in spruce forests. Also, as pine survives defoliation better, not as much attention has been paid to outbreaks in pine monocultures.

In summary, the nun moth seems to be a species that erupts inside a climatically well-defined zone, particularly in large old stands growing on meager drought-sensitive sites. Outbreaks on these sites seem to be triggered by climate acting mainly indirectly through the host. On such sites inside the high-risk zone, moth reproduction can be so great that it "breaks through the environmental resistance" probably created by bird predation and host vitality. Nuclear polyhedrosis virus and starvation are the main factors causing the collapse of outbreaks that have somewhat different durations and effects on the major host species, Norway spruce and Scots pine.

5. IMPLICATIONS FOR DEVELOPING MANAGEMENT STRATEGIES

5.1. Silvicultural Practices

Ideally, the goal of forest protection is to produce a forest that is stable against nun moth outbreaks. As outbreak zones are rather well defined in terms of temperature and precipitation, special attention should be paid to forests inside the zone. It is further evident that within this zone, some localities, especially the meager and drought-sensitive sites, are particularly susceptible.

To reduce the risk of large-scale outbreaks, spruce forests within the outbreak zone and on susceptible sites should either be well dispersed or removed, or the rotation should be shortened. This approach would diminish the areas at risk at any one time and place and would also impede moth expansion to some extent. However, areas as small as 10 ha are still at considerable risk.

Establishment of mixed stands rather than pure spruce would be another precaution to consider. Conifers like Scots pine and larch, and to some extent Douglas-fir, endure defoliation better than spruce. Theoretically, deciduous trees would be even better. The replacement of Norway spruce with provenances with later-flushing buds might also diminish the risk of nun moth outbreaks.

Not unexpectedly, these recommendations are not easy to implement. In Europe, conifer plantings have generally increased through the centuries. There are two main reasons for this. One contributing factor is the planting of many areas with poor unproductive soils typically heathland and moors. On these lands, conifers have hitherto been the only economical alternative, and sometimes the only biological alternative. The other reason is the relative economic competitiveness of conifers as compared to natural deciduous forest. This situation will only change in the event of a new economic balance between softwood and hardwood or by political dictate. Nun moth outbreaks can themselves interfere with the economic balance. With effective chemical control possible, the risk is probably considered relatively low, but the trend is in the direction of increased deciduous forestry, however.

European forestry attempts to develop a good spatial distribution of tree ages and species. On some soils, stands of mixed coniferous species (e.g., spruce-pine) are possible, and in some cases uneven-aged stands can be grown. There are strong forces against this, however: (1) many soils do not permit more than one economic tree species, (2) single old trees or small groups will rarely stand against the strong winds that are common on the continent, and (3) there is a technological and economical tendency toward mechanization and the consequent demand for large stands of the same tree species.

It appears that only moderate changes in silvicultural practice can be expected. As a result, nun moth risk will probably persist, although on meager soils some reduction can probably be achieved by fertilization with nitrogen or nitrogen combined with phosphorus and potassium. Single-tree nutrition achieved by heavy thinning might also have some influence. With the preference of the nun moth for old stands, fertilization and thinning might at the same time lead to shorter rotations and, again, a reduction of risk.

5.2. Outbreak Suppression

As the risk for nun moth outbreaks is likely to continue in the future, possibilities for controlling outbreaks takes on great importance. Practically all scientific effort against the moth during the past four decades has centered on aspects of suppression (and fertilization).

With most of the notorious European forest pest insects, continuous surveys have been carried out, especially in state-owned forests. In many cases, this is done by sampling the soil for pupae (e.g., in the case of *Bupalus piniarius* and *Panolis flammea*).[23] For the nun moth, which remains in the tree, the traditional, and best, method has been to count adult moths on the stems in selected old stands during the flight period. This is done every third day, and the accumulated average number of females per tree is then combined with experience on defoliation. The critical number must be modified according to stand age, yield, site-quality classes, and crown size. They can also be used for pine, which will tolerate somewhat higher moth numbers.

In the case of large outbreaks, the whole forest is scanned by adult counting. Unfortunately, this process may be too time consuming or too late. Instead, winter-sampling of egg masses or pupal holsters or, in the spring, young larvae beneath glued rings on stems has been used. All approaches are very time consuming. Counting pupal holsters gives a good estimate of fecundity, when their diameter is also measured.[18]

During the larval period, frass droppings can be sampled with sticky plates; this gives a good indication of expected defoliation from the second instar on.

Pheromone-trap monitoring using *disparlure* has recently attracted much attention. Apart from the well-known problems of identifying trapping area and a practical trap design, efforts have been directed toward establishing critical trap data. Experience is accumulating[7,25] in nonoutbreak populations with standards of five small traps on a line at 50-m intervals (or in square design). Dispersal of males varies with weather and their abundance is diluted in forests composed of mixed species. Calibration is therefore difficult. Even in the vicinity of outbreaks, catches may decline to background levels within 500 m.[12] This shows the necessity of using short intervals between monitoring traps as is done in orchards.

The main merit of trapping may be to follow the general nun moth fluctuations rather than delimiting actual outbreak risk areas. The general trend combined with climatic data from preceding years should provide an alert for moth flight.

Chemical control became a realistic possibility with the advent of the airplane. The first control operations were carried out against forest pests, and on a great scale against the nun moth, during the mid-1930s in Germany using dosages of about 100 kg of calcium arsenate per ha. This had great impact on wildlife, while dinitro-*o*-cresole, which was another possible choice, had much less effect on wildlife.[31] It was soon found that the operation often had to be repeated in the outbreak epicenters because even a 95% kill still left enough moths to be a threat the following year.

After World War II, DDT was commonly employed against nun moth outbreaks. It was followed by endosulfane and lately, in the enormous outbreak in Poland in 1982, 159 airplanes and 23 helicopters were actively spraying synthetic pyrethroids by both conventional and ultra-low-volume technology.[26]

Recently another type of insecticide, difluron (Dimilin) has been used against *Lymantria monacha* with good results. As a stomach poison, it has advantages in relationship to the tachinids and other parasites. However. it is slow acting (about 5–15 days), which may not be tolerated when larval populations are very high.[4]

The choice of insecticide depends on the degree of defoliation already present from the previous season and defoliation expected in the current season. In high-risk areas, very little feeding can be tolerated, which means that an instantly acting insecticide such as pyrethroids or endosulfane must be used. The former should be preferred due to the strong faunal side effects of endosulfane.[3] In low-risk situations, a slow-acting but effective stomach poison with few side effects, such as difluron, is acceptable.

It is extremely important to achieve control with as little loss of new (current-year) foliage as possible, because tree survival depends to a high degree on this. On the other side, because the small larvae eat only current foliage, it is important to wait for a major part of the new shoots to open. Only if the new shoots are sprayed can a decisive effect be attained. Thus, timing is a compromise between spraying early and not being effective, and spraying too late when too much needle loss has occurred. However, an early spraying with an insecticide (e.g., difluron), which has a residual effect lasting through larval hatching, has proved effective on trees such as larch, which will have flushed before most larvae have hatched, as well as pines, which flush later and where small larvae are forced to feed on old foliage.

Nun moth outbreaks usually expand in area during their first years (Fig. 10). As areas outside the spraying zone usually suffer some defoliation, they are already next year in a higher-risk class, involving the use of stronger insecticides and difficulties with spray timing. It is therefore relevant to consider a preventive spraying outside the risk zone proper even though populations are below the critical defoliation level. Such decisions might be correct in the early outbreak stage and wrong in the late stage, which again calls for refined prognosis. In the outer zone, difluron, but also biorational methods like virus and *Bacillus thuringiensis* might come into use.

5.3. Biological Methods

The nun moth NPV virus has been known for a long time, but it was not until the 1970s that attempts were made to use it in large-scale control operations.[37] There are some major obstacles to this. One is that the virus must be cultivated in living nun moth larvae. These larvae must be raised on artificial diets or started on acorns or wheat bran and ended on spruce foliage. Because this is labor-intensive, production costs are high and practical spraying operations have therefore been restricted. Another problem is that virus-treated larvae usually develop to large larvae or pupae before they die. Thus they consume nearly the normal amount of foliage. Attempts to use stressors such as copper sulfate, *B. thuringiensis,* difluron in conjunction with virus have shown some promise.[33]

The strategic question of whether nun moth larvae are as susceptible in the increase as in the collapse phases has not been answered.

Several strains of *B. thuringinesis* have been tried, especially in the U.S.S.R. and, although most seem to be satisfactory, the data are very inconclusive. Similarly, the effectiveness of commerical *B. thuringiensis* varies from about 75% reduction to practically nil; the latter case probably due to low temperatures.[4] Recent large-scale applications in Poland were relatively ineffective.[26] There are several indications that stresses such as low dosages of insecticides (e.g., pyrethroids, difluron) enhance the effect appreciably.[14]

Mating disruption using pheronomes is another possibility for suppressing nun moth populations. The pheromone of *L. monacha* is estimated to be about 10% (+)-disparlure and 90% (−)-disparlure, while that of *L. dispar* is nearly 100% (+)-disparlure.[10] Commercial disparlure attracts both species and has been used successfully in some instances for mating disruption. A dosage of about 20 g microencapsulated disparlure per hectare has resulted in up to 100% disorientation of males and at least 60% reduction of eggs.[11,30] Although the mating disruption method provides some promise, it is not clear whether it will be effective or economically feasible on large areas. Experience with the related gypsy moth is also inconclusive.[8]

6. CONCLUSION

A major part of the nun moth problem seems to lie in silviculture. More specifically, the creation of monotonous plantations of conifers on dry and meager soils seems to create high-risk conditions, even though outbreaks do occur in almost natural coniferous forests.

It is hoped that ecology, silviculture, and forest entomology will amalgamate in creating a forestry less dependent on (chemical) control of nun moth outbreaks. It is also hoped this will also turn out to be economical in the long run.

ACKNOWLEDGMENTS. The review of earlier drafts of this manuscript by Alexander Isaev and Michael Montgomery is gratefully appreciated. The chapter heads and Figs. 2 and 5 are from original pen and watercolor drawings done by C. Cordts under the supervision of J. E. V. Boas in 1899. I also acknowledge the Paul Parey Publishing House for permission to reproduce and redraw the other figures.

REFERENCES

1. Altenkirch, W., 1986, Die Nonne (*Lymantria monacha* L.) in Nordwestdeutschland 1977–80, *Anz. Schädlingskde., Pflanzenschutz, Umweltschutz* **59:**67–74.
2. Arnim, Frhr. von, 1936, Zur Kenntnis der Temperaturabhängigkeit der Nonne (*Lymantria monacha* L.), *Z. Angew. Entomol.* **22:**533–557.
3. Bejer, B., 1985, Nun Moth (*Lymantria monacha* L.) Outbreaks in Denmark and Their Association with Site Factors and Climate, For. Comm. Res. and Devel. Paper **135:**21–26.
4. Bejer, B., 1986, Outbreaks of Nun Moth (*Lymantria monacha* L.) in Denmark with remarks on their control, *Anz. Schädlingskde., Pflanzenschutz, Umweltschutz* **59:**86–89.

5. Boas, J. E. V., 1899, Nonnen, *Tss. Skovv.* **11**:73–85.
6. Boas, J. E. V., 1923, *Dansk Forstzoologi*, Copenhagen, 1923.
7. Bogenschütz, H., 1982. Monitoring of moth populations harmful to forests, *Z. Pflanzenkrankh.* **89**:586–594.
8. Cameron, E. A., 1979, Disparlure and its role in gypsy moth population manipulation, *Bull. Soc. Entomol. Suisse* **52**:333–342.
9. Carter, D. J., 1984, Pest Lepidoptera of Europe, with special reference to the British Isles, Junk, Dordrecht.
10. Hansen, K., 1984, Discrimination and production of disparlure enantiomers by the gypsy moth and the nun moth, *Physiol. Entomol.* **9**:9–18.
11 Jensen, T. S., and Bejer, B., 1985, Registration and control of the Nun Moth (*Lymantria monacha* L.) in Denmark 1978–84. *Da. Skovf. Tss.* **70**:182–205.
12. Jensen, T. S., and Nielsen, B., 1984, Evaluation of pheromone catches of the nun moth, *Lymantria monacha* L. Effect of habitat heterogenity and weather conditions in the flight period, *Z. Angew. Entomol.* **98**:399–413.
13. Klimetzek, D., 1976, Insektenvermehrungen und Sonnenflecken, *Z. Angew. Entomol.* **95**:226–338.
14. Kruchev, L. T., and Marchenko, Y., 1981, Against the nun moth, *Zashcita Rastenii* **11**:35.
15. Maercks, H., 1935, Der Einfluss der Nahrung auf die Entwicklung der Nonnenraupe, *Arb. Physiol. Angew. Entomol.* **2**:175–195.
16. Mors, H., 1942, Der Nonnenfalter während einer Massenvermehrung, in: *Die Nonne in Ostpreussen* (G. Wellenstein, ed.), pp. 176–206, Paul Parey, Berlin.
17. Mors, H., 1942, Aktivität und Frass der Nonnenraupe in den verschiedenen Jahren ihrer Massenvermehrung, in: *Die Nonne in Ostpreussen* (G. Wellenstein, ed.), pp. 126–175, Paul Parey, Berlin.
18. Mors, H., 1942, Untersuchungen zur Nonnenprognose Wellensteins, in: *Die Nonne in Ostpreussen* (G. Wellenstein, ed.), pp. 535–553, Paul Parey, Berlin.
19. Munster-Swendsen, M., 1987, The Effect of Precipitation on Radial Increment in Norway Spruce (*Picea abies*) and on the Dynamics of a Lepidopteran Pest Insect *J. Appl Ecol.* **24(2)**:551–561.
20. Niklas, O.-F., 1942, Schlupfwespen und Nonne in der Rominter Heide 1933–36, in: *Die Nonne in Ostpreussen* (G. Wellenstein, ed.,), pp. 389–404, Paul Parey, Berlin.
21. Sattler, H., 1939, Die Entwicklung der Nonne, *Lymantria monacha,* in ihrer Abhängigkeit von der Nahrungsqualität, *Z. Angew. Entomol.* **25**:543–587.
22. Schonherr, J., and Ketterer, R., 1979, Zur Frage der kombinierten Anwendung von Polyedervirus und *Bacillus thuringiensis* bei der Nonne, *Lymantria monacha* L. (Lepidoptera), *Z. Pflanzenkrankh.* **86**:483–488.
23. Schwerdtfeger, F., 1941, *Prognose und Bekämpfung forstlicher Grozschälinge,* Reichnärstand Verlag, Berlin.
24. Schwerdtfeger, F., 1963, *Ökologie der Tiere, Autökologie,* Paul Parey, Hamburg.
25. Skuhravy, V., and Zumr, V., 1978, Zur Migration der Nonnenmännchen (*Lymantria monacha* L.), untersucht durch Markierung und Fang an Pheromon-Lockfallen, *Anz. Schädlingskde., Pflanzenschutz, Umweltschutz* **51**:39–42.
26. Sliwa, E., and Sierpinski, Z., 1986, Gradation der Nonne (*Lymantria monacha* L.) in Polen von 1978 bis 1984. *Anz. Schädlingskde., Pflanzenschutz, Unweltschutz* **59**:81–86.
27. Stark, R. W., 1965, Recent trends in forest entomology, *Annu. Rev. Entomol.* **10**:303–324.
28. Steinfatt, O., 1942, Die Beziehungen zwischen Vogelwelt und Nonne in der Rominter Heide anlässlich der letzten Massenvermehrung, in: *Die Nonne in Ostpreussen* (G. Wellenstein, ed.), pp. 405–477, Paul Parey, Berlin.
29. Thalenhorst, W., 1951, Die Koinzidenz als gradologisches Problem, *Z. Angew. Entomol.* **32**:1–48.
30. Vrkoc, L., Skuhravy, V., Hochmut, R., and Zdarek, J., 1981, The disorientation of males with the pheromones of the grey larch moth (*Zeiraphera diniana* Gn.) and the nun moth (*Lymantria monacha* L.) in the conditions of Czechoslovakia, *Khemoretseptsyia Nasekomykh* **6**:130–133.
31. Wellenstein, G., 1942, *Die Nonne in Ostpreussen,* Paul Parey, Berlin.
32. Wellenstein, G., 1942, Zum Massenwechsel der Nonne, in: *Die Nonne in Ostpreussen* (G. Wellenstein, ed.), pp. 207–278, Paul Parey, Berlin.
33. Wellenstein, G., 1973, The Use of Insect Viruses for the Protection of Forests. *Eppo Bull.* **9**:43–52.
34. Wellenstein, G., 1978. *Lymantria monacha* L., in: *Die Forstschädlinge Europas* (W. Schwenke, ed.), pp. 349–368, Paul Parey, Hamburg.

35. Wilke, S., 1931, Über die Bedeutung tier- und pflanzengeographischer Betrachtungsweise für die Forstschutz, *Arb. Biol. Reichanst.* f. *L.* u. *Fw.* **18**:583–675.
36. Willkomm, M., 1864, Die Insectenverheerungen in Ostpreuszen und die durch dieselbe herbeigeführte Umgestaltung der ostpreuszischen Forsten und ihrer Bewirtschaftung. *Jahrb. kön. Sächs. Akad. Forst-Landwirth* z *Tharand*.**9**:161–215.
37. Zethner, O., 1976, Control experiments with the nun moth (*Lymantria monacha* L.) by nuclear-polyhedrosis virus in Danish coniferous forests, *Z. Angew. Entomol.* **81**:191–207.
38. Zwölfer, W., 1934, Studien zur Ökologie, insbesondere zur Bevölkerungslehre der Nonne, *Z. Angew. Entomol.* **20**:1–50.
39. Zwöfler, W., 1935, Die Temperaturabhängigkeit der Entwicklung der Nonne (*Lymantria monacha* L.) und ihre bevölkerungswissenschaftliche Auswertung, *Z. Angew. Entomol.* **21**:333–384.

CHAPTER 12

THE LARCH CASEBEARER IN THE INTERMOUNTAIN NORTHWEST

Garrell E. Long

Garrell E. Long • Department of Entomology, Washington State University, Pullman, Washington 99164-6432.

1. INTRODUCTION

The larch casebearer, *Coleophora laricella* Hubner (Lepidoptera: Coleophoridae), is a small silver-gray moth whose larvae mine needles of larches, *Larix* spp. Apparently of European origin, this defoliator was first recorded in North America near Northampton, Massachussetts, in 1886.[3] It had spread westward to Minneapolis by 1950[12] and, in the spring of 1957, was collected just south of St. Maries, Idaho. A survey of larch forests in the intermountain region that year indicated that about 44,000 ha (170 sq mi) were infested.[1] The population spread rapidly and uniformly until, in 1970, it occupied virtually all larch stands in the intermountain regions of eastern Washington, Oregon, and northern Idaho (Fig. 1).

2. BIOLOGY

Female larch casebearers (Fig. 2) lay about 50 eggs each, one-half of which produce female insects. Eggs are laid on larch needles in late June, usually no more than 1 per

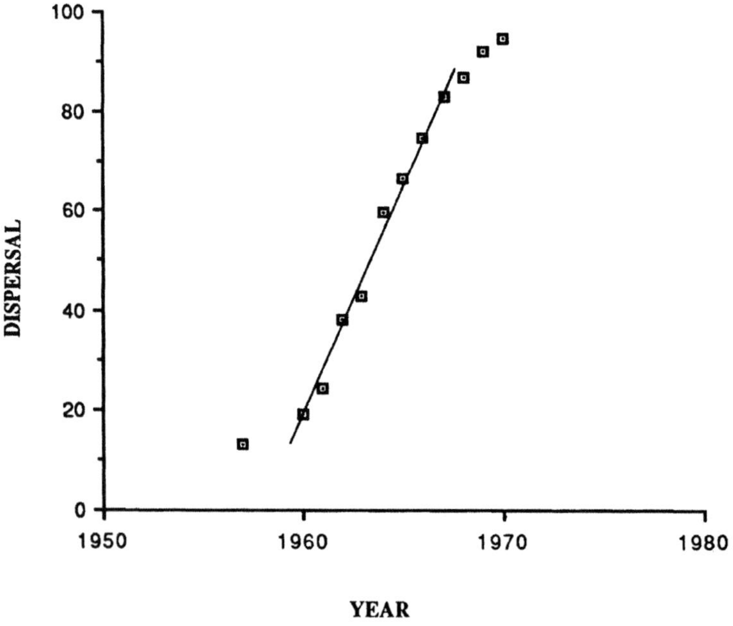

FIGURE 1. Spread of the larch casebearer over northern Idaho (dispersal = square root of hectares occupied). Values were obtained by planimetry on a hand-sketched map made by the USDA Forest Service, Moscow, Idaho. After 1968, larch casebearers were found in virtually all larch stands covered by this survey. The previous 9-year period was used to estimate an annual expansion of the infestation of about 4.6 mi/year. The linear relationship is characteristic of invading species.

FIGURE 2. (*Top left*) Adult casebearer resting on a larch needle. Moths are fairly quiescent during the day but at night, in areas of high infestation, large numbers are attracted to lighted windows. (*Top right*) Overwintering larch casebearer larvae. Under natural conditions, third instars diapause while attached, head down, to larch twigs. In this state, they are apparently less susceptible to low temperatures and dessication. (*Bottom left*) Spring-feeding larch casebearer larvae. Termination of diapause is closely related to larch bud burst. Because the larvae are large with respect to the new needles, feeding damage is most noticeable in early spring. (*Bottom right*) Pupating larch casebearers. Nonparasitized larch casebearers pupate within the foliage of larch spurshoots. Parasitized larvae tend to pupate elsewhere. Adults will emerge from the distal ends of the cases.

needle. When population densities are high, however, females may deposit eggs on needles already occupied, so that as many as five eggs can be found on a single needle.

On hatching, the first instar larva mines directly into the needle and feeds on the mesophyll. As it grows and molts to the second instar, it becomes too large to remain inside the needle and so forms a protective case by fastening pieces of needle epidermis together with silk. It then feeds externally, consuming the needle parenchyma through a hole cut in the epidermis.

As November approaches, the larva molts to the third instar, migrates toward the tip of a branch, and fastens itself to the base of a spurshoot with silk (Fig. 2). The larva overwinters in this stage. Diapause is not obligatory and is terminated at the time larch buds burst in April. The larva then molts to the fourth instar and begins to feed on new needles, consuming the distal portions first (Fig. 2). In light infestations, this leaves a large proportion of each needle intact, but in heavy infestations so much of the needle complement may be consumed that the tree generates a new crop, or second flush, of foliage.

In mid- to late May, fourth instars fasten their cases to the bases of needles in the interior of spurshoots and pupate (Fig. 2). Adults of unparasitized larvae begin to emerge about 2 weeks later.

3. INTERACTIONS WITH HOST TREES

Only 29 insect pests are recorded from western larch, as compared with 235 on Douglas-fir.[2] This finding suggests that western larch may have defenses or deterrents against insect attack. In no-choice feeding experiments, overwintering larch casebearer larvae survived to adult eclosion on western larch, Chinese larch, Douglas-fir, grand fir, Engelmann spruce, lodgepole pine and western white pine but failed to develop on ponderosa pine and western hemlock. Although larch casebearers have been observed on Douglas-firs growing next to heavily infested larches,[11,12] this is extremely rare in nature.

Larch is a deciduous conifer, shedding its needles in early November in the Pacific Northwest. Needle buds, or spurshoots, burst in April with needle elongation often continuing into May. Larch is a self-pruning species, with only those branches receiving sufficient light remaining alive. Heavy defoliation may also cause branches to die, and in severe infestations, tree death and a great deal of deformation due to epicormic branching has been known to occur, e.g., during the early infestations in northern Idaho. During the early 1960s, when the interaction between western larch and the casebearer was new,

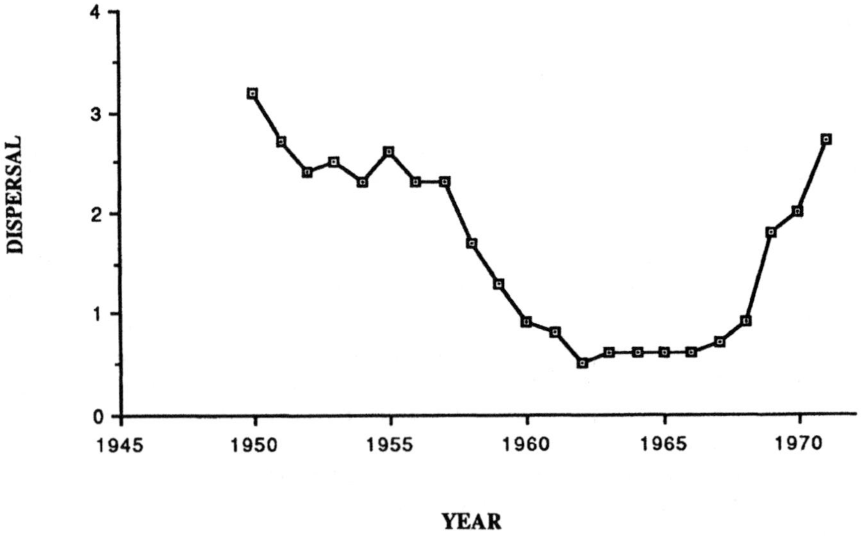

FIGURE 3. Impact of larch casebearer defoliation on the growth of larch trees in northern Idaho. Radial growth increments (mm) of larch trees near the site of discovery decreased significantly from 1957 (the year of discovery) to 1968, while that of nonhost trees from the same general area did not. Larch casebearer densities exceeded 3.5 larvae per spurshoot in 1957. Exotic parasitoids were first released in the area about 1960. By 1968, parasitism by *A. pumila* was 67%, and larch casebearer density had declined to 0.03 larvae per spurshoot on this site. By 1971, the growth rate of larch trees had recovered to normal.

larval densities often reached three to four per spurshoot. Since then, although defoliation has been severe at times, neither direct tree mortality nor casebearer densities in excess of 1.7 larvae per spurshoot have been noted. This pattern is similar to that reported in the eastern United States and suggests that compensatory mechanisms have begun to moderate the effects of the defoliator. Individual larch trees vary in susceptibility to casebearer defoliation, and it seems possible that highly susceptible individuals could have been killed by extreme defoliation during the early period of infestation.

When radial increments of larch trees are compared with those of other species growing nearby, it is apparent that significant growth reductions have occurred during the period of heavy casebearer defoliation (Fig. 3). Relating annual larch casebearer pupal counts to basal area increment growth in the current and succeeding years demonstrated that several year's defoliation were required to reduce basal area growth significantly.[10] For example, it is estimated that 0.01 larvae per spurshoot will decrease basal area increment by 30 sq mm, or about 1.5%, during the following growing season. However, tree-growth rates quickly return to normal when defoliation declines to low levels (Fig. 3).

It has been suggested that repeated defoliation may decrease height growth sufficiently to cause young larch trees to lose their competitive advantage.[9] Preliminary studies do not support this contention, however. Larch casebearer densities are usually low in the upper crown which bears most of the rapidly elongating shoots (lammas). However, the needles of lammas are not fed upon by larch casebearer until the expansion phase is over.[8] The single best predictor of height and basal area growth of western larch is tree vigor, as indicated by growth during the previous season. Casebearer defoliation is but one of the factors affecting this variable.

4. INTERACTIONS WITH NATURAL ENEMIES

As early as 1958, individuals of the native parasitoids *Bracon pygmaeus* Provancher, *Spilochalcis albifrons* (Walsh), and an unidentified member of the genus *Pristomerus* were reared from larch casebearer larvae. In 1960, the exotic braconid species, *Agathis pumila* (Ratzburg), was collected in Rhode Island and released at five locations near St. Maries, Idaho (Fig. 4). This host-specific European parasitoid was first released in Canada in 1938[7] and has been credited with natural control of the casebearer in Europe, the eastern United States, and Canada. Larch branches infested with parasitized casebearer larvae were later removed from these sites and taken to those where *A. pumila* had not yet been released. Since 1964, *A. pumila* has been reared in large numbers, and state and private foresters have participated in collecting and transporting parasitized larch casebearers to new locations in Idaho, Washington, and Oregon. Other releases of exotic parasitoids include *Chrysocharis laricinellae* (Ratzburg) and *Dicladocerus westwoodi* West, beginning in 1972.

Agathis females only parasitize needle-mining first and second instar casebearer larvae. One egg is inserted into each host, and female parasitoids seem to be able to determine whether a larva has previously been parasitized. The female parasitoid inspects larch needles closely, tapping them with her antennae. After a host is found and the egg is placed within it, the parasitoid tends to leave the immediate vicinity rather than searching for host larvae in adjacent needles. Each female can lay 50–80 eggs.

FIGURE 4. *Agathis pumila* inspecting a larch casebearer egg. Female parasitoids fly along branches until a host is detected. Then the spurshoot and needle containing the host are examined by drumming the antennae. An egg is inserted through the needle epidermis and into the young larch casebearer larva.

There is little development of parasitoid larvae within the larch casebearer until late spring. About the time that unparasitized larvae are getting ready to pupate, however, parasitoids begin to develop rapidly and the development of their host is arrested. Adult parasitoids start to emerge in early June with peak emergence occurring in early to mid-July.

In 1972, larch casebearer larvae and pupae parasitized by *C. laricinellae* were imported to the northwestern United States from Austria and England. This eulophid parasitoid only attacks casebearing larvae. It was not thought to be host specific, but here in the northwest it has three generations a year, and their timing is such that no alternate hosts seem to be required.[4] *C. laricinellae* appears to be more abundant than *A. pumila* in the northwest. Several years ago, the reverse was true, but I have found no evidence that competitive displacement was occurring between these two species.

About 35 species of native insect parasitoids have been recovered from larch casebearers in the northwest. Two of them, a chalcid wasp (*S. albifrons*), and a pteromalid wasp [*Mesopolobus verditer* (Norton)], are found consistently and in sufficient numbers to attract interest as potential regulating agents. However, only males of the chalcid emerge from larch casebearer pupae, possibly because the larger females may not be able to complete development within this host.

The parasitoid complex varies considerably between locations and seems to be influenced by site characteristics such as tree and understory composition.[4] Some of the parasitoids, such as *S. albifrons*, apparently require alternate hosts and these might not be available in or near all larch stands. Still others may require a food source, such as aphid honeydew, for the female to survive and perhaps even to oviposit. Lower parasitoid diversity occurs on sites with more homogeneous plant communities, and the disturbance of understory vegetation influences parasitism rates differently among parasitoid species.[6] There is some evidence that parasitism acts as a density-dependent mortality factor on larch casebearer populations.[5]

5. OTHER INFLUENCES

High populations of larch casebearer do not generally develop at elevations much above 1500 m or at latitudes beyond 50 degrees north. There are exceptions, however. For example, I have found single trees with larch casebearer densities as high as 1 larvae per spurshoot at 1800 m elevation in the Blue Mountains of Washington. Temperature does not seem to be a severe limiting factor as casebearer populations north of Moscow, Idaho, were not greatly affected by temperatures of −40°C during December 1968 or −35°C during January 1978. By contrast, samples collected near Boville, Idaho, on April 29. 1981 averaged 0.18 casebearers per spurshoot, but by 26 May only 0.11 remained. The cause, a late frost on May 4th. Although late frosts can certainly reduce spring larch casebearer populations, this effect differs with site aspect and elevation and cannot be considered a significant factor in the regulation of casebearer populations.

Another important mortality factor was observed on the more moist sites, especially in the vicinity of Priest River, Idaho. Larch needle cast diseases, such as *Hypodermella laricis* and *Meria laricis*, effectively eliminated casebearers from heavily infested trees by killing the new needles.

Unlike other insect populations, larch casebearers did not suffer extreme mortality in areas receiving ashfall from the May 18th eruption of Mt. St. Helens in 1980. Many of the larvae had begun to pupate and so were not exposed to the direct abrasive effects of the ash.

6. POPULATION DYNAMICS

In 1976 I began to monitor larch casebearer pupal densities and parasitism on specific trees at known locations in the St. Joe and Priest River National Forests of Idaho. Larch casebearer pupal densities were fairly high for a few years and then, in 1980 dropped to very low levels, where most have remained ever since. During the first years of the study, before the introduced parasitoid population had built up, casebearer densities varied independently on each site. When annual rates of increase are compared with density for these sites and years, a roughly linear scatter is apparent for initial densities above about 8 pupae per 100 spurshoots (Fig. 5a). This suggests that population growth was being regulated by the availability of a resource, such as food in the form of larch foliage. These populations fluctuated dramatically around an equilibrium level, or carrying capacity. of

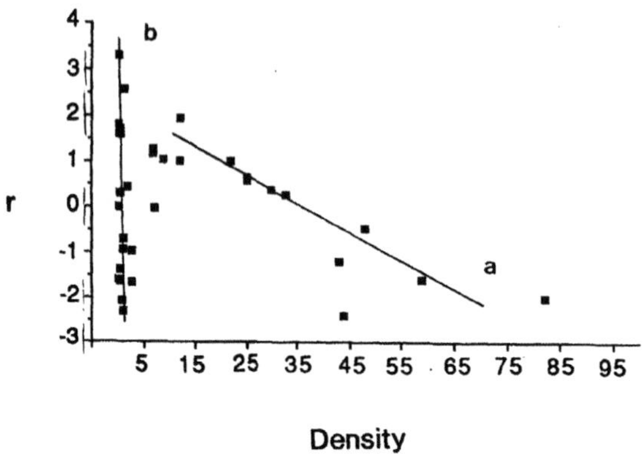

FIGURE 5. The annual per-capita rate of change r of larch casebearer pupal populations expressed as a function of initial pupal density per 100 spurshoots, N_t: $r = \ln(N_{t+1}/N_t)$. During the early stages of the infestation, before exotic parasitoids became established, r was inversely related to the initial pupal density N_t (line a), and an upper equilibrium, or carrying capacity, occurred at about 35 pupae per 100 spurshoots. Later, after exotic parasitoids had built up, the relationship between r and N_t was negative but much steeper, and the equilibrium occurred at a very low density of around 1 pupa per 100 spurshoots (line b).

about 35 pupae per 100 spurshoots. During this period, mortality from native parasitoids was fairly low, and the introduced species were just becoming established. After 1978, however, parasitism by *A. pumila* began to increase rapidly, and larch casebearer pupal densities plummeted. Since that time, casebearer densities have fluctuated around a much lower equilibrium density of about 1 pupa per 100 spurshoots (Fig. 5b).

Because exotic parasites did not build up simultaneously on all plots, it is necessary to examine each plot separately to see the details of the casebearer–parasitoid interaction (Fig. 6). We first note that the casebearer population varied wildly, from 7 to 83 unparasitized pupae per 100 spurshoots, over the period 1975–1980. These oscillations can be fit by a linear logistic model of the form $r = 2 - 0.06 \cdot N_t$, as shown graphically in Fig. 5a. Using this model, and without implying statistical rigor, the pupal density for 1981 was predicted to be $N_{t+1} = \exp(r) \cdot N_t = 9.8$, which is close to the observed value of 11 pupae per 100 spurshoots. The following year, however, the pupal density should have increased again to about 90 per 100 spurshoots. Instead it dropped to 1 pupa per 100 spurshoots and has remained at very low densities thereafter (<1.7/100 spurshoots).

The unexpected decline in the rate of increase of the larch casebearer occurred in the

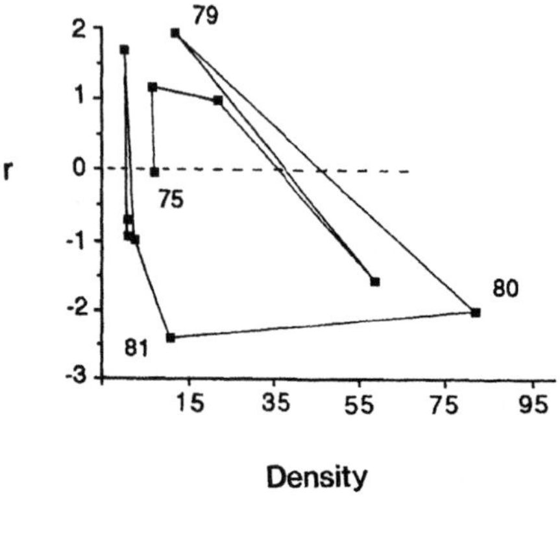

FIGURE 6. Time course of pupal densities N_t and annual per-capita rates of change, r, for larch casebearer pupal populations on a typical plot near Boville, Idaho. The points are connected in chronological order. The density value for 1981 is very nearly what would be predicted based on the sequence 1977–1980, but r was very much lower than expected.

same year that parasitism by the exotic species *A. pumila* and *C. laricinellae* reached peak levels of around 40% (Fig. 7). The obvious inference is that the specific exotic parasitoids reduced the casebearer population to a low density in 1982 and maintained it at this level thereafter by density-dependent responses. This inference is supported by data from other populations, all of which collapsed in different years but always in the year of maximum parasitism or the next year. Unfortunately, however, the data do not support the necessary hypothesis that exotic parasitoids respond in a density-dependent manner to variations in casebearer populations. Without such a feedback response, it is difficult to see how casebearers can be regulated at sparse densities by these parasitoids. Part of this problem may be due to errors inherent in sampling very sparse populations. In the end, it will only be possible to prove whether parasitoids are in fact responsible for controlling casebearer populations by experiments that remove or drastically disturb the parasitoid populations.

The larch casebearer in the intermountain northwest appears to be an example of an organism whose populations have, under natural conditions, two potential equilibrium densities (Fig. 5). The higher equilibrium, or carrying capacity, is about 35 larch casebearer pupae per 100 spurshoots, and is enforced by food limitations. At these densities, parasitoids, predators, and density-independent mortality factors are insufficient to reduce larch casebearer oviposition and survival rates below that required to decouple the resource–recruitment relationship. This equilibrium may be unstable in that the oscillations are undamped; in any event, the casebearer density fluctuates wildly around its equilibrium level (Fig. 6).

When larch casebearer densities are low, the two introduced parasitoids seem to be able to find and parasitize most of the larvae and thereby keep the densities low. This lower parasitoid-maintained equilibrium is about 1 larch casebearer per 100 spurshoots. By implication, there also exists an unstable third equilibrium, or escape threshold, that can be imagined by drawing a line from the bottom of line b in Fig. 5 to the top of line a. If, through immigration or other enhancing effect, casebearer densities exceed this value, the parasitoids will loose control and larch casebearer densities will again be limited by food availability. Whether such an escape threshold exists depends on the efficiency of the density-dependent responses of these parasitoids. If their responses saturate or break down at moderate to high casebearer densities, then an escape threshold is likely to occur.

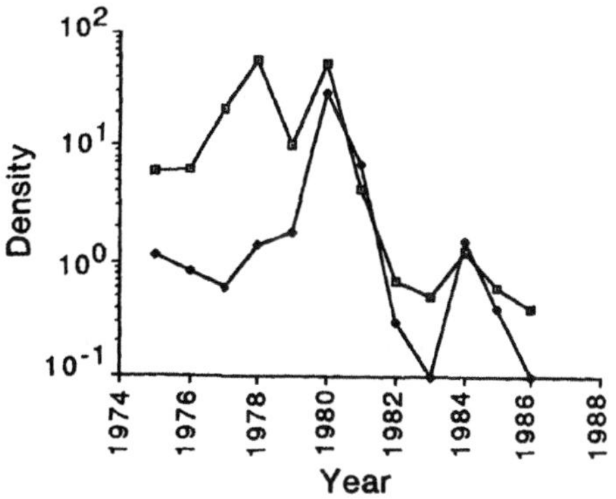

FIGURE 7. Time-series plot of the logarithms of density of living (unparasitized) larch casebearer pupae per 100 spurshoots (open symbols) and percentage parasitism by exotic species (closed symbols) on one plot near Boville, Idaho.

7. MANAGEMENT IMPLICATIONS

It appears that the larch casebearer will not continue to be a commercially significant problem in the intermountain northwest. If the natural controls that now seem to be operating fail, a return to higher oscillating densities can be expected. These will persist for only a few years, unless the disturbance that disrupted the natural control is allowed to continue. Larch, as a seral species, often develops in mixed stands with some natural complement of understory vegetation. Some of the other plant species on the site are important to the maintenance of at lease one of the introduced parasitoids.[6] If clear cultivation is desired, the precise ecological requirements of the introduced parasitoids should be determined, or they may be lost. It is probably unwise to manage larch in pure stands of plantations, unless the trees have been selected for larch casebearer resistance. Similarly, the use of spray programs should be considered carefully since some substances, pyrethrins for example, are particularly detrimental to small Hymenoptera.

Although heavy defoliation still occurs occasionally, individual trees and even stands are not subject to repeated defoliation episodes. While there is no doubt that heavy defoliation can lead to losses in height and radial growth, the apparent lack of concern for larch casebearer in Europe and the eastern United States suggest that the problem will not remain economically important in the Intermountain Pacific Northwest.

REFERENCES

1. Denton, R. E., and Tunnock, S., 1972, Larch casebearer in western larch forests, USDA For. Serv. Leafl., 8 pp., Washington, D.C.
2. Furniss, R. L., and Carolin, V. M., 1977, Western Forest Insects, USDA For. Serv. Misc. Publ. No. 1339.
3. Hagen, H. A., 1886, *Coleophora laricella* Hb, very injurious to *Larix europea* in Massachusetts, *Can. Entomol.* **18**:125–126.
4. Ismail, A., and Long, G. E., 1982, Interactions among parasites of the larch casebearer in northern Idaho, *Environ. Entomol.* **11**:1242–1247.
5. Long, G. E., 1977, Spatial dispersion in a biological control model for larch casebearer, *Environ. Entomol.***6**:843–852.
6. Nathanson, R. A., 1983, Observations on the influence of site characteristics on parasites of larch casebearer, M. S. thesis, Department of Forest Resources, University of Idaho.
7. Quednau, F. W., 1970, Competition and co-operation between *Chrysocharis laricinellae* and *Agathis pumila, Can. Entomol.* **102**:602–612.
8. Ryan, R. B., 1979, The larch casebearer and its parasites. II. The influence of different stages of *Larix occidentalis* foliage on casebearer development and survival, *Can. Entomol.* **111**:477–480.
9. Schmidt, W. C., Shearer, R. C., and Roe, A. L., 1976, Ecology and silivculture of waste larch forests, USDA For. Serv. Tech. Bull. 1520, 96 pp.
10. Tunnock, S., Denton, R. E., Carlson, C. E., and Janssen. W. W., 1969, Larch casebearer and other factors involved with deterioration of western larch stands in northern Idaho, USDA For. Serv. Res. Pap. INT-68, 10 pp, Ogden, Utah.
11. Voute, A. D., 1952, Aantasting van Douglas door de larikswot, *Ned. Boschboum Tijdschr* **24**:244.
12. Webb, F. E., 1953, An ecological study of the larch casebearer, *Coleophora laricella,* Ph.D. thesis, University of Michigan, Ann Arbor, 210 pp.

CHAPTER 13

THE PINE BEAUTY IN SCOTTISH LODGEPOLE PINE PLANTATIONS

Allan D. Watt and Simon R. Leather

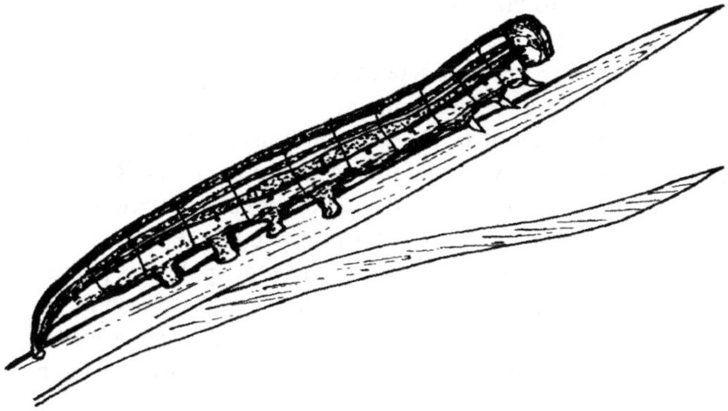

Allan D. Watt • Natural Environment Research Council, Institute of Terrestrial Ecology, Edinburgh Research Station, Penicuik, Midlothian EH26 OQB, Scotland. **Simon R. Leather** • Forestry Commission, Northern Research Station, Roslin, Midlothian EH26 9SJ, Scotland.

1. INTRODUCTION

The pine beauty moth, *Panolis flammea* Denis and Schiffermuller (Lepidoptera: Noctuidae), is a pest of lodgepole pine (*Pinus contorta*) in Scotland. This indigenous species has a widespread distribution throughout the United Kingdom and Europe on its natural host plant, Scots pine (*Pinus sylvestris*). In the British Isles, very low densities of *P. flammea* are normally found in Scots pine plantations, the average density being about 0.1 pupae m^{-2}, and densities above 0.3 pupae m^{-2} being rare.[6] However, serious outbreaks have been recorded on Scots pine in central Europe from 1810 onward,[20] and it is thought likely that *P. flammea* was the pest responsible for serious damage in Nurnberg during 1449–1450.[9] Pine beauty moth outbreaks have also occurred on Scots pine in Sweden,[29] Norway,[4] and Finland.[35]

In 1973, *P. flammea* was recorded on a new host, lodgepole pine, in the Highland Region of Scotland.[39] Lodgepole pine, although introduced into Scotland from Oregon in 1853 and planted on a small scale since 1912, was not planted extensively until the late 1950s and early 1960s.[30] Between 1961 and 1980, however, it was the second most frequently planted tree, so that by 1980 there were 127,000 ha of lodgepole pine in the United Kingdom.[3] The rise in popularity of this species was due to its ability to grow in wet, acidic, and infertile soils, too poor to sustain Sitka spruce without considerable addition of fertilizer.[30] More than 80% of the lodgepole pine in the United Kingdom is in Scotland, where it occupies 13.6% of the land planted with conifers. Moreover, 75% of the Scottish lodgepole pine grows on poor soils in the north and northeast, where it occupies 21% of the land planted to conifers.

The first serious outbreak of *P. flammea* on lodgepole pine occurred in 1976, when 120 ha of 10–19-year-old trees were defoliated and killed in the Rimsdale block of Naver Forest (Highland Region).[38] In 1977, a second outbreak occurred in the Strathy block of Naver Forest, 14 km north of Rimsdale. This prompted the Forestry Commission to carry out a systematic survey of the density of pupae in all lodgepole pine plantations more than 7 years old in the north Highland Region. Based on a control threshold of 15 pupae m^{-2}, 5000 of the 12,000 ha surveyed were found to be at risk and were sprayed with insecticide in 1978. In addition, three outbreaks occurred outside the survey area, one in the Highland Region, one in Bareagle Forest in the extreme southwest of Scotland, and one in Craigellachie Forest in the Grampian Region.

Between 1979 and 1981, a further 4500 ha were sprayed with insecticide (Fig. 1). Populations declined to low levels in 1982 and 1983, but in 1984 two severe outbreaks occurred. One of these, at Pollie Hill (Highland Region), led to the death of about 150 ha of lodgepole pine. In 1985 and 1986, *P. flammea* numbers were again very high in many parts of northern Scotland, and 5000 and 3000 ha were sprayed in those years, respectively. Most of the forest blocks sprayed in 1985 were experiencing their second outbreak in less than 10 years.

Thus, the pine beauty moth, previously an innocuous insect restricted to Scots pine, has become a serious threat to lodgepole pine in Scotland. Its depredations are restricted largely to the north, but this is where most of the lodgepole pine is grown in the United Kingdom and where most new planting is likely to take place.

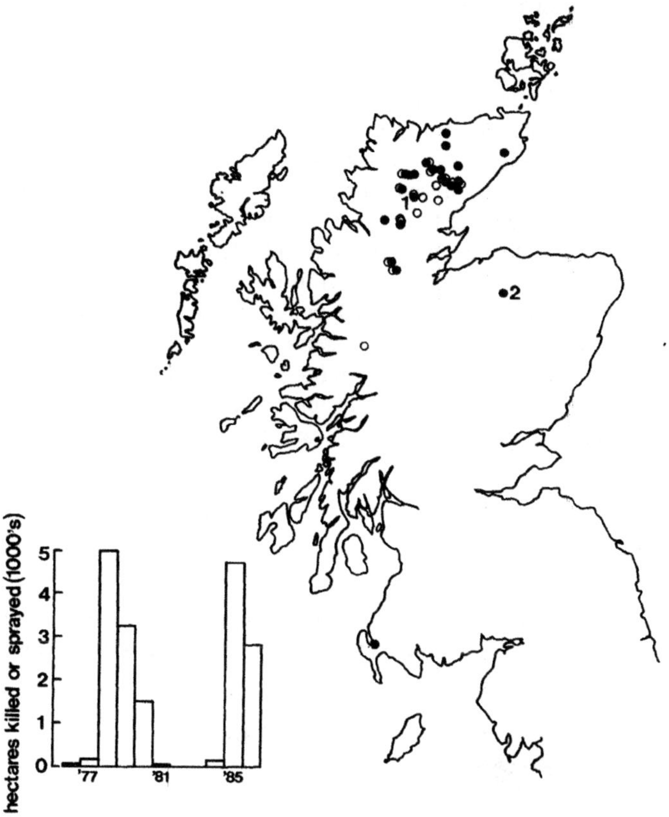

FIGURE 1. Distribution of pine beauty moth outbreaks in Scotland. (●) Outbreaks 1976–1979; (○) outbreaks 1984–1986. Locations of study sites: (1) North Dalchork, (2) Elchies, and the area destroyed or sprayed during 1976–1986.

Although the pine beauty moth is the most important pest of lodgepole pine in Scotland, the European pine sawfly (*Neodiprion sertifer*) also severely defoliates young plantations (prior to canopy closure). This does not cause tree death, only a temporary reduction in growth, but control measures using nuclear polyhedrosis virus appear to be economically justifiable.[11] An additional problem posed by *N. sertifer* is that it may be found at high densities during *P. flammea* outbreaks (e.g., in 1985).

The most serious defoliator of Scots pine in the United Kingdom is the pine looper (*Bupalus piniaria*) (see Chapter 15). Pine looper caterpillars feed on mature pine needles and much later in the year (July to December) than do pine beauty larvae. After a looper outbreak, Scots pine trees that are heavily defoliated will survive the winter but often succumb to secondary attack by the bark beetle, *Tomicus piniperda*. *B. piniaria* also feeds on lodgepole pine as does, among others, *Cinara* spp., *Rhyacionia* spp., *Zeiraphera diniana* (see Chapter 17), and *Diprion pini* (see Chapter 18). As yet, none of these species has caused serious problems to lodgepole pine.

2. BIOLOGY AND LIFE HISTORY

Adult pine beauty moths emerge from their overwintering pupal stages during late March and April (Fig. 2), females emerging slightly before males.[26] Eggs are laid in short rows on pine needles, predominantly in the upper part of the crown, during April and early May. The eggs hatch in late May or early June. There are five larval stages. The young larvae feed on the newly developing needles, and cannot survive on old needles.[43] Older larvae are not only able to feed on needles formed in previous years, they seem to prefer them, so that trees that are almost completely defoliated will often have patches of new needles remaining. At the end of the fifth instar, in late July or early August, the larvae descend from the tree canopy and pupate in the soil in a lightly constructed hibernaculum, normally at the soil–litter interface.

The precise timing of each stage in the life history of the pine beauty moth can vary greatly from year to year. Likewise, the fecundity of *P. flammea* and the mortality that acts on different stages of the life cycle can vary markedly from place to place and from year to year.

2.1. Adult Emergence

The timing of adult emergence is strongly dependent on temperature and, since spring temperatures in northern Scotland vary greatly from year to year, it is not surprising

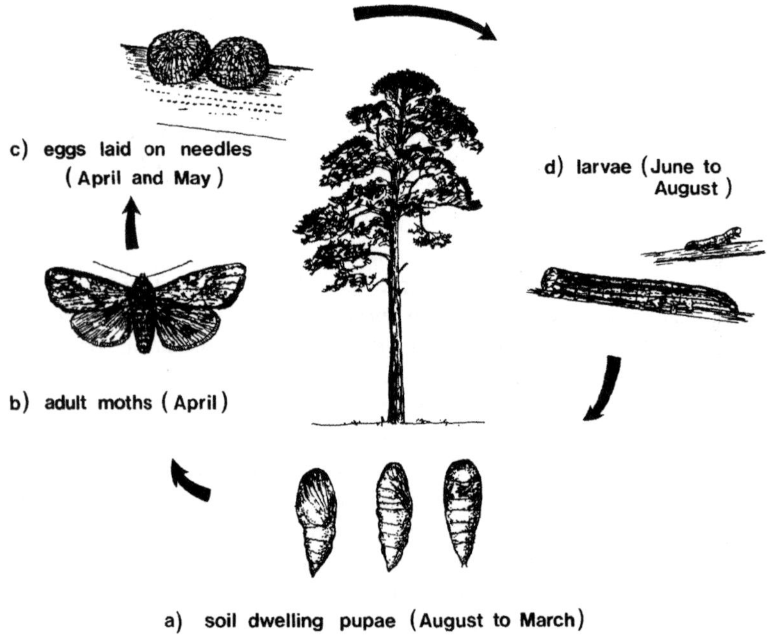

FIGURE 2. Life cycle of the pine beauty moth.

that adult emergence is extremely variable, with flights being seen as early as February 27 and as late as May 23. Pupae of *P. flammea* appear to have an obligatory diapause, which is largely independent of temperature and lasts 120–130 days.[22] Thereafter, the speed and duration of adult emergence are dependent on temperature. If, as seems likely, shallow litter layers warm up faster in the spring, adult emergence should be earlier in stands with sparse litter. This would in turn affect the synchronization of egg hatch and host plant phenology (see Section 2.3). There is conflicting evidence from German studies, with some reports suggesting that emergence is earlier from shallow litter layers, while others show no difference.[16]

2.2. Mating and Egg Laying

The number and pattern of egg laying are affected by temperature, adult feeding, host plant provenance, and the timing of mating. *P. flammea* females may lay up to 430–460 eggs.[13,14,21] Oviposition rates are maximal at temperatures of 15–20°C but drop by about 50% at 10°C and cease at around 6°C. If *P. flammea* adults are provided with food in experimental conditions, their fecundity is considerably higher than when they are starved. For example, in the presence of food, 50% more eggs are produced at 20°C and twice as many are laid at 10°C.[21] It is not known to what extent *P. flammea* actually feeds in the field. They have been observed feeding on willow catkins, but these are very localized within upland forests and generally do not occur in the outbreak areas of northern Scotland.

Periods of inclement weather could lead to considerable reduction in the number of eggs produced by *P. flammea* through their effect on mating (directly, or indirectly by delaying the emergence of males) and on longevity. This prediction is borne out by measurements of egg production in the field. The average number of eggs laid per female at two study sites in northern Scotland over a 5-year period varied between 30 and 123. This variation could be largely explained by spring temperatures[27,41] (Fig. 3).

Lodgepole pine grown in Scotland originates from a number of different areas in

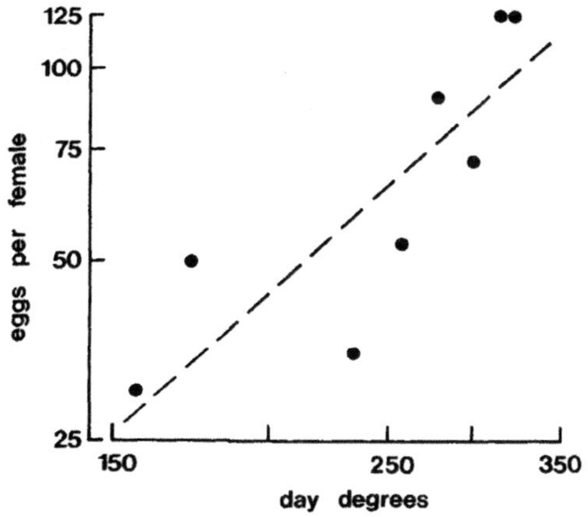

FIGURE 3. Effect of spring temperatures on the fecundity of pine beauty moth. (From Watt.[41])

North America. Each of these seed origins, or provenances, has foliage that is chemically distinct[17] and has characteristic growth forms and habitat requirements.[2] South Costal lodgepole pine, for example, grows vigorously in Scottish upland areas, whereas Alaskan and North Coastal provenances are slower growing in the same areas.[30] *P. flammea* lays more eggs on some provenances than on others.[23,25,27] In a series of laboratory experiments, for example, most eggs were laid on Skeena River, South Coastal, Central Interior, and Southern Interior provenances and least on North Coastal and Alaskan (Table I). This preference was repeated even when the females had no choice, and it appears that egg-laying behavior is strongly affected by the chemical composition of the host plant; specifically higher β/α pinene ratios are preferred.[24]

The previous defoliation history of the plant also affects its acceptability as an oviposition site. Young lodgepole pine that have previously been sublethally defoliated are less preferred than plants of a similar age that have never been defoliated. This has been related to changes in the monoterpene profiles of previously defoliated plants, which have lower β/α pinene ratios.[28]

2.3. Larval Development and Survival

The time of egg hatch depends on when adults emerged (see Section 2.1), the timing of egg laying (see Section 2.2), and egg development, which takes 6.3 days at a constant 20°C and 19.7 days at 10°C. Larval performance is affected by temperature, as well as host species, provenance, and phenology. Development takes 28 days at 20°C, and 71 days at 10°C.[52] The second to fifth instars each last approximately the same time while the first is about 25% longer than the others.

Laboratory tests show that some provenances are better than others for larval growth and survival.[23,25] It is notable that the provenances on which *P. flammea* larvae perform best are the same as those on which adults prefer to lay their eggs (Fig. 4).

Larval survival and growth are strongly affected by the phenology of their host

TABLE I
Influence of Lodgepole Pine Provenance on Number of Eggs Laid by *Panolis flammea* at 15°C

Provenance	Total fecundity[a]	Eggs/tree/moth/day[b]
Central Interior	—	2.6 ± 1.2
Southern Interior	—	2.5 ± 1.4
Skeena River	166.3	1.2 ± 0.6
South Coastal	131.5	1.1 ± 0.5
Southern Interior	114.3	—
Alaskan	72.7	0.7 ± 0.3
North Coastal	65.1	0.6 ± 0.3

[a]When confined to the particular host.
[b]When given a choice of hosts on which to oviposit.

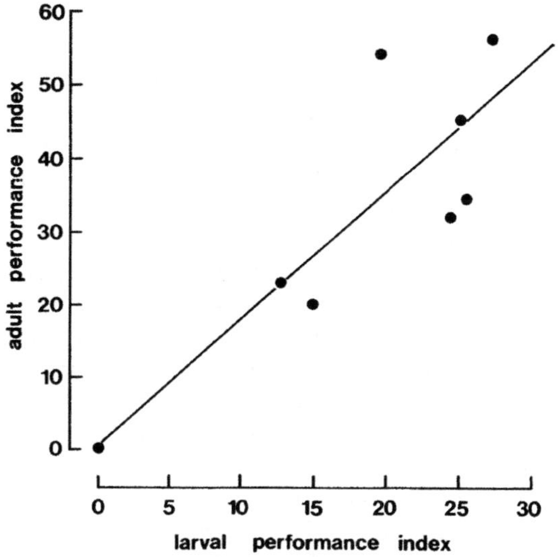

FIGURE 4. Relationship between the preference of adult pine beauty moths and the performance of larvae on different provenances of lodgepole pine.

plants. Newly emerged larvae can only survive on current year's needles, which remain suitable for several weeks, but survival is greatest when new shoots are 50% expanded and needles are just beginning to emerge. Shoot development of lodgepole pine precedes that of Scots pine; consequently, lodgepole becomes suitable for larval establishment before Scots pine.[43] On both hosts, larval survival exceeds 70% for a period of about seven weeks; this is the phenological window within which egg hatch must occur for maximal survival. Early- and late-hatching individuals suffer much greater mortality.

Larval growth and development are also affected by the previous history of defoliation, with growth being slower and survival lower on young trees that have been sublethally defoliated the year before. This apparent induced defense response does not appear to occur in older trees (i.e., >15 years) which may help to explain why outbreaks usually occur in lodgepole plantations more than 12 years old.

2.4. Pupal Survival

The pupal stage of *P. flammea* lasts two thirds of the year. During this time, it is vulnerable to waterlogging, desiccation, and very low temperatures. The severity of these factors is considerably modified by the pupation substrate. Prolonged and excessive waterlogging causes high mortality to pupae,[22] while exposure to −20°C, even for only 1 day, causes 100% mortality. At −10°C, pupal mortality is dependent on the degree of waterlogging. Desiccation is also a threat to pupae, while high humidity is favorable.[22,52]

Extreme temperatures are unlikely to affect *P. flammea* pupae significantly in Scotland, for even when ground surface temperatures fall to −20°C, soil temperatures rarely drop below −5°C. Moreover, snow cover can keep soil temperatures at 0°C regardless of air temperature.[19] Waterlogging is a more important threat to *P. flammea*, since lodgepole pine tends to be planted in surface water gleys and undrained deep peats. However, waterlogging is likely to occur intermittently and, in any case, will only affect certain areas, probably always leaving the ploughed ridges freely drained.

The most serious physical threat to *P. flammea* pupae is desiccation. Pupal survival is better in needle litter than in either soil or peat,[22] but the litter layer must be deep enough so that low humidity conditions, which are less favorable for *P. flammea* pupae, do not develop. Deep litter layers are associated with deep peats and poorly drained soils (in which lodgepole pine tends to be planted) and with younger plantations. Shallow litter layers are associated with sandy mineral soils (in which Scots pine tends to be planted) and with older stands. This may explain why *P. flammea* outbreaks in Scotland tend to occur on lodgepole pine growing in deep unflushed peat;[39] it may also explain, in part, why outbreaks do not occur on Scots pine. In addition, it suggests that *P. flammea* outbreaks, at present occurring on 12–25-year-old trees, may subside as the trees get older.

A similar association has been noted between *P. flammea* pupal survival and certain sites in Germany.[31] The relative humidity within thin litter sites on permeable soils fell below 100% during dry periods but the relative humidity within deep litter layers, also on permeable soils, never fell below 100%. The survival of *P. flammea* pupae was found to be greatest in deep litters. Nevertheless, all forest stands, irrespective of age or site type, can be defoliated during severe outbreaks in Germany, even Scots pine growing among hardwoods in mixed stands.[16]

2.5. Natural Enemies

The pine beauty moth is attacked by a number of parasitoids and predators, as well as by fungal and viral diseases. In the outbreak areas of Scotland, several ichneumonid, braconid, and tachinid parasitoids have been recorded. Most of these attack *P. flammea* larvae but do not kill their hosts until the prepupal or pupal stages. *Meteorus* spp., however, kill their hosts while still in the larval stages. *Aphanistes xanthopus* Schrank is the most abundant parasite of *P. flammea* in Scotland. *Zele albiditarsus* Curtis and *Ernestia* spp. are usually found in significant numbers; *Meteorus* spp. are occasionally very abundant. *Banchus hastator* Fabricius, *Ichneumon septentrionalis* Holmgren, and *Therion circumflexum* Linnaeus have also been recorded.

Such diverse bird species as tits (*Parus* spp.) and ravens (*Corvus corax*) have been seen taking *P. flammea* pupae, adults, and larvae. The fungus *Spiralia farinosa* is responsible for heavy mortality in some years, as is disease caused by a nuclear polyhedrosis virus.

It is probable that our knowledge of *P. flammea* predators, especially invertebrates, is incomplete. An extensive list of natural enemies has been compiled from German studies.[16,36] However, a number of parasites and predators found in Germany are absent either from Scotland or from commercial plantations of lodgepole pine. For example, wood ants (*Formica* spp.) are thought to be significant predators of larvae in central Europe[1,46,47] but, although found in Scotland, they are absent from lodgepole pine plantations.

3. PATTERNS OF NUMERICAL BEHAVIOR

3.1. Temporal Behavior of Pine Beauty Moth Populations

The pine beauty moth problem in Scotland is relatively new, stemming from 1976. During the past 11 years, however, there have been outbreaks in 32 different locations, and two outbreaks have occurred at 12 sites. Population data have been collected at all these locations, and more intensive work has been carried out at two of them: North Dalchork, part of Shin Forest in the Sutherland District of the Highland Region of Scotland, and Elchies, part of Speyside Forest in the Moray District of the Grampian Region of Scotland (see Fig. 1).

Population monitoring by autumn pupal sampling has continued in North Dalchork since 1977 (Fig. 5). In 1978 the pupal density was high enough in parts of the block to warrant control in 1979. The population declined further in 1980 but rose again in 1981 and 1982. The average density for North Dalchork as a whole dropped slightly in 1983 but increased to very high levels in the northcentral part of the forest. A massive increase in the pine beauty moth population occurred over the whole forest in 1984 and, as expected, there was severe defoliation in the northcentral part of the block, killing about 30 ha of lodgepole pine. An insecticidal spraying operation was carried out in 1985. but this was not completely successful and a further 50 ha of forest was destroyed.

In Elchies, complete defoliation and the death of clumps of trees occurred in 1978 and the block was, like North Dalchork, sprayed successfully in 1979. The pine beauty moth population then increased steadily until 1982 but has shown erratic behavior since then (Fig. 5).

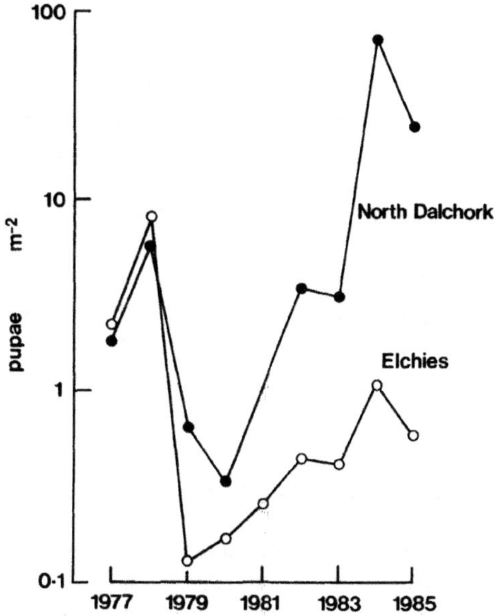

FIGURE 5. Population behavior of *Panolis flammea* at North Dalchork, 1977–1985 (insecticide-spraying operations were carried out in 1979 and 1985), and at Elchies, 1977–1985 (an insecticide-spraying operation was carried out in 1979).

The populations of *P. flammea* in North Dalchork and Elchies did not, therefore, follow the same pattern after 1979. In North Dalchork, two outbreaks have occurred in 7 years, but in Elchies, a second outbreak has yet to occur. The variable population development at these sites is representative of the situation over the whole of Scotland in 1986, with some areas experiencing a second outbreak and others with populations well below economic damage levels. In addition, at a number of sites, *P. flammea* populations have recently reached outbreak levels for the first time.

In addition to the pupal samples from North Dalchork and Elchies, there have been more intensive population studies at both sites, including (1) life-table investigations at North Dalchork from 1981 to 1984, Elchies 1983 to the present, and Borgie (Naver Forest, Caithness District, Highland Region) 1981[5,6,41]; (2) studies of manipulated and natural populations on trees growing in different soils and on different host species in Elchies 1983 to the present[40,41,44]; (3) studies on the effects of soil type, thinning, and fertilizer application in Elchies 1983 to the present; and (4) investigations into adult emergence and dispersal within North Dalchork 1985. These studies indicate that a number of factors are responsible for the temporal patterns of abundance of *P. flammea*, including host plant condition, natural enemies, intraspecific competition, and weather.

3.1.1. Host Plant Condition

The effect of host plant condition on the temporal dynamics of *P. flammea* has been approached in three ways: (1) by analyzing population data, (2) by studying larval survival, and (3) by chemical analysis of plant foliage. No correlation could be found between *P. flammea* density fluctuations and January–March waterlogging,[44] but population growth was found to be correlated with late spring and early summer drought, being highest in years when rainfall was low (Fig. 6). However, the exceptionally low rainfall in 1984 did not result in higher foliar nitrogen concentrations. Phosphorus levels were also the same in 1984 and 1985, but the soluble tannin content was higher in 1984 than in 1985. Moreover, larval survival in cages which excluded natural enemies was the same in 1984 and 1985, despite the differences in rainfall and plant chemistry, and a laboratory study of the performance of pine beauty larvae on artificially stressed trees showed that drought and waterlogging were detrimental to larval growth and survival.[42] Thus, our studies on the influence of host plant condition provide contradictory results and we must conclude that this variable plays no significant role in the temporal dynamics of *P. flammea* populations.

There is, however, one possible exception to this conclusion. Host plant quality has been shown, in the laboratory, to be significantly affected by defoliation and this, in turn, can affect the oviposition behavior and larval survival.[28] To date, there is no good field evidence on this topic but, given that induced chemical defenses have been implicated in a number of other studies, this aspect requires further study.

3.1.2. Natural Enemies

Insect parasitoids attack *P. flammea* at both low and high densities, but the proportion parasitized seems to be related to population density. If both North Dalchork and

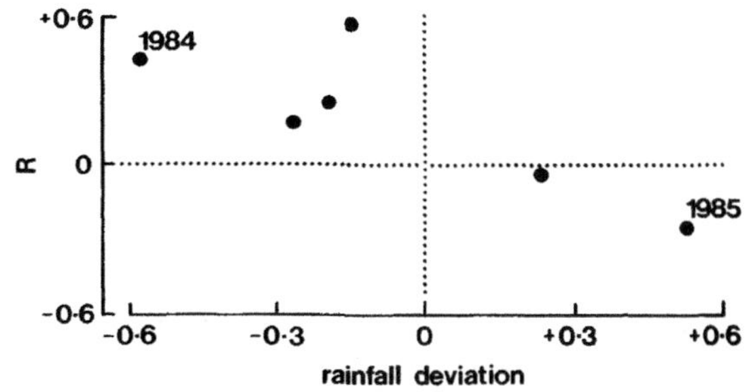

FIGURE 6. Relationship between the population growth of *Panolis flammea* at Elchies $R = \ln N_t - \ln N_{t-1}$) and the deviation from average rainfall in April, May, and June. (From Watt.[44])

Elchies are considered together, pupal parasitism seems to be more significant at lower *P. flammea* densities (Fig. 7); i.e., it acts as an inverse density-dependent factor.

Mortality caused by fungal and viral pathogens is negligible until there are approximately five pupae m^{-2} of soil (Fig. 7). Mortality then increases rapidly but is insufficient to prevent defoliation of the host plant.

The role of predation in the population dynamics of *P. flammea* is difficult to assess. Predators attack all stages of the pine beauty moth, but there is little direct evidence of their impact. Two approaches can be used to study mortality caused by predators: exclusion experiments and life-table data analysis. Exclusion experiments have only recently been initiated; thus, most of our information comes from life tables. Larval mortality data are difficult to interpret because predators usually remove the whole insect, and it is impossible to separate the mortality due to predation from that due to starvation. By contrast, the mortality acting on the pupa is much easier to evaluate.

The number of overwintering pupae lost to predation over winter was extremely variable in the six generations studies.[6] Much of this variability may be attributable to errors in sampling at low population densities. However, the level of predation shows no significant direct relationship to *P. flammea* density.

Predation of eggs and larvae is even more variable than that of pupae but can reach higher levels. Natural enemy exclusion experiments indicate that about 30% of pine beauty moth larvae die before they become established. If this figure is subtracted from

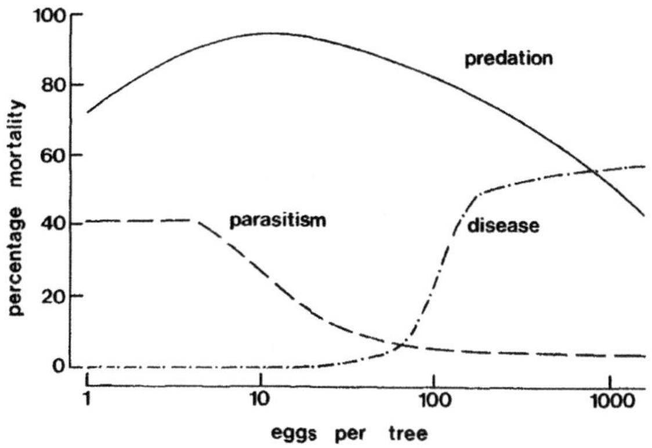

FIGURE 7. Relationship between predation of larvae and other larval mortalities, parasitism of pupae, and fungal and viral disease in pupae, and the density in the egg/young larva stage. (Based on Barbour[5,6] and unpublished data.)

estimates of egg and larval mortality obtained from life tables,[6] a measure of predation can be obtained. This indicates that predation of eggs and young larvae was about 20% in Elchies during 1985.[41] In the fourth instar alone, mortality rose to 61%; in the fifth instar and prepupal stages, it rose to 96%. The total mortality from the egg stage to the start of pupation was 99.3% compared with 56.5% at North Dalchork 1984 (Fig. 8). This suggests that larval predation is much more significant at low *P. flammea* densities. A similar relationship was also observed in 1981 at Borgie, where a small population crashed to a level too low to detect by conventional sampling.[6] However, observations in 1986 suggest that larval predation becomes less significant as *P. flammea* densities drop to very low levels (see Fig. 7).

3.1.3. Intraspecific Competition

Intraspecific competition is important in unsprayed *P. flammea* outbreaks, where complete defoliation of the host plant results in larval starvation. However, there is no indication that competition exerts a significant effect on the reproduction and survival of *P. flammea* in the years between outbreaks. In North Dalchork in 1984, mortality attributed to predation could have been partly caused by starvation, since there were 1500–2000 larvae per tree, but total mortality did not exceed 57%. Similarly, there seems to be no reduction in *P. flammea* fecundity as its population density increases, at least until widespread defoliation occurs. Even then, the number of eggs laid in a defoliated area is largely determined by moth dispersal.

3.1.4. Weather

The influences of temperature, humidity, and rainfall have been discussed in earlier sections. Weather affects fecundity directly or indirectly and it has an impact on several

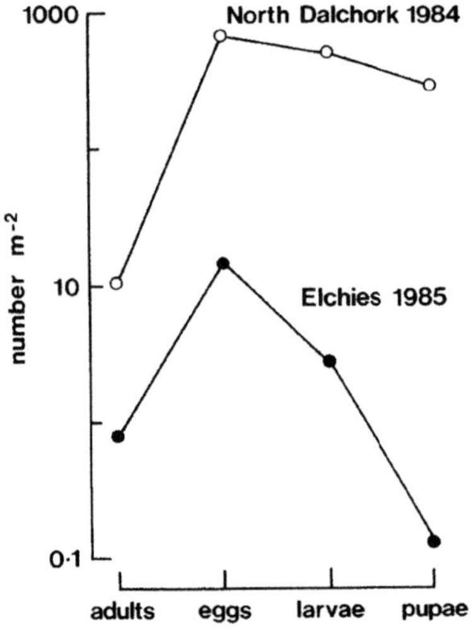

FIGURE 8. Densities of *Panolis flammea* developmental stages at North Dalchork 1984 and Elchies 1985. (From Watt.[41])

aspects of the phenology of *P. flammea*. Weather does not appear to affect the pine beauty moth through its impact on host plant chemistry, but it probably influences the degree of coincidence between egg hatch and the optimum stages of growth of the host plant. Weather may also influence parasitism or disease, and it appears that predation of older larvae and prepupae may be more severe during cool summers because of slower larval development and protracted exposure to predation. This is supported by the fact that unusually high predation on larvae and prepupae occurred during unusually cool summers of 1981 and 1985.

3.2. Spatial Behavior of Pine Beauty Moth Populations

There are two distinct aspects of the spatial dynamics of pine beauty moth populations. First, outbreaks appear to be more likely in some places than in others. In particular, there are associations between outbreaks and certain soil types, and they occur on lodgepole pine rather than Scots pine. Second, once outbreaks start, they may spread rapidly to other areas.

3.2.1. Soil Type

Many attempts have been made to identify factors that make some sites more susceptible to insect outbreaks than others.[8,34] Pine beauty moth outbreaks were also noted to be associated with lodgepole pine stands growing on deep unflushed peat, at least in their early stages.[39] This led to the extension of the Forestry Commission pine beauty moth surveys beyond the initial outbreak areas and was instrumental in discovering an imminent outbreak at Elchies in the Grampian Region.

The association between outbreaks and deep peat may be due to the greater susceptibility of the trees to water stress, so that their foliage is more nutritious for *P. flammea* larvae.[48,49,51] This argument was raised earlier in relationship to the effects of rainfall. Thus, there are both temporal and spatial aspects to the hypotheses that link herbivore population dynamics to plant stress. Some studies implicate both aspects but others suggest that only one is significant, presumably because either weather, on the one hand, or the site conditions, on the other, are always likely to promote stress.

In the case of the pine beauty moth, it has already been argued that variations in rainfall and host plant condition play no significant role in its population dynamics. However, spatial variability in soil type and topography might explain why outbreaks of pine beauty moth occur in certain areas only.

In the Elchies outbreak of 1979, a pupal survey prior to spraying showed that high numbers of *P. flammea* were associated with areas of deep unflushed peat. However, the experimental establishment of populations on trees growing on deep unflushed peat and iron pan soils showed that population growth was not higher on trees growing in deep peat. Other experiments involving the caging of larvae on trees growing on the two soil types also gave negative results. Thus, we failed to support the hypothesis that trees growing in deep peat are of better nutritional quality for *P. flammea* larvae. Other possibilities that might explain why *P. flammea* outbreaks occur in areas of deep unflushed peat are discussed in Section 4.2.

3.2.2. Host Plant Species

Pine beauty moth outbreaks frequently occur on lodgepole pine in Scotland, but never on Scots pine, a native host plant. To investigate this difference we monitored artificially manipulated and natural populations of *P. flammea* and showed that mortality during the egg and early larval instars was higher on Scots pine than on lodgepole pine. However, this difference disappeared in the later instars. Exclusion experiments showed that the survival of *P. flammea* in the absence of natural enemies was marginally (although not statistically) higher on Scots pine than on lodgepole pine. This implies that a larger proportion of *P. flammea* eggs or young larvae are killed by natural enemies on Scots pine or are simply lost during dispersion.

Scots pine and lodgepole pine have different phenologies and different structures, which may lead to differences in the number of young larvae lost while dispersing. Scots pine shoot development lags behind that of lodgepole and, although there is considerable variation in shoot development within individual trees, lodgepole pine becomes suitable for young *P. flammea* larvae before Scots pine and, similarly, ceases to be suitable before Scots pine. Since pine development and egg hatch vary from year to year, it is impossible to generalize about the importance of the different phenologies of the two pine species. On the other hand, the two pines differ consistently in their morphology, Scots pine having many more branches and shoots than lodgepole pine of the same age. All other things being equal, dispersing larvae should have less difficulty finding shoots of Scots pine than those of lodgepole pine. To investigate these factors, sticky traps were placed under lodgepole and Scots pines in Elchies in 1984. Small numbers of larvae fell onto these traps but significantly more were lost from the canopy of lodgepole pine (7.6%) than from Scots pine (4.0%).

We conclude that natural enemies cause greater mortality to the young larvae of *P. flammea* on Scots pine than they do on lodgepole pine. There was no difference in the population development of *P. flammea* during the later instars on the different hosts but, since the mortality during these stages was exceptionally high in 1984 (in the manipulated population) and in 1985, the importance of mortality to young larvae should not be underestimated. Clearly, there is a need to know more about the natural enemies that occur on Scots pine and their role in preventing *P. flammea* outbreaks on this host plant.

3.2.3. Spread of Pine Beauty Moth Outbreaks

Pine beauty moth outbreaks in Scotland are usually controlled by human actions, but on several occasions they seem to have been restricted at first to relatively small areas but then to spread over larger areas. For example, the outbreak in North Dalchork in 1984 destroyed about 30 ha, but the following year 1550 ha had to be sprayed. The development of such an outbreak can be explained in two ways. First, populations may develop in different parts of a forest at different rates, so that what appears to be a spreading outbreak is just a population reaching damaging levels at different times in different parts of the forest. Second, there may be dispersal from areas of high numbers into the surrounding parts of the forest, generating an outbreak in these new regions.

Both processes were investigated in North Dalchork. In 1979, an insecticide spraying operation was carried out in the older parts of the forest block. Thereafter, the density of

P. flammea rose and fell at slightly different rates in different parts of the forest but the overall pattern of change was much the same everywhere, especially in the years when the population increased (Fig. 9). A substantial population increase occurred throughout the area from 1980 to 1984. The rate of increase in 1984 was particularly significant because it was only marginally greater in the defoliated area than in the other parts of the forest. The reason the outbreak centred on a particular part of the forest in 1984 seems to be that pine beauty moth numbers were not reduced to such low densities there during the control projects of 1979 and 1980 (Fig. 9).

In 1985, pine beauty moth populations over most of North Dalchork caught up with that in the area defoliated in 1984. What appeared to be a spreading outbreak was, to a large degree, merely a delay in the buildup rather than spread by dispersal. There is also evidence, however, that dispersal by adult moths also occurred. For example, 181.5 moths emerged per m^2 of soil in the defoliated area of North Dalchork in 1985. but emergence at seven sites 1–3 km distant ranged from 12.4 to 21.9 moths m^{-2} (Fig. 10). There was much less variation in the number of eggs laid per tree, however, and individual females laid fewer eggs in the defoliated region. These results imply that female moths emigrated from the defoliated region, laying, on average, two thirds of their eggs in surrounding areas.

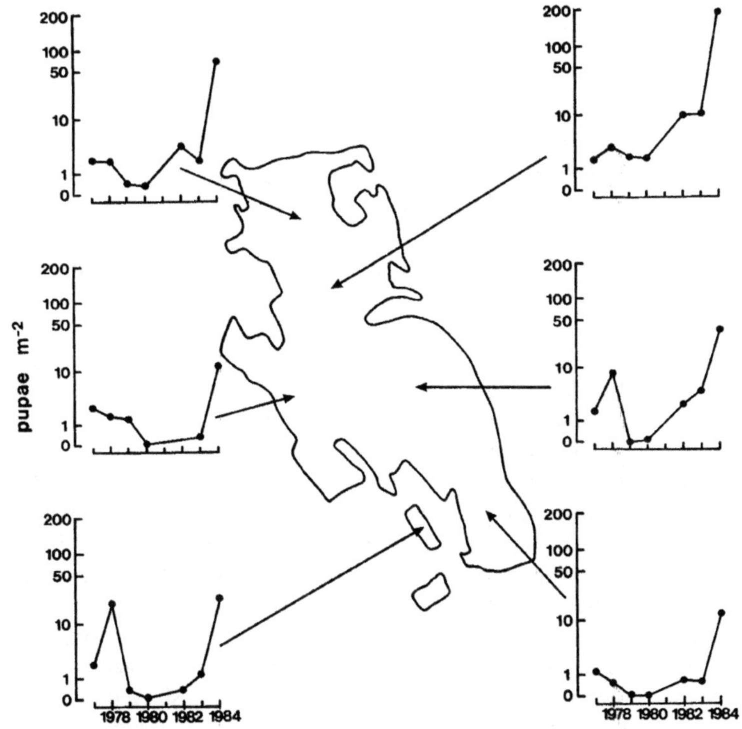

FIGURE 9. Population behavior of *Panolis flammea* in different areas of North Dalchork, 1977–1984.

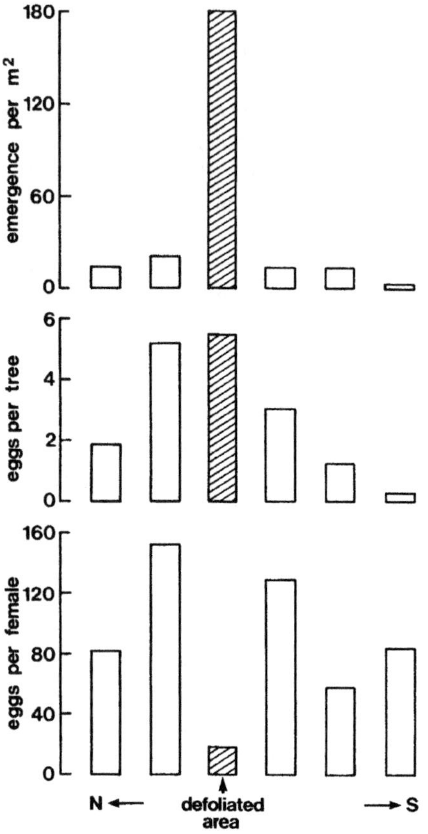

FIGURE 10. Emergence, oviposition, and fecundity of adult *Panolis flammea* in different parts of North Dalchork 1985.

4. HYPOTHESES FOR THE CAUSES OF OBSERVED POPULATION BEHAVIOR

The factors that have a major influence on the dynamics of pine beauty moth populations appear to be natural enemies. intraspecific competition, and weather and site factors, while host plant condition appears to be of minor importance, except through phenological coincidence. The evidence suggests that parasites and predators are more effective at low population densities and on certain sites, while fungal and viral pathogens cause significant mortality at high densities but not enough to prevent host plant defoliation. Pine beauty moth populations collapse as a result of intense intraspecific competition following extensive host plant defoliation. Weather is known to affect egg production and also appears to have an indirect effect on the intensity of predation.

There is no concrete evidence that host plant condition affects pine beauty moth populations. In particular. hypotheses relating to water stress and plant nitrogen are not supported by our experiments. This confirms the view that flush-feeders (e.g., the pine beauty moth) are less likely to be influenced by water stress than are insects feeding on mature foliage (e.g., the pine looper).[50] By contrast, our experiments demonstrate that survival of *P. flammea* is strongly affected by the growth stage of its host plants and laboratory studies have shown that induced plant defenses may be important.

The following sections, examine three hypotheses in an attempt to explain the observed dynamics of pine beauty populations.

4.1. Population Cycles

Pine beauty moth outbreaks appear to occur at fairly regular intervals. However, they are probably not true cycles as found in some other forest insects, such as larch bud moth (Chapter 17); autumnal moth (Chapter 9); Douglas-fir tussock moth (Chapter 10). In the years before an outbreak, for example, pine beauty moth populations do not show a steady increase but tend to rise and fall erratically (see Fig. 5). Population collapse usually results from human intervention, but even uncontrolled outbreaks exhibit dramatic declines following complete defoliation and dispersal of surviving moths.

Although true population cycles do not appear to occur on outbreak sites, it is possible that any tendency toward cyclic behavior is disrupted by spraying or defoliation at outbreak sites but not at nonoutbreak sites. If cycles do occur, there should be evidence of delayed density dependence, which could be caused by natural enemies (see Chapter 15), induced host-plant defenses (see Chapter 9), or a combination of both factors.[10] There is no evidence of delayed density-dependent factors acting on *P. flammea* populations in Scotland. On the other hand, the population behavior of this species on Scots pine in some parts of Germany does appear to be cyclical.[20]

4.2. Resource Limitation (Gradient Outbreaks)

At outbreak sites, pine beauty moth populations usually behave rather erratically. The only exception to this unpredictable behavior is that populations always collapse when densities reach 1000–2000 larvae per tree. The crash is not attributable to predators, parasites, or disease but simply to food shortage, an inevitable consequence of host–plant defoliation. Therefore, *P. flammea* outbreaks could be classified as resource limited,[12,33] with intraspecific competition for food being the only significant density-dependent factor.

According to this hypothesis, the erratic behavior of *P. flammea* populations between defoliation episodes may be seen to be caused by weather variations, moth migrations, or unpredictable variability in the activity of natural enemies. The evidence for the importance of weather variations is particularly good. First, the number of eggs laid by *P. flammea* is influenced by temperatures from March to early May. Second, predation of older larvae and prepupae appears to be correlated with summer temperatures. Although these are the only variables that have been studied intensively, key factor analysis has shown them to be the most variable factors affecting *P. flammea*.[6] A third way in which weather variations can influence populations is through the phenological coincidence of insect and foliage development.

4.3. Eruptive Population Behavior

An alternative hypothesis is that *P. flammea* is an eruptive species with endemic and epidemic domains of behavior (see Chapters 16, 22, and 23). Evidence for this comes from the population studies discussed earlier, which have shown that parasitism is only important at low prey densities, disease exerts significant mortality at high densities, and predation of larvae appears to be most marked at intermediate densities (see Fig. 7). These factors can be considered together with predation of pupae and fecundity to give a series of population change (or recruitment) curves (Fig. 11). The result is strongly affected by fecundity. At average and above-average fecundity, endemic behavior is absent, but population growth rates are notably lower at intermediate population densities. However. when fecundity is below average, an endemic region of population behavior appears. This endemic behavior is mediated by predation and is similar to that reported in a number of other studies.[7,32,37] In addition to the marked dip in recruitment rate caused by predation, there is another less significant decline in recruitment at higher densities of *P. flammea* caused by the combination of parasitism and disease. If the disease-causing pathogens were more efficient at lower host densities or if the pine beauty's parasites were more effective at higher densities, then these factors would combine to create a much more significant effect of the recruitment rate of *P. flammea*.

This model predicts that a period of a few years with average or above-average fecundity is required to trigger an outbreak. Once an epidemic starts, however, the population is unlikely to collapse before defoliating its host plant, unless fecundity is extremely low.

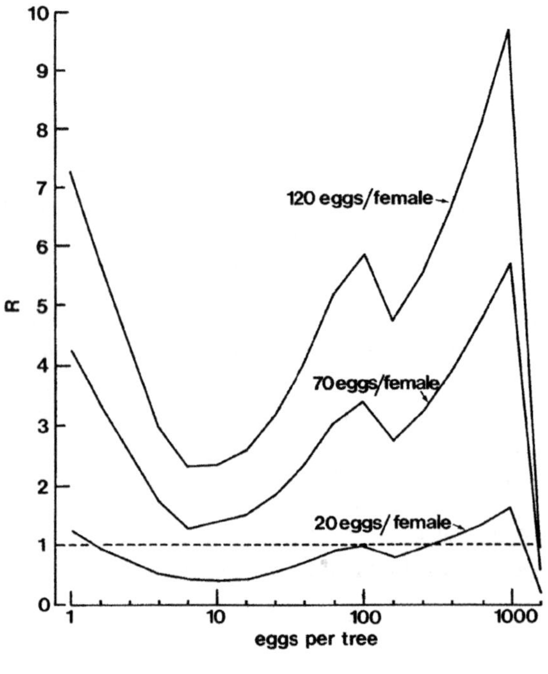

FIGURE 11. Relationship between population growth rate ($R = N_{t-1}/N_t$) and population size at outbreak sites, calculated from the relationships between parasitism, predation and disease, and the density of *Panolis flammea* (as shown in Fig. 7), the effect of larval starvation, and a constant level of mortality to overwintering pupae. The upper and lower lines represent the limits of fecundity in northern Scotland (observed 1981–1985); the middle line represents average fecundity. The large dip in population growth is caused by predation and the smaller dip by the combination of parasitism and disease. When fecundity is below average, a low-density stable equilibrium appears, plus an unstable equilibrium, or outbreak threshold, at prey densities above which natural enemies lose control.

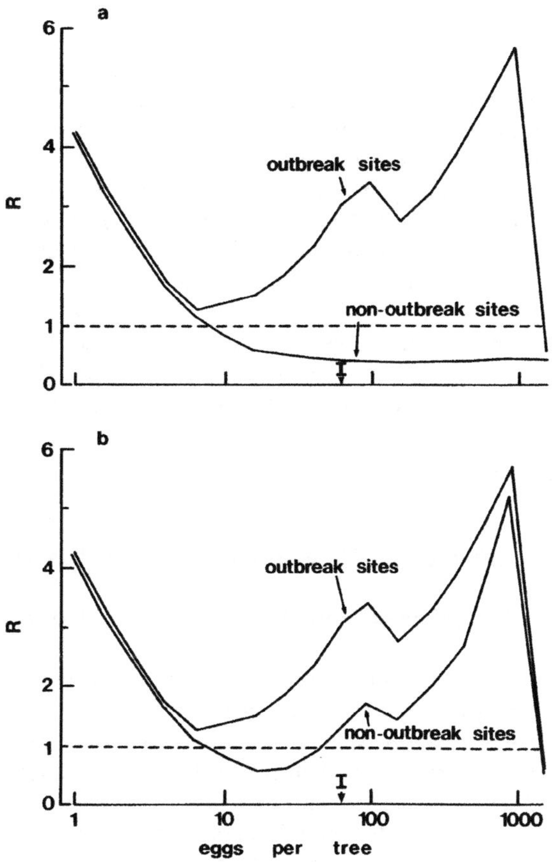

FIGURE 12. Alternative hypotheses for the population behavior of *Panolis flammea* at outbreak and nonoutbreak sites. (a) Resource limitation hypothesis. At outbreak sites, population growth is eventually limited by foliage availability, but at nonoutbreak sites, the population is limited by the availability of pupation sites. If immigration raises the population to *I*, an epidemic will still not occur. (b) Eruptive population behavior hypothesis. At outbreak sites, endemic population behavior is absent unless moth fecundity is very low (see Fig. 14), but at nonoutbreak sites, the action of natural enemies ensures that *Panolis flammea* is held at an endemic level. However, if immigration raises the population to *I*, an epidemic will occur.

4.4. Site Effects

Outbreaks are particularly prevalent in lodgepole stands growing on deep unflushed peat soils but do not occur in Scots or lodgepole pine growing on other soils. This could be because thick litter layers associated with deep peats may promote pupal survival or reduce predation by small mammals, or adult emergence from deep peat soils may be better synchronized with host phenology.

There are also some possibilities that do not relate directly to soil type. For example, some areas of deep peat might act as sinks, collecting moths from surrounding areas because their topography, associated temperature profiles, and wind patterns lead to net immigration of adult moths.[18] Another possibility is that areas of deep peat happen to be in areas in which the parasite and predator complex is less diverse or which are less suitable for the reproduction and survival of these organisms.

Concerning the absence of outbreaks on Scots pine, we have already argued that this is caused by the greater impact of natural enemies on this tree species. It is also possible that the practice of planting Scots pine on different soils leads to its immunity from attack (because, for example, these soils may be less suitable for pupation).

The possible reasons why Scots pine and lodgepole pine in certain areas are free from pine beauty moth attack can be considered within the theoretical framework discussed previously. Populations in Scots pine forests vary erratically. but within a fairly limited range of densities, and do not exhibit outbreak tendencies.[6] This behavior does not support the cyclic outbreak or the food resource limitation hypotheses. It is possible, however, that *P. flammea* numbers are limited by a shortage of another resource, pupation sites, on nonoutbreak sites and by food shortage on outbreak sites, i.e., a two-resource limitation model (Fig. 12a). The third possibility is that the eruptive population behavior model applies to all populations of *P. flammea* but that, under the influence of natural enemies, endemic behavior is more dominant at nonoutbreak sites (Fig. 12b).

Unfortunately, the available data do not allow us to separate the two most likely models. We are left with the conclusion that either a shortage of pupation sites or more effective natural enemies, or both, are responsible for preventing outbreaks on certain sites.

5. IMPLICATIONS FOR DEVELOPING MANAGEMENT STRATEGIES

5.1. Outbreak Prediction and Timing Control Procedures

There is a clear need to control pine beauty moth outbreaks on lodgepole pine, as they always threaten to kill trees over large areas. This contrasts with chronic pests such as the pine sawfly where control costs have to be more carefully weighed against the long-term effects of growth losses. Although pine beauty moth outbreaks are irregular, they are frequent enough to threaten lodgepole pine stands several times during a rotation.

The need for control can be judged by sampling the densities of pupae, adults, eggs, or a combination of these. Combined sampling of pupae and eggs provides the manager with advance warning (from pupal surveys) and accuracy (from egg samples). Another alternative is to assess the adult population by traps baited with synthetic pheromones. Although such systems are difficult to calibrate, they provide a cheap technique for monitoring populations at low densities. For a number of reasons, including accuracy, length of time available, and the need to organize and seek advance clearance for large-scale control operations, the Forestry Commission currently carries out autumn pupal surveys that are later verified by egg surveys. The level used as a threshold for economic damage is 15 pupae m^{-2}, followed by about 1500–2000 eggs per tree. Egg surveys are time consuming and expensive and have to be executed within a 3-week period. One way in which egg assessment could be avoided is to predict egg numbers from the relationship between fecundity and temperature (see Fig. 3), although this would require accurate counts of pupae.

Given our understanding of the effect of weather and other factors on the population dynamics of *P. flammea,* it may also be possible to calculate the risk of outbreaks for one or more years into the future. Although risk assessment obviously becomes less accurate as the forecast period increases, it should be feasible to suspend population monitoring for

1 or 2 years on sites at which the outbreak probability is low, so that greater effort can be concentrated on sites where the risk of an outbreak is high. There seems to be little prospect of developing definitive (rather than probabilistic) forecasts of pine beauty moth outbreaks because of the overriding influence of weather on its population dynamics.

The timing of insecticide application is critical: the chemical should be applied soon after egg hatch, but before significant damage has been done. Similar guidelines are even more relevant to the application of viral insecticides because the sensitivity of larvae to virus infection declines rapidly as they grow older.[15] Control measures are carefully timed by directly monitoring egg hatch, but in the future we may be able to predict egg hatch from phenological models.

5.2. Assessing Alternative Pest-Management Strategies

Pine beauty moth control can be achieved in two ways; (1) by treating outbreaks just before damage occurs; or (2) by creating conditions where outbreaks occur rarely, if at all. An example of the latter approach would be to use provenances of lodgepole pine that are least suitable for the pine beauty moth. The population behavior of *P. flammea* on different provenances has been explored by simulation, and the results suggest that outbreaks could be reduced substantially by planting North Coastal or Alaskan provenances.[45] A compromise between the ideal provenance silviculturally and one relatively resistant to pine beauty moth attack may have to be reached.

The absence of pine beauty moth outbreaks in Scots pine stands and in susceptible lodgepole pine provenances growing on certain sites suggests other ways in which this pest could be managed. For example, if outbreaks were found to be absent from Scots pine because of the action of certain natural enemies, ways should be sought to encourage those natural enemies into lodgepole pine plantations. perhaps by planting tree mixtures, e.g., lodgepole with Scots pine or birch. Any management plan will have costs that will have to be compared with the full costs of control by insecticides. Moreover. insecticides will continue to be needed while the currently growing stands continue to experience pine beauty moth outbreaks. Thus, research has to be undertaken on two fronts. First, it should be designed to seek better ways of predicting outbreaks and timing control measures. Second, it must identify practical alternatives to insecticides, not just because this may provide cheaper control, but also because of the possible side effects of insecticides. Sound pest management requires sound ecological knowledge—if we fail to achieve it, the considerable investment placed in Scottish forestry will be put at risk. Pine beauty moth is a pest only of lodgepole pine, but it would be rash to assume that there will be no new pest problems on this and other trees. The lessons learned on the pine beauty moth may prove to be of general benefit toward the problem of indigenous pests on exotic conifers in the United Kingdom and elsewhere.

ACKNOWLEDGMENTS. We thank D. A. Barbour, B. Bejer, J. P. Dempster, H. F. Evans, N. A. C. Kidd, and J. T. Stoakley for their comments on this chapter; G. M. L. Locke for information on lodgepole pine in Scotland; J. T. Stoakley for unpublished data on the outbreaks of the pine beauty moth; and H. Crick, R. H. Dennis, P. F. Entwistle, and M.

R. Shaw for information on various groups of natural enemies. Numerous others have contributed to and encouraged our work, but among these we must particularly thank D. A. Barbour.

REFERENCES

1. Adlung, K. G., 1960, A critical evaluation of the European research on use of red wood ants (*Formica rufa* Group) for the protection of forests against harmful insects, *Z. Angew. Entomol.* **57:**167–189.
2. Aldhous, J. R., 1976, Lodgepole pine seed zones with reference to British requirements, in: *Pinus contorta Provenance Studies,* (R. Lines, ed.), pp. 6–39, Forestry Commission Research and Development Paper 114, Edinburgh.
3. Anonymous, 1984, *Census of Woodlands and Trees 1979–1982,* Forestry Commission, Edinburgh.
4. Austarå, Ø, 1982, Survey of the pine beauty moth *Panolis flammea* in Norway in 1980 and 1981 using traps with synthetic pheromone analogues, *Fauna Norv. Ser. B* **29:**105–109.
5. Barbour, D. A., 1985, *Panolis* life-table study, Report to the Forestry Commission (unpublished).
6. Barbour, D. A., 1987, Pine beauty moth population dynamics: General considerations and the life table work, in: *Population Biology and Control of the Pine Beauty Moth (Panolis flammea)* (S. R. Leather, J. T. Stoakley, and H. F. Evans, eds.), pp. 7–13, Forestry Commission Bulletin 67, Edinburgh.
7. Berryman, A. A., 1982, Biological control, thresholds, and pest outbreaks, *Environ. Entomol.* **11:**544–549.
8. Berryman, A. A., and Stark, R. W., 1985, Assessing the risk of forest insect outbreaks, *Z. Angew. Entomol.* **99:**199–298.
9. Berwig, W., 1926, Die Forleule in Bayern. Historisch-statistisch-klimatologische Betrachtung, *Fw. Ctrbl.* **70:**165–181.
10. Van den Bos, J., and Rabbinge, R., 1976, *Simulations of the Fluctuations of the Grey Larch Bud Moth,* Wageningen, Pudoc.
11. Britton, R. J., 1985, Effects of defoliation by *Neodiprion sertifer* (Geoff.) on the growth of young crops of *Pinus contorta* Dougl., PhD thesis, University of Edinburgh.
12. Dempster, J. P., and Pollard, E., 1981, Fluctuations in resource availability and insect populations, *Oecologia (Berl.)* **50:**412–416.
13. Eidmann, H., 1928. Eizahl und eireifung einiger forstlich wichtiger Schmetterlinge, *Z. Angew. Entomol.* **13:**549–570.
14. Eidmann, H., 1929, Morphologische und physiologische untersuchungen am weiblichen genitalapparat der lepidopteren. I. Morphologischer teil, *Z. Angew. Entomol.* **15:**1–66.
15. Entwistle, P. F., and Evans, H. F., 1986, Trials on the control of *Panolis flammea* with a nuclear polyhedrosis virus, in: *Population Biology and Control of the Pine Beauty Moth (Panolis flammea)* (S. R. Leather, J. T. Stoakley, and H. F. Evans, eds.), pp. 61–68, Forestry Commission Bulletin 67, Edinburgh.
16. Esherich, K., 1931, *Die Forstinsekten Mitteleuropas,* 3. Paul Parey, Berlin.
17. Forrest, G. I., 1980, Geographical variation in the monoterpenes of *Pinus contorta* oleroresin, *Biochem. Syst. Ecol.* **8:**343–359.
18. Greenbank, D. O., Schaefer, G. W., and Rainey, R. C., 1980, Spruce budworm (Lepidoptera: Tortricidae) moth flight and dispersal: New understanding from canopy observations, radar and aircraft, *Mem. Ent. Soc. Can.* **110:**1–49.
19. Hart, G., and Lull, H. W., 1963, Some relationships among air, snow and soil temperatures and soil frost, U.S. Forest Service Research Note NE-3, 1–4.
20. Klimetzek, D., 1979. *Insekten-Grosschadlinge an Kiefer in Nordbayern und der Pfalz: Analyse und Vergleich 1810–1970,* Institut fur Forstzoologie, Freiburg.
21. Leather, S. R., 1984, The effect of adult feeding on the fecundity, weight loss and survival of the pine beauty moth. *Panolis flammea* (D&S), *Oecologia (Berl.)* **65:**70–74.
22. Leather, S. R., 1984, Factors affecting pupal survival and eclosion in the pine beauty moth, *Panolis flammea* (D&S), *Oecologia (Berl.)* 63:75–79.

23. Leather, S. R., 1985, Oviposition preferences in relation to larval growth rates and survival in the pine beauty moth, *Panolis flammea*, *Ecol. Entomol.* **10:**213–217.
24. Leather, S. R., 1987, Pine monoterpenes stimulate oviposition in the pine beauty moth, *Panolis flammea*, *Entomol. Exp. Appl.* **43:**295–296.
25. Leather, S. R., 1987, Lodgepole pine provenance and the pine beauty moth, in: *Population Biology and Control of the Pine Beauty Moth*, (S.R. Leather, J. T. Stoakley, and H. F. Evans, eds.), pp. 27–30, Forestry Commission Bulletin 67, Edinburgh.
26. Leather, S. R., and Barbour, D. A., 1983, The effect of temperature on the emergence of pine beauty moth, *Panolis flammea* Schiff. (Lep., Noctuidae), *Z. Angew. Entomol.* **96:**445–448.
27. Leather, S. R., Watt, A. D., and Barbour, D. A., 1985, The effect of host plant and delayed mating on the fecundity and lifespan of the pine beauty moth, *Panolis flammea* (Denis and Schiffermuller) (Lepidoptera: Noctuidae): their influence on population dynamics and relevance to pest management, *Bull. Entomol. Res.* **75:**641–651.
28. Leather, S. R., Watt, A. D., and Forrest, G. I., 1987, Insect-induced chemical defences in lodgepole pine (*Pinus contorta*): The effect of previous defoliation on oviposition and growth and survival of the pine beauty moth, *Panolis flammea* (D&S), *Ecol. Entomol.* **12:**275–281.
29. Lekander, B., 1954, Skoginsekternas upptradande i Sverige under tiden 1946–50, *Medd. Stat. Skogforskinst* **44:**1–46.
30. Lines, R., 1976, The development of forestry in Scotland in relation to the use of *Pinus contorta*, in: *Pinus contorta Provenance Studies* (R. Lines, ed.), pp. 2–5, Forestry Commission Research and Development Paper 114, Edinburgh.
31. Meyer, E., 1931, Beobachtungen und untersuchungen zur biologie und bekampfung der forleule, *Z. Angew. Entomol.* **18:**1–56.
32. Peterman, R. M., Clark, W. C., and Holling, C. S., 1979, The dynamics of resilience: Shifting stability domains in fish and insect systems, in: *Population Dynamics* (R. M. Anderson, B. D. Turner, and L. R. Taylor, eds.), pp. 321–341, Blackwell, Oxford.
33. Pollard, E., 1981, Resource limited and equilibrium models of populations, *Oecologia (Berl.)* **49:**377–378.
34. Rhoades, D. F., 1983, Herbivore population and plant chemistry, in: *Variable Plants and Herbivores in Natural and Managed Systems* (R. F. Denno and M. S. McClure, eds.), pp. 155–220, Academic, London.
35. Saalas, V., 1949, *Suomen Metsahyonteiset*, Werner Soderstrom OY, Helsinki.
36. Schwenke, W., 1978, *Die Forstschadlinge Europas*. Vol. 3: *Schmetterlinge*, Paul Parey, Hamburg.
37. Southwood, T. R. E., and Comins, H. N., 1976, A synoptic population model, *J. Anim. Ecol.* **45:**949–965.
38. Stoakley, J. T., 1977, A severe outbreak of the pine beauty moth on lodgepole pine in Sutherland, *Scott. For.* **31:**113–125.
39. Stoakley, J. T., 1979, The pine beauty moth—Its distribution, life cycle and importance as a pest in Scottish forests, in: *Control of Pine Beauty Moth by Fenitrothion in Scotland* (A. V. Holden and D. Bevan, eds.), pp. 7–17, Forestry Commission Occasional Paper #4, Edinburgh.
40. Watt, A. D., 1985, The influence of host plant species and soil type on the population ecology of the pine beauty moth, *Annu. Rep. Inst. Terr. Ecol.* 1984:28–30.
41. Watt, A. D., 1986, The ecology of the pine beauty moth in commercial woods in Scotland, in: *Trees and Wildlife in the Scottish Uplands* (D. Jenkins, ed.), pp. 79–87, Institute of Terrestrial Ecology, Abbots Ripton, Huntingdon.
42. Watt, A. D., 1986, The performance of the pine beauty moth on water-stressed lodgepole pine plants: A laboratory experiment, *Oecologia (Berl.)* **70:**578–579.
43. Watt, A. D., 1987, The effect of shoot growth stage of *Pinus contorta* and *Pinus sylvestris* on the growth and survival of *Panolis flammea* larvae. *Oecologia (Berl.)* **72:**429–433.
44. Watt, A. D., 1987, Pine beauty moth outbreaks: The influence of host species, plant phenology, soil type, water stress and natural enemies, in: *Population Biology and Control of the Pine Beautv Moth* (S. R. Leather, J. T. Stoakley, and H. F. Evans, eds.), pp. 21–26, Forestry Commission Bulletin 67. Edinburgh.
45. Watt, A. D., and Leather, S. R., 1987, Pine beauty moth population dynamics synthesis, simulation and prediction, in: *Population Biology and Control of the Pine Beauty Moth* (S. R. Leather, J. T. Stoakley, and H. F. Evans, eds.), pp. 41–45, Forestry Commission Bulletin 67, Edinburgh.

46. Wellenstein, G., 1957, Die Beeinflussung der forstlichen Arthropodenfauna durch Waldameisen (*Formica rufa* Gruppe). I. Teil, *Z. Angew. Entomol.* **41**:368–385.
47. Wellenstein, G., 1980, Auswirkung hugelbauender Waldameisen der *Formica rufa* Gruppe auf forstschadliche Raupen und das Wachstum der Waldbaume, *Z. Angew. Entomol.* **89**:144–157.
48. White, T. C. R., 1969, An index to measure weather-induced stress of trees associated with outbreaks of psyllids in Australia, *Ecology* **50**:905–909.
49. White, T. C. R., 1974, A hypothesis to explain outbreaks of looper caterpillars, with special reference to populations of *Selidosema suavis* in a plantation of *Pinus radiata* in New Zealand, *Oecologia (Berl.)* 16:279–301.
50. White, T. C. R., 1976, Weather, food and plagues of locusts, *Oecologia (Berl.)* **22**:119–134.
51. White, T. C. R., 1984, The abundance of invertebrate herbivores in relation to the availability of nitrogen in stressed food plants, *Oecologia (Berl.)* **63**:90–105.
52. Zwolfer, W., 1931, Studien zur okologie und epidemiologie der insekten: Die kieferneule, *Panolis flammea* Schiff, *Z. Angew. Entomol.* **17**:475–562.

CHAPTER 14

THE TEAK DEFOLIATOR IN KERALA, INDIA

K. S. S. Nair

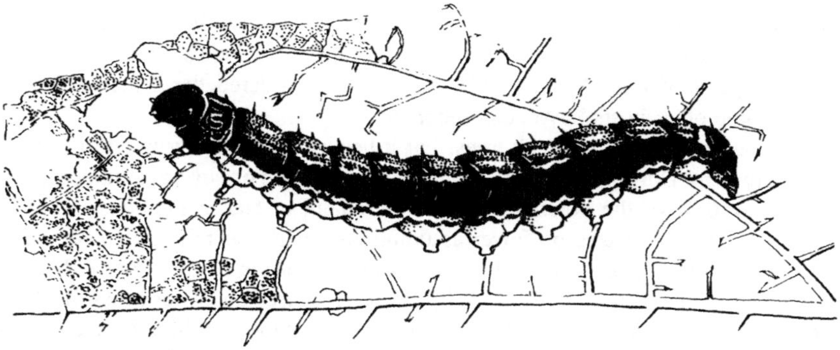

K. S. S. Nair • Division of Entomology, Kerala Forest Research Institute, Peechi 680 653, Kerala, India.

1. INTRODUCTION

Two species of insects are well-known pests of the teak tree (*Tectona grandis*) in India—*Hyblaea puera* Cramer (Lepidoptera: Hyblaeidae), popularly known as the teak defoliator, and *Eutectona machaeralis* (Walker) syn. *Pyrausta machaeralis* Walker (Lepidoptera: Pyraustidae), popularly known as the teak skeletonizer. Larvae of the former feed on the entire leaf, leaving only the major veins, while those of the latter feed only on the green matter, leaving all the veins intact, thus qualifying for the name, skeletonizer. Of the two, *H. puera* is the more serious because it feeds on young leaves during the early part of the growing season, compared with *E. machaeralis*, which feeds on old leaves not long before natural leaf fall. First described in 1794,[14] *H. puera*, originally included in the family Noctuidae, has been recognized as a pest of teak in India for almost a century. This is not surprising, since the first teak plantation in India was raised as early as 1842, in Nilambur, Kerala State. Preliminary information on the life history of *H. puera* and the nature of its damage were published from 1898 to 1903.[10,11,19,30] The species is fairly widely distributed in the tropics: in the Oriental and Australian regions (India, Burma, Sri Lanka, Java, Papua–New Guinea, Cape York Peninsula of northern Queensland in Australia, and the Solomon Islands); in Central America (West Indies); and in Africa (South Africa and parts of East Africa).[3,9,12,14] Teak is considered the principal host plant, although some other hosts have been recorded.

Defoliation does not kill teak trees, but it has usually been assumed to cause heavy loss of increment. During the 1930s, rough estimates based on several assumptions placed the loss at 6–65% of the potential volume increment of teak plantations,[21] but a 1941 estimate of 13% loss, based on fewer assumptions,[3] was generally accepted and quoted extensively in subsequent years. More recent studies at the Kerala Forest Research Institute[24] showed that natural defoliation by *H. puera* caused an average loss of 44% of the potential volume increment in 4- to 9-year-old teak plantations, while *E. machaeralis* had no significant impact on growth. Although it was not possible to quantify the benefit in terms of volume gain over the entire rotation (60 years), we demonstrated that *H. puera* can have a substantial impact on wood production.

Teak is a highly prized constructional and multipurpose timber species. In India, it occurs in nearly 9 million ha of natural forests (moist deciduous and dry deciduous), where it constitutes 10% to nearly 100% of the tree species present, depending on the forest subtype.[28] In addition, plantations have been raised in about 1.4 million ha.[25] It is obvious that control of the teak defoliator can lead to substantial economic gain due to the high value of teak timber and the large area under teak. No control methods are currently employed, although silvicultural-cum-biological methods have been advocated[3] and attempts have been made to standardize chemical methods.[1,29] In our search for lasting and environmentally acceptable strategies for management of this pest, increasing attention is now being focused on its population dynamics. From the scientific point of view, some unique features of the dynamics of this tropical forest pest[22] promise new insights into the theory of insect population ecology, nurtured thus far mainly by examples from temperate insects.

2. BIOLOGY

2.1. The Life Stages

The moths (Fig. 1a) are comparatively small, with a wing span of 3–4 cm, and have a characteristic resting posture that conceals the black-and-orange-yellow hindwings under the grayish-brown forewings. Newly emerged moths can sometimes be found resting on the surfaces of leaves of teak coppice or other shrubs. They are inactive during the day but, when disturbed, fly briskly to adjacent shrubs. Males and females emerge more or less simultaneously, and mating takes place within a couple of days. Eggs are laid over a week-long period starting the third or fourth day after emergence, the longest recorded oviposition period being 12 days.[3] In Burma, where winter temperatures are not cold enough for hibernation, maximum longevity has been recorded at 111 days for males and 48 for females. At Dehra Dun, in northern India, the moths are believed to hibernate for a period of about 3 months during the winter,[3] but no details of the hibernation behavior or of the place of hibernation have been reported.

Eggs are laid on tender new leaves, placed singly near the veins, and usually on the undersurface (Fig. 1b). They are oval, flat, and white and measure about 1 mm in length. About 500 eggs are laid per female with a recorded maximum of 1000.[3] Larvae hatch in about 2 days.

There are five larval instars. The neonate larva eats a shallow depression on the surface of the tender leaf and protects itself with strands of silk. The first and second instars feed mainly on the leaf surface. Starting with the third instar, the larva cuts out a leaf flap, usually at the edge of the leaf, folds it over, fastens it with silk, and feeds from within (Fig. 1c). Fourth- and fifth-instar larvae also feed from within the shelter of leaf folds. The entire leaf, excluding the major veins of tender leaves, is eaten, but more veins are left in older leaves. The early instars cannot feed successfully on old tough leaves and fail to establish when they are given no other food.[3] Under optimal conditions, the larval period lasts 10–12 days, but an average of 21 days has been recorded in November at Dehra Dun. The full-grown larva (Fig. 1d) measures about 3.5–4.5 cm, and there is considerable color variation in the fourth and fifth instars; the body may be either wholly black or dark grayish to black, with longitudinal colored bands that may include a dorsal orange or ocherous band and lateral white lines. The dark and light forms occur together in the same population, with the darker forms predominating during epidemics. The color difference is so prominent that the black form was originally designated as *H. puera* var. *nigra*.[30]

Following heavy defoliation, the mature larvae usually descend to the ground on silken threads and pupate under a thin layer of leaf litter or soil, within a loosely built cocoon made of dry or decayed leaves, or soil particles held together with silk. Pupation may sometimes occur within green leaves of other plants in the undergrowth, folded or juxtaposed with silk. The average pupal period lasts 6–8 days under optimal conditions,

a

b

FIGURE 1. (a) Female moth. (b) Eggs on leaf. (c) Third-instar larva inside leaf fold. (d) Mature larva. (e) Diseased larva.

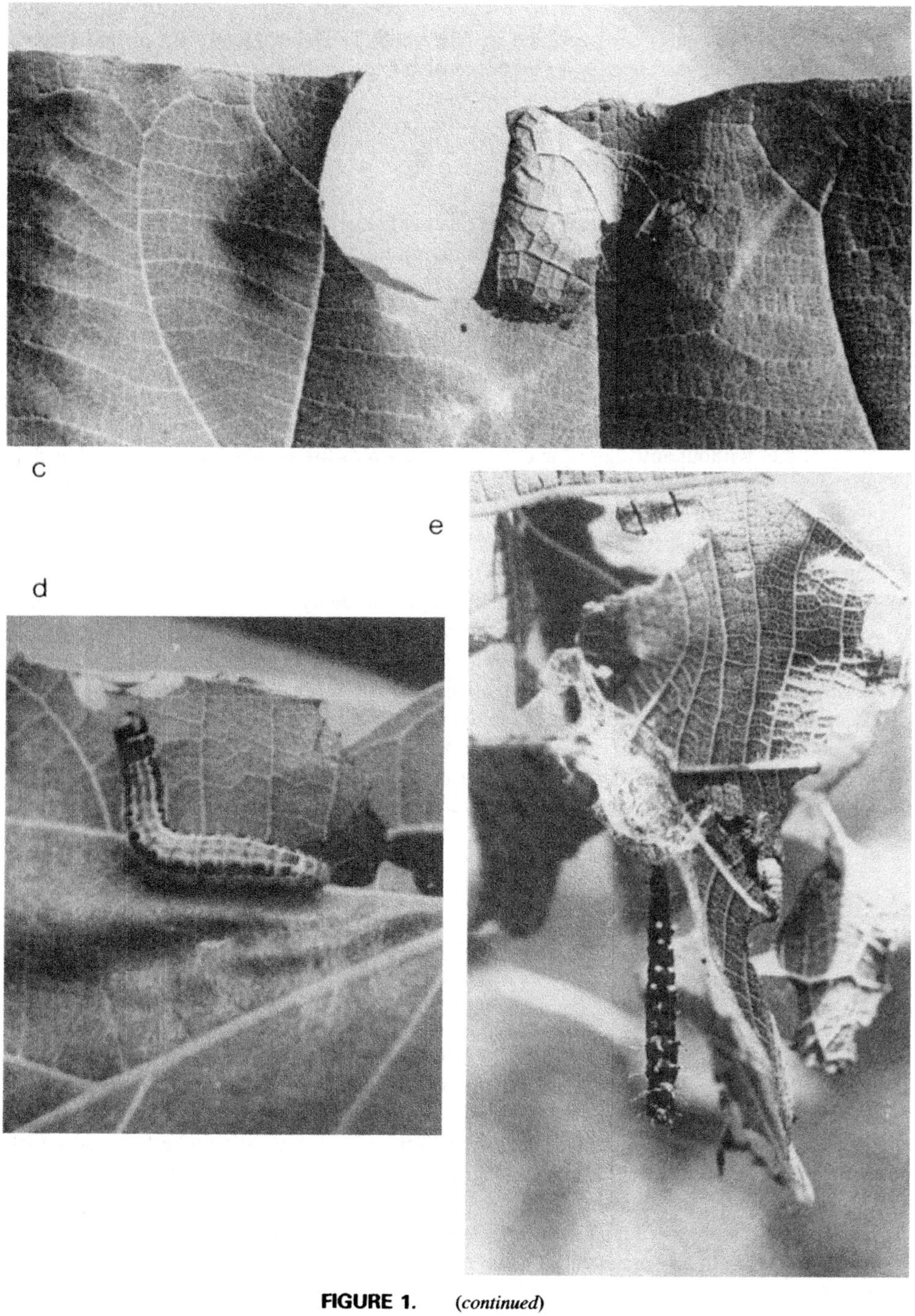

FIGURE 1. (*continued*)

but up to 20 days has been recorded in November at Dehra Dun, with an individual maximum of 25 days. There is no evidence of hibernation or aestivation of pupae.

2.2. Life Cycle and Voltinism

Populations of *H. puera* can complete their life cycles within a minimum of 19 days and a maximum of 36 days, depending mainly on the temperature. In field insectaries in southern India and Burma, generations are continuous, with a theoretical sequence of 14 complete generations and a partial 15th each year.[3] At Dehra Dun in northern India, outside the natural range of distribution of teak[33] and where there is a distinct winter season with chances of occasional frost, the number of generations is reduced to 10, with a partial 11th, and the moths may hibernate for a period of about 3 months from December to February. We have also been able to rear the insect continuously in our laboratory on an artificial diet without any sign of undue prolongation of life in any stage. Thus, throughout the natural range of teak, at least 13 continuous generations per year are theoretically possible, and there is no strong evidence of an estivation or hibernation period. This suggests that the insect should be prevalent in teak plantations throughout the year. Contrary to this expectation, however, is the almost total absence of the insect during the greater part of the year. Heavy defoliations begin suddenly, usually during the premonsoon season (May to June), and then spread rapidly over extensive areas during the next 2 months. In some years, visible but smaller-scale defoliation may occur again in August to October; thereafter, until the infestation appears again in May–June of the next year, the insect is not noticeable. Light traps at Jabalpur in Madhya Pradesh captured no moths over a period of 6 months (from January to June), leading to the conclusion that they do not breed in the same area on any host plant during this season.[34] This disappearance of huge defoliator populations is one of the unsolved mysteries of *H. puera,* to which we shall return later.

3. ECOLOGICAL RELATIONSHIPS

3.1. Host Plants

H. puera is oligophagous; of the 29 host plant species recorded, 13 including teak belong to the family Verbenaceae, 13 to the Bignonaceae, and one each to the Araliaceae, Juglandaceae, and Oleaceae.[3,20] Most are trees, others woody climbers or shrubs. It has been suggested[3] that some of them may sustain the insect population when teak is not in flush, but there are few field data. Although nine of the recorded alternative hosts occur in Kerala, only once in the many field trips we have made over the past several years has infestation been observed on a species (*Vitex altissima*) other than teak, and that too when infestation was already prevalent on teak.[20] Experimental studies in Burma indicated that teak is the most preferred among the 18 host species tested.[3] However, in another study in India,[3] teak, *Vitex negundo,* and *Premna latifolia* were equally acceptable to newly hatched larvae, but development was fastest on *Vitex* and survival was highest on *Premna,* teak being the poorest with respect to both the parameters.

Eggs are deposited only on tender leaves, and the newly hatched larva requires soft young leaf tissue for initial establishment. This dependence on young foliage can be expected to lead to population shifts among various host plants depending on their state of flushing, but there is very little published information on natural infestations on species other than teak. Another interesting consequence of the dependence of *H. puera* on tender foliage is that trees that reflush after defoliation (they do so within 2–3 weeks) are susceptible to reinfestation, while trees missed in the first attack continue to escape further defoliations.

Plantations of a variety of teak known as Teli (meaning "oily"), prevalent in some parts of Karnataka State, which flushes at least a month in advance of the normal variety, are known to escape defoliation.[15]

3.2. Relationships with Other Organisms

3.2.1. Parasites

A total of 34 species of insect parasites have been recorded on *H. puera* in India, including 15 Diptera and 19 Hymenoptera. Among the Diptera, one is a sarcophagid and the others are tachinids; among the Hymenoptera, eight families are represented.[31] Extensive surveys conducted during the 1930s demonstrated the highly complex web of interrelationships among these parasites and their various hosts.[16] Biological control recommendations were formulated as early as 1934[2] and included (1) retention of strips of natural forests between plantations as natural enemy reserves and their improvement by promoting desirable plant species and eliminating undesirable ones; (2) maintenance of a varied flora of desirable species under teak canopy to support parasites; and (3) introduction of selected species of natural enemies in areas where a deficiency exists in the natural enemy complex.

The quantitative aspects of the relationship between the parasites and *H. puera* has received attention only recently. In a study at Nilambur, Kerala[31] only five parasites were found: the tachinid, *Palexorista solennis;* an eulophid, *Sympiesis* sp., the chalcidid, *Brachymeria lasus;* and two unidentified ichneumonids. Overall parasitism by all species was about 9% during the year, the tachinid, *Palexorista* accounting for nearly 6%. Parasites were either absent or rare at the beginning of the epidemic, but their numbers increased subsequently, particularly in the case of *Palexorista* (see Fig. 3). The incidence of parasitism also varied widely between locations and between species. Parasitism as high as 60% was recorded for *Sympiesis* sp. at one location in October, when the host population was low.[31] In an earlier study at a different location,[22] near total parasitism by *Palexorista* and an unidentified ichneumonid was recorded in June, when the host population was high. In general, insect parasites do seem to have a significant influence on the dynamics of *H. puera* populations, but detailed quantitative studies are yet to be made. A mermithid nematode, *Mermis* sp., is also known to parasitize the larvae late in the season.[16]

3.2.2. Predators

Four species of birds have been observed feeding on *H. puera:* the jungle crow, the black drongo, the common mynah, and the jungle babbler;[31] many others are probably involved on occasion. These birds, particularly the crows, often visit the site of an epidemic in large groups that may number more than 100 individuals. Bird predation appears to cause a significant reduction in the number of larvae in certain locations during epidemics, but its impact has not been quantified; it would appear, in epidemics, that a sizeable population of larvae probably survives in spite of bird predation because of the sheer number of larvae.

Insect predators of *H. puera* include carabids (six spp.), mantids (five spp.), pentatomids (two spp.), and a syrphid, but again the information is only qualitative.[31] Many unidentified spiders also feed on *H. puera* larvae.

3.2.3. Pathogens

Moribund and dead larvae are a common sight in teak plantations, particularly during the declining phase of epidemics during the rainy season (Fig. 1e). Large-scale deaths of larvae characterized by cessation of feeding, flaccidity, and subsequent liquefaction of body tissues has been reported as early as 1903[30] and confirmed by subsequent observations.[16,31] Two bacteria, *Enterobacter aerogenes* and *Bacillus thuringiensis* var. *thuringiensis,* and a nuclear polyhedrosis virus have been identified recently from field collected diseased larvae.[31,32] These observations indicate that diseases are a major mortality factor during the collapse of epidemics, but this has not been confirmed quantitatively.

3.2.4. Interspecific Competition

A large number of leaf-feeding insects are associated with teak in the Indian region, including 78 Lepidoptera, 40 Coleoptera, and 18 Orthoptera.[17] Although visible defoliation may be caused by several species on occasion, only two, *H. puera* and *E. machaeralis,* have acquired pest status. Grasshoppers and chrysomelids (*Colasposoma* sp.) are usually the first to appear after the flushing of teak in March–April. Both feed on tender leaves and are only active for a short period, usually before the other species arrive. Grasshoppers may cause heavy damage in some patches, and chrysomelids cause damage to the lower canopy in some localities. An unnamed curculionid, which may cause up to 15% leaf loss in some plots, fed on a second flush following total defoliation by *H. puera* (see Fig. 2). Although these insects feed, like *H. puera,* on tender leaves, they exert little influence on its population dynamics because they are isolated in either time or space.

Hyblaea puera and *E. machaeralis* are also ecologically isolated from each other by their specialized feeding habits; the former feeding predominantly on tender leaves and the latter on old leaves. Some degree of temporal isolation also occurs, with *H. puera* being more prevalent in the early part of the season and *E. machaeralis* later. However, low-density populations of *E. machaeralis* are present in teak plantations throughout the year[24,26] and, when given the choice, the larvae will apparently select the younger and softer leaves.[3] The interaction between *H. puera* and *E. machaeralis* and its influence on their population dynamics is not understood.[23] In Kerala, only localized outbreaks of *E. machaeralis* occur in most years, but a very heavy and widespread outbreak was recorded in November 1976.[24]

4. POPULATION DYNAMICS

In spite of a century of accumulated data on the biology and ecological relationships of the teak defoliator, our knowledge of its population dynamics is meagre. Population dynamics, which deals with fluctuations in numbers, is basically a quantitative science but most data on the teak defoliator up until now have been of a qualitative nature. Although we can identify some of the components of its population system, we do not yet know how they are interconnected. However, with an understanding of the biological attributes and ecological relationships of the insect and within the theoretical framework developed to explain the behavior of insect population systems in general,[4,6] it should be possible to make intelligent guesses about the natural behavior of *H. puera* populations. The following example will serve to highlight the problem of teak defoliator population dynamics.

On the night of April 26, 1986, a dramatic event occurred in a teak plantation at Nilambur in Kerala State. A group of at least two million *H. puera* moths descended on about 20,000 newly flushed teak trees in a 30-ha patch within the plantation. Together, they deposited between 50 to 100 eggs per leaf on most of the tender leaves in the top canopy, and by morning all the moths had disappeared. Within a fortnight, the trees were stripped clean by the feeding caterpillars, and the falling frass sounded like mild rainfall on the dry leaves. Other nearby plantations remained untouched. No one knows where the moths came from or where they went. Such events have been happening all over the country for years, but no one has ever seen the airborne fleets of moths. The task of unravelling the origin, growth, and movements of populations of *H. puera* moths resembles the solving of a crime. We have a few clues, some direct and some circumstantial, with which we shall try to piece together a story of the sequence of events and conditions leading to the "crime". In the process, we will try to build a simple model to explain the behavior of *H. puera* populations in time and space. This model will be highly speculative, but we hope that it will illuminate the gaps in our understanding and help us ask the right questions. In order to avoid confusing facts with hypotheses, we will examine the facts first.

4.1. The Facts

4.1.1. Dynamics in Time

The seasonal abundance of *H. puera* in teak plantations at Nilambur, Kerala, over a 6-year period from 1979 to 1984 is shown in Fig. 2 and 3. Figure 2 depicts the defoliation intensity which is a faithful index of larval population size, except when defoliation reaches 100%. Figure 3 shows the actual number of larvae collected per unit effort as well as the number killed by three parasites. These data permit the following conclusions:

1. The insect is very abundant at certain times of the year and is not noticeable at other times: This conclusion is supported by general observations in literature[3] and light trap catches.[34] The first major population buildup usually occurs during the premonsoon season, anywhere from late March to May in Kerala but is delayed until July in Madhya Pradesh[34] and Uttar Pradesh.[3]
2. During the early phase of the outbreak, huge numbers of insects are produced from teak plantations: Epidemic infestations appear suddenly in areas where the insect was not previously seen. Trees with comparatively young leaves are heavily infested, with 50–100 eggs often being laid on each leaf in the top canopy (Fig. 4a). In trees with older leaves, egg laying is confined to the peripheral layer of tender leaves. The tender leaves are consumed within a few days, the larvae then hang down on silken threads and continue feeding on older leaves. In this way, all the leaves on a tree may be consumed (Fig. 4c). From heavily infested trees, the larvae then disperse to adjacent trees. Mature, even-aged larvae descend to the ground on silk threads, and crawl about in search of suitable pupation sites. Some continue to feed on teak saplings (Fig. 4b) before pupating in the ground.

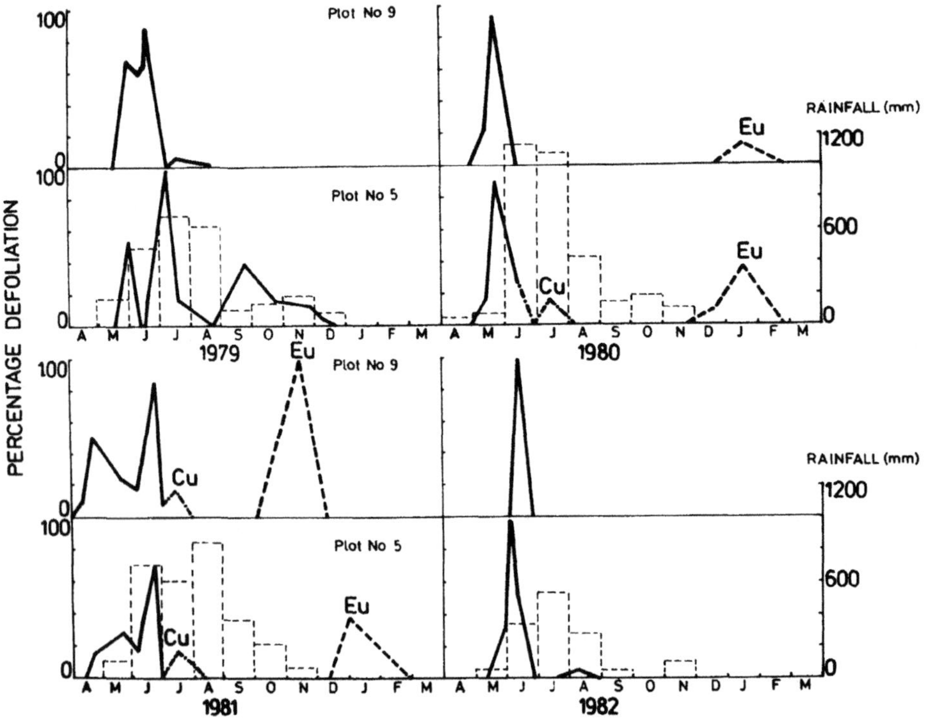

FIGURE 2. Seasonal incidence of defoliation over 4 years in two representative plots in 5–9-year-old teak plantations at Nilambur, Kerala. Defoliation caused by *Hyblaea* is shown by solid curves; other curves show defoliation caused by *Eutectona* (Eu) and by curculionids (Cu). The distribution of rainfall is presented as a histogram. (From Nair and Sudheendrakumar.[22,24])

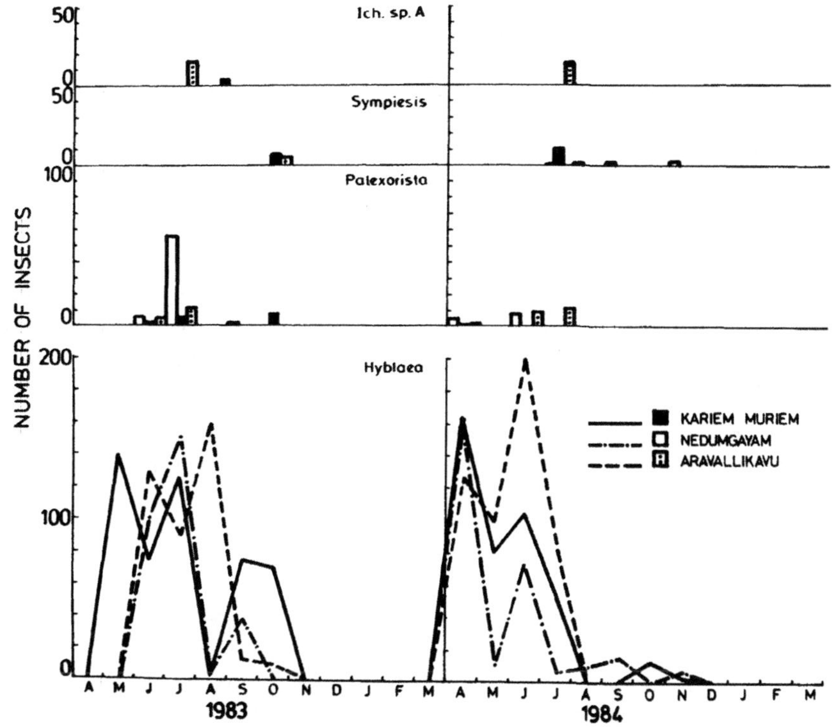

FIGURE 3. Changes in the populations of *Hyblaea* and its major parasites in representative plantations at Nilambur, over 2 years. Numbers represent host larvae collected per unit effort (1 hr). (Adapted from Sudheendrakumar.[31])

About 6 days after pupation, moths emerge in large numbers. Assuming 50 eggs per tender leaf and 1000 tender leaves per tree, it is roughly estimated that during epidemics each infested tree, of about 10 to 20 years of age, may support as many as 50,000 larvae. Therefore, even after allowing 25% mortality due to parasites and predators, a 20-year old plantation with about 400 trees per hectare, could produce about 15 million moths per hectare in one generation (20 odd days). A 30-ha infestation described earlier could generate 450 million moths in this time period—a staggering number, indeed.

3. Generally, two waves of heavy defoliation are observed in a given location, representing two population peaks: The time interval between these two peaks may vary from location to location. In some years only one peak is noticed (e.g., 1980, 1982: Fig. 2), and occasionally a third peak is seen in August[22] or September-October (e.g., 1979: Fig. 2; 1983: Fig. 3).
4. Very small numbers of uneven-aged larvae may be found during the nonoutbreak period: Although it is generally believed that the insect disappears from teak areas during the nonoutbreak period, in Kerala, small numbers of uneven-aged larvae have been sighted on teak saplings in the understory of plantations and/or natural

forests in most months of this period.[22] Such stray larvae are common in plantations immediately after the annual epidemic but have also been found, on rare occasions, prior to the outbreak, in areas in which epidemics did not develop.

5. Parasites and diseases are more prevalent during the second wave of infestation than in the first.

4.1.2. Dynamics in Space

Following the discovery that moths emerging from a particular location do not oviposit in the same place, a ground survey was made along roadside teak plantations and the distribution of infestations was mapped (Fig. 5). This and other observations led to the following conclusions:

1. During outbreaks, *H. puera* populations are highly aggregated: The defoliated patches may be very small, containing as few as 10 trees or may cover a few to

FIGURE 4. Scenario of an outbreak. (a) Tender teak leaf harboring about 80 young larvae within small leaf folds. (b) Older larvae feeding on mature leaf of sapling near ground. Note the fecal pellets and one larva within leaf fold at right. (c) Teak tree completely defoliated by *Hyblaea*. (d) Defoliated teak plantation. (a and b from Nair and Sudheendrakumar.[22])

c

d

FIGURE 4. (*continued*)

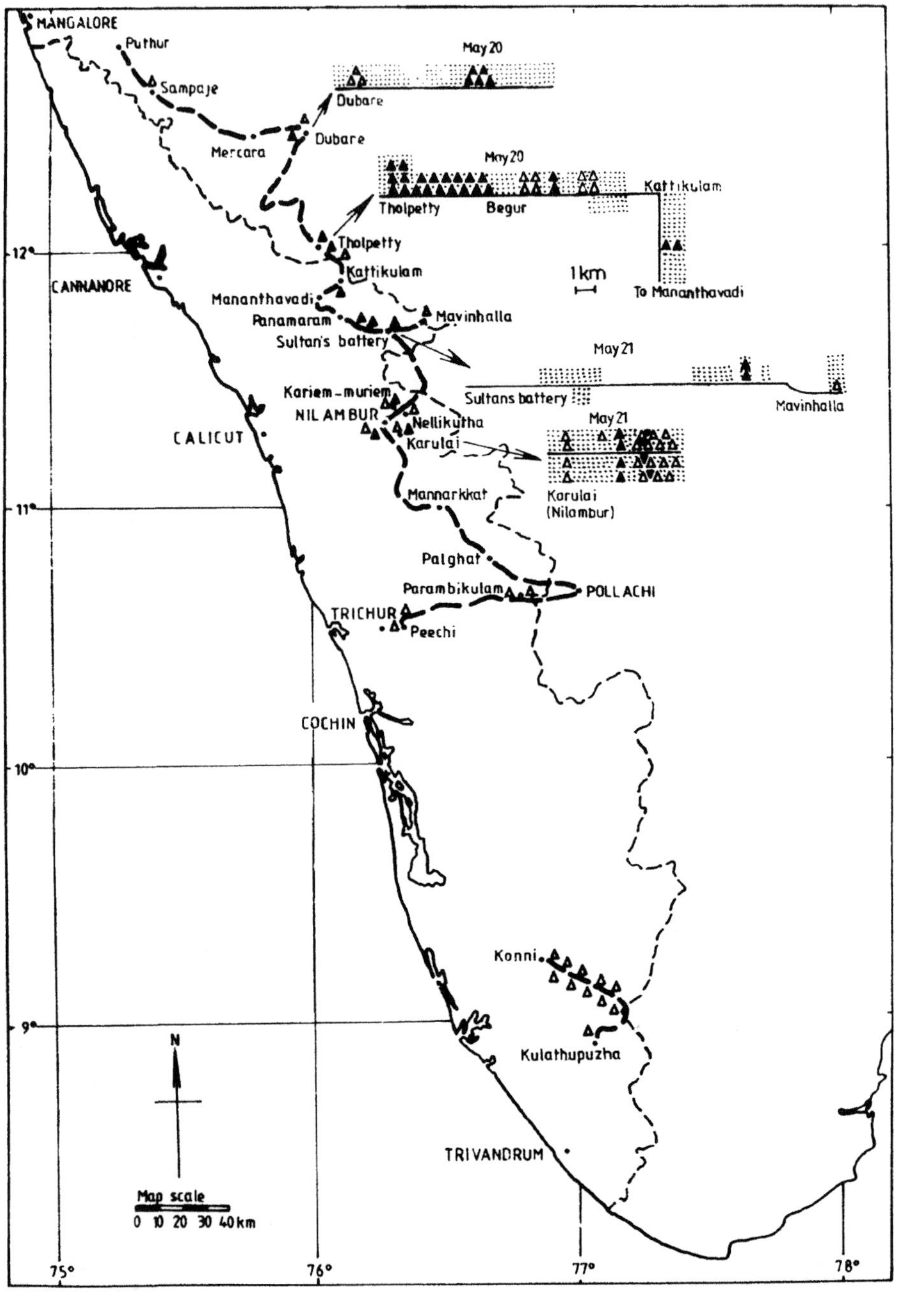

FIGURE 5. Map of Kerala showing the distribution of current (closed triangles) and old (open triangles) infestations of the teak defoliator in roadside teak plantations along the survey route. Inserts show details of the teak plantations (stipled areas) and infestations, in selected locations. (Adapted from Nair and Sudheendrakumar.[22])

many hectares (Fig. 4d). Small discrete patches of defoliation have been noticed in natural forests, where teak is mixed with other species, and also in the midst of plantations, particularly during the early outbreak period. However, larger patches covering a hectare or so are more common at this time, and areas as large as 30 ha may sometimes become infested. The second outbreak wave also creates discrete boundaries but tends to cover more extensive areas.

In some places, infestations are sometimes less intense and more diffuse but the boundaries are still distinct, even though the area covered may be large. In diffuse infestations, some trees or patches of trees are left totally uninfested in the midst of defoliated trees.

2. Successive generations of the insect do not inhabit the same place: Thus no significant reinfestation occurs in or around recently defoliated areas even when trees possessing suitable tender foliage are present. Instead, a small number of uneven aged larvae may be seen in the area, feeding mostly on tender leaves of teak coppice in the understory and occasionally the lower branches of bigger trees. This is quite unlike the original infestation which started from the tree tops. However, the same area may be reinfested subsequently, in the typical fashion, by a different population of insects. The unequal time intervals between successive defoliation peaks in different locations provide proof that successive defoliations are not caused by successive generations originating in the same place. Sometimes reinfestation may also occur before moths of the previous generation have emerged.
3. At any given time, populations in different patches within a large area may be in different developmental stages, but within the same patch the population is even-aged.
4. In the early stages of outbreaks, old and newly infested patches are generally separated by a distance of a few kilometers (Fig. 5).
5. Over wider geographical areas, outbreaks occur at different times: The first epidemic may occur as early as late March in southern India[22] but may come as late as July in central India.[34] Even within Kerala there is about a 1-month delay between outbreaks in the south and north. For example, in 1986, at Kulathupuzha in southern Kerala the first epidemic occurred around April 1, but about 270 km north at Nilambur (Fig. 5) it was not observed until April 26.
6. Not all plantations are affected every year: Although outbreaks occur every year, some plantations or patches within plantations may escape defoliation each year. The spatial distribution of such patches is not constant over years.
7. Aggregations of moths are sometimes observed in bushes near recently defoliated teak plantations: This observation was made at Parambikulam on May 29, 1986, about 1 week after moths of the first wave had emerged. A small group of about 100 moths was observed at one location near a defoliated plantation. In another area of natural forest adjoining the plantation, large numbers of moths were present, each sapling or shrub in the undergrowth harbouring dozens of moths. Both sexes were prevalent and dissections showed that the females had immature ovaries. The moths were observed resting on the leaves of a variety of plants including bamboo bushes and the shrubby weed, *Chromolaena odorata,* and appeared to choose cool, shaded areas with good undergrowth. When this area

was examined 11 days later, no moth aggregation was seen, although small numbers of moths were found scattered over the general area. A few young larvae were also present on teak saplings.

4.2. Dynamics in Space and Time: An Explanatory Hypothesis

A conceptual model of the population dynamics of *H. puera* is presented in Fig. 6. This model attempts to accommodate all the observed facts but also includes some assumptions and open questions. The model is based on two major assumptions: First, most of the adult moths emerging from a defoliated center migrate *en mass* to another site, at least 4 km away, even when suitable host trees are available close by. The strongest evidence for migration includes the sudden appearance of outbreaks, the disappearance of new moths from the emergence site without producing a new generation, and the unequal intervals between successive defoliation peaks in adjacent areas. The 4-km minimum displacement distance was prompted by measuring distances between old and new infestations during the early outbreaks (Fig. 5). Second, during the nonoutbreak period, a residual nonmigratory population with continuous generations survives in the natural forests. The only support for this assumption is the sighting of small numbers of uneven-aged larvae on teak saplings at different locations throughout the year.

Migration seems to be a major factor in the spatial distribution of *H. puera* outbreaks. Recent field observations suggest that newly emerged moths do not migrate immediately but move about in groups, feeding and resting in shaded forest undergrowth, away from teak plantations. In fact, teak plantations are usually devoid of an understory and therefore make poor resting and aggregation sites. Migration probably occurs after mating. This is unlike in the noctuid moth, *Spodoptera exempta* (Walker), which migrates while the females are immature and unmated.[27] The conditions that initiate migratory flights in *H. puera* are unknown. Recent studies on the dispersal of other moths indicate, however, that wind and other meteorological and physiognomic factors have important effects on the concentration and displacement of airborne moth aggregations.[8,27] The characteristic intense outbreaks of *H. puera* at places such as Nilambur, and the diffuse infestations often observed in other places such as Parambikulam and Konni, suggest that air movements are involved in concentrating or dispersing the airborne moths. In any case, given the observed facts, migration of *H. puera* moths appears to be an inescapable conclusion; the hypothetical part relates only to the range of the migratory flight, the role of air currents, and other details of migration.

The sudden appearance of large numbers of moths and the spatiotemporal pattern characteristic of early outbreaks can be explained by wind-aided unidirectional (or even multidirectional) movement of groups of moths from a source of origin, with egg laying occurring on successive nights at different sites (Fig. 7). Consider three separate groups of moths arriving at site 1 on the night of three different dates, April 12, 16, and 22. Consider also that after ovipositing, these groups continue to move, arriving at site 2 on the next night. At site 2, the number of eggs laid and the extent of the infested patch will be smaller than at site 1 because of depletion of eggs as well as mortality of some moths. After some time (about 27 days), moths emerging from infestations at site 1 also arrive at site 2 and start a larger infestation. This may encompass the area infested earlier if the

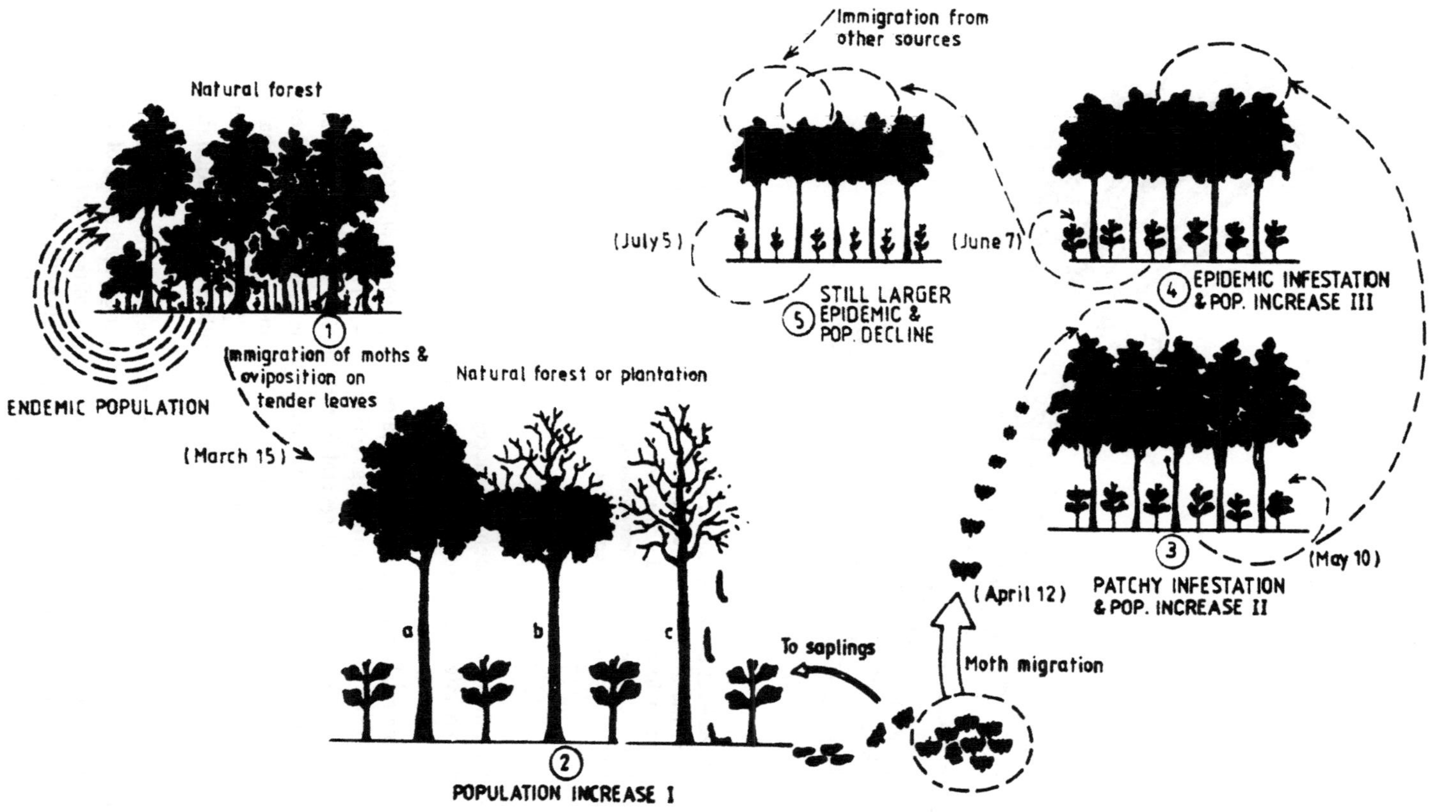

FIGURE 6. Diagram showing the proposed model of population dynamics of *Hyblaea puera*. (Modified from Nair and Sudheendrakumar.[22,23])

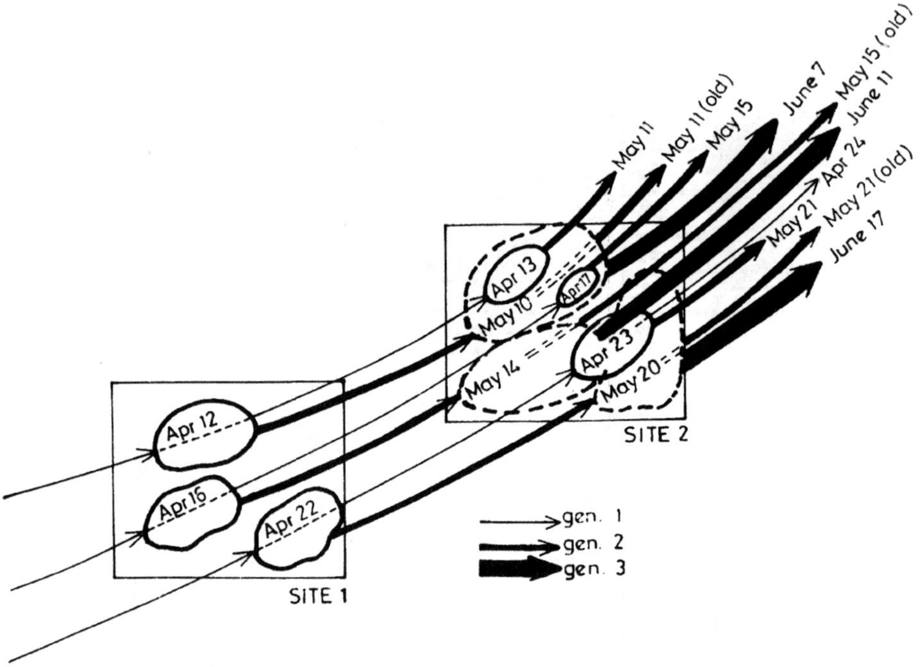

FIGURE 7. Hypothetical infestation patterns produced in two teak plantations when moths are assumed to migrate in groups and to oviposit on two consecutive nights. Patches with solid outline indicate the first infestation starting on the date marked within; larger patches with dashed outlines indicate infestation by the second generation, 28 days after the first oviposition. Although broad arrows are used to indicate a third generation of moths, these may never emerge, because of widespread disease incidence.

trees have reflushed. It can be seen that successive waves of small and large infestations as well as repeated infestations of the same area can result from this sequence of events (Fig. 7).

The geometrical increase of the defoliator population over three generations from mid-March to early July is shown in Fig. 6 (steps 2–4). The subsequent decline (step 5) may be caused by one or more of several factors but leaf maturity and natural enemies, including pathogenic microorganisms, are probably the most important. Leaf maturity by itself, however, cannot be the only cause of decline in population as small quantities of fresh young leaves are always available during this period. Although this continuous supply of tender leaves should be able to sustain a small larval population, in practice they do not.

Among the natural enemies, the tachinid parasite, *Palexorista solennis,* appears to respond numerically, but with a time lag, to increasing populations of *H. puera.* This density-dependent response of parasites is negated to some extent by migration of *H. puera* moths, which leaves behind the parasite population that built up on the previous generation. Although parasites may contribute to the decline of *H. puera* populations, their numbers do not seem sufficient to bring about a population collapse (Fig. 3). By contrast, disease epizootics are capable of causing population collapses. Diseased larvae are a familiar sight in teak plantations, especially during the declining phase of outbreaks during the rainy season; usually in July but sometimes at the end of June (Fig. 2). Disease epizootics may be triggered by (1) high inoculum loads being built up over repeated generations of dense moth infestations, (2) favorable weather brought about by the arrival of the monsoon, and/or (3) greater susceptibility of new generations of larvae due to genetic shifts or physiological stress. After the populations collapse, a small nonmigratory residual population may remain on understory saplings, or the populations may become locally extinct. In some years and on some areas, however, outbreaks may again flare up later in the year (e.g., Fig. 2, 1979, plot 5). It is not known, however, whether late outbreaks are due to local buildups or to immigrations.

Although *H. puera* populations may be locally exterminated or reduced to very low densities, they probably persist in larger numbers in some localities, perhaps inaccessible natural forests (Fig. 6, step 1). This condition seems necessary to account for the large influx of moths into plantations in spring. Over large geographical areas, outbreak populations tend to precede the monsoon, which starts first in the south (Fig. 5). Outbreaks then progress northward and only collapse after the arrival of the monsoon. Thus, in spite of local extinctions, the population of *H. puera* over a large geographical area, say India as a whole, remains high from late March to September. But what happens during the rest of the year; that is, for five months from October to February? According to our model, the population is assumed to maintain itself in the natural forest. Here, survival may be facilitated by (1) the absence of disease epizootics because of low inoculum loads in the dispersed *H. puera* population (there is a great deal of vegetational diversity in the natural forest) or genetic/physiological resistance in the heterogenous moth population, and (2) the presence of different species of alternative food plants as well as the greater phenological variability of natural teak which ensure the availability of tender foliage throughout the year. That the insects do survive in natural forests is indisputable, for, if they did not, the species would not be here today. The major question is whether populations of *H. puera* only survive in natural forests and then invade plantations every year, or whether outbreaks arise from residual populations within the plantations. Our current thinking is that populations build up in early spring (March–April) in the natural forest, coincident with the flushing of teak, and spill over to the teak plantations due to gypsylike migrations induced by a genetic or physiological shift (Fig. 8). In the plantations, the populations increase enormously and continue to invade extensive areas *en mass,* or more diffusely at the whim of air movement patterns, until deteriorating environmental conditions within the plantations cause their collapse.

There are counterarguments to this hypothesis. One is that the disappearance of larvae is the result of insect inactivity. Aestivation or diapause in summer and hibernation in winter have been proposed to explain the absence of insects during the leafless period in teak plantations.[3,34] However, there is little factual support for this hypothesis while at least two facts argue against it:[22] First, larvae are observed, albeit in small numbers, in October, November, December, and January in different areas in Kerala. Second, the insect is first observed at different times in unconnected patches, even within an area of a few hectares, while one would expect more synchronized emergence if the insects had been in diapause.

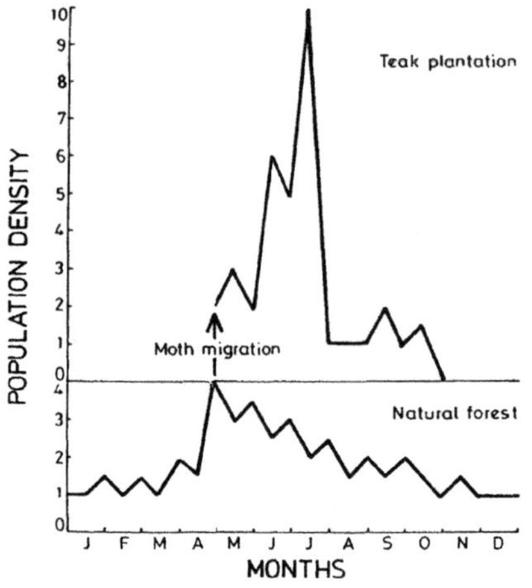

FIGURE 8. Hypothetical numerical dynamics of *Hyblaea puera* populations in natural forest and teak plantations. An alternative hypothesis is that the insect persists at low density in teak plantation itself.

Another hypothesis for the sudden appearance of millions of moths during spring is that the moths originate from a single or few major epicenters in which enormous numbers are produced in a specially favorable environment, as in the case of locusts.

If this discussion has posed more questions than it has answered, it reflects on the state of our current knowledge. Further research is needed to test these hypotheses concerning the origin and cause of *H. puera* outbreaks in teak plantations. The most characteristic feature of *H. puera* population dynamics is the gypsylike movement of populations, which helps the species exploit new sources of its ephemeral food, tender foliage, and to keep ahead of the parasites and pathogens they themselves help build up.

5. MANAGEMENT IMPLICATIONS

It is abundantly clear that knowledge of population dynamics is a necessary first step in the development of strategies for controlling this highly mobile pest. For example, the strategy will differ depending on whether the large invading spring population of moths originates from outbreak epicenters or locally from within plantations or nearby natural forests. If outbreak epicenters exist, it will be easy to prevent the population buildup by suitable operations in these centers, compared with the problems involved in preventing the buildup in more dispersed and extensive areas. The geographical location of the epicenters is also important, for it will be easy to eliminate epicenters in India but not in a far-off place such as Burma, because of intervening political boundaries.

If the hypothesis that outbreaks originate from natural forest proves true, it will have serious implications for management of the pest. Except for some misadventures with insecticides applied by air,[1,29] recommendations for managing teak defoliators have emphasized the retention of strips of natural forests between plantations and encouraging selected understory vegetation within the plantations to serve as refuges for parasites. In addition, the introduction of selected species of insect parasites in areas in which there is a deficiency in the natural enemy complex has been recommended. Although these recommendations were made as early as the 1930s, and lists of beneficial plants have been compiled, the recommendations have never been put to practice for various reasons. But the recommendations have been vigorously advocated by the entomologists, and the reluctance of forest managers to follow them up has often been criticized. In the light of our present knowledge, it is clear that the impact of a resident population of parasites on the millions of larvae that build up suddenly from immigrant moths will be small. If parasites are to be used, constant surveillance to locate infestation sites and inundative releases of large populations of parasites will be required.

If detailed population dynamics studies show pathogens to be important in the collapse of outbreaks, they could probably be used profitably in pest management. But here again, detailed knowledge of the conditions that influence population crashes are important. For example, if weather has an overriding influence on the buildup and effectiveness of pathogenic organisms, they may not be useful during the first wave of the epidemic before the monsoon arrives. The effectiveness of orally ingested virus is not related to weather, but weather may influence the persistence of infective virus polyhedra in the environment, and rain splashes may play a decisive role in dispersing them to cause a disease epizootic.[13]

Knowledge useful for managing *H. puera* will also come from detailed investigations of the underlying principles and mechanisms governing the spatial and numerical dynamics of its populations in natural forests and plantations. For example, knowledge about the conditions under which premigratory moth aggregations are formed can lead to methods for locating them in the field and destroying them or perhaps to the isolation and characterization of aggregation pheromones, which may then be deployed in various says.

From the insight already gained on the population dynamics of this insect, one of the most promising approaches to its management appears to be to attempt to break the synchrony between flushing of teak and spring arrival of moths, by breeding early flushing varieties of teak. These varieties would flush sufficiently early so that there would be a limited supply of tender leaves available for oviposition and larval establishment when the moths arrive. Sufficient variation exists in teak populations to suggest that breeding or cloning for this purpose is within practical reach. If the moths are dependent on premonsoon wind movements for their arrival, they will not be able to adapt to the altered phenology of teak plantations.

Our hypothesis for the population dynamics of *H. puera* suggests that outbreaks in plantations arise from the habitual gypsylike migration of the pest and the presence of a plentiful supply of tender foliage in spring. This is not in agreement with the widely accepted generalization that most forest insect outbreaks are associated with site and stand conditions that put trees under stress,[18] an idea that has led to the hope that outbreaks can be eliminated through silvicultural management practices aimed at increasing stand vigor.[5,7] In the case of *H. puera,* infestations do not seem to be associated with poor stand conditions.

ACKNOWLEDGMENTS. Thanks are due to Professor A. A. Berryman for stimulating discussions during his visit to India in May 1986 to see the teak defoliator in action. Some of the ideas discussed here originated during or as a consequence of discussions with him, but it does not mean that he endorses them. I also thank my colleagues Dr. V. V. Sudheendrakumar, Dr. R. V. Varma, Dr. George Mathew, and K. Mohanadas for their valuable contributions and for permission to quote from unpublished work. Thanks are due to Subash Kuriakose for preparing the diagrams and to The Indian Academy of Sciences for permission to reprint some of the figures.

REFERENCES

1. Basu-Chowdhury, J. C., 1971, Interim Report on Konny Aerial Spraying Project, in: *Working Plan for Nilambur Forest Division, 1967–68 to 1976–77* K. G. Vasudevan, ed., pp. 165–168, Government of Kerala.
2. Beeson, C. F. C., 1934, The biological control of teak defoliators. *Indian For.* **60:**672–683.
3. Beeson, C. F. C., 1941, *The Ecology and Control of the Forest Insects of India and the Neighbouring Countries,* 1961 reprint, Government of India, New Delhi.
4. Berryman, A. A., 1981, *Population Systems: A General Introduction,* Plenum, New York.
5. Berryman, A. A., 1981, Effects of site characteristics on insect population dynamics, pp. 541–549, in: *Proceedings of the Seventeenth IUFRO World Congress, Japan,* Div. 2.
6. Berryman, A. A., 1986, *Forest Insects: Principles and Practice of Population Management,* Plenum, New York.
7. Berryman, A. A., and Baltensweiler, W., 1981, Population dynamics of forest insects and the management of future forests, pp. 423–430, in: *Proceedings of the Seventeenth IUFRO World Congress, Japan,* Div. 2.
8. Berryman, A. A., and Safranyik, L., (eds.), 1980, *Dispersal of Forest Insects: Evaluation, Theory and Management Implications: Proceedings of IUFRO Conference at Sandpoint, Idaho,* Washington State University, Pullman, Washington.
9. Bigger, M., 1980, *Hyblaea puera* on teak. Forest pests of the Solomon Islands, No. 5, Forestry Div., Solomon Islands.
10. Bourdillon, T. F., 1898, Insects attacking teak in southern India, *Indian For.* **24:**126–129.
11. Bourdillon, T. F., 1898, Insects attacking teak in southern India, *Indian For.* **24:**424.
12. Browne, F. G., 1968, *Pests and Diseases of Forest Plantation Trees,* Clarendon Press, Oxford.
13. Entwistle, P. F., Institute of Virology, Oxford, Personal communication.
14. Hampson, G. F., 1894, *The Fauna of British India, Including Ceylon and Burma.* Vol. II: *Moths,* 1976 reprint, Today & Tomorrow's Printers and Publishers, New Delhi.
15. Kaushik, R. C., 1956, Teli variety of *Tectona grandis, Proc. 9th Silv. Conf., Dehra Dun, India* **2:**92–94.
16. Mathur, R. N., 1960. Pests of teak and their control, *Indian For. Rec.* (New Ser.) *Entomology* **10:**43–66.
17. Mathur, R. N., and Singh, B., 1960, A list of insect pests of forest plants in India and adjacent countries. 10. List of insect pests of plant genera T to Z, *Indian For. Bull. (New Ser.) Entomol.* **171(9):**1–116.
18. Mattson, W. J., and Addy, N. D., 1975, Phytophagous insects as regulators of forest primary production, *Science* **190:**515–522.
19. Menon, V. K. G., 1898, The life history of the insects which attacked the teak plantations at Perunthode, *Indian For.* **24:**127–129.
20. Mohanadas, K., 1986, A new host record for the teak defoliator *Hyblaea puera* (Lepidoptera: Hyblaeidae), *Curr. Sci.* **55**(23):1207–1208.
21. Nair, K. S. S., 1980, The problem of insect defoliation of teak—To spray or not to spray, in: *Proceedings of the Second Forestry Conference, Dehra Dun, India* (cyclostyled).
22. Nair, K. S. S., and Sudheendrakumar, V. V., 1986, The teak defoliator *Hyblaea puera:* Defoliation dynamics and evidences for short-range migration of moths, *Proc. Indian Acad. Sci. (Anim. Sci.)* **95:**7–21.
23. Nair, K. S. S., and Sudheendrakumar, V. V., 1986, Population dynamics of teak defoliators, pp. 673–684, in: Proceedings of the Eighteenth IUFRO World Congress, Yugoslavia, Div. 2.
24. Nair, K. S. S., Sudheendrakumar, V. V., Varma, R. V., and Chacko, K. C., 1985, Studies on the seasonal incidence of defoliators and the effect of defoliation on volume increment of teak, KFRI Res. Rep. 30, Kerala Forest Research Institute, Peechi, India.
25. Pande, D., 1983, *Growth and Yield of Plantation Species in the Tropics,* FAO, Rome.
26. Patil, B. V., and Thontadarya, T. S., 1983, Seasonal incidence of teak skeletonizer, *Pyrausta machaeralis* Walker (Lepidoptera: Pyralidae) in Prabhunagar forest, *Indian J. Ecol.* **10:**204–209.
27. Riley, J. R., Reynolds, D. R., and Farmery, M. J., 1983, Observations of the flight behaviour of

armyworm moths, *Spodoptera exempta*, at an emergence site using radar and infra-red optical techniques, *Ecol. Entomol.* **8**:395–418.

28. Seth, S. K., and Kaul, O. N., 1978, Tropical forest ecosystems of India: The teak forests, pp. 628–640 in: *Tropical Forest Ecosystems. A State-of-Knowledge Report Prepared by UNESCO/UNEP/FAO*, UNESCO, Paris.
29. Singh, P., 1980, Aerial spraying of chemicals to control teak defoliators, in: *Proceedings of the Second Forestry Conference, Dehra Dun, India* (cyclostyled).
30. Stebbing, E. P., 1903, Departmental notes on insects that affect forestry 2, Government of India Central Printing Office, Calcutta.
31. Sudheendrakumar, V. V., 1986, Studies on the natural enemies of the teak pests, *Hyblaea puera* and *Pyrausta machaeralis*, KFRI Res. Rep. 38, Kerala Forest Research Institute, Peechi, India.
32. Sudheendrakumar, V. V., Kerala Forest Research Institute, Peechi, India, personal communication.
33. Troup, R. S., 1921, *The Silviculture of Indian Trees*, Vol. II: *Leguminosae to Verbenaceae*, Clarendon, Oxford.
34. Vaishampayan, S. M., and Bahadur, A., 1983. Seasonal activity of adults of teak defoliator, *Hyblaea puera* and teak skeletonizer, *Pyrausta machaeralis* (Pyralidae: Lepidoptera) monitored by light trap catch, in: *Insect Interrelations in Forest and AgroEcosystems* (P. K. Sen-Sarma, S. K. Kulshrestha, and S. K. Sangal, eds), pp. 137–146, Jugal Kishore, Dehra Dun, India.

CHAPTER 15

THE PINE LOOPER IN BRITAIN AND EUROPE

David A. Barbour

David A. Barbour • British Forestry Commission, Forest Research Station, Wrecclesham, Farnham, Surrey GU10 4LH, England.

1. INTRODUCTION

The pine looper moth, *Bupalus piniaria* Linnaeus, has been a serious pest of European forestry for more than 200 years, particularly in Germany,[15,17,30,32,61] In the Netherlands, where the looper has never attained pest status, the population dynamics and general biology have been intensively studied.[10,20,37,38] The pine looper was not known as a pest in Britain until 1953, when larvae totally defoliated 40 ha of Scots pine (*Pinus sylvestris*) at Cannock Chase, Staffordshire. Subsequent secondary infestations by the bark beetle, *Tomicus piniperda* Linnaeus, resulted in death of the trees.[11] Since then, pupal samples have been taken annually in some 50 different Forestry Commission pine areas. Seven subsequent outbreaks on five different sites have been treated with insecticides. High populations have also been found in plantations of Corsican pine (*Pinus nigra*) and lodgepole pine (*Pinus contorta*). This chapter describes the dynamics of populations in Britain, but makes frequent reference to studies in Germany and the Netherlands.

2. ECONOMIC SIGNIFICANCE

In the two most severe outbreaks in Britain, a total of 110 ha of pines were killed. If insecticidal control had not been employed, however, the loss would probably have been 10–20 times greater. About 5000 ha of forest have been treated, using DDT prior to 1963, and diflubenzuron in the more recent outbreaks. The need for insecticide application is established, in advance of any visible damage to trees, by counting pupae in the forest litter each winter. Transects of ten 0.25 m^2 circular plots are examined, with about one transect every 40 ha of forest.[8] When pupal surveys reveal populations in excess of 30 per m^2, egg samples are taken the following summer before the final decision to spray is made. The expense of monitoring is justified by the high value of plantations, which consist of Scots and Corsican pine over 20 years of age.

The impact of *Bupalus* infestations on pine plantations in continental Europe has been on a much larger scale. The worst recorded outbreak, in Bavaria during 1892–1896, resulted in the complete or near-complete defoliation of 40,000 ha.[17] Since the late 1920s, outbreaks in Germany have been sprayed with insecticides, which may have forestalled the development of vast infestations comparable to that of the 1890s. For example, a total of 4668 ha were treated in North Bavaria from 1925 to 1970.[31] The reduced significance of *Bupalus* and other pine defoliators since 1930 has been attributed to underplantings of hardwoods (particularly beech) and to planting mixed stands of pine and hardwoods.[30]

3. BIOLOGY AND ECOLOGICAL RELATIONSHIPS

Adult moths first appear at the end of May and fly from June to early August. Male moths emerge slightly earlier than females, and there is strong sexual dimorphism in wing-markings: The males have white or cream-colored wings with a dark chocolate border, hence their common English name, the bordered white moth. Females are much more drab, of a more uniform rust-brown color. Both have a wingspan of around 35 mm.

During the day, female moths generally sit motionless in the undergrowth or tree crowns, while males take part in vigorous flights, particularly on still sunny days, when they are seen in swarms around tree crowns. Pairing takes place in the crowns during the middle of the day, but oviposition flights occur at night. Eggs are laid in rows of about 3–7 on the needles of pines. Females each lay between 100–200 eggs, and fecundity is closely correlated to the weight of individual females. Thus, fecundity can be predicted from the weight of female pupae.[9] Egg laying commences in June and may continue as late as mid-August. All parts of the tree crown may bear eggs, but the mid-levels are favored. Eggs hatch between late June and the end of August.

Bupalus larvae feed first on older foliage, biting a narrow channel in the epidermis of the needle and leaving a characteristic feeding groove, which may partly fill with resin. Older larvae consume the whole tissue of the needle, removing large and irregular-shaped sections along its length. The larvae typically pass through five instars, but development in the field may cease at the fourth instar or proceed to a sixth instar.

The young looper is a uniform gray-green color with a dull brown head, but from the third instar onward it attains its distinctive pattern of white or yellowish-white stripes along the full length of the green body and head. The full-grown larva is about 26 mm long. Larval development may take as long as 4 months. Pupation takes place from October onward, in a loosely constructed chamber in the litter. The pupal period lasts through winter and spring until adult emergence in late May or June.

Female pupae are notably fatter and heavier than males. It has also been established that larvae reared under crowded conditions pupate at smaller sizes than those reared singly.[40] This in turn results in lighter, less fecund female moths. A complicated relationship exists among pupal weight, the number of larval instars, temperature, photoperiod, and the sex of individual larvae.[20] Pupal weight is positively correlated with the number of larval instars, and crowded larvae go through fewer instars. The mechanism of growth reduction involves direct bodily contacts between larvae, during which regurgitated gut fluid is transferred. The number of encounters required for growth reduction is quite low, in accordance with field studies showing reduced larval growth at even moderate population densities. It has been suggested that larval crowding might increase the tendency of adults to disperse by reducing their wing loading.[20] However, no experimental evidence exists to support this contention.[10]

On average, both male and female moths travel about 150 m during their lifetimes,[10] so that females deposit most of their eggs very close to their points of origin. Release–recapture experiments in the forest show that differences in flight ranges are not related to pupal weight. The restricted dispersal of individuals also indicates that genetic drift might cause differences between subpopulations in some habitats.

3.1. Natural Enemies

The pine looper is attacked by numerous predators, including spiders, Hemiptera, Coccinellidae, Elateridae, and wood ants of the genus *Formica*.[56] The latter can reduce the number of looper pupae by 70% within 5 m of their nests.[1] Bird predation on *Bupalus* adults is also important,[19,45] but attempts to influence looper numbers by manipulating coal tit populations have not proved successful.[12]

As many as 56 ichneumonid and 9 tachinid parasitoids attack *Bupalus* larvae and pupae.[56] One particular ichneumon, *Cratichneumon viator* (Scopoli), seems to predominate during periods of low *Bupalus* density, but during outbreaks several other host-specific parasitoids predominate, particularly as populations decline.[22] *Cratichneumon viator* is also important in Britain,[13] with up to 75% parasitism caused by a second generation on looper pupae in early spring. *Poecilostictus cothurnatus* Gravenhorst is another ichneumonid that can have a major impact on *Bupalus* populations.[39,63]

A tachinid, *Eucarcelia rutilla* Villiers, has also been intensively studied,[25,26,34–36,54]) and its life cycle is well synchronized with that of its host. It is an important mortality factor during the collapse of outbreaks.

The chalcid *Trichogramma embryophagum* Hartig is a polyphagous parasitoid that produces five to six generations per year in eggs of different hosts.[33] The parasitoid deposits an average of six to seven eggs in each looper egg.[41]

3.2. Host Trees

Outbreaks of defoliating looper larvae have been hypothesized to result from summer drought and/or winter waterlogging, which causes stress to their host trees.[66] Water stress is thought to increase the concentration of mobile nitrogen compounds in the foliage, which in turn benefits the growth and survival of small foliage-feeding larvae and may release the populations to outbreak levels. Although a critical stage in the *Bupulus* life cycle appears to be the small larvae, whose mortality is highly variable and averages around 65–70%,[37] corresponding data on changes in the nitrogen content of pine foliage are not available.

Experimental studies seem to support the stress hypothesis. For example, larval mortality was found to be higher in pine stands treated with nitrogen fertilizer than in unfertilized stands.[48] Mortality was also higher in a stand in which lime had been worked into the soil 23 years previously but was not significantly higher in a stand given similar treatment only 6 years prior to the test. Foliage sugar concentrations have also been associated with increases in the proportion of females, the weight of pupae and hence fecundity, and larval survival.[55] Refractometer measurements of the foliage of forest trees show that sugar content increases progressively during periods of drought and is higher in trees growing on drier sites or affected by mechanical damage.

Needle monoterpenes have been shown to be toxic to *Bupalus* larvae, and hexenal is even more poisonous.[49] Although resin is not itself toxic, heavy resin flow from feeding wounds at the needle surface will repel feeding larvae. Resin exudation was found to be greater in young plantations growing on nutrient-rich or fertilized sites, but no such difference was found in middle-aged stands.

4. PATTERNS OF NUMERICAL BEHAVIOR

The numerical behavior of *Bupalus* populations has been recorded in two long-term studies in Germany and Holland, both widely quoted in the ecological literature.[37,38,57,58,61] The earlier data from Germany consist of 60 years of annual pupal counts showing violent density fluctuations exceeding the outbreak level on five different occasions (Fig. 1a). The later study in Holland spanned 17 years, during which the population never attained outbreak densities (Fig. 1b).

More recently, 25 years of British survey data from 47 populations have been analyzed.[4,5] Most of these populations were relatively stable and produced no outbreaks, but a few showed wide-amplitude oscillations (as in Germany), with outbreaks occurring at intervals of about 7 years (Fig. 1c). Between the most stable and the most oscillatory types was a continuous spectrum of population behavior that can be characterized in terms of immediate and delayed density-dependent components.[4,5]

The immediate density-dependent effect is characterized by high-frequency oscillations, in which deviations from the supposed equilibrium density tend to be corrected for by a compensatory swing in the following generation. The delayed density-dependent pattern shows a longer period of oscillation, with deviations from equilibrium only being corrected for after a time lag of one generation. The delayed effect is expressed in the recurring cycles of outbreak populations (Figs. 1a,c). These two feedback effects have also been termed noninertial and inertial[28] or first-order and second-order processes.[50,52]

The immediate and delayed density-dependent effects can be quantified by plotting the specific growth rate of the population (i.e., the net change in the natural logarithm of population density) against population density in the current and preceding years. When immediate effects dominate, the change from year t to year $t + 1$ will be negatively

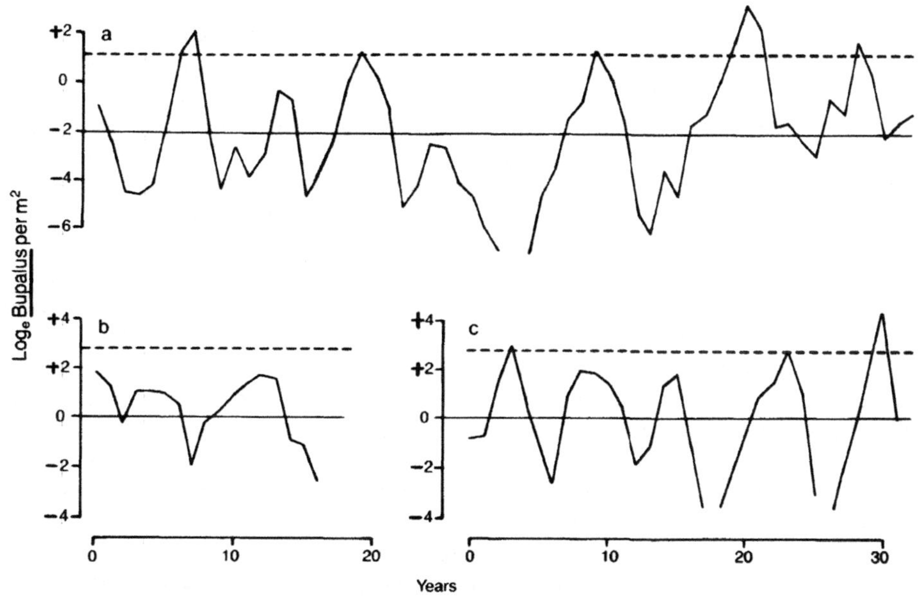

FIGURE 1. Annual counts of *Bupalus* pupae at three sites. (a) Letzlingen, Germany, 1881–1940. (After Schwerdtfeger[57,58] and Varley[61]) (b) de Hoge Veluwe, Netherlands, 1951–1967. (After Klomp.[38]) (c) Tentsmuir Forest, Scotland, 1954–1985. Insecticides were applied in 1958, 1977, and 1984. For each population, the approximate equilibrium density (——) and outbreak level (- - -) are shown.

correlated with density in the first year (t). For delayed effects, the change from year t to year $t + 1$ is negatively correlated with density in the preceding year (t-1). The two effects can be combined in a multiple regression equation, such as

$$r_t = r_0 + iN_t + dN_{t-1} \tag{1}$$

where r_t is the specific growth rate of the population from year t to $t + 1$, r_0 is the specific growth rate at $N = 0$, N_t is the logarithm of population density in year t, and i and d are the immediate and delayed feedback coefficients, respectively. This is mathematically equivalent to the second order autoregressive model used to describe cyclical and quasicyclical behavior of a wide range of animal populations, e.g., red grouse,[51] spruce budworm,[53] and Canadian lynx.[46]

The parameters of this second-order model have been calculated for the German, Dutch, and 47 British populations of *Bupalus* (Fig. 2). The results are as follows:

1. There is a significant tendency for populations to lie along an axis from large immediate–small delayed effects (top left) to small immediate–large delayed effects (bottom right).
2. Populations with small immediate and large delayed feedback effects are those with the most severe, repeated outbreaks, e.g., Cannock and Tentsmuir in Britain and Letzlingen in Germany.
3. The coefficients i and d define the expected period of the population cycles.[52]
4. The mean coefficients for the 47 British populations are $i = -0.509$ and $d = -0.267$, which corresponds to an expected cycle period of 5.84 years.

Recognizing a typical 6-year cycle period, we can classify representative cycles from different forests at different times into a common scheme so as to compare numerical and biological data directly. Twenty different cycles from 13 different forests were averaged to give the pattern shown in Fig. 3a. The cycle is approximately symmetrical, showing an S-shaped pattern of increase followed by an S-shaped decline of the same magnitude. The amplitude of the typical cycle is approximately 40-fold, from just over 0.1 pupae per m^2 (year 1) to just under 5.0 (year 4). The steepest increase, from year 2 to 3, represents a 6.8-fold multiplication of the population.

Records of female pupal weight, and the proportion of parasitized pupae, show characteristic changes at different stages of the population cycle (Fig. 3b,c). The interpretation of these changes is discussed in Section 5 on causal mechanisms.

An important population parameter is the maximum per-capita rate of increase, sometimes called the finite rate of increase λ. This parameter can be estimated from real *Bupalus* population data by averaging the maximum rates of increase observed in the data. Of the 47 populations, 10 rather stable populations without pronounced cycles had a λ value of <3. The highest value ($\lambda>10$), was recorded by the strongly cyclic Tentsmuir population (Fig. 1c). Thus, cycles tend to be most pronounced in environments that are most favorable for the looper, where λ is highest.[6]

Variations in λ between populations are probably due to site or stand characteristics or to climatic variables acting both directly on the insect and indirectly through the medium of the host tree. For instance, outbreaks of pine loopers are associated with pine stands growing on areas with poor and mostly sandy soils, with average to low yield

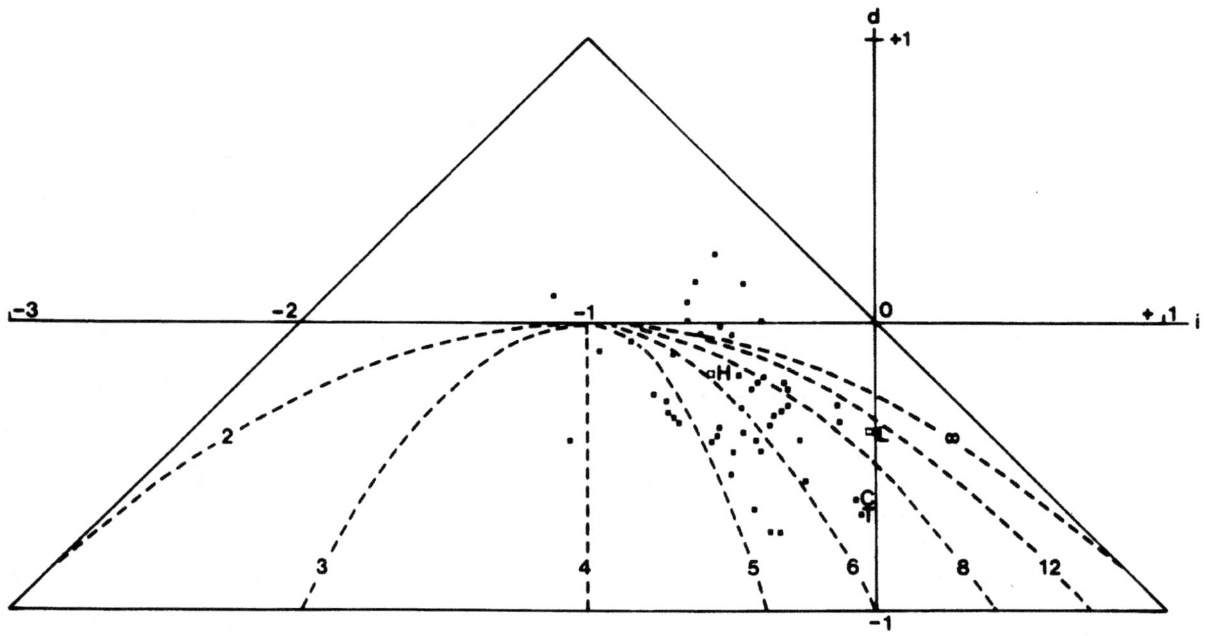

FIGURE 2. Parameters of the second-order model ($r_t = r_0 + iN_t + dn_{t-1}$) are calculated for 47 British (■) and 2 European (□) *Bupalus* populations. All but 10 populations fall below the parabola $(1+i)^2 + 4d = 0$, hence show a tendency to cycle. (---) show the expected cycle period. C, Cannock; T, Tentsmuir; L, Letzlingen; H, de Hoge Veluwe. (See Royama[52] for details and underlying theory.)

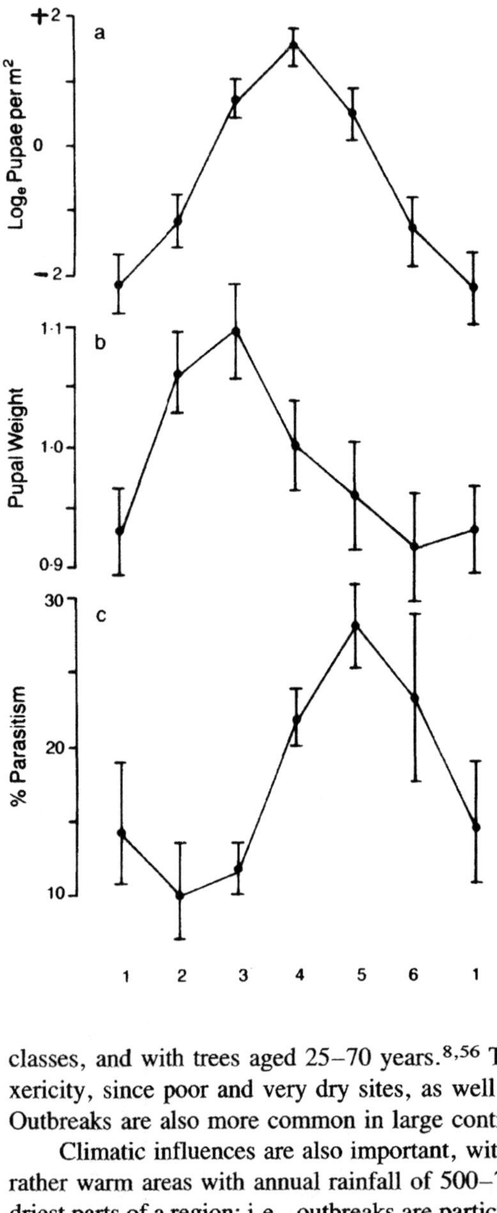

FIGURE 3. Analysis of a typical 6-year *Bupalus* cycle. The analysis is based on 20 recorded cycles in 13 different populations over the period 1954–1980. Mean ±95% confidence limits are shown for (a) $\log_e$ *Bupalus* pupae per m^2, (b) weights of female pupae as a proportion of the long-term mean weight for each population, and (c) percentage parasitism of pupae.

classes, and with trees aged 25–70 years.[8,56] They seem to prefer sites of intermediate xericity, since poor and very dry sites, as well as waterlogged ones, are little affected. Outbreaks are also more common in large continuous pine forests.[8]

Climatic influences are also important, with outbreaks occurring most frequently in rather warm areas with annual rainfall of 500–700 mm.[56] These are not necessarily the driest parts of a region; i.e., outbreaks are particularly prevalent in the wettest parts of the northeast German lowlands, but still in the range 600–700 mm rainfall.[16] The most serious outbreaks in Britain have occurred in a similar rainfall range, e.g., Tentsmuir (700 mm) and Cannock (750 mm). Thus, considerable areas of Scots pine forest growing in areas of more than 1000 mm annual rainfall are not considered at risk from *Bupalus* outbreaks and are not included in the routine pupal surveys.

The influence of weather on the population dynamics of *Bupalus* is more difficult to explain. Drought conditions are said to predispose stands to outbreaks and are usually associated with the preparation year, one or two seasons before the population peak.[15] The three outbreaks of *Bupalus* at Tentsmuir in Scotland (Fig. 1c) were each preceded by severe droughts in late summer in 1955, 1976, and 1983. However, outbreaks at other sites showed no such pattern, and an extensive correlation analysis of population trends in relationship to weather over the full range of survey sites showed little if any consistent influence of temperature or rainfall.[4]

5. HYPOTHESES FOR CAUSAL MECHANISMS

5.1. A General Hypothesis

A general inclusive hypothesis for the occurrence of cycles in population numbers states that they are caused by delayed negative feedback[6] or delayed density-dependence.[62] In other words, a time lag of some order operates in the negative feedback loop regulating population growth. The first step, therefore, is to identify the length of the time lag, after which it may be possible to deduce the causal factor or factors which are responsible.

The cyclical outbreaking populations of concern in both Britain and Germany are characterized by strong delayed density-dependent effects, measured with a time lag of exactly one generation. In these cases, density dependence of zero lag (the so-called immediate density-dependence) or a lag of two generations make negligible contributions to the determination of population trend. We can be confident that the factor(s) determining these cycles show what is essentially a 1-year time lag.

Any specific hypotheses for *Bupalus* cycles must explain this time lag, and most such explanations involve interactions between trophic levels. That is to say, interaction of *Bupalus* (the primary consumer level) with either its host plant (the producer level) or its natural enemies (the secondary consumer level). Explanations based on a single trophic level, such as intrinsic changes in the genetic structure of the population, seem implausible for purely theoretical reasons.[59]

5.2. Host Tree Responses

The first interaction we shall consider is that with the host plant. Certain characteristics of the host tree (e.g., age, nutritional state, water stress) are known to have important effects on *Bupalus*. However, I have not yet demonstrated that host tree condition changes in any systematic way in response to the *Bupalus* cycle. Fortunately, studies have been carried out on the growth of Scots pine under nutritional and climatic stress at Alltcailleach forest in Scotland, an area in which cyclical fluctuations in *Bupalus* numbers were also evident. These studies demonstrate that foliar nigrogen content is strongly and positively correlated with looper density after a lag of 2 or 3 years (Fig. 4). If foliar nitrogen is critical to looper survival and if trees under stress have higher levels of available nitrogen, it is possible to explain the cyclical population changes in terms of the tree's nitrogen economy. Thus, moderate defoliation could benefit trees by speeding up

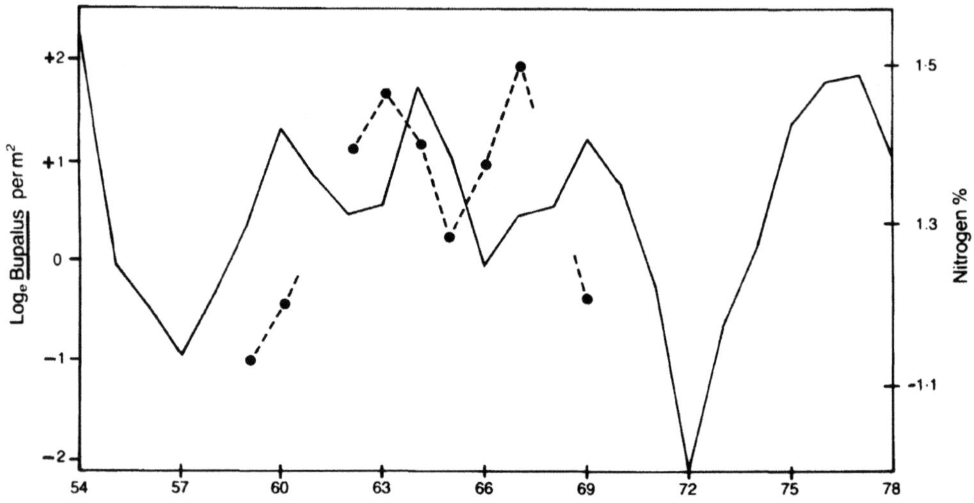

FIGURE 4. Annual estimates of *Bupalus* pupal densities at Dalliefour Wood, Alltcailleach Forest (——) and determinations of foliar nitrogen (---). (From Miller *et al.*[44])

the recycling of nutrients between foliage and roots,[43] or defoliation could be followed by increased concentrations of foliar nitrogen after a time lag, which could generate positive feedback to tree growth (equivalent to a relieving of nutritional stress).[18]

Closer examination of the Alltcailleach data exposes the fallacy of this explanation. For example, there is no correlation between nitrogen concentration and the rate of increase of *Bupalus* populations. Also, the time lag of at least 2 years between peak *Bupalus* numbers and enhanced nitrogen levels is too long, since we have already established that a feedback lag of 1 year is necessary to explain the observed cycles.

Other (indirect) evidence for the importance of food-plant quality comes from records of pupal weights, which are maximal at the end of the generation of greatest reproduction, and minimal following the steepest decline (Fig. 3b). Such a strong association ($r = 0.954$) between pupal weight and the rate of population change suggests that survival and weight may be determined by some common factor, such as the nutritional quality of the food. However, simple starvation effects have not been observed in field studies of *Bupalus*. The reduction in pupal size seems to be a more subtle effect, such as can result from increased contact between larvae in crowded populations[37] or from sublethal infections by a cytoplasmic polyhedrosis virus.[27]

Variations in female pupal weight lead, in turn, to changes in the fecundity of the resulting moths. The highest pupal weight observed in the third generation should be associated with a fecundity of 133 eggs per female.[9] The minimum pupal weight, in the sixth generation, would give a probable fecundity of 104. So the total reduction in fecundity (22%) is much too small to explain the transition from increasing to declining populations, where more than 97% total mortality is involved. It would only be possible if the progeny of the smaller females had greatly reduced viability, a hypothesis that has been disproved in laboratory and field experiments.[20,21]

5.3. Parasitoid Numerical Responses

As demonstrated by classical predator–prey models, host–parasitoid interactions can potentially generate population cycles, a hypothesis that has been proposed previously for *Bupalus* cycles.[47] Pupal parasitism has been shown to change systematically through the looper cycle in such a way that highest parasitism is recorded 1–2 years after peak numbers (Fig. 3c). This parasitism may influence the course of the cycle but can only account for a small part of the total mortality; i.e., in order to convert the population increase of generations 2–3 into a decrease of generations 5–6 requires a mortality of 97%, but maximal pupal parasitism is only 28%.

Parasitoids attacking other developmental stages (egg and larval parasitoids) may also play a role in population cycles. However, as these cannot be assessed from routine survey data, some other means of evaluating their impact must be used. The host-specific larval parasitoid *Dusona oxyacanthae* Boie kills pine looper larvae in the prepupal stage and moves away to spin its own cocoon in the forest litter. Comparison of independent estimates of *Dusona* abundance (cocoons in litter) with subsequent performance of the *Bupalus* populations[5] showed a highly significant negative correlation between *Dusona* abundance and *Bupalus* survival (Fig. 5).

Real data for *Bupalus* and *Dusona* populations can be used to develop parameters for

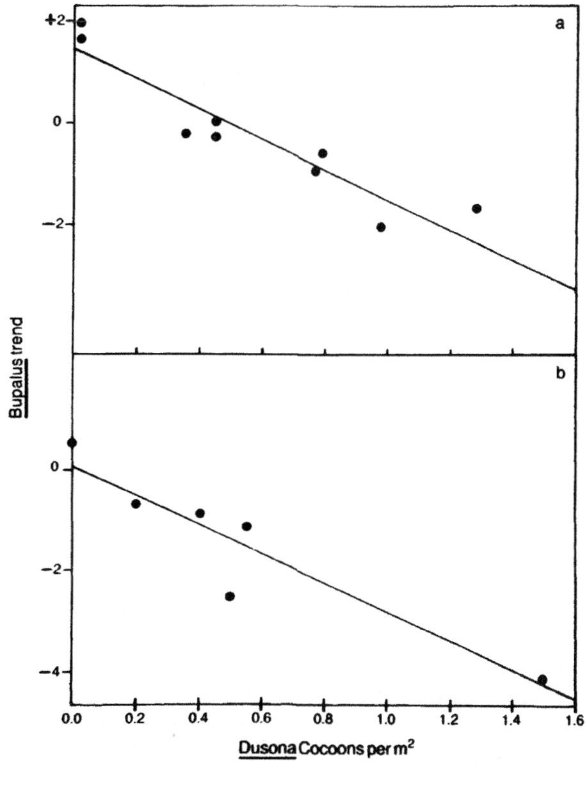

FIGURE 5. Correlations between *Bupalus* population trend, *r*, and the abundance of the parasite *Dusona oxyacanthae* at (a) the Cannock study plot in nine different years, and (b) six different sites in 1978–1979.

a simple host–parasitoid model.[24] The particular model used has three parameters: m, the mutual interference constant of the parasitoid; Q, the quest constant of the parasitoid; and λ the finite rate of increase of the host. Values of $m = 0.7$, $Q = 4.6$, and $\lambda = 25$ and 6, respectively, gave a realistic fit to the data shown in Fig. 5a,b.

Differences in the value of λ have important effects on the resulting population fluctuations.[23] For example, given $m = 0.7$, (1) for $\lambda<2$ one obtains a stable equilibrium with no tendency to cycle, (2) when $2<\lambda<25$ damped-stable cycles of abundance are produced, and (3) when $\lambda>25$ one obtains unstable cycles.[23]

Two 50-generation simulations with $\lambda = 3$ and 12, respectively, are shown in Fig. 6. The higher value causes the population to cycle with a period of about 7 years, while the lower value produces much less definite cycles. These differences may arise from a mortality factor acting on first-instar larvae.[5,37] Thus, in addition to providing a feasible explanation for *Bupalus* cycles, the host–parasitoid model also gives insight into factors that may change the system from a stable, noncyclical state to a wildly cycling condition, i.e., a reduction in first-instar mortality due, perhaps, to better quality pine foliage.

5.4. Diseases

Infectious diseases have also been proposed as causes of cycles in forest insect populations, particularly if the disease has high pathogenicity and long-lived infective stages.[2] A cytoplasmic polyhedrosis virus that has these properties can kill up to 70% of the loopers in declining populations,[29] and other types of virus and bacterial disease may sometimes be important.[7,37] The selection of broadly appropriate parameter values for *Bupalus* and its viral pathogen generates a cyclical solution of a pathogen–host model (Fig. 7). The dynamic pattern is characterized by a steady rise followed by a rapid fall in the host population, while the increase and decline in percentage infection tend to be more symmetrical. During the latent and increase phases, the infection virtually disappears from the host population, surviving only by virtue of its long-lived transmission stages.

This model, although superficially a feasible explanation for cycles in *Bupalus*, has weaknesses when applied to real forest insect populations. For example, virus epizootics by no means always manifest themselves in the decline phase of defoliator cycles.[4] In *Orgyia pseudotsugata*, it was found that realistic parameter estimates do not generate the expected cycles.[64]

5.5. Conclusion

Bupalus cycles seem to be induced by density-dependent factors acting with a one-generation time delay, the most plausible proximal factors being insect parasitoids. Host tree conditions due to stress may also be involved, particularly in determining areas in which outbreak cycles recur. It should be emphasized that the hypotheses need not be mutually exclusive. In principle, any regulatory factor that operates with a time delay of about 1 year will tend to cause population cycles within a period of around 6 years. Several such factors operating together would probably reinforce the cyclical pattern rather than disrupt it.

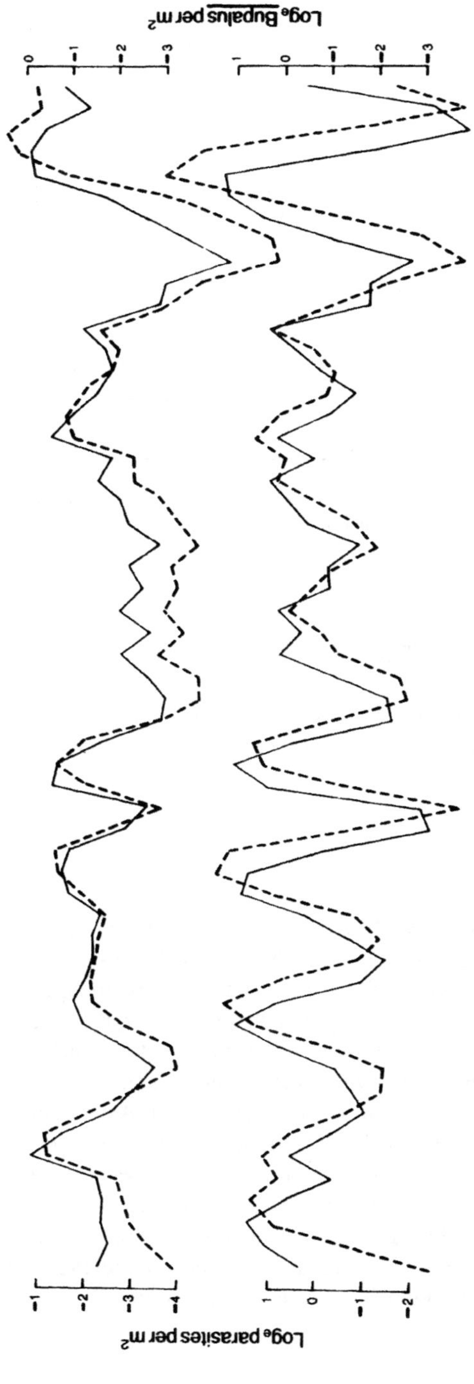

FIGURE 6. Simulations of the interaction between *Bupalus* and a hypothetical parasite population (using the model described by Hassell and Varley[24]), with λ = 3 (upper) and λ = 12 (lower). The random variations in parasite unrelated mortality are identical in both runs. Note the distinct cyclical pattern in the lower simulation, which is much less visible in upper one.

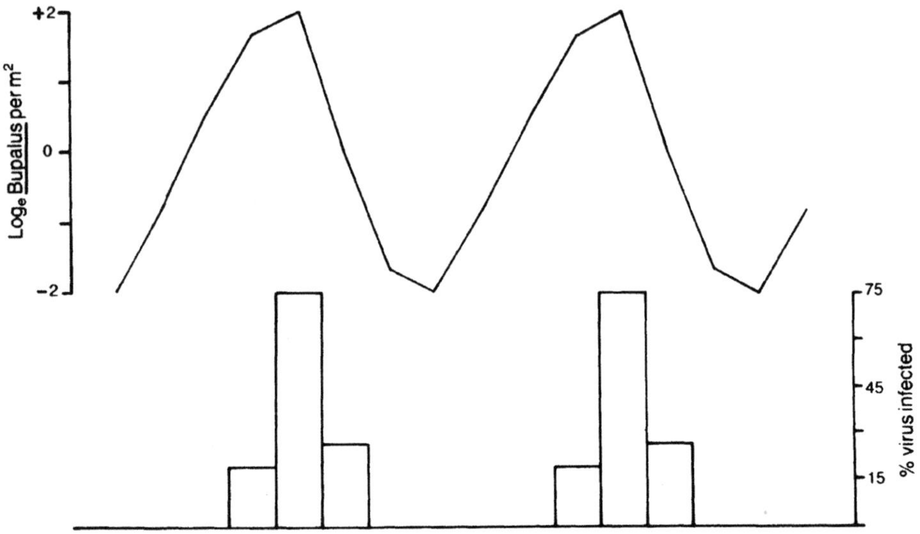

FIGURE 7. Simulated interaction between a *Bupalus* population and a cytoplasmic polyhedrosis virus (according to the model presented by Anderson and May.[2])

6. MANAGEMENT STRATEGIES

Understanding of the cyclical pattern of *Bupalus* outbreaks illuminates some new approaches for managing this important pest, as well as modifications to existing practices.

6.1. Surveillance

The delayed density-dependent processes that produce cyclical oscillations are not necessarily synchronized between individual forest blocks; thus, populations in some blocks may be increasing while others are declining. These independent population cycles in Britain are unlike those in Germany, which tend to be synchronized over wide areas.[32] Synchronized cycles are also common in other defoliators, such as *Panolis flammea*,[32] (see Chapter 13) *Zeiraphera diniana*,[3] (see Chapter 17), *Orgyia pseudotsugata*,[42] (see Chapter 10), and *Epirrita autumnata*,[60] (see Chapter 9). Asynchrony causes problems for surveillance because it is not possible to extrapolate from one population to another. Thus, suggestions that only a single block in an area should be monitored as an indicator for surrounding plantations should be resisted. Likewise, the extreme population oscillations make it impossible to predict the long-time susceptibility of forests to *Bupalus* from only a 1 or 2 year survey. In fact, it is necessary to monitor populations for a number of years before the potential amplitude of the cycle can be assessed.

6.2. Outbreak Prediction

If surveys show an increasing population over a 2- or 3-year period, we may wish to know whether a further increase is likely or conversely whether the population will stabilize or decline. Knowledge of pupal weight changes and parasitism during different phases of the cycle may help predict these trends. Specifically, a reversal of population increase is likely under any combination of the following conditions (Fig. 3):

1. Mean pupal weight has declined significantly from the preceding year.
2. Pupal parasitism exceeds 25%.
3. Cocoons of the parasite *Dusona oxyacanthae* are present in the litter at densities of two or more per m^2.

If none of these conditions is met, there is a high probability that the population will increase to outbreak levels in the coming season.

6.3. Control

The cyclical nature of *Bupalus* population dynamics has important implications for conventional insecticide control. It is essential that natural control mechanisms be allowed to develop, and to do this insecticide usage must be kept to a minimum. It would be a mistake to spray pre-emptively at an early phase of the cycle because natural controls, such as delayed density-dependent parasitism or disease, would be short-circuited. In that case, a rapid resurgence of the *Bupalus* population would be likely.

In addition, it is probably a mistake to apply insecticides so heavily that *Bupalus* populations are reduced to low levels over large blocks of forest. Simple models of the delayed density-dependent system show that the lower the trough, the higher the subsequent peak. This effect is counteracted if parts of a block containing moderate populations are left unsprayed. This is equivalent to deliberately leaving a reservoir of hosts to sustain small numbers of specific parasitoids and diseases. Thus, pressures from forest managers to spray more extensively and intensively should be resisted. Such reactions are motivated by a fear of increases in the surviving population next season; but, provided the control has not been applied too early, this is unlikely to be a great risk.

Recently, the selective insecticide diflubenzuron has been used for *Bupalus* suppression instead of the conventional organochlorine and organophosphorus insecticides. On the two occasions in which this insecticide was used, it proved highly effective, and the few surviving larvae were heavily parasitized. Unlike conventional insecticides, this substance has no effect on adult parasitoids, making it especially suitable for the type of minimal control strategy outlined above.

Finally, a useful insight for managers is the recognition of the effect of moderate changes in *Bupalus* survival or reproduction on the severity of outbreaks. Simulation models show that *Bupalus* populations with high rates of increase (λ values) have much more violent outbreak cycles. This makes it possible to consider cultural and other ameliorative measures that, by imposing a constant reduction of *Bupalus* reproduction and/or survival, would largely eliminate outbreaks, that is, would damp the cycles as shown in Fig. 6 (upper curve). Growing pines in mixed stands, either with other conifers or hardwoods, would be expected to have this effect. Hardwood mixtures are particularly useful for maintaining the complex of natural enemies, particularly general predators and nonspecific parasitoids.[30,65] Fertilizing and drainage can also reduce *Bupalus* survival.

A comprehensive understanding of the population dynamics of the pine looper would enable us to develop our techniques for managing the pest still further. The understanding that has been gained already has contributed to substantial improvements in our capacity to deal with the problem in just over 30 years' experience with it in Britain.

REFERENCES

1. Adlung, K. G., 1966, A critical evaluation of the European research on use of red wood ants (*Formica rufa* group) for the protection of forests against harmful insects, *Z. Angew. Entomol.* **57**:167–189.
2. Anderson, R. M., and May, R. M., 1981, The population dynamics of microparasites and their invertebrate hosts, *Philos. Trans. R. Soc.* B **191**:451–524.
3. Baltensweiler, W., 1964, *Zeiraphera griseana* Hübner (Lepidoptera: Tortricidae) in the European Alps. A contribution to the problem of cycles, *Can. Entomol.* **96**:792–800.
4. Barbour, D. A., 1980, Population dynamics of the pine looper moth *Bupalus piniaria* (L.) (Lepidoptera, Geometridae) in British pine forests, PhD. thesis, University of Edinburgh.
5. Barbour, D. A., 1985, Patterns of population fluctuation in the pine looper moth *Bupalus piniaria* L. in Britain, in: *Site Characteristics and Population Dynamics of Lepidopteran and Hymenopteran Forest Pests* (D. Bevan and J. T. Stoakley, eds.), pp. 8–20, Forestry Commission Research and Development Paper 135.
6. Berryman, A. A., 1985, Site characteristics and population dynamics: A theoretical perspective, in: *Site Characteristics and Population Dynamics of Lepidopteran and Hymenopteran Forest Pests* (D. Bevan and J. T. Stoakley, eds.), pp. 1–7, Forestry Commission Research and Development Paper 135.
7. Bevan, D., and Brown, R. M., 1961, The Pine looper moth *Bupalus piniarius* in Rendlesham and Sherwood Forests—1959, *Rep. Forest Res. (Lon.)* **1961**:172–179.
8. Bevan, D., and Brown, R. M., 1978, Pine looper moth, *For. Comm. Forest Rec.* 119.
9. Bevan, D., and Paramonov, A., 1962, Fecundity of the pine looper moth, *Bupalus piniarius, Rep. For. Res. (Lon.)* **1962**:174–176.
10. Botterweg, P. F., 1978, Moth behaviour and dispersal of the pine looper, *Bupalus piniarius* L. (Lepidoptera, Geometridae), *Neth. J. Zool.* **28**:341–464.
11. Crooke, M., 1959, Insecticidal control of the pine looper in Great Britain. I. Aerial spraying, *Forestry* **32**:166–196.
12. Crooke, M., 1970, Studies on tit and pine looper moth populations at Culbin Forest, *Rep. Forest Res. (Lon.)* **1970**:185.
13. Davies, J. M., 1962, The pine looper moth, *Bupalus piniarius,* at Cannock Chase in 1960, *Rep. Forest Res. (Lon.)* **1962**:176–182.
14. Ebert, W., 1967, Uber den Einfluss von Klima und Witterung auf den Massenwechsel des Kiefernspanners, *Arch. Forst.* **16**:787–792.
15. Ebert, W., 1968, Die Schadgebiete unserer wichtigsten Kiefern-bestandesschädlinge im Tiefland der DDR. I. Die Schadgebiete des Kiefernspanners und der Forleule, *Arch. Forst.* **17**:125–144.
16. Ebert, W., and Otto, D., 1969, Zur Populationsdynamik des Kiefernspanners (*Bupalus piniarius* Linné) in nordostdeutschen Tiefland, *Ber. 10 Wandervers Entomol.* (DAL Berlin) **80**:309–316.
17. Escherich, K., 1931, *Die Forstinsekten Mitteleuropas,* Vol. 3, Paul Parey, Berlin.
18. Fagerstron, T., Larsson, S., Lohm, U., and Tenow, O., 1978, Growth in Scots pine (*Pinus sylvestris* L.): A hypothesis on response to *Blastophagus piniperda* L. (Col., Scolytidae) attacks, *Forest Ecol. Man.* **1**:273–281.
19. Gibb, J. A., and Betts, M. M., 1963, Food and food supply of nestling tits (Paridae) in Breckland pine, *J. Anim. Ecol.* **32**:489–533.

20. Gruys, P., 1970, Growth in *Bupalus piniarius* (Lepidoptera: Geometridae) in relation to larval population density, *Agric. Res. Rep.* **742:**127.
21. Gruys, P., 1971, Mutual interference in *Bupalus piniarius* (Lepidoptera, Geometridae), in: *Dynamics of Populations. Proceedings of the Advanced Study Institute on "Dynamics of Numbers in Populations," Oosterbeek, The Netherlands, September 7–18, 1970,* pp. 199–207, PUDOC, Wageningen.
22. Haeselbarth, E., 1979, Zur Parasitierung der Puppen von Forleule (*Panolis flammea* [Schiff.]), Kiefernspanner (*Bupalus piniarius* [L.]) und Heidelbeerspanner (*Boarmia bistortata* [Goeze] in Bayerischen Kiefernwaldern, *Z. Angew. Entomol.* **87:**186–202, 311–322.
23. Hassell, M. P., and May, R. M., 1973, Stability in insect host–parasite models, *J. Anim. Ecol.* **42:**693–726.
24. Hassell, M. P., and Varley, G. C., 1969, New inductive population model for insect parasites and its bearing on biological control, *Nature (Lon.)* **223:**1133–1137.
25. Herrebout, W. M., 1966, The fate of the eggs of *Eucarcelia rutilla* VILL. deposited upon the integument of the host, *Z. Angew. Entomol.* **58:**340–355.
26. Herrebout, W. M., 1969, Some aspects of host selection in *Eucarcelia rutilla* VILL, *Neth. J. Zool.* **19:**1–104.
27. Hukuhara, T., 1985, Pathology associated with cytoplasmic polyhedrosis viruses, in: *Viral Insecticides for Biological Control* (K. Maramorosch and K. E. Sherman, eds.), pp. 121–162, Academic, London.
28. Isaev, A., and Khlebopros, R. G., 1977, Inertial and non-inertial factors regulating forest insect population density, in: *Proceedings of a Conference on Pest Management. Int. Appl. Systems Analysis CP-77-6, Laxenburg, Austria* (G. A. Norton and C. S. Holling, eds.), pp. 317–319, International Institute for Applied Systems Analysis, Laxenburg, Austria.
29. Jahn, E., and Sinreich, A., 1957, Zum Auftreten des Kiefernspanners, *Bupalus piniarius* L. im Burgenland in den Jahren 1952–1956, *Z. Angew. Entomol.* **41:**184–195.
30. Klimetzek, D., 1972, Die Zeitfolge von Ubervermehrungen nadelfrassender Kiefernraupen in der Pfalz seit 1810 und die Ursachen ihres Rückganges in neuerer Zeit, *Z. Angew. Entomol.* **71:**414–428.
31. Klimetzek, D., 1975, Umfang und Auswirkung von Begiftungsaktionen gegen Kiefernraupen in Nordbayern, *Allg. Forst. Z.* **146:**186–191.
32. Klimetzek, D., 1979, *Insekten-Grosschädlinge an Kiefer in Nordbayern und der Pflaz: Analyse und Vergleich 1810–1970,* Institut für Forstzoologie, Freiburg.
33. Klomp, H., 1953, Die Bedeutung der populations-dynamischen Forschung in der angewandten Entomologie, *Mitt. Biol. Zentr. (Berl.)* **75:**28–33.
34. Klomp, H., 1956, Die morphologischen Merkmale und die Bionomie der Kiefernspanner-tachine *Carcelia obesa* Zett. (=*rutilla* B.B.), *Z. Angew. Entomol.* **38:**288–294.
35. Klomp, H., 1958, On the synchronization of the generations of the tachinid *Carcelia obesa* Zett. (=*rutilla* B.B.) and its host *Bupalus piniarius* L., *Z. Angew. Entomol.* **42:**210–217.
36. Klomp, H., 1959, Infestations of forest insects and the role of parasites, in: *Proceedings of the Fifteenth International Congress on Zoology, London, 1958.*
37. Klomp, H., 1966, The dynamics of a field population of the pine looper *Bupalus piniarius* L. (Lep. Geom.), *Adv. Ecol. Res.* **3:**207–305.
38. Klomp, H., 1968, A seventeen-year study of the abundance of the pine looper *Bupalus piniarius* L. (Lepidoptera: Geometridae), *Symp. R. Entomol. Soc. (Lon.)* **4:**98–108.
39. Klomp, H., 1974, Population dynamics: A key to the understanding of integrated control, in: *Insects: Studies in Population Management* (P. W. Geiber, L. R. Clark, D. J. Anderson, and H. A. Nix, eds.), pp. 69–79, *Ecol. Soc. Aust.* (memoirs 1), Canberra.
40. Klomp, H., and Gruys, P., 1965, The analysis of factors affecting reproduction and mortality in a natural population of the pine looper, *Bupalus piniarius* L., *Proceedings of the Twelfth International Congress on Entomology, London, 1964.* pp. 369–372.
41. Klomp, H., and Teerink, B. J., 1962, Host selection and number of eggs per oviposition in the egg–parasite *Trichogramma embryophagum* HTG, *Nature (Lond.)* **195:**1020–1021.
42. Mason, R. R., and Luck, R. F., 1978, Numerical behaviour of populations, in: *The Douglas-fir tussock moth: A synthesis.* (M. H. Brooks, R. W. Stark, and R. W. Campbell, eds.), pp. 39–47, U.S. Forest Service Tech. Bull. #1585.
43. Mattson, W. J., and Addy, N. D., 1975, Phytophagous insects as regulators of forest primary production, *Science* **190:**515–522.

44. Miller, H. G., Miller, J. D., and Binns, W. O., 1977, Growth of Scots pine under nutritional and climatic stress, *Plant Soil* **48:**103–114.
45. Mook, J. H., Mook, L. J., and Heikens, H. S., 1963, Further evidence for the role of "searching images" in the hunting behavior of titmice, *Arch. Neerl. Zool.* **13:**448–465.
46. Moran, P. A. P., 1953, The statistical analysis of the Canadian Lynx cycle, *Aust. J. Zool.* **1:**163–173, 291–298.
47. Ohnesorge, B., 1963, Beziehungen zwischen Regulationsmechanisms und Massenwechselablauf bei Insekten, *Z. Angew. Zool.* **50:**427–483.
48. Oldiges, H., 1959, Der Einfluss der Waldbodendüngung auf das Auftreten von Schadinsekten, *Z. Angew. Entomol.* **45:**49–59.
49. Otto, D., and Geyer, W., 1970, Zur Bedeutung des Kiefernnadelhurzes und des Kiefernnadelöles für die Entwicklung Nadelfressender Insekten. *Arch. Forst.* **19:**151–167.
50. Poole, R. W., 1974, *An Introduction to Quantitative Ecology,* McGraw-Hill, New York.
51. Potts, G. R., Tapper, S. C., and Hudson, P. J., 1984, Population fluctuations in red grouse: Analysis of bag records and a simulation model, *J. Anim. Ecol.* **54:**21–36.
52. Royama, T., 1977, Population persistence and density dependence, *Ecol. Monog.* **47:**1–35.
53. Royama, T., 1984, Population dynamics of the spruce budworm *Choristoneura fumiferana, Ecol. Monog.* **54:**429–462.
54. Schoonhoven, L. M., 1963, Diapause and the physiology of host parasite synchronization in *Bupalus piniarius* L. (Geom.) and *Eucarcelia rutilla* Vill. (Tach.), *Arch. Neerl. Zool.* **26:**111–174.
55. Schwenke, W., 1968, Neue Hinweise auf eine Abhängigkeit der Vermehrung blatt und nadelfressender Forstinsekten von Zuckergehalt ihrer Nahrung, *Z. Angew. Entomol.* **611:**365–369.
56. Schwenke, W., 1978, *Forstchadlinge Europas,* Vol. 3: *Schmetterlinge,* Paul Parey, Hamburg.
57. Schwerdtfeger, F., 1935, Studien über den Massenweschsel einiger Forstschädlinge, *Z. Forst. Jagdw.* **67:**15–38, 85–104, 449–482, 513–540.
58. Schwerdtfeger, F., 1941, Uber die Ursachen des Massenwechsels der Insekten, *Z. Angew. Entomol.* **28:**254–303.
59. Stenseth, N. C., 1981, On Chitty's theory for fluctuating populations: The importance of genetic polymorphism in the generation of regular density cycles, *J. Theor. Biol.* **90:**9–36.
60. Tenow, O., 1972, The outbreaks of *Oporinia autumnata* Bkh. and *Operophthera* spp. (Lep. Geometridae) in the Scandinavian mountain chain and northern Finland 1862–1968, *Zool. Bidr. Uppsala.* (Suppl. 2)
61. Varley, G. C., 1949, Population changes in German forest pests, *J. Anim. Ecol.* **18:**117–122.
62. Varley, G. C., and Gradwell, G. R., 1970, Recent advances in insect population dynamics, *Annu. Rev. Entomol.* **15:**1–24.
63. Veen, J. C., Van, 1981, The biology of *Poecilostictus cothurnatus* (Hymenoptera, Ichneumonidae) an endoparasite of *Bupalus piniarius* (Lepidoptera, Geometridae), *Ann. Entomol. Fenn.* **47:**77–93.
64. Vezina, A., and Peterman, R. M., 1985, Tests of the role of a nuclear polyhedrosis virus in the population dynamics of its host, Douglas-fir tussock moth, *Orgyia pseudotsugata* (Lepidoptera: Lymantriidae), *Oecologia (Berl.)* **67:**260–266.
65. Voute, A. D., 1964, Harmonious control of forest insects, *Int. Rev. Forest Res.* **1:**325–383.
66. White, T. C. R., 1974, A hypothesis to explain outbreaks of looper caterpillars with special reference to populations of *Selidosema sauvis* in a plantation of *Pinus radiata* in New Zealand, *Oecologia (Berl.)* **16:**279–301.

CHAPTER 16

THE SPRUCE BUDWORM IN EASTERN NORTH AMERICA

William J. Mattson, Gary A. Simmons, and John A. Witter

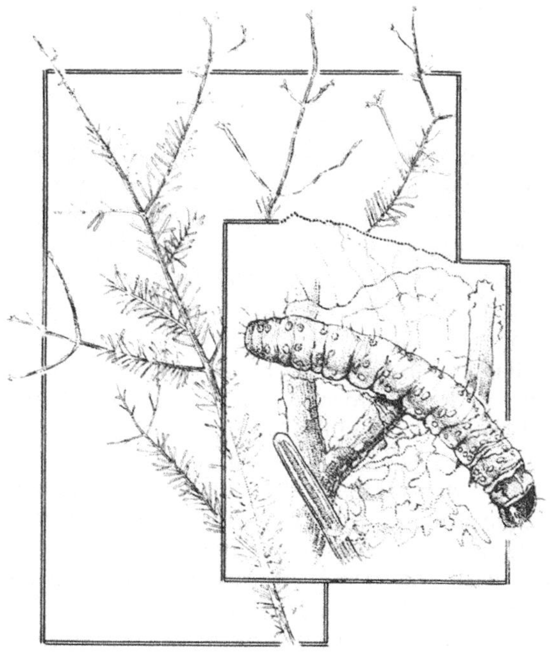

William J. Mattson • U.S.D.A. Forest Service, North Central Forest Experiment Station, Pesticide Research Center, Michigan State University, East Lansing, Michigan 48824. **Gary A. Simmons** • Department of Entomology, Michigan State University, East Lansing, Michigan 48824. **John A. Witter** • School of Natural Resources, University of Michigan, Ann Arbor, Michigan 48109.

1. INTRODUCTION

The spruce budworm, *Choristoneura fumiferana* (Clemens), (Lepidoptera: Tortricidae) was first described from specimens collected in Virginia,[17] but this native insect occurs primarily in the northern boreal forest from Newfoundland west to the McKenzie River near 66°N.[53] The most extensive and destructive outbreaks have occurred in the maritime provinces (New Brunswick, Nova Scotia, Newfoundland), Quebec, Ontario, Maine, and the Great Lakes states. This defoliator feeds primarily on the new growth of balsam fir (*Abies balsamea*), red spruce (*Picea rubens*), white spruce (*Picea glauca*), and black spruce (*Picea mariana*). Sometimes it feeds on other conifers, such as eastern larch (*Larix laricina*), eastern hemlock (*Tsuga canadensis*), Engelmann spruce (*Picea engelmannii*), subalpine fir (*Abies lasiocarpa*), and eastern white pine (*Pinus strobus*).[28,47]

1.1. Economic Impact

The spruce budworm poses an economic threat to more than 60 million ha of susceptible spruce–fir forests in Canada and the eastern United States. For centuries, the North American boreal forest has experienced natural periodic spruce budworm outbreaks. As an integral component of spruce–fir forests, the budworm is no ecological threat because it normally does not interfere with their continuity in space and time. In fact, fir and spruce usually regenerate rapidly after spruce budworm outbreaks. However, from the land manager's point of view, the impact of the spruce budworm can be considerable, including growth loss, top kill, cone and seed mortality, and widespread tree mortality (Table I). Radial growth usually declines by 30–90% in the first year following heavy defoliation.[35] Top kill and tree mortality typically begin during the third year of an outbreak, and seed production is virtually eliminated. The average annual wood loss to mortality and growth reduction was 42.5 million m^3 in eastern Canada from 1978 to 1982.[84] This is equivalent to about two thirds of the annual harvest and about one half the annual allowable cut of softwoods in eastern Canada.

Any spruce–fir stand in eastern North America is susceptible to spruce budworm feeding. There are no immune trees or stands. However, the size of the budworm population and the resulting growth loss and tree mortality varies directly with certain stand attributes. Factors contributing to chronic outbreaks, and severe damage are (1) a high composition of mature trees, (2) a large percentage of balsam fir, (3) open stands with spiked tops protruding above the forest canopy, (4) abnormally dry or wet sites, (5) extensive and continuous spruce–fir stands, (6) downwind position from current outbreaks, and (7) elevations lower than 700 m and latitudes south of 50°N.[84]

Spruce–fir forests of similar composition usually grow back after spruce budworm outbreaks because susceptible stands usually contain an abundance of spruce/fir seedlings and saplings in the understory. Although larvae may spin down in large numbers from the denuded overstory trees and feed on the reproduction, mortality from budworm feeding is low. Thus, major changes in forest composition do not usually occur after budworm outbreaks.

The status of the spruce budworm as a forest management problem depends primarily

TABLE I
Succession of Events Associated with a Spruce Budworm Outbreak on Balsam Fir without Human Intervention[a]

Years of severe defoliation[b]	Impact
1	Flowers and cone crops die.
	Radial growth loss occurs in the upper crown.
2–3	Small roots begin to die.
	Radial growth loss occurs over the entire stem.
	Height growth ceases.
	Some treetops die.
4–6	Suppressed trees in the understory and mature and overmature trees in the overstory begin to die.
	Tree growth and wood production nearly cease.
7–15	Budworm populations begin to collapse.
	More trees die, particularly balsam fir.
	Some seedlings and saplings die.
	Dead trees begin to deteriorate as a result of disease, secondary insect attack, and wind breakage.
	Protective cover in deer yards is diminished.

[a]Adapted from Witter *et al.*[84]
[b]Seventy-five percent or more of current year's growth.

on current social and economic conditions and the objectives of the landowner.[18] Thus, there are considerable regional variations in the objectives and strategies for managing the insect. In eastern Canada and Maine, for instance, the budworm was not considered a problem until the 1930s, when the development of a pulp and paper industry created an economic demand for fir and spruce fiber. Major economic, social, and political considerations in New Brunswick led to the use of insecticides for foliage protection when the spruce budworm outbreak started during the early 1950s. The situations in Quebec and Maine are similar, except that these regions have not used insecticides as intensively. In Ontario, land managers have emphasized protecting high value recreation areas and harvesting other susceptible and infested stands as markets become available. Managers in Michigan, Wisconsin, and Minnesota have primarily allowed outbreaks to run their course, but susceptible and heavily infested stands have been harvested when markets were available. The regional economies and the forests of Ontario and the Great Lakes states are far more diverse than those in New Brunswick, Quebec, and Maine. Thus, forest industries in the Great Lakes Region are less dependent on fir and spruce.

2. BIOLOGY AND BEHAVIOR

2.1. Life Cycle

The spruce budworm has a 1-year life cycle (Fig. 1). Moths emerge from pupae in late June to late July and mate soon afterwards. Female moths usually deposit about 170 eggs over a 1–2-week period, laying 5–50 eggs in each of 5–10 egg clusters, with an

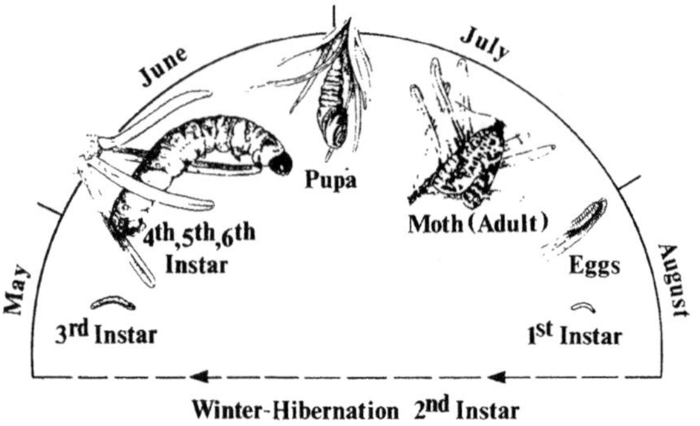

FIGURE 1. Life cycle of the spruce budworm.

average cluster containing 20 eggs. Egg clusters are found primarily on the new needles of peripheral shoots of fir and spruce at all crown levels. However, most eggs are deposited in the upper half of the crown, and usually on the tallest trees.[47,69]

Eggs hatch within 10–14 days and the first-instar larvae, reacting photopositively, move toward the branch tips. This movement can be interrupted by high temperatures that cause larvae to settle down and spin a cocoonlike shelter called a hibernaculum. Turbulent winds or crowded conditions on branch tips cause larvae to drop and to balloon outward on silken threads. Newly hatched larvae do not feed and, after spinning a hibernaculum, they molt to the second instar and then overwinter in diapause. Hibernacula can occur on both host and nonhost trees between bark scales, bark fissures, staminate flower bracts, or in lichens.

In late April to mid-May, second-instar larvae emerge from the hibernacula. Again responding to light, they move toward the branch tips, where wind dispersal can cause their redistribution (up to several kilometers) throughout the forest. After settling, they mine old needles, unopened vegetative buds, or feed on staminate flowers. As the buds open and the shoots expand, the caterpillars begin feeding on the succulent new foliage. A half-grown larva constructs a small silken tunnel on a developing shoot and feeds inside this shelter. Larger larvae prefer the current year's tender foliage but may be forced to consume old foliage when populations are high. The mature larvae are 20–30 mm long and weigh about 100 mg fresh weight. This represents at least a 1500-fold increase in mass since first feeding. Females are usually twice as large as males. Pupation occurs during mid-June to early July within the feeding shelter or in other protected sites. After 10–14 days, the adult moth emerges and the budworm has completed its 1-year life cycle.

2.2. Mating Behavior

Adult emergence is largely temperature-dependent, occurring primarily in mid-afternoon. Virgin females normally remain sedentary until darkness after which "calling" (extrusion of the sex pheromone producing gland) begins.[58,61] Mating, which varies from 3 to 6 hours depending on ambient temperature, may occur twice during the first 24-hr period. Up to one third of females will then move from their place of emergence while the other two thirds remain in place. The first and second egg masses are laid in early to mid-afternoon of the second day, usually on the same tree that the female fed on as a larva. The remaining egg masses are laid elsewhere in the same or other stands.

2.3. Adult Dispersal

Newly emerged females are not vigorous fliers until the second or third day and after having laid about half of their eggs. At this time, females often leave heavily defoliated stands in mass flights.[26] Weather conditions dictate the initiation, direction, and distance of flights. There are three potential mechanisms for moth dispersal: (1) prevailing wind dispersal with moths being transported above tree tops by wind; (2) cold front dispersal, where moths are lifted into the atmosphere by cold fronts and deposited en masse downwind; and (3) cold temperature sink dispersal, where local moths are trapped in and dispersing moths pulled into local pockets of cold air. Dispersal flights are regular, nightly occurrences when population densities are high. On the other hand, moths usually do not disperse from lightly or moderately defoliated stands. Dispersing females begin their long range flight by ascending to heights of 150 to 300 m above the forest canopy. Take off occurs around 1930 hours when light and temperature conditions are changing. Females make several flights on separate evenings with each flight covering about 80 km. Sometimes large numbers of moths can invade regions as far as 600 km from their point of origin.[20] There is evidence that moths are visually attracted to the spires of dominant host trees that break through the forest canopy and may use them as cues for landing.

3. ECOLOGICAL RELATIONSHIPS

3.1. Relationships with Host Plants

3.1.1. Variety of Hosts

The spruce budworm occurs naturally on all the native spruces and firs in eastern North America as well as on several exotic species that occur there as ornamentals. In fact, spruce budworm larvae are capable of feeding on practically any species in most genera of the family Pinaceae: *Abies, Larix, Picea, Pinus, Pseudotsuga,* and *Tsuga*.[28] This extremely wide host range suggests that budworm larvae are capable of accepting and detoxifying a wide range of host defensive compounds, because each genera has unique spectra of these materials.[79] This implies that the spruce budworm's apparent specialization on just two primary hosts (white spruce and balsam fir) is largely due to behavioral and/or ecological constraints. For example, the electromagnetic spectrum and volatile cues from white spruce and balsam fir may provide the optimal stimuli for host acceptance and oviposition, or the phenology of these species may be more closely matched to that of the spruce budworm's feeding cycle. Furthermore, these observations suggest that there is unlikely to be any intraspecies genetic differences in biochemical resistance/susceptibility to budworm because larvae are physiologically capable of dealing with even larger biochemical differences among the major genera of the Pinaceae.

3.1.2. Synchrony with Host Growth and Chemistry

Second-stage budworm larvae emerge from their overwintering hibernacula in late April to mid-May (at about 110 degree-days, using 2.7°C as a base and January 1 as a starting date). This is usually several days before bud break of balsam fir and white spruce which occurs at about 236 and 254 degree-days (dd) respectively. At about 40 days postemergence, or at about 700 degree-days, they finish feeding and enter the pupal stage. By this time, white spruce has already completed its annual shoot elongation (about 640 dd), while balsam fir shoots are still about 2 weeks away from finishing growth (about 875 dd). Black spruce, which is a much less common host, initiates bud break much later (about 330 dd) and likewise completes its shoot elongation substantially later (at about 1000 dd).

How important is it for the budworm to be closely synchronized with its host plants? No one yet knows for sure but there is reason to suspect that it is crucial.[66] For example, needle mining by second-instar larvae is much less common when buds have already begun expansion so that larvae can get between the normally impervious bud scales. This is especially true when there are abundant staminate flower buds because they normally break before foliar buds.[11] On the other hand, if after leaving mined needles, the third instars still cannot enter buds, owing to their retarded expansion, they usually disperse and commonly die.[21] This observation has been used to explain the apparent resistance of black spruce to serious budworm outbreaks.[66]

When second-stage larvae mine year-old needles, they face perhaps the most nutrient-impoverished food during their entire lives. These needles are low in nitrogen (less than 1%) and phosphorous (about 1500 ppm), both of which must be concentrated nearly 8-fold to attain the insect's usual tissue levels.[42,43] As soon as third-instar larvae begin feeding on expanding needles, they encounter the most nutrient-rich food of their lives. At this time, nitrogen and most mineral elements are at their seasonal maxima. These nutrients decline in concentration as needles and shoots elongate, typically following a hyperbolic decay function with time (Fig. 2a). By contrast, sugar concentrations are extremely low in the newly breaking buds but steadily increase up to the time that larvae reach the pentultimate stage (Fig. 2b). Crude fat levels likewise tend to rise steadily during budworm development.[64,65] In short, protein and mineral levels decline as needles age while the opposite is true for energy giving substrates such as crude fats and sugars.

At the initiation of budworm feeding on newly developing needles, the concentration of alleochemicals would be expected to be low because cells are using resources primarily for growth, not the production of secondary metabolites.[38] Several studies[78] confirm this, at least with respect to mono- and sesquiterpenes whose concentrations steadily increase in new foliage until the end of the growing season. For example, there is a four- to fivefold increase in volatile oils over this period. On the other hand, total phenolic concentrations show the opposite pattern. Highest values (4–6%) occur early in the season and then decay hyperbolically with time (Fig. 2c). This means that late season levels are about one half the early season maxima for both balsam fir and white spruce.

3.1.3. Effects of Leaf Chemistry on Spruce Budworm

The mixtures of nutrients and allelochemics in host foliage profoundly affect spruce budworm growth and survival, with nutrients having a greater apparent impact than

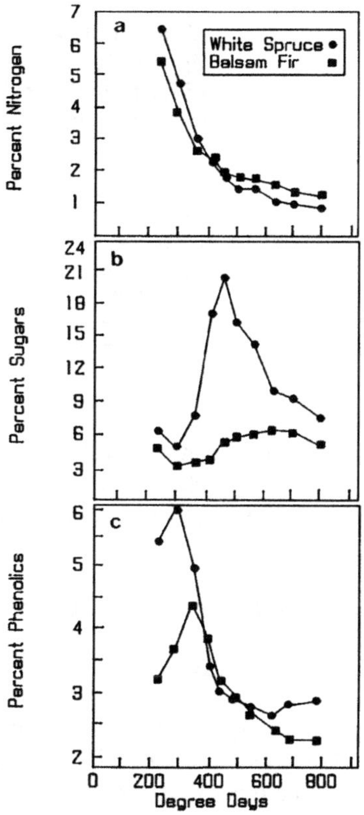

FIGURE 2. Seasonal changes in the foliar levels (% dry weight) of nitrogen (a), sugars (glucose, fructose, sucrose) (b), and total phenolics (c) in new growth of balsam fir and white spruce. Each point represents an average of three trees.

constitutive allelochemics.[42] In particular, variations in foliar nitrogen seem most important, probably because nitrogen is the most limiting nutrient. Variation in foliar sugars do not seem as important because spruce budworm growth responses are relatively insensitive to increases in sugars.[45] Budworm larvae can apparently detoxify large quantities of most monoterpenes because when individual compounds have been added to artificial diets, mortality and growth reductions occur only at abnormally high concentrations. Little is known about the budworm's ability to cope with the more complex terpenoids such as sesquiterpenes and resin acids. The same is true for the broad spectrum of phenolic compounds found in both firs and spruces.

There is limited information about inducible defenses in the two major host plants. Preliminary evidence suggests that small scale, branch by branch, defoliation of current and 1-year-old fir needles elicits a weak, immediate inducible defense that lowers budworm growth but not development or survival (Table II). Likewise, long-term and heavy foliage depletion, as occurs during outbreaks, also lowers growth because it causes budworms to consume older foliage, which is much less nutritious, tougher (especially in spruces), and higher in allelochemical content. Such feeding invariably results in smaller, less fecund insects (50% lower), reduced survival of small larvae, but increased survival of large larvae.[47] Whether a true induced defensive reaction occurs under these conditions is not known, but if it does, it is insufficient to prevent widespread tree mortality in

TABLE II
Mean Adult Weight, Development Time, and Survival of Spruce Budworm Fifth-Instar Larvae Placed on Branches of Balsam Fir Having Different Levels of Prior Defoliation

	Prior defoliation level		
Performance variable	0%	25%	50%
Weight (mg)[a]	22.1	19.3	19.9
Development time (days)	24.3	24.7	24.8
Survival (%)	63.3	61.7	62.0

[a]The weight of the 0% prior defoliation differs significantly ($p < 0.01$) from that of the 25% and the 50% prior defoliation.

susceptible, mature forests.[11] In immature fir forests, however, and in those dominated by white and/or black spruce, outbreaks often collapse while most of the trees are still alive and bearing sufficient foliage for oviposition and feeding.[28]

3.3. Relationships with Natural Enemies

3.3.1. Predators

The spruce budworm is attacked by an enormous number and variety of predators during its life cycle,[33,36,47,48] many of which attack only certain developmental stages. On the other hand, others attack all or most life stages by virtue of their extraordinary mobility and hunting and trapping skills. Birds, spiders, and ants belong to this latter class. Some of the largely life-stage specific predators are mites, plant bugs, lacewings, and beetles on eggs; phalangids, and beetles on small larvae; beetles, wasps, and small mammals on large larvae and pupae; and dragonflies and robber flies on adult moths.

The significance of predation in the population dynamics of the spruce budworm is a controversial subject. There is universal consensus, however, that the influence of predators is insignificant when budworm populations are high. During most other years, when densities are low, there is a longstanding hypothesis that predators can regulate budworm numbers at low densities by their rapid feeding (functional) responses and breeding (numerical) responses to changes in budworm abundance.[29] Numerical responses are derived not only from increased breeding but also from flocking (behavioral numerical response) that permits immediate increases in predation in response to localized increases in budworm density. However, some argue that predators, especially birds, are insignificant because they tend to feed on alternative more easily available foods when budworm numbers are low; later when budworm numbers are high birds may feed vigorously on them although to little avail because they do not substantially increase their breeding populations.[57] However, many birds (especially warblers) can increase their population densities in response to spruce budworm, but no one has yet unequivocally demonstrated their role in regulating budworm populations.[15,82] The best evidence for the regulatory role of bird predation comes from population studies of the closely related western spruce budworm, *C. occidentalis*.[36,75]

3.3.2. Parasites

There are close to 100 species of flies and wasps that parasitize budworms.[47,74] Only a handful, however, are common, and each is highly stage specific, e.g., *Trichogramma minutum* (on eggs), *Apanteles fumiferanae, Glypta fumiferanae, Synetaeris tenuifemur* (on small larvae), *Meteorus trachynotus, Actia interrupta, Aplomya caesar, Phryxe pecosensis* (on large larvae), and *Agria housei* (on pupae).[74]

As was true for predators, there is widespread consensus that parasites are ineffective at high budworm densities, but they may have some regulatory capacity at low densities. In order to be effective in regulating budworms, the parasites must respond in a density-dependent way to changes in budworm numbers. Recent studies in New Brunswick indicate that parasites may be among the most significant mortality factors affecting large larval survival, the life stage that determines population trends.[47,57] But their effect is significant only when coupled to a largely unknown source of mortality, the fifth agent.[57] Other studies in the western United States concluded that parasites probably have an insignificant impact on the dynamics of the western spruce budworm.[76]

3.3.3. Microorganisms and Unknown Factors

The spruce budworm is remarkably free of serious microorganism parasites although it is known to harbor at least 15 species in four diverse groups: three viruses, several species of bacteria and fungi, and three or more protozoans (Microsporidia).[49–51] Besides these, consistent mortality occurs in wild and laboratory populations due to some unknown factor(s). In fact, this unknown mortality often equals or exceeds that which can be attributed to disease, and fluctuates from year to year (range 2–30%).[49,54] This complex of diseases and unknown mortality factors has been called the fifth agent.[57]

Of all the diseases, the Microsporidia and fungi are usually most prevalent and therefore assumed to be most important. However, the disease organisms appear to be of rather low virulence,[49] and their impact on budworm survival and fecundity is highly dependent on coincidental environmental stresses (e.g., crowding, shortages of food, and nonoptimal temperatures and humidity). For example, it has been shown that budworm mortality from disease tends to change nonlinearly with temperature, as does unknown mortality.[49]

Long-term studies in New Brunswick support the hypothesis that the mortality caused by fifth agent factors may be largely responsible for driving the oscillations of spruce budworm. These factors are likely candidates for regulation of the budworm cycles because they appear to have the necessary strong numerical response that is weak or lacking in other mortality factors, such as parasites and predators. Moreover, their synergistic interaction with the host plant and the physical environment may constitute a fundamental triad that basically governs the budworm's population dynamics.

3.3.4. Competitors

The spruce budworm coexists with a host of other phytophagous insects in the buds, foliage, cones, and staminate flowers. For example, there are at least 20 other species of Lepidoptera associated with the budworm,[37] not to mention members of other taxonomic groups such as aphids, spittlebugs, and sawflies. Normally, one does not think about contemporaneous species of folivores competing with one another for food and space because food and space are usually abundant. However, there are some reports of competition and outright predation on spruce budworm by the larvae of various species of *Dioryctria*.[2,80] In addition, high populations of the balsam twig aphid (*Mindarus abietinus*) can lower the susceptibility of balsam fir to the spruce budworm by causing stunted shoots and distorted needles, probably rendering them less desirable for oviposition.[41] In addition, needles covered with the sticky honeydew and waxy secretions of the aphid are less suitable as feeding sites (Fig. 3). Because twig aphids are most abundant in young forests, they may interfere with the budworm's ability to successfully exploit such stands. Furthermore, they may promote populations of polyphagous predators, such as coccinellids, which could provide additional predator pressure on budworm populations.

3.3.5. Life Tables

The various mortality factors acting on the spruce budworm are summarized in a life table (Table III). Most individuals (85%) die between the egg and the third larval stage before beginning to feed on the opening buds. Of this mortality, only about 18% are killed by parasites and predators, the remainder being lost to unknown factors and dispersal. From the initiation of bud feeding to the final emergence of adults (the key age interval in determining budworm population trends), parasites kill 42%, unknown factors 25%, diseases 23%, and predators less than 1%.

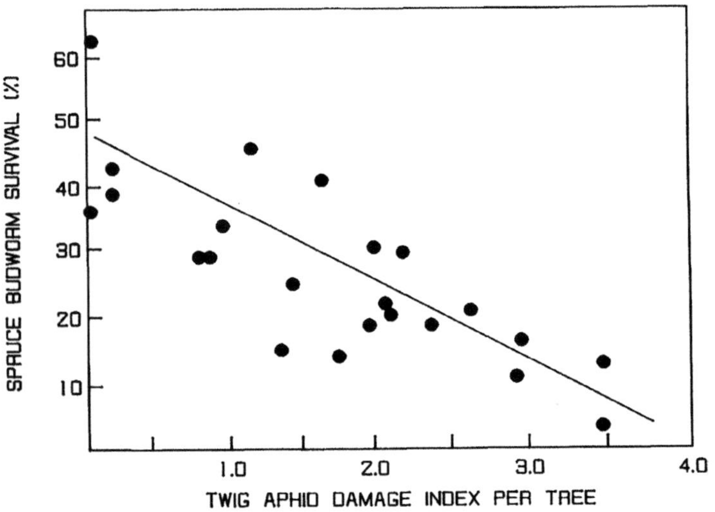

FIGURE 3. Relationship between spruce budworm survival rates (%) and aphid damage index per tree (0 index = no damage, 4 index = all needles and shoots severely infested).

TABLE III
Representative Life Table for the Spruce Budworm During an Outbreak[a]

x Age interval	N_x Number alive at beginning[b]	M_xF Mortality factors	100M_x: M_x Number dying	100M_x: N_x Percentage dying	S_x Survival within interval
Eggs	200	Parasites	18.0	9	
		Predators	12.0	6	
		Unknown	8.0	4	
			38.0	19	0.81
Instar I–II	162	Dispersal	132.8	82	0.18
Instar III–VI	29.2	Parasites	11.7	40	
		Disease	6.7	23	
		Unknown	6.7	23	
			25.1	86	0.14
Pupae	4.1	Parasites	0.5	13	
		Predators	0.2	4	
		Unknown	0.7	17	
			1.4	34	0.66
Moths	2.7				
Generation total	2.7	All factors	197.3	99	0.01

[a]Adapted from Morris.[47]
[b]Number of insects per 10 ft² of foliage.

4. OUTBREAK PATTERNS

4.1. Frequency of Outbreaks

The widespread recurrence of massive outbreaks of the spruce budworm, sometimes covering millions of hectares, has captured the imagination and curiosity of many. Outbreaks have two intriguing dimensions—the time period between them and the geographical area covered. The period between outbreaks ranges from 17 to 100 years and has averaged about 35 years.[10,11] Tree-ring analyses suggest that outbreaks have occurred more frequently in the twentieth century than in the previous century. For example, 23 outbreaks were recorded in the past 80 years as opposed to 9 in the preceding 100 years.[11] Furthermore, earlier outbreaks were more limited in size than those of the twentieth century. Three major outbreaks (1910, 1940, 1970) in northeastern North America have covered 10, 25, and 55 million ha, respectively.[11] The increases in frequency, extent, and severity of outbreaks are believed to have been caused by increased harvesting, improved fire protection, and the use of pesticides, all of which favor mature spruce–fir stands and create conditions more favorable for the budworm.[11] There is good evidence to support this contention; e.g., in western Ontario and eastern Quebec, where forests have not been harvested intensively or had consistent fire protection, the intervals between outbreaks averages about 100 years, while in the central portion of this territory, where harvesting and fire protection have been practiced more intensively, the outbreak intervals now average about 25 years (from a previous periodicity of 75–100 years).

4.2. Factors Triggering Outbreaks

Budworm outbreaks have long been linked to extensive areas of mature balsam fir.[73,77] The abundance of staminate flowers on mature trees was thought to be a major contributor to small larval survival and subsequent outbreaks.[7,9,25] Others concluded that early summer drought, 3–4 years prior, was correlated with budworm outbreaks.[24,27,32,52,81] Such weather conditions indirectly increase early larval survival through increased staminate flower production as well as directly enhance late larval survival.[11] Certain weather conditions were also linked to massive moth dispersal and the deposition of egg-carrying moths in new areas.[20,26] Such infusions may augment local populations enough to swamp their natural enemies and thereby initiate outbreaks.[16,48]

4.3. Spatial Patterns and Outbreak Spread

4.3.1. Epicenter Hypothesis

Defoliation records from Quebec and Ontario suggest that high populations erupt first in special locations (epicenters) and then spread elsewhere.[27,59,81] For example, between 1967 and 1974 in Quebec there was a large (32,000,000-ha) provincewide outbreak that apparently sprung from seven local epicenters.[27] Each epicenter ranged in size from 154 to 2873 ha and occurred from 50 to 600 km from the main outbreak. All were mixed stands containing a preponderance of temperate forest species such as sugar maple, yellow birch, and white pine. Balsam fir and white spruce were only minor components of the canopy (ranging from 10 to 37 percent). Host tree to nonhost tree ratios ranged from roughly 1 : 1 to 1 : 5. Moreover, each epicenter had been previously subjected to an ecological disturbance, such as selective harvesting of commercial species, or fire, which encouraged the establishment of spruce and fir where it otherwise would not have grown.

A similar analysis of infestation maps from Ontario over 1947–1983 revealed the following[71]: (1) the area of infestation exhibited extreme fluctuations from only 40 ha in 1964 to 18 million ha in 1981; (2) an infestation was present every year in Ontario or in an adjacent province or state; (3) between outbreaks, budworm numbers decreased with distance from known infestations; (4) new infestations appeared to result from moth influxes from outbreak areas; (5) infestations tended to spread following the track of prevailing winds; (6) infestations that developed in forest types of low vulnerability (river valleys, bogs, tundra) tended to be short-lived; (7) infestation severity seemed to be governed by stand character (high fir component) and climate; (8) collapse of outbreaks was dependent on host tree mortality; and (9) small infestation remnants persisted after outbreak collapse.

4.3.2. Simultaneous Outbreak Hypothesis

One of the major criticisms of the epicenter hypothesis is that observing defoliation is a low-resolution sampling technique that only recognizes population densities above some threshold of observed defoliation.[22,55,72] A number of authors argue that outbreaks may arise from simultaneous population increases over large geographical areas. For example, average annual egg mass densities appeared to fluctuate in unison over the entire Province of New Brunswick between 1952 and 1980, even though their mean levels and amplitudes varied among specific regions.[57] A major budworm outbreak (1969–1975) in eastern Canada took only 6 years to rise, spread, and encompass nearly all susceptible stands from central Ontario to Newfoundland. The areas covered by visible defoliation were 1, 10, 22, and 56 million hectares in 1969, 1970, 1971, and 1972, respectively. It is unlikely that such a rapid increase in population could result solely from moths dispersing from the three original epicenters in eastern Ontario, western Quebec, and New Brunswick.

4.4. Local Population Processes

4.4.1. The Double Equilibrium Theory

The tendency of budworm populations in a given forest stand to exist either at very low (endemic) or at very high (epidemic) density led to the double equilibrium theory of budworm dynamics[29,47] (Fig. 4a). In this model, two potentially stable equilibria (E_1 and E_2) are seen in the population growth curve; the plot of the per-capita rate of change ($\ln N_{t+1} - \ln N_t$) on the initial population density ($\ln N_t$) (Fig. 4b, center curve). Between these two points, the curve can also cross the replacement line ($\ln N_{t+1} - \ln N_t = 0$) from below creating an unstable equilibrium point or threshold (T) from which the population will either release and increase to the region of the higher equilibrium point or collapse to the lower equilibrium. The higher and lower equilibria represent epidemic and endemic population states, respectively. If populations are in the epidemic state for several years, the condition of the forest deteriorates and the population then recedes to the lower, endemic equilibrium region. The lower equilibrium is normally maintained through the combined action of predators and parasites, but if favorable weather conditions prevail, or if large numbers of migrants arrive, the population may escape their regulatory actions. Stand age also plays a role in this model because several stand properties (temperature, relative humidity, foliage quality, and natural enemies) change with time and cause the budworm's per-capita rate of change curve to substantially increase (Fig. 16b, upper curve). Similar, but more embellished, explanations can be found elsewhere.[16,29,34,39] These interpretations of population numerical behavior lead to eruptive-type outbreak systems.[5,6]

4.4.2. The Continuous Oscillation Theory

Some of the same data that led to the double equilibrium hypothesis have also been used to develop the continuous oscillation theory. Reexamination and reevaluation of the

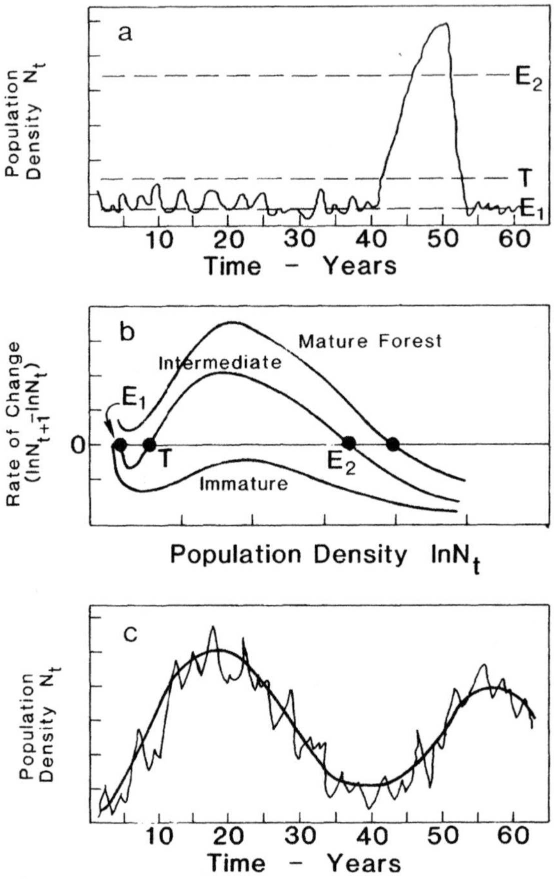

FIGURE 4. Spruce budworm population change in relationship to time and population density. (a) Double equilibrium theory of population pattern over time; E_1, E_2 are endemic and epidemic equilibria, respectively, and T is the unstable threshold from which populations either increase to E_2 or decrease to E_1. (b) Per-capita rate of budworm population change [$\ln(N_{t+1}/N_t)$] in relationship to population density ($\ln N_t$) for immature, intermediate, and mature forests. Rate of change is generally less than 0 (population declining) in immature forests, whereas it is usually greater than 0 (population growing) in mature forests. (c) Continuous oscillation theory of population pattern over time.

Green River Project data[47] and other historic records led to the conclusion that the population fluctuations within a given forest stand were governed by a basic cyclical oscillation and a secondary random fluctuation about this oscillation[57] (Fig. 4c). The oscillation was hypothesized to be driven by a combination of parasitism and mortality from disease and unknown factors affecting the large larvae (L3–L6)—the so-called fifth agent. The fluctuations about the basic oscillation were hypothesized to be caused by egg-bearing moths leaving stands (in the case of points below the basic oscillation) or entering stands (in the case of points above the basic oscillation). This interpretation leads to a cyclical outbreak population system.[6]

Analysis of annual egg count averages over New Brunswick support the explanation that populations oscillate in unison over large geographic areas, even though mean levels and amplitudes of individual populations were not the same.[57] Consequently, if one area is heavily invaded by egg-carrying moths, populations there might reach outbreak levels sooner than stands in the surrounding area, thereby giving the appearance of an epicenter even though no such epicenter existed. Populations can oscillate in unison over a large geographic area because density-independent factors such as weather, which are correlated among localities, will bring locally oscillating populations into synchrony even though weather has no oscillatory trend itself.[57] It should be noted, however, that the double equilibrium hypothesis also predicts constantly high densities or cycles of abundance if the forest is kept alive and in a susceptible state (Fig. 4b, upper curve), as has been done in New Brunswick by insecticide spraying. Hence, there is as yet no conclusive test of either hypothesis.

5. MANAGEMENT IMPLICATIONS

5.1. Forest Management Practices to Minimize Damage: Traditional Thinking

5.1.1. Management Strategies at the Stand Level

The land manager can influence the time, place, and quantity of tree mortality from the spruce budworm by certain practices that reduce spruce–fir vulnerability, e.g. (1) reducing the rotation age of balsam fir to 50 years or less; (2) breaking up the continuity of extensive areas of spruce-fir forests; (3) maintaining a mixed-species composition whenever possible; and (4) converting spruce-fir stands to less susceptible species such as aspen, pine, larch, or black spruce whenever feasible.[3,14,23,67,70]

There are numerous rating systems to assist land managers in predicting the vulnerability of spruce–fir stands to the spruce budworm.[83] One example is a short-term rating system that combines aerial photographic interpretation of stand defoliation, mortality, density, and proportion of host species into a simple system for predicting potential damage. The procedure ranks stands according to their damage probabilities to allow land managers to plan for immediate (next 1–3 years) protection or salvage.[44] Another example is a long-term rating system that can be used at the stand and regional levels. These systems are either quantitative empirical models that provide numerical estimates of potential loss of basal area or qualitative systems that provide a categorical index of vulnerability.[12,40]

Decisions to spray a stand using microbial or chemical insecticides are based on the present and future value of the stand.[8,31] Short-term spray strategies are designed to (1) save the most valuable mature spruce–fir stands under heavy budworm attack that would otherwise die within 2 or 3 years, or (2) suppress budworm populations before severe defoliation occurs (see refs. 1, 19, and 46, for additional information on the use of microbial and chemical insecticides against the spruce budworm).

5.1.2. Management Strategies at the Regional Level

Regional management strategies are often necessary to maintain the flow of raw materials to mills and thus help sustain state or provincial economies.[4,63] Potential long-term regional strategies also include various silvicultural techniques that produce less susceptible forest types.[13,14] In particular, two factors that can be manipulated at the regional level are the distribution of age classes within stands and the spatial distribution of the stands themselves. Land managers should aim for a patchwork of stands over the entire region that would include recent clearcuts, young stands of natural and artificial regeneration, and nonhost and nonvulnerable stands interspersed with vulnerable stands. This type of regional management should lessen the overall impact of a budworm outbreak on regional wood supply and would probably also decrease losses in individual stands. This strategy requires a long-term sustained effort between different ownerships and jurisdictions to plan the location of harvest cuts. This represents an ideal, however, that has not been accomplished anywhere to date.

5.2. Management Possibilities in Light of Population Theories

5.2.1. The Double Equilibrium Theory

The double equilibrium theory essentially generates an irregular eruptive type of outbreak.[6] What are its implications for innovative management approaches for the spruce budworm? If we assume that outbreaks can arise at the local level through the release of endemic populations from natural enemies (or other density-dependent regulating factors), by (1) an influx of moths that increases local populations enough to escape natural enemy pressure, and/or (2) the occurrence of weather conditions that favor the reproduction and survival of moth populations, we need only detect local population increases to affect management actions at that level. In the past, epicenter detection has been limited to the sighting of defoliation. Direct insect sampling techniques using foliage collections were too insensitive or too labor intensive to detect changes at population densities below those that create visible defoliation. Now, however, new tools for directly sampling budworm populations and new techniques for interpreting numerical changes at low densities have been developed. For example, synthetic pheromones and large-capacity pheromone traps are available at low cost,[60] and practical methods have been developed that permit more precise prediction of population release at the local level.[67,68]

With the ability to detect population release at a local level comes a potential regional management strategy.[27,71] When a network of pheromone traps, distributed over a large region, detects a surge of some locale, the population could be reduced with microbial or chemical insecticides at that locality, driving it back to its lower stable equilibrium and thereby preventing the spread of moths over a larger area. Taking this idea to its limits, it has been proposed that all outbreaks in North America could be entirely eliminated by focusing on epicenters at the western edge of the budworm's range, which are the sources of most dispersing moths.

Historically, the immediate management of a spatially confined and distantly removed insect population has been the approach taken by regulatory agencies when dealing with introduced pests. The gypsy moth, *Lymantria dispar,* is handled this way outside of its established range in northeastern North America. Thus, there is experience and precedent in using this management method. There is also an example of such an effort with the spruce budworm as well—an expanding outbreak at Burchell Lake in Ontario was successfully eliminated and the threat to surrounding forests removed by aerial pesticide application.[30]

The epicenter management approach is intriguing and exciting to contemplate. To be successful, the enormous logistical challenges of monitoring all spruce–fir forests, must be overcome. However, some epicenters[27] are made up of vegetation dominated by nonhost plants with only a small percentage of spruce–fir. These kinds of stands may harbor dangerous low-density populations of budworm because the normal regulatory factors found in pure spruce–fir stands may not be present, thus creating a kind of refuge for spruce budworm populations. This being the case, it might be possible to identify and monitor only these stands rather than having to monitor all spruce–fir forests within a region. One could also remove the budworm threat entirely by removing all host tree species so that the stands could no longer sustain budworm populations.

5.2.2. The Continuous Oscillation Theory

The continuous oscillation theory essentially generates cyclical outbreaks and the management implications differ considerably from the eruptive hypothesis. If populations cycle in unison from one stand to another over large geographical areas, local population management would not be practical, and true regional management efforts would be much more difficult to attain. Management focus would be on damage control rather than on budworm populations per se. Inventory management within individual ownerships would stimulate competitive rather than cooperative approaches. Management would focus more on developing a distributed age class inventory and preventing disruptions in the year-to-year flow of raw materials to the mills. This would appear much more like traditional management as discussed earlier. During a nonoutbreak period, annual harvests within the allowable cut would be conducted within the constraints of market demand and management would proceed as normally as possible. Silvicultural techniques would focus on reducing vulnerability through the methods discussed earlier. During outbreak periods, those stands scheduled for harvest within a few years and those unavailable for harvest (due to management objectives or inaccessibility) would receive no protection from budworm damage. The remaining stands within the inventory would be protected from significant tree mortality by insecticide spraying. The choice of protecting a given stand would depend on expected value losses and the cost of protective action. Large geographical areas within a given political jurisdiction might be sprayed in a given year, but such efforts would not be a true regional management strategy, but rather a cost-minimizing approach to protect individual stands or management parcels.

6. CONCLUDING REMARKS

Many facts about spruce budworm biology and behavior (particularly its extraordinary mobility) suggest that its long-term success is not closely tied to individual trees or even stands—as are some less mobile insects, such as scales and even sawflies. Instead, successful interaction of the spruce budworm with its hosts seems to depend on the quality and organization of the host plant landscape. Only at this level can one begin to understand the long-term population dynamics of the budworm and to interpret its biology. The quality of a landscape for the spruce budworm depends on the collective suitability of the various watersheds contained therein because they are all within the dispersal potential of moths ($\leqslant$ 600 km). From this perspective, an individual moth's success (fitness) depends not on selecting or arriving at the right tree, but rather on arriving at the right stand. In other words, moths select suitable stands or watersheds in the landscape and the ensuing early larvae select suitable trees.

This interpretation enables one to explain why moths lay eggs rather haphazardly within stands but are more discriminating when it comes to selecting stands—unless they are unable to complete dispersal because of unsuitable weather. For example, moths usually select mature forests with spike-top canopies as opposed to immature forests with more even-top canopies or those that have sustained serious defoliation. The ultimate reasons for differential stand selection must rest with the fact that immature and defoliated stands provide far less favorable environments for budworm larvae than do unattacked mature stands. The evidence suggests that immature stands are less favorable for late larval growth and survival because they have a less optimal microclimate, more natural enemies, and perhaps better plant defenses.

Differences of opinion about the basic numerical behavior of budworm populations may be more apparent than real. For example, the dual equilibrium theory probably applies to natural landscapes that contain a broad spectrum of age classes and species compositions. Such a checkerboard system would have a blend of young and otherwise weakly susceptible stands as well as highly susceptible ones. One would expect budworm populations there to erupt irregularly and sporadically—showing up first in the mature stands depending on when climatic and other conditions are most suitable for budworms reaching the higher (outbreak) equilibrium level. Spillover to younger and otherwise less susceptible stands may occur, but these populations are soon driven to near extinction by emigration, and unfavorable host, microclimatic, and other conditions. Population levels in these stands depend on influx from nearby populations in mature stands and may be correlated with them.

By contrast, if the landscape is composed largely of mature susceptible stands and is climatically uniform, and furthermore if tree mortality is prevented by spraying, one would expect a widespread and regular population cycle or oscillation. This, in fact, is merely population movement around the epidemic equilibrium position of the dual equilibrium model. As long as the individual stands are not allowed to die from severe defoliation, such a cycle could go on indefinitely.

REFERENCES

1. Armstrong, J. A., 1985, Tactics and strategies for larval suppression and prevention of damage using chemical insecticides, in: *Recent Advances in Spruce Budworms Research* (C. J. Sanders, R. W. Stark, E. J. Mullins, and J. Murphy, eds.), pp. 301–319, Canadian Forestry Service, Ottawa.
2. Barker, R. B., and Fyfe, H. A., 1947, A mixed infestation in the spruce woods forest reserve, Manitoba, *Can. Dep. Agric. Div. Ent. Bi-mon. Prog. Rep.* **3**(4):3.
3. Baskerville, G.L., 1975, Spruce budworm: The answer is forest management or is it?, *For. Chron.* **51**:157–160.
4. Baskerville, G. L., 1976, Report of the task-force for evaluation of budworm control alternatives, prepared

for the Cabinet Committee on Economic Development, Province of New Brunswick, Department of Natural Resources, Fredericton, New Brunswick, Canada.
5. Berryman, A. A., 1978, Towards a theory of insect epidemiology, *Res. Popul. Ecol.* **19**:181–196.
6. Berryman, A. A., 1986, *Forest Insects: Principles and Practice of Population Management*, Plenum, New York.
7. Bess, H. A., 1946, Staminate flowers and spruce budworm abundance, Can. Dep. Agric., Forest Insect Invest., Ottawa Bi-Mon. Prog. Rep. 2.
8. Bible, T. D., 1985, Economic considerations for spruce budworm management in North America, in: *Recent Advances in Spruce Budworms Research* (C. J. Sanders, R. W. Stark, E. J. Mullins, and J. Murphy, eds.), pp. 176–187, Canadian Forestry Service, Ottawa.
9. Blais, J. R., 1952, The relationship of the spruce budworm [Choristoneura fumiferana (Clem.)], to the flowering condition of balsam fir [*Abies balsamea* (L.) Mill.], *Can. J. Zool.* **30**:1–29.
10. Blais, J. R., 1983, Trends in the frequency, extent, and severity of spruce budworm outbreaks in eastern Canada, *Can. J. For. Res.* **13**:539–547.
11. Blais, J. R., 1985, The ecology of the eastern spruce budworm: A review and discussion, in: *Recent Advances in Spruce Budworms Research* (C. J. Sanders, R. W. Stark, R. J. Mullins, and J. Murphy, eds.), pp. 49–59, Canadian Forestry Service, Ottawa.
12. Blais, J. R., and Archambault, L., 1982, Rating of vulnerability of balsam fir to spruce budworm attack in Quebec, Inform. Rep. LAU-X-51. Canadian Forestry Service, Laurentian Forestry Research Center, Sainte-Foy, Quebec.
13. Blum, B. A., and Maclean, D. A., 1984, Silviculture, forest management and the spruce budworm, in: *Managing the Spruce Budworm in Eastern North America*, pp. 83–102, USDA Agric. Handb. #620.
14. Blum, B. A., and Maclean, D. A., 1985, Potential silviculture, harvesting and salvage practices in eastern North America, in: *Recent Advances in Spruce Budworms Research* (C. J. Sanders, R. W. Stark, E. J. Mullins, and J. Murphy, eds.), pp. 264–280, Canadian Forestry Service, Ottawa.
15. Campbell, R. W., 1985, Effects of birds and ants on pupae of the spruce budworm, in: *Recent Advances in Spruce Budworms Research* (C. J. Sanders, R. W. Stark, E. J. Mullins, and J. Murphy, eds.), p. 100, Canadian Forestry Service, Ottawa.
16. Clark, W. C., Jones, D. D., and Holling, C. S., 1979, Lessons for ecological policy design: A case study of ecosystem management, *Ecol. Model.* **7**:1–53.
17. Clemens, B., 1865, North American Micro-Lepidoptera, *Proc. Entomol. Soc. Phila.* **5**:139.
18. Coulson, R. N., and Witter, J. A., 1984, *Forest Entomology: Ecology and Management*, Wiley, New York.
19. Cunningham, J. C., 1985, Biorationals for control of spruce budworms, pp. 320–351, in *Recent Advances in Spruce Budworms Research*, Ref. 1.
20. Dobesberger, E. J., Lim, K. P., and Raske, A. G., 1983, Spruce budworm moth flight from New Brunswick to Newfoundland, *Can. Entomol.* **115**:1641–1645.
21. Eidt, D. C., and Little, C. H. A., 1970, Insect control through induced host–insect asynchrony: A progress report, *J. Econ. Entomol.* **63**:1966–1968.
22. Fleming, R. A., 1985, How should one view the historical record of spruce budworm outbreaks?, in: *Recent Advances in Spruce Budworms Research* (C. J. Sanders, R. W. Stark, E. J. Mullins, and J. Murphy, eds.), pp. 136–137, Canadian Forestry Service, Ottawa.
23. Flexner, J. L., Bassett, J. R., Montgomery, B. A., Simmons, G. A., and Witter, J. A., 1983, Spruce–fir silviculture and the spruce budworm in the Lake States, Mich. Coop. For. Pest Manage. Prog. Handb. 83-2.
24. Greenbank, D. O., 1956, The role of climate and dispersal in the initiation of outbreaks of the spruce budworm in New Brunswick. I. The role of climate, *Can. J. Zool.* **34**:453–475.
25. Greenbank, D. O., 1963, Host species and the spruce budworm, in: The dynamics of epidemic spruce budworm populations (Morris, R. F., ed.), pp. 219–223, *Mem. Entomol. Soc. Can.* 31.
26. Greenbank, D. O., Schaffer, G. W., and Rainey, R. C., 1980, Spruce budworm (Lepidoptera: Torticidae) moth flight and dispersal: New understanding from canopy observations, radar, and aircraft, *Mem. Entomol. Soc. Can.* 110.
27. Hardy, Y., Lafond, A., and Hamel, A., 1983, Epidemiology of the current spruce budworm outbreak in Quebec, *For. Sci.* **29**:715–725.
28. Harvey, G. T., 1985, The taxonomy of the coniferophagous *Choristoneura* (Lepidoptera: Tortricidae): A review, in: *Recent Advances in Spruce Budworms Research* (C. J. Sanders, R. W. Stark, E. J. Mullins, and J. Murphy, eds.), pp. 16–48, Canadian Forestry Service, Ottawa.

29. Holling, C. S., Jones, D. D., and Clark, W. C., 1976, Ecological policy design: A case study of forest and pest management, in: *Pest Management* (G. A. Norton, and C. S. Holling, eds.), pp. 13–90, Pergamon, New York.
30. Howse, G. M., and Sippell, W. L., 1975, Aerial control operations against the spruce budworm in Ontario, in: *Aerial Control of Forest Insects in Canada.* (M. L. Prebble, ed.), pp. 84–93, Department of the Environment, Ottawa.
31. Irland, L. C., and Runyon, K. L., 1984, Economics of spruce budworm management strategy, in: *Managing the Spruce Budworm in Eastern North America,* pp. 65–82, USDA Agric. Handb. 620.
32. Ives, W. G. H., 1974, Weather and outbreaks of the spruce budworm, Can. For. Serv. Nor-X-118.
33. Jennings, D. T., and Crawford, H. S., 1985, *Predators of the spruce budworm,* USDA Agric. Handb. 644.
34. Jones, D. D., 1976, The budworm site model, in: *Pest Management* (G. A. Norton and C. S. Holling, eds.), pp. 91–155, Pergamon, New York.
35. Kulman, H. M., 1971, Effect of defoliation on tree growth, *Annu. Rev. Entomol.* **16:**289–324.
36. Langelier, L. A., and Garton, E. O., 1986, *Management Guidelines for Increasing Populations of Birds That Feed on Western Spruce Budworm,* USDA Agric. Handb. 653.
37. Linquist, O. H., 1982, *Keys to Lepidopterous Larvae Associated with Spruce Budworm in Northeastern North America,* Environment Canada Distribution Center, Hull, Quebec.
38. Lorio, P. L., 1986, Growth-differentiation balance: A basis for understanding southern pine beetle–tree interactions, *Forest Ecol. Mgmt.* **14:**259–273.
39. Ludwig, D., Jones, D. D., and Holling, C. L., 1978, Qualitative analysis of insect outbreak systems: The spruce budworm and forest, *J. Anim. Ecol.* **47:**315–332.
40. Lynch, A. M., Fowler, G. W., and Witter, J. A., 1984, Development of empirical models to rate spruce–fir stands in Michigan's Upper Peninsula for hazard from the spruce budworm (Lepidoptera: Tortricidae): A case history, *Great Lakes Entomol.* **17:**163–174.
41. Mattson, W. J., Haack, R. A., and Lawrence, R. K., 1988, Do balsam twig aphids (Homoptera: Aphididae) lower tree susceptibility to spruce budworm. *Can. Entomol.* (in press).
42. Mattson, W. J., Slocum, S. S., and Koller, C. N., 1983, Spruce budworm performance in relation to foliar chemistry of its host plants, in: Proceedings, forest defoliator/host interactions: A comparison between gypsy moth and spruce budworms, pp. 55–65, USDA For. Serv. Gen. Tech. Rep. NE-85.
43. Mattson, W. J., and Scriber, J. M., 1987, The nutritional ecology of insect folivores of woody plants: Nitrogen, water, fiber, and mineral considerations, in: *Nutritional Ecology of Insects, Mites, and Spiders* (F. Slansky and J. Rodriguez, eds.), pp. 105–146, Wiley, New York.
44. McCarthy, J., Olson, C. E., Jr., and Witter, J. A., 1983, Assessing spruce budworm damage with small-format aerial photographs, *Can. J. For. Res.* **13:**395–399.
45. McLaughlin, B. M., 1986, Performance of the spruce budworm, *Choristoneura fumiferana,* in relation to dietary and foliar levels of sugar and nitrogen, M.S. thesis, Michigan State University, E. Lansing, Michigan.
46. Montgomery, B. A., Dimond, J. B., Witter, J. A., and Simmons, G. A., 1984, Insecticides for control of the spruce budworm, USDA Agric. Inf. Bull. 615.
47. Morris, R. F. (ed.), 1963, The dynamics of epidemic spruce budworm populations, Mem. Entomol. Soc. Can. 31.
48. Morris, R. F., Cheshire, W. F., Miller, C. A., and Mott, D. G., 1958, The numerical response of avian and mammalian predators during a gradation of the spruce budworm, *Ecology* **39:**487–494.
49. Neilson, M. M., 1963, Disease and the spruce budworm, in: The Dynamics of Epidemic Spruce Budworm Populations, (R. F. Morris, ed.), pp. 272–287, *Mem. Entomol. Soc. Canada* 31.
50. Perry, D. F., 1985, Summary-Discussion Group A-Diseases, in: *Recent Advances in Spruce Budworms Research* (C. J. Sanders, R. W. Stark, E. J. Mullins, and J. Murphy, eds.), p. 111, Canadian Forestry Service, Ottawa.
51. Perry, D. F., 1985, Epizootic development of Entomophthoralean fungi, in: *Recent Advances in Spruce Budworms Research* (C. J. Sanders, R. W. Stark, E. J. Mullins, and J. Murphy, eds.), p. 107, Canadian Forestry Service, Ottawa.
52. Pilon, J. G., and Blais, J. R., 1961, Weather and outbreaks of the spruce budworm in the province of Quebec from 1939 to 1956, *Can. Entomol.* **93:**118–123.
53. Powell, J. A., 1980, Nomenclature of Nearctic conifer-feeding Choristoneura (Lepidoptera:Tortricidae): Historical review and present status, USDA Gen. Tech. Rep. PNW-100.

54. Raske, A. G., 1985, Collapsing budworm populations, in: *Recent Advances in Spruce Budworms Research* (C. J. Sanders, R. W. Stark, E. J. Mullins, and J. Murphy, eds.), pp. 141–142, Canadian Forestry Service, Ottawa.
55. Regniere, J., 1985, Interpreting historical records, in: *Recent Advances in Spruce Budworms Research* (C. J. Sanders, R. W. Stark, E. J. Mullins, and J. Murphy, eds.), pp. 143–144, Canadian Forestry Service, Ottawa.
56. Ricker, W. E., 1954, Stock and recruitment, *J. Fish. Res. Bd. Can.* **11**:559–623.
57. Royama, T., 1984, Population dynamics of the spruce budworm *Choristoneura fumiferana*, *Ecol. Monog.* **54**:429–462.
58. Sanders, C. J., 1975, Factors affecting adult emergence and mating behavior of the eastern spruce budworm, *Choristoneura fumiferana* (Lepidoptera: Tortricidae), *Can. Entomol.* **107**:967–977.
59. Sanders, C. J., 1976, Pest management strategies of epicenter control. in: Proceedings of a symposium on the spruce budworm, pp. 61–63, USDA For. Serv. Misc. Publ. 1327.
60. Sanders, C. J., 1985, Sex pheromone traps for monitoring spruce budworm populations, in: *Recent Advances in Spruce Budworms Research* (C. J. Sanders, R. W. Stark, E. J. Mullins, and J. Murphy, eds.), p. 234, Canadian Forestry Service, Ottawa.
61. Sanders, C. J., and Lucuik, G. S., 1972, Factors affecting calling by eastern spruce budworm, *Choristoneura fumiferana* (Lepidoptera: Torticidae), *Can. Entomol.* **104**:1751–1762.
62. Sanders, C. J., Stark, R. W., Mullins, E. J., and Murphy, J., (eds.), 1985, *Recent Advances in Spruce Budworms Research,* Canadian Forestry Service, Ottawa.
63. Seymour, R. S., 1985, Sex pheromone traps for monitoring spruce budworm populations, in: *Recent Advances in Spruce Budworms Research* (C. J. Sanders, R. W. Stark, E. J. Mullins, and J. Murphy, eds.), pp. 200–213, Canadian Forestry Service, Ottawa.
64. Shaw, G. G., and Little, C. H. A., 1977, Natural variation in balsam fir foliar components of dietary importance to spruce budworm, *Can. J. For. Res.* **7**:47–53.
65. Shaw, G. G., Little, C. H. A., and Durzan, D. J., 1978, Effect of fertilization of balsam fir trees on spruce budworm nutrition and development, *Can. J. For. Res.* **8**:364–374.
66. Shepherd, R. F., 1985, A theory on the effects of diverse host-climate environments in British Columbia on the dynamics of the western spruce budworm, in: *Recent Advances in Spruce Budworms Research* (C. J. Sanders, R. W. Stark, E. J. Mullins, and J. Murphy, eds.), pp. 60–70, Canadian Forestry Service, Ottawa.
67. Simmons, G. A., Cuff, W., Montgomery, B. A., and Hardmann, J. M., 1984, Integrated pest management, in: *Managing the Spruce Budworm in Eastern North America*, pp. 11–20, USDA Agric. Handb. 620.
68. Simmons, G. A., and Elliott, N. C., 1985, Use of moths caught in light traps for predicting outbreaks of the spruce budworm, *J. Econ. Entomol.* **78**:362–365.
69. Simmons, G. A., and Fowler, G. W., 1984, Considerations when sampling spruce budworm egg masses on balsam fir in the Lake States: low to extreme population levels, *Great Lakes Entomol.* **17**:87–95.
70. Simmons, G. A. and Montgomery, B. A., 1985, Progressing toward integrated pest management of the spruce budworm in the eastern United States, in: *Recent Advances in Spruce Budworms Research* (C. J. Sanders, R. W. Stark, E. J. Mullins, and J. Murphy, eds.), pp. 418–442, Canadian Forestry Service, Ottawa.
71. Sippell, W. L., 1984, Planning now to reduce, postpone or prevent tne next spruce budworm outbreak, in: Proceedings: new and improved techniques for monitoring and evaluating spruce budworm populations, pp. 59–67, USDA For. Serv. Gen. Tech. Rep. NE-88.
72. Stehr, G. W., 1968, On some concepts in the population biology of the spruce budworm, *Proc. Entomol Soc. Ont.* **99**:54–60.
73. Swaine, J. M., and Craighead, F. C., 1924, Studies on the spruce budworm (*Cacoecia fumiferana* (Clem.)). Part I. A general account of the outbreak, injury and associated insects, *Ont. Tech. Bull. (New Ser.)* **37**:3–27.
74. Tilles, D. A., and Woodley, N. E., 1984, *Spruce Budworm Parasites in Maine: A Reference Manual for Collection and Identification of Common Species,* USDA Agric. Handbook 616.
75. Torgerson, T. R., and Campbell, R. W., 1982, Some effects of avian predators on western spruce budworm in North Central Washington, *Environ. Entomol.* **11**:429–431.
76. Torgerson, T. R., Campbell, R. W., Srivastava, N., and Beckwith, R. C., 1984, Role of parasites in the population dynamics of western spruce budworm (Lepidoptera: Tortricidae) in the Northwest, *Environ. Entomol.* **13**:568–573.

77. Tothill, J. D., 1919, The passing of the balsam budworm in New Brunswick, *Can. For. Mag.* **July**:306–307.
78. von Rudloff, E., 1972, Seasonal variation in the composition of the volatile oil of the leaves, buds, and twigs of white spruce (*Picea glauca*), *Can. J. Bot.* **50**:1595–1603.
79. von Rudloff, E., 1975, Volatile leaf oil analysis in chemosystematic studies of North American conifers, *Biochem. Syst. Ecol.* **2**:131–65.
80. Warren, G. L., 1954, The spruce needle worm, *Dioryctria reniculella* Gvt. as a predator of spruce budworm, *Can. Dep. Agric. For. Biol. Bi-Monthly Prog. Rep.* **10**(3):2.
81. Wellington, W. G., Fettes, J. J., Turner, K. B., and Belyea, R. M., 1950, Physical and biological indicators of the development of outbreaks of the spruce budworm and the forest tent caterpillar, *Can. J. Zool.* **30**:114–127.
82. Welsh, D. A., 1985, Preliminary studies of the relation between spruce budworm and birds in the boreal forest of Ontario, in: *Recent Advances in Spruce Budworms Research* (C. J. Sanders, R. W. Stark, E. J. Mullins, and J. Murphy, eds.), pp. 100–101, Canadian Forestry Service, Ottawa.
83. Witter, J. A., and Lynch, A. M., 1985, *Rating Spruce-Fir Stands for Spruce Budworm Damage in Eastern North America,* USDA Agric. Handb. 636.
84. Witter, J., Ostaff, D., and Montgomery, B., 1984, Damage assessment, in: *Managing the Spruce Budworm in Eastern North America,* pp. 37–64 USDA Agric. Handb. 620.

CHAPTER 17

THE LARCH BUDMOTH IN THE ALPS

Werner Baltensweiler and Andreas Fischlin

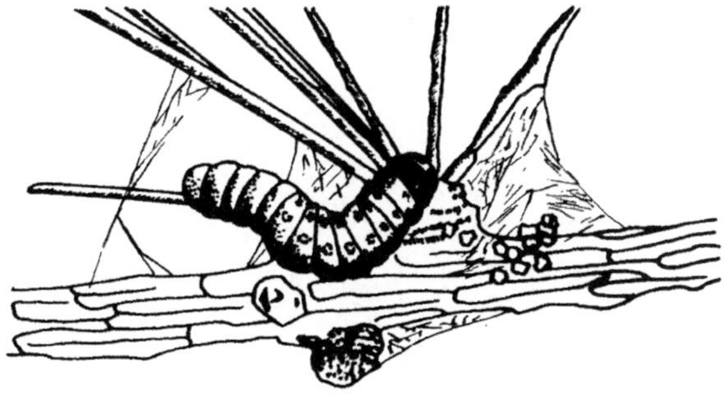

Werner Baltensweiler • Department of Plant Sciences, Swiss Federal Institute of Technology, CH-8092 Zurich, Switzerland. **Andreas Fischlin** • Department of Plant Sciences, Project Centre IDA/Celtia, Swiss Federal Institute of Technology, CH-8092 Zurich, Switzerland.

1. INTRODUCTION

During the late 1940s, immediately after World War II, the lush green forests of the Engadine Valley, high in the Swiss Alps, turned an ugly red-brown in the midst of the tourist season. This was due to a spectacular outbreak of the larch budmoth, *Zeiraphera diniana* Guenée (Lepidoptera: Tortricidae). Preparing for a revival of the tourist industry, and having the new insecticide DDT at hand, it seemed only appropriate that the tourist office urge the forest service to control the pest. This was the beginning of what was to become a 34-year study of the population dynamics of the larch budmoth (Fig. 1).

Zeiraphera diniana is known for its periodic devastation of the subalpine larch–cembran pine forests (*Larici pinetum cembrae*[41]) of the European Alps. The wasteful feeding and subsequent drying of needles, as well as the masses of frass and webbing, change the green forests in midsummer into a dirty red-brown, as if a crown fire had scorched them. This spectacular sight, and the regular reappearance of the phenomenon, has for centuries attracted the attention of laymen and foresters alike.

Larch budmoth outbreaks have occurred at regular 9-year intervals in the Alps as far

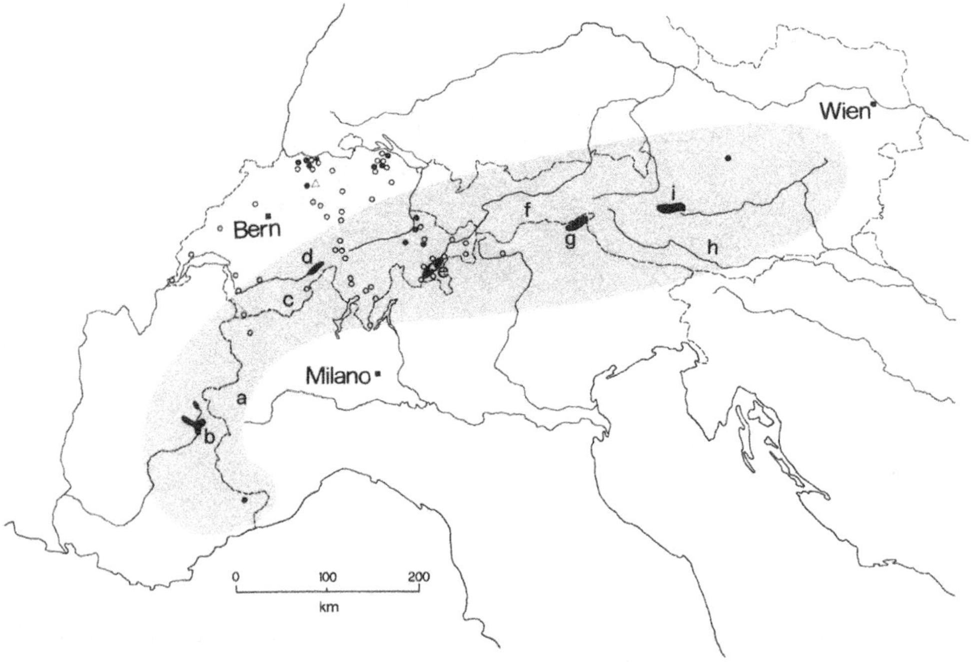

FIGURE 1. Locations of sites at which research on the larch budmoth has been conducted between 1949 and 1986 in the European Alps: large black locations, main study areas >2000 ha; black dots, local sites ~100 ha; open triangles, life-table studies; open circles, light trap or pheromone trap census sites; black squares, major cities. Approximate locations of regions mentioned in Fig. 2: a, = Piemonte; b, Hautes Alpes; c, Vispertäler; d, Goms; e, Oberengadin; f, Tirol; g, Trentino; h, Kärnten; i, Salzburg.

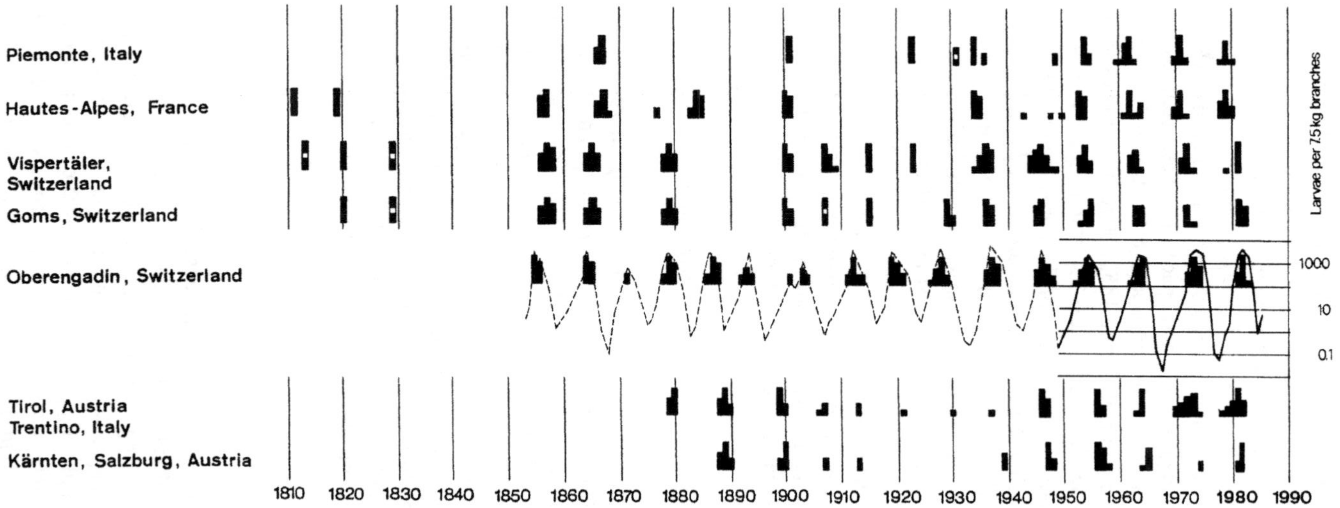

FIGURE 2. Pattern of apparent defoliation by the larch budmoth in seven major regions of the European Alps (see Fig. 1). The height of the columns represents an index of the area affected and columns with a white inset indicate fragmentary information only. The combination of results of the regional population census in the upper Engadine Valley from 1949 until 1986 (see Fig. 4) with the recorded defoliation periods allows us to extrapolate the fluctuations in larval density back to 1853.

back as recorded history, growth ring reductions with a 9-year frequency having been observed in fossil larch dating from Roman times and in the beams of fifteenth century farm houses. The first written account, in 1815, attributed the outbreak to spiders that covered the larch trees with their silvery webbing.[33] Since then periodic outbreaks have been reported in various parts of the Alps (Fig. 2). Although such historical records can never provide a completely accurate picture, they do illustrate the remarkably constant frequency of defoliation damage in time and space. The best-recorded sequence of outbreaks is from the Engadine valley in southeastern Switzerland, where foresters have observed and documented 16 outbreaks since 1855. Here, the heaviest defoliation occurs at altitudes of 1600–2100 m, lasts 2.93 ± 0.21 (standard error) years, and recurs at intervals of 8.47 ± 0.27 years. In other parts of the Alps, the outbreaks are less regular. Nevertheless, when considering the entire alpine arch, trends in the spatial sequence of outbreaks can be recognized[20]: For example, defoliation occurred first in the eastern parts of the Alps during the 1850s and 1860s; in the western parts during the 1940s, 1950s, and 1960s; and simultaneously in both east and west and last in the central Alps in all other instances. At lower altitudes of 1200–1600 m, defoliation occurs only half as often, and therefore rather irregularly, and also appears 1–4 years later than in the subalpine zone.

The species was once considered the most dangerous pest of larch forests,[29] but the results of our more recent research, as presented in this chapter, led us to develop a quite different appreciation of the larch budmoth problem.

2. BIOLOGY AND ECOLOGICAL RELATIONSHIPS

The larch budmoth was first described in 1845 as *Sphaleroptera diniana* Gueńee from specimens collected from larch trees in the French Alps and 1 year later as *Grapholita pinicolana* Zeller, from Scots pine in the Baltic and as *Poecilochroma occultana* Douglas from fir-trees (?) in England. These have since been synonymized and, after being included in several tortricid genera such as *Steganoptycha, Enarmonia, Semasia,* and *Eucosma,* the species was finally placed in the holarctic genus *Zeiraphera.*[22,30] The species has several ecological equivalents throughout the holarctic region, being described as *Z. diniana* var. *desertana* in Siberia, *Zeiraphera lariciana* in Japan,[55] and *Zeiraphera improbana* in North America.[62] With the exception of *Z. lariciana,* however, no detailed differentiation on taxonomic grounds has yet been provided.

2.1. Life Cycle

Unless explicitly specified, the following description applies to the univoltine larch budmoth living in its optimal habitat at 1700–2000 m (Fig. 3). Moth flight occurs from July to October, with adults living for more than 30 days and egg laying lasting around 20 days. Females lay 20–180 eggs, with a maximum of 350, depending on the nutritional status of the larval food supply. Eggs are deposited by means of a protractile ovipositor underneath the lichen *Parmelia exasperatula,* which covers 3-year-old and older larch branches at altitudes above 700 m. Eggs may also be laid between cone scales or beneath bark scales.

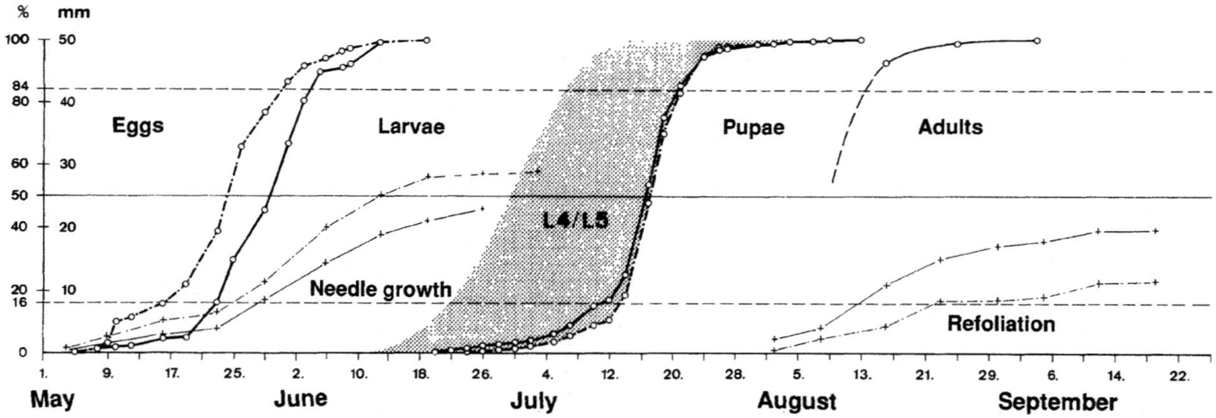

FIGURE 3. Phenological diagram of larch budmoth development and larch foliage growth in the Oberengadin (Upper Engadine); at the lower (1820 m —·—·—) and upper (1900 m ——) Sils sites, in 1972.

Eggs overwinter as diapausing embryos in the late gastrula stage, their supercooling point during this state being −51.3°C.[12] Diapause enables the species to synchronize egg development with the phenology of the host plant at altitudes of 500–2200 m. This flexibility is crucial because the coincidence of first instar larval emergence and the sprouting of the larch foliage is essential for the establishment of the population. Depending on weather, altitude, and exposure, the emergence period may take up to 4 weeks during May and June but may also vary by more than a month from one year to the next.

Optimal first-instar survival occurs when needles are 6–18 mm long. Once established, the first and second instars feed within the needle cluster, at the base of the needles, where they are protected from environmental hazards. During the fourth instar, they construct a characteristic needle case and in the fifth they live within webbing along the branch axis. The last larval instar exists for 10–14 days and is the longest and most destructive stage. On maturing, the fifth instar becomes negatively phototactic, leaves the larch tree in full sunshine, and immediately moves into the litter, where it constructs a pupation chamber at the base of the litter horizon. There it pupates, remaining in this stage for 25–36 days.[59]

Moths emerge early in the morning, with the males some days before the females. Flight activity starts at dusk, when temperatures are above 7°C, and lasts until midnight. Late in the season, however, when nights become too cold, flight activity shifts to midday. The female budmoth produces a powerful sex attractant.[23]

2.2. Relationship with the Host Plant

Each new generation of a univoltine herbivore has to cope with a food supply that may vary both quantitatively and qualitatively. The coincidence of first-instar larval emergence with the sprouting of larch foliage is crucial for the establishment of the new generation. This quantitative aspect, in itself rather variable due to site and weather factors, can be drastically modified by variability in food quality.

In the deciduous genus *Larix,* needle growth occurs before shoot elongation. Therefore, the chemical and morphological constitution of the new needles is strongly dependent on nutrient reserves stored during the previous fall and on the prevalent growth conditions, i.e., temperature and moisture in spring. The deciduous growth rhythm of larch probably evolved in regions in which late frosts frequently destroyed the new needles.[32] In response, the plant has evolved the capacity to flush a new set of needles during the same season. However, the second set of needles is unable to mature properly and is susceptible to early killing by fall frosts before the nutrients can be resorbed. The following spring, the needles of these trees, short of starch and nitrogen, grow more slowly and do not reach normal lengths.

This growth process can be modified to some extent by the prevailing weather conditions. Warm temperatures immediately after snow melt induce rapid needle growth, so that full-grown needles may measure more than 30 mm. Needle growth under these conditions is sigmoid and results in soft, lush needles. Conversely, needle growth during cold springs is slow and linear,[18] with mature needles being stiff and short (<20 mm). During the course of needle growth, the raw fiber content increases from about 8%

needle-fresh-weight to 16%, while the raw protein content falls from about 6% to 4%.[28,64] These values characterize the intrinsic variability of host traits to which budmoth populations are exposed, i.e., the constitutive defense capacities.[57]

Larch trees that have been defoliated to more than 50% refoliate in a similar manner to trees that have suffered from frost. Such trees are also depleted of nutrient reserves in the spring following defoliation and therefore grow short needles (<20 mm) with a raw fiber content of about 18%[28] (see Fig. 3). It may then take 2–7 years for the larch needles to recover their average quality.[46] Although the physiological processes are the same as those following frost damage, this change in foliage quality is induced by insect feeding.[50] As we shall see, this induced change in food quality has important effects on the dynamics of *Z. diniana* populations.

In dense populations, food waste also becomes an important factor. The late fourth- and fifth-instar larvae, living in loose webbing along the branch axis, interfere with each other during their feeding activities. The result is that many needles are only nibbled or partly consumed. In the dry alpine climate, such needles are lost as a nutritional resource within a day. This wasteful feeding is the cause of the ugly red-brown larch forests seen at the height of the outbreak. As a consequence of this destructive feeding, large larvae leave the defoliated trees in search of better food and often feed indiscriminantly on cembran pines, spruces, and firs in the understory. These larvae are robust enough to consume even the perennial needles on these hosts.

In contrast to larch trees, which refoliate within 3–4 weeks, young evergreens stripped of more than 70% of their needles generally die. Larches that have lost more than 50% of their needle biomass sprout quickly and uniformly. If by chance rain is lacking during this crucial refoliation period in July and August, larch trees can also die. However, mortality rarely exceeds 1% of the standing volume in any one outbreak.[26]

2.3. Ecotypes

The original descriptions of *Z. diniana* under three different names are symptomatic of its present status, for even today we are not sure of whether we are dealing with several ecotypes, several host races of one species, or separate species. In most populations, two sympatric color morphs can be distinguished in the fifth instar.[31] A dark morph that lives on larch is called the larch form, and an orange-yellow morph on cembran pine is called the cembran pine form. It has recently been established that these color traits are related to other morphological and ecological characteristics. For instance, the black larch form has a much faster postdiapause development than the cembran pine form, a trait that ensures synchronization of larval emergence with the sprouting of their respective host plants. Rapid development of the larch form usually results in larger adults but also leads to lower fitness on suboptimal foods. Thus, the larch form fails to survive on cembran pine, even when synchronized with foliage flushing, whereas the light cembran pine form thrives on larch but does not reach the size of the larch form. Intermediate color morphs are also found, and these have intermediate postdiapause developmental rates and are superior to the larch form in surviving on suboptimal foliage.[37] Finally, the two forms differ in their sexual communication.[23,49]

By crossing the two color morphs, it was found that the F_1 offspring exhibit the full range of intermediate colors. But even more importantly, the ecological traits are also transferred in a more or less correlated manner.

2.4. Relationship with Natural Enemies

A granulosis virus disease and a diverse parasitoid fauna lead to considerable mortality among larch budmoth larvae, particularly during the peak and subsequent crash of the population. Following a widespread virus epizootic at the end of the 1954–1957 outbreak, this pathogen was considered of great importance.[60] Subsequently, however, widespread virus epizootics have not been observed, yet the amplitude and period of the later cycles have not differed from the earlier one.

A total of 90 obligate primary and 13 obligate secondary parasitoids have been reared from the larch budmoth.[38] They attack all life stages except the adult. The most important parasitoids are the ichneumonid *Phytodietus griseanae* Kerrich and three eulophids, *Sympiesis punctifrons* Thomson, *Dicladocerus westwoodii* Westwood, and *Elachertus argissa* Walker. During the course of a budmoth cycle, a sequence of parasitoid species dominates and this leads to a predictable pattern of mortality. Apparent mortality amounts to about 10% in the population growth phase, rises to 10–20% at population peak, and reaches 70–80% during the collapse. However, the subsequent extremely low density of the budmoth brings about a dissociation between specific parasitoids and their host. Several studies on the most frequent parasitoid species demonstrate that they are regulated by the density of the budmoth and not vice versa.[39,52]

2.5. Flight Behavior and Dispersal

During studies in a 150-year-old larch stand located within a deciduous forest near Lenzburg, Switzerland, a surprising second moth flight was detected about 1 month after all local flights had ceased.[48] Since large masses of budmoths had been observed on many earlier occasions on glaciers, passes, and mountain tops,[42] the most plausible explanation was that the moths had emigrated en masse from areas in defoliated subalpine larch–cembran pine forests. To test this hypothesis, an extensive network of light and pheromone traps was installed on both sides of the Swiss Alps (see Fig. 1). Flight and oviposition behavior were also studied in relation to abiotic factors and defoliation intensity in the Engadine Valley.[25,71]

Regular diurnal breezes, characteristic of mountainous areas, induce moths to fly, by positive anemotactic stimuli, against the downslope and downvalley breezes toward timberline and, given certain weather conditions, even beyond this to the mountain ridges and passes. At these sites, conditions are optimal for airborne insects to be taken up by the gradient winds that can carry them for long distances. Thus, we recognize both local and long-range dispersal patterns. Local upwind dispersal occurs constantly throughout the flight period whenever temperatures exceed 7°C, while weather conditions favoring long-range dispersal are characterized by indifferent barometric pressure over central Europe with airmasses above 10°C at 2500 m.[21] Thus, dispersal on a local scale probably contributes to the regularity of the population cycle, while long-range dispersal, because of its stochastic nature, may be conducive to the conspicuous time sequence of defoliation throughout the Alps. In 1979, for example, masses of moths from the French Alps swamped the southern slope of the Alps and led to defoliation in the following years as far east as the Carinthian Alps in Austria, a distance of approximately 600 km.[17]

In later experiments, we released marked moths in defoliated areas and undamaged forests. The results showed that moths were attracted to green undamaged trees[71] and flew greater distances in defoliated forests. Caged females were found to lay 12.5 ± 8.2 eggs in defoliated stands, as compared with 116 ± 19.1 in green stands.

3. PATTERNS OF NUMERICAL BEHAVIOR

In order to discuss changes in animal numbers, we use the following terms:[69] Oscillation describes numerical changes occurring within a generation; fluctuation describes numerical variations during the course of many generations in sequence. Thus, in its optimum area, the larch budmoth exhibits a regular cyclic fluctuation; i.e., regular cycles of abundance and sparsity follow each other immediately without extended periods of low numbers in between (Fig. 4).

3.1. Oscillation

Although only incomplete life tables have been constructed for budmoth populations in the optimum area, they change dramatically between the growth and decline phases of the cycle. In the growth phase, mortality to the different life stages rarely exceeds 20–30% and total generation mortality is approximately 90%. This low mortality, together with high fecundity, gives rise to a 10-fold increase from one generation to the next. In the decline phase, total mortality rises to 99.98%, egg mortality doubles, small larval instars suffer high mortality, larval parasitism increases 10-fold, and fecundity decreases by a factor of 10.[16]

3.2. Fluctuation

When the forest service of the Canton of Grisons was asked to control the larch budmoth in 1948, it was faced with the problem of predicting when and where defoliation would reach damaging levels. The obvious answer was to design a census method to provide information on the spatial pattern of population fluctuations in successive generations over the 6200 ha of larch–cembran pine forests of the Engadine Valley.[54] This task was facilitated by the forest inventory system, which records the numbers and sizes of trees by species for each stand. Individual larch trees were used as the sample unit, more than 400 of them being selected in a stratified random manner each year. Sample trees are stratified in relationship to topographical features (altitude, exposure) and are weighted

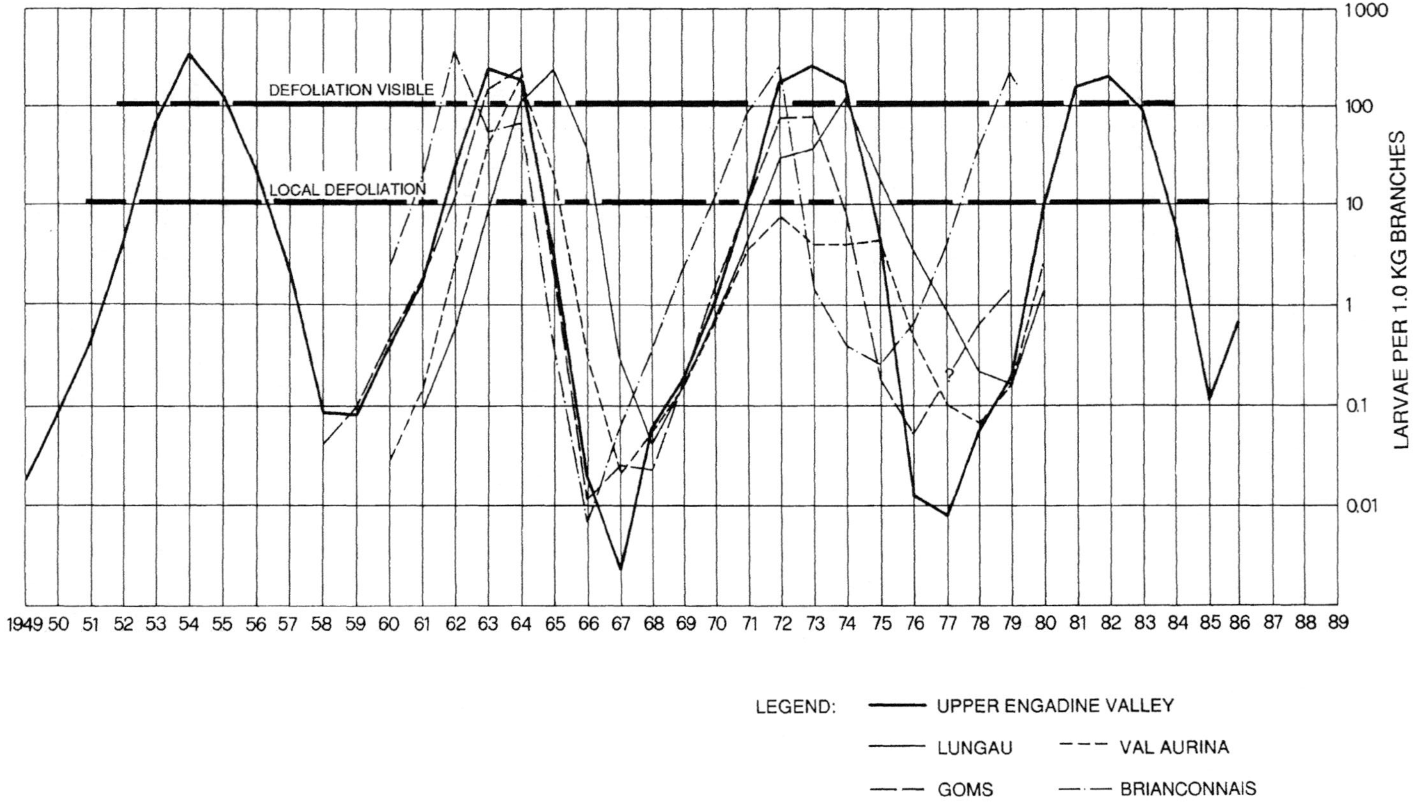

FIGURE 4. The cyclic fluctuation of larch budmoth populations at the five major study areas (see Fig. 1) expressed by the weighted mean of large larvae per 1.0 kg of larch branches (1949–1976 data from Auer,[11] 1977–1981 data from Baltensweiler,[20] 1982–1986, unpublished data).

according to host-tree density per unit area.[6] The sample, 1 kg of twigs and foliage, is extracted from three branches cut at three levels within the crown. This annual census provides an average density estimate for the larval population and can be converted to an absolute population estimate for the entire Engadine Valley.[4,11]

The population census was initiated in the Engadine Valley in 1949 and, with minor modifications, continued until 1976.[4,9,11] From 1977 to 1986 the census was continued on a reduced scale,[20] but from 1959 to 1980 it was extended to four additional outbreak areas spread along the entire alpine arch from the Brianconnais to Salzburg. In addition, 20 sample plots containing 4–20 larch trees were selected systematically along an altitudinal profile from high-elevation areas, characterized by large-amplitude cyclic fluctuations, to low-elevation stands within the deciduous forests of the colline region, where populations fluctuate much less extremely.[10]

This census for the high Alps is summarized in Fig. 4. Budmoth larval densities vary with remarkable regularity and by more than 10,000-fold over four or five generations. Altogether 13 complete cycles have been described, 11 of them causing general defoliation throughout the area, while only one cycle caused no visible damage (Valle Aurina, 1967–1977).[64] By superimposing the density curve of the Engadin on the defoliation records (Fig. 2), it is easy to visualize how populations have fluctuated in this subalpine region for more than 100 years.

At altitudes of 1200–1600 m, populations only occasionally reach densities that cause defoliation, and at altitudes below 1000 m they never do. However, it is interesting to note that these lower populations also exhibit cyclical fluctuations, albeit with a much smaller amplitude, which reach their peaks 1–3 years later than do subalpine populations.[10] Still farther away, along the Rhine River between the Lake of Constance and Basel, the budmoth fluctuates irregularly and, in small larch stands within the deciduous forest, it may become locally extinct.

4. HYPOTHESES FOR THE CAUSES OF OBSERVED POPULATION DYNAMICS

The question of central importance is: What causes larch budmoth populations to cycle so regularly and with such little variation in amplitude and periodicity? Many hypotheses have been erected to explain this rather common phenomenon in animal populations, including weather or sun-spot activity; interactions with parasites, predators, pathogens, and host plants; genetic factors; and finally mass dispersal. The more plausible of these hypotheses are discussed below:

4.1. The Food Quality Hypothesis

Changes in food quality are considered an important factor in regulating larch budmoth populations. The raw-fiber content of needles increases considerably during an outbreak and has a strong negative effect on larval survival and female fecundity.[28,46,63] In addition, increases in raw-fiber content can be induced by budmoth feeding in the previous generation, and it then takes several years for these changes to relax.[28,45] On theoretical grounds alone, this delayed negative feedback mechanism could be sufficient to generate regular population cycles.[28,43]

This hypothesis was tested by constructing a detailed mathematical model of the larch–larch budmoth system (Fig. 5).[43] Simulations of budmoth population dynamics resulting from this model closely mimic the real population cycles (Fig. 6). In addition, *a priori* model predictions of larch needle raw-fiber content correspond closely to later experimental data.[46] These results led us to believe that food quality change induced by previous budmoth feeding is the most plausible explanation for the 9-year population cycle.

4.2. The Self-Regulation Hypothesis

The self-regulation hypothesis proposes that cycles can be induced in populations consisting of several genotypes correlated with ecologically different phenotypes, if mechanisms for density-dependent selection are present.[34,56] The data on larch budmoth ecotypes are consistent with such a hypothesis. Not only have color morphs been observed,[13–15,35–37] but their proportions are correlated with the population cycles.[14] How-

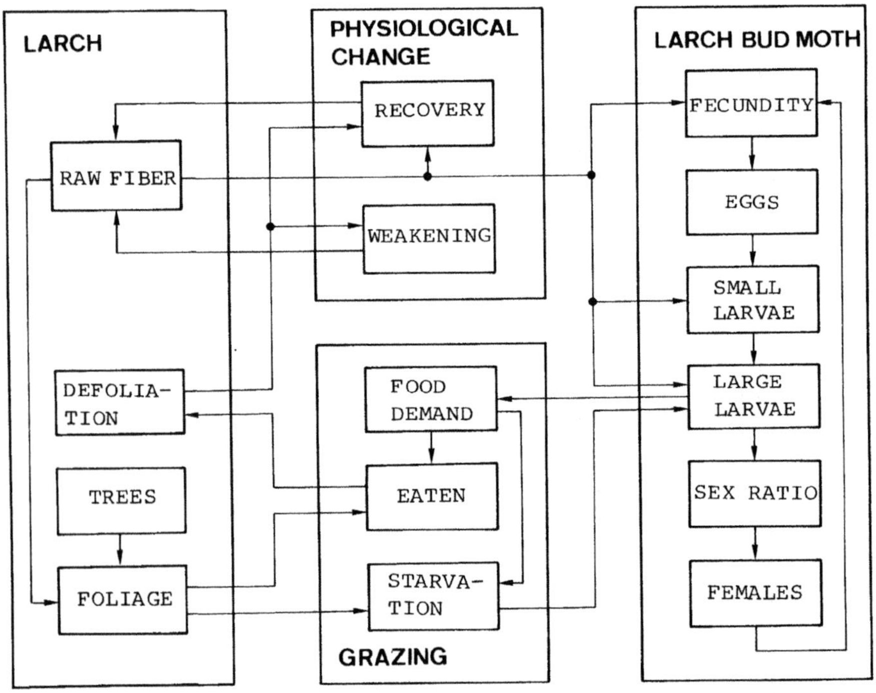

FIGURE 5. Structure of the larch–larch budmoth relationship. The depicted functional dependencies are based on observed and quantified correlations. (From 43).

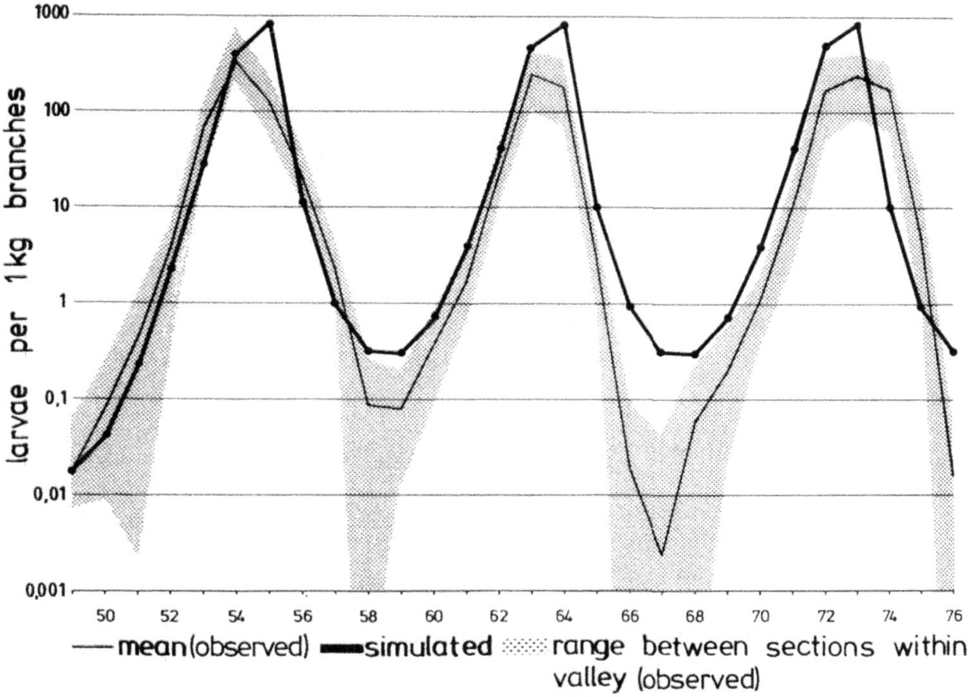

FIGURE 6. Simulated population fluctuations of the larch budmoth in the Upper Engadine Valley. The model is based on the larch–larch budmoth relationship (see Fig. 5) and excludes dispersal. (From 43, 45).

ever, theoretical studies suggest that polymorphism–selection cannot alone induce sustained cycles.[47,70] For this reason, the classical self-regulation hypothesis has to be modified by combining it with the food-quality hypothesis or by including assortive mating. In the first case, changing food quality becomes the selective process that induces varying genotypic frequencies. The different ecological traits then cause populations to increase up to the point where defoliation alters food quality.[19] This is essentially equivalent, however, to the food-quality hypothesis because the mechanism inducing the cycles is a delayed negative feedback between host plant and budmoth populations.

In a second form of this hypothesis, sex-attractant polymorphism is assumed to separate subpopulations temporarily, by means of assortive mating, during the phases of low population density when pheromone communication is essential for reproduction.[19,43,49]

4.3. The Antagonism Hypothesis

On the basis of classical theory, we might expect interactions with antagonists (predators, parasitoids, pathogens) to induce cyclic population dynamics.[3,5,7,53,58,72,73] In 1957, immediately after the first cycle we had studied intensively, it seemed obvious to everybody that the granulosis virus disease played a crucial role in suppressing the outbreak.[27] But since substantial numbers of diseased larvae could not be found in subsequent outbreaks, this idea had to be abandoned. It is unfortunate that this fact was not taken into account in recent theoretical papers that revive the pathogen–host relationship hypothesis for budmoth cycles.[3] It is worth noting that if the larch budmoth project had been terminated after the first 10 years of research, we would not have discovered that disease epizootics were not a regular phenomenon,[27] and most scientists would now be convinced that the cycles were caused by the interaction between virus and budmoth.

The impact of several species of parasitoids was studied carefully in an attempt to determine their roles in budmoth population dynamics.[1,38,52,66] These studies generally indicate that parasitism merely tracks the larch budmoth population; i.e., budmoth population fluctuations regulate the numbers of parasitoids and not vice versa.

4.4. The Dispersal Hypothesis

The dispersal hypothesis cannot be considered independently from the other hypotheses because budmoth population dynamics are determined by a combination of local factors as well as dispersal processes. The influence of dispersal has been postulated in the form of four hypotheses: the *Herdtheorie,* the conglobation hypothesis, the translocation hypothesis, and the cycle synchronization hypothesis.[43,44]

Forest pest fluctuations cause defoliation to appear and disappear in distinct patterns in time and space. This observation led to the formulation of the *Herdtheorie* during the 1930s.[40] This theory proposes that populations spread out from hot spots (*Herd*), or epicenters, and swamp other unpopulated areas. In the case of the larch budmoth, this theory was abandoned, however, when the annual census demonstrated that budmoth larvae were everywhere all the time.[4] Another observation that went against the *Herdtheorie* was that budmoth populations grew at similar rates in all alpine valleys, a fact that is most consistent with autochthonous population growth.

The conglobation hypothesis proposes that locally high budmoth populations occur because they tend to aggregate in preferred habitats.[68] This is based on the observation that budmoths tend to fly from suboptimal montane levels (1000–1700 m) to optimal subalpine levels (1700–2000 m).[25,38] This behavioral characteristics of flying moths, together with the topology of the European Alps, are assumed to be responsible for the observed conglobation.[25,43] According to this hypothesis, budmoth populations are exposed to a regular and more or less constant aggregated influx of moths that replenishes subpopulations that have become locally extinct following severe outbreaks.

The translocation hypothesis refers to the emigration of a substantial proportion of the population to new, less hostile habitats.[69] Since subpopulations of the larch budmoth regularly deplete their food resource, they are forced to lead a nomadic life. The increased mobility of moths in defoliated stands, their attraction to green larch foliage,[71] the conspicuous mass flights on mountain passes and mountain tops,[21,42,59] and defoliation patterns in neighboring valleys[13] support this hypothesis. According to the translocation hypothesis, outbreaks collapse because of overexploitation of the food resources and mass exodus of the adult population.[43]

Finally, the synchronization hypothesis attempts to explain the synchronous pattern of conspicuous defoliation along the alpine arch (Fig. 2). Population cycles are sometimes

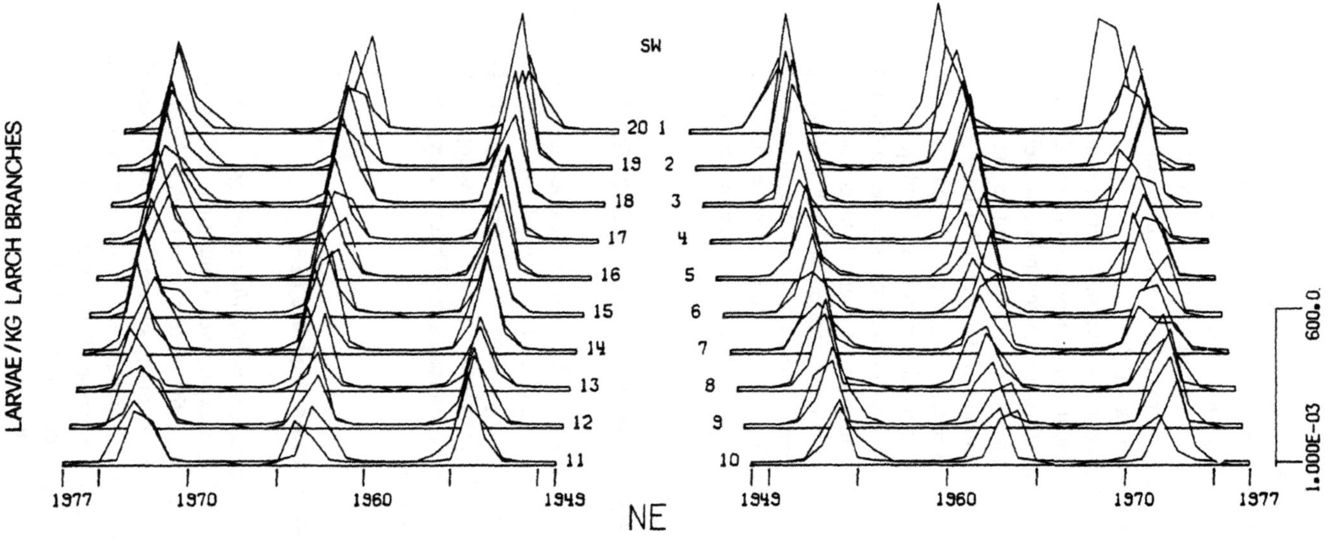

FIGURE 7. Simulated population fluctuations of the larch budmoth in the Upper Engadine Valley based on a model of the larch–larch budmoth relationship (see Fig. 5) that includes dispersal processes (translocation as well as conglobation). The simulated and observed larval densities (linear scale) are shown for each of the 20 sites within the Engadine Valley (From 43).

slightly out of synchrony, e.g., the delayed defoliation sequence from the western to eastern Alps during the 1960s (Fig. 4). However, these delays are usually small and follow a distinct pattern. Otherwise, the cycles show remarkable synchrony, even from one end of the Alps to the other, and the closer the valleys the greater their synchrony. The synchronization hypothesis explains these patterns by the exchange of relatively small numbers of moths between neighboring valleys that would otherwise be fluctuating independently of each other. Cycles initially out of synchrony can be synchronized in our model by the exchange of surprisingly few moths, provided the valleys are arranged in a line, as is the case in the alpine arch.[44]

The various dispersal hypotheses (except for the *Herdtheorie*) have been studied by means of simulation. Their inclusion generally increases the realism of the simulation (Fig. 7) but does not substantially alter the general cyclic behavior and therefore does not refute the food quality hypothesis. By contrast, migration is critical in stabilizing the system, guaranting re-establishment of the cycle after local extinctions[43] and synchronizing population fluctuations over large geographical regions.

As a result, our studies generally support the food-quality hypothesis. We should note, however, that our model of induced host responses to defoliation is based on rather limited data, and new information could change our conclusions. If additional mechanisms need to be included in our explanation of larch budmoth population cycles, we suspect that parasitism or genetic polymorphism are the most likely candidates. By contrast, dispersal is responsible for the stability of the cycles, the resilience of the larch budmoth system, and the synchronization of fluctuations between alpine valleys.

5. IMPLICATIONS FOR DEVELOPING MANAGEMENT STRATEGIES

A diverse palette of techniques has been developed for controlling larch budmoth populations, including chemical[8] and microbiological[61] insecticides, mating disruption by means of sex attractants,[24] biological control by releasing parasitoids,[2] releasing budmoths in order to enrich the parasitoid potential at low population densities,[51] and silvicultural measures such as the afforestation with less susceptible tree species or various stand-management practices.[26]

5.1. Feasibility of Management Strategies

Spraying chemical insecticides or *Bacillus thuringiensis* with helicopters in alpine areas is technically feasible but expensive. Simulations on our model indicate that this strategy may fail, however, since the protection of larch foliage is equivalent to eliminating the key regulating mechanism in the system. Without the negative feedback between foliage and insect, the budmoth population would remain high, and the application of insecticides might have to be repeated at frequent intervals.[43] It is also feared that spraying would result in high tree mortality due to continuous stress from high populations near the threshold of defoliation.

Biological control by inundative releases of parasites is extremely expensive and poses some almost insurmountable difficulties. Rearing parasites is even more difficult than rearing the budmoth. Mating disruption using sex pheromones seems to offer some potential. Model simulations over large units (e.g., a whole alpine valley such as the Upper Engadine) suggest that this technique could succeed.[43] However, it would again become necessary to repeat the treatment almost on a continuous basis, rendering the method very expensive. In addition, immigration due to conglobation or long-range dispersal might soon devaluate the control effort.[43]

Changes in forest management practices take a long time to have an impact on subalpine forests. Not only are growing conditions poor, but few other tree species are available or can compete with larch. Within the Engadine Valley, for example, native inhabitants favor larch because of its rapid growth, its capability to reproduce naturally, and its aesthetic properties.

5.2. Economic Aspects of Population Management

Only recently has it become possible to make a realistic estimate of the diverse impacts of larch budmoth outbreaks on the forest ecosystem and forestry economy. Proponents of tourism have often claimed that budmoth outbreaks cause tourists to leave or avoid the area. However, no data have been gathered to support this view. It could well be that tourism is less affected than has been feared, since visible defoliation rarely affects a whole valley during the same season. Also, not all outbreaks coincide with dry weather, so the conspicuous defoliation patterns are not always visible.

Tree mortality due to budmoth defoliation is negligible ($<1\%$ of the total volume).[26] However, local impacts may be significant; e.g., salvage operations often disrupt the management and cutting plans of the district. Within the realm of a single community, therefore, large numbers of dead larch trees can cause severe economic hardship by disrupting sustained harvesting.

Natural reproduction of host trees is not directly affected by the larch budmoth outbreaks and, although trees are weakened by defoliation, this has never been observed to reduce reproductive output.

Larch budmoth outbreaks do not threaten either the existence or composition of the subalpine larch–cembran pine forest. On the contrary, the components of this forest ecosystem seem to be well adapted to each other. Dendrochronological data show that larch and budmoth coexisted at least since Roman times. A hypothesis could even be proposed that without the larch budmoth the subalpine larch–cembran pine forest would change to a less desirable state. For example, the consumption of 50–100% of the larch needle biomass every 9 years over large areas probably accelerates nutrient cycling, and the defoliation of vulnerable cembran pine in the understory can cause considerable mortality and postpone the succession to the climax community, i.e., a subalpine larch–cembran pine forest with a much higher percentage of pines.[26] Larch does not reproduce well under a tight pine canopy and survives only in relatively open stands. Thus, the interaction between the larch budmoth and its host, the larch, may even be considered mutualistic.

5.3. Conclusions

Since control would have to be applied almost continuously to entire alpine valleys, even in the case of the most promising pheromone disruption technique, and since economic losses due to the budmoth turn out to be rather minor, we no longer consider the larch budmoth a pest worth controlling. The decision to abolish control measures is made even easier when the risks of increasing larch tree mortality are considered. The larch budmoth research project not only exemplifies how easily wrong decisions can be made, but that long-term research on a scale appropriate to the organism is necessary to provide the basis for correct management decisions. Close scrutiny has shown that this forest defoliator is, in fact, not a pest. In this respect, the story of the larch budmoth provides a lesson for the administration of future ecological research.[27]

ACKNOWLEDGMENTS. We thank Dr. D. Rubli for drawing and preparing the figures and the innumerable scientists who have made possible this review by their research or who have contributed with their stimulating ideas during many discussions. All this has helped us tremendously to clarify our own ideas.

This chapter represents contribution no. 120 of the Working Group for the Study of the Population Dynamics of the Larch Budmoth of the Swiss Federal Institute of Technology, Zürich, Switzerland.

REFERENCES

1. Aeschlimann, J.-P., 1969, Contribution à l'étude de trois espèces d'eulophides (Hymenoptera: Chalcidoidea) parasites de la tordeuse grise du mélèze, *Zeiraphera diniana* GUENÉE (Lepidoptera: Tortricidae) en Haute-Engadine, *Entomophaga* **14:**261–320.
2. Aeschlimann, J.-P., 1975, Biologie, comportement et lacher expérimental de *Triclistus pygmaeus* Cresson (Hym, Ichn.), *Mitt. Schweiz. Entomol. Ges.* **48:**165–171.
3. Anderson, R. M., and May, R. M., 1980, Infectious diseases and population cycles of forest insects, *Science* **210:**658–661.
4. Auer, C., 1961, Ergebnisse zwölfjähriger quantitativer Untersuchungen der Populationsbewegung des Grauen Lärchenwicklers *Zeiraphera griseana* Hübner (=diniana GUENÉE) im Oberengadin (1949/1960), *Eidg. Anst. Forstl. Versuchswes. Mitt.* **37:**173–263.
5. Auer, C., 1968, Erste Ergebnisse einfacher stochastischer Modelluntersuchungen über die Ursachen der Populationsbewegung des grauen Lärchenwicklers *Zeiraphera diniana* Gn. (= *griseana* Hb.) im Oberengadin, 1949/66, *Z. Angew. Entomol.* **62:**202–235.
6. Auer, C., 1971, Some analyses of the quantitative structure in populations and dynamics of larch bud moth 1949/1968, in: *Statistical Ecology* (G. P. Patil, ed.), Vol. 2, pp. 151–173, Pennsylvania State University Press, University Park, Pennsylvania.
7. Auer, C., 1971, A simple mathematical model for "key-factor" analysis and comparison in population research, in: *Statistical Ecology* (G. P. Patil, ed.), Vol. 2, pp. 33–48, Pennsylvania State University Press, University Park, Pennsylvania.
8. Auer, C., 1974, Ein Feldversuch zur gezielten Veränderung zyklischer Insektenpopulationsbewegungen, *Schweiz. Z. Forstwes.* **125:**333–358.
9. Auer, C., 1975, Dendrometrische Grundlagen und Verfahren zur Schätzung absoluter Insektenpopulationen, *Eidg. Anst. Forstl. Versuchswes. Mitt.* **50:**88–131.
10. Auer, C., 1975, Jährliche und langfristige Dichteveränderungen bei Lärchenwicklerpopulationen (*Zeiraphera diniana* Gn.) ausserhalb des Optimumgebietes, *Mitt. Schweiz. Entomol. Ges.* **48:**47–58.
11. Auer, C., 1977, Dynamik von Lärchenwicklerpopulationen längs des Alpenbogens, *Eidg. Anst. Forstl. Versuchswes. Mitt.* **53:**71–105.

12. Bakke, A., 1969, Extremely low supercooling point in eggs of *Zeiraphera diniana* (GUENÉE), *Norsk. Entomol. Tidsskr.* **3**:81–83.
13. Baltensweiler, W., 1964, *Zeiraphera griseana* HÜBNER (Lepidoptera: Tortricidae) in the European Alps. A contribution to the problem of cycles, *Can. Entomol.* **96**:792–800.
14. Baltensweiler, W., 1971, The relevance of changes in the composition of larch bud moth populations for the dynamics of its numbers, in: *Dynamics of Populations*. Proceedings of the Advanced Study Institute of Dynamics Numbers Popul. Oosterbeek. (P. S. DenBoer and G. R. Gradwell, eds.), pp. 208–219, Pudoc, Wageningen.
15. Baltensweiler, W., 1977, Colour-polymorphism and dynamics of larch bud moth populations (*Zeiraphera diniana* Gn., Lep. Tortricidae), *Mitt. Schweiz. Entomol. Ges.* **50**:15–23.
16. Baltensweiler, W., 1978, Insect Population Dynamics, pp. 381–394 in Proc. XVI IUFRO World Congress, Norway 1976, Div. 2.
17. Baltensweiler, W., 1982, Zur Entwicklung der gegenwärtigen Massenvermehrung des Grauen Lärchenwicklers (*Zeiraphera diniana* Gn.). Inst. Phytomedizin, ETH Zürich, unpublished report 91.
18. Baltensweiler, W., 1983, The impact of weather on the life system larch–larch bud moth in the European Alps. in: *The role of insect–plant relationship in the population dynamics of forest pests,* (A. S. Isaev, ed.), pp. 73–83, Proc. Symp. IUFRO/MAB, Irkutsk 1981.
19. Baltensweiler, W., 1984, The role of environment and reproduction in the population dynamics of the larch bud moth, *Zeiraphera diniana* Gn. (Lepidoptera, Tortricidae), in: *Advances in Invertebrate Reproduction* (W. Engels, W. H. Clark, Jr., A. Fischer, P. J. W. Olive, and D. F. Went, eds.), Vol. 3, pp. 291–302, Elsevier, Amsterdam.
20. Baltensweiler, W., 1985, On the extent and the mechanisms of the outbreaks of the Larch Bud Moth (*Zeiraphera diniana* Gn., Lepidoptera, Tortricidae) and its impact on the subalpine Larch-cembran pine Forest ecosystem, in: *Establishment and Tending of Subalpine Forest: Research and Management,* Proc. 3rd IUFRO Workshop P 1.07-00, 1984, *Eidg. Anst. Forst. Versuchswes., Ber.* 270 (H. Turner and W. Tranquillini, eds.), pp. 215—219.
21. Baltensweiler, W., and G. von Salis, 1975, Zur Dispersionsdynamik der Falter des Grauen Lärchenwicklers (*Zeiraphera diniana* Gn., Tortricidae). *Z. angew. Entomol.* **77**:251–257.
22. Baltensweiler, W., Benz, G., Bovey, P., and Delucchi, V., 1977, Dynamics of larch bud moth populations, *Annu. Rev. Entomol.* **22**:79–100.
23. Baltensweiler, W., Priesner, E., Arn, H., and Delucchi, V., 1978, Unterschiedliche Sexuallockstoffe bei Lärchen- und Arvenform des Grauen Lärchenwicklers (*Zeiraphera diniana* Gn., Lep., Tortricidae), *Mitt. Schweiz. Entomol. Ges.* **51**:133–142.
24. Baltensweiler, W., and Delucchi, V., 1979, The study of larch bud moth migration in the Engadine valley by means of a parapheromone, *Mitt. Schweiz. Entomol. Ges.* **52**:291–296.
25. Baltensweiler, W., and Fischlin, A., 1979, The role of migration for the population dynamics of the larch bud moth *Zeiraphera diniana* Gn. (Lep., Tortricidae), *Mitt. Schweiz. Entomol. Ges.* **52**:259–271.
26. Baltensweiler, W., and Rubli, D., 1984, Forstliche Aspekte der Populationszyklen von *Zeiraphera diniana* Gn. (Lep., Tortricidae), *Eidg. Anst. forstl. Versuchswes., Mitt.* **60**:1–148.
27. Baltensweiler, W., and Fischlin, A., 1987. On methods to analyze ecosystems: lessons from the analysis of forest-insect systems, in: *Potentials and limitations of ecosystem analysis* (E. D. Schulze and H. Zwölfer, eds.), pp. 401–415, Springer, Berlin.
28. Benz, G., 1974, Negative Rückkoppelung durch Raum- und Nahrungskonkurrenz sowie zyklische Veränderung der Nahrungsgrundlage als Regelprinzip in der Populationsdynamik des Grauen Lärchenwicklers, *Zeiraphera diniana* (GUENÉE) (Lep., Tortricidae), *Z. angew. Entomol.* **76**:31–49.
29. Bovey, P., 1958, Le problème de la tordeuse grise du mélèze *Eucosmu griseana* (Hübner) (Lepidoptera: Tortricidae) dans les forêts alpines, Proc. 10. Congr. Internat. Entomology, 1956, **4**:123–131.
30. Bovey, P., 1978, *Spilonota* und *Zeiraphera,* in: *Die Forstschädlinge Europas.* Band 3: Schmetterlinge (W. Schwenke, ed.), pp. 153–177, Verlag Paul Parey, Hamburg.
31. Bovey, P., and Maksymov, J. K., 1959, Le problème des races biologiques chez la tordeuse grise du mélèze *Zeiraphera griseana* (HB), *Vierteljahrsschrift d. Naturforsch. Ges. Zürich* **104**:264–274.
32. Burger, H., 1926, Untersuchungen über das Höhenwachstum verschiedener Baumarten, *Eidg. Anst. Forstl. Versuchsw., Mitt.* **14**:29–158.
33. Chaix, B., 1845, *Préoccupations statistiques, géographiques, pittoresques et synoptiques du Département des Hautes Alpes,* Allier, Grenoble.

34. Chitty, D., 1977, Natural selection and the regulation of density in cyclic and non-cyclic populations, in: *Evolutionary Ecology* (B. Stonehouse and C. Perrins, eds.), pp. 27–31. MacMillan Press Ltd., London.
35. Day, K. R., 1976, An ecological study of *Zeiraphera diniana* GN. (=griseana HB.) on conifers in Britain, Ph.D. thesis, University of London, London.
36. Day, K., 1984, Phenology, polymorphism and insect-plant relationships of the larch budmoth, *Zeiraphera diniana* (Guenée) (Lepidoptera: Tortricidae), on alternative conifer hosts in Britain, *Bull. Entomol. Res.* **74:**47–64.
37. Day, K. R., and Baltensweiler, W., 1972, Change in the proportion of the larval colourtypes of the larchform *Zeiraphera diniana* when reared on two media, *Entomol. Exp. Appl.* **15:**287–298.
38. Delucchi, V., 1982, Parasitoids and hyperparasitoids of *Zeiraphera diniana* GN. (Lep., Tortricidae) and their rôle in population control in outbreak areas, *Entomophaga* **27:**77–92.
39. Delucchi, V., and Renfer, A., 1977, The level of abundance of *Phytodietus griseanae* Kerrich (Hym. Ichneumonidae) determined by its host *Zeiraphera diniana* Guenée (Lep., Tortricidae) at high altitude, *Mitt. Schweiz. Entomol. Ges.* **50:**233–248.
40. Eidmann, H., 1934, Untersuchungen über Herdtheorie und Überflug, in: *Die Forleule in Preussen im Jahre 1933* (H. Eidmann, ed.), pp. 28–34. Sonderheft, Verlag M. und H. Schaper, Hannover.
41. Ellenberg, H., and Klötzli, F., 1972, Waldgesellschaften und Waldstandorte der Schweiz, *Eidg. Anst. forstl. Versuchswes., Mitt.* **48:**589–930.
42. Escherich, K., and Bär, W., 1909, Einiges über den grauen Lärchenwickler, *Naturw. Z. Land- u. Forstw.* **7:**188–194.
43. Fischlin, A., 1982, Analyse eines Wald-Insekten-Systems: Der subalpine Lärchen-Arvenwald und der graue Lärchenwickler *Zeiraphera diniana* GN. (Lep., Tortricidae), Ph.D. Diss ETHZ Nr. 6977.
44. Fischlin, A., 1983, Modelling of alpine valleys, defoliated forests, and larch bud moth cycles: The rôle of moth migration, in: *Mathematical models of renewable resources* (R. Lamberson, ed.), Vol. II, pp. 102–104. Proceedings of the 2nd Pacific Coast Conference on Mathematical Modeling of Renewable Resources, University of Victoria, Victoria, B.C., Canada.
45. Fischlin, A., and Baltensweiler, W., 1979, Systems analysis of the larch bud moth system. Part I: the larch-larch bud moth relationship, *Mitt. Schweiz. Entomol. Ges.* **52:**273–289.
46. Fischlin, A., Baltensweiler, W., Delucchi, V., Francesci, G., and Seiffert, F., Is the rawfiber content really correlated with the population cycles of larch bud moth? Some results from chemical analysis of larch needles from the Upper Engadine Valley, unpublished.
47. Fischlin, A., Chitty, D., and Krebs, C. J., *Modelling the real Chitty Hypothesis,* Institute of Animal Resource Ecology, University of British Columbia, Vancouver, B.C., Canada, unpublished.
48. Graf, E., 1974, Zur Biologie und Gradologie des Grauen Lärchenwicklers, *Zeiraphera diniana* Gn. (Lep., Tortricidae) im schweizerischen Mittelland, *Z. angew. Entomol.* **76:**233–251, 347–379.
49. Guerin, P. M., Baltensweiler, W., Arn, H., and Buser, H.-R., 1984, Host race pheromone polymorphism in the larch budmoth, *Experientia* **40:**892–894.
50. Haukioja, E., 1980, On the role of plant defenses in the fluctuation of herbivore populations, *Oikos* **35:**202–213.
51. Herren, H. R., 1976, Manipulation d'une population de la tordeuse grise du Mélèze en vue d'augmenter l'efficacité de ses parasitoides, *Mitt. Schweiz. Entomol. Ges.* **49:**307–308.
52. Herren, H. R., 1977, Le rôle des eulophides dans la gradation de la tordeuse grise du mélèze, *Zeiraphera diniana* GUENÉE (Lep., Tortricidae) en Haute-Engadine, Ph.D. Thèse, ETHZ No. 6037.
53. Howard, L. O., and Fiske, W. F., 1911, The importation into the United States of the parasites of the Gipsy Moth and Brown-Tail Moth: A report of progress with some considerations of previous and concurrent efforts of this kind, *U.S. Dept. Bur. Ent. Bull.* **91:**105–109.
54. Kälin, A., and Auer, C., 1954, Statistische Methoden zur Untersuchung von Insektenpopulationen, *Z. angew. Entomol.* **36:**241–282, 423–461.
55. Kawabe, A., 1980, Descriptions of fourteen new species of the micro moths of Japan, *Tinea* **11:**17–31.
56. Krebs, C. J., 1978, A review of the Chitty Hypothesis of population regulation, *Can. J. Zool.* **56:**2463–2480.
57. Levin, D. A., 1976, The chemical defences of plants to pathogens and herbivores, *Ann. Rev. Ecol. Syst.* **7:**121–159.
58. Lotka, A. J., 1925, *Elements of physical biology,* Williams and Wilkins, Baltimore.
59. Maksymov, J. K., 1959, Beitrag zur Biologie und Oekologie des Grauen Lärchenwicklers *Zeiraphera*

griseana (Hübner) (Lepidoptera, Tortricidae) im Engadin, *Mitt. schweiz. Anst. forstl. Versuchsw.* **35**:277–315.

60. Martignoni, M. E., 1957, Contributo alla conoscenza di una granulosi di *Eucosma griseana* (Hübner) (Tortricidae, Lepidoptera) quale fattore limitante il pullulamento dell'insetto nella Engadina alta, *Mitt. schweiz. Anst. forstl. Versuchsw.* **32**:371–418.
61. Martouret, D., and Auer, C., 1977, Effets de *Bacillus thuringiensis* chez une population de tordeuse grise du mélèze, *Zeiraphera diniana* (Lep.: Tortricidae) en culmination gradologique, *Entomophaga* **22**:37–44.
62. Mutuura, A., and Freeman, T. N., 1966, The North American species of the genus *Zeiraphera*. *J. Res. Lepid.* **5**:153–176.
63. Omlin, F. X., 1977, Zur populationsdynamischen Wirkung der durch Raupenfrass und Düngung veränderten Nahrungsbasis auf den Grauen Lärchenwickler *Zeiraphera diniana* Gn. (Lep.: Tortricidae), Ph.D. Diss. ETHZ Nr. 6054.
64. Omlin, F. X., and Herren, H. R., 1976, Zur Populationsdynamik des Grauen Lärchenwicklers, *Zeiraphera diniana* Guenée (Lep., Tortricidae), im Ahrntal (Südtirol, Italien): Lebenstafeluntersuchungen und Nahrungsverhältnisse während der Vegetationsperiode 1975, *Mitt. Schweiz. Entomol. Ges.* **49**:203–228.
65. Priesner, E., 1979, Specificity studies on pheromone receptors of noctuid and tortricid Lepidoptera, in: *Chemical Ecology: Odour Communication in Animals* (F. J. Ritter, ed.), pp. 57–71, Elsevier, Amsterdam.
66. Renfer, A., 1974, Caractéristique biologiques et efficacité de *Phytodietus griseanae* Kerrich (Hym., Ichneumonidae) parasitoide de *Zeiraphera diniana* Guenée (Lep., Tortricidae) en haute montagne, Ph.D. Thèse, ETHZ No. 5278.
67. Schweingruber, F. H., 1979, Auswirkungen des Lärchenwicklerbefalls auf die Jahrringstruktur der Lärche, *Schweiz. Z. Forstwes.* **130**:1071–1093.
68. Schwerdtfeger, F., 1963, *Ökologie der Tiere. I. Autökologie,* Paul Parey, Hamburg.
69. Schwerdtfeger, F., 1968, *Ökologie der Tiere. II. Demökologie,* Paul Parey, Hamburg.
70. Stenseth, N. C., 1981, On Chitty's theory for fluctuating populations: the importance of genetic polymorphism in the generation of regular density cycles, *J. Theor. Biol.* **90**:9–36.
71. Vaclena, K., and Baltensweiler, W., 1978, Untersuchungen zur Dispersionsdynamik des Grauen Lärchenwicklers, *Zeiraphera diniana* Gn. (Lep., Tortricidae): 2. Das Flugverhalten der Falter im Freiland, *Mitt. Schweiz. Entomol. Ges.* **51**:59–88.
72. Van den Bos, J., and Rabbinge, R., 1976, *Simulation of the fluctuations of the grey larch bud moth,* Simulation monographs, Centre for Agricultural Publishing and Documentation, Wageningen.
73. Volterra, V., 1926, Variazioni e fluttuazioni del numero d'individui in specie animali conviventi, *Memorie della R. Accademia dei Lincei* **2**:31–113.

CHAPTER 18

THE GYPSY MOTH

A Westward Migrant

Michael E. Montgomery and William E. Wallner

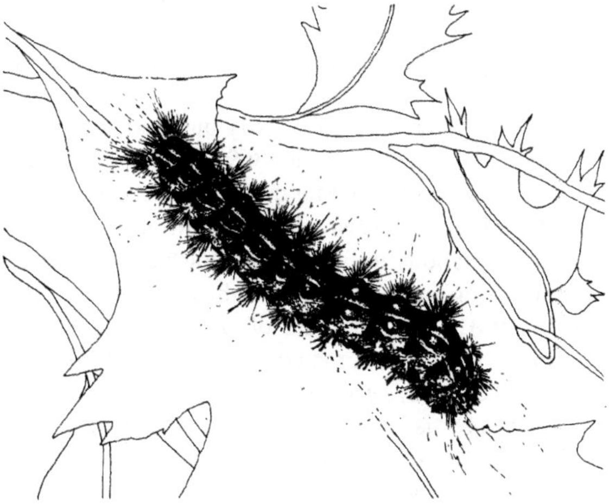

Michael E. Montgomery and William E. Wallner • U.S.D.A. Forest Service, Northeastern Forest Experiment Station, Center for Biological Control of Northeastern Forest Insects and Diseases, Hamden, Connecticut 06514.

1. INTRODUCTION

The gypsy moth, *Lymantria dispar* (Linnaeus), is one of the most serious pests of hardwood forests in temperate regions. A cartographical analysis of gypsy moth[35] shows it to occur within the latitudes 20°–60°N, where annual rainfall is 25–100 cm and temperature isotherms are 15–27°C for July and −18–12°C for January. The genus probably originated in East Asia,[36] which is where *L. dispar* exhibits its greatest variability[39] (Fig. 1). Proceeding westward from East Asia to western Europe, adult and larval coloration patterns become less variable and females progressively lose the ability to fly.[36,65] The range of the gypsy moth was extended through a series of founder events westward into Europe and from France to North America.

Currently, the gypsy moth is a minor pest in most of Eurasia. Outbreaks in Europe have been recorded periodically since the 1600s[48]; those during the mid-1950s and 1960s were especially widespread and devastating. For example, 70% of Yugoslavian hardwood forests were defoliated in 1957.[81] Within the past decade, however, outbreaks in Europe have been minor and localized. The historical record is not as well documented for Japan and China, where it is recognized as a pest insect but where severe widespread outbreaks are apparently rare/[64,65]

In North America, the gypsy moth has slowly but gradually spread from its point of introduction near Boston, Massaschusetts, in 1869. Infestations currently extend west to Ohio, south to Virginia, and north to Montreal, Canada. In this area, it is the most important defoliator of deciduous hardwoods. Long-distance inadvertent spread by man has resulted in isolated infestations in California, Washington, Oregon, and British Columbia, but attempts are being made to eradicate these soon after discovery. At the peak of the last outbreak in 1981, 6 million ha of mixed oak forest were defoliated. In Pennsylvania alone, timber loss was estimated to be more than $72 million and more than $9 million was expended for spray programs.

While the forest manager is justifiably concerned with timber loss (mortality and reduced growth increment) due to gypsy moth defoliation, the impact on urban environments creates a much more volatile problem. Homeowner concerns include defoliation of ornamental shrubs and trees, hordes of migrating larvae, and allergenic reactions to caterpillar hairs. The public pressure to control the problem is exemplified by personal accounts from Forbush and Fernald[30] of the first outbreak in North America in 1889:

> The place simply teemed with them (gypsy moth caterpillars), and I used to fairly dread going down the street to the station. . . . One morning, in particular, I remember that I was completely covered with caterpillars inside my coat as well as out. The street trees were completely stripped down to the bark. . . . The fronts of the houses were black with caterpillars, and the sidewalks were a sickening sight, covered as they were with crushed bodies of the pest. . . . Many got into the house. . . . [I'd even] find them in the beds when I turned down the blankets.

and by letters to the Pennsylvania Department of Environmental Resources 1982 of the invasion into that state:

> For three weeks the destruction wore on, and the incessant munching almost led me to insanity. . . . All vegetation up here is being denuded by the gypsy moth. As a result, all life is being threatened, even human.

FIGURE 1. Examples of size and color variation in gypsy moth males (a) and females (b) from Japan (top row) and from the United States (bottom row).

2. BIOLOGY

2.1. Life Cycle

Eggs are laid in clusters or masses covered with buff-brown hair from the female abdomen. Each mass contains 100–1200 eggs, with an average of 700 for sparse populations feeding on good hosts and 250 for dense declining populations. Egg masses tend to be aggregated in the forest.[49] Eggs are laid in early summer, enter diapause after completing embryonic development, and remain in diapause through winter. There is no evidence that diapause is controlled by photoperiod, but there is a requirement of 60–150 days of low temperature.[55] In spring, egg hatch begins at the time red oak buds open, continuing for 2–3 weeks. Accurate prediction of egg hatch and early larval growth[47,78] is important, since most pesticides are applied as soon as possible after all larvae have hatched and begun feeding.

Larvae disperse within 24 hr of hatching, but dispersal may be delayed by precipitation or temperatures below 7°C. In order to disperse, the larvae climb to the top of the tree and spin down on silken threads that eventually fracture. The small hairy larvae and attendant silk are then carried by the wind, but usually no more than 200 m.[56] In mountainous terrain, strong vertical wind fields may lift larvae above the tree canopy and carry them as far as 16 km.[72] However, under average conditions, only 3% of the larvae that become airborne would gain an altitude of more than 500 m above treetops, and only 0.1% of the total larval population would travel more than 5 km. Thus, dispersal may be important in the spread of new outbreaks within a region[12,14] but would seem unlikely to cause outbreaks in areas remote from infestations.

There are normally five larval instars in males and six in females. Full-grown females are three times heavier than males. Complete development at 20°C takes 44 and 50 days, respectively, for male and female larvae.[53] Larvae generally remain on or near the leaves for the first three instars but older larvae leave the foliage during daylight and seek resting places on the bole or in ground litter. This behavior is changed in dense populations, where larvae often remain in the canopy and feed intermittently over the entire day.

Consumption of oak foliage is about 10 mg for each mg gain in larval biomass.[75] Each larva in a dense population consumes about 1270 mg of foliage (dry weight) and attains a weight of 114 mg. This is about 170 cm^2 foliage or about 3 red oak leaves. However, a large (750 mg) female caterpillar typical of a sparse population would remove a total of 1000 cm^2 (1 sq ft) of foliage.

Pupation occurs at or near the daytime resting locations, which are usually at the tree base, on the ground-litter, or on the underside of rocks, but may be in the tree crown if population density is high. Male pupae appear about a week before females but spend 15–20% longer in this stage and thus eclose 1–3 days before females.

Mating and egg laying occur soon after adults emerge. This occurs within 1 or 2 m of the female pupation site in North America and western Europe, where females are flightless. In Japan, females preferentially fly to and oviposit on trees with light-colored bark, such as *Betula platypyhlla*.[63]

2.2. Host Plants

The gypsy moth has an exceptionally broad range of hosts. In Massachusetts, 41 tree species are considered favored food, 21 intermediate, and 35 unfavored.[57] Approximately half of the 185 native tree species in Europe are attacked.[51] In China, the host range may be even broader; larch, elm, persimmon, and several species of *Prunus* are frequently defoliated as well as oak, birch, poplar, willow, linden, maple, and apple, which are also fed on heavily in North America and Europe.[65]

TABLE I
Host Plant Preferences of the Gypsy Moth Relative to *Q. rubra*

	Laboratory[a] rearing	Defoliation[b] level	Defoliation[c] level	Larval[d] distribution
Quercus rubra	++	1.00	1.00	1.00
Q. velutina	++	1.13	1.35	−
Q. prinus	++	1.11	1.47	−
Q. alba	++	1.11	0.83	−
Populus spp.	++	1.18	1.10	0.96
Tilia americana	++	1.24	0.56	0.52
Betula papyrifera	++	0.76	0.56	0.28
Fagus grandifolia	++	0.50	0.54	0.44
Acer rubra	+	0.42	0.42	0.10
A. saccharum	+	0.20	0.68	0.69
Carya spp.	+	0.33	0.76	0.36
Prunus serotina	+	0.44	0.29	0.05
Pinus strobus	+	0.34	0.22	0.64
Tsuga canadensis	+	0.24	0.01	−
Robinia pseudoacacia	−	0.20	0.15	−
Fraxinus spp.	−	0.07	0.20	0.22

[a]++, favored host; +, acceptable; −, avoided; Massachusetts.[37]
[b]190 plots surveyed for 20 or 30 years in New England.[19]
[c]575 plots surveyed 1 year in Pennsylvania.[33]
[d]Larvae under bark flaps on 944 trees surveyed 2 years in Quebec.[52]

In North America, oak and poplar are consistently the most preferred hosts (Table I). Species such as black locust and ash are usually avoided, but considerable variability exists in the relative preference for intermediate species such as *Betula papyrifera* and *Acer saccharum*. This variability may reflect differences in population density[19] but could also be a consequence of differences in site conditions such as stand age or soil moisture.

3. NATURAL ENEMIES

3.1. Parasitoids

More than 90 species of parasitoids are known to attack *L. dispar* in Eurasia.[60] Most have been introduced into the United States in what has been one of the most extensive efforts in the history of biological control.

The first importations occurred between 1905 and 1914 and were from Japan and Europe. Four of the six most important parasitoids in North America were quickly established: *Ooencyrtus kuvanae* (Howard) that attacks eggs, *Cotesia melanoscelus* (Ratzeburg) and *Compsilura concinnata* (Meigen) that attack small larvae, and *Blepharipa pratensis* (Meigen) that attacks large larvae. The establishment of *Parasetigena silvestris* (Robineau-Desvoidy), which attacks large larvae, and *Brachymeria intermedia* (Nees), which attacks pupae, was not confirmed until 1937 and 1965, respectively. These parasitoids are all host specific except *C. concinnata,* which attacks more than 200 species of Lepidoptera. Four other species have been established in North America, but they are rare.

Establishment of other parasitoids has been unsuccessful despite the importation of over 200,000 individuals representing at least 80 species during the past 20 years. Lack of suitable alternate hosts is a major reason why many of these species have not become established in North America.[46]

3.2. Predators

Bird predation may have a substantial impact on gypsy moth populations. In Europe, both orioles and nuthatches are important predators of larvae and eggs[27] and birds seem to help maintain populations at low levels.[73] In Japan, birds decimate small larvae during outbreaks but have little impact on sparse populations;[31,32] and bird predation on overwintering egg masses averaged 34% overall and up to 90% in sheltered areas.[64] In North America, bird predation of egg masses is uncommon, but some birds consume considerable numbers of larvae. For example, cuckoos, although seldom numerous, concentrate in outbreak areas and feed voraciously on gypsy moths and other hairy caterpillars.[30]

Small mammals are considered crucial in maintaining gypsy moth populations at innocuous levels in North America. Mesic well-stocked forests with a deep layer of litter have higher populations of mice and shrews and lower rates of survival of gypsy moth pupae than drier open stands.[7] In a sparse stable population, 80% of the total mortality was attributed to vertebrates.[16] When small mammals were trapped out, a six-fold increase in gypsy moth survival resulted, whereas excluding birds had no effect.[20] Interestingly, there is little mention of mammalian predators of gypsy moth in the Eurasian literature, except that gypsy moth remains have been reported in the stomachs of wood mice and the forest dormouse.[61] Small mammals in the Ukraine occur at densities of over 100/ha or 1.5 times higher than the highest estimate from the northeastern United States. Gypsy moth larvae stay on the tree in the Ukraine and this behavior could have evolved to avoid the numerous litter-dwelling mammals. The tendency for North American larvae to aggregate in the litter is believed to be an adaptation that evolved in Europe in response to birds and parasites.[18] It may be that Western Europe, the origin of North American gypsy moth, lacks the predation pressure on the forest floor that is present in North America and Eastern Europe.

The impact of invertebrate predators on the gypsy moth has not been documented as thoroughly as for vertebrates, but several species of insects and spiders are known to attack larvae.[30,71] The shiny green beetle, *Calosoma sycophanta* Linnaeus, was imported 1906–1926 and can be very abundant during outbreaks. It may account for as much as 25% of the pupal mortality;[11] the increase in beetle populations, however, is insufficient to prevent gypsy moth overpopulation and tree defoliation.

3.3. Disease

Whereas predation may be important in sparse populations, a nuclear polyhedrosis virus or wilt disease is the mortality agent prevalent in dense outbreak populations.[26]

Other pathogens that occur naturally in North America are a cytoplasmic polyhedrosis virus, two bacteria, and two fungi.[17] These infectious diseases often kill later stage larvae in dense populations, but their incidence is usually much less than the nuclear polyhedrosis virus.[17]

Microsporidia are absent in North America, but they are ubiquitous in Europe and natural infection rates of 5–25% are common.[10] Four microsporidian species are major mortality agents in Russia with infection rates of 12–40% in low-density gypsy moth populations and 30–80% in outbreak populations.[80] In Iran, where no serious gypsy moth outbreaks have been recorded, *Nosema* sp. is the most prevalent mortality agent.[41]

4. NUMERICAL PATTERNS

4.1. Temporal Variations

European gypsy moth populations have been described as cyclical with populations increasing from latent to outbreak densities every 8–11 years.[37,48] One area in Yugoslavia has had density peaks at approximately 9-year intervals for more than 80 years (Fig. 2).

Europeans use the terms latency, progradation, culmination, and retrogradation to describe the major phases of these cycles. These terms correspond to innocuous, increase, outbreak (including year of collapse) and postoutbreak phases.[15] In Europe, outbreaks usually persist for 3 years, with seldom more than two successive years of extensive defoliation; however, the Iberian Peninsula has harbored sustained high populations on cork oak, *Quercus suber*.[3,9]

In North America, the gypsy moth has been described as bimodal[21,22] with two stable population phases, innocuous and outbreak, both of which can last for several successive years. These conclusions were based on three studies (Fig. 3):

1. An example of a sustained outbreak is the Melrose population. The gypsy moth was first noted as a problem in this area of Massachusetts, New Hampshire, and

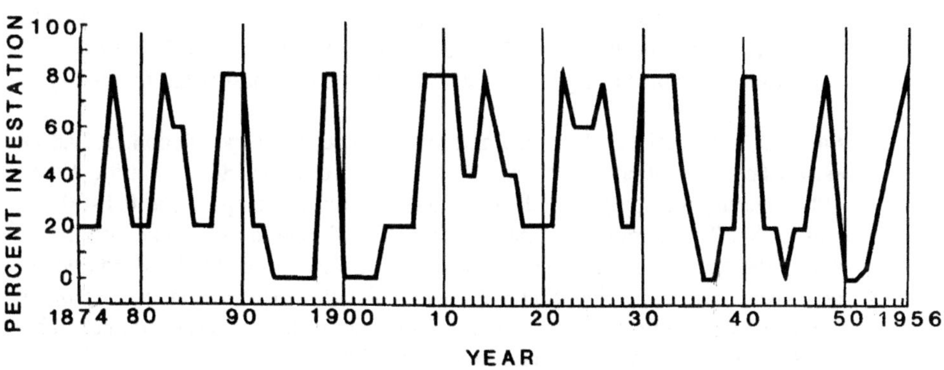

FIGURE 2. Percentage of trees infested with egg masses in Croatia, Yugoslavia. Values are relative with the highest infestation for each cycle set at 80%, hence do not represent severity but show when population peaks occurred. (From Kovacevic.[50])

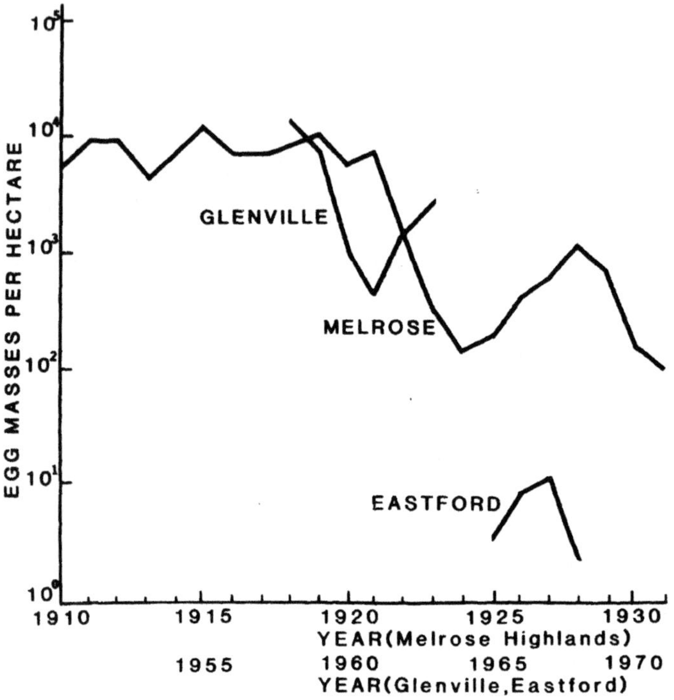

FIGURE 3. Three long-term studies of gypsy moth populations in New England. (Data from Campbell.[15])

Maine in 1889; its wrath was first felt about 1904–1910. From the first year of monitoring, 1910, egg mass densities were 7000/ha and did not subside until 1922. From then until 1930, populations remained at much lower levels, at 100–1000/ha.

2. Glenville, New York, is another example of a sustained, initial outbreak. The first outbreak began there in 1956 and was monitored from 1958 to 1963. The populations remained at high levels, at 800–10,000 egg masses/ha, for the 6-year study.
3. Eastford, Connecticut, is an example of a sparse stable population. It had an established population for at least 50 years before it was studied during 1965–1968. During the study period, the population remained sparse, at 0.1–20 egg masses/ha. This population finally increased and caused noticeable defoliation during the outbreaks in 1972 and 1980 (Fig. 4). Thus, while the Eastford populations were not stable indefinitely, they depict the extended innocuous phase prevalent in established populations. Populations that undergo similar stable sparse phases are not uncommon in Eurasia.[31,67]

Annual defoliation surveys of towns in Connecticut (Fig. 4) illustrate the periodicity of outbreaks during the past 20 years in an area in which gypsy moth has been endemic for

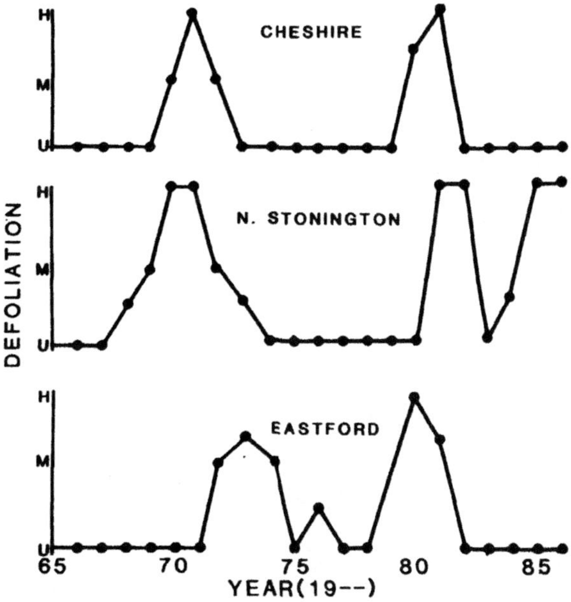

FIGURE 4. Defoliation in three Connecticut towns classified as heavy (H) when it was 75–100% of the maximum ever recorded, moderate (M) when 25–50% of the maximum, and unnoticeable (U) when no defoliation was recorded. (Data from Anderson and Gould[1] and from Anderson and Weseloh.[2])

at least 50 years. While each of the three areas exhibited different population behaviors, each had outbreak populations that coincided during the early 1970s and 1980s. These are also the years in which record-breaking levels of defoliation occurred over the entire Northeast. Gypsy moth populations in North America appear to progress through a colonization phase and then to a cyclical pattern similar to that in Europe.

4.2. Spatial Variations

Population behavior is very much influenced by site or habitat conditions. Susceptible forests are defoliated more often and for longer periods of time than are resistant forests (Fig. 5). Resistant stands typically occur on mesic sites that support vigorous tree growth where *Quercus rubra* is the predominant oak species. Susceptible forests have a high percentage of preferred food species, usually *Quercus prinus, Q. alba, Q. velutina,* or *Betula populifolia.*[45,76] Susceptible stands also have a higher proportion of trees with bark flaps and fissures that allow gypsy moth larvae to remain on the tree bole, hence escape litter-dwelling predators.[7] Susceptible stands usually occur on xeric soils, such as rocky ridgetops and coastal–sand plains, hence are more likely to be effected by drought.

Defoliation of susceptible forests on coastal-sand plains occurs at egg mass densities that are much lower than the densities that cause defoliation of mesic-sited resistant stands. Defoliation will exceed 60% of foliar biomass when egg masses/ha are above 1000 on xeric sites and 4000 on mesic sites; but egg mass densities range up to 14,000 on both sites.[29] A major difference between the two types of forest is the threshold density at which significant defoliation can occur.

FIGURE 5. Examples of a susceptible forest on a xeric ridgetop (a) and a resistant forest on a mesic lowland (b).

4.3. Within Generation Survival

Life tables suggest that parasitism is more important in European than in North American populations (Table II). Conversely, predation of pupae and large larvae seems relatively more important in North America than in Europe. On both continents, disappearance of first instars, presumably through dispersal, was the greatest source of loss. The other category for the fourth to sixth instars in North America also represents disappearance, probably due to bird predation.[14]

The three North American populations show the influence of population density and intrinsic site-related factors. The Eastford population was sparse and stable, the Glenville population (with 450 eggs/mass) was sparse and increasing, whereas the Glenville population (with 250 eggs/mass) was dense and stable.[14] Egg survival was constant in the three populations and did not seem to be affected by density. Survivorship of instars I–III appeared to be density dependent, but windborne dispersal of larvae into and out of plots complicates the interpretation; e.g., survivorship was greater than 1.0 in the sparse Glenville subpopulations near the outbreak region. In Glenville, survivorship of late instar larvae and pupae was higher at low densities. As the life table shows, however, survivorship of pupae at Eastford was much lower, even though pupal density was similar.

TABLE II
Numbers and Percentage Mortality of Gypsy Moth Life Stages in North American[15] and European[80] Populations

Stage	Mortality agent	Europe Ukraine	North America Eastford	North America Glenville	North America Glenville
Eggs[a]		1023	550	450	250
	Parasites	3	15	15	20
	Birds	29	—	—	—
	Microsporidia	6	—	—	—
	Other	—	15	15	15
Instars 1–3		661	385	315	162.5
	Dispersal, etc.[b]	82	37	50	70
Instars 4–6		120	242.5	157.5	48.7
	Parasites	38	2	5	5
	Predators	12	20	—	—
	Disease	26	5	5	60
	Other	—	69	75	25
Pupae		30	14.6	23.6	4.9
	Parasites	70	—	—	22
	Predators	10	87	20	18
	Disease	11	—	—	14
	Other	—	3	13	18
Adult females		1.3	0.4	10.4	0.4
Generation		99.74	99.93	97.69	99.84

[a]Eggs/mass for North America; eggs/tree for Europe.
[b]The term "etc." was not specified for the United States but includes 18% disease and predation for the U.S.S.R.; for U.S.S.R., 163 larvae began instar 2–3, of which 17%, 6%, and 3% died from microsporidia, predators, and disease, respectively.

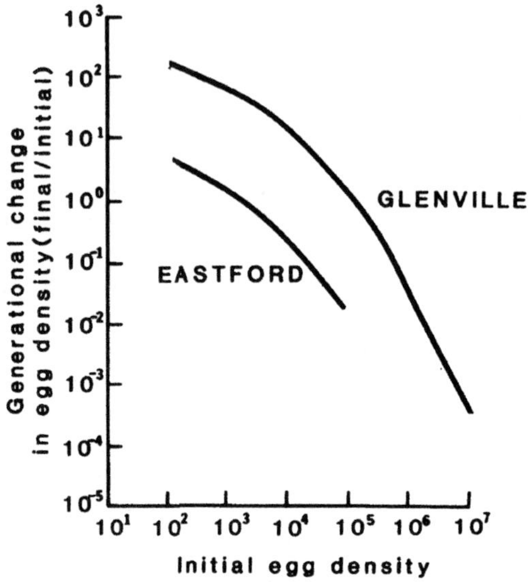

FIGURE 6. Change in egg density from one generation to the next as related to initial egg density for an area susceptible to outbreaks (Glenville) and an area resistant to outbreaks (Eastford). (From Campbell.[15])

This may be why this population remained at stable innocuous densities, while the sparse Glenville population increased to outbreak levels. Overall survivorship was density dependent but this relationship was distinctly different for the two areas (Fig. 6). Thus, while both populations exhibited density-dependent trends, the equilibrium density about which Eastford populations fluctuated was low and populations remained innocuous, whereas fluctuations around a higher equilibrium density in Glenville resulted in defoliating outbreak populations.

5. FACTORS AFFECTING NUMERICAL DYNAMICS

The gypsy moth life system is complex and is influenced by many interacting factors, but the numerical behavior of populations seems to be especially affected by the following forces.

5.1. Tachinid Parasitoids

Tachinid flies are often abundant during outbreaks both in Europe and in North America. The relationship between parasitism and host density has been measured for several years in Yugoslavia.[68,69] The most abundant species at high host densities are *Blepharipa pratensis* and *Parasetigena silvestris;* both are univoltine species specific to the gypsy moth and diapause as puparia in the soil. There are also seven polyphagous species present. These have two or more generations per year and diapause as young larvae in an alternate host.

Both of the univoltine species respond to changes in host density but lag behind the host population (Fig. 7). *B. pratensis,* which can deposit up to 5000 eggs on the foliage, responds more rapidly than *P. silvestris,* which can lay up to 300 eggs directly on the host. Heavy defoliation discourages egg laying by *B. pratensis,* however, and the lower humidity of defoliated stands increases the adult mortality of both species.[68] By contrast, the polyphagous species do not respond to change in host density but are usually abundant when host densities are low. Their abundance is determined more by availability of alternate hosts, which often increase as gypsy moth population densities decrease.[68] Except for *C. concinnata,* these polyphagous species are not present in North America.

5.2. Vertebrate Predators

Vertebrate predators, although poorly understood, undoubtedly play an important role in maintaining the gypsy moth at sparse population levels.[18] Predation by vertebrates is not only related to predator density but is also affected by the presence of alternative foods and by habitat. For instance, most birds prefer to eat smooth-skinned caterpillars rather than hairy gypsy moth larvae.[70] Mice must learn to bite off the head capsule of the larva and then turn the skin back in order to avoid the hairs;[13] thus, it is not surprising that they disregard gypsy moth larvae if other foods are available.

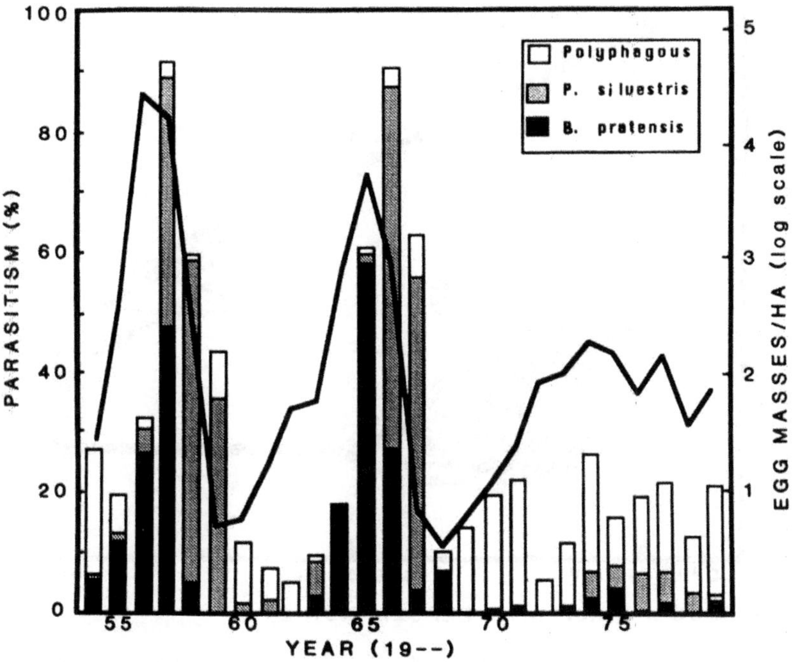

FIGURE 7. Cycles of abundance of gypsy moth egg masses/ha in Yugoslavia (solid line) and corresponding percentage parasitism by polyphagous tachinids and two host-specific tachinids (bars). (From Sisojevic.[69])

On xeric susceptible sites, the overstory is open, leaf litter is sparse, and many trees have deep bark fissures and bark flaps.[7,45] This encourages gypsy moth larvae to rest and pupate on the tree bole, where they are less likely to be found by ground-foraging predators.[16] Moreover, the diversity of mammal and bird species, as well as their total numbers, is lower on the xeric, susceptible sites than on mesic, resistant ones (Fig. 8).

It may seem incongruous that gypsy moth larvae should migrate to the forest floor, where they are more vulnerable to predation. It has been suggested that this behavior evolved in Europe as an escape from birds and parasites.[18] Both the life table and the literature indicate that bird and parasite pressures are greater in Europe than in North America.

The numerical responses of mammalian predators to gypsy moth density seem to be greatest at intermediate densities.[70] This is similar to the initial rise and subsequent decline of small mammal predation on sawfly cocoons as prey density increased.[43] This type of response would tend to segregate populations into two phases, an innocuous (endemic) phase and an outbreak (epidemic) phase. A large increase in prey numbers or an abundance of alternative, preferred food would result in the gypsy moth escaping control by vertebrates and a shift from the endemic to the epidemic phase.

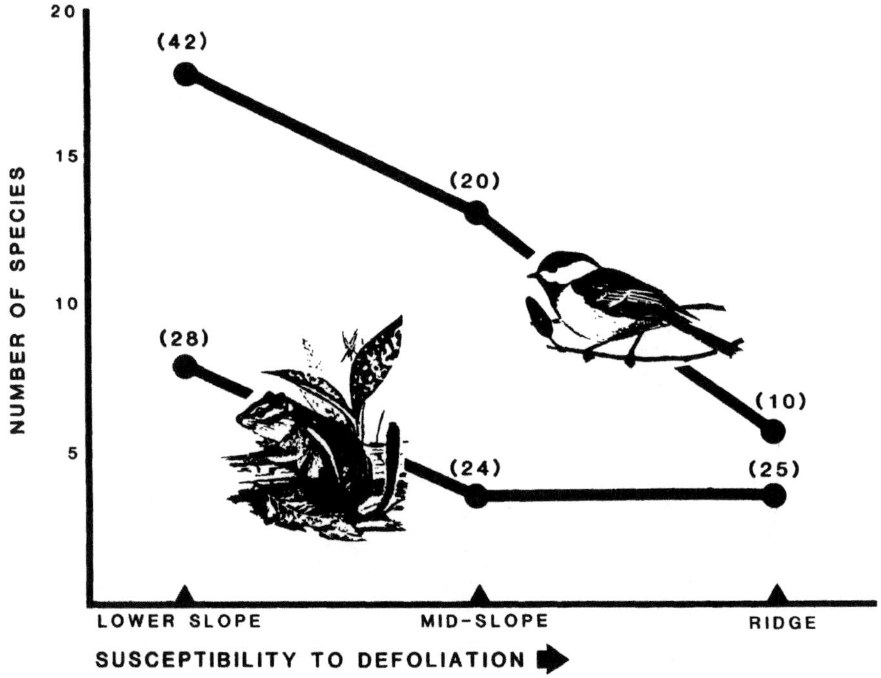

FIGURE 8. Avian and mammalian diversity and abundance (in parentheses) along a gradient of defoliation susceptibility on Bryant Mountain, Vermont. (Modified from Smith.[70])

5.3. Weather

Weather may act directly on the insect itself or indirectly through the effects on natural enemies or host plants. Several reports from Europe suggest that weather variations may trigger outbreaks. For example, three successive years of cold winters followed by dry springs preceded outbreaks in the Soviet Union.[5] Similarly, in Bulgaria, temperatures 3–9% above normal and rainfall 50% below normal in April, May, and June preceded the last two outbreaks.[48] Cold weather in May also inhibited the development of an outbreak in Czechoslovakia,[58] but the specific rate of change in gypsy moth populations in the European part of the Soviet Union was positively correlated with the minimal temperature in May.[82] In America, relationships between weather and population fluctuations were examined by regression analysis. In one model, precipitation and temperature in May of the previous year was positively related to egg mass density,[8] while in the other, a negative relationship was found with May precipitation in the previous year.[12]

5.4. Food

The ability of a forest to support gypsy moth populations can change in response to defoliation in two ways: (1) long-term changes in species composition and stand structure that follow a severe outbreak, and (2) short-term changes in foliage quality that occur during an outbreak.

Susceptible forests are characterized by a high proportion of oak and other favored species, such as poplar and birch.[42] In the initial outbreak, which may last for several years, oaks and other preferred hosts have higher mortality rates than do less favored species, such as red maple.[19] Vigorous dominant trees are also more likely to survive than are suppressed trees and this creates single-story stands. These changes reduce the susceptibility of the stand to defoliation in subsequent years.

Gypsy moth feeding can also cause changes in the chemistry of remaining foliage, and this may cause populations to decline.[66] Larvae grow more slowly and gain less weight on defoliated oak trees,[79] the leaves of which have lower levels of foliar nutrients[40,77] and more secondary defensive compounds.[66]

In addition to changes in the nutritional quality of foliage, the quantity of foliage available may also become limiting at high gypsy moth densities. These effects of defoliation on the numerical behavior of gypsy moth populations have been examined with the aid of a population model.[74] When the nutritional quality of the foliage is held constant, population densities peak at 7–8-year intervals and are followed by catastrophic declines due to food shortage. When defoliation induced changes in food quality are included in the model, population growth is retarded but not enough to shift the cycle more than one year. If the defoliation-induced effect is increased, the population becomes more stable but defoliation remains unexpectedly high. Although these models are but caricatures of reality, they do imply that density-related changes in food quality, by themselves, are insufficient to explain the dynamics of gypsy moth populations.

5.5. An Integrated Analysis

Having examined the various factors involved in gypsy moth population dynamics, we shall now attempt to show how they function together to regulate populations. We will do this by plotting natural logarithms of gypsy moth egg mass densities in the Spring of the current year (N_t) or the previous year (N_{t-1}) against the specific (per-capita) rate of change of the egg mass population over a generation ($N_{t+1} - N_t$).

When this is done with the current year density as the abscissa (Fig. 9a), the result is a phase trajectory that goes through two cycles of decreasing amplitude (* and ●) and then settles into a more or less stable pattern of low-amplitude oscillation (▲). An orbital trajectory on the phase plane indicates that second-order density-dependent processes are operating[6,62]; i.e., time delays, probably due to interactions between the subject popula-

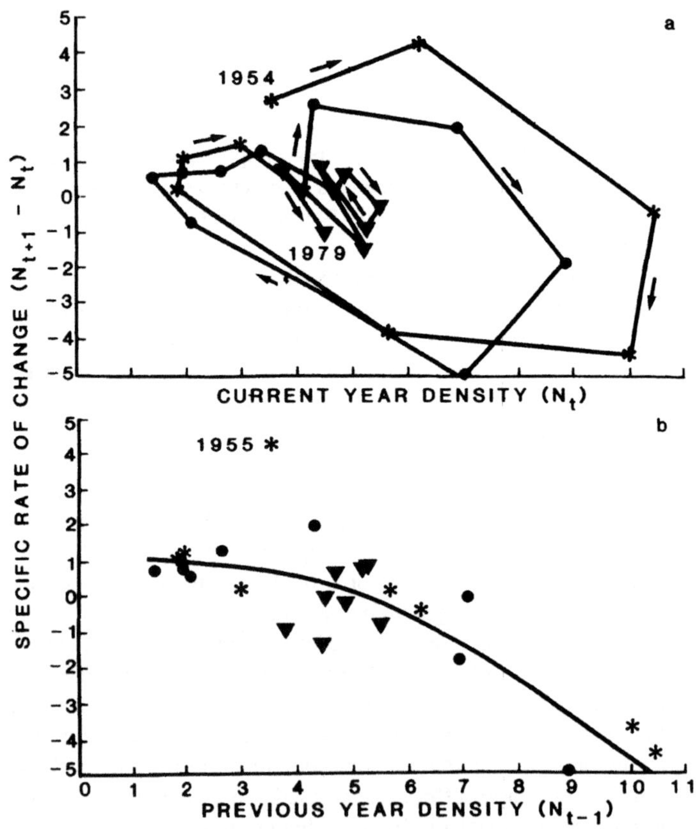

FIGURE 9. Rate of change in population density from one generation to the next plotted on the natural logarithm of egg masses/ha in (a) the current generation, and (b) the previous generation, through which a density-dependent function is visually fitted. (See Fig. 7.) (See Berryman[6] for theoretical details; data from Sisojevic.[69])

tion and its food supply or its natural enemies, are present in the density-dependent feedback response. In the case of the gypsy moth, the most plausible hypothesis is that specific tachinid parasitoids, with generation spans approximately the same as that of their host, are responsible for the cyclical dynamics. This hypothesis is based on the fact that specific parasitoids show an obvious delayed numerical response, becoming very numerous during the collapse of outbreaks (see, e.g., Fig. 7). During the low-density phase, however, they either die out or move away.

If this hypothesis is true, the cyclical orbit on the phase plane should be suppressed when the specific rate of increase is plotted on egg mass density in the previous generation (N_{t-1}); i.e., specific parasitoids should respond with a one-generation time lag because their numbers depend on prey density in the previous generation. That the cyclical trajectory is almost completely eliminated by this manipulation (Fig. 9b) lends some support to this hypothesis.

Figure 9a also indicates that other more rapidly responding feedback processes could become important at intermediate population densities (data marked ▲). These oscillations do not have such obvious cyclical orbits which suggests that density-dependent feedback occurs much more rapidly. Polyphagous parasitoids and predators could be responsible for such responses if they rapidly switch their attacks from one prey species to another. This would tend to hold prey populations in a fairly stable state.

Weather disturbances could play an important role in destabilizing this steady state, however. For instance, the specific rate of change of the gypsy moth population was unusually high in 1955 (Fig. 9b), which could have been due to extremely favorable weather. This high reproductive rate could have enabled the gypsy moth population to escape from its polyphagous natural enemies and set off the cyclical interaction with its specific parasitoids.

Population patterns in North America may not be the same as those observed in Europe. Parasitism by tachinids, for example, is much lower in North America, with maximum rates around 40%[59] as compared with 90% in Yugoslavia.[69] Lower parasitism during the decline of outbreaks in North America could result in prolonged infestations that rebound more quickly (e.g., see Fig. 4, North Stonington, 1980–1985)). By contrast, predation by small mammals seems to have more influence on gypsy moth dynamics in North America. Where conditions are favorable for predators they can, apparently, maintain gypsy moth populations at densities below 25 egg masses/ha.[13,20] This is well below the levels at which parasitoids stabilized gypsy moth populations in Yugoslavia (i.e., 50–200/ha) (Fig. 9). The outcome of these interactions between parasites and predators in North America could be more rapid shifts from endemic to epidemic behavior, with longer periods in each phase than is observed in Europe. Such bimodal numerical behavior has been proposed for North American populations[21] but has yet to be verified by long-term studies.

6. MANAGEMENT IMPLICATIONS

Historically, gypsy moth population management has been limited to the importation of natural enemies and to the use of chemical and biological insecticides to suppress outbreak populations. The population dynamics of the insect provides some insight about current practices and suggests some approaches that could be used to maintain populations at socially and economically acceptable levels.

6.1. Predicting and Controlling Outbreaks

In the United States, detection of gypsy moth outbreaks is often based on aerial defoliation surveys with follow-up ground egg-mass counts to estimate population densities. Since there is little risk of tree mortality from a single defoliation,[23] this method protects the timber crop while keeping survey costs low. The only management option, however, is to treat with insecticide because gypsy moth populations will already be high when defoliation is observed from the air. Because insecticides are applied after the larvae disperse, the treatment area may have to be enlarged to take care of dispersing larvae. If this fringe area is not treated, populations may either increase to defoliating levels or be held at innocuous levels by natural processes. The outcome depends on the numerical variability of subpopulations in the entire region; if the variability is low, the infestation is likely to decline.[22]

Prevention of gypsy moth outbreaks is preferable to "putting out the fire after it's raging," but this requires considerable investments in monitoring populations. Annual egg mass surveys would be necessary to detect population increases in time to implement preventive measures. These costs are prohibitive over a large area; about 12 hr of labor being required to obtain an accurate estimate from a population of 600 egg masses/ha on 50 ha.[49] Costs could possibly be reduced by limiting annual surveys to susceptible areas and expanding them to more resistant forests when outbreaks seem imminent. Another way to reduce costs would be to monitor population density with pheromone-baited traps. While pheromone traps have proved useful in detecting new infestations, further development is necessary before they can be used to accurately estimate gypsy moth densities.[28]

Cost is not the only thing that hinders development of preventive treatments. Much more critical is the lack of information about the behavior of gypsy moth populations as they increase to outbreak densities. Studies in North America have been conducted on populations that were at outbreak levels and remained there (Melrose, Glenville) (Fig. 3), were collapsing from outbreak levels,[22] or were at extremely low levels over the entire study period (Eastford) (Fig. 3). Without more information about the transition phase, from endemic to epidemic, such techniques as augmenting sparse gypsy moth populations in order to increase parasitism (see Section 6.2), do not seem to be practical.

6.2. Biological Control and Outbreak Prevention

Elaborate, prolonged efforts to establish gypsy moth parasitoids in North America have only been partially successful.[25] Of the 80 natural enemies that have been introduced, only 10 have become established and, although these are often abundant at high gypsy moth population densities, they do not seem important in preventing outbreaks.[14,59] Parasitoids that are more significant at low host densities are multivoltine[68] and require alternate hosts not found in North America. Although considerable effort is still made to import parasites, this is unlikely to result in the elimination of the gypsy moth problem, since it is also a pest in its native habitat, where both insect and parasites have long coexisted.

Microsporidian protozoa are natural enemies that are missing in North America but are present at both high and low densities in Eurasia.[10,67] These parasites do not usually cause direct mortality but result in decreased fecundity, egg sterility, and increased susceptibility to other natural agents.[67,80] Establishment of microsporidia in North America would likely be beneficial, but the many species differ greatly in virulence and persistence, and they should be evaluated thoroughly before release.

A unique control strategy that has been tested in Europe is the addition of gypsy moth eggs to sparse populations in order to sustain specific parasitoids.[54] Eggs collected from outbreak areas were stocked in low-density plots at a level of about 100 egg masses/ha. One half the eggs are released 1 month later to provide food for a second generation of *Cotesia melanoscelus*. Although egg augmentation increased the number of gypsy moth caterpillars, egg mass densities in the fall were lower than in check plots because fewer larvae survived to adults.

An innovative variation of this technique is to supplement feral populations with eggs having lethal genes. Male moths from irradiated pupae are mated with normal females in the laboratory to obtain eggs that produce sterile adults. The addition of these eggs with defective sex cells to isolated infestations has successfully eradicated such populations in North America. However, the outcome is less predictable in more dense populations. For example, natural enemies were apparently overwhelmed in one instance and feral populations increased, whereas in another case the addition of eggs with lethal genes increased rates of parasitism dramatically and feral densities did not increase. Augmenting gypsy moth populations to enhance natural enemies shows promise, but the host densities at which these treatments are effective will need to be established first.

Vertebrate predators seem to play a unique role in gypsy moth population regulation, which is related to their learning abilities.[44] As a result, very sparse populations may be less vulnerable to predation by birds and mammals than are populations of intermediate density.[32,70] Control by predators can fail, however, when (1) populations exceed the capacity of predators to regulate them through functional or numerical responses, or (2) the presence of alternative preferred food shifts the interest of predators away from the gypsy moth. For example, berries are a favorite food of white-footed mice,[38] and an abundant crop would likely result in less predation on the gypsy moth. Numerical responses of vertebrate predators to increases in gypsy moth densities are probably weak since larvae and pupae are present only for a few weeks and not during winter when food is likely to be limiting. Providing nest boxes and winter feeding can increase mammal populations but, because these predators are opportunistic, this does not necessarily result in increased gypsy moth mortality.[71]

6.3. Silvicultural Management

The concept of managing gypsy moth populations through silvicultural treatments was proposed in 1917,[24] and considerable activity occurred during the 1930s.[4] With the advent of synthetic pesticides, this approach waned but has emerged again with the current concern about and restrictions on the use of these chemicals.[36]

Early treatments stressed the removal of preferred susceptible species from the stand,[4,24] but this may not be practical, since oaks are often the most valuable timber species. Trees with structural defects (e.g., bole wounds, bark flaps, crooks) have little value and are removed in most stand-improvement thinnings. This should enhance predation since these trees provide refugia from litter-dwelling predators. However, it will also reduce nesting sites for cavity-dwelling birds.[7,16] Piling the brush from cuttings near, but not at the base of, preferred species should not only attract gypsy moths to the litter but should also improve the habitat for predators.

Current silvicultural prescriptions focus on the timing of thinnings or cuttings in relation to the imminence of a gypsy moth outbreak.[34] Unless the forest is mature and ready for harvest, thinnings should not be made if defoliation is expected within 5 years. Stresses created by removing trees could combine with stress from defoliation and result in high tree mortality. Thinning may also reduce predation by *Peromyscus leucopus* by stimulating berry production and creating a more open habitat.[7,38] However, after 6–10 years, the thinned stand will be more vigorous and canopy reclosure will reduce berry production. Thus, by timing thinnings and cuttings to the population phase of the gypsy moth, both tree vigor and predator effectiveness can be enhanced.

ACKNOWLEDGMENTS. We are grateful to Ivan Spaic of Zagreb University and to Harvey Smith of the Northeastern Forest Experiment Station for sharing with us observations on the gypsy moth in Yugoslavia and Russia, respectively. Especially appreciated is Pelagija Sisojevic of the Institute for Biological Research, Belgrade, for providing us with her extensive data on tachinids and the gypsy moth. Thomas ODell, of the Northeastern Forest Experiment Station, provided information on recent field work on inherited sterility. We also thank Robert Campbell and David Leonard for reviews of earlier drafts of this manuscript, David Mikus for preparing the figures, and Katherine McManus for typing the manuscript. Permission of the Wildlife Society to reprint copyrighted material is gratefully acknowledged.

REFERENCES

1. Anderson, J. F., and Gould, S. W., 1974, Defoliation in Connecticut 1969–1974, Conn. Agric. Exp. Sta. Bull. 749.
2. Anderson, J. F., and Weseloh, R. M., 1981, The gypsy moth in Connecticut, Conn. Agric. Exp. Sta. Bull. 797.
3. Baeta Neves, C. M., 1947, The principal insects that attack the cork oak in Portugal, *Montes* **3:**140–147 (in Spanish).
4. Behre, C. E., Cline, A. C., and Baker, W. L., 1936, Silvicultural control of the gypsy moth, Mass. For. Park Assoc. Bull. 157.
5. Benkevich, V. I., 1964, Information on the forecast of widespread appearance of *Ocneria dispar* (Lepidoptera: Liparidae). VII and VIII. The widespread appearance of *Ocneria dispar* and their forecast in the forests, *Tr. Orekhovo-Zuevskogo Pedagog. Inst.* **3:**67–95. (in Russian).
6. Berryman, A. A., 1986, On the dynamics of blackheaded budworm populations, *Can. Entomol.* **118:**775–779.
7. Bess, H. A., Spurr, S. H., and Littlefield, E. W., 1947, Forest site conditions and the gypsy moth, Harvard For. Bull. 22.
8. Biging, G. S., Giese, R. L., and Nordheim, E. V., 1980, Gypsy moth population simulation for Wisconsin, *For. Sci.* **26:**710–724.

9. Burgess, A. F., and Crossman, S. S., 1929, Imported insect enemies of the gypsy moth and browntail moth, U.S. Dept. Agric. Tech. Bull. 86.
10. Cabral, M. T., 1977, The role of diseases in the natural control of populations of *Lymantria dispar* L. (Lepidoptera: Lymantriidae), *An. Inst. Superior Agron.* **37:**153–177. (in Portuguese).
11. Campbell, R. W., 1967, The analysis of numerical change in gypsy moth populations, *For. Sci. Monog.* 15.
12. Campbell, R. W., 1973, Forecasting gypsy moth egg-mass density, U.S. Dept. Agric. For. Ser. Res. Pap. NE-268.
13. Campbell, R. W., 1974, The gypsy moth and its natural enemies, U.S. Dept. Agric. Info. Bull. 381.
14. Campbell, R. W., 1976, Comparative analysis of numerically stable and violently fluctuating gypsy moth populations, *Environ. Entomol.* **5:**1218–1224.
15. Campbell, R. W., 1981, Population dynamics, in: *The Gypsy Moth: Research toward Integrated Pest Management,* (C. C. Doane and M. L. McManus, eds.), pp. 65–86, U.S. Dept. Agric. Tech. Bull. 1584.
16. Campbell, R. W., Hubbard, D. L., and Sloan, R. J., 1975, Location of gypsy moth pupae and subsequent pupal survival in sparse, stable populations, *Environ. Entomol.* **4:**597–600.
17. Campbell, R. W., and Podgwaite, J. D., 1971, The disease complex of the gypsy moth. I. Major components, *J. Invert. Pathol.* **18:**101–107.
18. Campbell, R. W., and Sloan, R. J., 1976, Influence of behavioral evolution on gypsy moth pupal survival in sparse populations, *Environ. Entomol.* **5:**1211–1217.
19. Campbell, R. W., and Sloan, R. J., 1977, Forest stand responses to defoliation by the gypsy moth, *For. Sci. Monog.* 19.
20. Campbell, R. W., and Sloan, R. J., 1977, Natural regulation of innocuous gypsy moth populations, *Environ. Entomol.* **6:**315–322.
21. Campbell, R. W., and Sloan, R. J., 1978, Numerical bimodality among North American gypsy moth populations, *Environ. Entomol.* **7:**641–646.
22. Campbell, R. W., and Sloan, R. J., 1978, Natural maintenance and decline of gypsy moth outbreaks, *Environ. Entomol.* **7:**389–395.
23. Campbell, R. W., and Valentine, H. T., 1972, Tree condition and mortality following defoliation by the gypsy moth, U.S. Dept. Agric. For. Serv. Res. Pap. NE-236.
24. Clement, G. E., 1917, Control of the gypsy moth by forest management, *U.S. Dept. Agric. Bur. Entomol. Bull.* **484:**1–54.
25. Debach, P., 1964, Success, trends and future possibilities, *Biological Control of Insect Pests and Weeds* (P. Debach, ed.), pp. 673–713, Reinhold, New York.
26. Doane, C. C., 1970, Primary pathogens and their role in the development of an epizootic in the gypsy moth. *J. Invertebr. Pathol.* **15:**21–33.
27. Dobrivojevic, K., 1963, Some observations on the mass occurrence of gypsy moth in the district of Valjevo from 1947–1961, *Zast. Bilja* **14:**677–692 (in Serbo-Croatian, English summary).
28. Elkinton, J. S., and Cardé, R. T., 1984, Effect of wild and laboratory-reared female gypsy moths, *Lymantria dispar* L. (Lepidoptera: Lymantriidae), on the capture of males in pheromone-baited traps, *Environ. Entomol.* **13:**1377–1385.
29. Etter, D. O., 1981, Pest management systems development, in: *The Gypsy Moth: Research toward Integrated Pest Management* (C. C. Doane and M. L. McManus, eds.), pp. 697–727, U.S. Dept. Agric. Tech. Bull. 1584.
30. Forbush, E. H., and Fernald, C. H., 1896, *The Gypsy Moth.* Wright and Potter, Boston.
31. Furuta, K., 1976, Studies on the dynamics of the low density populations of gypsy moth and Todo-fir aphid. Analysis of the environmental resistance factors by artificial host increase method, Bull. Govt. For. Exp. Sta. Tokyo **279:**1–85.
32. Furuta, K., 1977, The mortality and distribution pattern of the low density population of gypsy moth, *J. Jpn. For. Soc.* **59:**428–430.
33. Gansner, D. A., and Herrick, O. W., 1985, Host preferences of gypsy moth on a new frontier of infestation, U.S. Dept. Agric. For. Serv. Res. Note NE-330.
34. Gasner, D. A., Herrick, O. W., Mason, G. N., and Gottschalk, K. W., 1987, Coping with the gypsy moth on new frontiers of infestation, *South. J. Appl. For.* **11:**201–209.
35. Giese, R. L., and Schneider, M. L., 1979, Cartographic comparisons of Eurasian gypsy moth distribution (*Lymantria dispar* L.; Lepidoptera: Lymantriidae), *Entomol. News* **90:**1–16.

36. Goldschmidt, R., 1934, Lymantria, *Bibl. Genet.* **11**:1–186.
37. Hadzistevic, D., and Hadzihalilovic, H., 1959, Gypsy moth gradations on the territory of Pr Bosnia and Herzegovnia, *Zast. Bilja* **52/53**:155–159 (in Serbo-Croatian, English summary).
38. Hamilton, W. J., 1941, The food of small forest mammals in eastern United States, *J. Mammal* **22**:250–263.
39. Harrison, R. G., Wintermeyer, S. F., and ODell, T. M., 1983, Patterns of genetic variation within and among gypsy moth, *Lymantria dispar* (Lepidoptera: Lymantriidae) populations, *Ann. Entomol. Soc. Am.* **76**:652–656.
40. Heichel, G. H., and Turner, N. C., 1976, Phenology and leaf growth, in: *Perspectives in Forest Entomology* (J. F. Anderson and H. K. Kaya, eds.), pp. 31–40. Academic, New York.
41. Herard, F., Mercadier, G., and Abai, M., 1979, The status of *Lymantria dispar* (Lepidoptera: Lymantriidae) and its parasite complex in Iran, in 1979, *Entomophaga* **24**:371–384 (in French, English summary).
42. Herrick, O. W., and Gansner, D. A., 1986, Rating forest stands for gypsy moth defoliation, U.S. Dept. Agric. For. Serv. Res. Pap. NE-583.
43. Holling, C. S., 1959, The components of predation as revealed by a study of small mammal predation of European pine sawfly, *Can. Entomol.* **91**:293–320.
44. Holling, C. S., 1965, The functional response of predators to prey density and its role in mimicry and population regulation, *Mem. Entomol. Soc. Can.* No. 45.
45. Houston, D. R., and Valentine, H. T., 1977, Comparing and predicting forest stand susceptibility to gypsy moth, *Can. J. For. Res.* **7**:447–461.
46. Hoy, M. A., 1976, Establishment of gypsy moth parasitoids in North America: An evaluation of possible reasons for establishment or nonestablishment, in: *Perspectives in Forest Entomology* (J. F. Anderson and H. K. Kaya, eds.), pp. 215–232, Academic, New York.
47. Johnson, P. C., Mason, D. P., Radke, S. L., and Tracewski, K. T., 1983, Gypsy moth, *Lymantria dispar* (L.) (Lepidoptera: Lymantriidae), egg eclosion: Degree-day accumulation, *Environ. Entomol.* **12**:929–932.
48. Keremidchiev, M. T., 1972, Dynamics of outbreaks of the gypsy moth (*Lymantria dispar* L.) in the People's Republic of Bulgaria, *Proc. 13th Int. Congr. Entomol. Nauk* **3**:51–54. (in Russian).
49. Kolodny-Hirsch, D., 1986, Evaluation of methods for sampling gypsy moth (Lepidoptera: Lymantriidae) egg mass populations and development of sequential sampling plans, *Environ. Entomol.* **15**:122–127.
50. Kovacevic, Z., 1956, *Applied Entomology*, Poljaprevredni Nakladni Zavod, (Zagreb) (in Serbo-Croatian).
51. Kurir, A., 1953, The food plants of the gypsy moth (a contribution to the knowledge of its feeding habits), *Z. Angew. Entomol.* **34**:543–586 (in German).
52. Lechowicz, M. J., and Jobin, L., 1983, Estimating the susceptibility of tree species to attack by the gypsy moth, *Lymantria dispar, Ecol. Entomol.* **8**:171–183.
53. Leonard, D. E., 1966, Differences in development of strains of the gypsy moth (*Porthetria dispar* L.), Bull. Conn. Agric. Exp. Sta. No. 680.
54. Maksimovic, M., and Sivcev, I., 1984, Further studies on the numerical increase of natural enemies of the gypsy moth (*Lymantria dispar* L.) in forests, *Z. Angew. Entomol.* **98**:332–343.
55. Masaki, S., 1956, The effect of temperature on the termination of diapause in the eggs of *Lymantria dispar* Linne, *Jpn. J. Appl. Zool.* **21**:148–157.
56. Mason, C. J., and McManus, M. L., 1981, Larval dispersal of the gypsy moth, in: *The Gypsy Moth: Research toward Integrated Pest Management* (C. C. Doane and M. L. McManus, eds.), pp. 161–202, U.S. Dept. Agric. Tech. Bull. 1584.
57. Mosher, F. H., 1915, Food plants of the gypsy moth in America, U.S. Dept. Agric. Bull. 250.
58. Patocka, J., and Capek, M., 1971, Population changes of certain oak defoliators (Lepidoptera) in Slovakia, *Acta Inst. For. Zvolen.***2**:461–485.
59. Reardon, R. C., 1976, Parasite incidence and ecological relationships in field populations of gypsy moth larvae and pupae, *Environ. Entomol.* **5**:981–987.
60. Reardon, R. C., 1981, Parasites, in: *The Gypsy Moth: Research toward Integrated Pest Management* (C. C. Doane and M. L. McManus, eds.), pp. 299–320, U.S. Dept. Agric. Tech. Bull. 1584.
61. Rothschild, E. V., 1958, Extermination by rodents of the gypsy moth in the area of mass reproduction, *Byull. Mosk. O-Va. Ispyt. Prir. Otd. Biol.* **63**:129–130 (Trans. from Russian).
62. Royama, T., 1977, Population persistence and density dependence, *Ecol. Monog.* **47**:1–35.

63. Schaefer, P. W., 1978, *Betula platyphylla:* the preferred oviposition host of *Lymantria dispar japonica* in Hokkaido, Japan, *Environ. Entomol.* **7:**168–170.
64. Schaefer, P. W., 1980, Natural enemies of gypsy moth (*Lymantria dispar*) in Japan and Korea, especially new and potentially useful species, XVI Int. Congr. Entomol., Kyoto. Abstr. 9P2, p. 297.
65. Schaefer, P. W., Weseloh, R. M., Sun, X. L., Wallner, W. E., and Yan, J. J., 1984, Gypsy moth, *Lymantria* (=*Ocneria*) *dispar* (L.) (Lepidoptera: Lymantriidae), in the People's Republic of China, *Environ. Entomol.* **13:**1535–1541.
66. Schultz, J. C., and Baldwin, I. T., 1982, Oak leaf quality declines in response to defoliation by gypsy moth larvae, *Science* **217:**149–150.
67. Semevsky, F. N., 1973, Studies of the dynamics of gypsy moth, *Porthetria dispar* L. (Lepidoptera: Lymantriidae) at low population density levels, *Entomol. Rev.* **52:**25–29.
68. Sisojevic, P., 1975, Population dynamics of tachinid parasites of the gypsy moth (*Lymantria dispar* L.) during a gradation period (in Serbo-Croatian), *Zast. Bilja* **26:**97–170.
69. Sisojevic, P., 1979, Interactions in the host-parasite system, with special reference to the gypsy moth-tachinids (*Lymantria dispar* L.-Tachinidae), in: *Papers of the Sixth Interbalcanic Plant Protection Conference, Izmir, Turkey,* pp. 108–111, Turkish Ministry of Food, Agric., and Anim. Husb., Research Section, No. 13.
70. Smith, H. R., 1985, Wildlife and the gypsy moth, *Wildl. Soc. Bull.* **13:**166–174.
71. Smith, H. R., and Lautenschalger, R. A., 1981, Gypsy moth predators, in: *The Gypsy Moth: Research Toward Integrated Pest Management* (C. C. Doane and M. L. McManus, eds.), pp. 96–125, U.S. Dept. Agric. Tech. Bull. 1584.
72. Taylor, R. A. J., and Reling, D., 1986, Density/height profile and long-range dispersal of first instar gypsy moth (Lepidoptera: Lymantriidae), *Environ. Entomol.* **15:**431–435.
73. Turcek, F. J., 1950, The bird population in some deciduous forests during a gypsy moth outbreak, *Bull. Inst. For. Res. Czech. Repub.* **3:**108–131 (in Czechoslovakian, English summary).
74. Valentine, H. T., 1983, The influence of herbivory on the net rate of increase of gypsy moth abundance: A modeling analysis, in: *Proceedings of Forest Defoliator–Host Interactions: A Comparison between Gypsy Moth and Spruce Budworms,* pp. 105–111, U.S. Dept. Agric. For. Serv. Gen. Tech. Rep. NE-85.
75. Valentine, H. T., and Talerico, R. L., 1980, Gypsy moth larval growth and consumption on red oak, *For. Sci.* **26:**599–605.
76. Valentine, H. T., and Houston, D. R., 1984, Identifying mixed-oak stand susceptibility to gypsy moth defoliation: An update, *For. Sci.* **30:**270–271.
77. Valentine, H. T., Wargo, P. M., and Wallner, W. E., 1983, Nutritional changes in host foliage during and after defoliation, and their relation to the weight of gypsy moth pupae, *Oecologia* **57:**298–302.
78. Waggoner, P. E., 1984, The hatching of gypsy moth eggs, a phenological model, *Agric. For. Meteorol.* **33:**53–65.
79. Wallner, W. E., and Walton, G. S., 1979, Host defoliation: A possible determinant of gypsy moth population quality, *Ann. Entomol. Soc. Am.* **72:**62–67.
80. Zelinskaya, L. M., 1980, The role of microsporidia in the population dynamics of the gypsy moth (*Porthetria dispar* L.) in forest plantations in the lower Dneiper region, *Vestn. Zool.* **1:**57–62 (in Russian).
81. Zivojinovic, S., 1957, Introduction. Gypsy-moth—Results of the activity on its study and control in our country in the course of 1957, *Zast. Bilja* **52/53:**3–6.
82. Znamiensky, V. C., 1984, Population dynamics of gypsy moth in oak stands of the forest steppe, *Akad. Nauk SSR (Seriia) Lesoviedienye* **4:**12–20 (in Russian).

CHAPTER 19

THE PINE SAWFLY IN CENTRAL FRANCE

Claude Geri

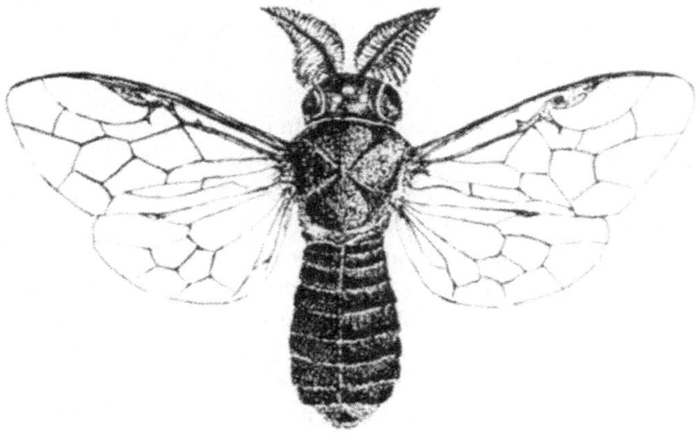

Claude Geri • Institut National de la Recherche Agronomique, INRA-CRF-Station de Zoologie Forestière, Ardon 45160, Olivet, France.

1. INTRODUCTION

The pine sawfly, *Diprion pini* Linné (Hymenoptera, Diprionidae), is widespread from the British Isles to the Soviet Union and from North Africa to Finland. The family Diprionidae includes about 85 species living exclusively in the coniferous forests of the Northern Hemisphere. The species are assumed to be very close to primitive forms derived from panorpoid ancestors in the Permian.[70,53] Many of these species belong to the genera *Diprion* and *Neodiprion* that differentiated in the subfamily Diprioninae, in the Palaearactic and Nearctic continents, respectively. In the Old World, the subfamily is represented by the genera *Diprion, Gilpinia, Microdiprion, Macrodiprion* (essentially in Europe), *Prionomeion* (North Africa), and *Nesodiprion* (Asia) and by a few species of American origin, such as *Neodiprion sertifer* Geoffroy.

The most serious outbreaks in North America are due to species introduced from Europe: *D. similis* Hartig, *G. frutetorum* Fabricius, *G. hercyniae* Hartig, and *N. sertifer,* reintroduced in 1925. In Europe, heaviest damage is caused by *D. pini,* an indigenous species. This sawfly chiefly attacks *Pinus silvestris* Linné, but in France *Pinus nigra* Arnold ssp. *laricio* and *Austriaca* can be damaged at the end of outbreaks.[13,37] In Yugoslavia, *P. nigra* is the preferred host.

The hosts of *D. pini* include, in order of preference[25] *Pinus silvestris, P. uncinata, P. nigra, P. cembra, P. griffithi,* and *P. strobus.* In France, *P. strobus* and *P. pinaster* are rarely attacked, and *P. uncinata* suffers some damage in the Alps. In nurseries, *P. contorta* and, to a lesser extent, *P. radiata* are infested, making this sawfly a threat to reforestation in Europe and a potential pest to North American and Australian pine forests.

Diprion pini outbreaks are sudden and spectacular throughout Europe. Thousands of hectares can be totally defoliated in less than 2 years. However, outbreaks subside just as rapidly and are followed by periods of low population levels, which may last several decades. Occasionally, the outbreaks may last 5 years of more, especially when they occur as successive attacks on several forest stands. Figure 1 summarizes some of the outbreaks reported in the literature and shows no clear relationship between the spatial distribution of the infested areas and the succession of infestations. Outbreaks appear to occur at random, although some periods appear more favorable for their development (e.g., the years around 1920, 1930, 1960).

Infestations generally occur in the lowland plains, while at higher elevations populations remain at fairly low levels without successions of rarity and abundance.[18] In the warmer Mediterranean regions, however, outbreaks occur in the mountains such as the Sierra de Albaracin in Spain and Maljen in Yugoslavia.

In spite of dramatic defoliations, the effects of *D. pini* infestations are frequently limited to loss of growth without tree mortality. As established during the past century, only one out of five outbreaks causes severe damage.[3] In the southern Paris Basin, for instance, the 1935 outbreak had only limited impact on the Fontainebleau forest[46]; the 1963–1964 infestation, which was widespread in the forests of Orleans and Fontainebleau, also had limited effects, while the latest outbreak starting in the 1980 resulted in serious damage only

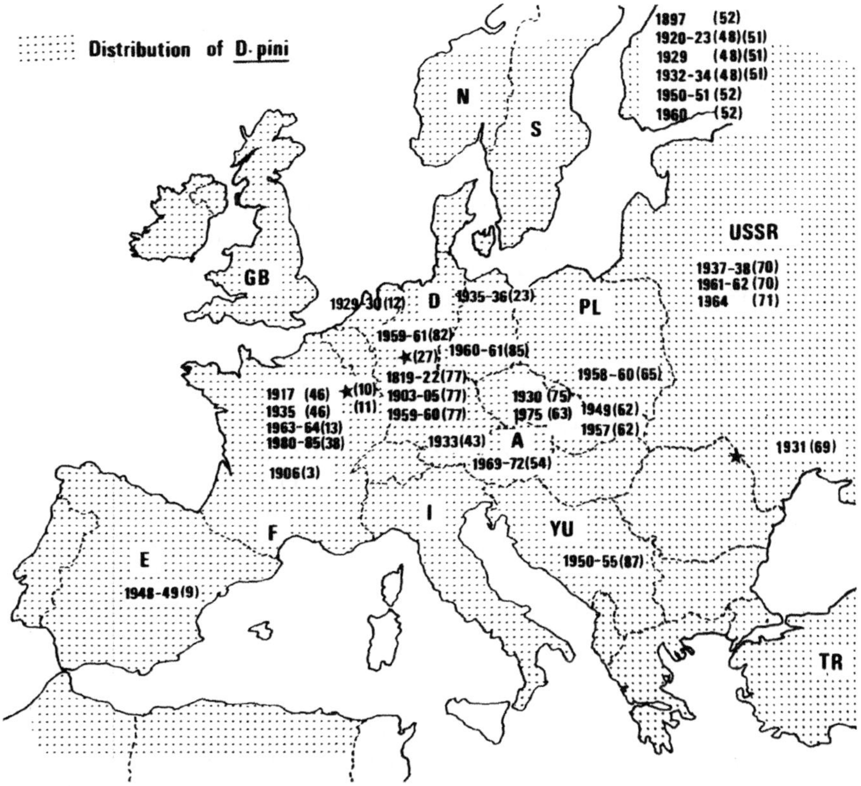

FIGURE 1. Distribution of *Diprion pini* and outbreaks cited in the literature. Numbers in brackets refer to references cited. *(10), (11) = outbreaks in 1839, 1888, 1905, 1856–1957. *(27) outbreaks in Germany in 1781–1989, 1819–1920, 1834, 1840–1943, 1903–1905, 1920–1922, 1927–1929, 1935–1937.

in the Lorris stands of the Orleans forest. These stands consisted of old pines growing on hydromorphic soils and the outbreak was prolonged by a second infestation in 1983–1984. Tree mortality was observed as early as 1982 and was aggravated by attacks of the bark beetles *Ips sexdentatus* BOERMER and *Tomicus piniperda* Linné (35.000 m^3 of salvage cuttings were carried out in 1983 and 70.000 m^3 in 1984).

Dendrochronological studies, with maritime pine as a control, showed that growth-increment losses ranged from 36.6. to 49.5% in Fontainebleau after the 1963 outbreak and reached 54.8% in Orleans on surviving trees after the 1982 peak infestation.[55] In the former forest, growth reduction was 20% throughout the life of the old pines, while in the latter, growth reduction averaged 15% during the past 30 years. Although defoliation by *D. pini* does not usually cause tree mortality, economic impacts are of no little importance and may be severe in forest stands of low vigor planted on poor soils.

2. BIOLOGY AND BEHAVIOR

2.1. Life Cycle

Diprion pini is usually bivoltine in the lowland plains and univoltine in mountain regions. On the Atlantic plain (France, Belgium, Netherlands, northern Germany and Poland), it has two generations per year,[13,17,32,65,81] while in Central and Eastern Europe (Germany, Austria, Czechoslovakia, Russia, Eastern France) two generations occur at lower altitudes and one in the mountain regions.[17,22,43,69,70] In the northern part of its distribution (Finland), it is also univoltine,[48] while in the mountain regions of Spain and Yugoslavia it is usually bivoltine.[9,88]

These variations in the life cycle enable the insect to adapt its development to various climates. At moderate elevations (200–500 m) in Central Europe, a part of the population may produce a second generation during warmer years.

In the Paris Basin, a typical bivoltine cycle occurs. The first generation develops from April to July (Fig. 2a). Eggs are laid as early as mid-April and hatch in about 1 month. Larvae, known as false carterpillars, feed on old pine needles; their growth generally ends at the beginning of July (five instars for the male, six for the female). Larvae then disperse and transform to eonymphs, which spin cocoons on vegetation or in the duff and then metamorphose successively into pronymphs, pupae, and adults.

Some adults emerge from their cocoons at the end of July and give birth to a second generation. Larvae of this second generation develop from late August to October (exceptionally November) and eonymphs overwinter within the cocoon, generally as enomymphs. Adults then emerge in April to initiate the first generation.

A proportion of the eonymphs of each generation may undergo a prolonged diapause, ranging from a few months to 3 years; the emergence of adults varies during this period (Fig. 2b). However, these adults always emerge in June and August. Both flights may be very important, but populations produced by the June flight are usually small because of poor egg deposition and hatch, high larval mortality, and heavy parasitism. These larvae complete their development between late August and early October and enter diapause; adults emerge in June or August, the following years. Populations produced by the August flight are often confused with the second generation, even though their development is slightly later. A high percentage of the adults of these populations emerge the following April.

In order to clearly distinguish the origins of the populations observed in nature, we use the following definitions:

1. *Spring cohort:* The set of all individuals from the first generation
2. *Summer cohort:* The set of all individuals from the June flight, resulting from eonymphs with prolonged diapause formed during the previous three years
3. *Autumn cohort:* The set of all individuals from the second generation plus those derived from the August flight of the long diapausing insects

This life cycle seems to be similar in other lowland countries.[17,18,27,30,65,71,88] For example, larvae brought from Eidlitz, Austria (200 m), to Delemont (500 m), produced adults the following year at periods similar to those observed in the Paris Basin, while larvae from other regions (Wallis, 850 m; Unterbrand, 1000 m) showed an altitudinal

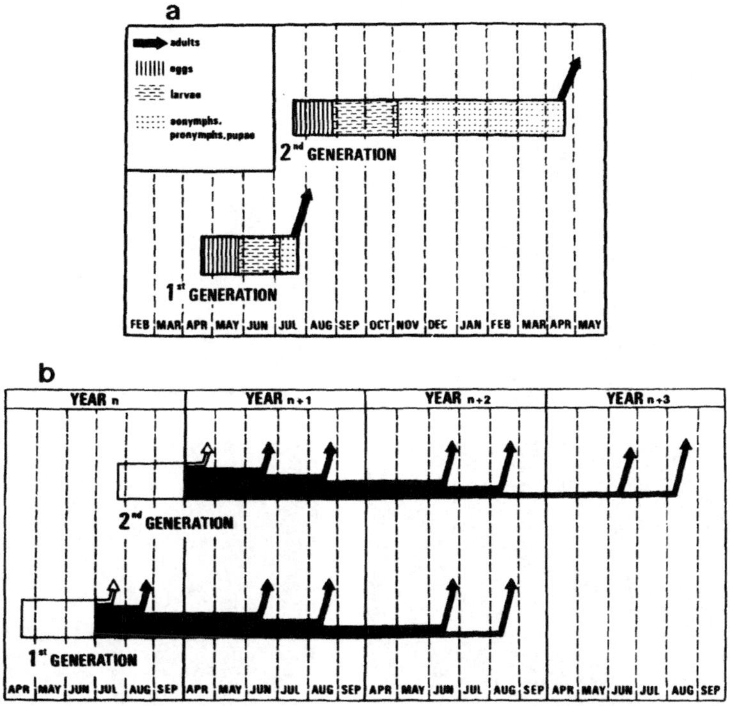

FIGURE 2. Bilvotine biological cycle of *D. pini* (a), and possibilities of delayed adult emergence under prolonged diapause (b).

progression from the bivoltine to the univoltine cycle, including increases in the rate of individuals entering diapause and in the length of diapause, and decreases in the number of annual flights.[17,18] At high elevations, the first and second generations disappear and only flights of long-diapausing insects remain, usually June flights. This cohort, which contributes little to populations existing at low elevations, allows populations to survive in mountain areas with an obligatory diapause. An exceptional case, of larval diapause, has also been discovered in upper Franconia.[77] Here populations from the first two flights stop developing and spin cocoons simultaneously with those originating from the third flight.

It is commonly assumed that *D. pini* outbreaks are related to the bivoltine cycle, permitting the insect to multiply rapidly and shortening the time during which cocoons are exposed to mortality agents. This theory is supported by the fact that outbreaks almost always start in lowland regions and that heaviest infestations occur in autumn.

2.2. Adults and Eggs

Sawfly adults are 7–10 mm in length, with the larger female having a thick yellow abdomen and serrate antennae and the smaller black male having pectinate antennae (Fig.

FIGURE 3. *Diprion pini* life stages: (a) adult male, (b) adult female, (c) eggs, (d) larvae, and (e) cocoons.

3a,b). They remain up to 7 days within the cocoon[25] and then cut a hole at one end. Emergence occurs at 1200–1400 hr, and the sexually mature adults do not feed and are usually considered poor flyers. In fine weather, however, flights over some hundreds of meters may be observed, and the insects may be carried longer distances by the wind.

Females lay eggs in pine needles, cutting the needle along its length with their ovipositors, depositing the eggs in successive rows, and covering them with a spumous coating (Fig. 3c). Eggs of the first generation are deposited on the foliage of the previous year, while those of the second generation are laid on new foliage. Females from the June flight oviposit at the border between old and new foliage, alternatively on top of one or the other. Fecundity varies from 50 to 160 eggs with a mean value of 100 to 120 per female.[9,13,37,74,81,88] During heavy infestations, several females may oviposit on the needles of the same twig.

Mating occurs only once and lasts 25–75 min. Females that do not mate produce only males. Females live 7–11 days, males about 10 days. Adults are resistant to cold temperature,[48] but in summer they are susceptible to strong exposure to sun.

The white translucent eggs increase in size from 1.4 × 0.45 mm to 1.60 × 0.90 mm during incubation.[9] As in *N. sertifer*, this swelling requires water absorbtion from the leaf parenchyma.[8] Thus, eggs are sensitive to parenchyma dessiccation, probably explaining the poor egg hatch of the June cohort in lowland regions. Eggs can withstand high moisture conditions,[43] but mortality can be caused by rising sap following heavy rainfall.[70] Egg mortality can also be extensive when temperatures decrease below 10°C or rise above 35°C.[24,43]

The embryo develops in 40 days at 10°C, 18 days at 18°C, 13–14 days at 20°C, 12 days at 25°C, and 7 days at 32°C,[13,24,43] and variable temperatures accelerate incubation.[57] On average, incubation lasts 25–35 days in the Paris Basin and is more extended in the spring cohort. Under adverse climatic conditions, incubation may take up to 45 days.

2.3. Development of Larvae and Free Eonymphs

False caterpillars have red head capsules, are greenish in the first instar, and then become yellowish-white with rows of gray-black spots that disappear in the fifth instar and darken by late autumn. The larvae grow from 5 mm in the first instar to 26 mm in the final stage.

Larvae live in colonies on pines (Fig. 3d). First-generation larvae feed on the needles of the previous year and die if their only food is new foliage.[13] The autumn cohort essentially feeds on current foliage, while the summer cohort can also partially consume these needles.

First-instar larvae feed on needles close to the site of oviposition, consuming all but the vascular bundles, so that foliage takes on a reddish color. Later, larvae move to other branches where they consume the entire needle. Some colonies remain in groups, while others separate. In overpopulated pine stands, feeding aggregations often break up as larvae migrate to other trees during the fourth instar.

False caterpillars develop normally at temperatures at 12–25°C and can withstand temperatures below 0°C and above 30°C for only short periods of time. They are quite susceptible to short periods of sun exposure and, in summer, often cluster on the shaded side of trees. Larval development accelerates with temperature and day length and is more extended for the female, which has a sixth instar. At 20°C and 16 hr daylight, the first stage lasts about 5 days, the second through fourth stages 4 days each, and the fifth and sixth stages 16 days together. In nature, these values vary according to climatic conditions and may be much higher in autumn. Individuals that undergo diapause appear to take longer to develop, even under the same photoperiod and temperature conditions.[13] At the end of the feeding period, larvae disperse on the ground and vegetation and transform after 2–5 days to free-moving eonymphs. After dispersing for another 1 or 2 days, the eonymphs, whose head caspules still remain white, spin cocoons on trees in litter, or on grass.

2.4. Development within the Cocoon

Cocoons are cylindrical in shape, 8–12 mm long for females and 6–9 mm for males (Fig. 3e). Cocoons are first white but become colored by the second day. Their color varies from white to dark brown, passing through yellowish, ochre, reddish, or gray gradations. The brown color, which appears to be due to 3,4-dihydroxyphenylalanine (DOPA),[25] intensifies with ambient humidity. Cocoons found on the ground are generally darker and more of them contain diapausing eonymphs than those spun on vegetation.[13,33]

Although the distribution of cocoons appears to be random in nature, some locations are preferred, e.g., rock cavities, border vegetation, and heaps of dry leaves. First-generation cocoons are often spun on trees or plants, while those from the autumn are often found on the ground. One or 2 days after spinning its cocoon, the eonymph coils up, and its false legs regress. It then grows into a true eonymph.[88] This stage may be prolonged considerably by diapause. Nondiapausing spring cohort eonymphs develop in 8–10 days. The development of autumn cohort eonymphs lasts at least 6 months over-winter (April flight). The eonymphs can survive within the cocoon for at least 1 month at −20°C.

After further shortening of its body, the eonymph transforms to a pronymph, which shows pupal eye prefiguration. Its development takes 3–18 days, depending on climatic conditions, but can also be delayed by winter. Pronymphs may also enter diapause.[80] Pronymphs from the June and August flights mature by late May or early June and late June to late July, respectively. Those from the April flight appear in November[77] or after winter.[25,88] The pupal stage usually lasts 6–12 days, depending on temperature.[9,88]

3. RELATIONSHIPS WITH NATURAL ENEMIES

3.1. Parasitoids

More than 120 species are known to be parasitic on *D. pini,* including chalcid egg parasites, tachinid and ichneumonid larvae parasites, ichneumonid parasites of free eonymphs, and ichneumonid and chalcid cocoon parasites. The main species are reviewed below.

3.1.1. Egg Parasites

The biology of chalcid parasitoids specific to diprionid sawflies, i.e., *Achrysocharella ruforum* Krausse, *Achrysocharella ovulorum* Ratzeburg, and *Dipriocampe di-*

prioni Ferriere, has been described.[4,22,60,67] The most important parasitoid *A. ruforum* may destroy 80–100% of the eggs of *D. pini*.[4,14,36,37,73,82,83,87]

The immature stages of these parasites are found within the host eggs, which turn black (*A. ruforum, A. ovulorum*) or brown (*D. diprioni*). Chalcids are bivoltine, and diapause depends on photoperiod. *A. ruforum* parasitizing the first generation produces adults by July, and these attack the second generation (Fig. 4). Individuals that parasitize the second generation diapause until May–June of the following year. Thus, parasitoids are closely synchronized with the summer and autumn *D. pini* cohorts, but not with early first generations or with *D. pini* populations emerging after long diapause.

3.1.2. Parasites of Larvae and Free Eonymphs

Two tachinids *Drino gilva* Hartig and *Drino inconspicua* Meigen, attack these stages[19,47,58,67,72] but *D. gilva*, which is specific to Diprionidae, is most commonly observed. The other species, which also attacks Lepidoptera, becomes abundant during outbreaks. Superparasitism and multiparisitism are frequent, but tachinid flies generally survive when competing with other parasites.

Eggs of *D. gilva* are sometimes deposited on second instars but more frequently on later instars. The full-grown maggots emerge from larvae in the first case, from cocoons in the second, and then pupate. If sawfly cocoons are spun in July, *D. gilva* adults emerge in August and attack the second generation. If cocoons are spun later in the year, larvae enter diapause in increasing numbers—100% in September.[19] The life cycle of *D. inconspicua* is similar, and adults of both species emerge from diapause in June and August. Tachinid emergence is therefore closely synchronized with the host, with the exception that prolonged diapause rarely occurs.

By contrast, the ichneumonids *Holocremus cothurnatus* Holmgreen and *Lamachus cohalitorius* Thunberg may have their development arrested for several years depending on the diapause state of *D. pini* (Fig. 4). The former lays its eggs on young larvae, the latter on older larvae. Parasitoid larvae penetrate their host and develop internally, and adults emerge from cocoons 4–8 days after the unparasitized sawflies.

Ichneumonids of the genus *Exenterus* oviposit on free eonymphs. The most important species, *E. amictorius* Panzer and *E. adspersus* Hartig, are specific to Diprionidae.[59] Eggs hatch within the cocoon, in which the *Exenterus* larvae develop as ectoparasites. *E. amictorius* is most abundant species and has been introduced to Canada, where it parasitizes several diprionid sawflies.[29]

Exenterus species are bivoltine, with adults emerging from the first-generation host cocoons and ovipositing on the second-generation host cocoons. From this second-generation host adults emerge by May–June simultaneously with a few diapausing *Exenterus* from the first-generation host.[14,20] The long life-span of the adults (over 70 days) ensures good synchronization with all host cohorts, except those delayed by long diapause.

3.1.3. Parasites of Eonymphs within Cocoons

The major parasites during this stage are two Ichneumonids, *Pleolophus basizonus* Gravenhorst and *Agrothereutes adustus* Gravenhorst.[20,29,44,45,59,66,67,86] The first spe-

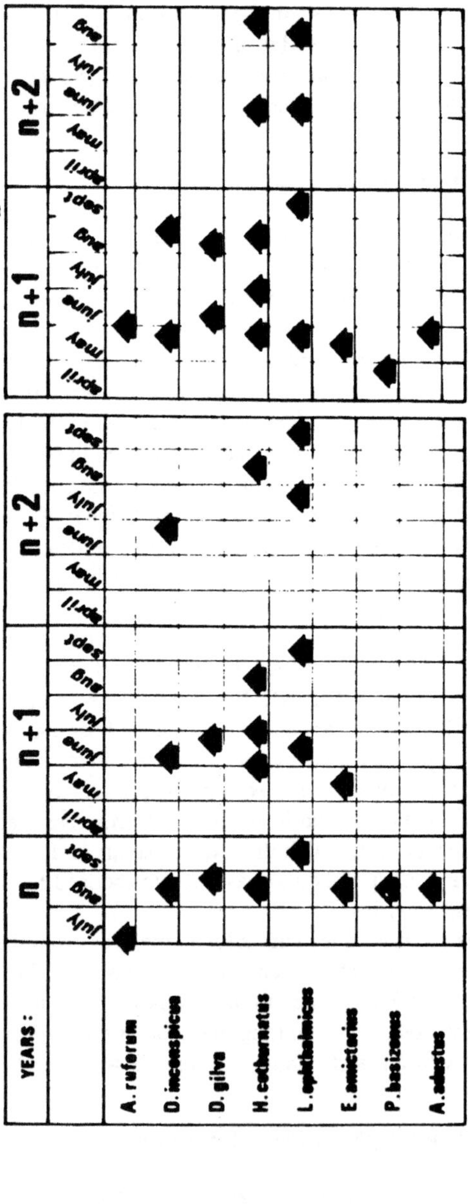

FIGURE 4. Adult emergence of the main parasitic species in the Paris Basin from the first (a) and second (b) sawfly generations.

cies, which is also parasitic on other Diprionidae and has been successfully introduced to Canada, is effective on *D. pini* all over Europe.[14,37,48,59,71,81,86] The second species is very active during some outbreaks. Adults feed and live a long time—40–50 days for *P. basizonus*. The adult female paralyzes the eonymph in the cocoon and then deposits an egg on it. Larval development is endophagous and superparasitism is frequent, but only one adult emerges from each host.

The emergence of parasitoids from a given generation is closely synchronized with the development of *D. pini* (Fig. 4). Diapause of *P. basizonus* does not occur between 10°C and 20°C[20] and, under favourable weather conditions, a new generation can thus be produced every 27 to 30 days on diapausing cocoons. This rapid development and early emergence in spring makes *P. basizonus* an effective parasite, particularly during the years following an outbreak.

The chalcid *Dahlbominus fuscipennis* Zetterstedt is an important parasite on *D. pini* in Southern Europe.[10,58,87] It lays up to 50 eggs per cocoon and completes its development within 21 days. It is frequently bivoltine, with both generations synchronized with those of the host.

3.2. Predators and Pathogens

Reduvidae and Pentatomidae attack *D. pini* eggs and larvae, while many birds feed the on false caterpillars. More important, however, is predation by tits, nuthatches, and magpies,[73,76] as well as by mice and shrews[13,62,63] on the cocoons. It is not easy to determine the effects of predators on sawfly populations because cocoons from several generations are mixed and are often displaced or piled up. Predator activity is particularly evident during outbreaks in which cocoons are available all winter.[27,37,48,69,73,85,86] However, predators also destroy diapausing sawflies after the outbreak is over.

Diseases often appear by the end of sawfly outbreaks, but it is difficult to distinguish between mortality due to pathogens and that due to physiological stress. Mortality by pathogens is seldom greater than 20%, and *D. pini* appears to be more resistant to disease than may other *Diprionidae*.[21,86,87]

4. RELATIONSHIPS WITH THE PHYSICAL AND FOREST ENVIRONMENT

4.1. Effects of Soil and Climate

According to numerous investigators, outbreaks frequently occur in warm locations with shallow infertile soils and deep ground water.[21,50,73,75,78] Some consider that *Diprion* populations are more dense or more vigorous and cause more severe damage on trees growing on poor dry soils, especially dune soils, than on those growing on rich, wet soils.[23,73,77] It is also suggested that infestations occur during dry, warm years, referring, in particular, to the outbreaks that followed the exceptionally dry and warm summer of 1959.[65,71,77,82,84]

A clear link between climate and outbreaks cannot be obtained from the meteorological data in France. In addition, outbreaks have been observed in forests with very different soils, e.g., those of the Fontainebleau forest, which are dry and sandy, those of the Ingrannes stands, which are rich and deep, and those of the Lorris stands, which are hydromorphic. Although recent attacks have been moderate in the Ingrannes area, heavy infestations have been recorded in the past in these three areas, with extremely severe damage in the Lorris region.

Certainly, higher temperatures increase sawfly development rates, encouraging the bivoltine cycle, but its effect is only important in regions in which bivoltinism can only occur during warm years. Damage is also more severe when trees are weakened by dry or hydromorphic soil, but the influence of soil properties on insect numbers is less obvious.

4.2. Effects of Stand Structure

Contrary to other defoliators, whose attacks are closely linked to exposure, tree density, and species,[33–35,42] the abundance of *D. pini* does not seem to be greatly affected by the local environment in the Paris Basin.[38] However, populations tend to be higher in locations in which they had been important in previous years, while damage is low on black pines and nonexistent on maritime and white pines (Fig. 5). Attack on black pines, however, becomes heavier as the outbreak progresses. In the same way, large older trees are more heavily infested at the beginning of an outbreak, but damage to young pines often occurs later on.

Although local variations in infestation intensity do not seem be associated with the

FIGURE 5. Defoliated Scots pines (right) growing next to black pine (a) and maritime pine (b).

species composition of stands, when the whole forest is considered, areas of pure *P. silvestris* have experienced the most severe, the earliest, and the longest infestations. Other regions have only been subjected to moderate infestations of short duration.

The impact of *D. pini* infestations on the host tree is quite variable and depends on the appearance of secondary pests, such as *Tomicus piniperda, Ips sexdentatus,* and *Pissodes notatus* Fabricius,[51,56,57] as well as direct mortality resulting from severe defoliation and other stress factors (old pines, hydromorphic soils, improper silviculture, and maintenance). A study of these factors in the Lorris area showed that mortality rates are lower in mixed stands than in pure Scots pine stands and also on soils that are slightly acid and hydromorphic.[38] Thus, the influence of environmental factors is more clearly related to damage severity than to infestation levels.

5. NUMERICAL POPULATION DYNAMICS

Quantitative data concerning the dynamics of *D. pini* populations over long periods of time are generally lacking, largely because the insects are so rare during interoutbreak periods. However, our observations in the South Paris Basin seem to be representative of situations reported elsewhere in Europe.

Observations were first made in late 1963 after dramatic defoliation of the Fontainebleau forest but ended in 1969 because of the scarcity of insects. A new outbreak at the end of 1980, however, resulted in the resumption of observations, which have since consisted mainly of the following:

1. Simple notations of the relative abundance of *D. pini* (nonexistent, rare, low, high, total defoliation) determined while driving throughout areas south of Paris containing stands of *P. silvestris* (Fig. 6)
2. Intensive surveys of infestations and forest environmental conditions in Sologne stands and in the Orleans and Fontainebleau forests
3. Detailed observations on each sawfly cohort on l-ha plots in these forests, including phenological observations, assessment of fecundity, mortality, parasitism, diapause rates, and the importance of the different adult flights (Samples were either examined immediately or reared in forest shelters

These studies provided information on the numerical dynamics of sawflies in the Sologne, Fontainebleau, and Orleans (Lorris and Ingrannes areas) forests during both outbreaks (Fig. 7). Summer cohorts, present in very low numbers, are not taken into account. For the Fontainebleau forest, the level of infestation during the early outbreak of 1935 is assessed from observations by foresters.

Figure 7 shows that outbreaks are rare, without any obvious periodicity, and that they are separated by long periods of latency during which sawflies are rarely seen. Outbreaks generally develop and collapse in 2 years and occur almost synchronously across the entire southern Paris Basin. However, outbreaks seem to occur earlier in the Southern forests (Sologne) and to spread northward (Figs. 6 and 7).

Maximum infestations occur in autumn from much smaller spring populations and collapse, usually within one generation, in the year following peak infestation. One or 2

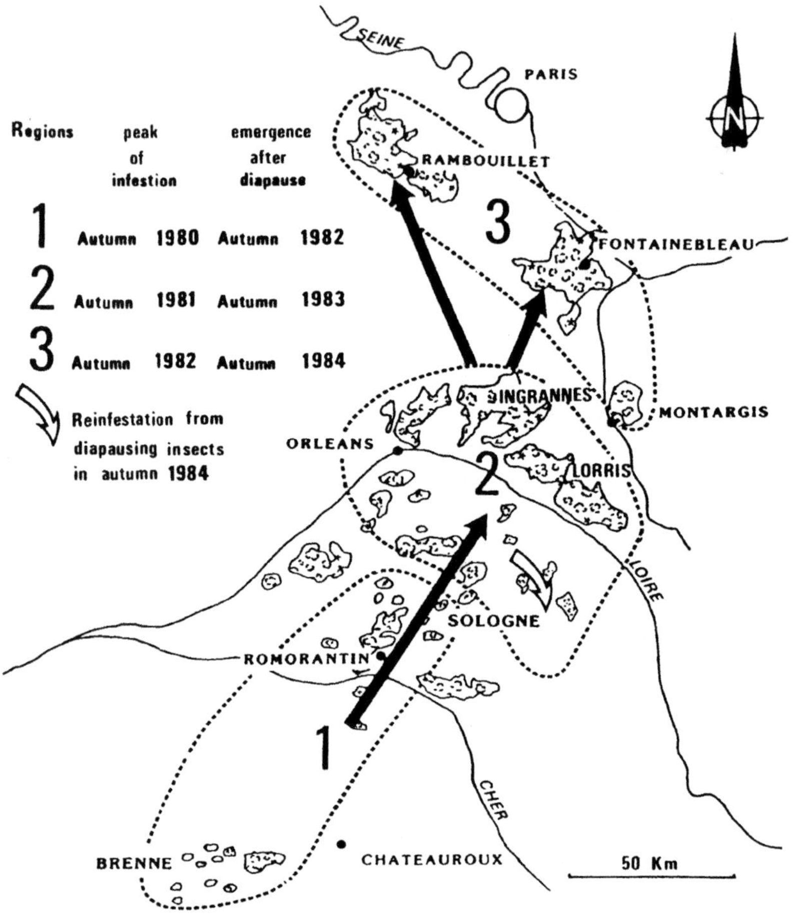

FIGURE 6. *Diprion pini* study areas in the South Paris Basin, and the progress of the outbreak since 1980.

years later, a new infestation may appear from diapausing populations, in some cases, two outbreak episodes may be observed (e.g., Sologne and Orleans 1980–1981 and 1983–1985).

The first outbreak in the South Paris plain (1963) probably started previously in southern Sologne, reaching Orleans and Fontainebleau stands in spring 1963. This infestation reached its peak in fall 1963 and caused severe defoliation in both forests. The North Sologne and the Rambouillet and Montargis forests were also affected, while the infestation spread eastward as far as Burgundy and southern Champagne and westward to Normandy. Populations in the Fontainebleau forest, which were already reduced by fall starvation, parasites, and rodents in 1963, were decimated by parasitoids and rodents by spring 1964; 60% of individuals were parasitized by prior cocoon spinning, and many surviving eonymphs were destroyed by hyperparasites and rodents after cocooning. Almost all the eggs of the autumn cohort were killed by chalcids.[36]

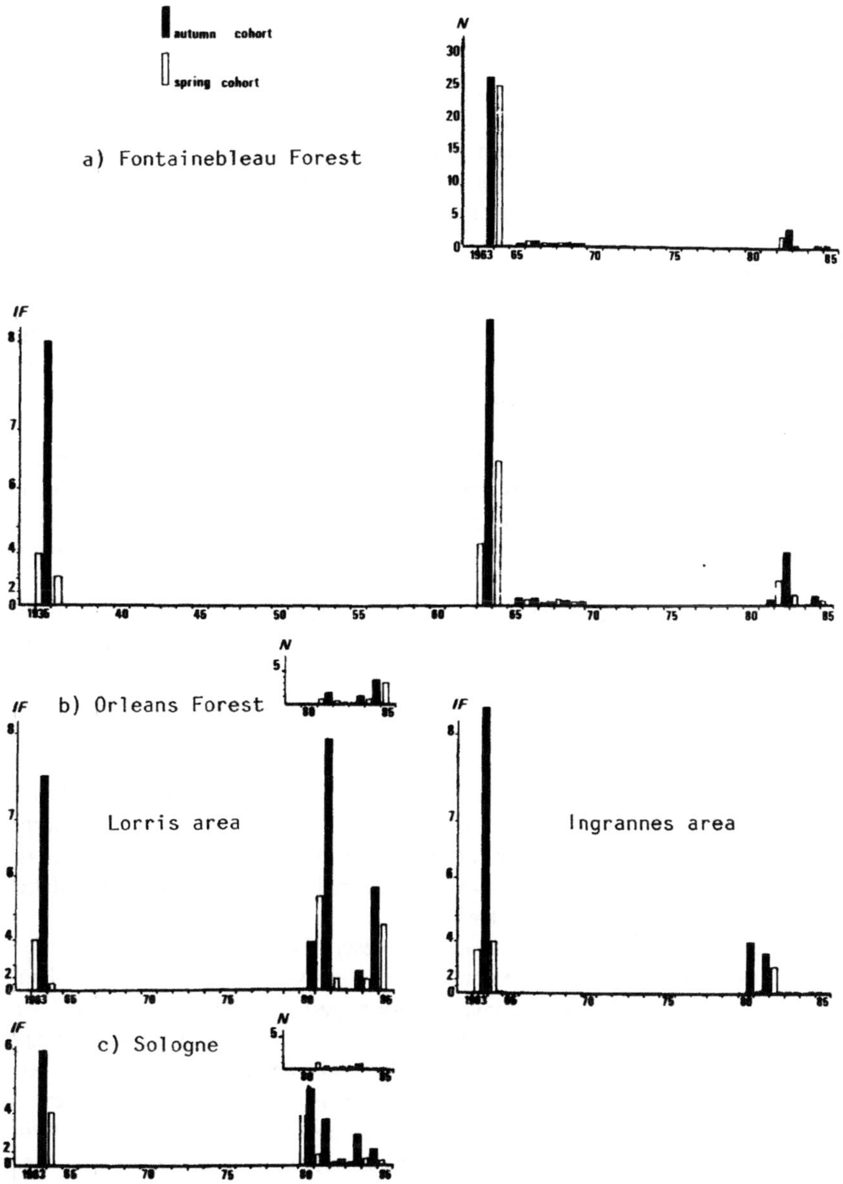

FIGURE 7. Mean infestation index (*IF*) and mean number of colonies (*N*) from plots in Fontainebleau forest (a), Orleans forest (b), and Sologne (c). Scale of infestation index is built according to corresponding mean population levels.

Populations samples from the various forests showed widespread regression of the outbreak attributable to the same causes. By spring, only very small populations were present in the Orleans forest, and large populations were only later observed in Normandy and near Rambouillet.

Small populations reappeared in Fontainebleau in autumn 1965,[14] arising from individuals entering diapause in 1963 and 1964 (50% and 90%, respectively, of the surviving eonymphs). These populations remained at low levels for the next 4 years. Diapause and parasitism were very low in 1965 but increased rapidly thereafter. The population rose again in 1967, as more diapausing insects emerged, but was rapidly extinguished by parasites, so that only minor infestations could be found in 1969. Thus, the outbreak seemed to collapse in a series of damped oscillations.

The second outbreak began in Brenne and southern Sologne and spread to the northeast, reaching successively the northern Sologne and Orleans forests, then Fontainebleau and Rambouillet forests. Peaks were observed in 1980, 1981, and 1982 in these three areas (Fig. 6). New infestations from diapausing individuals also occurred in all these areas 2 years later.[37,38]

Three infestation foci were detected in autumn 1980 in Brenne, southern Sologne, and the Lorris stands of the Orleans forest. Populations at the first two localities had built up earlier and caused heavy defoliation. Then, after 99% of the individuals had entered diapause, the progeny of the few nondiapausing sawflies were eradicated by parasites in spring and autumn 1981.

By contrast, the Lorris populations had low levels of diapause and were not heavily parasitized in autumn 1980 or in 1981. These populations multiplied and infested the Orleans forest and northern Sologne in the fall of 1981, while, at the same time, parasitism and diapause increased in the Lorris populations. The infestation also spread to the Rambouillet and Fontainebleau forests, where low levels of parasitism and diapause were recorded.

In spring 1982, sawfly populations in northern Sologne and the Orleans forest were decimated, with 60% of eggs and 94% of cocoons being parasitized, and 97% mortality of the nonparasitized larvae, probably resulting from nutritional stress. Later, in autumn, 80% of the second-generation eggs were parasitized. Only a few local populations remained in the east of the Orleans stands.

In the Fontainebleau and Rambouillet forests, the attack reached higher levels in 1982 because the spring cohort was only weakly parasitized and few insects diapaused. However, parasites quickly limited this expansion, particularly in the Fontainebleau forest and, in the fall, 80% of eonymphs entered diapause. Sawfly populations also reappeared in southern Sologne at this time from insects that had entered diapause in 1980. Although weakly parasitized in the egg stage and lightly diapausing, these populations were quickly decimated by tachinids. By spring 1983, residual populations only survived in the Fontainebleau, Rambouillet, and southern Sologne areas, but these disappeared by the end of the year.

By contrast, sawflies that had entered diapause in 1981 appeared in autumn 1983 in the Orleans forest; 60% of the adults of this autumn cohort emerged in April, giving a first generation in 1984. As these populations were only weakly parasitized in 1983 and 1984, and because 70% of the eonymphs of the first generation of 1984 developed immediately, a severe infestation developed again in autumn 1984, causing total defoliation at Lorris and in eastern Sologne. Many eonymphs of the second 1984 generation entered diapause, but this new infestation was aggravated and prolonged by the emergence of insects that had entered diapause in the remaining foci of 1982. In the Rambouillet and Fontainebleau areas, populations reappeared in fall 1984 after a 2-year diapause; they were weakly parasitized, and their numbers were high at Rambouillet and low at Fontainebleau.

Although a very cold winter in 1984–1985 did not cause significant mortality, cocoons were actively preyed upon by rodents and birds. In spring 1985, high population levels were only found in the Rambouillet and eastern Lorris areas. These populations were weakly parasitized, but climatic conditions caused them to develop later than normal: 60% of the eonymphs entered diapause.

During these outbreaks, the most active entomophagous insects were *A. ruforum, P. basizonus,* and *E. amictorius,* and to a lesser extent *D. gilva, D. inconspicua,* and *E. adspersus,* while *H. cothurnatus* and *L. cohalitorius* were mainly observed during the interoutbreak (latent) period. Fecundity remained normal (90 to 130 eggs) and no disease symptoms were observed.

6. EXPLANATION OF POPULATION BEHAVIOR

The previous discussion illustrates the importance of parasites, predators, diapause, and nutritional factors in the populations dynamics of *Diprion pini.* In particular, the evolution of parasitism and diapause shows similar patterns during outbreaks and latent periods (Fig. 8); i.e., small, weakly diapausing and weakly parasitized populations build up in specific localities, but increasing parasitism and diapause cause the population to subside after three generations. One or 2 years later, populations emerging from diapausing insects may develop in a similar way but often remain at low levels.

Migration, although poorly understood, also seems to play an important role in *D. pini* population dynamics. Sawfly populations showed a similar spatial progression during both outbreaks, suggesting that flights spread the infestation from one forest to the next.

6.1. Role of Parasitism and Predation

The importance of parasites, especially *A. ruforum, P. basizonus, D. gilva,* and *E. amictorius,* in regulating sawfly populations is widely recognized in Europe.[9,23,48,62,65,69,82,85,87] Parasitism was appraised in the Paris Basin by dissecting and rearing insects collected during a given developmental stage. This enabled us to determine the rate of parasitism as populations developed over time, while the emergence of parasitoids reared under natural conditions enabled us to establish their synchrony with the host.

Parasitism appears to act as a delayed density-dependent factor, being very low at the beginning of an outbreak, increasing gradually during the development of the outbreak, and reaching very high levels as populations collapse. Similar delayed density-dependent responses also occurred as host populations oscillated during the latency period.

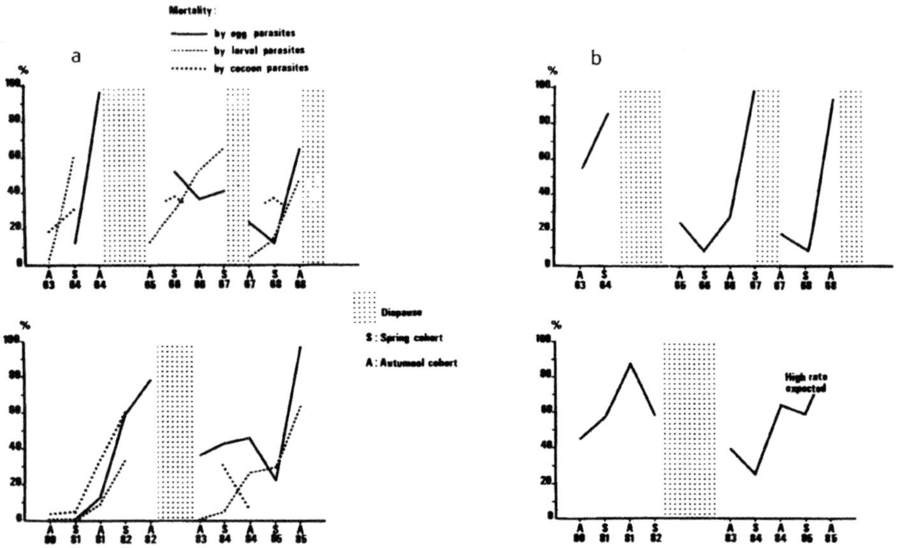

FIGURE 8. Evolution of parasitism (a) and prolonged diapause (b) from 1960 to 1968 in Fontainebleau forest and from 1980 to 1985 in Orleans forest.

Decreasing parasitism following outbreaks probably results from a loss of synchrony with long-diapausing sawfly populations. Thus, after the 1963–1964 outbreak, *E. amictorius* and *D. inconspicua,* which parasitized the first 1964 generation, emerged in 1964 or too early in 1965 to parasitize the sawfly populations arising in the fall from diapausing insects. This inability of parasites to synchronize their life cycle with diapausing *D. pini* increases with the length of the sawfly diapause and was even more obvious during the second outbreak, when the host was subject to a 2-year diapause. Lack of coincidence disrupted the parasitoid–prey interaction in several other instances, such as the renewed attack in the Orleans forest in 1984.

Variation in the parasitic species complex is also important. Tachinids and ichneumonids, having little capacity for long diapause, are mainly effective during outbreaks, while ichneumonids, having high capacity for extended diapause, are more important during the latency period. The impact of these ichneumonids is probably reduced during outbreaks by competition with tachinids, but during the latency period they are more efficient because of their adaptability to host diapause. Disruption in this host–parasitoid interaction by climatic events, host scarcity, and/or sawflies arising from diapause or immigration can give rise to a new outbreak epicenter that can then spread by immigration into adjacent forests.[5,6]

The activity of vertebrate predators was mainly observed during sawfly outbreaks and during the regression phase, e.g., in spring 1964. They probably also contribute to density-dependent regulation of low-density populations because of their rapid functional responses.[5] However, when sawfly populations become relatively dense, an unstable outbreak threshold can be breached as predator-functional responses saturate.[5,6]

6.2. Role of Diapause

Diapause is avoided by exposure of the last larval stages to long photophases, whereas short photophases induce diapause.[15–18,21] However, threshold photophases vary according to the origin of the population, with insects from northern regions where summer days are longer needing longer photophases to prevent diapause. In the Paris Basin, the critical photophase usually lies between 14 and 16 hr daylight.[39] Spring cohorts, which are exposed to the longest photophases of the year, are weakly diapausing unless their development is retarded, in which case later instars experience shorther photophases and enter diapause (Fig. 9). Summer cohorts, which experience even shorter photophases, all enter diapause. Moreover, photophase acting on first instars modifies the diapause response of the later instars.[39] The greatest number of prompt emergences is observed after passage from short to long photophases and vice versa. Besides, photophase can also act on eonymphs during the first weeks following cocoon formation and low temperatures also increase the number of diapausing individuals.

The autumn cohorts, which are exposed to very short days during late larval instars, produce adults in April. This is confirmed by experiments in which cocoons are replaced in the field after treatment; adult emergence in April only occurs if the larvae have been continually exposed to short photophases[39]; this flight can be induced in the spring cohort by rearing larvae under decreasing photophases that simulate autumn conditions.

Termination of diapause depends both on increasing photophase[17,18] and temperature. However, experimental responses to these factors are quite variable. Eonymphs react to very small increases in photophase and, when warmed to 20–25°C in natural daylight, can resume development in January. Under constant photophase, some individuals remain in diapause until they die. Cold temperatures stimulate reactivation, and eonymphs have successive periods of susceptibility to relatively short and then long photophases.[80] In nature, the reactivation of individuals might be due to the cumulative effects of variations in photophase and temperature during successive years.

By contrast, it is difficult to prevent sawflies that have developed without interruption for several generations from entering diapause, even when larvae are reared under the most favorable photoperiodic conditions. Furthermore, the eonymphs derived from diapausing individuals seem to be reactivated more readily than do those that have developed for several generations without diapause. Until now we have failed to obtain more than three of four generations without diapause by rearing insects under natural or laboratory conditions.

These results provide some insights into the population dynamics of *D. pini* (Fig. 9). In spring, larvae develop under increasing photophases and eonymphs cocoon during long days, so that few enter diapause. However, the larvae of late spring and summer cohorts hatch on long days, and develop under decreasing photophases, and eonymphs cocoon in shorter days. Thus, most of the eonymphs from these generations enter diapause. Because larvae of spring and summer cohorts were exposed to long periods of daylight, however, they cannot produce adults in April of the next year, but only in June and August, and in the same months during subsequent years.

Autumn cohorts develop on increasingly shorter days and cocoon on very short days, entering diapause. This diapause is more labile because the larvae were exposed to shorter

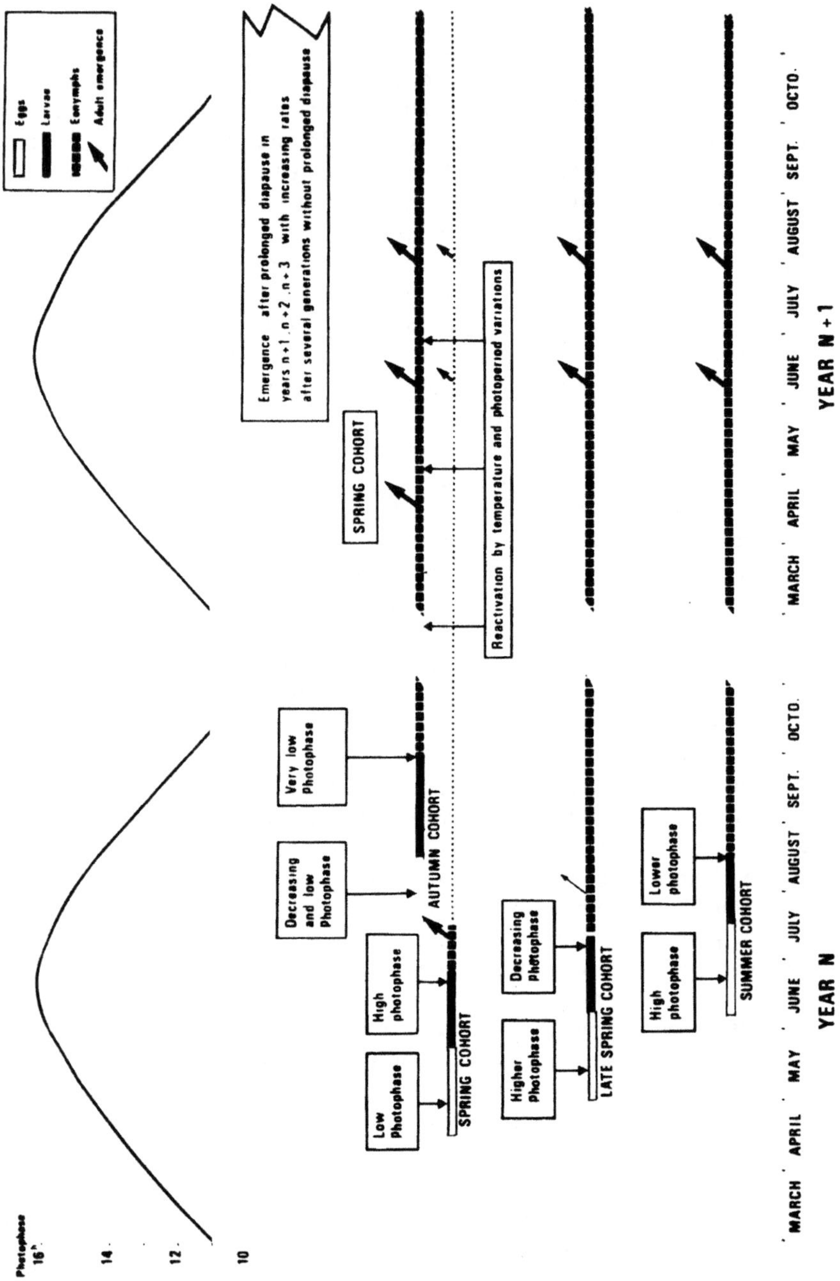

FIGURE 9. Effects of diapause on *Diprion pini* population patterns.

photophases and may be interrupted by slight increases in day length, so that some adults appear in April. Insects remaining in diapause for more than 1 year are apparently reactivated by the cumulative effects of temperature and photoperiod variations during successive years.

By contrast, variations in diapause during outbreaks, i.e., low diapause rates when populations are first observed and high rates during the collapse, agree with observations that an obligatory diapause occurs when three generations have developed continuously, whatever the photophase and food conditions. At the present time, this mechanism cannot be explained by external factors and probably involves genetic processes or some other form of transfer of information between generations. Nevertheless, some external factor, such as foliage quantity or quality, may affect in this increasing rate of diapause during the outbreak.

6.3. The Influence of Foliage

Both foliage quantity and quality can affect the reproduction and survival of the sawfly during outbreaks. Competition for food and starvation are observed mainly in autumn during periods of heavy infestation. This lack of food is felt even more strongly the following spring because needles become very scarce after autumn defoliation, which removes the current year's foliage. There is now no experimental evidence that larval density or lack of food affects diapause rates.

Foliage quality was investigated because young needles have been shown to be distasteful and toxic to neodiprionid larvae.[1,2,61] The active ingredients are diterpene resin acids, particularly 13-keto-8(14)-podocarpen-18-oic acid.[49] Concentrations of this diterpene decreases quickly as foliage ages, which would explain why sawflies prefer old foliage.

Recent experiments show that young foliage of Scots pine is repellent to *D. pini* larvae, and that this repellency decreases with time (Table I). New foliage is accepted in late July, which explains its acceptability to the larvae of the second generation. Prior to

TABLE I
Percentage of *D. pini* Larvae on the Distal Half of *P. silvestris* Twigs Having Only Mature Foliage and Young and Mature Foliage at Four Times after Release at Five Different Dates (10 Larvae per Replicate)

	Mature foliage					Young and mature foliage				
		% of larvae					% of larvae			
Dates	Number of replicates	1 hr	2 hr	3 hr	4 hr	Number of replicates	1 hr	2 hr	3 hr	4 hr
14.06.83	6	49	48	49	49	12	34	20	13	8
28.06.83	6	49	49	45	47	18	37	23	16	10
12.07.83	6	52	56	61	57	6	39	35	30	29
26.07.83	/	/	/	/	/	4	49	46	47	50
29.08.83	6	47	45	48	48	6	49	48	48	49

this, feeding on new foliage induces heavy mortality, slower development, reduced weight and fecundity, and increased diapause rates.[40] The latter effect may be a consequence of slower development induced by food quality, causing last instars to develop under shorter days, but it could also result from the direct effects on the initiation of diapause.

Variations in foliage quality are important in the population dynamics of *D. pini*. After heavy defoliation in autumn, larvae born the next spring eat the old residual needles and are then forced to feed on new foliage. This causes heavy mortality, slower development, and higher rates of diapause as the populations collapse.

6.4. General Conclusions

Diprion pini populations in the Paris Basin appear to exhibit a general pattern of behavior as described in Fig. 10. At the beginning of the outbreak, a few population epicenters develop in autumn under favorable environmental conditions (e.g., undamaged foliage, low parasitism) and exhibit low diapause rates. The following spring, parasitism and diapause remain low and foliage is only slightly damaged and, if climatic conditions are favorable, the bivoltine cycle and the natural fecundity of the sawfly result in a dramatic population increase. Overpopulation is usually evident by autumn, when intense competition for food causes starvation and the incidence of parasitism and diapause become high. Increased predator activity is also observed at this time. As a consequence, the number of eggs is greatly reduced the following spring.

Migration to other forest areas seems to result from the numerous adults that generate the autumnal cohort, since new foci are observed at this time. These migrations presumably raise local populations above the outbreak threshold inducing a new outbreak cycle.[6]

The collapse of the outbreak usually occurs in the spring, following heavy defoliation, and is brought about by a shortage of old foliage, very high activity by outbreak parasites and predators, and numerous individuals entering diapause. If residual populations remain in autumn, they are completely exterminated by parasites, mainly the egg parasite *A. ruforum*.

One or 2 years later, small weakly diapausing populations reappear from the diapausing sawflies. By this time, the outbreak parasites have been reduced by a lack of hosts, and so the sawfly populations begin to grow. After a few generation, however, they are rapidly suppressed by the more synchronized parasites (ichneumonids such as *Lamachus* and *Holocremnus*), predators, and increasing diapause rates. Nevertheless, a reinfestation may occur if a large number of individuals emerge simultaneously from long diapause (2 years or more) to find good foliage conditions and few parasitoids.

In most cases, the temporary reappearance of the sawflies has little effect on the forest, and the populations now enter a relatively long period of latency, in which local increases in density are quickly suppressed by the responses of parasite and predator populations (Fig. 10). Latent populations, widely scattered throughout the forest, become sparse, while environmental conditions become progressively more favorable for the sawfly; i.e., old foliage accumulates and predators become disinterested in this rare food source. These dispersed and scarce populations may also get out of synchrony with their parasites. As time progresses, therefore, it becomes more likely that environmental distur-

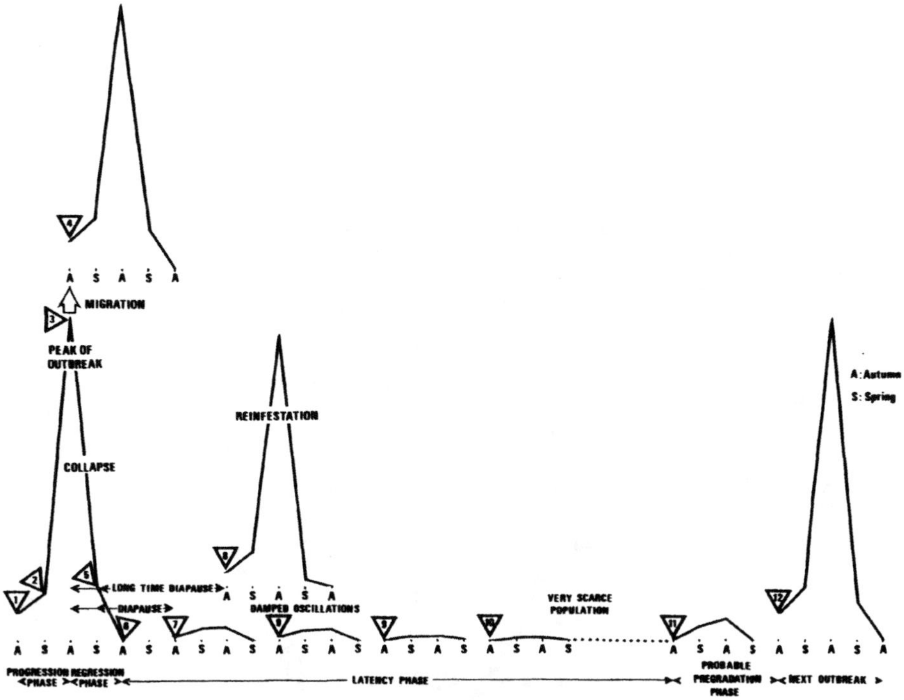

FIGURE 10. General pattern of *Diprion pini* population fluctuations. 1, Undamaged foliage, low diapause, scarce parasites, disinterested predators. Supply by immigration or emergence from diapause; 2, little damage to foliage, low diapause, and parasitism. In good climatic conditions multiplication by fecundity—increasing interest of predators and "outbreak" parasite populations; 3, overpopulation, starvation, high diapause, and parasite and predator activity, migration to other forests; 4, migrating insects raise the populations above the outbreak threshold and trigger an outbreak as in 1; 5, collapse occurs in the first infested forest due to high mortality the previous autumn and to competition for food, poor quality of foliage, very high parasitism, predation, and diapause; 6, diapausing insects excepted, populations are destroyed, mainly by egg parasites; 7, reappearance of small populations from diapause regulated by remaining "outbreak" parasites, predators, diapause, and perhaps inadequacy of foliage; 8, long diapause results in a more disruption of "outbreak" parasites and better foliage. Reinfestation may occur; 9, reappearance of insects from diapause is suppressed by "latency" specific parasites, predators, and return to diapause. Populations oscillate with decreasing amplitude; 10, very scarce and scattered populations, foliage is now good and predators become disinterested, scarcity and dispersion are favorable for disruption of parasites and predators; 11, lack in host–parasite synchrony and probably high rates of emergence from diapause cause in favorable environmental conditions an increase in populations entering diapause; 12, new outbreak starts with emergence of large numbers of diapausing sawflies or immigration and saturation of parasite and predators responses.

bances, such as climatic factors that induce emergence from diapause, will trigger another outbreak episode. If this reasoning is correct, *D. pini* populations apparently exhibit eruptive outbreaks that spread, following a disturbance, from epicenters.[7] It also implies that the sawfly system possesses at least three coincident equilibria, one of which is an unstable outbreak threshold (Fig. 11).

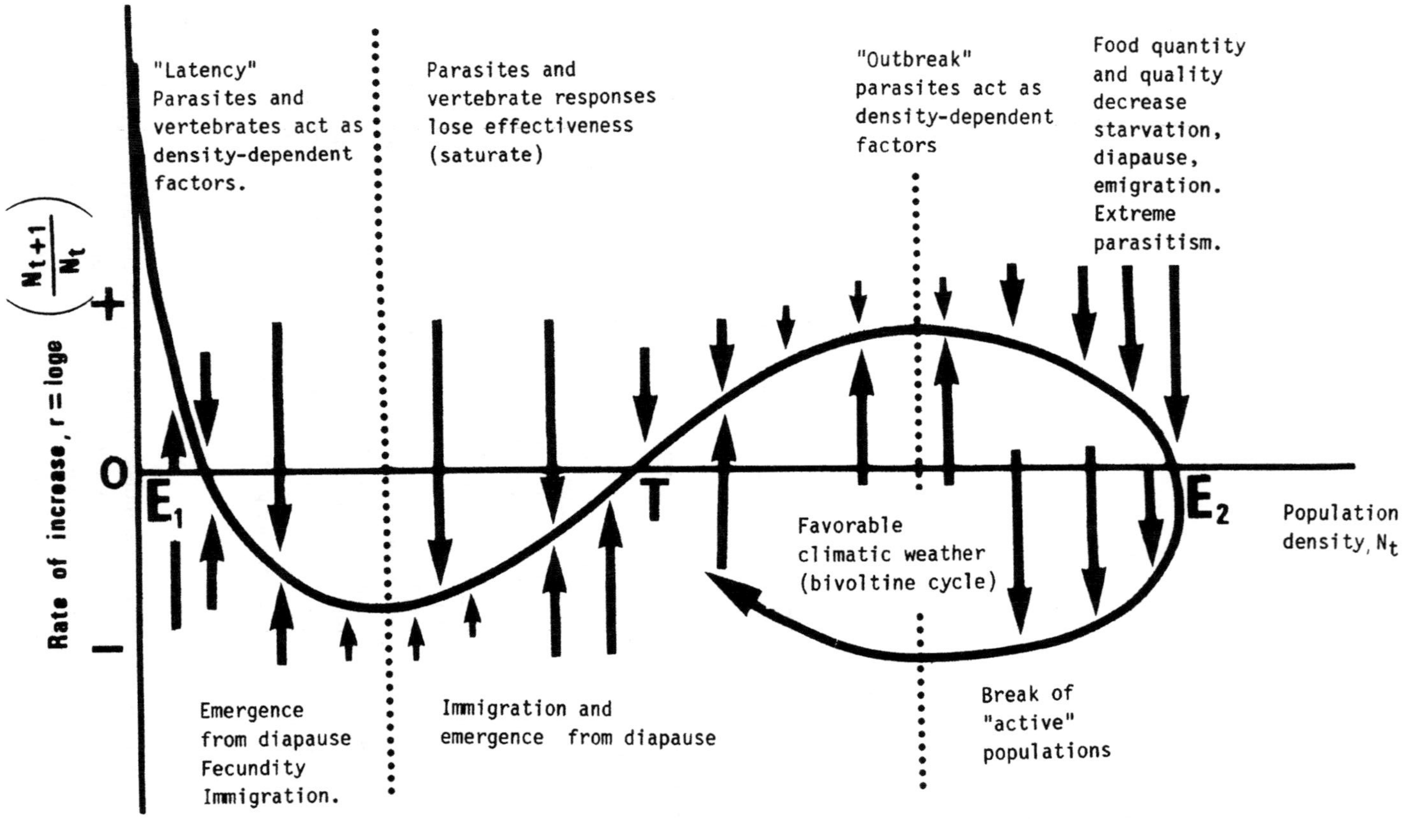

FIGURE 11. Proposed relationship between the per-capita rate of increase of *D. pini* and population density. E_1, low-density or latent population equilibrium; T, outbreak threshold; E_2, unstable (cyclical) outbreak equilibrium.

7. IMPLICATIONS FOR PEST MANAGEMENT

Pine sawfly larvae are susceptible to a number of insecticides but, because of strict legislation, few, e.g., diflubenzuron (Dimilin), are permitted in French forests. To prevent damage, Dimilin must be applied at the beginning of larval development because it causes larval mortality only during ecdysis.

Our analysis, however, suggests that control of *D. pini* is essentially a biological problem. Obviously, it is dangerous to use chemicals against an insect that is regulated by a rich parasitic and predacious fauna and whose diapausing eonymphs escape treatment. This is especially true during heavy infestation periods, when intervention can severely reduce the number of parasites at a time when many insects are in an extended diapause. By contrast, it may be sensible to spray populations in the early outbreak phase because they have low diapause and parasitism rates at this time and can possibly be suppressed below the outbreak threshold. Detection of such scarce populations is difficult, however, requiring intensive surveys and accurate prognosis techniques.

Local chemical treatment of small areas is often more appropriate than widespread application because populations of parasites can be preserved. These parasites can build up on populations in untreated areas to prevent reinfestation of treated areas.[72] Treatment of the most heavily infested areas at the beginning of an outbreak will usually limit its expansion.

Manipulation of indigenous parasites may also be practical for *D. pini* control.[41] Outbreaks usually follow predictable patterns that culminate in heavy parasitism while new foci without parasites appear in nearby forests. Thus, it may be feasible to transport parasitized eggs and cocoons from heavily infested areas to these new foci. The technical feasibility of this approach was tested in 1985 by transporting several thousand parasitized cocoons and more than 60,000 parasitized eggs to a plot in the Rambouillet forest. Rearing efficient parasites such as *A. ruforum* and *P. basizonus* could permit intervention at an earlier stage, soon after populations are first detected. Both species can be reared rather easily, for it only requires regular supplies of eggs or cocoons. These parasitoids could be used in conjunction with local insecticide spraying, in an effort to enrich the parasitic fauna in unsprayed areas.

There seems to be little hope of reducing oubreaks by local alterations of stand structure and species composition. However, large areas of pure pine are very favorable for outbreak development and should be avoided. In general, forest management should attempt to mix pines with deciduous trees and to keep important areas of deciduous stands. The less susceptible black pine may also be encouraged, but its monoculture must be avoided in countries in which this species is attacked by the pine processionary caterpillar. These actions would reduce the severity of sawfly attacks and the seriousness of its damage. Treatments should be carried out principally on pure stands, which favor outbreaks. In the long run, research may open new avenues, such as the breeding of resistant Scots pine varieties according to the chemical properties of their needles or their ability to synthesize repellent and antifeeding substances.

REFERENCES

1. All, J. N., and Benjamin, D. M., 1975, Deterrent(s) in jack pine, *Pinus banksiana* influencing larval feeding behavior and survival of *Neodiprion swainei* and *N. rugifrons*, *Ann. Entomol. Soc. Am.* **68**:495–499.

2. All, J. N., and Benjamin, D. M., 1975, A feeding preference bioassay for *Neodiprion* Sawflies on jack pine *Pinus banksiana, Ann. Entomol. Soc. Am,* **68:**553–556.
3. Barbey, A., 1925, *Traité d'entomologie forestière,* Berger-Levrault, Paris.
4. Besemer, A. F. H., 1942, Die Verbreitung und Regulierung der *Diprion pini*. Kalamität in der Niederlanden in den Jahren 1938 bis 1941, *Ned. Boschb. Toijdschr.* **15:**135–165; 198–242; 262–302.
5. Berryman, A. A., 1978, Towards a theory of insect epidemiology, *Res. Pop. Ecol.* **19:**181–186.
6. Berryman, A. A., 1981, *Population Systems. A General Introduction,* Plenum, New York.
7. Berryman, A. A., 1986, *Forest Insects: Principles and Practice of Population Management,* Plenum, New York.
8. Breny, B., 1954, *Neodiprion sertifer:* l'oeuf et le végétal, *Bull. Inst. Agron. Gembloux* **22:**179–188.
9. Ceballos, G., and Zarco, E., 1952, Ensayo de lucha biologica contra une plaga de *Diprion pini* L. en masas de *Pinus silvestris* L. de la Sierra de Albarracin, *Madrid Inst. Esp. Entomol.*
10. Chatelain, L., 1957, Le lophyre du pin en Champagne, *Phytoma* **91:**32–33.
11. Chatelain, L., 1958, Le lophyre du pin en Champagne, *Phytoma* **98:**39–40.
12. De Fluiter, H. J., 1934, Over het tijdstip, wasdrop de gewone deunenblaswesp, *Diprion* = *Lophyrus pini* L. bestreden dient te worden, *Ned. Boschb. Tijdschr.* **7**(3):70–81.
13. Dusaussoy, G., and Geri, C., 1966, Etude d'une population de *Diprion pini* L. (Hym. Symphites) en forêt de Fontainebleau. I. Données biologiques sur *Diprion pini* L. et sur les principaux parasites, *Ann. Soc. Entomol. Fr.* (NS), **2**(3)**:**503–534.
14. Dusaussoy, G., and Geri, C., 1971, Etude des populations résiduelles de *Diprion pini* après la gradation de 1963–1964, *Ann. Sci. For.* **28**(3)**:**297–322.
15. Eichhorn, O., 1976, Dauerzucht von *Diprion pini* L. (Hym. *Diprionidae*) in laboratorium unter Berücksichtigung der Fotoperiode, *Auz. Schadlingskd. Planz. Umweltsch.* **49:**38–41.
16. Eichhorn, O., 1976–1977, Autökologische Untersuchungen an Populationen der gemeinen Kiefern-Buschhornblattwespe *Diprion pini* (L.) (Hym. *Diprionidae*). I. Herkunftsbedingte Unterschiede im Schlüptverlauf und Diapauseverhasten, *Z. Angew. Entomol.* **82:**394–414.
17. Eichhorn, O., 1977, Autökologische Untersuchungen an Populationen der gemeinen Kiefern-Buschhornblattwespe *Diprion pini* (L.) (Hym. *Diprionidae*). III. Laborzuchten, *Z. Angew. Entomol.* **84:**264–282.
18. Eichhorn, O., 1979, Autökologische Untersuchungen an Populationen der gemeinen kiefern-Buschhornblattwespe *Diprion pini* (L.) (Hym. *Diprinidae*). IV. Generations and Schlüpfwellenfolge, *Z. Angew. Entomol.* **88:**378–398.
19. Eichhorn, O., 1980, Autökologische Untersuchungen an Populationen der gemeinen Kiefern-Buschhornblattwespe *Diprion pini* (L.) (Hym. *Diprionidae*). V. Untersuchungen über die larvenparasitischen Tachinen, insbesondere über *Drino gilva* Htg. *Z. Angew. Entomol.* **89:**455–470.
20. Eichhorn, O., 1981, Autökologische Untersuchungen an Populationen der gemeinen Kiefernbuschhornblattwespe *Diprion pini* L. (Hym. *Diprionidae*). VI. Zur Kenntnis der Kokonparasiten und *Exenterus* Arten, *Z. Angew. Entomol.* **92:**252–285.
21. Eichhorn, O., 1982, Untersuchungen zur Ökologie der Gemeinen Kiefern-Buschornblattwespe *Diprion pini* L. (Hym. *Diprionidae*). VII. Populations dynamische Faktoren, *Z. Angew. Entomol.* **94:**271–300.
22. Eichhorn, O., and Pschorn-Walcher, J., 1976, Studies on the biology and ecology of the egg-parasite (Hym. *Chalcidoidae*) of the pine sawfly *Diprion pini* L. (Hym. *Diprionidae*) in Central Europe, *Z. Angew. Entomol.* **80**(4)**:**355–381.
23. Eidmann, H., 1936, Zur Frage der Blattwespen-Prognose, *Arb. Physiol. Angew. Entomol. (Berl.)* **3:**229–234.
24. Elens, A., 1953, Etude écologique des Lophyres en Campine (Belgique). II. Incubation des oeufs et adaptation à la température chez *Diprion pini* L. *Diprion pallidum* KL. et *Diprion sertifer* GEOFFR, *Agriculture* **1:**19–29.
25. Eliescu, G., 1932, Beiträge zur Kenntnis der Morphologie, Anatomie und Biologie von Lophyrus *pini* L. *Z. Angew. Entomol.* **19:**22–67, 188–206.
26. Elliger, C. A., Zinkel, D. F., Chan, B. G., and Waiss, A. C., Jr., 1976, Diterpene acides as larval growth inhibitors, *Experientia* **32**(11)**:**1364–1365.
27. Escherich, K., 1942, *Die Forstinseketen Mitteleuropas,* Paul Parey, Berlin.
28. Finlayson, L. R., and Finlayson, T., 1957, Influence of adult food on viability of early stages of *Aptesis basizonia* GRAV. (Hym. Ichneum). A parasite of pine sawflies (*Diprion*), *Can Entomol.* **89:**507–509.

29. Finlayson, T., 1963, Taxonomy of cocoons and puparia and their contents of Canadian parasites of some native *Diprionidae* (*Hymenoptera*), *Can. Entomol.* **95**:475–507.
30. Fintelmann, L., 1839, Bieträge zur näheren Bestimmung und Naturgeschichte einiger auf der Kiefer (*Pinus sylvestris* L.) lebender Lophyren, *Verh. Kaiserl. Leopoldinisch—Carolinischen Akad. Naturforsch* **11.**
31. Fransen, J. J., 1937, De bestrigding van de dennebladwesp *Diprion pini* L. *Tijdschr. Ned. Heidenlij* **49**(11):395–410.
32. Fransen, J. J. 1938. De Plagen van de Dennenbladwesp, *Diprion* (*Lophyrus*) *pini* L. *Landbouwkd. Tijdschr.* **50**:609–610.
33. Geri, C., 1980, Application des méthodes d'études démécologiques aux insectes défoliateurs forestiers. Cas de *Diprion pini* L. (Hyménoptère, *Diprionidae*). Dynamique des populations de la Processionnaire du Pin *Thaumetopoea pityocampa* Schiff. (Lepidoptère *Thaumetopoeidae*) dans l'île de Corse, *Thèse d'état, Université de Paris Sud, Centre d'Orsay,* 260 p.
34. Geri, C., 1983. Répartition et évolution des populations de la Processionnaire du Pin (*Thaumetopoea pityocampa* Schiff.) dans les montagnes corses. I. Régimes d'apparition de l'insecte et dynamique des populations, *Acta Oecol. Oecol. Appl* **4**(3):247–268.
35. Geri, C., 1984, Répartition et évolution des populations de la Processionnaire du pin (*Thaumetopoea pityocampa* Schiff.) dans les montagnes corses. II. Relations entre les populations et le milieu forestier. *Acta Oecol. Oecol. Appl.* **5**(1):3–22.
36. Geri, C., and Dusaussoy, G., 1966, Etude d'une population de *Diprion pini* en forêt de Fontainebleau. II. Etablissement d'une table de mortalité. *Ann. Soc. Entomol. Fr.* (NS) **2**(3):535–548.
37. Geri, G., and Goussard, F., 1984, Evolution d'une nouvelle gradation du Lophyre du pin (*Diprion pini* L.) dans le sud du Bassin Parisien. I. Développement de la gradation et relation avec les facteurs du milieu, *Ann. Sci. For.***41**(4):375–404.
38. Geri, C., and Goussard, F., 1986, Observations sur l'incidence du mélange résineux feuillus au cours d'une gradation de lophyre du pin, *Diprion pini* (Hym. *Diprionidae*) dans le sud du Bassin Parisien, in *Conséquences de la monoculture des résineux et alternatives possibles* (M. Bonneau, ed.), C.N.R.S., Paris.
39. Geri, C., Goussard, F., and Liger, A., 1984, Influence de la photopériode sur les arrêts de développement de *Diprion pini* L. (*Hymenoptera, Diprionidae*), *C.R. Acad. Agric. Fr.* **70**(2):245–251.
40. Geri, C., Allais, J. P., Goussard, F., Liger, A., and Yart, A., 1985, Incidence de l'âge du feuillage sur le développement de *Diprion pini* L. (Hym. *Diprionidae*). Conséquences sur l'évolution des populations, *Acta Oecol. Oecol. Appl* **6**(4):349–365.
41. Geri, C., Goussard, F., and Reveillon, M., 1986, Réflexion sur l'utilisation des auxiliaires entomophages contre *Diprion pini* (Hym. *Diprionidae*), *Entomophaga* **31**(1):27–38.
42. Geri, C., and Millier, C., 1985, Mésure des populations de Processionnaire du Pin (*Thaumetopoea pityocampa* Schiff. *Lepidoptera Thaumetopeidae*) au Mont-Ventoux, *Ann. Sci. For.* **42**(2):143–184.
43. Gösswald, K., 1935, Physiologische untersuchungen über die Einwirkung ökologischer Faktoren, besonders Temperatur und Luftfeuchtigkeit, auf die Entwiklung von *Diprion* (*Lophyrus*) *pini* zur Festellung der Ursachen des Massenwechsels, *Z. Angew. Entomol.* **22**:331–384.
44. Green, T. U., 1938, A laboratory method for the propagation of *Microcryptus basizonius* GRAV, *Annu. Rep. Entomol. Soc. Ontario* **69**:32–34.
45. Griffiths, K. J., 1961, The life history of *Aptesis basizonia* GRAV. on *Neodiprion sertifer* GEOFFR. in Southern Ontario, *Can. Entomol.* **93**:1005–1010.
46. Grison, P., and Jacquiot, C., 1964, Observations sur une invasion de *Diprion pini* L. en forêt de Fontainebleau, *C.R. Acad. Agric.* **50**:998–1006.
47. Herting, G., 1960, Biologie der westpaläarktischen Raupenfliegen. Dipt. *Tachinidae, Monog. Angew. Entomol.* **16.**
48. Hertz, M., 1933, Tutkimuksia tavallisesta mânnynneulaspiskiäisestä (*Lophyrus* L.) ja sen metsätaloudellisesta merkityksestä, *Commun. Inst. For. Fenn.* **18.**
49. Ikeda, T., Matsumura, F., and Benjamin, D. M., 1977, Mechanism of feeding discrimination between matured and juvenile foliage by two species of pine sawflies, *J. Chem. Ecol.* **3**:677–694.
50. Jahn, E., Koller, F., and Schedl, W., 1980, Zum Auftreten von *Diprion pini* L. in einer Extremläge bei Schönwies, Tirol, 1975–1979 (Hym. *Diprionidae*), *Ber. Nat. Med. Ver. Innsbruck.* **67**:157–172.
51. Kangas, E., 1934, Huomioita mäntypistiäistuhoita. II. *Metsät. Aikak.* **51**(1):9–12.
52. Kangas, E., 1963, Über das schädliche Auftreten der Diprion-Arten (Hym. *Diprionidae* in finnischen Kiefernbeständen in diesem Jahrhundert, *Z. Angew. Entomol.* **51**:188–194.

53. Knerer, G., and Atwood, C. A., 1973, Diprionid sawflies: Polymorphism and speciation, *Science* **179:**1090–1099.
54. Kurir, A., 1977, Einspinnen und überwinterung der l-Generation der Gewöhnlichen Kiefernbuschhornblattwespe, *Diprion pini* L. (Hym., *Diprionidae*) in der Baumkrone der Weisskiefer (*Pinus sylvestris* L.), *Z. Angew. Entomol.* **84:**47–52.
55. Laurent-Hervouët, N., 1986, Mésure des pertes de croissance radiale sur quelques espèces de *Pinus* dues à deux défoliateurs forestiers. II. Cas du lophyre du pin dans le Bassin Parisien, *Ann. Sci. For.*
56. Lieutier, F., Geri, C., Goussard, F., and Rousseau, G., 1984, Problèmes entomologiques actuels du pin sylvestre en région Centre, *For. Priv.* **155:**25–36.
57. Maksimovic, M., 1954, Cycle of development of *Diprion pini* L. and its mass appearence on the Maljen mountain in 1951–1952. *Zast. Bitja* **(Beograd) 24:**20–40 (in Serbo-Croatian).
58. Morris, K. R. S., and Cameron, E., 1935, The biology of *Microplectron fuscipennis* Zett. (*Chalcid*), a parasite of the Pine sawfly (*Diprion sertifer Geoffr.), Bull. Entomol. Res.* **26:**407–418.
59. Morris, K. R. S., Cameron, E., and Jepson, W. F., 1937, The insect parasites of the spruce sawfly (*Diprion polytomum* HTG.) in Europe, *Bull. Entomol. Res.* **28:**341–393.
60. Niklas, O. F., 1956, Die Erzwespe *Tetracampe diprioni* FERRIERE als Eiparasit der Kiefernblattwespe *Neodiprion sertifer* GEOFFR, *Beitr. Entomol.* **6:**320–332.
61. Niemelä, P., Mannila, R., and Mantsälä, P., 1982, Deterrent to scots pine, *Pinus silvestris,* influencing feeding behavior of larvae of *Neodipiron sertifer* (*Hymenoptera, Diprionidae*), *Ann. Entomol. Fenn.* **48**(2):57–59.
62. Novak, K., 1957, Vliv prirozenych nepratel na prubeh premnozeni hrebenule borové (*Diprion pini* L.) na zapadnim Slovenska, *Acta Soc. Entomol. Cech.* **54**(4):356–362.
63. Obrtel, R., Zedja, J., and Holisova, V., 1978, Impact of small rodent predation on an overcrowded population of *Diprion pini* during winter, *Folia Zool.* **27**(2):97–110.
64. Ohigashi, H., Wagner, M. R., Matsumura, F., and Benjamin, D. M., 1981, Chemical basis of differential feeding behavior of the larch sawfly *Pristiphora erichsonii* (Hartig), *J. Chem. Ecol.* **7:**599–614.
65. Pilawa, J., and Piskor, M., 1962, Gradacja borecznika sosnowca na Gornym Slasku w latach 1958–1960, *Sylvan, Warszawa* **106**(6):57–76.
66. Price, P. W., 1970, Dispersal and establishment of *Pleolophus basizonus* (*Hymneoptera: Ichneumonidae*), *Can. Entomol.* **102:**1102–1111.
67. Pschorn-Walcher, H., 1964, Zur Kenntnis der gesellig lebenden Kiefernbushhornblattwespen (Hym. *Diprionidae*) der Ostalpen, *Pflanzenschutzberichte* **31:**49–66.
68. Pschorn-Walcher, H., 1967, Biology of the *Ichneumonid* parasites of *Neodiprion sertifer* (Geoff.) (Hym. *Diprionidae*), in Europe, *Technol. Bull. Comm. Inst. Biol.* **8:**7–51.
69. Rudnew, D. F., 1933, Materialen zur ökologie von *Lophyrus pini* L. in zusammenhang mit den Besonderheiten seiner Verbretung in den Waldbeständen, *Zb. Pratzsekt. Ekol. Nazemm.* **1:**135–144 (in Ukrainian with a German summary).
70. Ryvkin, B. V., 1963, Zur Kenntnis der Biologie von Kiefernblattwespen in Bjelorussland, *Beitr. Entomol.* **13:**455–464.
71. Ryvkin, B. V., 1969, Die *Diprionidae* und die Komplexe ihrer natürlichen Feinde, *Beitr. Entomol.* **19:**595–605.
72. Sachtleben, E., 1942, Beiträge zur Kenntnis der Diprion-Parasiten. I. Einleitung und Bemerkungen über einige Diprion-Tachinen, *Arb. Physiol. Angew. Entomol.* **9:**88–107.
73. Scheld, K. E., 1938, Zur blattwespen-Prognose, *Mitt. Forstwirtsch. Forstwiss.* **9:**192–241.
74. Scheld, K. E., 1939, Blattwespenkokongrosse und Fruchtbarkeit der shlüpfenden wespen, *Anz. Schädlingskd.* **3:**25–29.
75. Schimitschek, K., 1941, Die übervermehrung von *Diprion pini* L., im Westslowakishen Kieferngebiet, *Z. Pfanzenkr. Pflanzenschutz* **51:**257–278.
76. Schwenke, W., 1960, Uber die Wirkung der Waldbodendüngung auf die Massenwermehrung der Kiefernbuschhornblattwespe (*Diprion pini* L.) 1959 in Mittelfranken und die hieraus abletibaren gradologischen Folgerungen, *Z. Angew. Entomol.* **46:**371–378.
77. Schwenke, W., 1964, Grundzüge der Populationsdynamik und Bekämpfung der gemeinen Kiefernbuschhornblattwespe *Diprion pini* L., *Z. Angew. Entomol.* **54:**101–107.
78. Schwenke, W., and Steger, O., 1961, Uber auftreten und Bekämpfung der kiefernbuschhornblattwespen in Nordbayern 1959/60, *Allg. Forstz.* **16:**145–147.

79. Sharov, A. A., 1979, Effect of the spatial structure of interacting populations of the fox colored sawfly (*Neodiprion sertifer*) and its parasites *Exenterus abruptorius* on the dynamics of their numbers, *Zool. Zh.* **58**(3):356–365 (in Russian).
80. Sharov, A. A., and Safonkin, A. F., 1982, Spontaneous and inductive processus of the mecanism of the reactivation of the common pine sawfly: *Diprion pini. Zool. Zh.* **61**(8):1171–1179 (in Russian).
81. Thalenhorst, W., 1942, Der Zusammenbruch einer Massenvermehrung von *Diprion pini* L. und seine Ursachen, *Z. Angew. Entomol.* **29**:368–411.
82. Thalenhorst, W., 1963, Das Massenauftreten der Kiefernbuschhornblattwespe *Diprion* pini L. in Niedersachsen 1959–1961, *Allg. Forst. Jagdztg.* **134**:76–82.
83. Urban, S., 1962, Auftreten, Verlauf und Zuzamenbruch der "Übervermehrung der Kiefernbuschhornblattwespe *Diprion pini* L. in der D.D.R., *D. Soz. Forstwirstch.* **12**(1).
84. Urban, S., 1964, Erfahrungen mit Acrosprühmitteln bei der prophylaktiishën Bekämpfung der Kiefernbuschhornblattwespe *Diprion pini* L. in Jahre 1981, *Tagungber. Dtsch. Akad. Landwirthsch.* **62**:197–204.
85. Urban, S., 1965. Die Begrenzungsfaktoren der Ubervermehrung der kiefernbuschhornblattwespe *Diprion pini* L. in den Jahren 1960 un 1961, *Arch. Forstwes.* 14(11/12):1123–1134.
86. Vasic, K., and Sisojevic, P., 1955, Paraziti obicne borove zolje (*Diprion pini* L.) i njihova uloga u regulaciji projnosti ove stetocine na Maljenu 1951–1952 godine, *Zast. Bilja, (Beograd)* **27**:3–43.
87. Vasic, K., and Petrovic, M., 1957, Kvantitativna analiza uloge parasitskih insekata i entomolagnih gljivica u redukciji prenamnozenih populacija obicne borove zloje (*Diprion pini* L.) na Maljenu 1954 i 1955, *Zast. Bilja (Beograd)* **43**:3–28.
88. Vasic, K., and Zivokinovic, D. 1960, The common pine sawfly *Diprion pini* L. in Maljen in years 1954 and 1955. Life history, ecology and population dynamics, *Zast. Bilja* **57–58**:31–71 (in Serbo-Croatian).

CHAPTER 20

SIREX IN AUSTRALASIA

John L. Madden

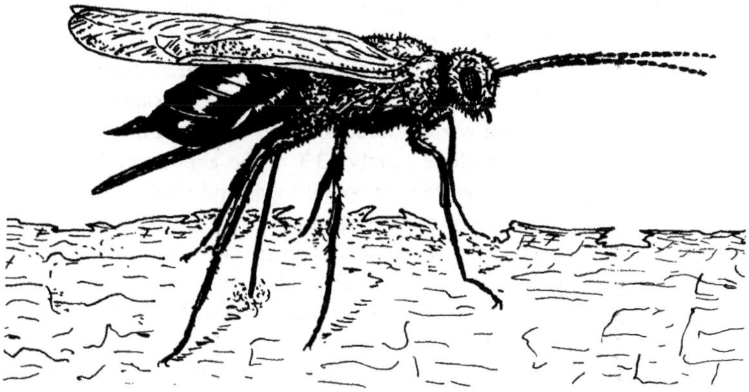

John L. Madden • Faculty of Agricultural Science, University of Tasmania, Hobart, Tasmania 7000, Australia.

1. INTRODUCTION

Woodwasps or horntails (Siricidae) are primitive phytophagous hymenopterans that naturally infest a variety of coniferous and hardwood trees throughout North America, Eurasia, North Africa, and Japan.[10,13,27,36,42,43,55,75,78] Although woodwasps have been reported to be responsible for economic degradation of otherwise marketable timber,[5,35,55,73,77,85] such infestations are more symptomatic of a prior pathological condition than a primary cause of tree mortality.[19] Therefore, in their natural habitat, woodwasps are secondary to other predisposing agents, which may include defoliation or debilitation of trees by insect and/or fungal attack,[6,19] fire or smog damage,[15,16] and mechanical injury.[46]

An exception to the secondary status of woodwasp attack was first recognized in New Zealand during the 1940s, when the woodwasp, *Sirex noctilio* Fabricius, was found attacking and killing apparently healthy Monterey pines, *Pinus radiata* D. Don.[66] In contrast to the northern hemisphere, where *S. noctilio* infestations were rare and localized, the primary nature and toxicity of the attack on *P. radiata* resulted in severe tree loss in New Zealand and subsequently in Australia.[33,70]

Both Australia and New Zealand lack extensive areas of fast-growing indigenous softwoods, but the introduction of *P. radiata* more than compensates for this deficiency with exceptional rapid growth and wood production over a wide range of climates and site conditions. Australian plantations alone supply 33% of the annual production of saw and veneer logs, pulpwood, preservative-treated round logs, and sleepers.[2]

Historically, *S. noctilio* became established in New Zealand around 1900,[56] but it was not until some 45 years later that severe and extensive losses occurred. During 1946–1954, approximately 30% of 120,000 ha of *P. radiata* was destroyed.[68] In 1952, *S. noctilio* was discovered in a 1100-ha *P. radiata* plantation at Pittwater, Tasmania[33] and by 1958 had destroyed 40% (range 30–80%) of trees aged 23–28 years.[48] In 1961, the wasp was discovered in the mainland state of Victoria and has since spread throughout that state and to forests in the adjacent states of South Australia and New South Wales, and the Australian Capital Territory.[4]

The capability of *S. noctilio* to kill individual *P. radiata* posed a potential threat to a rapidly expanding postwar timber industry. Initial control attempts in New Zealand and Tasmania involved the introduction and release of parasitoids collected in Europe by the Commonwealth Institute of Biological Control (CIBC), but a concerted program of research and control in Australia had to await the discovery of the wasp in mainland plantations in 1961–1962.[59]

In 1962, the National *Sirex* Trust Fund and National Sirex Fund Committee were established to finance immediate containment and eradication procedures in Victoria, and research into all facets of the *Sirex* problem. All field research was initially undertaken in Tasmania and included investigations on (1) biological control and the biology, behavior, and ecology of *S. noctilio* and its parasitoids by the Division of Entomology, Commonwealth Scientific and Industrial Research Organisation (CSIRO); (2) insect–host tree interactions, including susceptibility and resistance to attack, tree breeding, and silvicultural practice by the Forestry and Timber Bureau; and (3) insect–fungus relationships and evaluation of chemical control by the Department of Entomology, Waite Agricultural Research Institute, University of Adelaide. Subsequent research was conducted at the CSIRO *Sirex* Biological Control Unit established at Silwood Park, England, and from the mid-1970s by state forestry departments, notably the Victorian Forests Commission.[59]

2. LIFE HISTORY AND ECOLOGICAL RELATIONSHIPS

Woodwasp larvae cannot feed, grow, and develop without prior alteration and conversion of host–tree tissue into readily assimilable forms. This conversion is performed by *Amylostereum* and *Stereum* spp., basidiomycetous fungi that are mutualistically associated with, and vectored by, siricid females.[14,17,20,78] Female siricid larvae acquire and store the fungus as arthrospores within abdominal hypopleural organs,[65] which are recontaminated after each moult; ultimately, in the adult stage, arthrospores are lodged and nurtured within paired intersegmental, mycangial sacs at the base of the ovipositor.[30] Inoculation of new hosts with arthrospores is associated with oviposition. The relationship between the fungi and woodwasps is species-specific, with the exception of *Xeris spectrum*, which carries no fungus but deposits eggs into other siricid-infested wood.[31] The fungus associated with *Sirex noctilio* is *Amylostereum areolatum* Fries (Boidin).[32]

The general life cycle of *Sirex* in Australia takes 12 months but may range from 3 to 36 months, depending on tree size (volume) and the degree of insolation and/or shading.[58] An adult sexual dimorphism in color exists, with females being steel-blue and males possessing orange abdomens.[10] Adult emergence commences early in summer (early December) and continues to April, with peak emergence in February.[60] On emergence both sexes fly to tree tops, where they aggregate in predominantly male swarms. Mating occurs on the upper branches and the original photopositive response of the female is replaced by a host-location response. Mated females leave the swarm to seek trees as suitable oviposition sites. In the absence of mating the host-location response becomes increasingly dominant and unfertilized eggs are deposited. Such haploid eggs develop into males. Flight activity and mating frequency is favoured by temperatures greater than 20°C.

Sirex females are attracted to physiologically stressed trees. High girdling of pruned trees results in the region below the girdle becoming attractive some 10 days after the operation. At this time, the permeability of the bark tissue of the attractive region to water vapour and monoterpenes is twice that of the upper, nonattractive region.[18,50] The major monoterpenes, α- and β-pinene, elicit the greatest electroantennogram response[74] and arrestment in flight mill olfactometers, and trees baited with either solvent extracts of bark or the two monoterpenes alone, or in combination, have been preferentially attacked. That attraction actually occurs was evidenced by the flight to and searching behaviour of female wasps on fly wire screen surrounding but 6 cm from attractive stem sections.

Following the location of host trees, *Sirex* females walk over the bark surface and assess the tree as a potential oviposition substrate by probing the phloem with the ovipositor. Acceptance or rejection can occur after one or two probes. Eggs are deposited within drill shafts bored into the outer sapwood by the abrasive action of the ovipositor and removal of wood particles by a pair of reciprocating valvifers[5] (Fig. 1a).

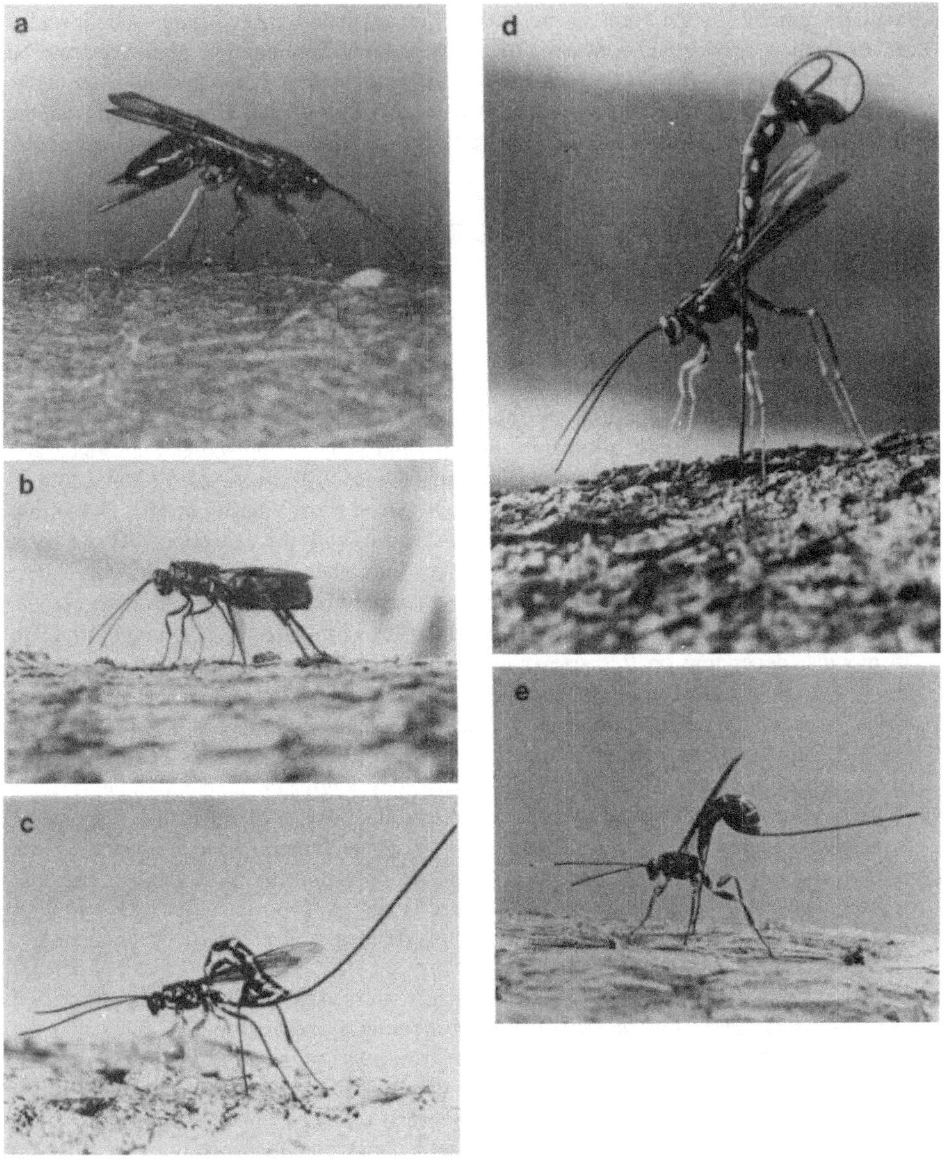

FIGURE 1. *Sirex* and parasitoids. (a) *Sirex noctilio*. (b) *Ibalia leucospoides*. (c) *Rhyssa persuasoria*. (d) *Megarhyssa nortoni nortoni*. (e) *Certonotus tasmaniensis*.

Initial acceptance does not necessarily result in immediate oviposition, as eggs are placed only in situations that maximize the probability of survival.[47] In the early stages of an attack, a high proportion of single drills are made, but the incidence of eggs in these drills is low. However, *A. areolatum* arthrospores and secretions derived from a large mucous gland reservoir are inoculated into the lumen of the drill holes.[24] When two drills are made, eggs only occur in the first drill and arthrospores and mucus are secreted into the second. As many as six drills may be made in the sapwood through an outer single drill and, irrespective of number, the last drill is always egg free but contains mucus and arthrospores. The capacity to vary the number of drills and hence eggs, is related to the vitality of the tree as expressed by the osmotic status of the phloem tissues.[47] Single drills are made when the tree is suboptimal, and increasing numbers of eggs are deposited as the tree's vitality and capacity to resist attack decline in time following successful attack. Aspects of tree susceptibility and resistance to attack are considered below.

Eggs usually hatch within 10–15 days, but viable eggs and first-instar larvae have been recovered from the base of large trees 10 months after attack. Such delayed hatching is attributed to the existence of low temperatures beneath the canopy for most of the year, inhibiting fungal growth as well.[47] On hatching, first-instar larvae feed on yolk fragments and fungal mycelia occupying the lumen of the drill hole. Second instars and subsequent stages feed through fungus-infested wood tunneling with the grain to create galleries of increasing diameter and whose general characteristics vary according to conditions of moisture content, aeration, and fungal activity. The number of instars varies from 6 to 12, but a minimum of 3 larval moults has been reported to result with very small adults.[63]

Greater numbers of instars are associated with larvae that experience suboptimal conditions during their development. Such factors as high moisture and consequently poor fungal growth or physical barriers such as resin patches result in more frequent moults and changes in gallery direction. Final adult size is directly related to the volume of fungus-infested wood displaced during development. Pupation occurs in the spring, when air temperatures increase and the moisture content of the larval habitat limits fungal growth.[47]

2.1. Tree Mortality and Resistance Mechanisms

Sirex has been reared successfully from *Pinus, Abies, Picea, Larix, Pseudotsuga,* and *Auricaria* spp., and treatment of *Pinus sylvestris* and *Picea sitchensis* trees with *Sirex* arthrospores and mucus, or oviposition, has been reported to cause death.[76] However, the interaction of the *Sirex* and Radiata pine is the only one in which the primary role of the insect in causing tree death has been substantiated.

The development of short-term periods of physiological stress within individual host trees appears sufficient to result in attraction and attack by *Sirex*. Stressors may involve damage by wind or felling operations, defoliation, herbicides, fire, cone harvesting, or thinning, particularly when it occurs during the *Sirex* flight season.[46,48,50,52,54]

Irrespective of cause, the injection of mucus into the sapstream, followed by its translocation to the foliage, results in the inhibition of translocation and a reinforcement and extension of the original stress condition. The chemical nature of the mucus is unknown, but its stability at high temperatures suggests that it has a low molecular weight, while its mode of action is to induce premature senescence.[29] The effect of the mucus is to alter the water balance of needles drastically through tissue desiccation and collapse of the phloem elements.

Consequently, the duration of a temporary deviation from normal physiological conditions is extended to increase the probability of additional attack and inoculation of the tree with arthrospores and mucus. At the same time, the rupture of xylem elements by ovipositor probes results in the invasion of air into the tracheids creating a suitable milieu for germination and growth of the fungus. Progressive coalescence of drying patches originating from individual points of inoculation results in the invasion of the fungus throughout the xylem and heartwood.[21,22]

Although the fungus was originally considered the sole pathogenic entity, it is now thought that tree death results from the combined effects of mucus and fungus. The mucus reinforces the original stress condition by impairing translocation, which in turn favors vigorous establishment of mycelia, which then occlude the xylem of the tree. Acting together, the insect and fungus create a suitable habitat for continued fungal growth and larval feeding. Trees successfully attacked by *Sirex* are characterized by a generalized chlorosis, or yellowing, of the crown, followed by needle fall and death that may occur 2 weeks to 8 months after attack. Such variation may be associated with locality and site characteristics and their influence on tree resistance mechanisms.

Resistance to *Sirex* wasp attack can be summarized under three traditional categories: nonpreference, antibiosis, and tolerance.[64] Nonpreference is observed when an initially attractive tree is rejected by the attracted wasp. This is commonly observed when a tree is damaged by localized wounding such as the removal of a phloem sample. Individual wasps may be attracted to these sites but, following a number of probes of the phloem with its ovipositor, often reject the site and leave the tree. Similarly, apparently normal trees attacked by three to four females on 1 day may be followed by a second attack some 10–12 days later or not be reattacked at all.[50]

Antibiosis in response to *Sirex* attack involves resinosis and polyphenol formation[23] (Fig. 2). Resinosis, a phenomenon similar to the hypersensitive reaction[71] and dynamic wound response[11] of bark beetle attack, is a nonspecific wound reaction that is dependent on water status and maximum cell turgor for complete expression. This expression includes both flooding of the drill and impregnation of surrounding wood with resin. Polyphenol formation involves the transformation of current photosynthate into two fungistatic polyphenols, pinosylvin and pinosylvin monomethyl ether.[37] These two polyphenols occur, along with three other nonfungistatic polyphenols, in the heartwood of radiata pine. Polyphenol production does not occur in response to a physical wound and is therefore a specific response to the invasive agent, *Amylostereum areolatum*. Furthermore, the dependence on current photosynthate as opposed to stored starch as the carbon source for these polyphenols emphasizes the importance of the effect of mucus in the pathology syndrome, as it directly inhibits translocation of photosynthate.

Tolerance to sustained *Sirex* attacks has not been investigated but undoubtedly entails the suppression of the mucus effect. This can be achieved in either of two ways. First, transpiration may be so reduced that the rate of mucus transport is insufficient to elicit a toxic effect in the needles. Such a condition could apply during drought periods when transpiration may fall to less than 10% of the predrought value.[72] Second, premature senescence and shedding of affected needles has been interpreted as a hypersensitive reaction resulting in the elimination of toxic mucus from the tree.[29] Extreme variability in response, which ranged from rapid needle chlorosis and wilt to no apparent effect, was observed when either detached shoots were placed in mucus solutions or mucus was injected into seedling trees.[22] Breeding for such resistance traits was not successful, even though needles from different trees retained their order of sensitivity to mucus when tested throughout the year.[12] Under field conditions, rapid development of chlorosis and wilt of the youngest needles is almost always associated with susceptibility to attack, whereas

FIGURE 2. Transverse section of Radiata pine tree (right) partially killed by *Sirex* attack 6 years before felling. Note resinosis and polyphenols associated with drilling and concentrated in killed sector, which contains emergence holes (right).

chlorosis and needle fall of more mature foliage and stem needles, but not meristematic needles, occurred on those trees that most often tolerated and resisted attack. Elongation of the leading shoot of Radiata pine is positively correlated with diurnal changes in diameter, and shrinkage of the leader shoot does occur in actively growing trees under high transpirational load.[44] By contrast, diurnal variation in transpiration declines from 600 ml/m^2 day to 200 ml/m^2 night predrought to a constant 50 ml/m^2 34 days after final watering.[72] Therefore, under drought conditions, it can be inferred that the supply of mucus to the needles is reduced and, of equal importance, that the target cells are metabolically dormant at this time. The response of droughted, watered, and untreated control trees to forced *Sirex* attack emphasized the importance of transpiration rate. Watered trees succumbed most rapidly and droughted trees took 2–3 months to die, while control trees were intermediate in their response.[12]

2.2. Parasitoids, Predators, and Pathogens

Prior to the establishment of the National *Sirex* Fund Committee program, attempts at biological control of *Sirex* were limited to the introduction of *Ibalia leucospoides* (Hochenwarth) and *Rhyssa persuasoria* Linnaeus into New Zealand and Tasmania.[80] The former parasitoid attacks *Sirex* eggs and early-stage larvae and the latter mature larvae in the spring. Both parasitoids are commonly associated with siricids in the Northern Hemisphere (Fig. 1b,c).

Concerted exploration in Europe by CSIRO officers from 1963–1971, supplemented by collections in North America, Japan, and Nepal by both CIBC and CSIRO, resulted in the introduction of 21 species and subspecies of parasitoids into Australia. Following laboratory rearing, testing, and quarantine, 10 species were released, 5 of which became established.[81]

In Tasmania, the most effective parasitoid was *Megarhyssa nortoni nortoni* Cresson (ex. California), which, with *R. persuasoria,* caused a significant reduction in a *Sirex* population at Pittwater between 1965–1974 (Fig. 1d). Key factor analysis indicated that these two parasitoids, combined as a guild, acted in a delayed density-dependent manner, whereas *I. leucospoides* apparently acted in a density-independent way.[82,83] By contrast, *Ibalia* spp., *I. leucospoides,* and *I. ensiger* Norton were the most effective parasitoids in Victoria, causing an average mortality of >20%. *Megarhyssa nortoni* (strains *nortoni* and *quebecensis* Provancher) averaged less than 12% mortality, and *R. persuasoria* has failed to establish in Victoria. Differences in the efficiencies of *Ibalia* spp. between Pittwater and Victoria appear to be related to differences in the rate of mortality of trees in different climatic zones. At Dulverton, northern Tasmania, *Ibalia* spp. caused 40% mortality during 1979–1980. Resinosis, a density-independent mortality factor, is the most likely factor obscuring a functional relationship between *Ibalia* and *Sirex* at Pittwater.

Indigenous parasitoids, particularly *Certonotus tasmaniensis* Turner in Tasmania[38] and *Guiglia schauinslandi* Ashmead in New Zealand,[68] have successfully parasitized *Sirex* larvae.

In addition to parasitoids, avian predators have been observed attacking *Sirex*.[51] Sixteen of a total of 42 species of birds recorded at Pittwater were implicated in the predation of both male and female woodwasps. Three species, the Dusky Wood Swallow,

Spine-tailed Swift, and Raven, predominantly attacked the mating swarms, which not only affected a numerical reduction but also reduced mating frequency as many single-attacked trees generated exclusively male or significantly high male to female ratios.

Both parasitoids and avian predators are responsible for the significant destruction of immature and adult stages of *Sirex*. Another less apparent but extremely efficient natural enemy was discovered in New Zealand in 1962.[86] Dissection of some *Sirex* adults demonstrated hypotrophy of the ovaries and hypertrophy of the testes and, within the ovaries, apparently normal eggs were found to be full of juvenile nematodes. The nematode was subsequently described as *Deladenus siricidicola* Bedding (Neotylenchidae),[8] which, together with a second species, *D. wilsoni* Bedding, has a unique life cycle involving extreme female dimorphism.[7] The neotylenchid form is free living and mycetophagous, feeding exclusively on *Amylostereum areolatum*, while the other possesses many diagnostic features of the Allantonematidae and parasitizes *Sirex* larvae. Hormones associated with insect pupation apparently trigger the rapid development of the reproductive system of the nematode, the ovoviviparous release of nemas into the hemocoel of its host, and their migration to the reproductive organs. The female *Sirex* is sterilized by the infection but is still able to locate host trees and oviposit normally, depositing arthrospores and mucus, but nematode-filled egg cases. The nematodes escape from the egg to feed on the fungus for many generations within the tree until, in the presence of an active siricid larvae, adult infectives are formed. These mate and then penetrate the larvae to reinitiate the cycle. Many parasitic strains of *Deladenus* spp. have been introduced into Australia, but only *D. siricidicola* has been released, for it only attacks *Sirex* larvae. *D. wilsoni* attacks both *Sirex* and the rhyssine parasitoids.

The remarkable life cycle of *D. siricidicola* has unique features that facilitate its use as a biological control agent. The mycetophagous form can be produced in the laboratory on fungus cultures and inoculated into *Sirex* infested trees in the field.[9] Artificial inoculation of felled infested trees results in virtually 100% parasitism, and flight and oviposition by infected females rapidly disperses the nematode throughout the forest.

3. PATTERNS OF NUMERICAL BEHAVIOR

The *Sirex* wasp in Australasia is associated predominantly with Radiata pine, which is grown on sites of variable quality that frequently experience hot, dry summers. Despite such conditions, woodwasp populations usually remain relatively low, with an annual turnover of 5–10 trees/ha.

Sirex outbreaks are characterized by an eruptive increase in tree mortality, followed by a rapid decline to endemic levels.[67] The duration of outbreaks ranges from 4 to 10 years, in the absence of parasitic nematodes.

3.1. New Zealand Outbreaks

Despite the woodwasp's presence in this country for >70 years, significant tree losses, due directly to *Sirex* attack, did not occur until the mid-1940s. Outbreaks in forests within the Rotorua region were attributed to overstocking and drought. Summer rainfall from December 1945 through February 1947 was about 50% of the long-term average, with 1945–1946 being the driest year on record at Kaingaroa.[66] Paradoxically, greater tree mortality occurred in thinned stands adjacent to overstocked and, presumably, more drought-affected trees. This preferential attack in thinned compartments was attributed, at the time, to "the 'effect' of solar radiation on the bark (phloem) of trees with thin crowns."[70] In areas of 9-year-old regeneration, it was observed that trees pruned and thinned during January–March 1946 were extremely vulnerable. This phenomenon was witnessed again in 1947–1948, and it was postulated that fresh resin elicited attraction and mass *Sirex* attack. High mortality in 20-year-old trees was also associated with pruning operations during *Sirex* flight, and a decreasing tendency for the wasp to attack trees at greater spacing intervals was observed.[66]

A severe drought occurred in 1945–1946, and *Sirex* populations increased through 1946–1947. Higher rainfall (90 mm/month) was experienced in the summer of 1947–1948, however, and *Sirex* activity declined. Rains were moderate (75 mm/month) in the summer of 1948–1949, with generally overcast and cool conditions prevailing. Unexpectedly a massive increase in *Sirex* activity occurred in this year and 25–30% mortality of trees was estimated over 300,000 acres of forest.[70] Surveys at Rotoehu Forest showed wide variation in tree susceptibility from year to year, as expressed by differences in the numbers of trees initially thought to be dying and those actually killed (Fig. 3a).[40] In subsequent years, *Sirex* numbers stabilized at subeconomic levels.

In one decade, official forest policy changed from terminating the Radiata pine planting program in favor of species not attacked by *Sirex* to its reintroduction. This about-face was attributed to the recognition of certain facts about the consequences of thinning and pruning operations during the growing season and the importance of appropriate stocking intensity with tree age (size). Furthermore, alternative softwood species failed to compare with the superior qualities of Radiata pine. Indeed, *Sirex* wasp was referred to as a beneficial insect through its thinning effects.[40,70] However, the unpredictability of its outbreaks in space and time serve more as a reminder of poor or delayed management practices.

3.2. Australian Outbreaks

3.2.1. Tasmania

Sirex was first discovered in the Pittwater plantation in 1952. However, evidence of its presence since 1947–1948 was found when the course of the outbreak was reconstructed by determining the frequency and age of *Sirex*-killed trees.[48] This analysis suggested that the outbreak developed from random foci within the compartment and that maximum tree mortality occurred some 10 years after the year of introduction. Resinosis and polyphenols within attacked but surviving trees indicated that a marked increase in resistance to attack commenced 8 years after *Sirex* introduction (Fig. 3b). Insect survival within trees declined through the outbreak reflecting the greater vigor and resistance of surviving trees. Numbers of oviposition drills per unit area of bark and the proportion of multiple drills also increased during summers experiencing above-average rainfall.

The Pittwater plantation was situated on sand dunes with a water table in the study

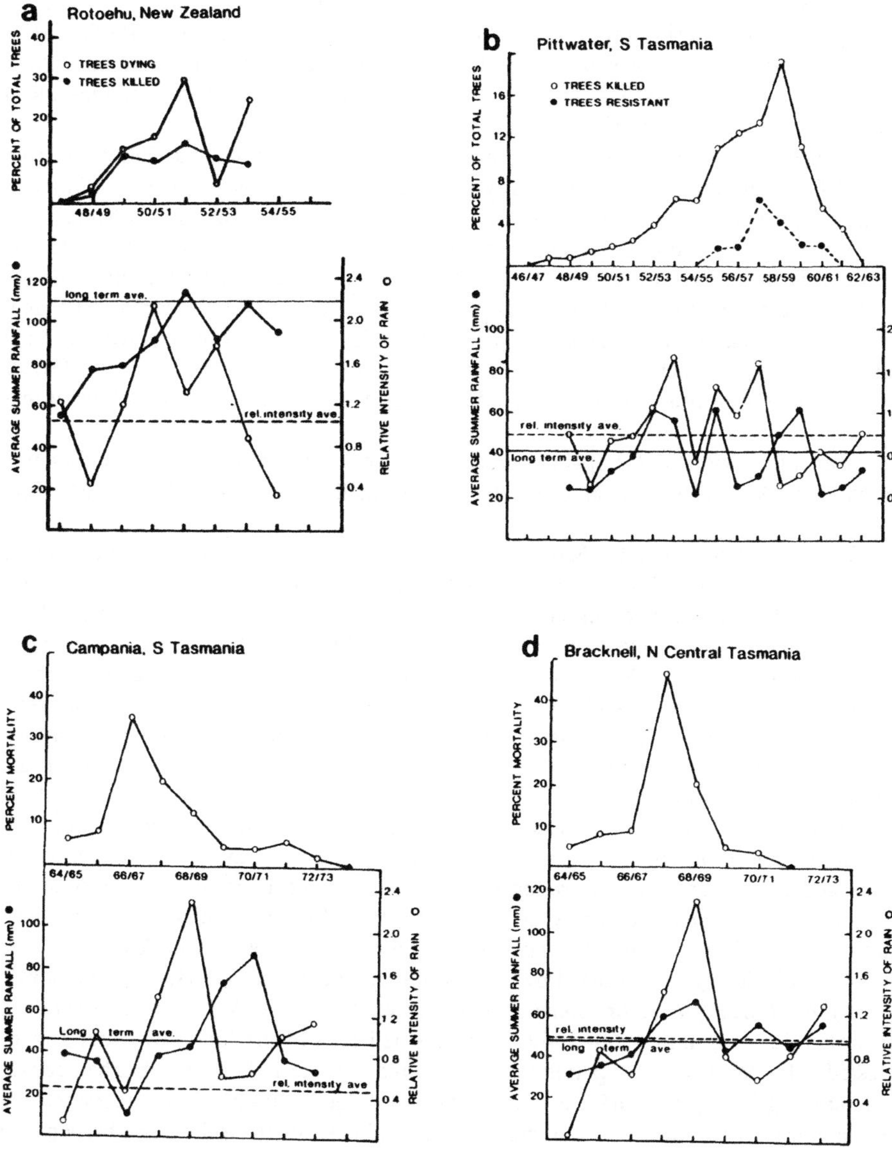

FIGURE 3. Relationship between the percentage of *P. radiata* trees attacked by *Sirex* and average summer rainfall (December–March) and the intensity of February rainfall relative to the summer average at locations in New Zealand (a) and Tasmania (b–d).

area at 2.4 m. Tree mortality was 68% at this site, with mortalities ranging from 30 to 80% in areas with deeper and shallower water table levels, respectively.

In common with the New Zealand experience, Pittwater was also disturbed by forestry operations during the *Sirex* flight season. Commencing in 1948–1949, and in response to postwar shortages of softwoods, the largest trees at Pittwater were selectively harvested. This practice of high grading caused damage to understory trees and produced large quantities of slash, in which the *Sirex* population increased.

A separate investigation of *Sirex* in Radiata pine regeneration at Pittwater from 1965–1974[82] showed that peak mortality occurred in 1968–1969 and that tree kill was influenced significantly by severe defoliation by the larvae of a geometrid, *Chlenias* sp.[52]

A small plantation at Campania, located 25 km north of Pittwater, was infested by *Sirex* during the early 1960s but in 1966–1967 a sixfold increase occurred, followed by a rapid decline to preoutbreak levels[83] (Fig. 3c). The only significant environmental features at this plantation were an elevated site with shallow soils overlaying parent rock and excessive hot air temperatures (approximately 40°C for ~6 hr) on February 7, 1967, a date on which disastrous fires destroyed large areas of southern Tasmania with great loss of life.

Coincident with the Campania outbreak, a fivefold increase in tree mortality was observed in 1967–1968 in a small plantation Bracknell, north-central Tasmania[83] (Fig. 4d).

Three plantations with the Scottsdale district of northeastern Tasmania were attacked by *Sirex* during the late 1960s and early 1970s. All plantations consisted of multiple age plantings and were in need of thinning. The Mt. Helen outbreak was monitored from 1968 to 1975. Growth-ring analysis on dead trees indicated that small numbers of trees were killed as early as 1965. Peak tree mortality occurred in 1971–1972, and the greatest frequency of resistance was observed during the preceding season. Increases in *Sirex* populations at Mt. Helen and to a lesser extent at the Cuckoo plantation, some 10 km east, was influenced by forestry operations and wind (Fig. 4a). Thinning during the *Sirex* flight season occurred at both plantations, and the impact of localized damage on tree attractiveness and susceptibility was evidenced in the death of 14 elite trees chosen for their growth rate and form and damaged during the flight season by the selective removal of cones and cone-bearing branches for seed. Levels of tree mortality were less at the Cuckoo plantation and it was recorded that Mt. Helen had approximately 5% of stems with evidence of crown damage or breakage and limb run compared with less than 1% at Cuckoo. This damage was caused by wind in the more exposed Mt. Helen plantation.[50]

At the Retreat plantation, some 14 km southwest of Mt. Helen, no real evidence of *Sirex* was detected until 1968–1969, when a small area of the forest was found to be infested. This locality had experienced severe wind damage in February that had caused stem breakage, lodging, and excessive crown damage. During this particular storm, dominant trees were observed to rotate violently and damage the foliage and branches of adjacent codominant and subdominant trees.[50]

In north and northwestern Tasmania, *Sirex* populations remained at low densities in a number of government- and company-owned plantations during the 1970s, but significant tree mortality occurred on two major plantations in 1980–1981 (Fig. 4b). Branches Creek plantation, which consisted of unthinned >10-year-old trees, experienced 30% mortality within two compartments from 1980 to 1984. In the other plantation, at Dulverton,

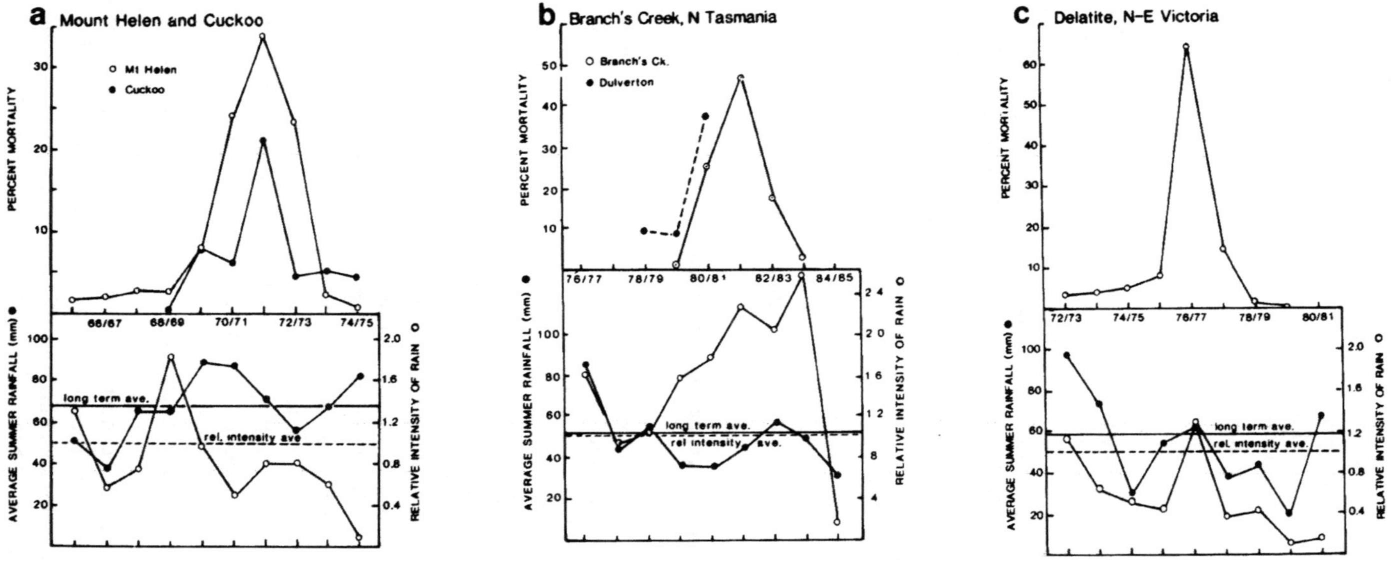

FIGURE 4. Relationship between the percentage of *P. radiata* mortality due to *Sirex* attack and average summer rainfall (December–March) and the intensity of February rainfall relative to the summer average at locations in Tasmania (a–b) and Victoria (c).

mortality increased gradually from 10% in 1977–1978 to 30% in 1980–1981. These increases in *Sirex* activity were attributed to drought, but the Dulverton outbreak was aggravated by thinning operations during the flight season of 1980–1981. One feature of the infestation at Branches Creek was that trees on poor sites with chlorotic needles were not attacked. Suppressed, subdominant, and then codominant trees growing on moister sites were particularly susceptible.[3]

3.2.2. Victoria

Sirex was discovered on mainland Australia in 1961 in the state of Victoria; until the early 1970s, control attempts consisted of the detection and eradication of infested trees and the release of parasitoids. In 1972, *Sirex* was discovered at Delatite plantation and in the subsequent 7 years destroyed approximately 12% of the marketable volume, notably in 12–15-year-old unthinned stands. Mortality ranged from 63 to 77% in the most severely affected stands. The remarkable feature of this outbreak was that 41–54 of the total mortality occurred during the 1976–1977 season alone, and tree losses were significantly greater in unthinned stands.[45] In addition, this outbreak occurred despite the presence of both parasitoids and nematodes in the plantation.

A sequence of progressively lower summer rainfalls occurred from 1972 to the end of the decade. Despite this and the large number of *Sirex* emerging from trees killed in 1976–1977, overall mortality declined markedly after 1978 (Fig. 4c).

4. HYPOTHESES FOR CAUSE OF OBSERVED POPULATION BEHAVIOR

Factors that seem to have contributed to the observed changes in *Sirex* populations, as reflected by annual tree mortality in the areas described, have been drought, thinning, and pruning during the *Sirex* flight season, and damage by any one of a number of agencies, including wind, fire, excessive dry heat, and mechanical wounds associated with selective harvest.

One hypothesis suggests that physiological stress due to any cause promotes damage responses characterized by increased transpiration and stem respiration, suppressed photosynthesis, and translocation. Such changes affect overall water relationships and lead to increased permeability of phloem to water and essential oils.[50] Experimentally girdling, which physically blocks translocation, leads to enhanced phloem respiration for 10 days, at which time phloem shrinkage occurs through direct loss of water. *Sirex* attacks occur coincidentally with these changes. A similar phenomenon is observed when trees are conditioned by the drilling activity of caged wasps. In this case, the mucus inhibits translocation and causes enhanced phloem respiration and stem shrinkage at 10 days and coincident attack by wild females.

Reduction of photosynthesis and labile and stored carbohydrates favours multiple drilling and increases the number of eggs laid and their survival because resinosis and polyphenol formation are dependent on cell turgor and terpenoids on the supply of labile carbohydrates, respectively. The rupture of xylem vessels experiencing high water tensions by drilling with subsequent invasion of air favours optimal conditions for growth of *Amylostereum areolatum*. The immediate and delayed effect of mucus on tree physiology and water status plus the progressive occlusion of the xylem vessels by the fungus results in tree death.[21,22,50,53]

All experimental treatments that result in predictable *Sirex* attack on individual trees involve a rapid change in the physiological status of the host tree.[46] This situation contrasts with the effects of drought, commonly regarded as a major cause of outbreaks.[3,45,66]

Growth in Radiata pine is closely correlated with rainfall and limited by extremes of temperature and evaporation during summer. Consequently growth may cease prior to the *Sirex* flight season (Fig. 5a) or may be progressively reduced during summer (Fig. 5b), primarily according to variations in water availability. In both instances, photosynthesis and transpiration decline, root growth ceases, and levels of starch and monosaccharides fall.[72]

Most Radiata pine plantations in Australia experience drought or shortages of water to varying degrees every summer. Such shortages progressively develop from late spring to early summer, with cessation of growth occurring before *Sirex* emergence occurs. Radiata pine trees adapt in time to tolerate periods of extreme water shortage and this tolerance to drought is expressed in zero growth and minimal levels of respiration and transpiration.[72] This quiescent stage may be sustained in seedling trees for 30 weeks, during which time, in the absence of watering, transpiration declines from 400 g to about 4 g/day.[25]

In contrast to the gradual adaptation to water shortage, Radiata pine resumes a high level of efficiency in most physiological functions within 24–48 hr of rewatering.[25,72] Absorption of rainwater by the foliage of water-stressed Radiata pine trees has also been demonstrated.[41]

In the light of the capacity of the host tree to adapt to, and tolerate, extended periods of drought, it is hypothesized that other influences occur during the flight season, which act catastrophically to alter host tree physiology rapidly. Such influences could be provided by unseasonal above-average rainfall, which would break the tree's quiescence or tolerance to drought. Subsequent to this, and in the absence of further rain, a rapid transition from a photosynthetically active to an acutely water deficient tree would occur. Cessation of growth is accompanied by active shrinkage of phloem tissue, which signifies an increase in tissue permeability. Thus, during the shrinkage phase water vapor and monoterpenes are essentially squeezed out of the tree, thereby increasing the probability of attack, which in turn depends on the degree of physiological impairment of individual trees. The role of the *Sirex* mucus in the initial stage of attack is to sustain the impairment and thereby increase the duration of attractiveness and attack and probably susceptibility.

Such a hypothesis would explain the results of droughting and watering experiments in which watered trees were more susceptible to *Sirex* attack than droughted trees. Such a result is dependent on rates of transpiration which, following *Sirex* attack, determine the rate at which mucus enters the foliage. In truly droughted trees, transpiration is minimal, and noneffective dosages would occur.

Although thinning, pruning, and felling during the *Sirex* flight season undoubtedly aggravated tree susceptibility, outbreak periods often coincide with periods of below-average but increasing summer rainfalls; greater tree mortality often occurs when dry summers were interrupted by intense, above-average rainfall of short duration. These two associations are shown in Figs. 3a–d and 4a–c: (1) average summer rainfall (mm) for the period December through March, and (2) the intensity of either February or March

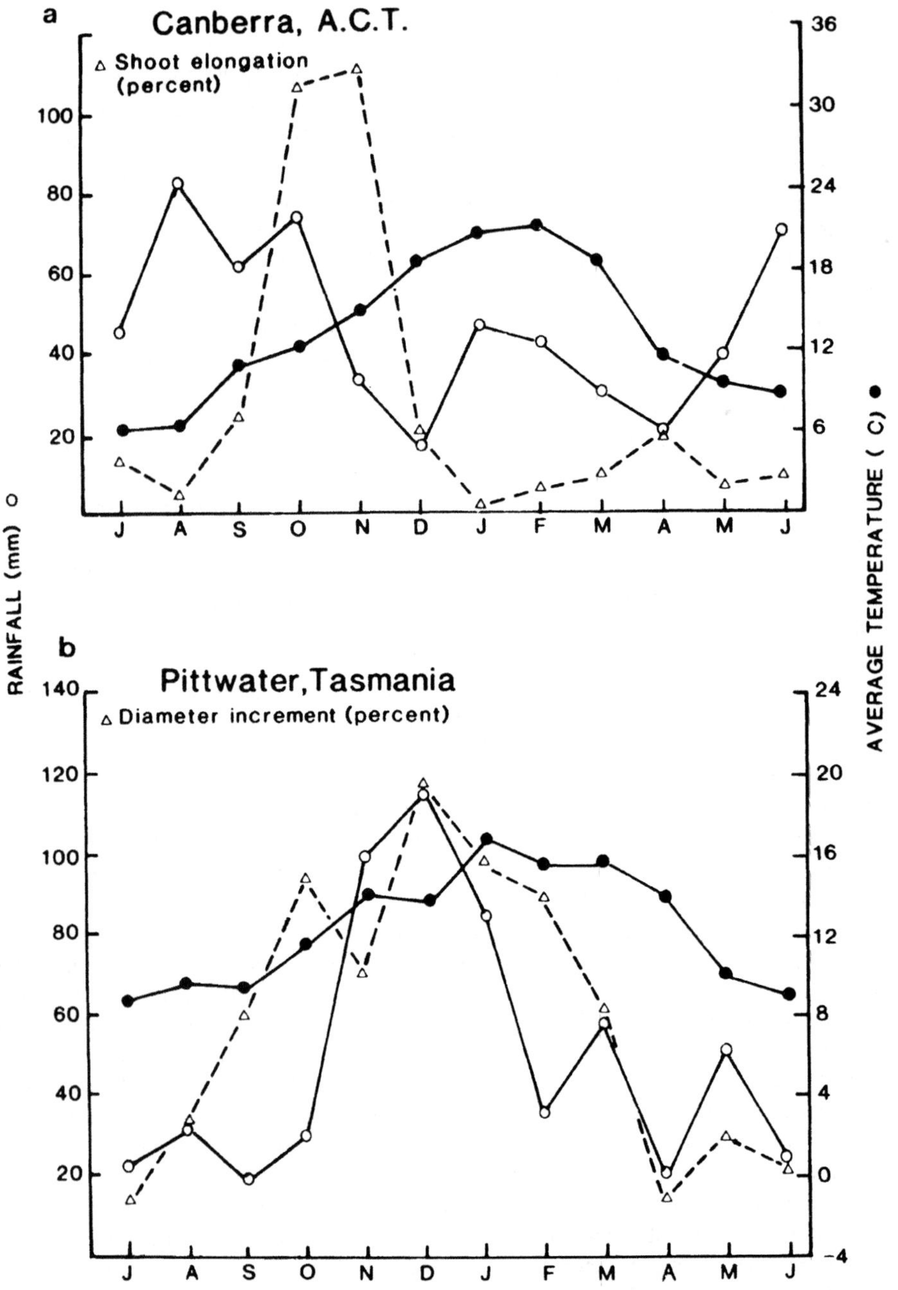

FIGURE 5. Growth of *Radiata* pine at (a) Canberra (A.C.T.) and (b) Pittwater (Tasmania) relative to rainfall and temperature. Growth monitored at the two sites by shoot elongation and diameter increment, respectively.

precipitation relative to the summer monthly average. The February–March period coincides with the peak emergence of *Sirex* in southeastern Australia, Tasmania, and New Zealand. Earlier rains in December or January may act to extend the growing period of Radiata pine, in contrast with those of February and March, which may break drought tolerance and reinitiate growth.

The series of graphs indicate the following data:

1. After a period of drought years at Rotoehu, New Zealand, a greater number of trees were attacked during wetter summers and subsequent tree mortality was greatest in 1950–1951 and 1952–1953, years with drier summers interrupted by above average rains in February. In 1951–1952, a season of more uniform and average precipitation, comparatively fewer trees were killed relative to the high proportion of trees initially observed to be dying (Fig. 3a).
2. The Pittwater outbreak coincided with a period of fluctuating summer rainfall but relatively intense February rains. Reduced tree mortality coincided with a year of below average rains (1954–1955), while high mortalities and increased frequencies of resistance to attack were associated with a period of dry summers with intense February rains (Fig. 3b).
3. A relatively dry summer with high February rains preceded a more intense summer drought and high tree mortality at Campania in 1966–1967. During this period, February rainfall was 5 mm, followed in early March by 11 mm. Tree mortality declined with increasing summer and less intense intermittent rains (Fig. 3c).
4. After three seasons of below-average rains, a pattern of ascending but still below-average rains was evident at Bracknell, with significant tree mortality occurring in a year marked by relatively intense February rains. As with Campania, tree mortality declined as summer rains became more average and February rains less intense (Fig. 3d).
5. Trends in tree mortality at the Mt. Helen and Cuckoo plantations in northeastern Tasmania were similar, although tree losses were more severe at Mt. Helen due to more severe wind damage in February 1969. The winds were accompanied by above-average rains, and together the two forces most probably acted to initiate the outbreak. Prior to this event, no change in *Sirex* status was observed despite below average rains (Fig. 4a).
6. At Branches Creek, a significant mortality of ~9% of all trees was recorded in 1980–1981 despite no evidence of *Sirex* presence previously. Summer rains during this and the preceding season were below average and the relative intensity of March rains was significantly high throughout the outbreak period (Fig. 4b).
7. More than 65% of the total mortality of unthinned Radiata pine at Delatite plantation, northeastern Victoria, occurred in one season, 1976–1977. Examination of rainfall data indicated that although preceded and immediately followed by dry summers, the 1976–1977 season experienced average precipitation but intense above-average February rains (Fig. 4c).

These examples, and notably 6 and 7 support the proposed hypothesis that intermittent drought during the *Sirex* emergence season contributes significantly to woodwasp outbreaks by increasing tree attractiveness and susceptibility through rapid physiological

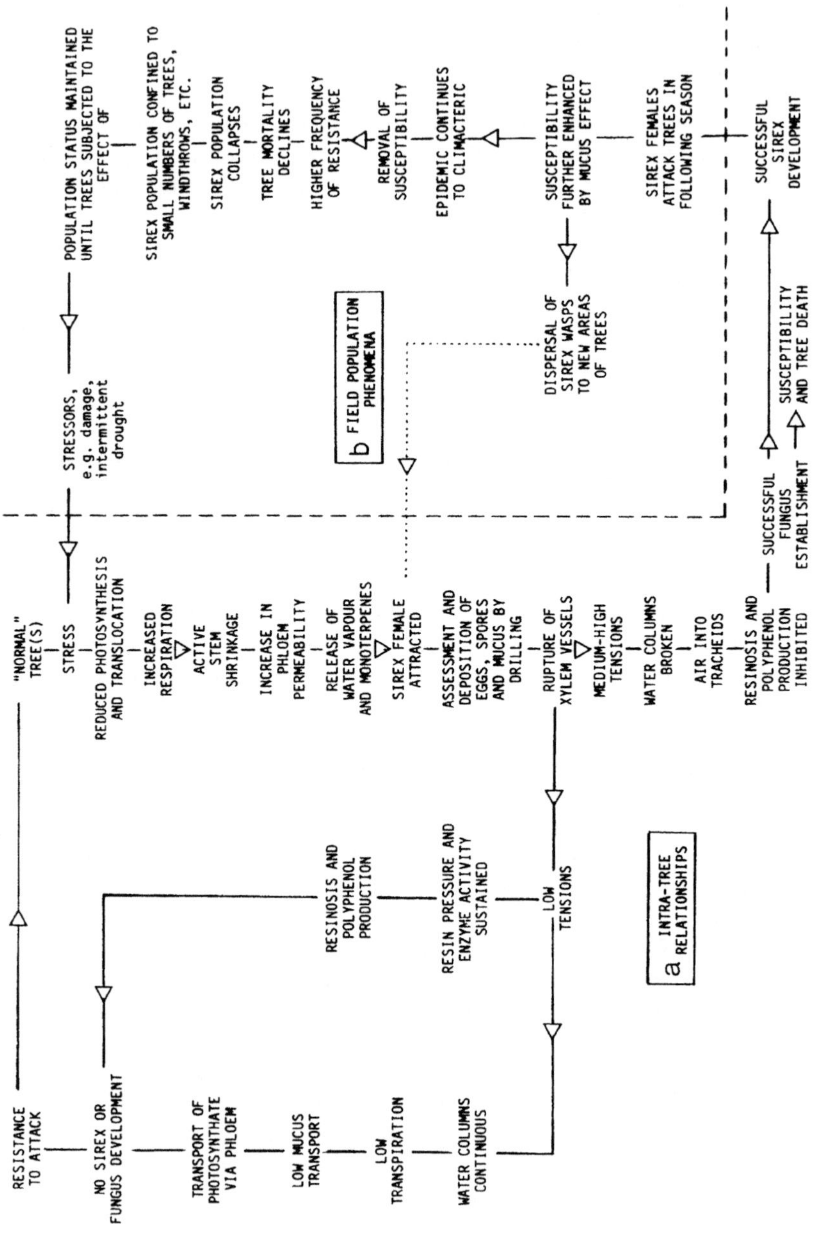

FIGURE 6. Proposed mechanism to explain (a) factors affecting susceptibility and resistance of *Radiata* pine to *Sirex* attack, and (b) changes in *Sirex* populations as reflected by tree mortality.

changes following rains of short duration. The proposed mechanism is summarized in Fig. 6.

5. MANAGEMENT IMPLICATIONS

Given the presence of the *Sirex* wasp and its capacity to exploit short-term disturbances within plantations, new plantations should be established in areas of low risk from fire and wind damage and in areas that experience dry summers, due consideration should be given to the probability of rain during the *Sirex* flight season. In established plantations, thinning and pruning schedules should be adhered to, but only during the late autumn and winter, and never in the summer months. Similarly, if felling during the summer is unavoidable, then clear felling with appropriate disposal of slash would be less harmful to remaining trees than selectively logging. Unprocessed logs also require protection from *Sirex* degrade through application of insecticide sprays.[34,39]

Attention to site conditions and timing of forestry operations would help prevent *Sirex* outbreaks. However, because of the unpredictability in the occurrence and intensity of stress agents, natural enemies should be introduced and established in order to suppress and stabilize resident populations. Distribution of nematodes through plantations by either female wasps or by planned artificial inoculation of *Sirex* infested trees contributes significantly to population control by reducing the reproductive capacity of the population. However, nematode parasitism does not affect the ability of female wasps to kill trees and, unless high levels of parasitism are sustained, high tree mortality can occur. The spectacular Delatite outbreak, in which 70% of the trees in unthinned areas were killed in 1 year, occurred despite the presence of both parasitoids and 40% nematode infection rates.[61]

The observation that high girdling of pruned trees during the flight season results in highly predictable attack 10–12 days later suggested that lure trees can be used to detect the presence of *Sirex* and avert attack from more valuable trees.[54] In a plantation of 12,000 trees, 80% of the girdled trees ($N = 55$) were attacked, compared with less than 1% of untreated trees, most of which were in the vicinity of girdled trees.

Tree girdling has also been used in New Zealand to increase the incidence of nematode parasitism,[7] but variability in the frequency and success of attacks, plus high establishment costs, limited the continued deployment of this method. A practical solution to this problem was discovered during field evaluation of different herbicides as thinning agents.[57] Trees, injected basally with 20% Dicamba (3,6-dichloro-2-methoxybenzoic acid) at 1-ml/10-cm circumference attracted wasps and, although not all treated trees were attacked, treatment during spring was 80% effective and was not detrimental to the development of parasitoids, fungus, or nematodes.[62]

The management strategy currently being used in high-risk plantations in Australia is to treat batches of low-quality trees with Dicamba in the spring. Attacked trees are then felled in autumn and inoculated with nematodes to maximize infection and attain sterilization rates in excess of 90% after 2 years.[63] This strategy concentrates the *Sirex* population in known locations and at the same time maximizes parasitism by insect parasitoids and nematodes.

In exceptional circumstances, significant tree mortality can still occur. Proper surveillance of forests during the flight season is required to ensure effective control. Unseasonal and unexpected events, such as wind or rain storms, should be considered potential stressors and initiators of *Sirex* outbreaks. Proper documentation of local plantation meteorological events should be accompanied by continual monitoring of tree growth to truly establish and identify the causal mechanisms of the observed effects of outbreaks, reduced growth and the resistance and death of forest trees.

ACKNOWLEDGMENTS. I wish to acknowledge the assistance of M. Nuttall, D. Rook, and R. Zondag (FRI, Rotorua, New Zealand), Ms. D. Campbell (Forestry Commission, New South Wales), H. Elliott, and R. Bashford (Forestry Commission, Tasmania), Bureau of Meteorology (Hobart, Tasmania), and Rural Water Commission of Victoria (Armadale, Victoria), for infestation and meteorological data and advice. I am especially grateful to K. L. Taylor and J. E. Dolezal (CSIRO, Hobart, Tasmania) for their comment, advice, and friendship over many years, and to F. G. Neumann (Division of Forests, Department of Conservation, Forests and Lands, Victoria).

Mrs. S. A. Jones patiently typed the manuscript. The chapter header was drawn by A. A. Berryman.

REFERENCES

1. Anonymous, 1980, Scientific and Common Names of Insects and Allied Forms Occurring in Australia, CSIRO, Melbourne, Australia.
2. Anonymous, 1984, Australian Forestry Resources 1984, Table 3, p. 9, Bureau of Agricultural Economics, Canberra.
3. Anonymous, 1984, Current Status of *Sirex* in Forestry Commission Plantations, Tas. For. Comm. Silv. Res. Rep. 3.
4. Anonymous, 1985, Forest and Forestry Products. Pests and Diseases Committee minutes, Canberra, A.C.T.
5. Ass, M., and Funtikow, G., 1932, On the biology and technical importance of wood wasps, *Z. Agnew. Entomol.* **19:**557–578 (in German).
6. Basham, J. T., and Belyea, R. M., 1960, Death and deterioration of balsam fir weakened by spruce budworm defoliation in Ontario, *Forest Sci.* **6:**78–96.
7. Bedding, R. A., 1967, Parasitic and free living cycles in entomophagous nematodes of the genus *Deladenus, Nature (Lond.)* **214:**174–175.
8. Bedding, R. A., 1968, *Deladenus wilsoni* n. sp. and *D. siricidicola* n. sp. (Neotylenchidae) entomophagous–mycetophagous nematodes parasitic in siricid woodwasps, *Nematologica* **14:**515–525.
9. Bedding, R. A., and Akhurst, R. J., 1974, Use of the nematodes *Deladenus siricidicola* in the biological control of *Sirex noctilio* in Australia, *J. Aust. Entomol. Soc.* **13:**129–135.
10. Benson, R. B., 1951, Handbook for the identification of British insects. Hymenoptera 2. Symphyta. Section a. *R. Entomol. Soc. (Lond.)* **vi**(2)(a):1–49.
11. Berryman, A. A., 1969, Response of *Abies grandis* to attack by *Scolytus ventralis* (Coleoptera: Scolytidae), *Can. Entomol.* **101:**1033–1042.
12. Bowling, P. J., and Dolezal, J. E., 1970, Initial note on the development of a "mucus" test to determine resistance of *P. radiata* to attack by *Sirex noctilio, Aust. Forest. Res.* **5:**57–62.
13. Bradley, J. C., 1913, The Siricidae of North America, *J. Entomol. Zool.* (Claremont) **5**(1)**:**1–35.
14. Buchner, P., 1928, *Holznahrung und symbiose,* Springer, Berlin.
15. Cameron, E. A., 1962, North American Survey for natural enemies of the Siricidae, May–October 1962, C.I.B.C. Report, Jamaica.
16. Cameron, E. A., 1968, *Sirex juvencus californicus* in smog-killed trees in Southern California (Hymenoptera: Siricidae), *Pan-Pac. Entomol.* **44:**168.

17. Cartwright, K. St. G., 1929, Notes on the fungus associated with *Sirex cyaneus*, *Ann. Appl. Biol.* **16:**182–187.
18. Chesterman, R., 1982, The sampling and identification of components in volatile organic mixtures, M.Sc. thesis, University of Tasmania, Hobart, Tasmania.
19. Chrystal, R. N., 1928, The Sirex woodwasps and their importance in forestry, *Bull. Entomol. Res.* **19:**219–247.
20. Clark, A. F., 1933, The horntail borer and its fungal association, *N. Z. J. Sci. Technol.* **15:**188–190.
21. Coutts, M. P., 1969, The mechanism of pathogenicity of *Sirex noctilio* on *Pinus radiata*. I. The effects of the symbiotic fungus *Amylostereum* sp. (Thelophoraceae), *Aust. J. Biol. Sci.* **22:**915–924.
22. Coutts, M. P., 1969, The mechanism of pathogenicity of *Sirex noctilio* on *Pinus radiata*. II. Effects of *S. noctilio* mucus, *Aust. J. Biol. Sci.* **22:**1153–1161.
23. Coutts, M. P., and Dolezal, J. E., 1966, Polyphenols and resin in the resistance mechanism of *Pinus radiata* attacked by the woodwasp, *Sirex noctilio*, and its associated fungus, *Leafl. Commonw. For. Timb. Bur.* **101:**10–19 (leaflet).
24. Coutts, M. P., and Dolezal, J. E., 1969, Emplacement of fungal spores by the woodwasp *Sirex noctilio* during oviposition, *Forest Sci.* **15:**412–416.
25. Cremer, K. W., 1973, Immediate resumption of growth by Radiata pine after five months of minimal transpiration during drought, *Aust. Forest. Res.* **6:**11–16.
26. Ermolenko, V. M., 1956, Ecologico-faunal survey of horntails (Hymenoptera: Siricidae) in the Soviet Carpathians, *Trudy Biol. Fak. Kiev Univ.* **13:**83–91 (in Russian).
27. Escherich, K., 1942, *Die Forstinsekten Mitteleuropas*, Vol. V: *Hymenoptera und Diptera*, Paul Parey, Berlin (in German).
28. Fielding, J. M., 1955, The seasonal and daily elongation of the shoots of the Monterey Pine and the daily elongation of roots, *Leafl. Commonw. For. Timb. Bur.* **75:**1–22.
29. Fong, L. K., and Crowden, R. K., 1973, Physiological effects of mucus from the woodwasp *Sirex noctilio* F. on the foliage of *Pinus radiata* D.Don, *Aust. J. Biol. Sci.* **26:**365–378.
30. Francke-Grosmann, H., 1939, Uber das Zusammen leben von Holzwespen (Siricinae) mit Pilzen, *Z. Angew Entomol.* **25:**647–680 (in German).
31. Francke-Grosmann, H., 1954, Tierische Holzschadlinge, Holzwespen, *Holz Roh Werkst.* **12:**35–38 (in German).
32. Gaut, I. P. C., 1969, Identity of the fungal symbiont of *Sirex noctilio* in Tasmania, *Aust. J. Biol. Sci.* **22:**905–914.
33. Gilbert, J. M., and Miller, L. W., 1952, An outbreak of *Sirex noctilio* in Tasmania, *Aust. J. For.* **16:**63–69.
34. Grujic, D., 1968, Preventative control of woodwasps (Fam. Siricidae, *Zast. Bilja* **99:**187–196 (in Serbo-Croatian).
35. Grujic, D., 1970, Contribution to the study of horntails (Hymenoptera: Siricidae) on conifers in mountain regions of Servia, *Zast. Bilja* **107:**51–61 (in Serbo-Croatian).
36. Gusakovsky, V. V., 1935, *Faune de l'URSS* (N.S.). No. 1. Insectes Hymenopteres T11 Vol. 1, Chalastogastra. Pt. I.
37. Hillis, W. E., and Inoue, T., 1968, The formation of polyphenols in trees. IV. The polyphenols found in *Pinus radiata* after Sirex attack, *Phytopathology* **7:**13–22.
38. Hocking, H., 1967, A native ichneumonid *Certonotus tasmaniensis* TURNER parasitizing *Sirex noctilio* F. (Siricidae) in Tasmania, *J. Aust. Entomol. Soc.* **6:**57–60.
39. Horwood, D. W., Morgan, F. D., and Stewart, N. C., 1970, Effectiveness of contact insecticides against the woodwasp *Sirex noctilio*, *Aust. For.* **34:**49–59.
40. Jackson, D. S., 1955, The *Pinus radiata–Sirex noctilio* relationship at Rotoehu Forest, New Zealand, N.Z. J. For. 7:26–40.
41. Johnston, R. D., 1964, Water relations of *Pinus radiata* under plantation conditions, *Aust. J. Bot.* **12:**111–124.
42. Krombein, K. V., 1958, Hymenoptera of America north of Mexico. Synoptic Catalogue; Suppl. 1. U.S. Government Printing Office, Washington, D. C.
43. Maa, T. C., 1949, A synopsis of Asiatic Siricoidae with notes on certain exotic and fossil forms, *Notes Ent. Chin.* **13**(2):76–163.

44. Macdougal, D. T., 1925, Reversible variations in volume, pressure and movement of sap of trees, *Publs. Carnegie Instn.* **365:**1–90.
45. McKimm, R. J., and Walls, J. W., 1980, A survey of damage caused by the Sirex woodwasp in the radiata pine plantation of Delatite, north eastern Victoria between 1972 and 1979, For. Comm. Vic. **28:**3–12.
46. Madden, J. L., 1971, Some treatments which render Moneterey Pine (*Pinus radiata*) attractive to the woodwasp *Sirex noctilio* F. *Bull. Entomol. Res.* **60:**467–472.
47. Madden, J. L., 1974, Oviposition behaviour of the woodwasp *Sirex noctilio* F., *Aust. J. Zool.* **22:**341–351.
48. Madden, J. L., 1975, An analysis of an outbreak of the woodwasp *Sirex noctilio* F. (Hymenoptera: Siricidae) in *Pinus radiata, Bull. Entomol. Res.* **65:**491–500.
49. Madden, J. L., 1977, Direct evaluation of cambial activity in trees, *Aust. Forest Res.* **7:**145–150.
50. Madden, J. L., 1977, Physiological reactions of *Pinus radiata* to attack by woodwasp *Sirex noctilio* F. (Hymenoptera: Siricidae), *Bull. Entomol. Res.* **67:**405–426.
51. Madden, J. L., 1982, Avian predation of the woodwasp, *Sirex noctilio* F. and its parasitoid complex in Tasmania, *Aust. Wildl. Res.* **9:**135–144.
52. Madden, J. L., and Bashford, R., 1977, Population biology of *Chlenias* sp., a geometrid defoliator of *Pinus radiata* in Tasmania, *J. Aust. Entomol. Soc.* **16:**379–388.
53. Madden, J. L., and Coutts, M. P., 1977, Role of fungi in the biology and ecology of woodwasps (Hymenoptera: Siricidae), in: *Insect–Fungus Symbiosis* (L. R. Batra, ed.), pp. 165–174, John Wiley & Sons, New York.
54. Madden, J. L., and Irvine, C. J., 1971, The use of lure trees for detection of *Sirex noctilio* in the field, *Aust. Forest* **35:**164–166.
55. Middlelkauff, W. W., 1965, The siricid woodwasps of California (Hymenoptera: Symphyta), *Bull. Cal. Ins. Surv.* **6**(4):59–77.
56. Miller, D., and Clark, A. F., 1935, *Sirex noctilio* (Hym.) and its parasites in New Zealand, *Bull. Entomol. Res.* **26:**149–154.
57. Minko, G., 1981, Chemicals for non-commercial thinning of *Pinus radiata* by basal stem injection, *Aust. Weeds* **1:**5–7.
58. Morgan, F. D., 1968, Bionomics of Siricidae, *Annu. Rev. Entomol.* **13:**239–256.
59. Neumann, F. G., 1979, Insect pest management in Australian Radiata pine plantations, *Aust. Forest* **42:**48–63.
60. Neumann, F. G., and Minko, G., 1981, The *Sirex* woodwasp in Australian Radiata pine plantations, *Aust. Forest* **44:**46–63.
61. Neumann, F. G., Harris, J. A., Kassaby, F. Y., and Minko, G., 1982, An improved technique for early detection and control of the *Sirex* woodwasp in Radiata pine plantations, *Aust. Forest* **45:**117–124.
62. Neumann, F. G., and Morey, J. L., 1984, Influence of natural enemies on the *Sirex* woodwasp in herbicide-treated trees of Radiata pine in north-eastern Victoria, *Aust. Forest* **47:**218–224.
63. Neumann, F. G., Morey, J. L., and McKimm, R. J., 1987, *The Sirex wasp in Victoria,* Bulletin of the Department of Conservation, Forests and Lands, No. 29, Melbourne.
64. Painter, R. H., 1951, *Insect Resistance in Crop Plants,* Macmillan, New York.
65. Parkin, E. A., 1942, Symbiosis and siricid woodwasps, *Ann. appl. Biol.* **29:**268–274.
66. Rawlings, G. B., 1948, Recent observations on the *Sirex noctilio* population in *Pinus radiata* forests in *N.Z. J. Forest* **5:**411–421.
67. Rawlings, G. B., 1953, Rearing of *Sirex noctilio* and its parasite *Ibalia leucospoides, Forest Res. Notes* **1:**28–34.
68. Rawlings, G. B., 1955, Epidemics in *Pinus radiata* forests in New Zealand, *N.Z. J. Forest* **7:**53–58.
69. Rawlings, G. B., 1957, *Guiglia schauinslandi* (ASHMEAD) (Hym. Orussidae) a parasite of *Sirex noctilio* (Fabricius) in New Zealand, *Entomologist* **90**(1125):35–36.
70. Rawlings, G. B., and Wilson, N. M., 1949, *Sirex noctilio* as a beneficial and destructive insect to *Pinus radiata* in New Zealand, N.Z. J. Forest Sci. **6:**20–29.
71. Reid, R. W., Whitney, H. S., and Watson, J. A., 1967, Reactions of lodgepole pine to attack by *Dendroctonus ponderosae* HOPKINS and blue stain fungi, *Can. J. Bot.* **45:**1115–1116.
72. Rook, D. A., Swanson, R. H., and Cranswick, A. M., 1976, Reaction of Radiata pine to drought, in: *Proceedings of the Soil and Water Symposium, DSIR, Palmerston North,* pp. 55–68.
73. Schimitschek, E., and Wienke, E., 1968, The susceptibility of tree species to secondary pests. IV. *Abies alba, Z. Angew. Entomol.* **62:**294–306.

74. Simpson, R. F., 1976, Bioassay of pine oil components as attractants for *Sirex noctilio* (Hymenoptera: Siricidae) using electro-antennogram techniques, *Entomol. Exp. Appl.* **19**:205–213.
75. Sonan, J., 1938, Siricidae of Formosa, *Trans. Nat. Hist. Soc. Formosa* **28**(175):88–94.
76. Spradbery, J. P., 1973, A comparative study of the phytotoxic effects of siricid woodwasps on conifers, *Ann. Appl. Biol.* **75**:309–320.
77. Stillwell, M. A., 1960, Decay associated with woodwasps in Balsam fir weakened by insect attack, *Forest Sci.* **6**:225–231.
78. Stroganova, V. K., 1968, *Horntails of Siberia,* Nauka, Novosibirsk (in Russian).
79. Talbot, P. H. B., 1977, The *Sirex–Amylostereum–Pinus* association, *Annu. Rev. Phytopathol.* **15**:41–54.
80. Taylor, K. L., 1967, The introduction, culture, hibernation and recovery of parasites of *Sirex noctilio* in Tasmania, 1962–1967, Aust. CSIRO Div. Entomol. Tech. Pap. No. 8.
81. Taylor, K. L., 1976, The introduction and establishment of insect parasitoids to control *Sirex noctilio* in Australia, *Entomophaga* **21**:429–440.
82. Taylor, K. L., 1978, Evaluation of the insect parasitoids of *Sirex noctilio* (Hymenoptera: Siricidae) in Tasmania, *Oecologia (Berl.)* **32**:1–10.
83. Taylor, K. L., 1980, Studies with *Sirex noctilio* (Hymenoptera: Siricidae) and its parasites that illustrate the importance of evaluating biological control attempts, *Acta Oecol./Oecol. Appl.* **1**:181–187.
84. Taylor, K. L., 1981, The *Sirex* Woodwasp: Ecology and control of an introduced forest insect, in *The Ecology of Pests—Some Australian case histories* (R. L. Kitching and R. E. Jones, eds.), pp. 231–248, CSIRO, Melbourne.
85. Wolf, F., 1967, Les Siricides en Belgique, leurs moers et leur importance en sylviculture, *Bull. Soc. R. For. Belg.* **76**:281–301 (in French).
86. Zondag, R., 1969, A nematode infection of *Sirex noctilio* (F.) in New Zealand, *N.Z. J. Sci.* **12**:732–737.
87. Zondag, R., 1971, Control of *Sirex noctilio* with *Deladenus siricidicola* BEDDING, *N.Z. J. Forest Sci.* **1**:5–14.

CHAPTER 21

THE JAPANESE PINE SAWYER

Fujio Kobayashi

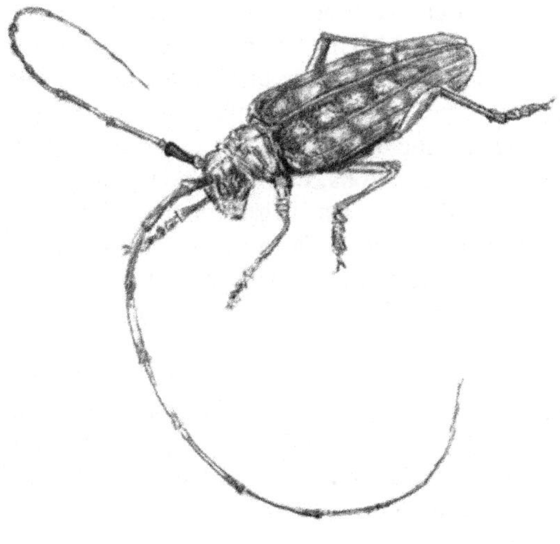

Fujio Kobayashi • Forestry and Forest Products Research Institute, Tsukuba, Ibaraki, 305 Japan.

1. INTRODUCTION

The major pine species in Japan, Japanese red pine (*Pinus densiflora*) and Japanese black pine (*P. thunbergii*), have suffered heavy mortality for several decades. Many beetle species are found under the bark of dead trees, including the Japanese pine sawyer, *Monochamus alteratus* Hope. On the basis of field observations, entomologists assumed that trees had been diseased before beetle attacks. A research project begun in 1968 proved that the causal agent of the disease was the pine wood nematode, *Bursaphelenchus lignicolus* Mamiya et Kiyohara, which was subsequently reclassified as *B. xylophilus* (Steiner and Buhrer) Nickle. Intensive inspection of the insects associated with dead pine trees demonstrated that the principal vector of the nematode was *M. alternatus*.[12,20]

Subsequently, the existence of the nematode was reported in the United States, where extensive surveys have demonstrated its widespread distribution. This does not mean, however, that an epidemic of pine tree mortality will occur as in Japan.[2,42] In the United States, only exotic pine plantings are severely damaged, a fact that suggests that the nematode could have been accidentally introduced into Japan from the American continent. Nagasaki, the area in which the first pine wilt outbreak was recognized in 1905, was the most prosperous international trade port at that time.

After a latent period of two decades, the infestation spread rapidly through Kyushu, and into western Honshu. During the war, annual tree losses continued to rise up to 1.3 million m^3 in 1948. Control operations over the whole country in the 1950s resulted in decreased infestations. This decrease was helped by the utilization of killed trees as fuel after the war. At the beginning of the 1970s, however, the infestation started to rise again because of the change to petroleum as the major energy source and the lack of labor for control operations. In 1978 and 1979, largely because of hot and dry summers, annual tree losses amounted to 2 million m^3/year (Fig. 1).

Pine forests in Japan occupy about 2.5 million ha, about 10% of the forested land. Pine trees have been economically important from ancient times mainly as fuel for domestic and industrial uses, and lumber for buildings. In recent times, extensive pine plantations have been established, but the importance of pine is declining because of cheap imported timber. On the other hand, pine forests contribute significantly to such public uses as watershed, erosion control, and outdoor recreation, and the pine tree is closely connected with Japan's heritage and culture (Fig. 2). For these reasons, the current epidemic attracts considerable public attention.

A large-scale national control project was started in 1977 when a federal law was passed implementing an insecticide spray program by either federal or local government. About $15 million is spent annually by the federal government for aerial and ground spraying of standing trees and about $8 million for the treatment of infested logs. Annual tree losses have gradually decreased, especially in the area continuously treated by aerial spray. However, infestations have spread into northern Honshu and into forests at higher elevation.

FIGURE 1. Japanese black pine forest completely killed by the pine wood nematode in two years, 1978 and 1979.

2. BIOLOGY AND ECOLOGICAL RELATIONSHIPS

2.1. Life History of the Pine Sawyer

Adult emergence occurs in early summer and is strongly influenced by weather, being accelerated by sunny, dry conditions. In Okinawa, the southernmost habitat in Japan, adults begin to emerge in mid-April, whereas in Akita, the northernmost habitat, emergence begins in late June. The span of emergence, which is usually 2 months, tends to be shorter in more northerly districts. Diurnal emergence activities are mainly from dusk to midnight.

Following emergence adult beetles walk upward and fly from the extremity of the object on which they are situated. Young adults are active until they reach a host tree, where they settle to feed on the bark of young branch tips (Fig. 3). Adults become active again after maturation feeding for 2 or 3 weeks. At this time, both sexes are attracted to weakened pine trees, such as those infected by nematodes, on which they copulate and oviposit. Females lay eggs on the bark of diseased trees in specially prepared slits on the main trunk, branches, or twigs as small as 2 cm in diameter. On an average, females lay 1.7 eggs a day and 60–100 eggs during their life. Oviposition reaches its peak about 45 days after emergence.

Eggs incubate for 6–9 days; after eclosion, the first-instar larva enters the phloem to feed. The three subsequent instars feed mainly in the outer sapwood, and the full-grown

FIGURE 3. Female Japanese pine sawyer feeding on a pine tree twig.

larva bores into the sapwood in late autumn to construct a pupal chamber. Most larvae then plug the opening of the tunnel with wood borings. The last or fourth-instar larva hibernates in the pupal chamber (Fig. 4).

Larval development is completed in 625 degree-days above a 12.5°C developmental threshold. The requirement for overwintering larvae to reach 50% adult emergence ranges

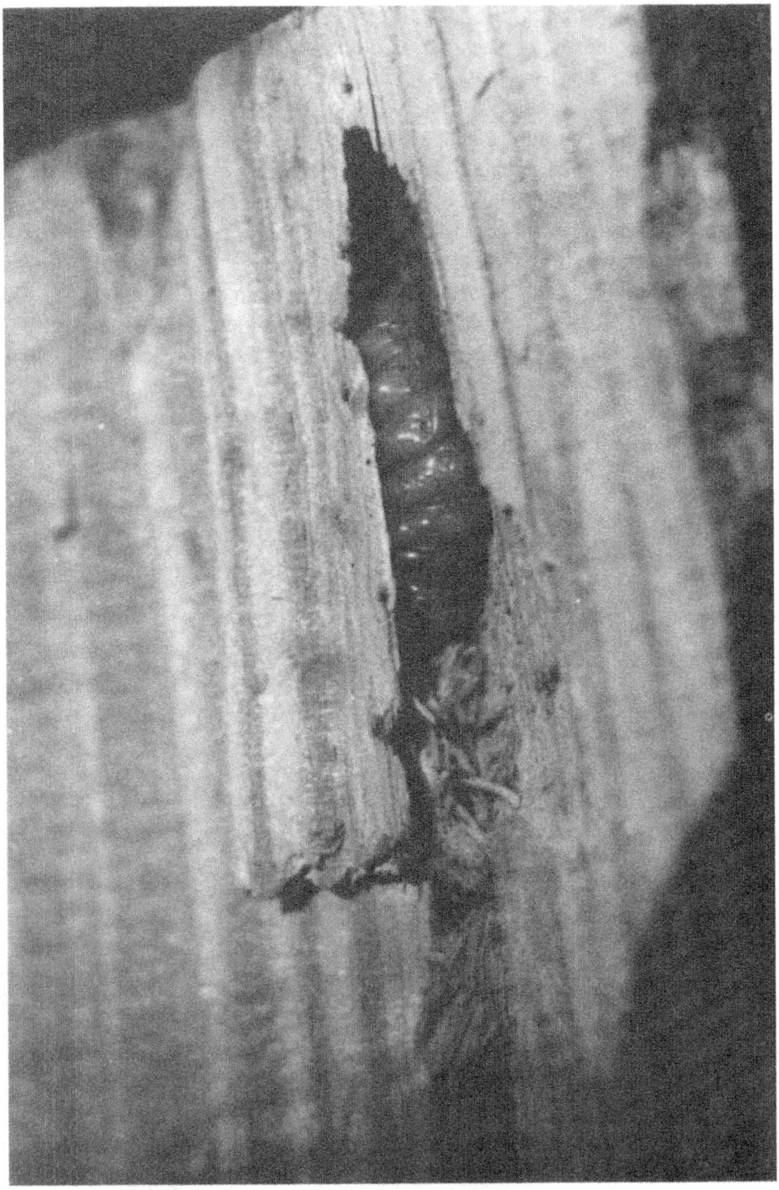

FIGURE 4. Larva of the Japanese pine sawyer in its pupal chamber.

from 500 to 540 degree-days, with larvae from warmer localities usually requiring more degree-days to complete development than those from cooler localities. Experience of low winter temperatures, from 10 to 15°C, is considered necessary for pupation. The pupa and callow adult stages last about 18 and 7 days, respectively.

Monochamus alternatus has been recorded from 18 species of *Pinus,* three species of *Picea,* and one species each of *Abies, Cedrus,* and *Larix,* but extensive tree mortality and normal larval development are only observed in three native pine species, *Pinus densiflora, P. thunbergii,* and *P. luchuensis,* with lesser effects in the exotic species, *Pinus pinaster* and *Cedrus deodora.*

Monochamus alternatus usually has one generation per year, but occasionally a single generation is completed in 2 years. In cooler areas, 20–30% of the beetles experience 2-year life cycles. In northern Honshu, larvae hatching later than mid-August overwinter, as young instars in the sapwood and high mortalities are common. Those that survive take 2 years to complete their life cycle.[39]

2.2. Insect–Tree Interactions

Adult maturation feeding occurs on pine species and on some other conifers. The degree of preference is roughly proportional to the hatching success of the eggs on that tree species. Although there may be a host influence on the maturation of beetle ovaria, the tree preferred by adults is not necessarily the best for larval development. For example, fresh logs of *Pinus elliottii* are strongly preferred for oviposition, but larval survival is significantly lower than in another preferred host, *P. densiflora.*

Adult beetles initially feed on current- and 1-year old twigs but eventually accept those that have aged for several years. Extracts from the bark of young twigs, including sucrose, fructose, β-sisterol, and glycosyl-flavonol, are known to stimulate beetle feeding.[29] Thin-barked regions of the tree are preferred as oviposition sites, but eggs may also be laid on somewhat thicker bark or wedged among old oviposition slits. The spatial distribution of oviposition slits is under-dispersed, suggesting that the female tends to avoid former oviposition slits. As suitable sites on a tree become scarce, and before the saturation point is reached, females fly to other trees.[10]

While attraction to host trees for maturation feeding has not been demonstrated, recently diseased trees or fresh logs strongly attract mature beetles. Behavior relating to oviposition and attraction is initiated by odors released from the diseased tree or log. Responses of adults to attractants vary with physiological age, with immature beetles having noticeably lower responses than those of mature individuals. Beetles usually begin to respond to host odors about 3 weeks after emergence.[43]

Volatiles collected from nematode-infected trees or fresh logs were identified as a mixture of terpene hydrocarbons and anaerobic fermentation products. Among them, α-pinene and ethanol are major attractive compounds for both sexes.[6,7] It is inferred that ethanol works as a synergist, with beetles recognizing diseased trees by the volatized ethanol. Ethanol or acetone injected into pine trees also attracts *Monochamus* beetles. The fact that paraquat-treated pine trees attract the beetles could be explained by the same mechanism.

2.3. Insect–Nematode Association

Monochamus alternatus is the major vector of the pine wood nematode, *Bursaphelenchus xylophilus,* in Japan. However, the nematode is occasionally isolated from adults of *Monochamus saltuarius,* which is limited to cooler localities. In the United States, the major vector is *Monochamus carolinensis,* which, on the average, carries more than 10,000 nematodes per individual.[17,18]

Most nematodes are found in the respiratory organs of *M. alternatus* adults, usually clumped within the metathoracic spiracle, but may also be found within the leading tracheae as far distant as the antennae and legs. Data on the density of nematodes isolated from beetles of both sexes just after emergence are presented in Table I. The frequency distribution of nematodes per beetle is strongly contagious with more than 90% of the total nematode population found in 20% of the insects. This means that many beetles have few or no nematodes.[16] It is commonly observed that more nematodes are present in beetles from heavily infested forests and that those emerging early in the season tend to have more nematodes than those emerging late.

Nematode larvae aggregate around the pupal chamber of *M. alternatus* in early spring and molt there to the dispersal fourth stage, called a dauerlarva. At the time of

TABLE I
Abundance of Pine-Wood Nematodes in Pine Sawyer Adults[a,b]

Locality	Year	Percentage of nematode-containing beetles	Number of nematodes per beetle	
			Average	Maximum
Miyagi Prefecture Ishinomaki	1978	65	6,900	88,200
Ibaraki Prefecture				
Iwai	1977	89	19,500	168,000
Mito	1974	95	8,100	52,000
Mito	1975	87	4,500	29,000
Yatabe	1972	57	4,200	135,000
Naka	1975	89	3,700	27,000
Naka	1976	91	3,000	23,100
Naka	1977	79	1,700	46,700
Chiba Prefecture				
Tateyama	1972	83	8,000	132,000
Ichihara	1972	84	10,100	150,000
Wakayama Prefecture				
Mirozu	1973	70	1,200	19,700
Shionomisaki	1973	27	170	5,100
Shionomisaki	1974	64	1,800	79,800
Shionomisaki	1975	46	1,100	12,300
Hidaka	1974	56	1,300	31,900
Tottori Prefecture Fukube	1975	61	2,200	23,800
Ehime Prefecture	1973	66	4,100	142,000
	1974	73	6,900	289,000

[a]From Kobayashi *et al.*[15]
[b]Samples limited to those exceeding 100 adults.

beetle emergence, the dauerlarvae move onto the callow adult.[19] Some unsaturated fatty acids extracted from the beetle have been reported as stimulants for nematode aggregation to the pupal chamber,[22] and CO_2 gas may also stimulate the nematodes to migrate.[23]

During maturation feeding, dauerlarvae appear from the spiracles and move down the body setae to the terminal abdominal sclerites, and then drop off onto pine twigs. Exodus of dauerlarvae is uncommon within a week of beetle emergence, reaching its peak in 2 or 3 weeks, and then gradually declining and terminating by the fourth or fifth week. The proportion of nematodes that invade pine trees after dropoff is estimated to be about 10% under field conditions and about 20% in the laboratory.[9]

2.4. Tree–Nematode Interactions

Pines native to Japan are highly susceptible to nematode infection, whereas many other exotic species seem to be resistant (Table II).

The pathology and physiology of the disease on susceptible pines are as follows:[19] At the early stage of infection, nematodes inhabit the resin canals of both xylem and cortex (Fig. 5). Death of ray and axial parenchyma cells surrounding the axial resin duct epithelium is the first pathological response. Cell death is extensive prior to the detection of nematodes in wood tissue. This indicates that some biochemical factors are involved in

TABLE II
Resistance or Susceptibility of Pine Species to *B. xylophilus*[a]

Resistant *Pinus* sp.	Susceptible *Pinus* sp.
banksiana	*densiflora*
brutia	*engelmannii*
bungeana	*koraiensis*
caribaea	*leiophylla*
contorta	*luchuensis*
echinata	*monticola*
elliottii	*mugo*
excelsa	*muricata*
halepensis	*nigra*
massoniana	*oocarpa*
palustris	*pentaphylla*
pungens	*pinaster*
resinosa	*ponderosa*
rigida	*radiata*
strobus	*rudis*
tabulaeformis	*strobiformis*
taeda	*sylvestris*
taiwanensis	*thunbergii*
rigida × *taeda*	*yunnanensis*
thunbergii × *massoniana*	

[a]From Mamiya.[19]

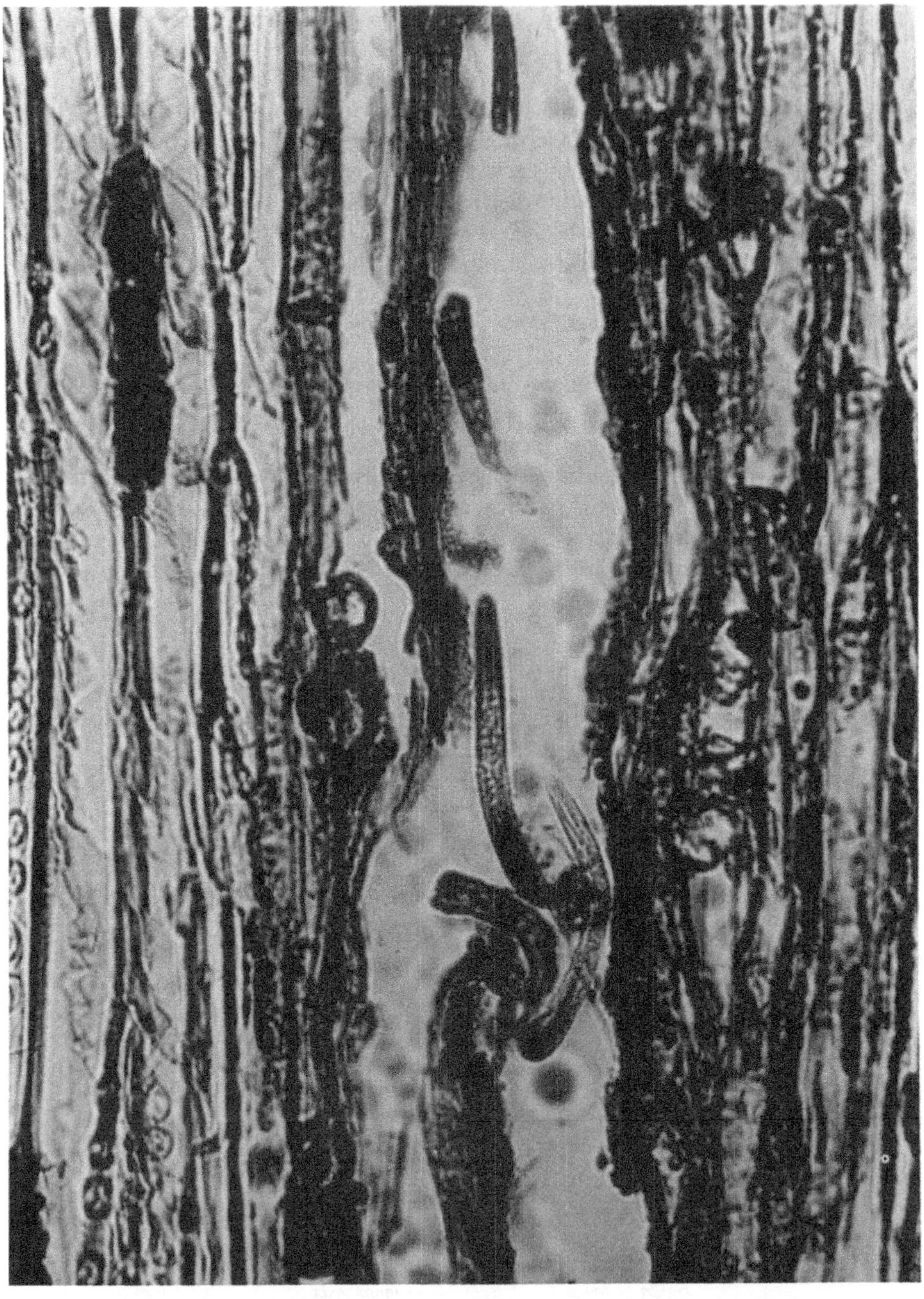

FIGURE 5. Pine wood nematodes in resin canals. (Photographed by Dr. M. Tamura.)

the pathological reaction of pine tissues. Candidates for the toxic substance were isolated from nematode infected trees,[32,33] and cellulase produced by the nematode was reported to be responsible for the early symptoms of the disease.[30]

Reduction and cessation of oleoresin exudation is the first obvious symptom, before tree death is apparent, and an increase in nematode populations is usually observed after the cessation of resin flow. *B. xylophilus* is easily propagated on cultures of many kinds of fungi, and is also successfully cultured on pine callus tissues.[40] This finding suggests the possibility of nematodes feeding on parenchyma cells, especially at the early stages of disease development, and feeding on fungi at the later stages when fungi become abundant in the wood.

The foliage of infested trees quickly fades to a reddish-brown. Most trees naturally infected with nematodes die within a year of infestation, usually by late autumn, in warm temperature zones. Disease development progresses more slowly in northern parts of Japan, where more than 40% of the trees show external symptoms of death in the spring of year following infection.[9]

3. POPULATION BEHAVIOR

3.1. Population Estimation

Several sampling methods have been employed to estimate adult beetle populations in the field:[15] (1) direct counting of adults on trees in a young forest by observing them along a footpath, (2) counting beetles that dropped on sheets or pathways after aerial spraying of mature forests, (3) measuring the area of maturation feeding wounds on twigs, (4) mark–recapture techniques, and (5) counting emergence holes on infested trees.

Using a mark-recapture techniques, the adult population in a heavily infested pine plantation was estimated to be 24,675 beetles per ha.[34] Counts of emergence holes on infested trees in a heavily infested young stand varied from 1000 to 10,000 per ha.[16]

Measurement of the population inside the tree is usually accomplished by removing and dissecting bark and wood or by soft X-ray photography.[11] For estimating population changes, 1-m logs are used providing relative precision of estimation, expressed by the ratio of standard error to mean, of 0.1–0.2, when 10–20 logs are sampled.[28] Large within-tree variation in larval density is mainly attributed to the effect of bark thickness. Bark thickness in *Pinus densiflora* changes sharply at about 2 m above the ground, with the upper thin-barked regions being more densely populated. Larger variations are found among trees, with some dead trees having no *M. alternatus* larvae, probably because of dissynchrony between the presence of weakened trees and adult beetles. As may be expected, intertree variation in sawyer population density declines as the stands become more heavily infested, and vice versa.

3.2. Within-Tree Populations

Seasonal changes in within-tree populations of *M. alternatus* may be divided, for convenience, into two developmental stages: (1) from egg to larval chambering in the sapwood, and (2) from larval chambering to emergence. Egg survival ranges from 80 to 90%, with most mortality due to ant predators (e.g., *Monomorium nipponense*), disease, and destruction of eggs by the larvae of other boring insects.[28] Mortality of first and second-instar larvae is low, tut high mortality (60–80%) occurs from the second to fourth larval instars (Fig. 6). Mortality in these stages is density dependent and may be a key factor in regulating within-tree populations.[24,25,28] More than 50% of the mortality at high density is caused by intraspecific competition between larvae for food or space. Interspecific competition with weevils and bark beetles does not strongly affect population change. Larvae are also killed by 18 species of predacious and parasitic insects, of which two predacious beetles, *Temnochila japonica* (Ostomidae) and *Thanasimus lewisi* (Cleridae), are the most abundant.[27]

Mortality in the second stage, after chambering, is caused by different factors and does not appear to be density dependent. Survival rates, as expressed by the ratio of emergence holes to larval tunnels, ranges from 30 to 70%, with more than 50% due to pathogenic microorganisms.[28] Intensive surveys throughout Japan revealed the presence of yeasts, eight genera of fungi, and a few bacteria associated with dead larvae, and inoculation tests proved *Beauveria bassiana, Verticillium* spp., and *Serratia marcescens*

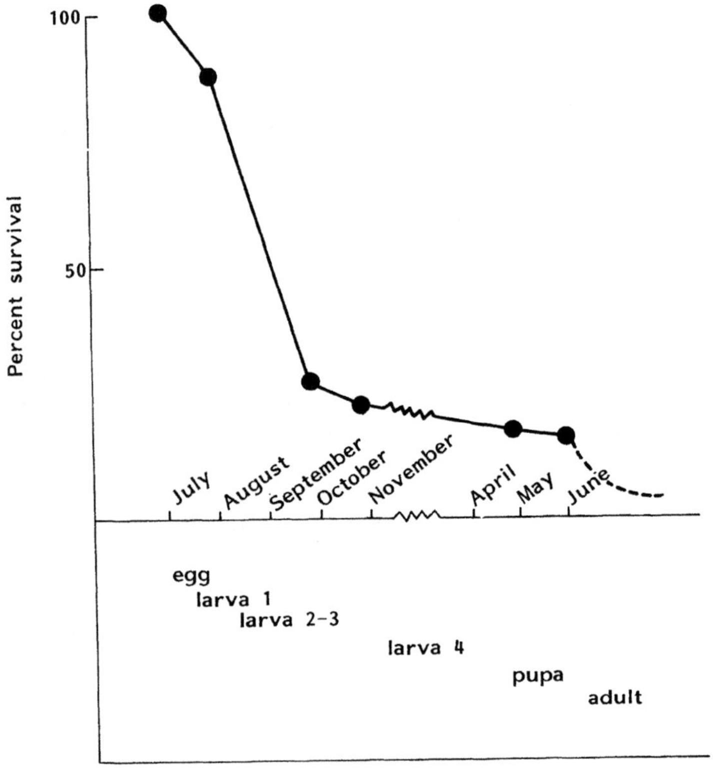

FIGURE 6. A schematic survivorship curve for *M. alternatus*.[13]

to be the major pathogens.[35] Predation by woodpeckers, such as *Dendrocopos major,* is sometimes important in reducing overwintering larval populations, especially in low-density *M. alternatus* populations in northern districts.[4,45]

The relationship between the number of eggs laid and adults emerging from dead trees indicates that density-dependent mechanisms maintain the number of adults emerging from dead trees at a constant value of about 20/m^2 (Fig. 7). However, the peak reproduction rate varies from year to year and from stand to stand.

3.3. Adult Populations

Female adults live about 100 days, a little longer than males. Mortality in field cages averages between 10 and 30% within a week of emergence, and this is higher than for any other period of adult life.[9] Mortality of females occasionally increases 2 or 3 weeks after emergence. This might be related to oviposition activity, which starts at this time. Early-emerging adults tend to live longer and to lay more eggs.

3.4. Infestation Patterns in Time and Space

When trees infested by beetles and nematodes are introduced into a pine stand, mortality is first noticed to nearby trees, and then gradually spreads from this center (Fig.

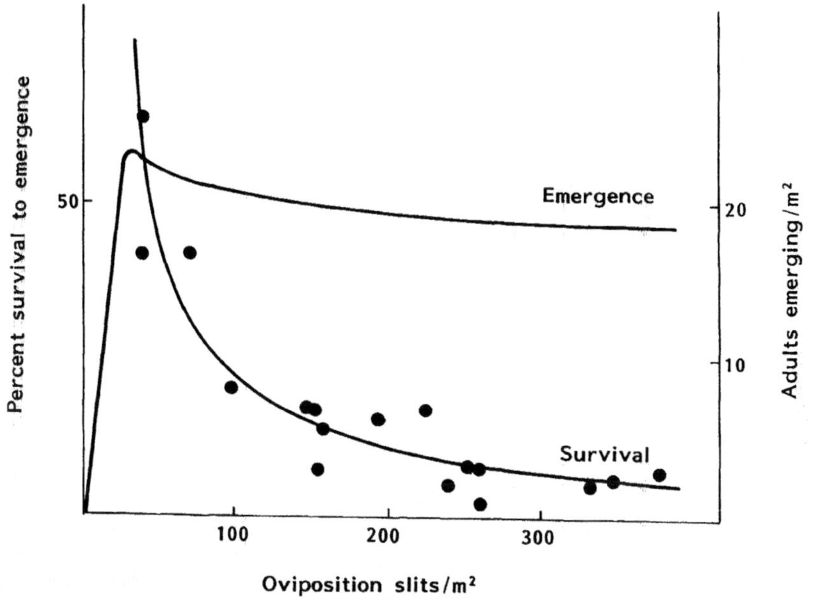

FIGURE 7. Relationship between the densities of oviposition slits and the adults derived from them. Density-dependent mechanisms keep adult emergence at a constant value, irrespective of the number of eggs laid.[25]

8). Most dead trees were found within 700 m of the center, but one dead tree was seen at a distance of 2.5 km. The number of trees killed in 1 year is usually two to seven times those killed the previous year.[31]

There are many reports of the spread of infestations over time. An extensive survey covering the entire infested area[9] showed that tree mortality increased very slowly in stands with 1% or less mortality, but infection increased suddenly in those stands with 2–9% mortality. The infection rate is accelerated by unusual weather conditions such as hot, dry summers and dry soil conditions. After the infection threshold is passed, the rate of increase in tree mortality is generally inversely related to the number of dead trees and to elevation.[26]

3.5. Reproduction and Dispersal Potential

The survival of within-tree populations of *M. alternatus* varies from 0.1 to 0.7 according to initial densities. The net reproductive rate from one generation to the next, calculated from age-specific survival rates and the fecundity of three caged adult populations, varied between 12 and 27.[41] These are fairly high values as compared with other tree-boring insects.

The potential for dispersal is also high. In mark–recapture experiments, 7% of released beetles were recaptured; of those recaptured, 75% were caught within 100 m of

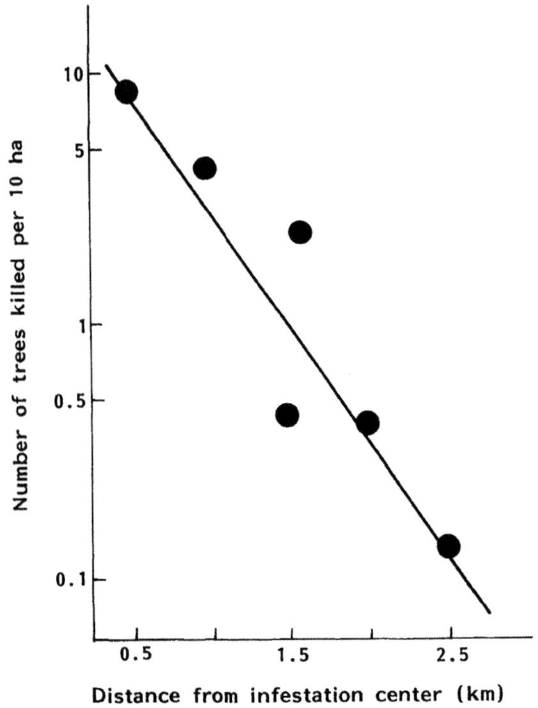

FIGURE 8. Distribution of trees killed in relationship to the distance from an infestation center, a large pile of infested logs. Most dead trees were found within about 1 km of the center, and one dead tree was found at 2.5 km.[31]

the point of release. However, a few beetles flew as far as 1–2.4 km (see Fig. 8). Repeated observation of small isolated groups of infested trees showed that beetles often moved 800 m from their origin and some dispersed up to 3.3 km.[15]

4. CAUSES OF OBSERVED POPULATION BEHAVIOR

4.1. Effects of Climatic and Site Factors

Climatic factors, particularly temperature and precipitation, have significant influences on tree death. Pine seedlings inoculated with *B. xylophilus* and kept at 25°C and 30°C became diseased, whereas those kept at 20°C and 15°C showed no disease symptoms. However, symptoms appeared in seedlings maintained at 18°C for 1 month after inoculation and then moved to warmer temperatures. On the contrary, seedlings kept at 30°C for about 1 week and then moved to 18°C did not become diseased. These results provide evidence that delayed disease development in naturally infected pine trees in northern Japan is due to the effect of low temperature.[19]

Larval development of *M. alternatus* is completed in about 625 degree-days above the threshold of 12.5°C. In cooler localities, where sufficient day-degrees are not accumulated, beetles suffer high mortality, and those that survive take 2 years to complete development. Feeding and dispersal activities of adults also cease as temperatures fall below 18°C. They never fly on rainy days. The amount of pine bark ingested by adults also increases geometrically with temperature.[15]

A flight index—days capable of flying activity—a combination of the number of days above 18°C daily temperature and below 10 mm daily precipitation—has been proposed as a measure of infestation potential.[37] Severe infestations occur mostly when this index exceeds 80 days, and no infestation were observed under 60 days. Another index, the *Monochamus–Bursaphelenchus* index (MB index), combines temperatures conducive to sawyer development, nematode infection, and tree weakening.[38] This is the annual summation of residues of mean monthly temperature exceeding 15°C. Epidemic pine infestations nearly always correspond to areas exceeding a MB index of 40. This region lies mainly along the coast and into the inland along river valleys up to about 200 m above sea level.

Elevation influences outbreak behavior through its relationship to temperature. In Kyushu, tree mortality decreases sharply above 300 m, except on south-facing slopes, and no infestations occur above 700 m. In the northern districts of Honshu, severe infestations may occur below 200 m, but no infestations are observed above 400 m.

Water stress also favors disease development. Experiments in altering the water content of the soil showed that dry conditions hastened the progress of the disease and caused higher mortality to nematode-inoculated pine trees.[36] Destructive infestations are also observed, sometimes in areas with dry soil.

In general, both temperature and precipitation are important factors that affect the severity of damage. High temperatures and low precipitation in summer cause accelerated damage because of effects on *M. alternatus* adult feeding, the propagation of *B. xylophilus*, and stress on the trees. Increases in pine tree mortality can therefore be expected in unusually hot, dry summers, such as those of 1978 and 1979 (Table III).

TABLE III
Pine Trees Killed in Ibaraki Prefecture as Related to Summer Temperature and Precipitation

	Temperature (°C)		Precipitation (mm)		
	Sum total of monthly means from June to August	One-month deviation from the average	Sum total from June to August	Deviation from the average (%)	Volume of trees killed (m^3)
Average	67.4	—	454.8	100.0	—
1977	65.4	−0.7	620.5	136.4	26,500
1978	74.2	+2.3	159.0	35.0	742,000
1979	71.0	+1.2	285.5	62.8	712,500
1980	64.7	−0.8	444.0	97.6	394,200

It should also be noted that only those trees that become infested by *M. alternatus* play a role in the spread of the disease. Adults lay eggs from July to early September on weakened trees, but the season of tree weakening does not necessarily coincide with that of beetle attacks. Unless tree weakening is synchronized with *M. alternatus* attack, dead trees may appear that do not become infested by the sawyer.

4.2. Causes for the Spread of Infestations

Pine wood nematode infestations usually spread from infection centers (Fig. 8). The radius of spread is normally within 2 km but occasionally exceeds 5 km. This type of spread can be explained by the dispersal of adult beetles contaminated by nematodes.

Another pattern of spread is sometimes observed when the disease appears unexpectedly in uninfested isolated areas, such as islands far from the mainland. This pattern can not be explained by beetle flight. In Okinawa Island, for example, the disease was first reported in a *Pinus luchuensis* stand in 1973. The infestation has since spread throughout the island and caused severe damage. The same pattern of infestation was repeated in the Ogasawara Islands. In both cases, the first infestations were due to pine logs introduced for public construction. Likewise, on the mainland, there are many cases of initial infestations caused by the movement of pine logs.

4.3. The Decisive Factor: Sawyer–Nematode Contact

Monochamus alternatus is native to Japan but was a rather scarce insect until it teamed up with the pine wood nematode. Since *M. alternatus* is a secondary insect that only lays eggs on weakened trees, populations were presumably maintained at a low level (in spite of their high net reproduction rates) by the limited amount of available food.

For example, annual tree mortality in an experimental forest located in northern Honshu, prior to the introduction of the nematode in 1970, remained low for many years, with a temporary increase for about 3 successive years after typhoons struck the area in 1959 and 1966 (Fig. 9). Typhoons acted in two ways to increase the beetle population: (1) by suppling food in the form of wind-thrown trees, and (2) by weakening standing trees and making them susceptible to attack by the beetle. However, since the nematode was introduced into this area, tree mortality has increased gradually until most of the pines have been destroyed.

The destruction following introduction of the nematode is exemplified by a 10-year-old plantation of *pinus thunbergii* located in central Honshu (Fig. 10). Heavy tree mortality first occurred in late summer, and most dead trees were occupied by *M. alternatus* larvae. This indicated that the trees were weakened in early summer, probably by nematode infections. Thus, the population explosion of *M. alternatus* in Japan is caused primarily by its association with the nematode, *B. xylophilus*. Although the origin of the nematode is not yet clear, it seems to have been introduced from outside Japan. The fact that certain pine species from areas of the world other than Japan are resistant to the nematode, while pines native to Japan are very susceptible, seems to confirm this theory. The recent discovery of the widespread distribution of the nematode in North America, where no severe mortality is observed in the native pine forests, suggests that it originated on this continent.

These ideas are summarized in Fig. 11. Figure 11a illustrates the feedback loop between pine and sawyer populations before introduction of the nematode. Sawyer adults succeed in attacking pine trees weakened by competition or random stresses. Thinning attributable to tree death reduces competition and strengthens the remaining trees. Thus, an increase in stand density leads to an increase in the abundance of sawyers (+ interaction) which causes stand density to decrease because weakened trees are killed (− interaction). The overall effect of the loop is negative feedback, and a stable system is maintained.

Once the nematode is introduced into this system, however, a new feedback loop is introduced that temporarily becomes dominant in the system (Fig. 11). Beetle maturation

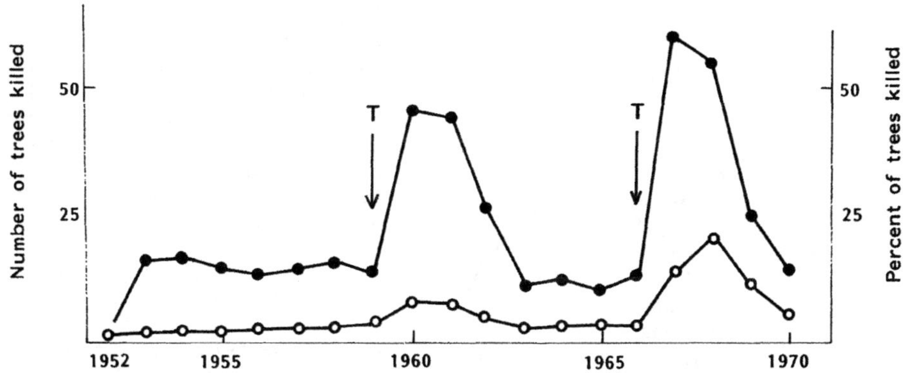

FIGURE 9. Annual tree mortality in Shidasan Experimental Forest prior to nematode introduction in 1970. *T* indicates the years when typhoons struck the forest.[44] ○ represents percent of trees killed. ● represents number of trees killed.

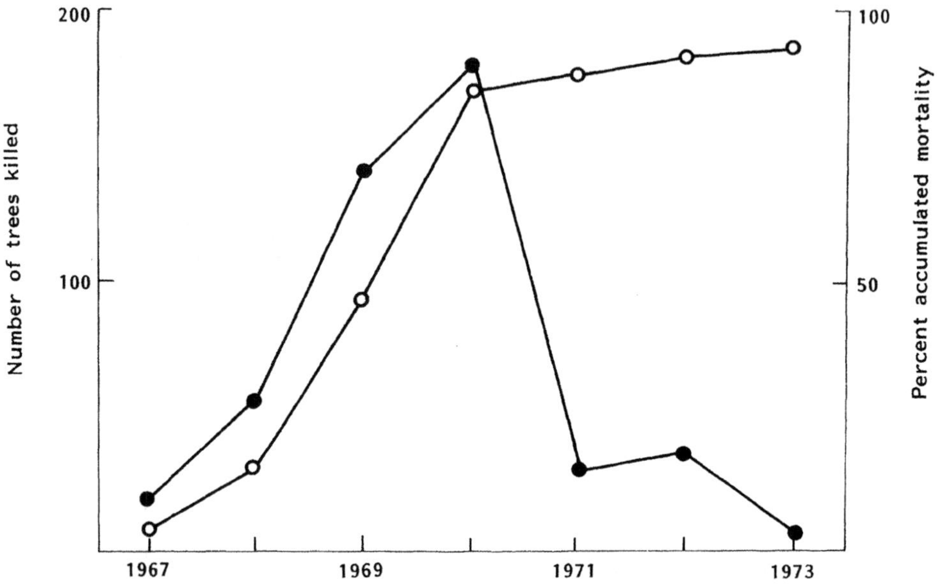

FIGURE 10. Annual tree mortality in a nematode-infected pine stand. Pine wood nematode caused complete destruction of a young pine plantation in 4–5 years.[12] ○ represents percent accumulated mortality. ● represents number of trees killed.

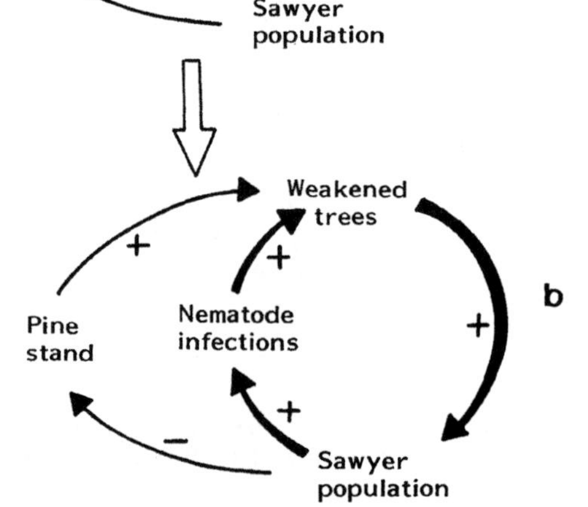

FIGURE 11. Change from stable system dominated by negative feedback (a) to an unstable system dominated by positive feedback (b). (a) An increase in stand density produces weakened trees (+ interaction), with which sawyer population increases (+ interaction). A decrease of stand density caused by the death of weakened trees promotes resistance to sawyer attack (− interaction). The overall effect is (+) × (+) × (−) = (−) feedback. (b) One more feedback loop is added to (a) by the introduction of nematode. Weakened trees provide the sawyer with breeding substance (+ interaction). An increased sawyer population produces more nematode infection (+ interaction), leading to an increase of weakened trees (+ interaction). The overall effect is (+) × (+) × (+) = (+) feedback. (Modified from [1])

feeding now increases its own food supply, creating an unstable positive feedback loop. This instability prevailing in Japanese pine forests was created by the accidental introduction of a new pathogen, the pine wood nematode.

5. MANAGEMENT STRATEGIES

Since the pine wilt in Japan is dominated by an unstable positive feedback loop, the continued operation of this structure will eventually destroy the pine forests. Theory suggests that evolutionary time is required to attain a stable equilibrium. Efforts should be directed toward the promotion of this trend, such as conversion to resistant pines. At the same time, urgent countermeasures are required.

Pine forest infestations occur through the linkage of three components: the pine tree, the pine wood nematode, and the Japanese pine sawyer beetle. A plan for control can therefore be divided into two approaches, one aimed at the individual components, the other at the linkages between components.[14] The various methods that have been tried are shown in Table IV.

5.1. Treatment and Utilization of Infested Trees

Felling infested trees and killing the *M. alternatus* larvae within them by burning, insecticide spraying, or rapid utilization is widely practiced in an attempt to reduce the rate of tree mortality.

Burning is the major control measure employed by some local governments because of its lethal effect on both beetles and nematodes. A newly devised portable charcoal kiln makes it possible to produce charcoal in the forest. Chipping trees for pulp or chip-board production is now being recommended. There is no effect on the quality of pulp material if trees are chipped within 6 months of death.

Spraying felled trees with insecticides is widely used[5] because logs do not need to be transported from the forest. Twenty chemical formulations with fenitrothion, carbaryl, fenthion, or diazinone as the active ingredient are registered to date and have been applied to more than 500,000 ha. Field tests showed 95–100% mortality of beetle larvae when logs were treated in autumn, but not as high in winter or spring when larvae have bored into the sapwood and formed pupal chambers. Fumigation with methyl bromide is also effective and the combination of chemical sprays and covering logs with vinyl sheets produces 100% mortality to the insects.

5.2. Preventive Chemical Sprays

Chemical spraying of tree crowns has been employed to prevent maturation feeding by *M. alternatus* adults and to reduce their populations. Eight chemicals, with fenitrothion, carbaryl, fenthion, or prothiofos as active ingredients, have been registered for aerial spraying, and 13 for ground spraying.[5]

Spraying from the air is considered more efficient than from the ground because

TABLE IV
Control Measures Judged from the Viewpoint of Current Practice[a]

Method	Technique	Use
A. Promotion of pine resistance	Genetic	
	Resistant exotic pines	□
	Resistant native pine clones	○
	Hybridization	○
	Environmental	
	Fertilization, nutrients	△
	Watering	□
	Thinning	□
	Mixed stands	△
B. Destruction of beetle populations	Chemical treatment of infested trees	
	Spraying	●
	Fumigation	□
	Physical treatment of infested trees	
	Debarking	□
	Burning	●
	Vynil sheet covering	□
	Water soaking	△
	Soil burying	□
	Chipping, charcoal making	●
	γ-radiation, radio heating	▲
	Natural enemies	○
	Attractants	□
	Ecdysis inhibitors	△
	Sterilization	△
C. Destruction of nematode populations in infested trees	Chemical treatment	△
	Radio heating	△
	Natural enemies	▲
A–B. Interruption between pines and beetles	Tree netting	□
	Preventative chemical spraying	●
	Repellents, antifeedants	▲
A–C. Interruption between pines and nematodes	Injection of nematicides	●
	Soil treatment of nematicides	○
	Induced resistance	▲
B–C. Interruption between beetles and nematodes	Application of nematode mounting mechanism	△
	Application of nematode dropoff mechanism	▲

[a]From Kobayashi.[14]
[b]Symbols: ●, widely practiced; ○, highly practiced; □, partially practiced; ▲, unknown practicality; △, low practicality.

maturation feeding is concentrated on the upper crown surface. Aerial spraying is usually conducted by helicopter from a height of about 10 m above the forest crown and at flying speeds of 35 mph. Spraying is usually applied twice so as to coincide with the beginning and the peak of adult emergence.

The protective effect of aerial spraying is generally satisfactory in reducing pine mortality. However, large fluctuations in efficacy are observed due to the rainfall or poorly timed application. Aerial spraying in heavily infested areas does not suppress tree mortality as effectively as in lightly infested areas.

5.3. Biological Control

Among eight fungi and two bacteria found on dead sawyer beetles, *Beauveria bassiana* is the most pathogenic. Inoculation of *B. bassiana* at 10^7 spore concentration in the laboratory resulted in 100% larval mortality. In field experiments, combined application of *B. bassiana* and a bacterium, *Serratia marcescens,* to infested logs causes higher mortality (90%) than does *B. bassiana* alone.[35]

An introduced entomogenous nematode, *Steinernema feltiae,* caused 80% or more mortality of *M. alternatus* larvae in infested logs.[8] This nematode seems to promise a new control technique because of its host-searching ability. Its successful use for biological control overseas also encourages further work. Biological control agents could be substituted for chemicals sprayed on logs.

5.4. Attractants

A commercially marketed attractant composed of benzoic acid and eugenol has been used for studying *M. alternatus* adult behavior and reducing its populations. Another lure composed of α-pinene and ethanol has also been developed into a commercial product. Since these attractants work on the matured adults from which most nematodes were already released, mass trapping is of no use in preventing tree death in the current year. Mass trapping may not be adequate as an independent control technique but may be useful as a supplementary one in sprayed stands where the beetle population level is low. They could also be used to monitor beetle populations.[5]

5.5. Nematicides

Laboratory and field experiments have led to the registration of three systemic nematicides: mesulfenhos, morantel tartalic acid, and levamysole hydrochloride. These chemicals, particularly morantel, are characterized by their low toxicity to mammals. The solution is injected into the tree trunk at the breast height about 3 months before infection by nematodes. Prevention lasts 1–2 years, depending on tree size and the dosage of chemicals.[21] The present method using these nematicides, however, is feasible for protection of valuable single trees or small stands but not for forests.

5.6. Resistant Pines

Exotic pines resistant to the nematode have been introduced but cannot be a permanent solution because many plantations of exotic pines have failed in the past due to their poor adaptability to local conditions. The production of resistant hybrids between *P.*

thunbergii and a Chinese pine (*P. massoniana*) that is resistant to the nematode, has recently started.

Breeding resistant cloning of *P. densiflora* and P. thunbergii is in progress as a national project. As a first step, candidate trees were selected from those that had survived in heavily infested stands. Grafts raised from their scions have been checked by nematode inoculation; of 23,000 candidates, 940 trees (748 *P. densiflora* and 192 *P. thunbergii*) passed the test. From these, 96 resistant clones (81 *P. densiflora* and 15 *P. thunbergii*) were selected through a second inoculation test.[3] Seeds of resistant clones will be supplied from scheduled seed orchards by 1995.

5.7. Necessity for a Nationwide Management Strategy

Since extensive infestations of pines in Japan is caused by the mutual interaction among the pine tree, Japanese pine sawyer, and pine wood nematode, management strategies should aim not only at one component but at their interactions as well. It is now certain that no single approach can be relied on to regulate the epidemic completely. Aerial spraying used to be thought of as a powerful tool but cannot cover all pine forests, even if financing could be found, because of technical and environmental problems.

Considering the background of the problem, we come to two essential conclusions, one biological, the other socioeconomic. First, native pines are highly susceptible to the nematode, probably because it is an introduced pest. The opportunity to eradicate this nematode at the initial stage of invasion has, however, been missed. The management strategy at present should be to reduce tree mortality in nematode-infested areas to a tolerable level and to stop the invasion into virgin areas. Second, neither treatment nor utilization of infested trees makes economic sense because of the low price of pine timber and labor shortages in forest lands, both of which are influenced by the nation's economic development. Having recognized these factors, a nationwide management strategy is a necessity.

A strategy to mitigate the impact of the present epidemic should be based on a reevaluation of pine forests from various viewpoints. They should be classified according to resource values, infestation levels, and the possibility of their replacement with other vegetation. The strategy should also be based on a reevaluation of available control techniques according to their effectiveness, cost, impact, and acceptability by forest managers. Direct control measures and their combinations should concentrate on forests where no other vegetation than pine is acceptable. Active convertion to resistant trees, other than pines, as well as to resistant pines, should be undertaken in heavily infested and easily converted areas.

ACKNOWLEDGMENTS. I thank Professor Alan A. Berryman, for his critical reading of the manuscript, and Dr. Mitsuru Maeda, for his illustration of the insect.

REFERENCES

1. Berryman, A. A., 1986, *Forest Insects: Principles and Practice of population management,* Plenum, New York.

2. Dropkin, V. H., Foudin, A., Kondo, E., Linit, M., and Smith, M., 1981, Pine wood nematode: A threat to U.S. forests?, *Plant Disease* **65:**1022–1027.
3. Fujimoto, Y., 1985, Project for breeding pines resistant to the pine wood nematode—Results obtained till the present. *Rinboku-no-Ikushu (Forest Tree Breeding)* **137:**11–14 (in Japanese).
4. Igarashi, M., 1980, Woodpecker predation on hibernating larvae of *Monochamus alternatus* HOPE, in: *Transactions of the Ninety-first Annual Meeting of the Japanese Forestry Society*, pp. 363–364 (in Japanese).
5. Ikeda, T., 1984, Integrated pest management of Japanese pine wilt disease, *Eur. J. For. Pathol.* **14:**398–414.
6. Ikeda, T., Enda, N., Yamane, A., Oda, K., and Toyoda, T., 1980, Attractants for the Japanese pine sawyer, *Monochamus alternatus* HOPE, *Appl Entomol. Zool.* **15:**358–361.
7. Ikeda, T., Miyazaki, M., Oda, K., Yamane, A., and Enda, N., 1981, The chemical ecology of *Monochamus alternatus* HOPE on the relationship with pine wood nematodes and host tree, in *Proceedings of the Seventeenth IUFRO World Congress, Division II*, pp. 297–303.
8. Katagiri, K., Mamiya, Y., Shimazu, M., and Tamura, H., 1984, A spray application of *steinernema feltiae* on pine logs infested with the pine sawyer, *Monochamus alternatus*, and its mortality induced by the nematode, in: *Transactions of the Ninety-fifth Annual Meeting of the Japanese Forestry Society*, pp. 479–480.
9. Kishi, Y., 1980, Mortality of pine trees caused by *Bursaphelenchus lignicolus* in Ibaraki Prefecture and its control, *Bull. Ibaraki For. Expt. Sta.* **11:**1–83 (in Japanese).
10. Kobayashi, F., 1975, Studies on the methods for estimating density and distribution of forest insect populations (I), *Bull. Gov. For. Expt. Sta.* **274:**85–124 (in Japanese with English summary).
11. Kobayashi, F., 1977, Distribution of coleopterous borers in pine, in: *Studies on Methods of Estimating Density, Biomass, and Productivity of Terrestrial Animals* (M. Morishita, ed.), pp. 84–96, JIBP Synthesis 17, Tokyo University Press, Tokyo, Japan.
12. Kobayashi, F., 1978, Pine bark beetle problem in Japan, referring to the discovery of the pine wood nematode, *Anz. Schädl. Pflanzensch. Umweltsch.* **51:**76–79.
13. Kobayashi, F., 1984, Biology of insects associated with the pine mortality in Japan, in: *Proceedings of the U.S.–Japan Seminar: The Resistance Mechanism of Pines against Pine Wilt Disease* (V. Dropkin, ed.), pp. 64–75, University of Missouri, Columbia, Missouri.
14. Kobayashi, F., 1984, Strategies for the control of pine wilt disease, in: *Proceedings of the U.S.–Japan Seminar: The Resistance Mechanism of Pines against Pine Wilt Disease* (V. Dropkin, ed.), pp. 171–177, University of Missouri, Columbia, Missouri.
15. Kobayashi, F., Yamane, A., and Ikeda, T., 1984, The Japanese pine sawyer beetle as the vector of pine wilt disease, *Annu. Rev. Entomol.* **29:**115–135.
16. Kobayashi, K., 1975, The number of pine sawyers in relation to pine mortality, *Shinrin Boeki (Forest Pests)* **24:**206–208 (in Japanese).
17. Kondo, E., Foudin, A., Linit, M., Smith, M., Bolla, R., Winter, R., and Dropkin, V., 1982, Pine wilt disease—Nematological, entomological, and biochemical investigations, University of Missouri, Columbia Agricultural Experiment Station SR282.
18. Linit, M. J., Kondo, E., and Smith, M. T., 1983, Insects Associated with the pinewood nematode, *Bursaphelenchus Xylophilus* in Missouri, *Environ. Entomol.* **12:**467–470.
19. Mamiya, Y., 1983, Pathology of the pine wilt disease caused by *Bursaphelenchus Xylophilus, Annu. Rev. Phytopathol.* **21:**201–220.
20. Mamiya, Y., 1984, The pine wood nematode, in: *Plant and Insect Nematodes* (W. R. Nickle, ed.), pp. 589–626, Dekker, New York.
21. Matsuura, K., 1984, Control of the pine wilt disease caused by pine wood nematode with trunk injection. *Shokubutsu Boeki (Plant Protection)* **38:**27–31 (in Japanese).
22. Miyazaki, M., Oda, K., and Yamaguchi, A., 1977, Behavior of *Bursaphelenchus lignicolus* toward unsaturated fatty acids, *J. Jpn. Wood Res. Soc.* **23:**255–261 (in Japanese with English summary).
23. Miyazaki, M., Yamaguchi, A., and Oda, K., 1978, Behaviour of *Bursaphelenchus lignicolus* in response to carbon dioxide released by respiration of *Monochamus alternatus* pupa, *J. Jpn. For. Soc.* **60:**249–254 (in Japanese with English summary).
24. Morimoto, K., 1976, Factors affecting population fluctuation of the pine sawyer, *Shinrin Boeki (Forest Pests)* **25:**195–198 (in Japanese).

25. Morimoto, K., and Iwasaki, A., 1975, Factors affecting the population fluctuation of *Monochamus alternatus. Shinrin Boeki* (*Forest Pests*) **24:**202–204 (in Japanese).
26. Morimoto, K., Iwasaki, A., Taromaru, I., and Takishita, K., 1977, Studies on *Monochamus alternatus* HOPE (XXV). Trend of pine tree mortality in Kumamoto Pref, in: *Transactions of the Eighty-eighth Annual Meeting of the Japanese Forestry Society*, 281–282 (in Japanese).
27. Nobuchi, A., 1980, Natural insect enemies of the pine sawyer, *Shinrin Boeki* (*Forest Pests*) **29:**23–28 (in Japanese).
28. Ochi, K., and Katagiri, K., 1979, Population dynamics and their factors of the pine sawyer larvae in dead pine trees, *Bull. Gov. For. Expt. Sta.* **303:**125–152 (in Japanese with English summary).
29. Oda, K., Hayashi, Y., Ohara, S., Enda, N., and Yamane, A., 1985, Pine-bark constituents relating to the maturation-feeding behavior of Japanese pine-sawyer adults (V). Isolation of a glycosylflavonol, in: *Transactions of the Ninety-sixth Annual Meeting of the Japanese Forestry Society*, pp. 519–520 (in Japanese).
30. Odani, K., Sasaki, S., Nishiyama, Y., and Yamamoto, N., 1985, Early symptom development of the pine wilt disease by hydrolytic enzymes exuded as a possible candidate of the pathogen, *J. Jap. For. Soc.* **67:**366–372.
31. Ogawa, S., and Hagiwara, Y., 1980, Expansion of the pine infestation caused by pine wood nematodes, *Shinrin Boeki* (*Forest Pests*) **29:**115–117 (in Japanese).
32. Oku, H., 1984, Biological activity of toxic metabolites isolated from pine trees naturally infected by pine wood nematode, in: *Proceedings of the U.S.–Japan Seminar: The Resistance Mechanism of Pines against Pine Wilt Disease* (V. Dropkin, ed.), pp. 110–118, University of Missouri, Columbia, Missouri.
33. Shaheen, F., Winter, R. E. K., and Bolla, R. I., 1984, Phytotoxin production in *Bursaphelenchus xylophilus* infected *Pinus sylvestris, J. Nematol.* **16:**57–61.
34. Shibata, E., 1981, Seasonal fluctuation and spatial pattern of the adult population of the Japanese pine sawyer, *Monochamus alternatus* HOPE, in young pine forests, *Appl Entomol. Zool.* **16:**306–309.
35. Shimazu, M., and Katagiri, K., 1981, Pathogens of the Japanese pine sawyer, *Monochamus alternatus* HOPE, and possible utilization of them in a control program, in: *Proceedings of the Seventeenth IUFRO World Congress*, Div. II, pp. 291–295.
36. Suzuki, K., and Kiyohara, T., 1978, Influence of water stress on development of pine wilting disease caused by *Bursaphelenchus lignicolus, Eur. J. For. Pathol.* **8:**97–107.
37. Takeshita, K., Hagiwara, Y., and Ogawa, S., 1975, Environmental analysis to pine damage in western Japan, *Bull. Fukuoka Pref. For. Expt. Sta.* **24:**1–45 (in Japanese).
38. Taketani, A., Okuda, M., and Hosoda, R., 1975, The meterological analysis on the epidemic mortality of pine trees, with special reference to the effective accumulated temperature, *J. Jap. For. Soc.* **57:** 169–176 (in Japanese with English summary).
39. Takizawa, Y., Igarashi, M., Yanbe, T., Shoji, T., and Saho, H., 1979, Studies on the pine sawyer in Tohoku District based on the rearing experiment, *Shinrin Boeki* (*Forest Pests*) **28:**84–89 (in Japanese).
40. Tamura, H., and Mamiya, Y., 1979, Reproduction of *Bursaphelachus lignicolus* on pine callus tissues, *Nematologica* **25:**149–151.
41. Togashi, K., and Magira, H., 1981, Age-specific survival rate and fecundity of the adult Japanese pine sawyer, *Monochamusa alternatus* HOPE, at different emergence times, *Appl. Entomol. Zool.* **16:**351–361.
42. Wingfield, M. J., Blanchette, R. A., Nicholls, T. H., and Robbins, K., 1982, The pine wood nematode: A comparison of the situation in the United States and Japan, *Can. J. For. Res.* **12:**71–75.
43. Yamane, A., 1980, Dispersal of the Japanese pine sawyer and its role in the spread and invasion of the pine wood nematode, in: *Proceedings of the Second IUFRO Conference on Dispersal of Forest Insects: Evaluation, Theory and Management Implications* (A. A. Berryman and L. Safranyik, eds.), pp. 153–163, Washington State University, Seattle, Washington.
44. Yamane, A., 1981, Problems of Scolytidae in Japan, in: *Proceedings of the Seventeenth IUFRO Congress Div. II*, pp. 469–473.
45. Yui, M., Suzuki, M., Yanbe, T., and Igarashi, M., 1985, Density of woodpeckers and predation rate of *Monochamus alternatus* larvae in pine-tree trunks, in: *Transactions of the Eighty-sixth Annual Meeting of the Japanese Forestry Society*, pp. 525–526 (in Japanese).

CHAPTER 22

THE GREATER EUROPEAN SPRUCE BEETLE

Jean-Claude Grégoire

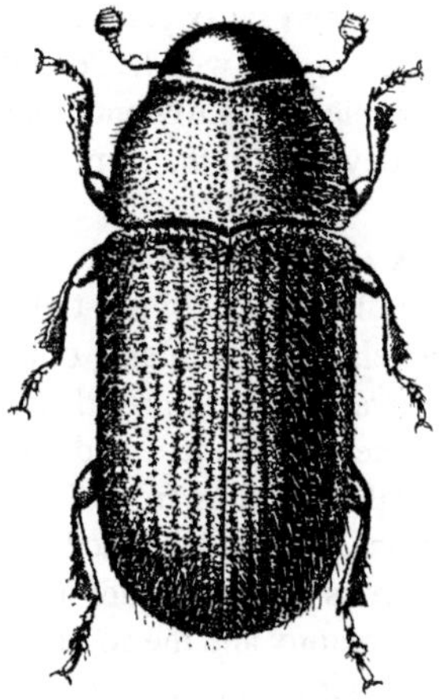

Jean-Claude Grégoire • Biologie Animale-C. P. 160, Université Libre de Bruxelles, B-1050 Brussels, Belgium

1. INTRODUCTION

The greater European spruce bark beetle, *Dendroctonus micans* (Kugelann) (Coleoptera: Scolytidae), is, with the Chinese species *Dendroctonus armandi* Tsai and Li, the only palaearctic representative of its genus. It appears to have moved to Eurasia in relatively recent times, and its closest relative is the Alaskan *D. punctatus* LeConte, the two species being doubtfully distinct.[79]

Dendroctonus micans, a large, jet black beetle (females 7–9 mm long, males slightly smaller), is a primary pest of spruce (*Picea*) but is sometimes found in other conifers (*Pinus, Abies, Larix, Pseudotsuga*). It usually attacks mature, apparently healthy trees, does not appear to be associated with pathogenic fungi, and does not need to kill the host by mass attack as the other primary species do (see Chapter 23). On the contrary, it inhabits a living tree for its entire life cycle.

In endemic conditions, *D. micans* is widespread but rare. Small numbers of trees are usually attacked, and these generally bear one or very few brood chambers. Local wounds due to single brood chambers can often be occluded by the host with no great consequences, except for a local defect in the wood. Few if any trees are killed. During outbreaks, however, *D. micans* is able to destroy entire stands. The trees are killed either by girdling, when several broods develop on the same host, or by secondary pests taking advantage of the weakened host.

The history of *D. micans* is one of continuous spread westward, with outbreaks occurring on the edges of its expanding range. In most of the interior parts of its range, *D. micans* is currently at endemic levels, and is causing only minor damage. By contrast, at the edge of its range, outbreaks covering an estimated 200,000 ha of spruce are occurring in Georgia, northeastern Turkey, central France, and the United Kingdom.

The present range of the beetle stretches from Siberia, the island of Sakhalin, and Japan (Hokkaido), to central France, with a localized spot in the United Kingdom. The Northern limit is that of the coniferous forest (including artificial plantations). The Southern boundary passes through Mt. Ventoux and the southern slopes of the Massif Central in France, Switzerland, the Austrian Alps, northern and eastern Yugoslavia, northern and central Rumania, and farther eastward, following the southern limit of Norway spruce to the Pacific coast. The insect is also present in some areas farther south, such as the parks around Kiev, the Soviet Socialist Republic of Georgia and northeastern Turkey.

2. BIOLOGY AND BEHAVIOR

The life cycle of *D. micans* is characterized by several peculiar habits. First, the females are fertilized in their birth chambers, usually by their own brothers. They then emerge to attack green trees; if the flow of primary resin does not expel them, they bore a gallery beneath the bark and lay batches of eggs in a groove. Immediately after hatching, the larvae start feeding collectively and remain gregarious until pupation. Pupation occurs in individual niches.

After metamorphosis, the young adults gather again, mate, and continue to feed under the bark until emergence. The entire life cycle takes 1–3 years, during which *D. micans* must face the defense reactions of the living host.

In accordance with its endogamous nature, the sex ratio of *D. micans* favors the female, e.g., in Germany, a 5 : 1 female sex ratio in autumn, and 20 : 1 to 48 : 1 in spring,[19] and ratios of 10 : 1 in Soviet Georgia.[80] Some possibilities for outbreeding do exist, however, because during outbreaks, the density of brood chambers per tree increases, and broods often merge, permitting extrafamilial mating. Beetles from such outbred matings tend to be heavier and more aggressive.[69] In addition, males do fly and respond to chemical signals,[78,70] so they may search for and enter mature brood chambers and fertilize pre-emergent females foreign to their own kin. Mating has never been convincingly reported in new egg galleries.

The extent of true heterosis provided by such outbreeding is questionable. Because of the normal kin-mating habits of the species, most of the insects inhabiting a particular area are likely to come, through one or several generations, from the same mother.

Fully mature fertilized females create their own solitary galleries. Each attacked tree bears only one or a very few brood systems, except under outbreak conditions. This strategy of individual colonization, whereby each female has to take her own chances, is favored by the kin-mating habits that permit larger proportion of females (potential colonizers) per brood. The loss of genetic diversity from inbreeding might be of little consequence due to the pest's relationship to the host; i.e., *D. micans* usually lives as a true parasite and does not kill its host, at least not immediately. Thus, it would not exert strong selective pressure on trees to evolve resistance mechanisms.

Dispersal of young females may occur in several ways. Some individuals do not emerge but simply bore new galleries at the edge of their own birth chamber. Others emerge but remain on the same tree, often establishing galleries in the vicinity of their own birth chambers, and some fly or possibly walk to colonize new trees. The factors governing this choice have not been systematically analysed, but temperature is probably critical. The flight threshold lies between 20° and 23°C,[16,20,78] temperatures that are rather uncommon in most of the *D. micans* range. Flight, however, does not necessarily lead to the colonization of new trees; females have been observed taking off, flying around, and coming back to their natal tree.[35] By contrast, colonization of new trees has been observed at temperatures well below the critical flight threshold. This phenomenon could be due to colonization of neighboring trees by walking insects or to flight occurring after brood chambers have been heated by sunlight.[31]

After probing the host, females eventually enter the bark. It should be stressed that, in Norway spruce at least, there is no evidence that *D. micans* is associated with any pathogenic fungus. Trees often have green crowns even after several successive beetle generations, and after having been almost girdled by the brood systems. It is not at all unusual for badly maimed trees to recover eventually. On reaching the xylem, a narrow channel, more or less oblique, is bored by the female and, after a few centimeters, this channel widens into a groove in which batches of eggs are deposited. The groove is gradually extended as oviposition continues. As they are laid, eggs are covered with sawdust and frass (Fig. 1a). This process takes slightly more than 20 days at room temperature[26] but is often protracted over much longer periods in the field. The average

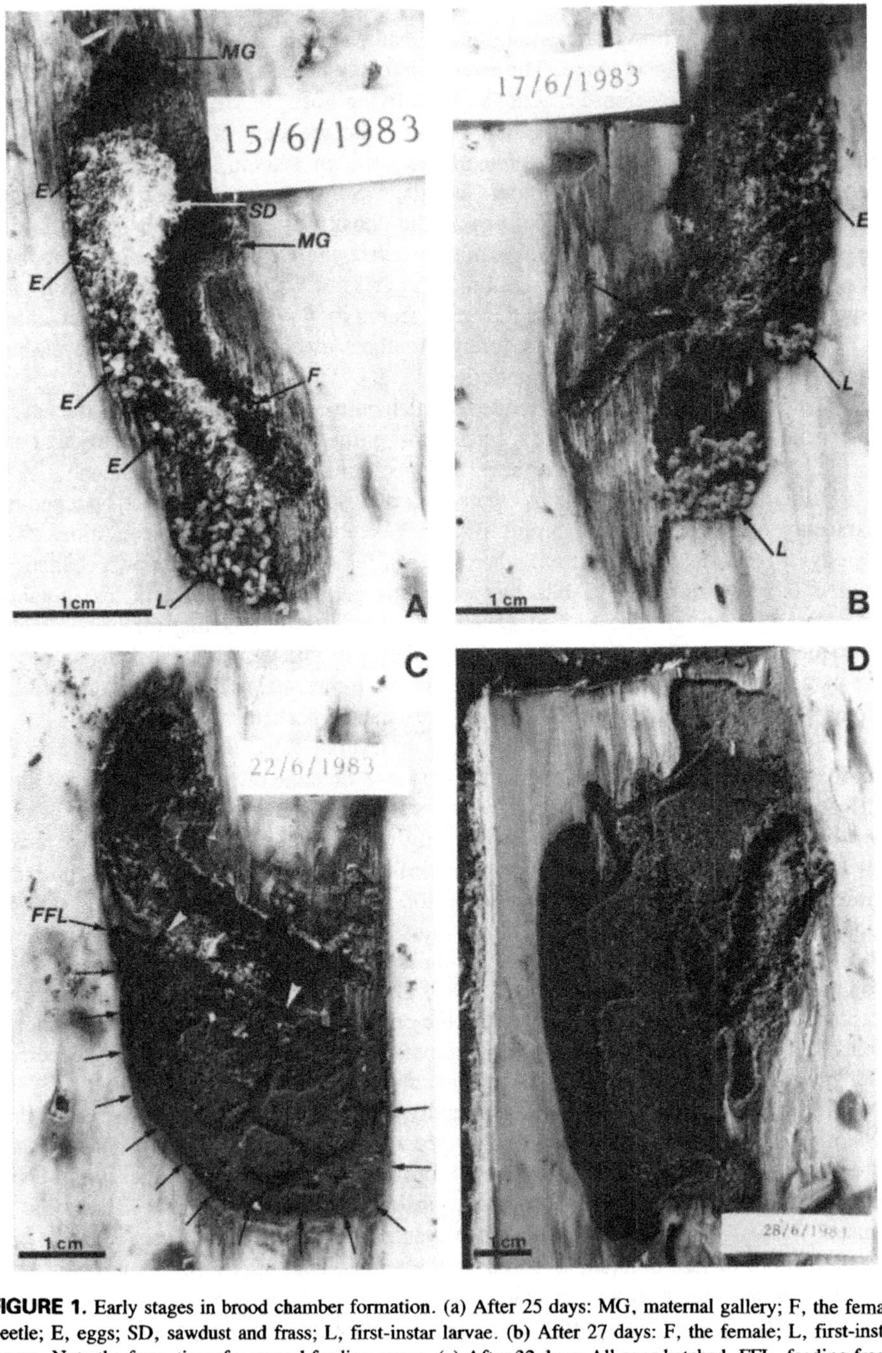

FIGURE 1. Early stages in brood chamber formation. (a) After 25 days: MG, maternal gallery; F, the female beetle; E, eggs; SD, sawdust and frass; L, first-instar larvae. (b) After 27 days: F, the female; L, first-instar larvae. Note the formation of a second feeding group. (c) After 32 days: All eggs hatched. FFL, feeding front-line probably resulting from the coalescence of two earlier groups coming from two single openings in the maternal gallery wall (white arrows). (d) After 38 days.

numbers of eggs laid in the field varies from 100 to 150 with maxima of 283 on Sitka spruce[22] and 293 on Norway spruce.

Female flight muscles degenerate once the egg gallery has been started and do not reconstitute afterward.[78] Dispersal of re-emergent females, if any, is probably limited to the neighborhood of their first egg gallery.

Attacks are usually located on the lower parts of the stems, root collars, or even underground roots. Under outbreak conditions, however, most of the galleries are usually higher on the trees.[4,16]

Egg-gallery formation and oviposition occur from April–June to August–November, according to latitude and elevation. Although there is no real swarming, a peak of new attacks often occurs at the start of the growing season, when the bulk of the overwintering adults emerge. Eggs laid too late in the season do not overwinter successfully.[20,22]

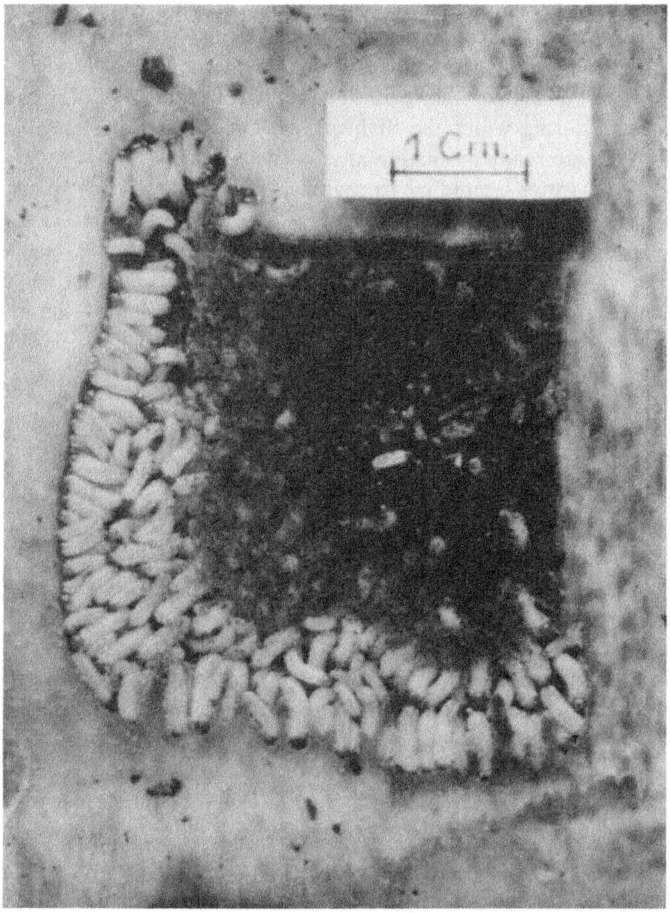

FIGURE 2. A group of fourth- to fifth-instar larvae feeding gregariously in a piece of phloem pressed between two glass plates. Frass and dead individuals have been left behind in the rectangular cavity from which the larva started.

The growth of the brood system has been followed in logs kept at room temperature.[26] Egg incubation takes 10–15 days at 20°C. The first larvae to hatch immediately bore a narrow channel through the discolored phloem surrounding the egg gallery. As more eggs hatch, the new larvae use the same channel, and a widening feeding front is formed in the green phloem (Fig. 1a). The larvae feed side by side, leaving the front only to defecate or molt, and frass is tightly packed to the rear, enclosing dead or diseased individuals. As hatching progresses, other channels are created, and these join the maternal gallery to the larval feeding chamber. Newly hatching larvae may join the initial group or start a new one; groups may split and gather, or fuse with other groups, according to the local topography of the system (Figs. 1 and 2). Gregarious larval groups appear to be maintained by aggregation pheromones (*trans*- and *cis*-verbenol, verbenone and myrtenol) produced by larvae during feeding or when exposed to the vapors of α-pinene.[25] Under laboratory conditions (19–23°C), larvae take 50–60 days to mature through five instars,[26] but in the field, this period may take a few months to more than a year.

Before pupation, the larvae move back into the frass and create individual niches, partly in the frass itself and partly in the underlying sapwood. Pupation lasts 4–6 days in the laboratory. At the end of this period, young adults gather again and proceed to maturation feeding during which the flight muscles and ovarioles slowly develop.[18,78] The young females are ready to emerge after 44 days at 20°C.[78]

The life cycle varies according to regions, and with the time of oviposition. In Turkey and Soviet Georgia, 12–15 months is sufficient,[40,64] but in Scandinavia 2 or 3 years may be required.[49] In France, Britain, and Belgium, the cycle lasts 1 year when the eggs are laid in May–June, but when oviposition occurs later, the first winter is spent in the larval stage, and one more winter may be necessary to complete the cycle.[16,24]

3. ECOLOGICAL RELATIONSHIPS

3.1. Relationships with the Host Tree

Throughout its vast range, *D. micans* attacks a number of indigeneous or introduced conifers. The usual hosts are spruces, mainly *Picea abies, P. sitchensis,* and *P. orientalis,* but also, on occasion, *P. ajanensis, P. breweriana, P. engelmanni, P. glauca, P. jezoensis, P. mariana, P. obovata, P. omorika,* and *P. pungens.* Scots pine (*Pinus sylvestris*) was also considered an occasional host, but recent outbreaks on this species in the Baltic area[76] and in Siberia[43] suggest that local shifts in host preference may occur. Sporadic attacks have also been observed on other pines (*Pinus contorta, P. nigra* var. *austriaca, P. sosnowskyi, P. strobus, P. uncinata*), firs (*Abies alba, A. holophylla, A. nordmanniania, A. pectinata,* and *A. sibirica*), larch (*Larix decidua*), and Douglas-fir (*Pseudotsuga menziesii*). Several investigators have noticed that Norway spruce is preferred to Sitka when both species coexist.[3,6,62] In Denmark, the following preference scale was observed[3]: *P. abies* > *pungens* > *sitchensis* = *glauca* > *omorika*. However, although less preferred, *P. sitchensis* has repeatedly been observed to be more susceptible to attack,[2,15,16,20] as is *P. orientalis.*[4] The susceptibility of several spruce species in Denmark was *P. pungens* = *orientalis* > *sitchensis* = *glauca* > *abies* > *omorika.*[3]

Extensive studies of *D. micans* attacking Norway spruce throughout the Massif

Central (France), showed that trees under 30 years of age were never attacked and that 55–85-year-old trees were preferred.[23] In Britain, however, Sitka and Norway spruces of all age classes over 13 years are attacked,[16] while in Turkey, any Oriental spruce older than 15 years and with a stem diameter greater than 7 cm can be attacked.[4] Considerable damage to young trees (21–25 years) was also reported in the recent outbreaks on *Pinus sylvestris* in Estonia and Siberia.[44,76] It is possible that bark thickness is of critical importance in determining the age at which a given species becomes susceptible to *D. micans*.[9] By contrast, the beetles seem to be less selective under outbreak conditions, attacking debarked timber and saplings.[69]

Dendroctonus micans colonizes green, standing trees; attacks on dying trees or on trees attacked by other insects are abnormal events that occur only at very high population densities.[69] Some attempts have been made to characterize the physiological state of susceptible trees. Bark sap osmotic pressure was similar in attacked and unattacked Norway spruces[12,46] but varied both along and around the trunk as well as with the season.[53] This heterogeneity within single trees provides support for the observation that beetles repeatedly probe a potential host until a favorable spot is discovered.[37] Oleoresin exudation pressure (OEP) and the rate of oleoresin flow (ROF) of Norway spruces in which *D. micans* adults had been artificially introduced were similar to those of sound trees (0–9 atm OEP and 3–4 ROF).[74] In trees in which larvae had been introduced, however, the observed values were much lower (0 atm OEP and 0–2 ROF).[74] Although these figures reflect the reaction of the tree to the bark beetle rather than its susceptibility, they suggest that adult *D. micans* are able to withstand the normal resin flow of its host.

By contrast, attacks are often noted on trees wounded by felling, logging, or pruning or damaged by frost, snow, sun, wind, lightning, or game, often at the wounds themselves. For example, of 386 Norway spruces attacked in the Morvan (France), 342 had been previously wounded.[12] This may perhaps be related to results obtained in the laboratory, where *D. micans* was strongly attracted to spruce resin and several monoterpenes.[49,75] High numbers of attacks have also been observed on the forks of double or multiple stems[4,62] and branch nodes.[16]

The occurrence of *fungal pathogens* has been found by some authorities to favor beetle establishment, e.g., trees infected by the root rot *Fomes annosus* appeared to sustain more attacks.[20,37] Similar associations between *Fomes* and *D. micans*, however, have not been observed by other workers, who generally found the organisms in different stands.[2,15] In Soviet Georgia, *Armillaria mellea* was thought to be responsible for weakening trees, making them susceptible to *D. micans*.[40] Attacks have also been observed at cankers of *Dasiscipha resinaria*.[37] Apart from their physiological effects on the host, however, cankers may simply provide easier penetration of the bark.[37] This might also explain the significant association between cankers of *Nectria cucurbitula* and attack by *D. micans*.[24]

Observations at the site level provide another type of information on host susceptibility to *D. micans*. Although attacks are often observed on good sites,[3,9,23,47] outbreaks usually occur on soils with poor mineral resources[23,30,32] or in stands with a deficient water economy, i.e., on shallow soils or well-drained soils with a remote or changing water table,[3,4,9,24,30,47] or on waterlogged soils, bogs, and more generally all poorly drained soils with a superficial impervious layer, which permits limited root development.[9,47] These observations suggest thus that, although *D. micans* attacks apparently healthy trees, its success is increased by factors adverse to the host, especially those affecting its water balance. We shall see later how this may explain the successive outbreaks marking the pest's progression through Europe.

Stand density is another variable that appears to influence the success of *D. micans*. Trees around openings in the stand and forest edges are often more heavily attacked,[23,63] perhaps because exposed edges create a warmer habitat for brood development or more stress on the tree. Several workers have also observed an inverse relationship between stand density and the percentage of attacked trees.[22,69] Others, however, found that the absolute number of attacked trees per hectare to be rather constant or even to increase with stand density.[23] Some infestations seem to start in sparsely stocked stands, but subsequent attacks are similar in all plots regardless of density.[3]

Stand composition may also affect the beetle population. Norway spruce is attacked more frequently when mixed with the more susceptible Sitka spruce.[62] Infestations also tend to be lighter in stands containing usually nonhost species (*Abies* spp., *Pinus* spp.) than in pure spruce stands.[23,69]

The resistance of a particular stand or species to *D. micans* attack is probably reflected by the proportion of aborted resin-flooded galleries. For example, 86% of the galleries on Norway spruce and 62% on Sitka were aborted in Britain,[16] 73% failed on Norway spruce in Belgium,[24] and 7–55% failed on Norway spruce in France.[23]

Abortive attacks, however, may prepare the way for future successful galleries. Abortive galleries often form horizontal tunnels, sometimes as long as 20 cm long,[20,37,47] and successful brood systems are often found a few centimeters under these aborted galleries.[47] Finally, dead beetles are found infrequently in resin-filled galleries, suggesting that the species is well adapted to escape from the defensive reaction of its host.

3.2. Relationships with Other Species

Although several bark beetles (e.g., *Ips typographus, Hylurgops palliatus, Pityogenes chalcographus, Polygraphus polygraphus, Cryphalus abietis, Pityophthorus pityographus*) may attack trees initially colonized by *D. micans,* none is a true competitor, since they do not bore in the living bark where *D. micans* lives and they usually attack higher up on the stems.[69] On occasion, *Ips typographus* incidentally attacks and kills a tree colonized by *D. micans,* causing the death of its slowly developing brood. The larvae of another xylophagous beetle, *Tetropium fuscum* (Cerambycidae), are occasionally observed to compete with and prey on *D. micans* larvae.[20]

The natural enemies of *D. micans,* although often represented by numerous individuals, are also few in species. For example, 71 species are associated with the sympatric spruce bark beetle, *Ips typographus,* but only 30 species are found with *D. micans,* 22 being common to both bark beetles.[56] This paucity of associated organisms is probably due in part to the solitary, often cryptic (underground), colonies of *D. micans* and the thick bark at the base of the trees, which may protect the beetles from attacks by some parasitoids. *D. micans* may also be protected by the large quantities of resin in its galleries that repel, deter, or intoxify some potential associates. The vapors of spruce monoterpenes α-pinene, limonene, and myrcene kill the generalist associate *Rhizophagus dispar,* whereas the specific predator *R. grandis* appears to be very resistant.[17]

In apparent contradiction to the above remarks, the list of all invertebrates ever found associated with *D. micans* includes about 130 species of insects, mites, and nematodes.[11,14,21,39,45,56,57,61,63] Most of these species, however, probably consist of inquilines, scavengers, fungus, or nematode feeders or of generalist predators living in the relatively innocuous shelter provided by the older parts of the brood systems. True predation and parasitism upon *D. micans* itself seems restricted to a few species, the most common of which are *Rhizophagus grandis* Gyllenhal (Coleoptera: Rhizophagidae), *Dolichomitus terebrans* Ratzburg (Hymenoptera: Ichneumonidae), and several *Lonchaea* spp. (Diptera: Lonchaeidae).

Rhizophagus grandis is a highly specific predator, attacking only *D. micans* (Fig. 3). The adults of this beetle locate the brood systems of their prey using chemical cues.[54,71] In Belgium, *R. grandis* was found with *D. micans* in all spruce stands sampled, even although the bark beetle population was generally low (1–5 brood chambers/ha). Predator colonization of brood chambers averaged 60%, increasing with the age of the brood chambers and reaching 85–90% at the third larval instar.[24] Predation rates were even higher in Soviet Georgia, where samples following releases of *R. grandis* showed 70–80% of the brood systems being colonized.[73]

Once in an egg gallery or brood chamber, adult *R. grandis* feed upon the eggs, larvae, and pupae of their prey and oviposit 50–230 eggs in the frass.[27,28,38,41] The larvae attack prey larvae and pupae in groups (Fig. 3c) and, on reaching the prepupal stage, leave the brood chambers to pupate in the litter. The predator may undergo one to three generations a year; as each brood develops independently, all stages are present throughout the year.

A complete assessment of the impact of *R. grandis* on field populations of *D. micans* has yet to be made. However, observations in the field show that it can completely consume *D. micans* broods[5]; laboratory experiments show that individual predators can consume 1.4–9.5 prey larvae during their lives.[42] Feeding experiments[54] established that each predator larva requires one fully grown prey larva to complete development and that pairs of predators introduced into brood systems when the prey were at the third to fourth larval instars devoured two thirds of the prey larvae.

Dolichomitus terebrans is not a specific parasitoid, attacking several xylophagous Coleoptera as well as various Microlepidoptera.[1] It attacks the last larval instar or the pupae of *D. micans*. The female parasitoid lays 200–500 eggs,[66] one per host, and the larvae develop as ectoparasites. Pupation occurs in a cocoon in the pupal niches of *D. micans*. The species is univoltine and overwinters in the pupal stage. Some *D. micans* broods may be nearly 100% parasitized,[20,30] but the effectiveness of the parasite is reduced by its univoltinism, specific host stage requirement, and inability to penetrate the thick bark where *D. micans* generally lives.

Lonchaea larvae are often found in the brood systems, especially among the late larval instars and pupae. These flies do not appear to be very specific, the larvae being found in the galleries of several bark beetles and xylophagous weevils.[63] The larvae of *Lonchaea* may be partly predaceous upon *D. micans*, and partly coprophagous, presumably feeding on beetle frass.[20,63] Older larvae seem capable of completing their development on frass only, but young larvae require dead *D. micans* larvae and live or dead pupae.[63] *Lonchaea* larvae feed gregariously, and 10 larvae will consume one *D. micans* pupa per day. One or two generations may be produced per year in Western Europe.[20,63]

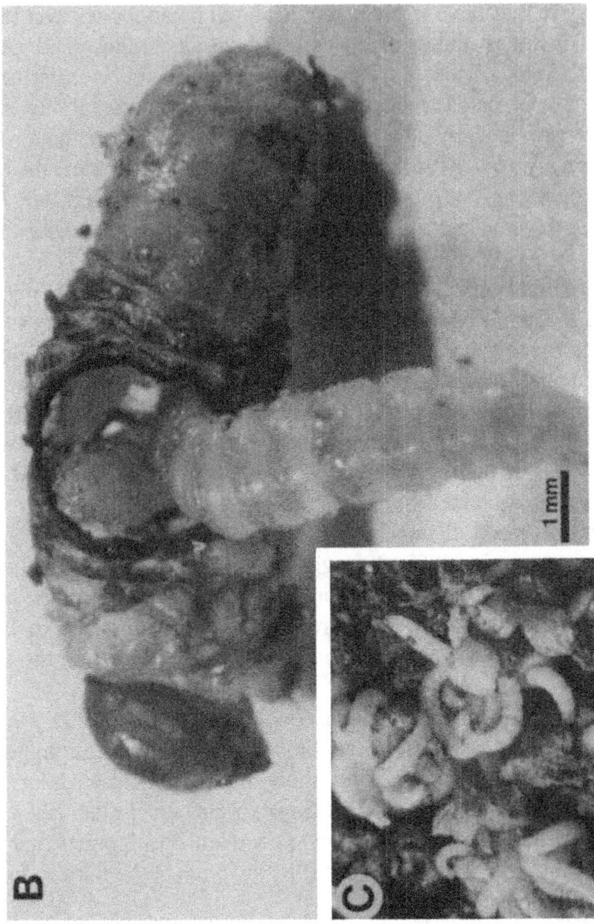

FIGURE 3. *Rhizophagus grandis* adult (a), larvae feeding inside a fifth-instar *D. micans* larva (b), and larvae feeding gregariously (c). Once a prey is wounded, *R. grandis* larvae aggregate rapidly and consume it in a few hours.

In addition to these invertebrates, several species of woodpeckers also attack the broods of *D. micans,* especially when they are high on the stem. However, as *D. micans* usually attacks the lower parts of the trunks and as the higher broods are often attacked by *Dolichomitus terebrans* or by *Lonchaea* spp., woodpeckers probably have little impact.[20]

4. PATTERNS OF NUMERICAL BEHAVIOR

4.1. Progression of Outbreaks in Eurasia

The history of *D. micans* outbreaks is clearly associated with its geographical expansion (Fig. 4). In many cases, there appears to be a gradual, inconspicuous spread into new areas followed, after a variable delay, by a sudden population increase, during which serious damage is inflicted to spruce stands. Outbreaks then decline to low and relatively stable population levels (Fig. 5). The speed of *D. micans* movement into Europe and neighboring countries since the middle of the nineteenth century is probably underestimated because of the cryptic nature of sparse populations. The pest was often acknowledged only after severe damage had occurred in a new area.

Dendroctonus micans was originally described in 1794, but there was no further mention of it until 1852.[36] Captures of *Rhizophagus grandis* in Bavaria in 1845 and in the Harz around 1850, however, provide indirect testimony to the inconspicuous earlier spread of *D. micans* into Germany.[33]

The first localized outbreak was observed in 1852 in Saxony.[36] Following this, the beetle increased its establishment in Germany, but still below outbreak levels; at a foresters' meeting in 1867, its presence was reported in the Harz, Thueringerwald, Anhalt, and Brunswick regions, but no serious damage was mentioned.[36] In 1872, an outbreak occurred in a 10-ha stand of 35-year-old spruce in the Harz Mountains.[36] Afterward, a series of outbreaks occurred in West Germany, particularly in the Rheinpfalz and the Eiffel between 1896 and 1900.[65,66] Similar damage occurred in the Eggebirge (Westphalia) and in Hesse.[55]

It was also during this period that Belgium was invaded from the Eiffel and Trier areas, although the pest might have been present in the country before.[51] Shortly after 1900, the outbreaks subsided in these parts of Germany and in Belgium and, except for a few very minor upsurges, no further outbreaks have been reported. Today, *D. micans* is present in almost all Belgian spruce stands but remains at low populations levels (1–5 brood systems/ha) and causes no serious damage.[24]

In Denmark, *D. micans* was first found in North Zealand in 1861 and was supposedly present throughout most of this region by 1898.[50] Serious damage was first recorded during 1907–1921 on *Picea abies* and *Pinus sylvestris.*[7] *D. micans* was then observed in South Jutland around 1920, causing some damage in 1923.[7] In 1943 and 1945, outbreaks were reported on *P. sitchensis* and *P. glauca* in Zealand and Jutland, respectively, and severe outbreaks occurred over the whole of Denmark in 1947–1949, especially in Sitka spruce stands[59]; e.g., 80–100% of the trees in some stands were attacked within 2–4 years.[3,22] New outbreaks occurred in later years, but the impetus of the beetle was reduced and, in 1966, the infested area had decreased to about one half that attacked in 1949.[3] Nowadays *D. micans* can be found in almost all Danish coniferous forest, but only about 50 ha suffer any damage.[3]

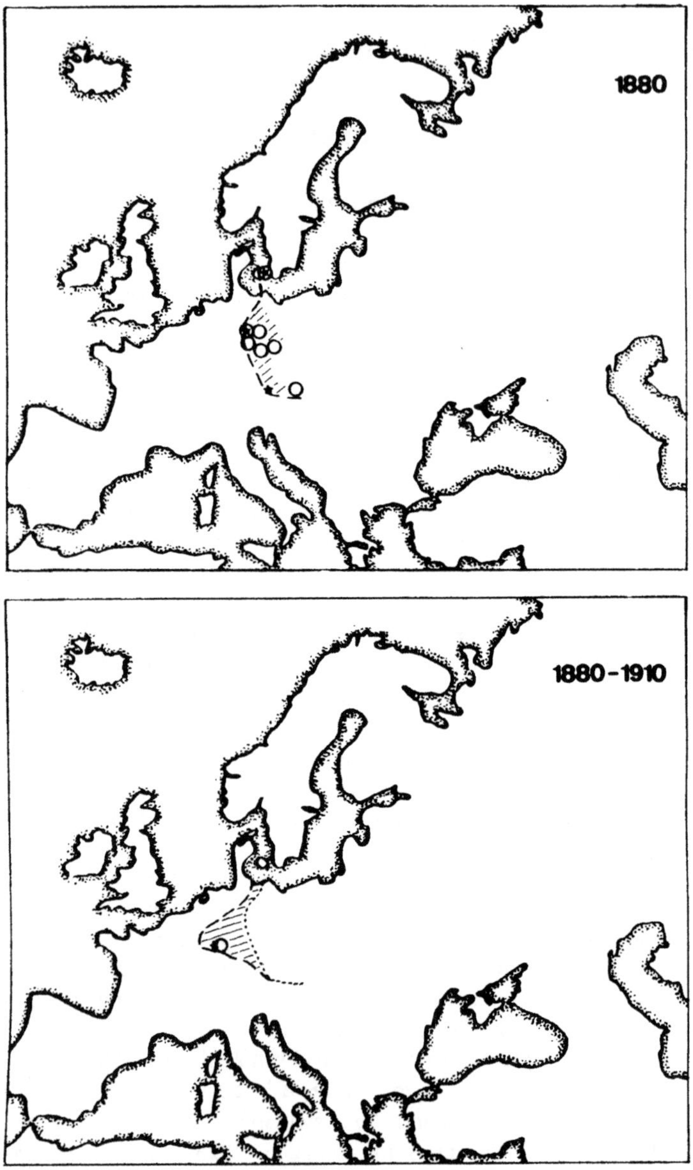

FIGURE 4. Progress of *Dendroctonus micans* in Europe and neighboring countries. (Shaded areas) outbreak zones; (○) first *D. micans*-infested stands observed; (★) *Rh. grandis* observed before, or the same year as *D. micans*; (●) outbreaks on *Pinus sylvestris*.

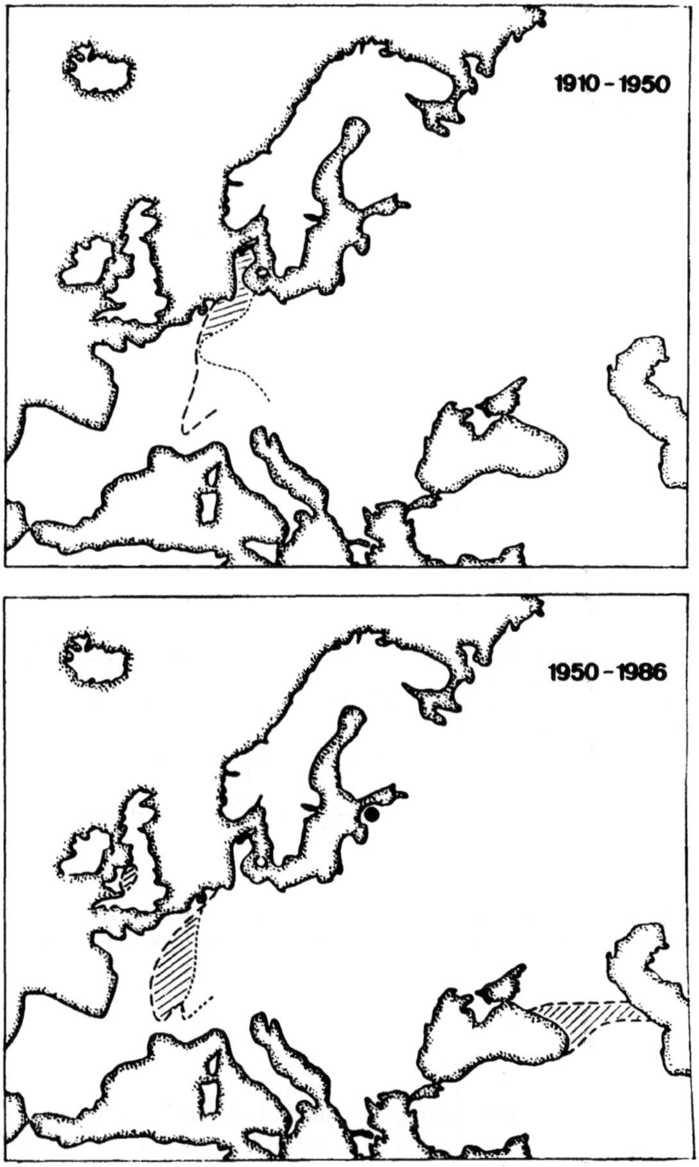

FIGURE 4. *(continued)*

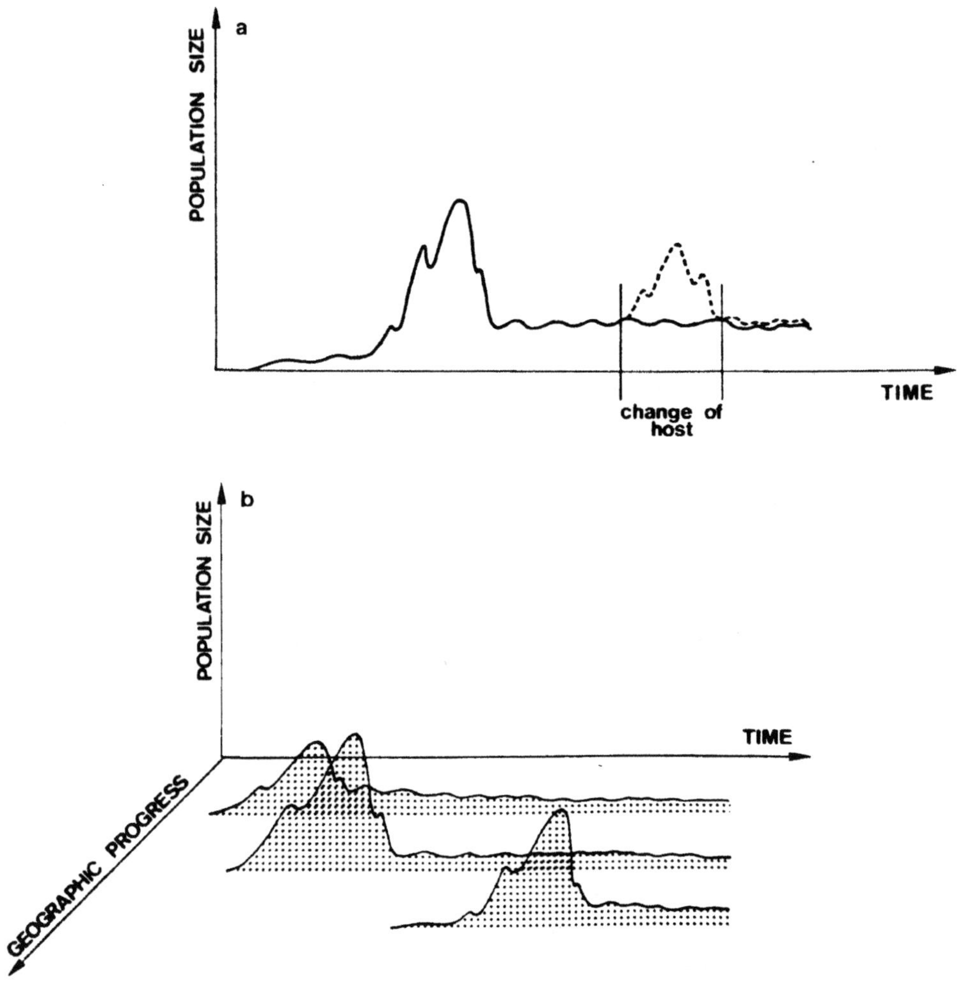

FIGURE 5. Hypothetical population changes of *D. micans* following arrival and establishment in a new area (a); population changes as the pest invades new territories (b). Broken line indicates shift to a new host.

Schleswig Holstein, a German region adjacent to Jutland, was invaded around 1938; however, *D. micans* had been previously observed sporadically.[20] As in Denmark, outbreaks occurred during 1947–1949, mainly in Sitka spruce plantations.[19,62] Afterward, there are no more records concerning important new outbreaks.

A localized outbreak occurred during 1952–1954 in the Saar but rapidly subsided.[34]

Dendroctonus micans was first found in the Netherlands in 1935, where it destroyed 0.77 ha of Sitka and oriental spruce in the province of Gelderland. Since that time, the beetle has been regularly reported from the few existing Dutch spruce stands but remains at very low densities. The total acreage of trees killed since 1935 is estimated at less than 1 ha.[30]

The first record of *D. micans* in France, near Metz, is dated 1897,[10] but it was not until 1950 that it was regularly observed in the Vosges, Jura, and Alps.[34] In 1959–1960, an outbreak occurred in Norway spruce forests of the Morvan region,[13] but it subsided soon afterward, with only 2–4% of the trees infested in 1976–1977.[23] Then, in 1972, outbreaks began on several thousand hectares of Norway spruce in the Massif Central.[10] In 1986, new outbreaks are still appearing as the pest continues to spread southward and westward.

Dendroctonus micans appeared in Soviet Georgia in 1956, probably coming in with imported timber from the North.[40] The bark beetle spread rapidly, and in 1976 about 135,000 ha of *Picea orientalis* were invaded and seriously damaged. The outbreaks gradually subsided and, at the end of the 1970s, the proportion of infested trees had fallen to 2.5–2.8%.[68]

Dendroctonus micans was discovered in Shropshire (Britain) in 1982 and, from tree-ring analysis of occlusions surrounding beetle attacks, it was determined that is must have arrived around 1972 with timber from the European mainland.[6] Extensive surveys found 53,000 Norway spruces and 10,100 Sitka spruces infested in 1982–1983,[16] and an estimated 24,000 infested trees were found in 1984.

Dendroctonus micans invaded the Pontic Mountains region of northeastern Turkey, near the border with Georgia (U.S.S.R), in 1966. From 1967 to 1972, 57,682 trees, mostly oriental spruce, had to be felled, and 268,562 young trees were burned.[64] During the early 1980s, the infested area reached 55,000 ha and is still increasing.[4]

It is interesting that some of the recent outbreaks have developed in Scots pine stands. In Estonia, where *D. micans* had been observed sporadically since 1912,[77] severe outbreaks were recorded in young (20–30 year) pine stands after 1971. In some places, more than 80% of the trees were damaged, and more than 55% killed.[76] Attacks on pine had also occurred earlier (~ 1950–1960) in other regions of the Baltic area, such as Leningrad, Belorussia, and the Bryansk region.[47] In the Lake Baikal region of Siberia, where *D. micans* had been recorded occasionally on overmature spruce, outbreaks occurred during 1974–1979 in young (22–25 year) Scots pine plantations[43] as well as in larch stands.[47]

4.2. Responses of Natural Enemies

Rhizophagus grandis, the major predator of *D. micans,* has tenaciously followed its prey wherever possible. In some instances, the two species have progressed very closely, with *R. grandis* being discovered before *D. micans* in Bavaria, the Harz Mountains, and eastern France.[52] Both species were found together in the first brood chamber discovered in Denmark.[50]

In most cases, however, there is a time lag between prey and predator movements. For example, *R. grandis* was first observed in the Rheinland in 1911,[33] 15 years after outbreaks of *D. micans* had begun in this area. In Belgium, *R. grandis* was caught only in 1947, some 50 years after the outbreaks had started, and in Holland, it was reported for the first time in 1950, 15 years after its prey had arrived.[15]

It is noteworthy that the mobility of *R. grandis* within stands appears to be much greater than its ability to spread from stand to stand. Predators artificially introduced into new areas expand no more than a few kilometers each year, but they colonize these areas very efficiently. In one case, predators were not recovered more than 3 km away from the release site 6 years after a release but were found in 48% of the brood chambers within this distance.[73] In another case, *R. grandis* expanded over 1.5 km in a year and colonized 42% of the available brood systems.[73]

In France, the predators expanded no more than 200 m after 7 months, but they had colonized about 17% of the suitable brood systems. Eighteen months later, the colonized area had extended to about 2 km, and 48% of the brood chambers contained *R. grandis*.

Although the territorial expansion of *R. grandis* from stand to stand appears to be rather slow, total colonization is nonetheless eventually achieved. For example, surveys made in Belgium during 1977–1983 showed the presence of *R. grandis* in all the *D. micans*-infested stands sampled.[24]

There is no precise information on the numerical response of *R. grandis* to its prey. However, the data suggest that density-dependent numerical responses do occur. For example, an average of 60% of *D. micans* brood chambers are colonized by R. grandis under endemic conditions in Belgium,[24] whereas up to 78% of the brood chambers can be colonized during outbreaks[73] (see Section 3.2).

5. HYPOTHESES FOR THE CAUSES OF OBSERVED POPULATION BEHAVIOR

As with other bark beetle species, the population behavior of *D. micans* may be related to changes in host susceptibility, depending on such influences as host species and origin, climate, stand and site conditions, and silvicultural practices. Unlike most other bark beetles, however, intra- and interspecific competition is not as significant in limiting population growth because of the solitary and cryptic habits of *D. micans* and possibly because it lives in a self-protected host.

In addition, *D. micans* has an unusually close relationship with its specific predator *R. grandis*. Under certain conditions, this predator might be a major regulating factor for the pest. In the following hypotheses, special emphasis is tentatively given to its possible role.

Finally, population changes of *D. micans* appear to depend strongly on its geographical expansion, characterized by periods of latency, rapid population growth, and collapse to low and relatively stable levels. These features should be included in any attempt to analyze the population behavior of this insect.

5.1. Population Spread and Establishment

The spread of *D. micans* is undoubtedly related to the extensive areas planted to spruce since the second half of the nineteenth century. In addition, the development of the timber trade provided a means for *D. micans* to cross otherwise impassable barriers such as the North Sea to Britain and the vast area devoid of spruce that isolates Georgia from the beetle's original range. Timber movements may have also contributed to its dispersal over less hazardous routes.

Because of the peculiar ecological habits of *D. micans,* the first steps of colonization were often quite furtive. Single fertilized females may establish sporadically, wherever they succeeded in infesting a tree, creating a loose network of small population foci. These small isolated spots often escape the notice of foresters but, as demonstrated by the records from Bavaria, Harz, and the Vosges, not always of *R. grandis.* Some *D. micans* spots, however, undoubtedly escape from *R. grandis.* This ability to escape from the predator is partly the result of the endogamous mating of *D. micans,* which produces fertilized emergent females able to start a new colony all by themselves. In respect to the selective pressure exerted by *R. grandis,* it may constitute an adequate counterweight to the loss of genetical diversity due to inbreeding.

5.2. Initiation of Outbreaks

The extensive spruce plantings in Europe over the past 150 years have not necessarily occurred in optimal conditions for these trees. Sites were often drought susceptible, waterlogged, or deficient in nutrients; elevations were sometimes too low; and silvicultural practices sometimes exacerbated these problems. These conditions set the stage for *D. micans* outbreaks, which often started after climatic factors had further weakened the stands. The critical influence of drought, particularly summer rainfall deficit, has been observed by several investigators[9,59,69]; the additional influence of very cold and dry winters has also been noted.[3] The outbreaks in northwestern Europe around 1935–1940 may also reflect the fact that the first Sitka spruces, which are more susceptible than Norway spruces, had been planted at the end of the nineteenth century and had reached susceptible age by the mid-1930s.[15]

Following their initiation by favorable conditions, outbreaks have often lasted for years, even after the restoration of normal climatic conditions, and have sometimes expanded over vast acreages of less susceptible spruce. Demographic pressure and increased vigor in the beetles, perhaps due to outbreeding in fused brood chambers, may have increased the capacity of the insect to overcome the defenses of the more resistant host trees.[69]

Regrettably, interest in *R. grandis* only arose during the 1960s, so there is little information on the role of the predator in outbreak initiation and expansion. It is a fact, however, that although *R. grandis* was present in many regions in which outbreaks had started, it was unable to prevent these outbreaks or to impede the growth of bark beetle populations during outbreak expansion. Two hypotheses have been proposed to explain this phenomenon. The first hypothesis proposes that, although *R. grandis* may have been present in stands in which outbreaks started, the increased susceptibility of the trees gave such a reproductive advantage to *D. micans* that the predator was unable to keep pace with it. The second hypothesis proposes that, as *R. grandis* is unable to locate every *D. micans* female that founds an infestation focus, and as site or/and weather conditions unfavorable to the trees could stimulate rapid growth of these predator-free foci, the subsequent arrival of *R. grandis* would be too late to prevent the outbreak from expanding and from generating new predator-free foci.

5.3. Outbreak Subsidence

Many factors have been associated with population declines of *D. micans*. The restoration of normal climatic conditions was probably a necessary prerequisite, and human actions such as sanitation thinning and clearfelling, chemical control, and quarantine measures have probably contributed. Intraspecific regulatory mechanisms may also have come into play at high population densities, i.e., intraspecific competition between broods on the same tree and unsuccessful attempts to establish attacks on unsuitable hosts. Natural enemies, especially *R. grandis,* were also important, although perhaps not the decisive factor at this stage.[5,20] For example, the mortality due to *R. grandis* was only about 15% during a recent outbreak on pines in Siberia, although the predator was abundant (up to 12 larvae/dm^2).[45]

5.4. Postoutbreak Behavior

The recent history of *D. micans* in Europe shows that outbreaks are followed by long periods of relative stability with populations remaining at low levels (Fig. 5). The absence of significant resurgences in stricken regions suggests that this stability may be more or less permanent. For example, the population decline in Denmark seems to have started after the drought of 1959, and *D. micans* has not caused serious damage since, in spite of the severe droughts in the 1970s.[3] In Belgium, 10 years or so after the outbreaks started in 1897, there are no further reports of *D. micans* as a noticeable problem. Surveys during the past 14 years show that *D. micans* is present in nearly every spruce stand of susceptible age in Belgium but remains at very low densities[24] irrespective of climatic stress (e.g., the severe droughts in 1947 and 1976 and in spite of reduced human care (e.g., sanitation practices back to normal). This reduced virulence cannot be attributed solely to the elimination of susceptible trees, because many plantations of the same genetic origins or on the same sites have reached susceptible ages since the last outbreaks.

Several hypotheses have been proposed. One hypothesis relies primarily on climatic and critical age influences on stand vigor. For example, the 1947 and 1956 outbreaks in Denmark were attributed to the coincidence of very low winter temperatures and annual rainfall deficiencies.[3] Rainfall deficiencies observed in later years did not coincide with very cold and dry winters and, therefore, would be less harmful to the trees. Furthermore, shorter rotations and several severe storms have reduced the numbers of Sitka spruces of susceptible ages.[3]

Another hypothesis includes the effect of climate and stand factors but places more emphasis on the role played by *R. grandis*. This predator has obviously taken advantage of outbreak periods to complete its spread, while the pest population has declined to a much lower level. The simultaneous occurrence of these two phenomena may have set the stage for effective long-term regulation by *R. grandis*. This hypothesis is supported by experiments in Georgia (U.S.S.R.), where the artificial introduction of *R. grandis* is followed by effective regulation if fewer than 3% of the trees are infested.[73]

Except for some local but minor resurgences, there seems to be one major exception to the generalization that *D. micans* populations remain at low stable levels following outbreaks. During the 1970s, severe infestations occurred in pine stands in Estonia and

Siberia, in regions occupied by *D. micans* for a long time but that had not suffered serious outbreaks on spruce. Although sporadic attacks on pine had been observed long ago,[58] pine stands usually constitute discontinuities in *D. micans*-infested areas. When outbreaks began in these areas, either because the bark beetle adapted to living on pine (one single fertilized female would have been enough) or because the trees themselves became more susceptible, *R. grandis* probably arrived too late to exert immediate control. Possible changes in the allelochemics produced by *D. micans,* due to the shift to another host, may have further delayed the predators that use chemical cues to locate their prey.

6. IMPLICATIONS FOR DEVELOPING MANAGEMENT STRATEGIES

The management of *Dendroctonus micans* has always been hindered by difficulties in detecting early infestations because attacked trees remain green, and population changes cannot be monitored with trap trees or pheromone trapping as they are with other bark beetles. Thus, prevention or suppression measures were rarely initiated during the initial invasion of the pest. Most control actions were taken under the most unfavorable circumstances, during the course of outbreaks. The effectiveness of these measures could not always be estimated unequivocally, because of the simultaneous restoration of stand vigor.[2] Similarly, the effectiveness of measures taken after the outbreaks could not be fully appreciated for the same reasons as well as the additional influence of biological control by *Rhizophagus grandis.*

Examples of preventive or early control are rare. An exception is in the United Kingdom, where *D. micans* has been expected for years. Quarantine procedures were initiated in 1961 and amended in 1980 and 1983. A special party visited infested stands in Schleswig-Holstein, Denmark, and Holland in 1964 and assessed the risks posed to British forestry.[9] The British were thus prepared when the insect was first discovered in 1982. Extensive surveys delineated a scheduled area subject to internal quarantine. Within this area, all the infested trees found in the annual extensive surveys were felled and peeled and the bark burned or sprayed with insecticide. As a result, the pest population was reduced by 20% between 1982 and 1983.[16] Then, in 1983, the Forestry Commission began mass producing *Rhizophagus grandis* and inoculative releases were made, in first priority, at the edges of the infestation areas; e.g., 32,000 predators were released in 961 sites in 1984,[38] and about 40,000 insects were released in 1985. The hope is that these early predator releases will establish a stable *D. micans* population without passing through the outbreak phase. A rather similar release strategy has been implemented during the same period by a Belgian-French group in areas surrounding the outbreak zone in central France. Differing from the British method, however, more massive releases were made (50 pairs/attacked tree; 500 pairs/ha when more than 20 infested trees/ha), but in a much smaller number of sites.[29]

Several types of control measures have been taken during the outbreak phases. Sanitation felling is still widely used in France and Britain.[16] However, the brutal modification of stand structure that results from heavy sanitation felling often exacerbated the problem because openings in the stand favor increased infestation.[8] In the German Eiffel and the Belgian Hertogenwald, the rule became to spare all trees not attacked over more than two thirds of their circumference, to clean and cover the wounds with a preservative, and to destroy the bark and excised insects. This labor-intensive method was later replaced in other regions by treatments with HCH or γ-HCH preparations sprayed directly on infested bark areas.[2,4,60,62] This method is generally effective, although failures have occurred on Sitka spruce growing on poor sandy soils and infected by root rot.[2] Biological control with *R. grandis* began in Soviet Georgia in 1963. By 1970, 54,000 predators had been released,[72] and by 1976, 20 rearing units were producing *R. grandis* throughout the country. Biological control was most effective, however, after populations had been reduced by insecticide treatment.[73] Finally, a report claims that soil fertilization halted a local outbreak in Schleswig-Holstein.[32]

Practices implemented during the postoutbreak periods have usually involved stand manipulation, such as growing Sitka spruce on shorter rotations together with heavy early thinning.[2] However, many forest districts no longer find it necessary to take special precautions against *D. micans*.[3] In Soviet Georgia, the observation that spruce in mixed stands (30% or less spruce with beech, fir, and pine) was less susceptible led to the creation of mixed forests on areas previously infested.[67]

The peculiarities of the population dynamics of *D. micans* call for an adaptive approach to pest management. During the outbreak phases, control is necessarily expensive and slow. Under these conditions, a combination of silvicultural, chemical, and biological measures, supplemented by the planting of mixed stands, as practiced in Soviet Georgia, would probably constitute the best approach. In the early establishment phase, a combination of extensive survey, sanitation thinning, and predator releases, as adopted in Britain, would seem appropriate although expensive. It may be that, following predator establishment, surveys could be partly replaced by kairomone trappings of *R. grandis*. It is also possible that some still unexploited potentials lie in anticipating *D. micans* with routine preventive predator releases in uninfested areas surrounding the infested districts.

ACKNOWLEDGMENTS. I am grateful to Broder Bejer, Colin J. King, Jacques M. Pasteels, Daniel Schvester, and David Wainhouse, who critically read the manuscript and provided valuable comments, suggestions, information, and references. Thanks also to Marianne Baisier and Joël Merlin for stimulating discussions. Alan Berryman went through the manuscript with great accuracy, and I owe much to him for this final version. Any errors and omission, however, are my own. The drawing of *D. micans* that heads this chapter is reproduced by kind permission of the British Forestry Commission.

REFERENCES

1. Aubert, J. F., 1969, *Les Ichneumonides Ouest-Paléarctiques et leurs hôtes*. I. Pimplinae, Xoridinae, Acaenitinae, Quatre Feuilles Editeur, Alfortville.
2. Bejer-Petersen, B., 1976, *Dendroctonus micans* KUG in Denmark. The situation 25 years after a "catastrophe," *Z. Pflanz. Pflschutz* **83**(1/2/3):16–21.
3. Bejer, B., 1984, *Dendroctonus micans* in Denmark, in: Proceedings of the EEC Seminar on the Biological Control of Bark Beetles (*Dendroctonus micans*), Brussels, pp. 2–19.
4. Benz, G., 1984, *Dendroctonus micans* in Turkey: The situation today, in: Proceedings of the EEC Seminar on the Biological Control of Bark Beetles (*Dendroctonus micans*), Brussels, pp. 43–47.
5. Bergmiller, F., 1903, *Dendroctonus micans* und *Rhizophagus grandis.*, *Zentralbl. Ges. Forstw.* **29**:252–256.

6. Bevan, D., and King, C. J., 1983, *Dendroctonus micans* KUG—A new pest of spruce in U.K., *Commonw. Forest Rev.* **62**(1):41–51.
7. Boas, J. E. V., 1923, *Dansk Forst Zoologi*, 2nd ed., Gyldendalske Boghandel, Copenhagen.
8. Brichet, O., and Severin, G., 1903, Le *Dendroctonus micans*. Dégâts, moyens préventifs et destructifs. *Bull. Soc. Centr. for. Belg.* **10**:244–261.
9. Brown, J. M., and Bevan, D., 1966, The Great Spruce Bark-Beetle *Dendroctonus micans* in North West Europe, *For. Comm. Bull.* 38, 41 pp. London, Her Majesty's Stationery Office.
10. Carle, P., Granet, A.-M., and Perrot, J.-P., 1979, Contribution à l'étude de la dispersion et de l'agressivité chez *Dendroctonus micans* KUG. en France, *Bull. Soc. Entomol. Suisse* **52**:185–196.
11. Ceianu, I., and Istrate, G., 1976, Observatii asupra entomofagitor gindacului de scoarta *Dendroctonus micans* KUG. (Coleoptera, Scolytidae) in Carpatii orientali, *Muzeul de Stiintele Naturii Bacau Studii si Comunicari*, pp. 107–121.
12. Chararas, C., 1960, Variations de la pression osmotique de *Picea excelsa* à la suite des attaques de *Dendroctonus micans* KUG. (Coleoptera, Scolytidae), *C. R. Acad. Sci.* **251**(18):1917–1919.
13. Chararas, C., 1961, Les Scolytides de l'épicéa dans la forêt domaniale de St-Prix (Morvan) et mesures de protection des peuplements, *Rev. Pathol. Vég. Entomol. Agric. Fr.* **40**(2):40–129.
14. Domocos, M., and Istrate, G., 1972, Mites (Acarina) from the spruce fir bark galleries of *Dendroctonus micans* KUG and other habitats from Cimpulung Moldovenesc zone, *Studii si comunicari de ocrotirea naturii suaceava*, pp. 221–229 (in Rumanian with English summary).
15. Elton, E. T. G., 1950, *Dendroctonus micans* KUG, a pest of sitka spruce in the Netherlands, VIII Int. Congr. Entomol., Stockholm, pp. 759–764.
16. Evans, H. F., King, C. J., and Wainhouse, D., 1984, *Dendroctonus micans* in the United Kingdom: The results of two years experience in survey and control, in: Proceedings of the EEC Seminar on the Biological Control of Bark Beetles (*Dendroctonus micans*) Brussels, pp. 20–34.
17. Everaerts, C., Grégoire, J.-C., and Merlin, J., (1988), Toxicity of spruce monoterpenes against bark beetles and their associates, in: *Mechanisms of Woody Plant Resistance to Insects and Pathogens* (W. J. Mattson, J. Lévieux and C. Bernard-Dagan, eds.), pp. 335–344.
18. Francke-Grossmann, H., 1949, Ueber Kopulation, Eiablage und Gelbkoerperbildung beim Riesenbastkaefer *Dendroctonus micans* KUG, *Mitt. Dtsch. Ges. Angew. Entomol.* **5**:142–153.
19. Francke-Grossmann, H., 1950, Die Gefaehrdung der Sitkafichte durch Rotfaule (*Fomes annosus*) und Riesenbastkaefer (*Dendroctonus micans* KUG.) in Aufforstungsrevieren Schleswigs, VIII Int. Congr. Entomol., Stockholm, pp. 773–780.
20. Francke-Grossmann, H., 1954, Populations-dynamische Faktoren bei der Massenvermehrung des *Dendroctonus micans* KUG. an der Sitkafichte in Schleswig-Holstein, *Verh. Dtsch. Ges. Angew. Entomol.*, pp. 108–117.
21. Gaprindashvili, N. K., Gumberidze, O. D., and Kharazishvili, K. V., 1967, A faunistic study of the natural enemies of *Dendroctonus micans* KUG. in Georgia, USSR, *Soobshch Akad. Nauk Gruz. SSR* **74**(1):167–172 (in Russian).
22. Gohrn, V., Henriksen, A. M., and Bejer-Petersen, B., 1954, Iagttagelser over *Hylesinus (Dendroctonus) micans. Forst. Forsogv. Danmark* **180**(XXI):380–433.
23. Granet, A.-M., and Perrot, J.-P., 1977, *Dendroctonus micans* Kug. dans le sud-est du Massif Central. Aire d'extension et premier essai d'interprétation des dommages. Mémoire de 3e année, Ecole Nationale des Ingénieurs des Travaux des Eaux et Forêts.
24. Grégoire, J.-C., 1984, *Dendroctonus micans* in Belgium: The situation today, in: Proceedings of the EEC Seminar on the Biological Control of Bark Beetles (*Dendroctonus micans*), Brussels, pp. 48–62.
25. Grégoire, J.-C., Braekman, J.-C., and Tondeur, A., 1982, Chemical communication between the larvae of *Dendroctonus micans* KUG (Coleoptera: Scolytidae), pp. 253–257, in: Les Colloques de L'INRA, 7. *Les Médiateurs chimiques.*
26. Grégoire, J.-C., and Merlin, J., 1984, *Dendroctonus micans:* The evolution of a brood system, in: Proceedings of the EEC Seminar on the Biological Control of Bark Beetles (*Dendroctonus micans*), pp. 80–86.
26. Grégoire, J.-C., Merlin, J., and Pasteels, J. M., 1984, Mass-rearing of *Rhizophagus grandis* for the biological control of *Dendroctonus micans:* An interplay between technical requirements and the species biological characteristics, *Med. Fac. Landbouww. Rijksuniv. Gent.* **49/3a**:763–769.
28. Grégoire, J.-C., Merlin, J., Pasteels, J. M., Jaffuel, R., Vouland, G., and Schvester, D., 1984, Mass-rearings and releases of *Rhizophagus grandis* in Lozère, in: Proceedings of the EEC Seminar on the Biological Control of Bark Beetles (*Dendroctonus micans*), Brussels, pp. 122–128.

29. Grégoire, J.-C., Merlin, J., Pasteels, J. M., Jaffuel, R., Vouland, G., and Schvester, D., 1985. Biocontrol of *Dendroctonus micans* by *Rhizophagus grandis* in the Massif Central (France): A first appraisal of the mass-rearing and release methods, *Z. Angew. Entomol.* **99:**182–190.
30. Grijpma, P., 1984, *Dendroctonus micans* KUG. in the Netherlands: The situation today, in: Proceedings of the EEC Seminar on the Biological Control of Bark Beetles (*Dendroctonus micans*), pp. 35–47.
31. Haarlov, N., and Bejer-Petersen, B., 1952, Temperaturmaalinger i bark og ved af Sitkagran, *Forst. Forsogsv. Danmark* **170**(XXI)**:**43–91.
32. Hansen, J. E., 1957, Kann der Riesenbastkaefer *Dendroctonus micans* KUG. in Schleswig-Holstein erfolgreich bekaempft werden, *Forst. Holzw.* **12**(10)**:**167–168.
33. Horion, A., 1960, Faunistik der Mitteleuropaischen Kaefer, 7(1) *Rhizophagus*, pp. 136–146. Ueberlingen, Bodensee.
34. Husson, R., and Stauder, F., 1954, A propos du *Dendroctonus micans* KUG. Coleoptera Ipidae, *Rev. For. Fr.* **6**(66)**:**355–359.
35. Istrate, G., 1972, Observatii asupra activitatii in perioada de zbor a lui *Dendroctonus micans* KUG. (Coleoptera, Scolytidae). Studii si comunicari de ocrotirea naturii, Consiliul judetean de indrumare pentru ocrotirea naturii suceava, pp. 247–256.
36. Judeich, J. F., and Nitsche, H., 1895, *Lehrbuch des Mitteleuropaeischen Forstinsektenkunde*, Vol. I, Paul Parey, Berlin.
37. Kangas, E., 1952, Ueber die brutstaetenwahl von *Dendroctonus micans* KUG. auf Fichten, *Ann. Entomol. Fenn.* **18**(4)**:**154–170.
38. King, C. J., and Evans, H. F., 1984, The rearing of *Rhizophagus grandis* and its release against *Dendroctonus micans* in the United Kingdom, in: Proceedings of the EEC Seminar on the Biological Control of Bark Beetles (*Dendroctonus micans*), pp. 87–97.
39. Kleine, R., 1944, Die europaeischen Borkenkaefer und ihre Feinde aus den Ordnung der Coleopteren und Hymenopteren, *Entomol. Blaetter* **40**(3/4)**:**68–83; **40**(5/6)**:**125–133.
40. Kobakhidze, D. N., 1967, Der Riesenbastkaefer (*Dendroctonus micans* KUG.) in Georgien (UdSSR), *Anz. Schaedlingsk.* **40**(5)**:**65–68.
41. Kobakhidze, D. N., Tvaradze, M. S., Yashvill, G. V., and Kraveishvili, I. K., 1968, Artificial rearing of *Rhizophagus grandis* Gyll. for the control of *Dendroctonus micans* KUG. in Georgia, *Soobshch. Akad. Nauk Gruz. SSR* **51**(2)**:**435–440 (in Russian).
42. Kobakhidze, D. N., Tvaradze, M. S., and Kraveishvili, I. K., 1970, Preliminary results of introduction, study of bioecology, development of methods of artificial rearing and naturalization of the effective entomophage, *Rhizophagus grandis* GYLL., against the European spruce beetle, *Dendroctonus micans* KUG., in spruce plantations of Georgia, *Soobshch. Akad. Nauk Gruz. SSR* **60:**205–208 (in Russian).
43. Kolomiets, N. G., 1981, Interrelations between the European spruce beetle (*Dendroctonus micans* KUGEL., Coleoptera, Scolytidae) and Scots pine (*Pinus sylvestris* L.) in western Siberia, in: *Rol'vzuimootnoshenii rastenie-nasekomoe v dinamike chislennosti populayatsi lesnykh vreditelei* (Tezisi dokladov sovetskikh uchastnikovk simposiumu IYUFRO/MAB, August 24–28, 1981, Irkutsk, SSSR), (A. S. Isaev, ed.), pp. 28–29, 68–69,
44. Kolomiets, N. G., and Bogdanova, D. A., 1976, Outbreaks of *Dendroctonus micans, Lesnoe khozyaistvo* **12:**71–73 (in Russian).
45. Kolomiets, N. G., and Bogdanova, D. A., 1980, Parasites and predators of xylophagous insects of Siberia, Siberian Branch of the USSR Academy of Sciences, Sukachev Institute of Forest and Wood, Novosibirsk (in Russian).
46. Kraemer, G. D., 1953, Die kritischen Grenzen der Brutbaumdisposition für Borkenkaeferbefall an Fichte, *Z. Angew. Entomol.* **34**(4)**:**463–512.
47. Krivosheina, N. P., and Aksentev, S. L., 1984, *Dendroctonus micans* in spruce stands, *Lesovedenie* **5:**63–68 (in Russian).
48. Lekander, B., Bejer-Petersen, B., Kangas, E., and Bakke, A., 1977, The distribution of bark-beetles (Coleoptera, Scolytidae) in the Nordic Countries, *Acta Entomol. Fenn.* **32:**1–100.
49. Lemperiere, G., 1979, Contribution à l'étude biologique et biochimique du comportement et de l'attractivité de *Dendroctonus micans* KUG. (Coleoptera: Scolytidae), ravageur de l'Epicéa, thesis of the 3e cycle, Pierre & Marie Curie University, Paris.
50. Lovendal, E. A., 1898, *De danske Barkbiller*, Det Schubotheske Forlag, Copenhagen.
51. Mathieu, C., 1859, Faune entomologique belge. Coléoptères, *Ann. Soc. Entomol. Belg.* **3:**2–92.

52. Méquignon, A., 1909, Révision des *Rhizophagus* paléarctiques, *L'Abeille* **31:**103–119.
53. Merker, E., 1956, Der Wiederstand der Fichten gegen Borkenkaeferfrass, *Allg. Forst- Jagdztg.* **127**(7):12–145.
54. Merlin, J., Grégoire, J.-C., Baisier, M., and Pasteels, J. M., 1984, Some new data on the biology of *Rhizophagus grandis,* in: Proceedings of the EEC Seminar on the Biological Control of Bark Beetles (*Dendroctonus micans*), Brussels, pp. 107–121.
55. Metzger, D., 1897, Forstentomologische Mitteilungen 1. *Hylesinus micans* KUG. und *Pimpla terebrans* RTZB, *Mündener Forstl. Hefte* **12:**59–68.
56. Mills, N. J., 1983, The natural enemies of scolytids infesting conifer bark in Europe in relation to the biological control of *Dendroctonus* spp. in Canada, *Commonw. Inst. Biol. Control Biocontrol News Inform.* **4**(4):305–328.
57. Nuorteva, M., 1956, Ueber den Fichtenstamm-bastkaefer, *Hylurgops palliatus* GYLL., und seine Insektenfeinde, *Acta Entomol. Fenn.* **13:**1–118.
58. Palm, T., 1948, Ett angrepp av *Dendroctonus micans* KUGEL. pa tall (Col. Scolytidae), *Entomol. Tidskr.* **69**(4):212–214.
59. Petersen, B., 1952, *Hylesinus micans,* artens udbredelse og en oversigt over dens optraeden i Danmark, *Dansk Skovforen. Tidsskr.* **37**(6):299–322.
60. Rudnev, D. F., and Khramtsov, N. M., 1963, The control of *Dendroctonus micans* in Gruziya forest, *Zaschita Rast. Vreditelei Boleznei* **7:**28–30 (in Russian).
61. Ruehm, W., 1956, Die Nematoden der Ipiden, *Parasitol. Schriftenreihe* **6:**1–437.
62. Ruehm, W., 1968, Zur mechanisch-chemischen und oekologischen Bekaempfung des Riesenbastkaefer *Dendroctonus micans* KUG. *Z. Angew. Entomol.* **43**(3):286–325.
63. Sanguignol, M., 1981, Recherches sur la biologie de *Lonchaea albitarsis* et *Dolichomitus terebrans,* deux insectes associés à *Dendroctonus micans.* Travail de fin d'études en Agronomie, Université libre de Bruxelles.
64. Serez, M., 1979, Der Riesenbastkaefer (*Dendroctonus micans* KUG.) in der Turkei (Coleoptera: Scolytidae), *Turk. Bitki Koruma Derg* **3**(1):17–24.
65. Severin, G., 1902, L'invasion de l'Hylésine géante, *Bull. Soc. Centr. For. Belg.* **9:**145–152.
66. Severin, G., 1908, Le genre *Dendroctonus, Bull. Soc. centr. For. Belg.* **15:**239–248; 312–320.
67. Shavliashvili, I. A., Mukhashavria, A. L., and Zharkov, D. G., 1976, *Dendroctonus micans* KUG. Population and integrated control in Georgia, XVI IUFRO World Congress, Dv. **II:**421–428.
68. Shavliashvili, I. A., and Murusidze, B. V., 1978, About dynamics of numbers of European spruce bark beetle (*Dendroctonus micans* KUG.) in Georgia. Abstract presented at the IUFRO Conference, *Dispersal of Forest Insects: Evaluation, Theory and Management Implications,* Zurich-Zuoz, Switzerland.
69. Shavliashvili, I. A., and Zharkov, D. G., 1985, Effects of ecological factors on the interactions between populations of *Dendroctonus micans* and *Ips typographus* (Coleoptera: Scolytidae), in: L. Safranyik (Ed.), *Proceedings of the IUFRO Conference The Role of the Host Plant in the Population Dynamics of Forest Insects, Banff, Canada,* pp. 227–232.
70. Tømmerås, B. A., Mustaparta, H., and Grégoire, J.-C., 1984, Receptor cells in *Ips typographus* and *Dendroctonus micans* specific to pheromones of the reciprocal genus, *J. Chem. Ecol.* **10**(5):759–769.
71. Tondeur, A., and Grégoire, J.-C., 1980, Chemical orientation of *Rhizophagus grandis* towards mates and towards prey: *Dendroctonus micans,* in: *Animals and Environmental Fitness* (R. Gilles, ed.), pp. 93–94, Pergamon, Oxford.
72. Tvaradze, M. S., 1976, On the acclimatisation of *Rhizophagus grandis* for control of *Dendroctonus micans, Sb. Nauch. rabot Izuch. B.E. Luboeda Gruzii, Tbilisi* **2:**76–90 (in Russian).
73. Tvaradze, M. S., 1977, Using *Rhizophagus grandis* to control *Dendroctonus micans, Sb. Nauch. rabot Izuch. B.E. Luboeda Gruzii, Tbilisi* **3:**56–61 (in Russian).
74. Vasechko, G. I., 1976, Host selection and colonization by some spruce bark beetles, in: *The Host Plant in Relation to Insect Behaviour and Reproduction, Symp. Biol. Hung.* **16:**287–290.
75. Vasechko, G. I., 1978, Host selection by some bark-beetles (Col. Scolytidae). I. Study of primary attraction with chemical stimuli, *Z. Angew. Entomol.* **85**(1):66–76.
76. Voolma, K. K., 1978, Distribution and harmfulness of the European spruce beetle, *Lesnoe Khozyaistvo* **31**(4):90–91 (in Russian).
77. Voolma, K. K., 1980, The distribution and ecology of the European spruce beetle *Dendroctonus micans* KUG. (Coleoptera, Scolytidae) in Estonia, *Lesovod. Issled.* **16:**44–51.

78. Vouland, G., Giraud, M., and Schvester, D., 1984, La période ténérale et l'envol chez *Dendroctonus micans* KUG. (Coleoptera, Scolytidae), in: *Proceedings of the EEC Seminar on the Biological Control of Bark Beetles (Dendroctonus micans)*, Brussels, pp. 68–69.
79. Wood, S. L., 1963, A revision of the bark-beetle genus *Dendroctonus erichson* (Coleoptera: Scolytidae), *Great Basin Natur.* **23**(1–2):1–117.
80. Zharkov, D. G., and Kobakhidze, T. D., 1976, Sex ratio in the *D. micans* populations of the Borjomi stands, *Sb. Nauch. rabot Izuch. B.E. Luboeda Gruzii, Tbilisi* **2**:121–124 (in Russian).

CHAPTER 23

THE SPRUCE BARK BEETLE OF EURASIA

Erik Christiansen and Alf Bakke

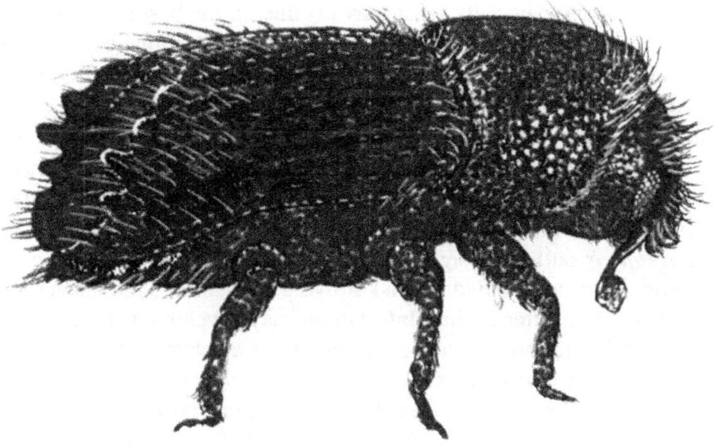

Erik Christiansen and Alf Bakke • Norwegian Forest Research Institute, N-1432 Ås-NLH, Norway.

1. INTRODUCTION

The spruce bark beetle, *Ips typographus* Linnaeus (Coleoptera: Scolytidae), is the most destructive scolytid in the coniferous forests of the palaearctic region. In Europe, outbreaks may lead to the destruction of millions of Norway spruces [*Picea abies* (Linnaeus) Karsten], its principal host tree. In Northeastern Asia a subspecies, *Ips typographus* Linnaeus f. *japonicus* Niijima, may cause severe damage to spruces in the group *P. jezoënsis* (Siebold and Zuccarini) Carrière *sensu lato*.

Norway spruce reproduces and grows well under shady conditions. On better sites, therefore, it outcompetes other forest trees (firs excepted) and frequently forms large continuous tracts of single-species climax forest. In the absence of human exploitation, several agents of disturbance, such as wind, snow, fungi, and insects, ensure periodic renewal of the climax forest. Among the insects, *I. typographus* plays the main role.

Norway spruce is not a long-lived species. Although it may survive for some 400 years in high altitudes or latitudes, its life span is generally much shorter. Even as far north as southern Scandinavia, its vigor may decline on the best sites after 50–60 years.

From the role of *I. typographus* as benefactor in the natural ecosystem, the species becomes the villain when humans claim all timber for their own use. The introduction of systematic forest utilization into Central Europe was followed by reports of calamities or catastrophies caused by the spruce bark beetle. The records from the Harcynian mountains of North Germany offer an example. Here, recurring outbreaks were reported during the seventeenth and eighteenth centuries, to be curbed by the introduction of clean management at the beginning of the nineteenth century.[55]

Ips typographus outbreaks signify a drastic change in the behavior of the species. From a low population maintained in dead or dying trees, slash, and unbarked timber, the species switches to active tree killing. Infestations may develop rapidly over large tracts of forest when climatic and forest stand conditions create an abundance of susceptible hosts.

Left unchecked by human interference, epidemics may eventually encompass tens of thousands of square kilometers. The epidemic in Central Europe after the World War II killed some 30 million cubic meters of Norway spruce.[56] This outbreak, the largest from which accurate loss data exist, lasted some 7 years in the southern areas[67] and about 5 in the northern regions.[55] A previous epidemic in the Harcynian Mountains lasted for 30 years, and in Southern Norwegian forests the pest was recorded as a serious problem for 50 years (1850–1900). Other outbreaks have been less extensive and persistent, depending on the degree of stress-causing external factors, as well as human control efforts.

Although the bulk of the killed trees are normally salvaged, the quality of the timber is reduced by cracking from desiccation, by wood-boring secondary insects (e.g., *Trypodendron, Tetropium, Siricidae,* etc., and by fungal blue stain. In the recent *I. typographus* epidemic in Norway, the death of 5 million m^3 of spruce timber supposedly led to a reduction of the gross national product (GNP) by some 500 million Norwegian kroner.[45]

European forests have a long history of systematic management for sustained yield. An unpredictable event such as a bark beetle outbreak may upset elaborate management plans or even eliminate the economic returns from a small holding for several decades.

2. BIOLOGY AND ECOLOGY

2.1. Hibernation

The spruce bark beetle generally hibernates in the adult stage, mainly in the forest litter close to the tree where it developed (Fig. 1). It may also overwinter under the bark of standing or felled trees or in logs and slash, particularly in the southern parts of its range. In Finland, larvae and pupae have supercooling points of −13°C and −17°C, while adults can tolerate temperatures close to −30°C.[2] In spring, the species undergoes a temperature-dependent maturation process prior to taking flight, the adaptive value of which may be the synchronization of attacks.[26]

2.2. Dispersal

The first spring flights occur when air temperatures rise to about 20°C.[2,7] Depending on latitude and altitude, this generally occurs during April–June in different parts of Europe.[58] Like most bark beetles, the flight period consists of two stages: (1) a dispersal phase, followed by (2) a search phase.[70]

The length of the disperal period, hence the distance covered during flight, is extremely variable among *I. typographus* individuals of both sexes, ranging from a few meters to several kilometers.[17,18,26,28] Laboratory experiments suggest that this variation is dependent on the energy (fat) reserves of the individual beetle. Apparently, some of this energy has to be used before a beetle responds to aggregation pheromones or reproductive instincts.[28,30] Therefore, the individual variation in energy appears to explain the variable flight capacity of *I. typographus*.[28] Starved beetles may fly readily, but their prolonged migration is contingent upon food intake. If a suitable breeding substrate is not found, beetles are able to continue flying day after day for a long time.[26]

When the dispersal flight is over, the beetles begin to search for suitable breeding material on which males make initial or pioneer attacks. After this, their powerful aggregation pheromones play a key role in host finding. This "spread out, then search" behavior seems to be an efficient mechanism for finding suitable breeding material scattered sparsely throughout the forest.

2.3. Chemical Communication

During the phase of host selection, volatiles from the spruce may guide the beetles to areas containing breeding materials.[5] Better known, however, is the chemical communication system involving pheromones. Aggregation pheromones as well as antiaggregation pheromones have been identified (Fig. 2).

The major components of the aggregation pheromone are (S) *cis*-verbenol and 2-methyl-3-buten-2 ol.[8] (S) *cis*-verbenol is produced by several *Ips* species,[63] while methylbutenol is specific to *I. typographus*. Ipsdienol, which is also shared by most *Ips* species, also occurs, but seems to play a minor role.

Two components inhibit the response to the aggregation pheromone, and act as

FIGURE 1. Life cycle in *Ips typographus* in southern Norway. (a) Adult beetles overwinter in the litter and under bark. (b) They emerge and attack trees, logs or windthrows in spring. (c) Parent beetles may re-emerge in summer and attack new trees or logs. (d) Brood beetles emerge in autumn and enter overwintering sites. (From Berryman.[15])

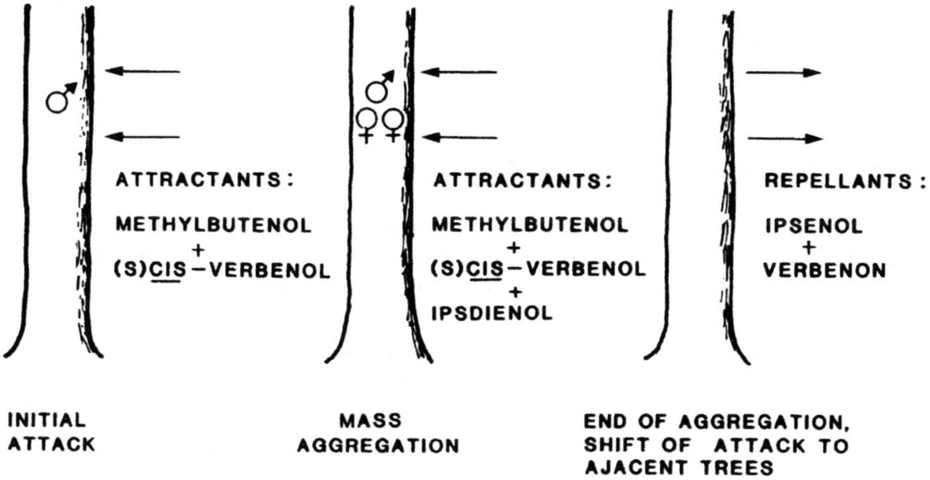

FIGURE 2. Principle of chemical communication in *Ips typographus*.

antiaggregation pheromones. These are verbenone and ipsenol.[6,16] They are released after the females have entered the gallery and seem to regulate gallery density and cause the shift of attack to new bark areas or neighboring trees. Recent experiments show that small quantities of ipsdienol added to the two major attractants increase the attraction, while larger amounts of ipsdienol or ipsenol reduce the attraction, especially when combined.[75]

Male beetles are the major producers of all pheromone components, and monoterpenes from host trees are the precursors for their biosynthesis.[34,63] Dispensers containing synthetic aggregation pheromones of *I. typographus* are marketed, mainly for use as baits in traps for the purpose of monitoring beetle population or for mass trapping.

2.4. Attack and Reattack

During endemic (nonoutbreak) periods, beetles breed in wind-felled trees, slash and unbarked timber, and occasionally in weak trees. During outbreaks, the beetles conquer healthy trees. Like other *Ips* species, but unlike *Scolytus* (Chapter 26) or *Dendroctonus* (Chapters 22,24,25) bark beetles, the male initiates gallery construction by excavating a nuptial chamber in the phloem. In doing so, it emits aggregation pheromones that draw conspecifics of both sexes to the site. While other males initiate their own nuptial chambers in the same tree, one to four females are welcomed to the completed ones. Here they mate with the resident male and begin excavating their longitudinal egg galleries in the phloem. The male helps in the excavation by pushing out boring dust. During construction, the phloem and cambium are contaminated by spores of several blue-stain fungi, one of which has an active role in killing healthy trees (see Section 2.6).

Parental beetles may leave successfully attacked host trees after a short period of time to produce a second "sister" brood in other trees and even reemerge again to start a third brood.[7,39] Parents reemerge sooner under crowded conditions in the laboratory.[1]

Whereas only one generation is produced per year at high altitudes and latitudes,[2,4] *I. typographus* is generally bivoltine in the lowlands of Central Europe,[59] where three generations per year may occur on warmer sites.[68]

2.5. Development and Emergence

In the field, high temperatures are generally favorable for *I typographus*, shortening the time for development. However, temperatures exceeding 47°C have proved fatal in the laboratory.[2] Breeding material exposed to direct insolation frequently reaches this temperature.

Dense beetle infestations in a tree cause intense intraspecific competition, so that the number of offspring produced per unit area of bark surface reaches a maximum at fairly low attack densities.[1,47,59] Furthermore, offspring from high density broods are smaller, have less energy reserves, and produce fewer eggs. Brood density can therefore influence the dynamics of *I. typographus* populations for more than one generation.[1]

In the northern areas, where *I. typographus* is univoltine, beetles emerge from July to October,[2,19,50] depending on the time of brood establishment, microclimate, and weather. An unusual second generation established during a particularly long and warm summer in Norway proved disadvantageous because the population entered winter in the more vulnerable larval and pupal stages and froze to death under the bark of standing trees.[4]

In areas of Central Europe in which bivoltinism is normal, emergence of the second generation may take place as late as November, but more often the adult beetles hibernate in the brood tree and emerge the following spring.

2.6. Host Relations

Throughout their long life span, coniferous trees have a high likelihood of being attacked by a variety of parasitic and pathogenic organisms. These heterotrophic organisms profit by obtaining access to a large body of carbon and nutrients bound up in the tree. As in other plants, a system of defense has evolved which normally protects the tree from its enemies.

2.6.1. The Constitutive Host Defense

Some genera of *Pinaceae*, spruces included, possess systems of ducts in the bark and wood that hold stores of pressurized oleoresin. When a duct is severed oleoresin is exuded and, in adequate quantities, will cleanse and later seal the wounds.[12]

Oleoresin has long been regarded as a defense of Norway spruce against *I. typographus*.[27] Resin exudation in this species is extremely variable both between trees, and between places on a single tree (Fig. 3). Within a given stand, large trees generally yield more primary resin than the small ones[55] and older trees more than young ones.[23]

Because the resin ducts are directly influenced by the turgor pressure of their adjoining parenchyma cells, the oleoresin pressure fluctuates with daily and seasonal changes in tree water potential.[61,62] However, the amount of oleoresin exuded probably depends less

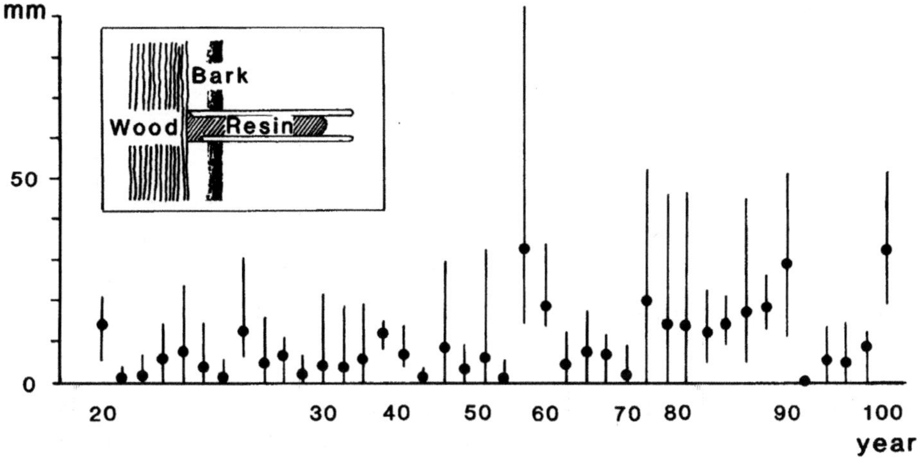

FIGURE 3. Yield of duct resin in 24 hr from Norway spruce trees of variable age (mean and range of 8 observations per tree). Insert shows measurement technique. (From Christiansen and Horntvedt.[23])

on pressure than on the storage capacity of the interconnected duct system (cf. ref. 32). Studies on the oleoresin exudation from artificial wounds show a distinct seasonality[55]: a spring increase, high values during the main growing season, and an autumn decrease. This fluctuation is closely correlated to ambient temperatures. By comparison, pines in the southern United States exhibit a close association between oleoresin yield and potential evapotranspiration of their environment.[43]

Different year-rings of Norway spruce often hold highly variable numbers of resin ducts. Ducts may be formed in response to external factors such as frost or drought, and to attack by microorganisms and insects. During a recent severe drought in South Norway, unusually high numbers of resin ducts were formed in spruce trees.

When the number of punctures to this interconnected duct system increases, the amount of resin exuded per wound rapidly tapers off (Fig. 4). *I. typographus* males may thus encounter a constitutive defense which is very effective, or practically non-existent. Hence, host characteristics and sequence in aggregation will determine whether a beetle will succeed in entering the phloem and start excavating its gallery.

2.6.2. The Inducible Host Defense

When the oleoresin exudation flow from a bore hole is insufficient, either for anatomical/physiological reasons or because a mass attack has reduced the flow, gallery construction may be successfully initiated. During gallery excavation, beetles contaminate the phloem and cambium tissues with a number of microorganisms, including several species of blue-stain fungi of the genus *Ceratocystis (or Ophiostoma)*. One of these fungi, *C. polonica* (Siemaszko) C. Moreau (=*O. polonicum* Siemaszko) plays a particularly important role: When inoculated into the phloem and cambium of Norway spruce trees in adequate quantities, this fungus may kill healthy trees.[33] As the hyphae make their way

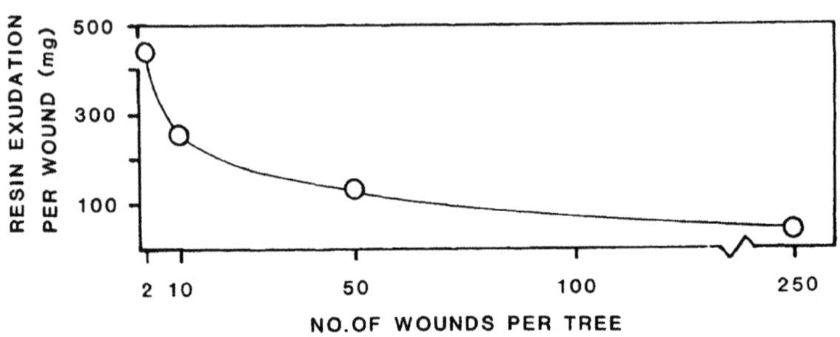

FIGURE 4. Yield of duct resin as a function of number of wounds to a tree. (Abridged from Schwerdtfeger.[55])

through the tracheids and ray parenchyma cells toward the central heartwood, the sapwood dries and can no longer transport water to the crown (Fig. 5). Given high transpiratory demands, the foliage rapidly dries, discolors, and drops within a few weeks.

The relationship between beetle and fungus appears truly mutualistic; the fungus helps the beetle kill the tree and in return is transported to new hosts by its winged symbiont. Spores of *C. polonica* and other fungi are carried in pits on the head, prothorax, and elytra (M. M. Furniss, H. Solheim, and E. Christiansen, in preparation). Possibly, the importance of the associated fungi could vary during the course of an epidemic, as indicated by a recent study of the Southern pine beetle, *Dendroctonus frontalis,* and its blue-stain fungus, *C. minor.*[20]

Against this combined assault, coniferous trees have developed a second, inducible defense.[11,12,53] Within a distinct zone surrounding the place of infection, a dynamic hypersensitive wound reaction takes place. The living cells of the reaction zone die rapidly, forming a necrosis depleted of nutrients essential to the fungus.[69] This is a general response of plant tissue to infections by fungi, bacteria or vira (cf. ref. 51). In coniferous trees, the necrotic area becomes impregnated with a resinous material produced by dying parenchyma cells within the reaction zone or by newly formed callus cells.[11] The resinous material has been shown to hold volatile components with a fungistatic effect.[52,57] Carbohydrates are translocated to the reaction zone for this energy-demanding process of allelochemic synthesis.[24,41]

The induced hypersensitive defense reaction is observed in Norway spruce.[23] As in pines [51] and firs,[11] the intensity of the reaction determines failure or success of the intruding organisms. Experiments with artificial inoculation of *C. polonica* reveal that, as the number of infection sites increases, the resin contents of individual reaction zones drops, and the fungus gradually succeeds in penetrating the sapwood (Fig. 6). Thus, this secondary defense is also exhausted as the number of attacks on a tree increases.

The relative importance of the two components of the defensive system appears to be very variable with no close association between their performance.[23] For example, beetle attacks on pheromone-baited trees may be judged successful at first because only dry frass is expelled from the entrance holes. However, if the tree is not overwhelmed by a mass attack, resin exudation may seep within a few weeks from formerly dry holes, indicating a wound reaction. Distinctive reaction zones may then surround extensive beetle galleries, spelling a sticky death to the parents and their offspring.[22]

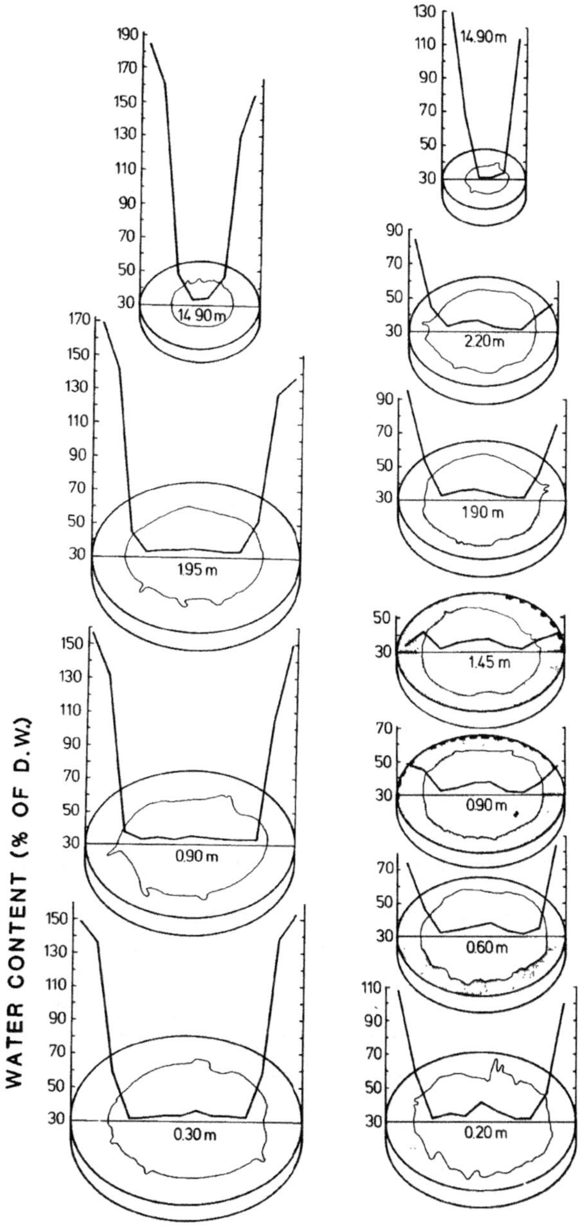

FIGURE 5. Water content of stem discs cut at different heights from two trees. The tree on the left was inoculated with sterile agar, and the one on the right with agar holding *Ceratocystic polonica*. The sapwood-heartwood borderline, blue stain (shaded), and resinous reaction zones (black) are indicated. (From Horntvedt *et al.*[33])

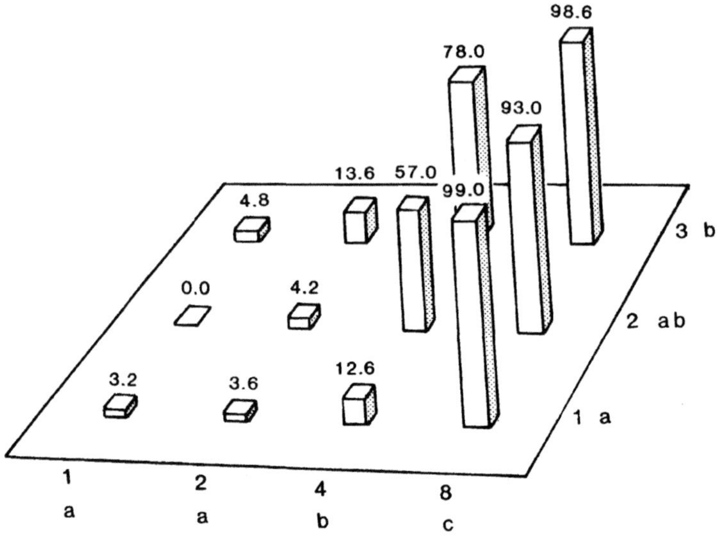

FIGURE 6. Percentage blue staining of the sapwood periphery in Norway spruce trees inoculated with *Ceratocystis polonica* as influenced by relative inoculation load (1, 2, 4, 8), and tree vigor (1 = high, 2 = medium, 3 = low vigor). Inoculation loads or tree-vigor categories with different letters have significantly different degrees of staining. (From Christiansen.[21])

Together, the two defense systems determine the "threshold of successful attack"[59] needed to kill a Norway spruce tree. The constitutive defense may "buy time" for the wound reaction to become effective. When a rapidly developing mass attack occurs under epidemic conditions, however, no signs of a wound response can be detected. This indicates that the main importance of the defense system is to protect trees during non-epidemic conditions, thereby preventing an epidemic.

2.6.3. Factors Influencing Host Tree Susceptibility

The effectiveness of the defensive systems are likely to depend on host tree physiology.[21,51,73] Several factors are known or suspected to be connected with susceptibility to *I. typographus* outbreaks. The site obviously has a strong influence on tree physiology. For example, there is evidence that spruces growing on certain geological formations suffer more from bark beetle attack. Such formations may produce a soil nutritious enough for spruce to become established and even to grow well under normal weather conditions, yet too well drained and drought prone during unusually dry periods. During the recent *I. typographus* epidemic of South Norway stands growing on permian Oslo-felt lavas were more susceptible to beetle damage than those on surrounding precambrian basement or quarternary deposits (Fig. 7). In drought years the soil on these lavas dries out because of the high degree of fissuring in the bedrock. In Baden, South Germany, limestone and sandstone areas have likewise been observed to be drought-prone and, consequently, the forest growing on them are vulnerable to *I. typographus* attack.[31]

Stands growing on more productive sites have been reported to suffer greater mor-

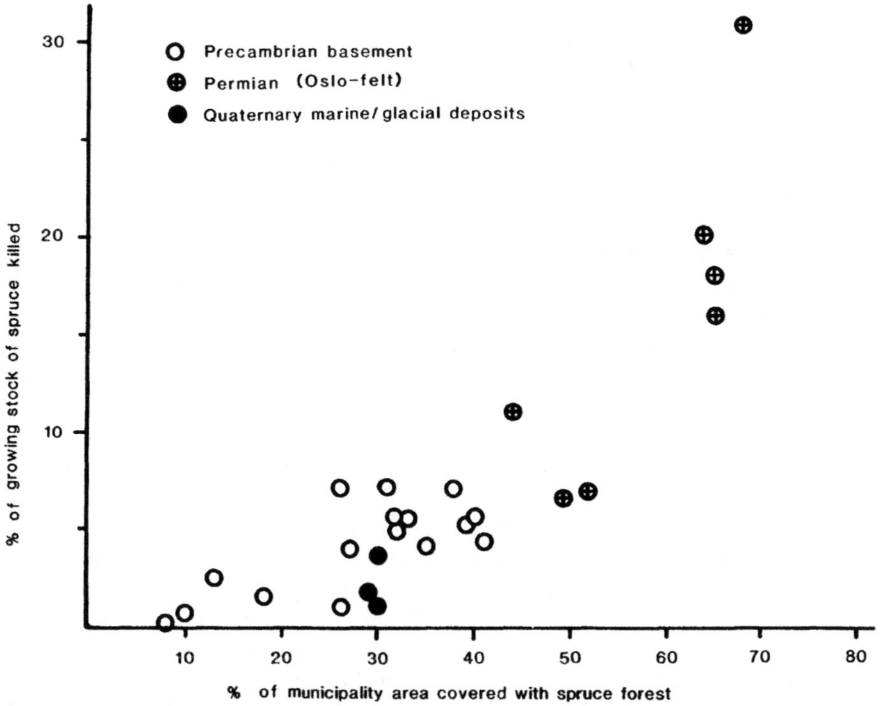

FIGURE 7. Degree of damage caused by *Ips typographus* in Southeastern Norway 1975–1980 in relation to local spruce forest cover and geological conditions. (From Worrell.[71])

tality than those on poorer soils.[55,71,72] Spruces which mature more rapidly may reach a state of susceptibility sooner. Valley slopes in South Norway suffered more damage than valley floors and hilltop sites during the recent epidemic.[71] This could be due to higher site quality on slopes, or to neglect because of their inacessibility, i.e., reduced thinning, harvesting, sanitation and salvage cutting.

Site aspect has also proved a significant variable in drought-affected areas during the Norwegian epidemic. East and north facing (shady) slopes generally had more damage than did south- and west-facing (sun-exposed) valley sides.[71] The explanation for this odd observation could be that shady valley slopes have larger continuous areas of spruce forest, or that trees growing on the shady slopes have a more shallow root system due to a (normally) high water table, hence are more vulnerable to prolonged drought. During epidemics in South Germany, however, the effect of aspect has been quite different, with more severe attacks recorded on south and west slopes.[40]

I. typographus attacks during the 1970s in South Norway showed a distinct decrease with elevation so that negligible damage occurred on sites over 600 m.[71] Likewise, during the postwar epidemic in Northern Germany, indigenous spruce forests suffered decreased mortality with elevation.[55] This pattern appears to be closely linked to humidity.

The genus *Picea* is generally adapted to humid situations, thriving in climates in which precipitation is high and evapotranspiration low. In many areas, however, Norway spruce has been planted far beyond its natural range, often in much drier situations. This is the case in many lowland areas of Central Europe. On such sites, it appears more susceptible to attack by *I. typographus*.[31,48,55]

In Southeast Norway, where spruce has occurred naturally for several centuries, growth is often restricted by summer drought. Unlike pines, spruce trees of the lowland areas exhibit reduced growth during warm (and dry) summers, a feature which was particularly striking during the prolonged drought of the 1970s. Spruce trees apparently exert less control over their water economy than do pines[36] because of their shallow root systems and less efficient stomatal control.

Windstorms are a frequent feature in the prehistory of *I. typographus* outbreaks. Strong winds favour beetle propagation directly by supplying downed breeding material and indirectly by leaving standing trees in a susceptible state, their fine roots torn by swaying and their water uptake impeded. Both effects are likely to facilitate the beetles' transition from endemic to epidemic behavior.

Heavy snows may also lead to felling in younger stands, but more frequently leave the trees with broken tops. Thin-barked tops are a less valuable breeding material than are whole fallen stems.[60]

Young stands of Norway spruce are rarely destroyed by *I. typographus*. In the Harcynian Mountains of Germany, few stands under 40 years of age were damaged in the 1946–1950 outbreak, while damage increased proportionally in older forest, being worst in stands of 120 years or more.[55] In several other bark beetle/conifer systems, dense stands have been shown to suffer more from beetle attack than sparsely stocked ones.[66] With *I. typographus* several factors interact to complicate this relationship. In one study in Norway the most extensive damage was found in the denser stands.[72] In northern Germany, on the other hand, attacks were worst in the more open forest of the cooler areas, probably due to the higher temperatures in such habitats during flight and larval development. In warmer areas, this was less important.[55]

Large continuous tracts of spruce trees in the critical age classes exhibit much higher attack frequencies and mortality rates than do areas in which more scattered stands occur.[55,72] This was the major factor associated with the local attack frequency during the recent *I. typographus* epidemic in Norway (Fig. 7). Beetle captures in pheromone traps may be as high in mixed stands as in pure spruce stands,[10] and spruce trees may readily be killed in mixed stands. If flight temperature is critical in an area, the presence of deciduous trees may discourage attacks by lowering the temperature of a stand.[55] In drought years, the deep roots of broadleaved trees such as birch and oak may drain deep soil layers and lower the water table significantly, exposing the shallow-rooted spruce to water stress.[55]

2.7. Relationship with Natural Enemies

Numerous parasites have been reared from *I. typographus*, among which the braconids and chalcids are the most abundant.[29,46,49,59] Among the predaceous insects, *Thanasimus* beetles are common. These predators use the aggregation pheromones of *I. typographus* for kairomones.[9] Birds, mice, shrews, ants, and wasps have also been listed as predators.[59] However, even the most effective parasites or predators rarely cause high mortality, and moderate predation or parasitization may merely relieve intraspecific competition between beetle broods.[59]

Many of the parasites and predators overwinter as immature stages under the bark of standing trees, in contrast to the majority of *I. typographus* beetles which hibernate in the litter. Thus, woodpeckers, which may significantly reduce other bark beetle populations during winter, are probably beneficial to *I. typographus* by devouring their natural enemies. Entomophagus fungi have been observed but are not judged important controlling agents.[59]

3. POPULATION PATTERNS: CASE HISTORIES

The Eurasian spruce bark beetle exhibits a typical eruptive population behavior.[15] For decades the population may remain below the epidemic threshold, surviving in fallen or weak trees, unbarked logs, slash, and stumps. Outbreaks occur following external events that either (1) lead to a population increase by supplying abundant, weakly defended, or undefended breeding material (e.g., fallen trees); or (2) lower the threshold of successful attack by reducing the resistance of standing trees over large areas.

The following discussion examines some *I. typographus* epidemics and evaluates the external changes that may have led to the outbreaks. These case histories form the basis for our later development of a hypothesis for the causes of *I. typographus* population behavior. Other epidemics may have been as important as the ones discussed below. For instance, in Hokkaido, Japan, some 2.4 million cubic meters of *Picea jezoënsis* were killed from 1956 to 1958 following a severe gale in 1954.[35,74] Swedish losses 1971–1982 were of the same magnitude as in the Norwegian epidemic.[25,54] Other examples are numerous (ref. 56).

3.1. Central Europe: 1944–1951

The postwar epidemic in Central Europe can be divided into a northern outbreak of moderate magnitude and a southern large-scale devastation. A schematic representation of the conditions that led to successful propagation of *I. typographus* in the Harcynian Mountains of North Germany is presented in Fig. 8. Practically all these conditions could be traced to human actions.

The prehistory of the outbreak is important. In 1933 the Forester-General ordered that harvesting be carried out by selecting the larger trees for cutting instead of the usual practice of clear-cutting. Combined with persistent over-exploitation, this decree led to large-scale wind-felling as the old and well-stocked stands were opened up. Abundant breeding substrate became available for endemic *I. typographus* populations.

The partial breakdown of civilian infrastructures towards the end of the world war also created insurmountable obstacles to effective forest management. The principle of *saubere Wirtschaft* (clean forest management), the *conditio sine qua non* of German forestry for more than a century, had to be abandoned due to shortages of labor. Sanitation/salvage cutting became impractical. Hence, no effective barriers could be erected against the upcoming epidemic.

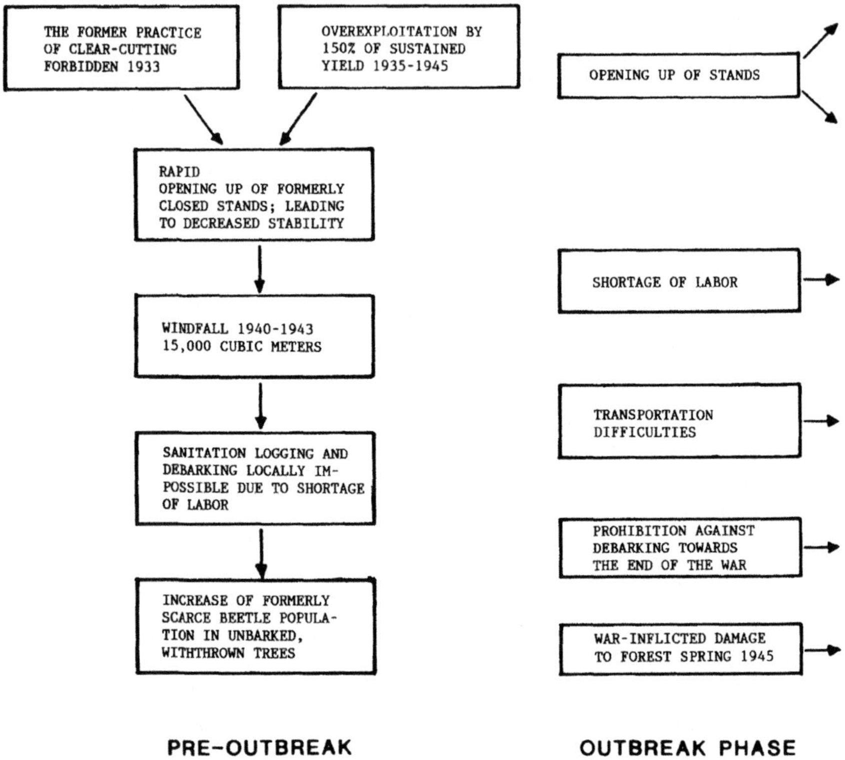

FIGURE 8. Conceptual model of the *Ips typographus* epidemic of Northwestern Germany 1946–1950. (Abridged from Schwerdtfeger.[55])

Weather conditions apparently played a minor role in the this outbreak. Apart from the extraordinarily dry year of 1947, there was no severe drought in the northern region during the 1940s; the average rainfall of the decade deviating by only 5% from the norm.

Several features of this northern outbreak can be recognised in the contemporary epidemic in Southwest Germany and adjoining areas of Switzerland, Austria and Czechoslovakia. There were, however, important differences.

Owing to recurring wind and snow damage, the *I. typographus* population of the southern areas was locally high in the early 1940s. Until 1942 weather conditions had been favorable for the forest, and no serious beetle outbreak had occurred. But during 1942–1949 the summers were unusually hot, and generally very dry.[31]

In Southern Baden, outbreaks started in the higher elevations of the Black Forest, where the beetle population was high, but became more severe and persistent in the drier lowlands (e.g., on the Bodenzee).[40] Heavy forest harvesting in 1947–1948 aggravated the situation; these extensive cuttings combined with a shortage of labor made an effective bark beetle campaign next to impossible.[31]

Despite these shortcomings of forest management, it was the extreme drought in the southern areas that brought the epidemic to the crisis point.[31,40] Out of 30 million cubic

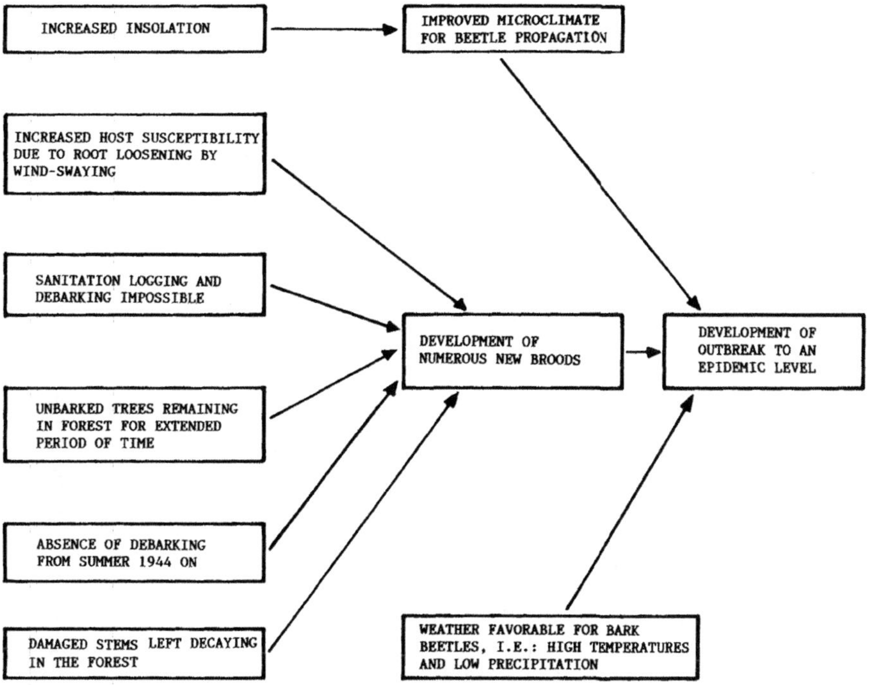

OUTBREAK PHASE

FIGURE 8. (*continued*)

meters of beetle-killed spruce in the whole of Central Europe, the vast majority was found in the drought-stricken southern areas.[56] By comparison, the northern outbreak claimed a "modest" 2 million cubic meters.

The factor which finally brought the devastating Southern epidemic to an end was a tremendous human effort. In a large-scale campaign, every thinkable control measure was taken, and the epidemic actually abated in the still hot and dry summer of 1949.[67]

Both in the northern and southern provinces, Norway spruce had been widely planted outside its natural range, frequently in lowland situations with unfavorable hydrology. These artificial stands suffered more severe damage than did those of the indigenous areas.[31,40,55]

3.2. Southeastern Norway: 1971–1982

Apart from a minor outbreak in the 1940s, the spruce forests of Southeastern Norway were by and large free from *I. typographus* damage for the first seven decades of this century. Before that, large-scale epidemics are known from the 1700s and 1800s.

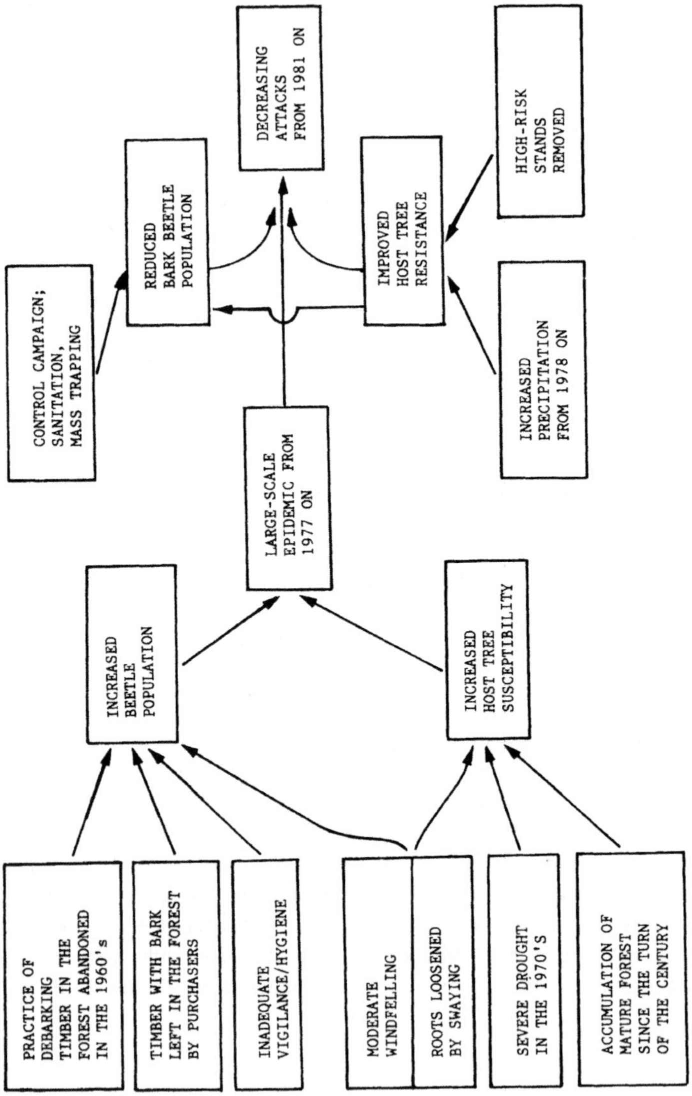

FIGURE 9. Conceptual model of the recent large-scale *Ips typographus* epidemic in southeastern Norway.

As in Central Europe, attention should be paid to the prehistory of the epidemic (Fig. 9). Since the turn of the century, the forests of Norway had undergone significant changes. After the long bark beetle outbreaks and heavy exploitation of the 1800s, the forests were generally sparsely stocked. But for most of the present century the standing volume has increased by nearly 1% per year, much of this increment being laid down in the larger tree dimensions.[37] This development continued after the introduction of clear-felling (around 1950). Although old stands have been extensively clearcut, large tracts of mature forest still remain and, overall, harvesting has been well under the level of sustained yield. By the onset of the *I. typographus* epidemic, therefore, Norwegian forests held excessive amounts of mature spruce.

Other changes included the abandonment in the 1960s of the practice of debarking all timber in the forest. Also, purchasers commonly left timber in the forests throughout the summer, instead of bringing it into timberyards. No legal regulations required the removal of logs from the forest. Sloppy forest hygiene added to the accumulation of breeding material, resulting in a gradually increasing beetle population.

In 1969, a heavy autumn gale left some 2 million cubic meters of fallen spruce and pine timber widely scattered throughout the eastern districts of South Norway (Fig. 10), with even much more damage done to Swedish forests across the border. The local bark beetle populations invaded the fallen and weakened trees in inaccessible areas, and rose to outbreak levels in 1971. Owing to intensive sanitation efforts, tree-killing decreased sharply after 1972, and in 1974 was reduced to some 30,000 m^3. The year of 1974, however, was to be the first of three extreme drought years, strongly affecting the lowland spruce forests of Southeastern Norway, and setting the stage for a far more extensive *I. typographus* epidemic.

The drought years of 1974–1976 was the climax of a decade of subnormal precipitation. In these three summers, large districts received less than half the normal rainfall, and for the whole period 1969–1978 the deficit equalled one whole year's precipiation. This led to a considerable drop in the water table of the generally shallow forest soils.

Several million trees died from drought alone, which means that many more must have been brought to the verge of death. Such trees dropped large quantities of foliage,

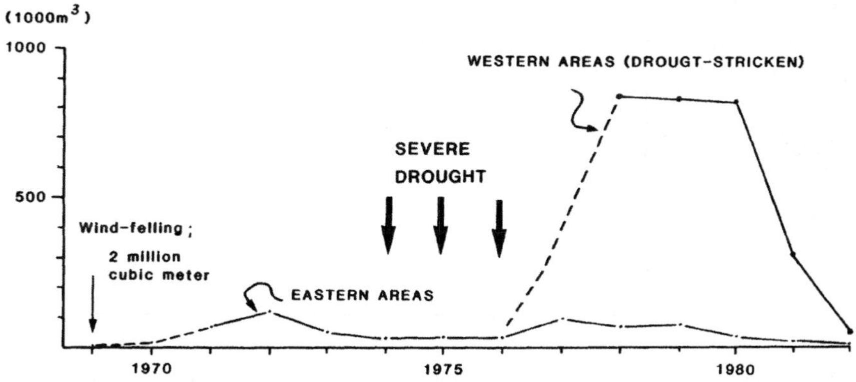

FIGURE 10. Development of the *Ips typographus* epidemics in two areas of southeastern Norway 1969–1982.

and ten years later they still exhibit sparse crowns. Salvage operations were often extremely difficult in roadless and rugged terrain, and severely damaged stands were left to the beetles. Minor windfellings in 1975 and 1976 added more fuel to the conflagration.

A large-scale epidemic started in the drought-affected areas in 1977. The devastation reached its maximum in 1978–1980, some 1 million cubic meters of spruce being killed each year. The epidemic declined in 1981 and by 1983 was largely finished. The factors contributing most to the decline were probably (1) an increase in host resistance brought about by the elimination of the most susceptible stands, plus abundant rainfall from 1978 onwards, and (2) an extensive control campaign which included salvage logging and mass trapping beetles in pheromone-baited traps (Fig. 11).

3.3. Southeastern Norway: 1850–1900

The large-scale *I. typographus* epidemic that ravaged the forests of southeastern Norway in the 1800s offers interesting similarities and differences to the more recent one.[3] In 1836, forest authorities lifted the prohibition against root-sale of timber, under which logging had to be done by the forest owner. Now, timber could be bought, cut and extracted by foreigners, and large quantities were harvested, commonly by high-grading (cutting the larger trees). Closed stands were opened and subjected to wind damage. Heavy storms in the 1830s left large numbers of fallen spruce trees scattered in the roadless terrain. Then, in the 1840s, an extremely dry period occurred with the deficit in Oslo, over the decade, equalling one year's normal precipitation.

Around 1850, a large-scale *I. typographus* epidemic broke out. Due to generally low

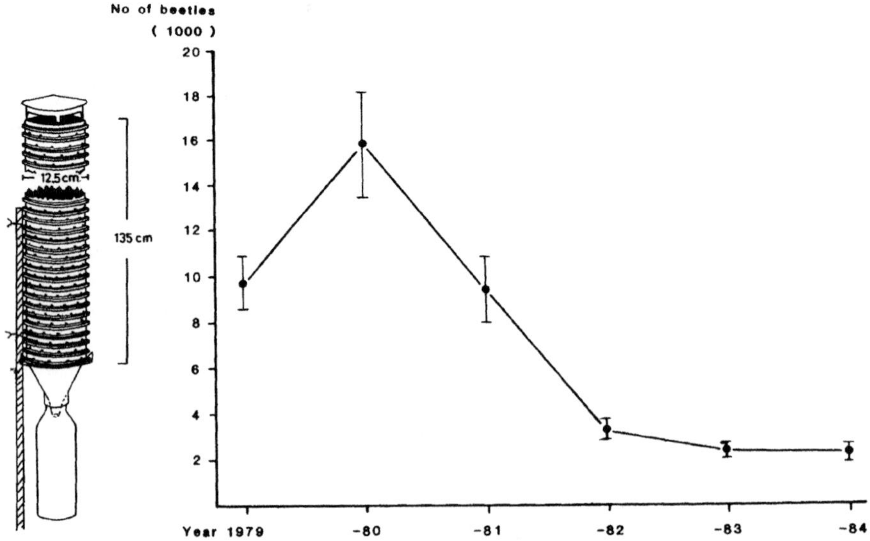

FIGURE 11. Mean catches of *Ips typographus* per pheromone trap (depicted) on fresh clearcuts in districts of southeast Norway with severe beetle damage.

standards in forest administration and inadequate knowledge on the part of forest managers, no effective suppression campaign was organized. High-ranking forest officials strongly advocated the "harmlessness-theory" for *I. typographus*. According to this view, the beetles only acted as the police force of the forest, clearing away diseased and worthless trees but being absolutely harmless to healthy ones. Thus, the best option for the foresters would be to do nothing at all. Consequently, the epidemic was left unchecked in large areas, and although it must have "burnt out of fuel" in many places, it repeatedly ignited elsewhere, the afterglow lasting to the end of the century.

3.4. Causes of Outbreak Behavior: A Hypothesis

A general theoretical model may serve as a starting point for our hypothesis concerning *I. typographus* outbreaks (Fig. 12). The theory recognizes an epidemic population threshold, above which active tree-killing occurs, which is directly related to stand vigor. This assertion is strongly supported by empirical data.[38,42,44,65,66] In the following, we examine the array of factors which apparently influence the two parameters of the model: (1) the size of the beetle population, and (2) the resistance of the host trees to beetle/fungus attack. We pinpoint key factors in the population dynamics which determine whether an endemic population will exceed the epidemic threshold.

3.4.1. The Size of the Beetle Population

In an endemic situation, the supply of weakly defended or nondefended host material (e.g., decaying and fallen trees, logs, slash,) determines the size of the beetle population. An increase in the amount of breeding substrate leads to an *I. typographus* population increase. If the resulting population exceeds the epidemic threshold, then an outbreak will occur.

The supply of breeding substrate depends to a large extent on local spruce abundance, human activities (e.g., logging), and wind damage. The manager's role is crucial in determining the standard of forest hygiene. However, extensive wind-throw in inaccessible areas may leave even the most skilled and dedicated manager with an overwhelming problem. On the other hand, long-term management for stable forest stands may reduce the impact of a gale. Whether due to sloppy forest hygiene or to insurmountable salvage problems after a storm, the amount of nondefended and weakly defended breeding substrate is a key factor in the dynamics of *I. typographus* populations.

3.4.2. Stand Resistance

The performance of the dual defense system determines the resistance of Norway spruce trees to the combined attack of *I. typographus* and its mutualistic fungus, *C. polonica*. Damage to the fine roots brought about by wind-swaying influences the uptake of water and nutrients and this may impair primary resin flow and reduce energy translocation for the secondary wound reaction.

Drought will have a similar effect. Reduced assimilation caused by water stress, and

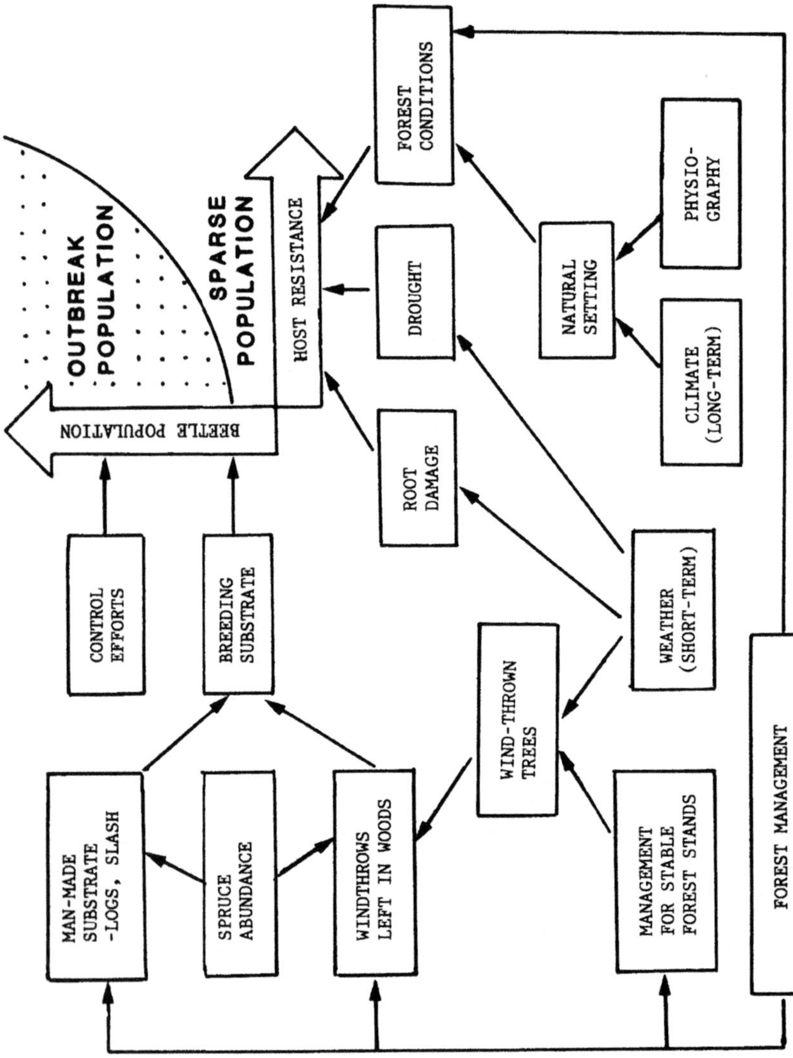

FIGURE 12. Conceptual model of interaction between factors that influence the threshold of successful attack by *Ips typographus*. (Adapted from Berryman.[13,14])

increased respiration will reduce the energy (carbohydrate) available for defensive reactions. However, increased formation of resin canals during droughty periods may strengthen the first line of defense. Whatever mechanisms are involved, drought appears to be the factor that exerts the strongest impact on host tree resistance. Drought should therefore be regarded as a key factor in the dynamics of *I. typographus* populations.

4. IMPLICATIONS FOR FOREST MANAGERS

Looking back at the model developed above in Fig. 12, we note that in spite of factors that remain beyond human control, others may be manipulated to the disadvantage of the spruce bark beetle. We therefore recommend the following management practices:

1. The stability of forest stands should be increased by thinning regimens that improve wind firmness by creating well-developed root systems. Thinning should preferably be implemented at an early age and should be avoided in older stands, particularly when an outbreak is imminent.
2. The principle of *saubere Wirtschaft* is no less important today than it was a century ago, particularly in areas of bivoltine *I. typographus* populations. Clean forestry is, however, contingent on accessibility and may be impractical or even undesirable in certain areas, such as wilderness.
3. In regions made up of many small forest holdings, clean forestry by one owner may be totally wasted if a neighbor mass-produces bark beetles because of sloppy forest hygiene. In such cases, strict enforcement of effective laws seems the only solution.
4. A particular aspect of forest hygiene is the timely removal of harvested timber from forest areas. If this cannot be done, the logs should be protected from *I. typographus* colonization. Effective forest hygiene depends on knowledge and vigilance on part of forest owners and advisers and is ultimately a question of education.
5. In low-lands and drought-prone areas, Norway spruce stands should not be maintained until old age. Overaged unstable stands represent a hazard to adjacent forest by serving as breeding sites for *I. typographus* beetles. Interspersion of other age groups or tree species will reduce habitat continuity, and may have a preventive effect against *I. typographus* proliferation.
6. Stand edges should be made with regard to maximum stability, taking advantage of natural windbreaks and well-rooted trees, as *I. typographus* frequently attacks the wind-affected and sun-exposed fringes of old stands.
7. Although by no means immune, indigenous forests appear more resistant to *I. typographus* outbreaks. Caution should thus be taken not to increase host susceptibility by extending the range of tree provenances beyond reasonable limits. The maintenance of high genetical variability in spruce trees seems a worthwhile precaution against future *I. typographus* ravages.
8. The development of synthetic aggregation pheromones, and an effective trapping technology opens new roads to population control.[10,64] Systematic monitoring of local beetle abundance offers a platform for rational population management

decisions. Catch trend indices may indicate whether short-term control measures are needed to forestall local outbreaks. Replacing the old trap-tree technique with mass deployment of pheromone-baited traps may also be a valuable suppression tactic for keeping endemic populations low and for extinguishing spot infestations that can quickly spread into general outbreaks.

Management for maximum resistance to *I. typographus* outbreaks may collide with other demands in practical forestry. After all, the bark beetle problem is only one of a multitude of considerations plaguing the forest manager. In modern forestry, economics is the major concern of management. However, short-term economical considerations may obscure the beacon ahead that most foresters aim for—the future wealth of the forest owner. Thus, it may be tempting to deviate from the principle of clean forestry, or to leave stands unmanaged, because the necessary operations do not appear immediately profitable.

Because forgetfulness is one of the flaws of human nature, managers often pay less than adequate attention to the lessons of the past. A few years after a major *I. typographus* outbreak, forestry may slip back to the old habits of leaving abundant breeding material in the woods.

In summary, we suggest that (1) forest stands be managed for a higher degree of stability, (2) the principle of clean forestry be adhered to, and (3) a monitoring and control program be established on a permanent base. These deviations from present practices should provide the forester with an advantage over an old enemy. In spite of such measures, however, unpredictable events, such as large-scale windthrow and severe drought, will ensure that future foresters have to face the challenges of a worthy opponent, the spruce bark beetle.

ACKNOWLEDGMENTS. We would like to thank the Paul Parey Publishing House for permission to reprint Figs. 3 and 6, and Plenum Publishing Corporation for Fig. 1, drawn by A. A. Berryman.

REFERENCES

1. Anderbrant, O., Schlyter, F., and Birgersson, G., 1985, Intraspecific competition affecting parent and offspring in the bark beetle *Ips typographus*, *Oikos* **45:**89–98.
2. Annila, E., 1969, Influence of temperature upon the development and voltinism of *Ips typographus* L. (Coleoptera, Scolytidae), *Ann. Zool. Fenn.* **6:**161–207.
3. Asbjørnsen, P. C., 1861, *Om skovtørk og markaat,* Steensballe, Christiania.
4. Austarå, Ø., Pettersen, H., and Bakke, A., 1977, Bivoltinism in *Ips typographus* in Norway, and winter mortality in second generation, *Medd. Nor. Inst. Skogforsk.* **33:**269–281.
5. Austarå, Ø., Bakke, A., and Midtgaard, F., 1986, Response in *Ips typographus* to logging waste odours and synthetic pheromones, *J. Appl. Entomol.* **101:**194–198.
6. Bakke, A., 1981, Inhibition of the response in *Ips typographus* to the aggregation pheromone; field evaluation of verbenone and ipsenol, *Z. Angew. Entomol.* **92:**172–177.
7. Bakke, A., Austarå, Ø., and Pettersen, H., 1977, Seasonal flight activity and attack pattern of *Ips typographus* in Norway under epidemic conditions, *Medd. Nor. Inst. Skogforsk.* **33:**253–268.
8. Bakke, A., Frøyen, P., and Skattebøl, L., 1977, Field response to a new pheromonal compound isolated from *Ips typographus*, *Naturwissenschaften* **64:**98.

9. Bakke, A., and Kvamme, T., 1981, Kairomone response in *Thanasimus* predators to pheromone components of *Ips typographus*, *J. Chem. Ecol.* **7**:305–312.
10. Bakke, A., and Strand, L., 1981, Feromoner og feller som ledd i integrert bekjempelse av granbarkbillen. Noen resultater fra barkbilleaksjonen i Norge i 1979 og 1980. (English summary: Pheromones and traps as parts of an integrated control of the spruce bark beetle. Some results from a control program in Norway in 1979 and 1980), *Rapp. Nor. Inst. Skogforsk.* **5/81**:1–39.
11. Berryman, A. A., 1969, Responses of *Abies grandis* to attack by *Scolytus ventralis* (Coleoptera: Scolytidae), *Can. Entomol.* **101**:1033–1041.
12. Berryman, A. A., 1972, Resistance of conifers to invasion by bark beetle–fungus associations, *BioScience* **22**:598–602.
13. Berryman, A. A., 1978, A synoptic model of the lodgepole pine/mountain pine beetle interaction and its potential application in forest management, in: *Theory and Practice of Mountain Pine Beetle Management in Lodgepole Pine Forests* (A. A. Berryman, G. D. Amman, R. W. Stark, and D. L. Kibbee, eds.), pp. 98–105, College of Forest Resources, University of Idaho, Moscow.
14. Berryman, A. A., 1982, Biological control, thresholds, and pest outbreaks, *Environm. Entomol.* **11**:544–549.
15. Berryman, A. A., 1986, *Forest Insects: Principles and Practices of Population Management*, Plenum, New York.
16. Birgersson, G., Schlyter, F., Löfqvist, J., and Bergström, G., 1984, Quantitative variation of pheromone components in the spruce bark beetle *Ips typographus* from different attack phases, *J. Chem. Ecol.* **10**:1029–1055.
17. Bombosch, S., 1954, Zur Epidemiologie des Buchdruckers (*Ips typographus* L.), in: *Die grosse Borkenkäferkalamität in Südwestdeutschland 1944–1951* (G. Wellenstein, ed.), pp. 239–283, Forstschutzstelle Südwest, Ringingen.
18. Botterweg, P., 1982, Dispersal and flight behaviour of the spruce bark beetle *Ips typographus* in relation to sex, size and fat content, *Z. Angew. Entomol.* **94**:466–489.
19. Botterweg, P., 1983, The effect of attack density on size, fat content and emergence of the spruce bark beetle *Ips typographus* L., *Z. Angew. Entomol.* **96**:47–55.
20. Bridges, J. R., Nettleton, W. A., and Connor, M. D., 1985, Southern pine beetle (Coleoptera: Scolytidae) infestations without the bluestain fungus, *Ceratocystis minor*, *J. Econ. Entomol.* **78**:325–327.
21. Christiansen, E., 1985, *Ceratocystis polonica* inoculated in Norway spruce: Blue-staining in relation to inoculum density, resinosis and tree growth, *Eur. J. For. Pathol.* **15**:160–167.
22. Christiansen, E., 1985, *Ips/Ceratocystis*-infection of Norway spruce: What is a deadly dosage?, *Z. Angew. Entomol.* **99**:6–11.
23. Christiansen, E., and Horntvedt, R., 1983, Combined *Ips/Ceratocystis* attack on Norway spruce, and defensive mechanisms of the trees, *Z. Angew. Entomol.* **96**:110–118.
24. Christiansen, E., and Ericsson, A., 1986, Starch reserves in *Picea abies* in relation to defence reaction against a bark beetle-transmitted blue-stain fungus, *Ceratocystis polonica*, *Can. J. For. Res.* **16**:78–83.
25. Eidmann, H. H., 1983, Management of the spruce bark beetle *Ips typographus* in Scandinavia using pheromones, in: Proceedings of the Tenth International Congress on Plant Protection, Brighton, 1983, Vol. 3, pp. 1042–1050.
26. Forsse, E., and Solbreck, C., 1985, Migration in the bark beetle *Ips typographus* L.: Duration, timing and height of flight, *Z. Angew. Entomol.* **100**:47–57.
27. Gmelin, J. F., 1787, *Abhandlung über die Wurmtroknis*, Crusius'sche Buchhandlung, Leipzig.
28. Gries, G., 1985, Zur Frage der Dispersion des Buchdruckers *Ips typographus* (L.), *Z. Angew. Entomol.* **99**:12–20.
29. Hedqvist, K. J., 1963, Die Feinde der Borkenkäfer in Schweden. I. Erzwespen (Chalcidoidea), *Stud. For. Suec.* **11**:1–176.
30. Helland, I. S., Hoff, J. M., and Anderbrant, O., 1984, Attraction of bark beetles (Coleoptera: Scolytidae) to a pheromone trap. Experiment and mathematical models, *J. Chem. Ecol.* **10**:723–752.
31. Hodapp, W., 1954, Der Verlauf der Fichtenborkenkäfer-Kalamität in Baden seit 1942, in: *Die grosse Borkenkäferkalamität in Südwestdeutschland 1944–1951* (G. Wellenstein, ed.), pp. 11–93, Forstschutzstelle Südwest, Ringingen.
32. Hodges, J. D., and Lorio, P. L., 1971, Comparison of field techniques for measuring moisture stress in large loblolly pines, *For. Sci.* **17**:220–223.

33. Horntvedt, R., Christiansen, E., Solheim, H., and Wang, S., 1983, Artificial inoculation with *Ips typographus*-associated blue-stain fungi can kill healthy Norway spruce trees, *Medd. Nor. Inst. Skogforsk.* **38**(4):1–20.
34. Hughes, P. R., 1973, Effect of α-pinene exposure on *trans*verbenol synthesis in *Dendroctonus ponderosae* Hopk., *Naturwissenschaften* **60**:261–262.
35. Inouye, M., and Yamaguchi, H., 1955, Analysis of the increase in beetle population in the storm-swept areas in the national forest of Hokkaido. I. *Hokkaido Branch, Gov. For. Exp. Stat. Spec. Rep.* **4**:72–94 (English summary).
36. Jarvis, P. G., and Jarvis, M. S., 1963, The water relations of tree seedlings. IV. Some aspects of the tissue water relations and drought resistance, *Physiol. Plant.* **16**:501–516.
37. Landsskogtakseringen, 1970, Taksering av Norges skoger -Landsskogtakseringen 50 år 1919–1969, Oslo.
38. Larsson, S., Oren, R., Waring, R. M., and Barrett, J. W., 1983, Attacks of mountain pine beetle as related to tree vigor of ponderosa pine, *For. Sci.* **29**:395–402.
39. Martinek, V., 1956, Prispevek k osvetleni problemu sesterskeho pokoleni u kurovce *Ips typographus* L. (German summary: Beitrag zur Erklärung des Problems der Geschwisterbrut beim Fichtenborkenkäfer *Ips typographus* L.), *Sb. CSAZV Lesnictvi* **29**:615–644.
40. Merker, E., 1952, Das Wetter der Jahre 1939 bis 1950 und sein Einfluss auf die Massenvermehrung des grossen Fichtenborkenkäfers in Südbaden, *Allg. Forst. Jagdzeitung* **123**:213–240; **124**:1–22.
41. Miller, R. H., and Berryman, A. A., 1985, Energetics of conifer defense against bark beetles and associated fungi, in: *The Role of the Host in the Population Dynamics of Forest Insects,* IUFRO Conference, Banff, Sept. 1983 (L. Safranyik, ed.), pp. 13–23, Pacific For. Res. Centre, Can. For. Serv., Victoria, B.C.
42. Mitchell, R. G., Waring, R. H., and Pitman, G. B., 1983, Thinning lodgepole pine increases tree vigor and resistance to mountain pine beetle, *For. Sci.* **29**:204–211.
43. Moser, J. C., and Dell, T. R., 1980, Weather factors predicting flying populations of a clerid predator and its prey, the southern pine beetle, in: *Dispersal of Forest Insects: Evaluation, Theory and Management Implications,* IUFRO Conference, Sandpoint, Idaho, Aug. 1979 (A. A. Berryman and L. Safranyik, eds.), pp. 266–278, Cooperative Extension Service, Washington State University, Pullman, Washington.
44. Mulock, P., and Christiansen, E., 1986, The threshold of successful attack by *Ips typographus* on *Picea abies:* A field experiment, *For. Ecol. Mgmt.* **14**:125–132.
45. NOU, 1979, *Granbarkbillen,* Norges off. utredninger 1979(22), pp. 1–71, Universitetsforlaget, Oslo.
46. Nuorteva, M., 1957, Zur Kenntnis der parasitischen Hymenoptera der Borkenkäfer Finnlands, *Ann. Entomol. Fenn.* **23**:47–71.
47. Ogibin, B. N., 1973, Effect of population density on fertility in *Ips typographus* L., *Ekologiya* **5**:66–72.
48. Otto, H.-J., 1985, Silviculture according to site conditions as a method of forest protection, *Z. Angew. Entomol.* **99**:190–198.
49. Pettersen, H., 1976, Parasites (Hym., Chalcidoidea) associated with bark beetles in Norway, *Nor. J. Entomol.* **23**:75–78.
50. Pettersen, H., and Austarå, Ø., 1975, Overvintringsforhold hos stor granbarkbille, *Ips typographus* L. (Col., Scolytidae) (English summary: Overwintering conditions for *Ips typographus* L. (Col., Scolytidae)), *Medd. Nor. Inst. Skogforsk.* **31**:569–580.
51. Raffa, K. F., and Berryman, A. A., 1982, Physiological differences between lodgepole pines resistant and susceptible to the mountain pine beetle and associated microorganisms, *Environ. Entomol.* **11**:486–492.
52. Raffa, K. F., Berryman, A. A., Simasko, J., Teal, W., and Wong, B. L. 1985, Effects of grand fir monoterpenes on the fir engraver, *Scolytus ventralis* (Coleoptera: Scolytidae) and its symbiotic fungus, *Environ. Entomol.* **14**:552–556.
53. Reid, R. W., Whitney, H. S., and Watson, J. A., 1967, Reactions of lodgepole pine to attack by *Dendroctonus ponderosae* Hopkins and blue stain fungi, *Can. J. Bot.* **45**:1115–1126.
54. Risberg, B., 1985, Uppskattning av granbarkborreskador i Värmland 1971–1982 genom korrigering och komplettering av tidligare flyginventeringsresultat (English summary: Estimation of *Ips typographus*-damage in Värmland 1971–1982 by correcting and supplementing earlier results from sketch mapping), *Sveriges Skogsvförb. Tidskr.* **83**(2):21–31.
55. Schwerdtfeger, F., 1955, Pathogenese der Borkenkäfer-Epidemie 1946–1950 in Nordwestdeutschland. *SchrReihe forstl. Fak. Univ. Göttingen* **13/14**:1–135.
56. Schwerdtfeger, F., 1957, *Die Waldkrankheiten,* 2nd ed., Paul Parey, Hamburg.

57. Shrimpton, D. M., and Whitney, H. S., 1968, Inhibition of growth of blue stain fungi by wood extractives, *Can. J. Bot.* **46**:757–761.
58. Svihra, P., 1973, On population dynamics of the bark beetle *Ips typographus* L. in the region of the upper Hron valley.), *Vedecké Práce. Vyskumného Ustavu Lesného Hospondárstva. Vo Zvolene* 18:227–258. (In Slovenian with English, German and Russian summary).
59. Thalenhorst, W., 1958, Grundzüge der Populationsdynamik des grossen Fichtenborkenkäfers *Ips typographus* L. *SchrReihe forstl. Fak. Univ. Göttingen* **21**:1–126.
60. Trägårdh, I., and Butovitsch, V., 1935, Redogörelse för barkborrekampanjen etter stormhärjningarna 1931–1932 (German summary: Bericht über die Bekämpfungsaktion gegen Borkenkäfer nach den Sturmverheerungen 1931–1932), *Meddn. St. SkogsförsAnst.* **28**(1):1–168.
61. Vité, J. P., 1961, The influence of water supply on oleoresin exudation pressure and resistance to bark beetle attack in *Pinus ponderosa, Contrib. Boyce Thompson Inst.* **21**:37–66.
62. Vité, J. P., and Wood, D. L., 1961, A study on the applicability of the measurement of oleoresin exudation pressure in determining susceptibility of second growth ponderosa pine to bark beetle infestation, *Contrib. Boyce Thompson Inst.* **21**:67–78.
63. Vité, J. P., Bakke, A., and Renwick, J. A. A., 1972, Pheromones in *Ips* (Coleoptera: Scolytidae): Occurrence and production, *Can. Entomol.* **104**:1967–1975.
64. Vité, J. P., and Francke, W., 1985, Waldschutz gegen Borkenkäfer: Vom Fangbaum zur Falle, *Chem. Unserer Z.* **19**:11–21.
65. Waring, R. H., and Pitman, G. B., 1983, Physiological stress in lodgepole pine as a precursor for mountain pine beetle attack, *Z. Angew. Entomol.* **96**:265–270.
66. Waring, R. H., and Pitman, G. B., 1985, Modifying lodgepole pine stands to change susceptibility to mountain pine beetle attack, *Ecology* **66**:889–897.
67. Wellenstein, G. (ed.) 1954, *Die grosse Borkenkäferkalamität in Südwestdeutschland 1944–1951*, Forstschutzstelle Südwest, Ringingen.
68. Wild, M., 1953, Die Entwicklung des grossen Fichtenborkenkäfers *Ips typographus* L. im Hochschwarzwald (1000–1200 m.ü.M.) und ihre Abhängigkeit von Klima 1947–1950, *Ber. Naturf. Ges. Freiburg Br.* **43**:121–176.
69. Wong, B. L., and Berryman, A. A., 1977, Host resistance to the fir engraver beetle. 3. Lesion development and containment of infection by resistant *Abies grandis* inoculated with *Trichosporium symbioticum, Can. J. Bot.* **55**:2358–2365.
70. Wood, D. L., 1982, The role of pheromones, kairomones, and allomones in the host selection and colonization behavior of bark beetles, *Annu. Rev. Entomol.* **27**:411–446.
71. Worrell, R., 1983, Damage by the spruce bark beetle in South Norway 1970–80: A survey and factors causing its occurrence. *Medd. Nor. Inst. Skogforsk.* **38**(6):1–34.
72. Worrell, R., 1983, Skoglige bestandsfaktorer som påvirker omfanget av granbarkbilleskade i Sør-Norge. (Stand factors affecting the occurrence of damage by *Ips typographus* in South Norway.), *Rapp. Nor. Inst. Skogforsk.* **18**/83:1–9.
73. Wright, L. C., Berryman, A. A., and Gurusiddaiah, S., 1979, Host resistance to the fir engraver beetle, *Scolytus ventralis* (Coleoptera: Scolytidae). 4. Effect of defoliation on wound monoterpene and inner bark carbohydrate concentrations, *Can. Entomol.* **111**:1255–1262.
74. Yamaguchi, H., Hirasa, T., Koizumi, C., Takai, M., Inouye, M., Kosugi, K., and Nobuchi, A., 1963, Survey and population studies of beetles in the wind-swept areas in Hokkaido (III). Beetle attacks on standing trees during the epidemic period, 1956 to 1958, *Bull. Gov. For. Exp. Sta.* **151**:75–135.
75. Schlyter, F., Byers, J. A., and Lofquist, J., 1987, Attraction to pheromone sources of different quantity, quality, and spacing: Density-regulation mechanisms in the bark beetle *Ips typographus*, *J. Chem. Ecol.* **13**:1503–1523.

CHAPTER 24

THE MOUNTAIN PINE BEETLE IN WESTERN NORTH AMERICA

Kenneth F. Raffa

Kenneth F. Raffa • Department of Entomology, University of Wisconsin, Madison, Wisconsin 53706.

1. INTRODUCTION

The mountain pine beetle, *Dendroctonus ponderosae* Hopkins (Coleoptera: Scolytidae), is generally considered the most destructive of all Western forest insects.[46] Periodic outbreaks may devastate entire forests, comprising millions of hectares. Between 1979 and 1983, more than 79 million trees were killed by this insect in the United States alone.[57] Total North American losses to the mountain pine beetle are estimated at above 2 billion board-feet per year.[86,87] In addition to the severe timber losses, mountain pine beetle outbreaks may affect watershed quality, wildlife composition, and recreational value.[57] The accumulation of dead wood is a major fuel source for subsequent wildfires.[5]

The mountain pine beetle is distributed throughout the southern half of British Columbia and Alberta, all of the northwestern United States, eastward to the Black Hills region of South Dakota and the Cypress Hills of the Alberta–Saskatchewan border, and southward through the Rocky Mountains into northern Mexico.[46,114] The beetle occurs from sea level in British Columbia to about 4000 m in the Rocky Mountains.

The designation *D. ponderosae* includes the previously described species *Dendroctonus monticolae* Hopkins and has also included *Dendroctonus jeffreyi* Hopkins (from 1963 to 1968), but the latter species is now considered distinct.[53,114] *Dendroctonus ponderosae* exhibits some genetic differentiation between widely separated populations,[106] but classification into strict subspecies is precluded by local differences related to host species and phloem thickness.[105,114]

Adult mountain pine beetles are small (~5 mm long), cylindrical, dark insects with clubbed, elbowed antennae. The eggs are smooth, oval, and cream colored, averaging ~1 × 0.5 mm. The larvae are soft, grublike, and legless. They have pronounced amber headcapsules with prominent mandibles and cream-colored thoracic and abdominal sections. There are four larval instars. The pupae are white and soft, and the newly formed immature adults are yellowish to tan. Pigmentation and sclerotization continue up to, and after, emergence.[79]

The most commonly attacked host species is lodgepole pine, *Pinus contorta* Douglas, accounting for 95% of all trees killed by the mountain pine beetle in the United States.[57] Therefore, this chapter concentrates primarily on lodgepole pine forests. The second major host is ponderosa pine, *Pinus ponderosa* Lawson, which accounts for an additional 4%. Ponderosa pine can be preferred in some portions of the beetle's range, particularly in the southwest. Western white pine and sugar pine are sometimes attacked in significant numbers, and outbreaks have occurred in these species. Whitebark, Coulter, limber, pinyon, bristlecone, and foxtail pines may also be attacked, but these are not common hosts. During severe outbreaks, beetles have occasionally attacked Engelmann spruce, blue spruce, grand fir, and incense cedar, although brood production in these trees ranged from very low to none at all.[114] A number of species outside the normal geographical range are also susceptible, and sometimes highly preferred, such as Scots, Eastern white, jack, red, Austrian pitch, cher, bishop, and Japanese red pines, and Norway spruce.[7,56] These preferences for exotic species should warn managers of the danger of introducing the mountain pine beetle into other parts of the world.

2. BIOLOGY AND ECOLOGICAL RELATIONSHIPS

2.1. Life History

The life cycle of the mountain pine beetle can be divided into three major phases: migration, colonization, and development. Upon emerging from killed trees, adult beetles disperse and search for suitable hosts. Little is known about their biology during this period. For example, the extent to which flight may influence beetle responsiveness to host stimuli is not clear. Daily flight periods vary, depending on temperature and location, but most of the seasonal flight occurs in July and August and is concentrated within a 3-week period.[79]

Attacks on host trees are initiated by females. Initial landing on trees is primarily elicited by visual cues, particularly vertical profiles.[99] Host chemistry appears to be of little importance during this stage of the colonization sequence: Beetles land in equal frequencies on host and nonhost species, as well as on suitable and nonsuitable members of the host population.[51,64] Once adult beetles land on a tree, orientation is influenced by visual, tactile, and chemical cues. Light and physical stimuli to the pronotum orient the beetles, which usually enter beneath bark flakes. A rough surface is preferred.[98] Nonpolar compounds in the outer bark elicit a biting response by the walking females, and polar compounds can arrest them and result in continued excavation.[71] If these stimuli are unsuitable, the orientation sequence is interrupted, and the beetle resumes flight. If the cues are satisfactory, the female continues to bore into the tree. The precise chemicals by which females distinguish suitable from unsuitable trees are unknown, but none of the host monoterpenes is strongly repellent. Likewise, beetles feed equally on tissues treated with extracts from resistant and susceptible lodgepole pines. Yet when caged onto trees, female beetles show a greater propensity to enter certain individuals, suggesting that olfactory cues may be involved at this level.[71]

As the female beetle bores into the host, she shovels resin, frass, and shredded bark away from the entrance site. She also releases pheromones that attract both mates and additional female colonizers.[66–68] The principal aggregation pheromone is *trans*-verbenol,[68] an oxidized product of the host monoterpene, α-pinene.[112] *trans*-Verbenol is synergized by small amounts of the pheromone exo-brevicomin and by the host monoterpenes myrcene and α-pinene.[25,38,85] Initially, about 90% of the arriving beetles are female. This proportion progressively declines until about one half the later arrivals are males. This is partly because high ratios of host resin to *trans*-verbenol principally attract females, and the opposite ratios principally attract males.[50,84] During the colonization sequence, which usually lasts about 4 days under outbreak conditions, there is a progressive decline in *trans*-verbenol emission and a release of antiaggregative pheromones. Most of the attacks occur in the lower 10 m of the bole.[89,98] During the final stages of aggregation on an attractive tree, the focus of aggregation may expand as congregated beetles begin landing on nearby trees, a process termed switching.[39,47,48]

Mating occurs beneath the bark, with females mating once and males usually mating with several females. After mating, females begin boring vertical oviposition galleries (Fig. 1a). The length of these galleries varies with attack density and phloem quality (e.g., moisture), but they are generally about 30 cm.[79,80] Ventilation holes to the outside are formed at irregular intervals along the gallery, and the lower portion is usually packed

FIGURE 1. (a) Early gallery construction by a mountain pine beetle in lodgepole pine, showing oviposition gallery beneath bark (removed), mated pair, and eggs along the sides. (b) Penetration of blue stain fungi into the sapwood (dark color) of a successfully infested tree. (c) Pitch tube and expelled beetle. (Photograph courtesy of R. Miller.) (d) Resistant response in the phloem of a lodgepole pine. The beetle and microbial associates are entomed within a dark resin-impregnated lesion formed around the gallery.

with frass. The female oviposits into small egg niches cut into the sides of the gallery. About 75 eggs are usually laid per gallery, but this number depends on attack density, phloem thickness, and moisture; potential fecundity is much higher, ranging up to more than 260 viable eggs.[80]

Following oviposition, females may re-emerge and attack nearby trees.[77] This behavior is apparently less common among mountain pine beetles than in some other bark beetles, such as *Dendroctonus frontalis* Zimmermann and *Ips typographus* Linnaeus, and has not been thoroughly studied. In general, re-emergence is more common in trees that have thin phloem, that desiccate rapidly, and that are heavily attacked early in the season.[7,78]

Eggs hatch in about 7–10 days under favorable conditions. The larvae mine at right angles to the oviposition gallery, usually maintain contact with the cambium, and try to avoid intercepting other larvae. They feed on the phloem tissue, and their nutrition is facilitated by various microbial symbionts.[109] The larvae usually develop to the second or third instar before winter temperatures arrest development. The final instar constructs a small pupal cell and clears away frass. The nonfeeding prepupal and pupal stages require about 3 weeks for development within this chamber. The number of offspring that successfully develop is strongly related to phloem thickness.[2,7,88]

There is usually one generation per year in the northern United States, but a partial second generation may develop in warm years. However, these second-generation progeny do not usually reach the cold-tolerant mid-larval stage by the onset of cool temperatures, and so survival is low. In the southern part of its range, there may be two full generations, and in some cases a partial third. At high altitudes and in much of Canada, a complete generation normally requires 1.5–2 years.[7,79] Emergence can occur between 16°C and 30°C, and flight is commonly initiated between 21°C and 38°C.[86] Temperatures below −18°C and −40°C are usually lethal to eggs and larvae, respectively.[86,88,90]

2.2. Microbial Associates

The mountain pine beetle is associated with a diverse flora of external and internal microbial symbionts that include fungi, bacteria, and yeasts. These microorganisms assist the beetles in overcoming host resistance and facilitate the nutritional physiology of the brood.[109] Their potential role in pheromone regulation under natural conditions is not well understood.[37] Some of the principal fungi and yeasts are transported in maxillary mycangiae.[110] Once inside the tree, fungi are spread by beetle larvae and mite associates. Adult beetles can also become recontaminated with the fungi during maturation feeding.

The principal phytopathogenic fungi associated with the mountain pine beetle are *Ceratocystis clavigera* and *Ceratocystis montia*.[109] These species rapidly invade the xylem and phloem tissues during the colonization sequence and are believed to be a key factor in both killing the tree and conditioning the substrate for brood development (Fig. 1b). Other microbial symbionts associated to varying degrees with the mountain pine beetle are summarized elsewhere.[109]

2.3. Host Relationships

2.3.1. Host Resistance

The above description of the mountain pine beetle life cycle presupposes that host colonization is successful. Successful reproduction, however, is contingent on host mortality,[10,70,73,88,111] and so there have been strong selective pressures acting on trees to evolve resistance mechanisms.

Lodgepole pine responds to bark beetle attack, as to other wounding, by exuding resin from the wound site. This resin flows under pressure from resin ducts in the sapwood and serves as an initial barrier to beetle entry.[74,83,102] As the beetle enters the tree, she must continuously remove this resin forming a pitch tube, or eventually be mechanically forced from her gallery or entombed (Fig. 1c).

The monoterpene fraction of these secretions has been studied most extensively. Monoterpenes have fungistatic properties against symbionts of *Scolytus ventralis* and *Ips pini* and so are presumably involved in the inhibitory effect of lodgepole pine resin on mountain pine beetle fungal symbionts. They are also ovidical,[74,82] which supports the field observation that egg hatch is prevented by strong resinosis.[73,83] Monoterpenes are not repellent to adults, however, which seems reasonable because mountain pine beetles can work within pitch tubes for long periods of time.[73] Although monoterpenes may be toxic to *D. ponderosae* adults at very high doses their total concentration in lodgepole pine phloem does not differ between resistant and susceptible trees.[73] Likewise, the relative proportions of monoterpenes are the same for resistant and susceptible trees.[72] Synergistic effects among monoterpenes and the effects of other compounds have been less studied. However, no constitutive differences have been found between susceptible and resistant trees in any resin fractions.[74]

In addition to the flow of preformed resin, pines also undergo an induced defense reaction.[72,74,83,100,101,103] This response commences as soon as beetles reach the living tissue and is triggered by the presence of microbial symbionts, mechanical wounding caused by beetle boring, and perhaps other elicitors such as chitosan and protein inhibitor-inducing factor (PIIF).[61] During this response, a necrotic lesion forms in advance of and around the beetle-fungus infection (Fig. 1d). A number of biochemical changes occur within the lesion. First, the concentration of allelochemicals such as monoterpenes, sesquiterpenes, and phenolics increases (Fig. 2). The total extent of monoterpene accumulation is related to levels of carbohydrates in the phloem, and is dependent on both localized conversion and transport of precursor compounds.[29,60] Second, the relative proportions of compounds within the monoterpene fraction change, with a disproportionately large increase in the more toxic and fungistatic compounds.[76] For example reaction tissue, but not constitutive tissue, inhibits the growth of *C. clavigera* and *C. montia*.[104] Simple mechanical injury, as opposed to simulated attack by fungal inoculations, does not induce these proportionate changes in monoterpene composition. Third, compounds that were not previously present in the phloem, particularly chemicals of higher molecular weight than the monoterpenes, appear in the reaction tissue.[74]

Almost all trees respond actively to invasion, but those trees that react most rapidly and extensively are most likely to survive.[72] The induced wound response is localized, in that no changes in terpene composition or quantity occur in the adjacent tissues, and is

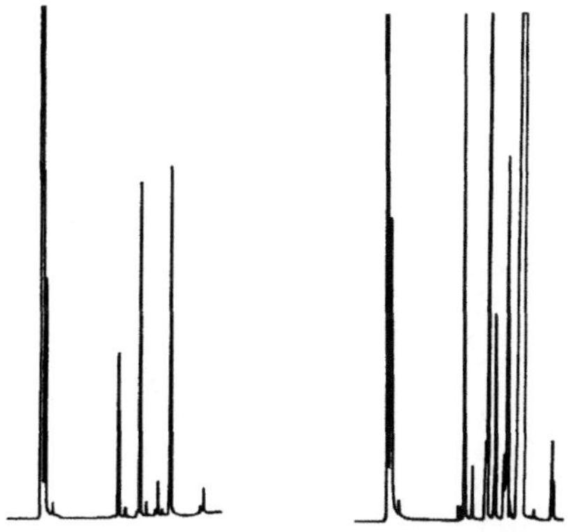

FIGURE 2. Compositional changes in the monoterpene fraction of lodgepole pine phloem in response to inoculation with fungal symbionts of the mountain pine beetle. The above chromatograms were made of extracts from the same tree before (a) and after (b) inoculation with *Ceratocystis clavigera*. Note the increase in total monoterpene content.

also energy conservative in that it only progresses as long as the beetle-fungal inclusion continues to invade new tissue. It is not known whether responsiveness changes dramatically during the course of the few weeks that comprise most of a typical flight season. However, there are pronounced seasonal and temperature-related differences. Responses are marginal in May, highest in July, and very weak by September.[88] The extent of the response is also related to some general indices of overall vigor, such as carbohydrate stores, periodic growth ratio (radial increment over the previous 5 years divided by increment during the 5 years before that), disease status, crown structure, and age.[60,88,101]

The preformed and induced defense systems operate in conjunction, rather than as independent processes.[10,72] The immediate resin flow can physically delay beetle progress, thereby providing time for the induced response to become effective. Resin flow can also interfere with beetle chemical communication. Even though host resin provides the precursors for aggregation pheromone synthesis, beetles engaged in shoveling a thick resin flow do not necessarily attract other beetles.[73] This is probably because pheromone synthesis by the beetle and pheromone emission from the entry site are distinct events. Regardless of the quantities of pheromone produced by the beetle, these volatiles do not necessarily permeate the tree's gummy secretions. The exact means by which resin flow interferes with chemical communication is unknown, but could also involve competitive inhibition of *trans*-verbenol at the site of chemoreception, metabolic inhibition of symbionts that produce the pheromones,[7]and/or autoxidation of α-pinene to *trans*-verbenone.[21,26]

2.3.2. Dynamics of Colonization

Although host resistance, beetle behavior, and fungal metabolism are discussed separately for descriptive purposes, they are best understood in terms of their interactions. Beetle density plays a critical role in determining the outcome of these interactions. Two physiological mechanisms are largely responsible for this relationship.

First, the defensive capacity of a tree is limited, and can be overwhelmed by mass attack (Fig. 3). Each attack depletes the remaining ability of the host to respond. This can

FIGURE 3. Lodgepole pine mass attacked by mountain pine beetles. The heavy flow of resin about each entry site indicates that this tree was relatively vigorous and resistant but that its defensive capacity was overcome by the high density of attacks.

be demonstrated by artificially interrupting the natural attack sequence, by screening the trees, to obtain a range of attack densities on different trees,[7] or by inoculating trees with varying densities of fungal symbionts.[74] The result in either case is a typical dose-response curve of the general form shown in Fig. 4. At very low attack densities, the tree's response is normally very effective, but the response declines with increasing beetle density, eventually reaching a point of inflection, and has no effect on brood development at high densities. At very high inoculation or attack densities there is no effective response, regardless of the tree's potential resistance. This physiological relationship be-

FIGURE 4. Relationship between defensive response of lodgepole pines and beetle attack density described by the equation $y = ae^{-bx^c}$, where y = host response and x = attack density. (a) Effect of attack density on host ability to limit brood production by mountain pine beetles. The host effect at any attack density is the difference in brood production between beetles in dead logs and living trees. (From Raffa and Berryman.[73]) (b) Monoterpene accumulation in response to controlled inoculation with *Ceratocystis clavigera*. (Data from Roffa and Berryman.[74]) (c) Between tree differences in the dose–response curve. Three hypothetical trees show different relationships with attack density. The threshold of resistance (T) for each tree is the attack density at which the combined allelochemical concentration falls below the effective dose needed to prevent brood and microbial establishment.

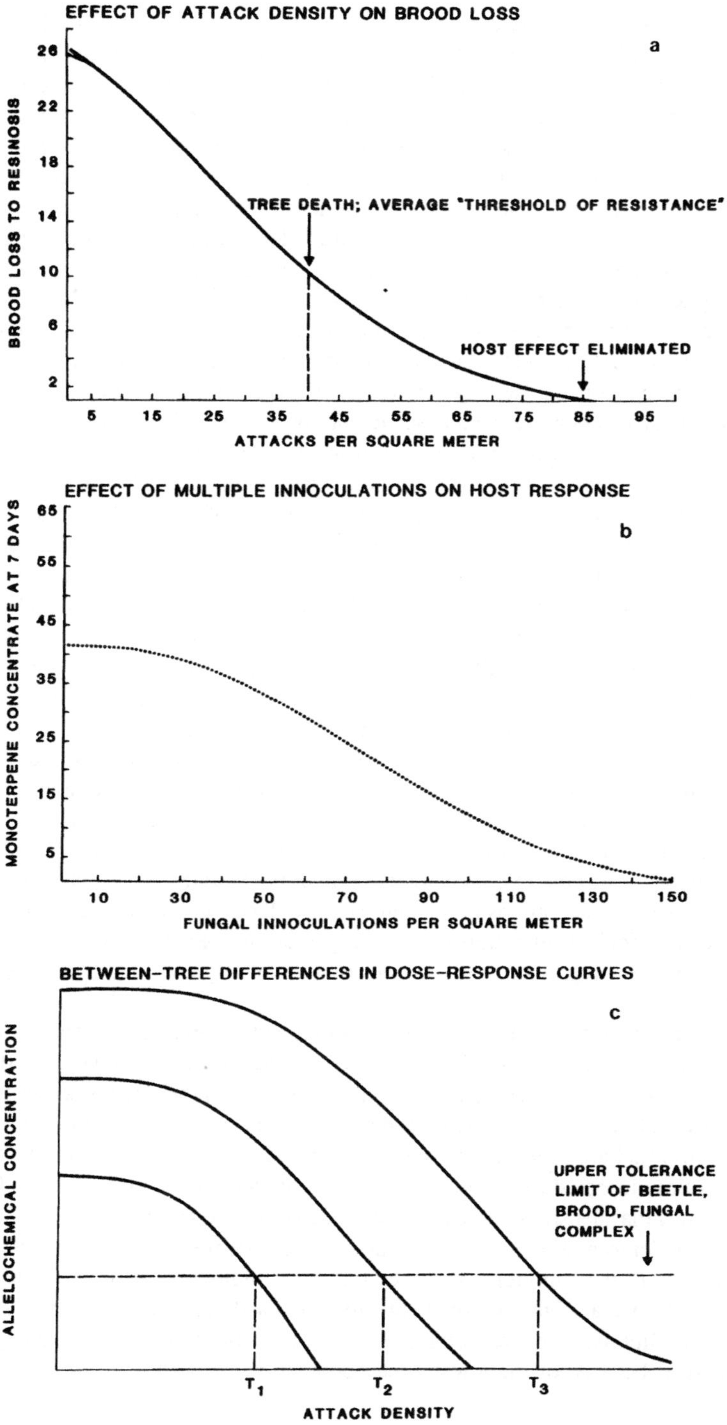
EFFECT OF ATTACK DENSITY ON BROOD LOSS
a
BROOD LOSS TO RESINOSIS
26
22
18
14
10
6
2
TREE DEATH; AVERAGE "THRESHOLD OF RESISTANCE"
HOST EFFECT ELIMINATED
5 15 25 35 45 55 65 75 85 95
ATTACKS PER SQUARE METER
EFFECT OF MULTIPLE INNOCULATIONS ON HOST RESPONSE
b
MONOTERPENE CONCENTRATE AT 7 DAYS
65
55
45
35
25
15
5
10 30 50 70 90 110 130 150
FUNGAL INNOCULATIONS PER SQUARE METER
BETWEEN-TREE DIFFERENCES IN DOSE-RESPONSE CURVES
c
ALLELOCHEMICAL CONCENTRATION
UPPER TOLERANCE
LIMIT OF BEETLE,
BROOD, FUNGAL
COMPLEX
T1
T2
T3
ATTACK DENSITY

tween beetle numbers and manifested host response is critical to understanding why brood intoxification is only weakly expressed in beetle life tables. Each tree is characterized by its own threshold of resistance, the density of beetle–microbial inclusions required for successful attack, and all successfully attack trees have had this threshold breached.

The second critical mechanism relating beetle behavior to host defense physiology concerns beetle arrival rates. Because *trans*-verbenol is an oxidized produce of host defensive compounds, the arrival of more beetles is assured for as long as the host is capable of secreting resin. So once a mass attack is initiated, aggregation does not cease until successful colonization is virtually assured. Thus if there are enough beetles within range of pheromone olfaction, almost any tree can be killed.

Despite this apparent ability of large beetle populations to be almost always successful, an offsetting defense mechanism enables some trees to survive. A rapid flow of resin about the early arrivals can prevent them from becoming attractive, and without effective aggregation the density remains well below the threshold of resistance. The expressed host responses are then sufficient for successful defense. Interference with beetle communication can itself be exhausted by multiple entries, however.[73] This is suggested both by field experiments and computer simulation.[69,73,75] Likewise, computer models were only able to simulate field observations of resistant and susceptible trees when this concept was included.[55] Therefore, a particular beetle density seems to be needed to exceed a threshold of aggregation above which entered trees become attractive.[27] The discrete outcome of a colonization attempt, then, is determined by a series of opposing rate reactions that govern beetle arrival and host responses. These reactions include resin secretion and allelochemical accumulation by the host, pheromone synthesis, pheromone emission, and arrival of beetles, and vascular penetration and perhaps phytotoxic metabolism by the associated microbial symbionts.

The dynamics of the colonization process are shown in mechanistic and conceptual terms in the upper and lower portions of Fig. 5, respectively. These models predict that the more vigorous trees have a lower likelihood of becoming foci of aggregation than stressed trees (upper pathway f–g–c), but if aggregation can be initiated the attraction will be more intense until the tree is overcome (upper pathway e–a–b). These predictions are supported by our observations on landing beetles. Trees with a high resin content prior to attack became foci of attraction more slowly than do trees with low resin content, but beetles eventually aggregate on these trees more intensely and for a longer period of time (Fig. 6).

The importance of competing rate processes in colonization dynamics can also be seen in the relationship between tree survival and the rate of oleoresin flow from a mechanical wound. The degree of correlation depends on the circumstances under which the initial attacks occur.[73,76] During epidemics, there is no difference in the oleoresin flow rates of trees that are subsequently killed or survive. However, the significance of a high oleoresin flow to tree defense may be obscured by beetles switching attacks from tree to tree so that initial densities can exceed the threshold of aggregation.[39,47,73] If only those trees that undergo attack by pioneer beetles are considered (i.e., the switching effect is eliminated from the sample), then trees that have higher oleoresin flow rates are more likely to survive a flight season than are trees with low oleoresin flow.[73,74] Under endemic conditions, then, we would expect oleoresin flow to be more strongly associated with survival because of the lower beetle population.

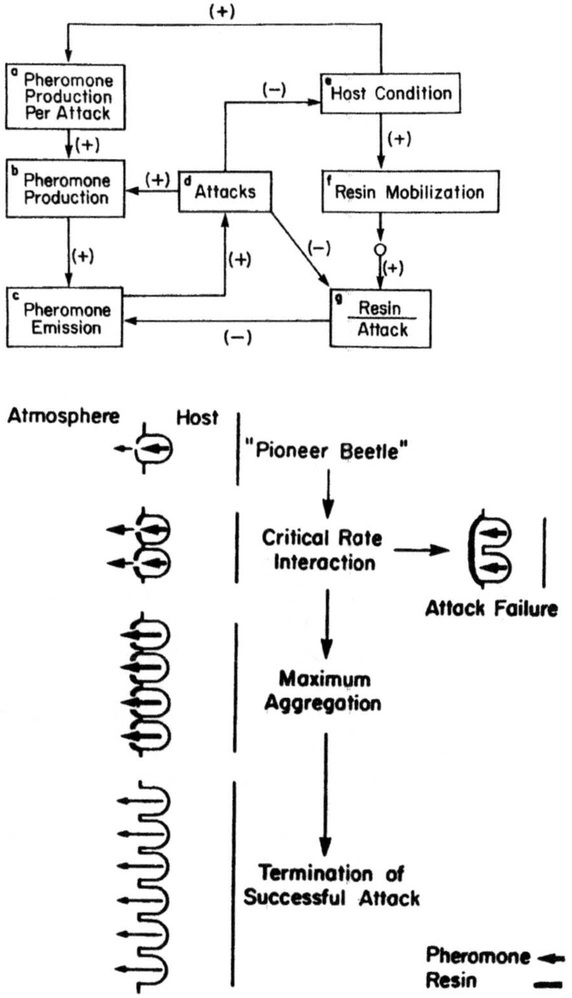

FIGURE 5. Dynamics of mountain pine beetle colonization of living lodgepole pines. (Upper) Feedback structure of the host–insect interaction. Direct and inverse relationships are indicated by + and −, respectively. Open circles indicate time lags. (Lower) Conceptual representation of the host insect interaction. The sequence of pheromone emission is depicted from the first pioneer beetle. Two outcomes are possible: successful colonization or interruption of attack by host defense secretions. Each arrow represents the pheromone emanating from one entrance tunnel. The quantity of aggregation pheromone production is depicted by the size of the arrows in the gallery. The pore diameter indicates the fraction of the pheromone which escapes into the environment.

2.4. Natural Enemies and Competitors

The mountain pine beetle is attacked by an array of natural enemies.[7,63] Woodpeckers are the most effective predators, not only consuming numerous larvae but also killing many more by bark removal that leads to desiccation.[6,7,34]

The most important insect predators are the clerid beetles *Enoclerus lecontei* (Wolcott), *Enoclerus sphegeus* (Fabricius), and *Thanasimus undatulus* SAY, which feed on both adults and larvae, and the long-legged fly *Medetera aldrichii* Wheeler (Diptera: Dolichopodidae), which feeds on eggs and larvae[93–95] The robber fly *Laphria gilva* Linnaeus (Diptera: Asilidae) feeds on adult beetles and can also be important in some

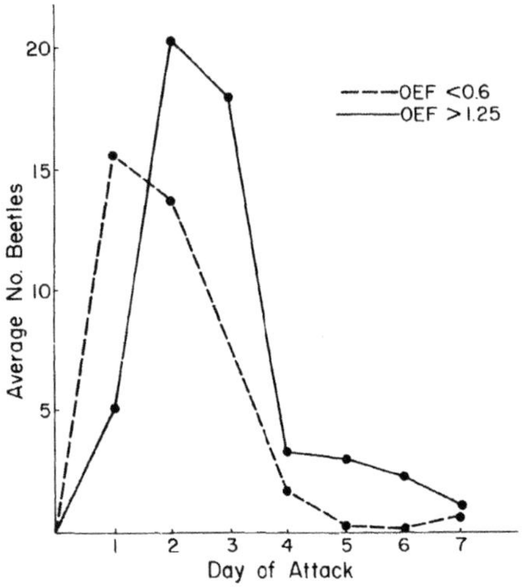

FIGURE 6. Relationship of aggregation pattern to pre-attack resin content. (- - -) Oleoresin flow (OEF) less than 0.2 ml/day. (——) Resin flow above 1.25 ml/day. (From Raffa and Berryman.[73])

areas.[92] The most effective insect parasitoid is *Coeloides dendroctoni* Cushing (Hymenoptera: Braconidae.[7,63] Many predators and parasites respond to aggregation pheromones and rapidly concentrate on trees early in the attack sequence. Acarine associates of *D. ponderosae* have not been as thoroughly studied as those of *D. brevicomis* and *D. frontalis,* but in general they appear to have a minor effect on brood mortality.[7,63]

Nematodes, such as *Mikoletzkya pinicola* (Thorne) (Rhabditidae) and various *Aphelenchoides* species may cause high egg mortality and adult sterility, respectively. Some pathogenic fungi can cause significant mortality to eggs and larvae.[7]

Several phloem-feeding insects colonize trees killed by the mountain pine beetle. These potential competitors include various secondary Scolytidae (e.g., *Ips*), Buprestidae, and Cerambycidae.

3. POPULATION PATTERNS

3.1. General Trends

Mountain pine beetle populations typically remain at low levels within a forest for several decades, and then build rapidly to outbreak proportions. One of the major differences between nonoutbreak and outbreak periods is the physiological condition of hosts that can be colonized. During periods of low density, beetles are most commonly found in individual trees or small groups of killed trees that had been previously weakened by severe stress factors such as lightning, disease, or injury.[9,12,44,114] During periods of high density, however, the population is more evenly distributed, and several fronts of killed trees may spread across a large region. Mountain pine beetles will successfully colonize almost every member of the host population during these periods, including trees that were previously healthy. After the host population has been severely depleted, food becomes a limiting factor, and beetles must colonize relatively small trees in which reproduction is very low.[35] Natural enemies and competitors contribute to the population decline. A few large trees may remain alive in the stand because of physiological immunity to attack, or chance.

3.2. Life-Table Parameters

Several life tables have been compiled to quantify the effects of major within-tree mortality agents.[6,34] Predators and parasites sometimes cause very high mortality, but their effectiveness is generally lower than with other forest pests, such as defoliators. For example, mortality from predation varies between 6 and 7% in pre-epidemic stands, 4% and 29% in epidemic stands, and 4 and 9% in postepidemic stands.[5,34] Likewise, parasitism and disease each kill an average of 1% of the brood. Losses to natural enemies generally show some density dependence but less so than for other forest insect pests. However, many of the predators, parasites, and pathogens show no consistent relationship between prey density and the level of mortality that they exert, and losses to woodpeckers and *Medetera* show inverse density dependence.[6,7,34] The generally low level of losses to invertebrate predators and parasites is probably due to the protected nature of the bark beetle habitat.

The role of interspecific competition varies, but it is generally minor in life tables. One of the major advantages of aggressive behavior in mountain pine beetles is that trees are attacked before they become available to other species. Therefore the mountain pine beetle can usually colonize most of the phloem of the lower bole before other species arrive. In addition, competing species are usually spatially and/or temporarily isolated from the mountain pine beetle brood; e.g., some smaller scolytid species colonize the tree above the height at which mountain pine beetles attack, and woodboring species usually invade after the brood has emerged.

Intraspecific competition can sometimes be a major source of mortality, averaging about 15%.[6,7,34,36] Factors that contribute to these losses are high attack densities, thin phloem, and narrow tree diameter.[2,33,34] The mountain pine beetle has several mechanisms that reduce intraspecific competition. First, attack density is partially related to the number of beetles required to kill the tree. Host compounds serve as substrate for conversion to *trans*-verbenol, so after the tree has been depleted of resin by mass attack, aggregation soon ceases.[73,84] This mechanism is enhanced by the release of antiaggregative pheromones, as well as by acoustic interactions.[85] Second, the number of eggs deposited per female is inversely related to the density of entered beetles.[33,69] Instead of laying all of their eggs, some females may re-emerge and attack new trees.[77] This behavior increases the likelihood of each larva developing to maturity. In general, within-tree intraspecific competition increases during the epidemic period, but this relationship is not pronounced or consistent.[6,34]

Mortality due to cold winter temperatures averages about 25.6% but can be much more at high elevations.[3,6,34] Cool weather also slows brood development and limits beetle flight. Outbreaks can only occur where environmental conditions do not delay beetle reproduction, cause high levels of mortality, or preclude flight during the normal periods of peak dispersal and colonization.[3,86,88] The effect of temperature on brood survivorship is modified by phloem and bark thickness, moisture, snow depth, and the opportunity for a gradual development of cold hardiness.[7] Mortality due to drying of the phloem averages 8% and is generally lowest under moderately moist conditions.

Losses directly attributable to host resistance are usually quite low. Resinosis, the flooding of beetle galleries with host resin, typically kills less than 1% of the brood.[6,34,73] Much of the brood mortality (about 40%) remains unexplained.[6,34]

3.3. Dispersal Losses

The major causes of mortality during dispersal are probably predation, environmental extremes, resinous responses during failed attacks where aggregation is not initiated, and simple energy exhaustion prior to locating a suitable host. The available number of suitable hosts probably plays a major role in affecting dispersal losses.[15] The more easily and quickly that beetles locate and enter susceptible trees, the shorter the duration of exposure to the elements. During nonoutbreak conditions, within-tree survivorship is around 7%,[6,34] and the between-generation replacement rate is presumably around 1. Assuming that around two thirds of the beetles are female,[79] and that each female has a fecundity of around 75 eggs (two rough approximations), about 60% adult dispersal mortality is needed to keep the population in equilibrium. During epidemics, when beetle population are able to overwhelm almost any tree, mortality during dispersal is conceivably quite lower.

4. HYPOTHESIS FOR THE CAUSES OF OUTBREAKS

The overall relationship between beetle survival and beetle population density is not clear, but it is generally concluded that none of the within-tree mortality agents can exert sufficient pressure to regulate beetle populations. It is therefore most likely that populations are food limited.[6,34]

Mountain pine beetle outbreaks commonly occur under three types of conditions. First, outbreaks often follow severe environmental stresses that reduce host vigor.[102] Second, mountain pine beetle outbreaks are much more likely in older and denser stands, particularly if radial growth has begun to slow down and if competition in the stand is severe,[9,16,44,102] Third, outbreaks most commonly start in stands with trees that have large DBH, thick phloem, and high basal area.[4,5] Because these host-related features are so frequently associated with mountain pine beetle outbreaks, and because within-tree life-table analyses show little evidence of any critical regulatory mechanisms, it has been proposed that the transition from nonoutbreak to outbreak behavior is due to changes in host availability.[6,11,13,14,16,17,34,73,75]

Host availability can be viewed at several levels. In its simplest sense it refers to the number of trees in a forest that are within the normal range of species colonized by the mountain pine beetle. However, host species composition is a relatively static component, at least until the later stages of an epidemic when the preferred species become rare. Within a population of a particular host tree species, however, availability can be considered more rigorously. It includes both the susceptibility and quality of individual trees. If each tree is characterized by its particular "threshold of resistance," then the number of susceptible trees will increase with increasing beetle population densities (Fig. 7). Each beetle population size is characterized by a level of tree resistance that can be overcome, and so the number of available trees depends on the distribution frequency of various resistance levels in the host population. Host quality, on the other hand, does not refer to susceptibility per se, but rather the capacity of the host to support viable brood if the tree can be successfully colonized. The most important factor affecting host quality is the thickness of the phloem.[2,5,12]

Host susceptibility and quality are to some extent inversely related.[12,16] In general, the more resistant trees are those which would support the most brood if they could be killed. This is because both defense and current growth are supported by the overall vigor (energy reserves) of the tree (Table I). So the trees of highest quality may not be available to nonoutbreak beetle populations because attack densities will be insufficient to kill them. That is they are not susceptible to low densities of beetles.

As host trees are used, beetles exert both negative and positive density-dependent effects on their own population.[13,18–20] As each tree is killed, it is eliminated from the potential food base and so adversely affects the beetle replacement rate in classical resource-depletion fashion. However, the resulting increase in beetle numbers also causes a corresponding increase in the available food base, i.e., a positive density-dependent effect.[13,14] In general, a small rise in beetle numbers will have a greater positive feedback effect in a homogeneous than heterogenous stand[16,17,41] (Fig. 7).

Because there are both positive and negative feedback processes operating on beetle populations, the effect of population density on its growth rate can show qualitatively different forms of behavior.[13,14,75] If population densities are low, and/or if host availability (susceptibility and/or quality) is low, an equilibrium state is maintained. Population gains during development are offset by population losses during dispersion. Outbreaks cannot occur under these conditions. However, if populations are sufficiently high

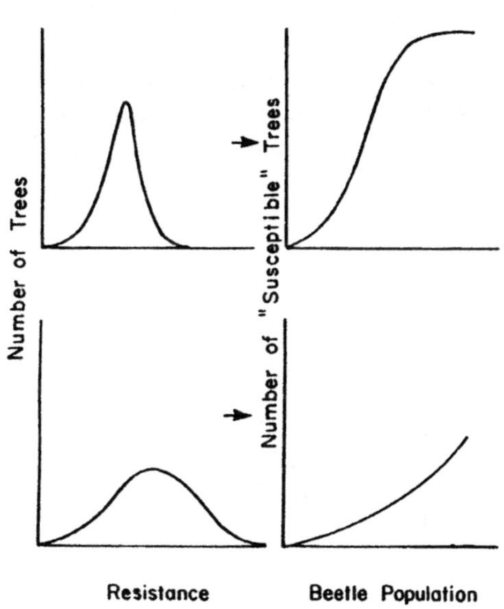

FIGURE 7. Relationship between the frequency distribution of resistant trees (left) and the number of susceptible trees at different beetle population densities (right). The relationship is shown for two stands, one with low between-tree variation in resistance (top), and one with high between-tree variation (bottom).

TABLE I
Association between Host Resistance Traits and Phloem Thickness in Lodgepole Pine[a]

	Thin phloem	Thick phloem
Primary oleoresin flow (ml/day)	5.73	12.85[b]
Radial growth (mm/year)	5.59	8.89[c]

[a]Data from Raffa.[69]
[b]$p < 0.05$.
[c]$p < 0.001$.

or host resistance is low, continuous population growth can occur in those stands where the quality (e.g., phloem thickness, bark thickness) of host trees is sufficient to yield high brood to parent ratios. Under these conditions, the population increases its own food base, expands rapidly, and devastating outbreaks can occur. Thus a critical threshold, determined both by beetle density and host availability, separates stable from outbreak populations.

The discrete separation of net positive or negative density-dependent effects can be displayed graphically by plotting population change against population size (Fig. 8). The metastable equilibrium[20] occurs at E in this figure and the outbreak threshold at U. Thus, if the population increases above U, or if U declines below E, an outbreak will erupt. This model provides some insights into how outbreak behavior can be generated. Factors such as old age that decrease host resistance, narrow the distance between E and U, while factors that enhance resistance broaden it. Thus much larger disturbances are required to initiate outbreaks in healthy stands than in decadent ones.

Based on this model, outbreaks can occur by immigration or within-stand changes in the beetle replacement rate, and various combinations thereof. Very little is known about mountain pine beetle flight, but studies on *D. frontalis*[40,43] suggest that dispersal can be responsible for the eruption of outbreaks in previously uninfested areas. Within stands, outbreaks can be initiated by widespread acute stress, localized acute stress, and aging. For example, a severe stress such as drought increases the number of susceptible trees, and beetle populations rise accordingly. Once the stress is removed, and if the population

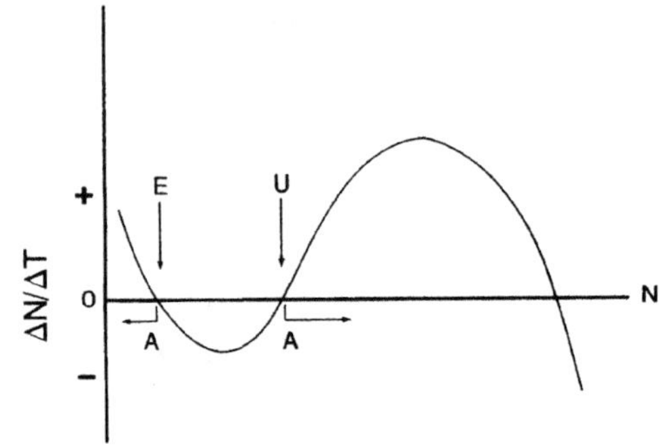

FIGURE 8. Relationship between beetle population growth rate $\Delta N/\Delta t$ and beetle population density N, generated by computer simulation. (From Raffa and Berryman.[75]) Potential equilibria occur where the growth rate is zero. A metastable endemic equilibrium occurs at the point E and an unstable outbreak threshold occurs at the point U.

is above the critical threshold density (U), positive density-dependent forces exceed the negative density-dependent effects, and an outbreak follows. If this critical threshold is not surpassed, negative density-dependent effects will predominate and the population will decline. Likewise, if a localized acute stress, such as root disease, develops among trees with very thick phloem, beetle reproduction in these trees is very high. Once the population has built up in such an epicenter, it can spread outward to neighboring trees or dissipate, again depending on the beetle population/host resistance manifold. A third type of eruption can occur following a gradual reduction in tree vigor over a large area. As lodgepole pine stands age, e.g., physiological resistance declines[102] and competition becomes more severe.[4,30,31] In addition, bark and phloem thickness can increase, at least until competition becomes very severe.[5,7,8] If the stand is even-aged, beetle reproductive success may progressively rise as more and more of the larger trees become decadent. Eventually, the entire stand can be destroyed. If the stand is uneven-aged, the older trees tend to be killed in smaller, more frequent outbreaks.

The ability of large beetle populations to overwhelm resistant trees cannot entirely explain their switch to healthy trees during outbreaks. Because selection operates at the level of individual beetles, the response of each female determines her fitness.[1] Mountain pine beetles show strict behavioral responses during host selection, and will not enter and/or remain in certain trees even when caged on them. If beetles enter trees that subsequently kill them (Fig. 1c), or conversely, if they never show positive orientation toward an encountered host, they fail to reproduce. The correct decision in each case is determined both by the relative resistance of the host and the available number of beetle recruits. Since beetles have no way of estimating the number of recruits available, it is unclear how host selection behavior varies between nonoutbreak and outbreak beetle populations.

There are at least three possible explanations (and combinations thereof):

1. There may be genetic differences between nonoutbreak and outbreak populations. That is, when the population is low, individuals that respond positively to both weakened and healthy trees would have a lower fitness than beetles that orient toward weakened trees only. Following initial entry, few additional beetles are likely to arrive, and so orientation to weakened trees alone is adaptive. However, when populations are high, individuals that enter both weakened and relatively healthy trees are likely to succeed because of the high number of subsequent arrivals. Thus, less discriminating beetles may be at an advantage because of their broader host range.
2. When populations are high, individual beetles may become less selective with time because they are unlikely to encounter weakened trees that have not already been killed. Therefore, a beetle might enter a tree that it previously would not have; That is, as its energy reserves are depleted, its threshold of acceptance may change. Such a mechanism would preserve discriminating genotypes during outbreaks, and be analogous to habituation by other insect species to feeding deterrents.[97]
3. Gustatory responses could be altered by olfaction of aggregation pheromones. That is, the presence of aggregation pheromones could conceivably cause a beetle to enter a tree that it would not normally enter. Mediation of gustatory responses

by pheromone would also preserve discriminating genotypes during outbreaks and, like (2), requires no genetic changes through time.

There is a lot of circumstantial evidence for and against all three of these hypotheses. However, no definitive tests have been made of these behavioral models.

5. MANAGING POPULATIONS

The available data suggest that the mountain pine beetle is governed by a variety of behavioral, physiological, and population thresholds that separate qualitatively different states (Table II) Identification and quantification of these various thresholds provides a framework for developing management strategies. By concentrating on the opposing processes that separate discrete forms of behavior, one can potentially predict and/or manipulate beetle population density. The strategies revolve around (1) determining the risk of outbreak in a given location; i.e., whether the relationship between population density and host availability is approaching the outbreak threshold; or (2) applying corrective measures that reduce the population, enhance resistance, or both.

TABLE II
Critical Thresholds in the Biology and Population Dynamics of the Mountain Pine Beetle

Threshold	Operating level of organization	Definition	Determining variables
Development	Physiological	Temperature (or nutrient, moisture, etc.) at which brood can develop from egg to reproductive adult	Energy demand versus nutrient uptake
Flight	Behavioral	Temperature (or wind, etc.) at which beetles fly	Innate threshold
Landing	Behavioral	Visual stimulus or semiochemical concentration required to arrest flying beetles	Innate threshold
Acceptance	Behavioral	Concentration of stimulants above which, or concentration of repellents below which, entry behavior is elicited	Innate threshold, energy depletion?, host semiochemicals
Aggregation	Within-tree population	Number of beetles required to elicit aggregation on an entered tree	Resin flow rate versus beetle entry and, pheromone synthesis rates
Colonization	Within-tree population	Density of beetles/microorganisms required to kill and physiological condition a tree so as to allow development	Host-defense reaction rate versus fungal pathogencity, and beetle arrival rate
Outbreak	Within-stand population	Density of beetles on an areawide basis, such that positive feedback effects predominate over negative feedback effects	Stand resistance versus beetle population density

Accurate prediction of mountain pine beetle outbreak risk requires measures of both beetle density and host resistance. The available tools for generating both indices are improving and include aerial surveys[52] to provide estimates of beetle population levels (Table II, row 7) as well as relatively general measures of tree vigor, stand resistance, and suitability (Table II, row 6) that are simple enough to be used on a routine basis.[55,107] Predictive ability is greatly enhanced by integrating these parameters with weather data (Table II, rows 1 and 2) in computerized risk-rating models.[42]

To be effective, direct control measures must reduce mountain pine beetle populations below the outbreak threshold (U of Fig. 8). However, no single method, when applied in the absence of an integrated approach, is economically practical, environmentally sound, or reliable.

Insecticide application is not economical or effective over large areas and also disrupts other management objectives, such as recreation, wildlife and livestock production, and watershed management. If the conditions that allowed the outbreak to occur are not removed, a small resurgence of the population will generate a new outbreak. Insecticides can temporarily reduce beetle populations, though, so that there is time for sanitation and salvage measures to be employed. Likewise, direct control by behavior-modifying compounds (Table II, rows 3–5) is not practical on a widescale basis.[113] Although beetles may be trapped and killed, these catches do not necessarily reduce tree mortality.[25,67] Also, the economic and logistic constraints for forest protection over large areas appear quite formidable.[113] Salvage has the direct beneficial effect of removing beetles.[87] Because debarking is required to kill the brood and prevent further spread, however, the economic feasibility of this approach varies with location and current market conditions. To date, there are no effective strategies for implementing biological control against the mountain pine beetle. Although predators, parasites, and pathogens can sometimes cause high beetle mortality under natural conditions, they are difficult to manipulate to an extent that will protect trees. The most efficient approach to biological control of the mountain pine beetle is to avoid deleterious effects on natural enemy populations that result from insecticide sprays and certain types of semiochemical and salvage treatments.

Judicious integration of the above tactics can provide effective direct control in many cases.[26,49,59,65] For example, sanitation logging followed by pheromone baiting of trap trees that are subsequently removed, or sanitation logging coupled with insecticide treatments to prevent emergence and pheromone baiting to attract the remaining population, can be very effective.[22–24,26,28,45] The advantages to these approaches are that semiochemicals and insecticides are used in smaller quantities and in a more focused fashion than in mass trapping, disrupting, or suppression programs, and that some of the more labor-intensive steps can be reduced.[22] The efficacy of such integrated approaches to direct control is likely to increase as knowledge of beetle behavior, antiaggregation pheromones, and optimal combinations of synergistic components, isomers, and enantiomers improves.[112]

Despite the need to control beetle numbers directly under certain conditions, the optimum approach to managing mountain pine beetle populations in commercial stands is by prevention, and the most effective technique is by silviculture.[31,32,62,91] Practices that enhance overall tree vigor raise the outbreak threshold and therefore provide the most reliable preventative measures.[62] For example, initial thinning at 15–25 years to a spacing of 3–4 m, followed by subsequent thinnings and thorough slash removal has been recommended.[30,31] Reduction of stand homogeneity is another useful silvicultural tactic for preventing outbreaks. Clear-cutting in small to moderate-sized blocks can create a mosaic of tree age, size, and species.[4,32] This mosaic can reduce the availability of old, large-diameter trees in which beetle reproduction is highest. It is also likely to increase beetle losses during the dispersal phase.

Because the silvicultural treatments recommended for mountain pine beetle control are aimed at increasing general tree vigor rather than a specific trait or phenological character[54,108] and at removing trees before they become susceptible,[4,32,87] these practices have an overall beneficial effect on the stand. They do not increase the susceptibility of stands to other insects or pathogens (provided there is proper slash removal and stump treatment), except for mistletoe under certain conditions. Mistletoe is not always a factor, however, and where present, adequate control measures have been developed.[32] Stand thinning and rapid rotation are also conducive to optimal wood growth because they are compatible with the natural ecology of lodgepole pine, a fast-growing, shade-intolerant, pioneer (usually) species. A major problem, however, is that lodgepole pine often colonizes poor sites[30] on which silvicultural treatment cannot always be justified economically.

Although silvicultural practices that reduce the threat to the mountain pine beetle also accomplish other intensive management goals, there may be added costs, particularly in areas that are relatively inaccessible, or if the current demand for the crop is low.[96] It is not yet possible to calculate the economic tradeoffs involved because the specifics of treatment costs, efficacy, stocking levels, timing of thinnings and rotations, site and weather predictive indices, and the numerous permutations thereof have not yet been derived. Moreover, the specifics and interactions of these parameters will have to be modified for local conditions wherever the mountain pine beetle is a threat.

Effective management tactics can only succeed if the objectives for each stand are clearly conceived, and are planned on a long-term basis. Intermittent changes in political designations can interfere with preventative tactics, often irrevocably. These tactics are based on gradual physiological changes that occur within the life of a tree. Where the management objective is to preserve wilderness in its natural state, the mountain pine beetle should be appreciated for what it is—a well-adapted resource manager that can use wood well in advance of its competitors. The natural cycle of lodgepole pine growth, beetle outbreaks, fire, and lodgepole pine reestablishment to the exclusion of competing late-successional trees represents a coadapted system that promotes long-term stability. Where the management objective is wood production, however, man is among the mountain pine beetle's competitors. The objective is to produce and remove trees before a large proportion of them is susceptible to attack. The mountain pine beetle will continue to survive at suboutbreak levels in lodgepole pine stands as it always has—an opportunistic scavenger of severely weakened trees. A major challenge in mountain pine beetle management in the coming years will be to develop the appropriate blend of control and preventive tactics at the interface between wilderness and commercial stands.

ACKNOWLEDGMENTS. The critical reviews of G. D. Amman, U.S.D.A. Forest Service, Ogden, Utah; J. H. Borden, Department of Biological Sciences, Simon Frasier University; E. Christiansen, Norwegian Forest Research Institute; S. G. Codella, Department of Entomology, University of Wisconsin; S. C. Krause, Department of Entomology, University of Wisconsin; and L. Safranyik, Canadian Forestry Service, Victoria, British Columbia, are greatly appreciated. The drawing heading this chapter was kindly provided by G. D. Amman.

REFERENCES

1. Alcock, J., 1981, Natural selection and communication among bark beetles, *Florida Entomol.* **65:** 17–32.
2. Amman, G. D., 1972, Mountain pine beetle brood production in relation to thickness of lodgepole pine phloem, *J. Econ. Entomol.* **65:**138–140.
3. Amman, G. D., 1973, Population changes of the mountain pine beetle in relation to elevation, *Environ. Entomol.* **2:**541–547.
4. Amman, G. D., 1976, Integrated control of the mountain pine beetle in lodgepole pine forests, in: Proceedings of the Sixteenth IUFRO World Congress, Div. II, Oslo, Norway, pp. 439–446.
5. Amman, G. D., 1977, The role of the mountain pine beetle in lodgepole pine ecosystems: Impact on succession, in: *The Role of Arthropods in Forest Ecosystem,* Proceedings of the Fifteenth International Congress on Entomology, Washington, D. C., (W. J. Mattson, ed.), pp. 3–18, Springer-Verlag.
6. Amman, G. D., 1984, Mountain pine beetle (Coleoptera: Scolytidae) mortality in three types of infestations, *Environ. Entomol.* **13:**184–191.
7. Amman, G. D., and Cole, W. F., 1983, Mountain pine beetle dynamics in lodgepole pine forests. Part II: Population dynamics. USDA Forest Service Gen. Tech. Rep. INT-145.
8. Amman, G. D., and Safranyik, L., 1984, Insects of lodgepole pine: Impacts and control, in: *Lodgepole Pine. The Species and Its Management* (D. M. Baumgartner, R. G. Krebill, J. T. Arnott, and G. F. Watman, eds.), pp. 107–128, Washington State University Cooperative Extension.
9. Beal, J. A., 1943, Relation between tree growth and outbreaks of the Black Hills Beetle, *J. For.* **41:**359–366.
10. Berryman, A. A., 1972, Resistance of conifers to invasion by bark beetle–fungal associations, *BioScience* **22:**598–602.
11. Berryman, A. A., 1974, Dynamics of bark beetle populations: Towards a general productivity model, *Environ. Entomol.* **3:**579–585.
12. Berryman, A. A., 1976, Theoretical explanation of mountain pine beetle dynamics in lodgepole pine forests, *Environ. Entomol.* **5:**1225–1233.
13. Berryman, A. A., 1978, Towards a theory of insect epidemiology, *Res. Popul. Ecol.* **19:**181–196.
14. Berryman, A. A., 1978, A synoptic model of the lodgepole/pine mountain pine beetle interaction and its potential application in forest management, in: *Theory and Practice of Mountain Pine Beetle Management in Lodgepole Pine Forests* (A. A. Berryman, G. D. Amman, R. W. Stark, and D. L. Kibbee, eds.), pp. 98–105, College of Forest Resources, University of Idaho, Moscow, Idaho.
15. Berryman, A. A., 1979, Dynamics of bark beetle populations: analysis of dispersal and redistribution, *Mitt. Schw. Entomol. Ges.* **52:**227–234.
16. Berryman, A. A., 1982, Mountain pine beetle outbreaks in Rocky Mountain lodgepole pine forests, *J. For.* **80:**410–419.
17. Berryman, A. A., 1982, Biological control, thresholds, and pest outbreaks, *Environ. Entomol.* **11:**544–549.
18. Berryman, A. A., 1983, Defining the resilience threshold of ecosystems, in: *Analysis of Ecological Systems State-of-the-Art in Ecological Modelling* (W. K. Lavenroth, G. V. Skogerboe, and M. Flug, eds.), pp. 57–60, Elsevier, Amsterdam.
19. Berryman, A. A., 1985, Threshold theory and its application to pest population management, in: *Pest and Pathogen Control: Strategy, Tactical and Policy Models* (G. R. Conway, ed.), pp. 40–57, Wiley (Interscience), Chichester, England.
20. Berryman, A. A., Stenseth, N. C., and Wollkind, D. J., 1984, Metastability of forest ecosystems infested by bark beetles, *Res. Pop. Ecol.* **26:**13–29.
21. Borden, J. H., 1984, Semiochemical-mediated aggregation and dispersion in the Coleoptera, in: *Insect Communication* (T. Lewis, ed.), pp. 123–149, Academic, London.
22. Borden, J. H., Chong, L. J., and Fuchs, M. C., 1983, Application of semiochemicals in post-logging

manipulation of the mountain pine beetle, *Dendroctonus ponderosae* (Coleoptera: Scolytidae), *J. Econ. Entomol.* **76:**1428–1432.

23. Borden, J. H., Chong, L. J., and Lacey, T. E., 1986, Pre-logging baiting with semiochemicals for the mountain pine beetle. *Dendroctonus ponderosae,* in high hazard stands of lodgepole pine, *For. Chron.* **62:**20–23.
24. Borden, J. H., Chong, L. J., Pratt, K. E. G., and Gray, D. R., 1983, The application of behaviour-modifying chemicals to contain infestations of the mountain pine beetle, *Dendroctonus ponderosae, For. Chron.* **59:**235–239.
25. Borden, J. H., Conn, J. F., Friskie, L. M., Scott, B. E., Cong, L. J., Pierce, H. D., Jr., and Oehlschlager, A. C., 1983, Semiochemicals for the mountain pine beetle, *Dendroctonus ponderosae* (Coleoptera: Scolytidae), in British Columbia: baited tree studies, *Can. J. For. Res.* **13:**325–333.
26. Borden, J. H., Hunt, D. W. A., Miller, D. R., and Slessor, K. N., 1986, Orientation in forest Coleoptera: An uncertain outcome of responses by individual beetles to variable stimuli, in: *Mechanisms in Insect Olfaction* (T. L. Payne, M. C. Birch, and C. E. J. Kennedy, eds.), pp. 98–109, Clarendon, Oxford.
27. Burnell, D. G., 1977, A dispersal–aggregation model for mountain pine beetle in lodgepole pine stands, *Res. Pop. Ecol.* **19:**99–106.
28. Cerezke, H. F., Borden, J. H., and Trott, T. N., 1984, Field tests with semiochemicals for the mountain pine beetle in the Cypress Hills, Alberta, *Can. For. Serv. Res. Notes* **4:**16–18.
29. Christiansen, E., and Ericsson, A., 1986, Starch reserves in *Picea abies* in relation to defense reaction against a bark beetle transmitted blue-stain fungus, *Ceratocystis polonica, Can. J. For. Res.* **16:**78–83.
30. Cole, D. M., 1978, Feasibility of silviculture practices for reducing losses to the mountain pine beetle in lodgepole pine forests, in: *Theory and Practice of Mountain Pine Beetle Management in Lodgepole Pine Forests* (A. A. Berryman, G. D. Amman, R. W. Stark, and D. L. Kibbee, eds.), pp. 149–157, University of Idaho, Moscow, Idaho.
31. Cole, D. M., 1985, Acceptable silvicultural systems in relation to desired stand character and successional roles of lodgepole pine, in: Integrating Management Strategies for the Mountain Pine Beetle with Multiple-Resource Management of Lodgepole Pine Forests (M. D. McGregor and D. M. Cole, eds.), pp. 45–46. USDA For. Serv. Gen. Tech. Rep. INT-174.
32. Cole, D. M., and McGregor, M. D., 1985, Silvicultural practices for lodgepole pine stands in commercial forests, in: Integrating Management Strategies for the Mountain Pine Beetle with Multiple-Resource Management of Lodgepole Pine Forests (M. D. McGregor and D. M. Cole, eds.), pp. 47–56, USDA For. Serv. Gen. Tech. Rep. INT-174.
33. Cole, W. E., 1962, The effects of intraspecific competition within mountain pine beetle broods under lab conditions, USDA For. Serv. Res. Note 97.
34. Cole, W. E., 1981, Some risks and causes of mortality in mountain pine beetle populations: A long-term analysis, *Res. Pop. Ecol.* **23:**116–144.
35. Cole, W. E., and Amman, G. D., 1980, Mountain pine beetle dynamics in lodgepole pine forests. Part I. Course of an infestation. USDA For. Serv. Gen. Tech. Rep. INT-89.
36. Cole, W. E., Amman, G. D., and Jensen, C. E., 1976, Mathematical models for the mountain pine beetle-lodgepole pine interaction, *Environ. Entomol.* **5:**11–19.
37. Conn, J. E., Borden, J. H., Hunt, D. W. A., Holman, J., Whitney, H. S., Spanier, O. J., Pierce, H. D., Jr., and Oeschlager, A. C., 1984. Pheromone production by axenically reared *Dendroctonus ponderosae* and *Ips paraconfusus* (Coleoptera: Scolytidae), *J. Chem. Ecol.* **10:**281–289.
38. Conn, J. E., Borden, J. H., Scott, B. E., Friskie, L. M., Pierce, H. D., Jr., and Oehlschlager, A. C., 1983, Semiochemicals for the mountain pine beetle, *Dendroctonus ponderosae* (Coleoptera: Scolytidae) in British Columbia: Field trapping studies, *Can. J. For. Res.* **13:**320–324.
39. Coster, J. E., and Gara, R. I., 1968, Studies on the attack behavior of the southern pine beetle. II. Response to attractive host material, *Contrib. Boyce Thompson Inst.* **24:**69–96.
40. Coster, J. E., Hicks, R. R., Jr., and Watterston, K. G., 1978, Directional spread of southern pine beetle (Coleoptera: Scolytidae) infestations in east Texas, *J. Georgia Entomol. Soc.* **13:**315–321.
41. Coulson, R. N., and Stark, R. W., 1982, Integrated management of bark beetles, in: *Bark Beetles in North American Forests: A System for the Study of Evolutionary Biology* (J. B. Mitton and K. B. Sturgeon, eds.), pp. 315–349, University of Texas Press, Austin, Texas.
42. Crookston, N. L., Roelke, R. C., Burnell, D. G., and Stage, A. R., 1978, Evaluation of management alternatives for lodgepole pine stands using a stand projection model, in: *Theory and Practice of Mountain*

Pine Beetle Management in Lodgepole Pine Forests (A. A. Berryman, G. D. Amman, R. W. Stark, and D. L. Kibbee, eds.), pp. 114–124, University of Idaho, Moscow, Idaho.
43. De Mars, C. J., and Hain, F. P., 1980, Bark beetle dispersal related to patterns of tree mortality and wind, in: *Dispersal of Forest Insects: Evaluation, Theory, and Management Implications*, Proceedings of the IUFRO Conference of Working Parties S2.07.05 and S2.07.06 (A. A. Berryman and L. Safranyik, eds.), pp. 66–78, Washington State University, Pullman, Washington.
44. Evenden, J. C., Bedard, W. D., and Struble, G. R., 1943, The mountain pine beetle, an important enemy of western pines, U.S. Dept. Agric. Circ. 664.
45. Fuchs, M. G., and Borden, J. H., 1985, Pre-emergence insecticide application for control of the mountain pine beetle, *Dendroctonus ponderosae* (Coleoptera: Scolytidae), *J. Entomol. Soc. Br. Col.* **82**:25–28.
46. Furniss, R. L., and Carolin, V. M., 1977, Western forest insects, USDA Forest Service Misc. Publ. No. 1339.
47. Gara, R. I., and Coster, J. E., 1968, Studies on the attack behavior of the southern pine beetle. III. Sequence of tree infestation within stands, *Contrib. Boyce Thompson Inst.* **24**:69–79.
48. Geiszler, D. R., Gallucci, V. F., and Gara, R. I., 1980, Modeling the dynamics of mountain pine beetle aggregation in a lodgepole pine stand, *Oecologia (Berl.)* **46**:244–253.
49. Hall, P. M., 1985, Background and techniques used in the British Columbia bark beetle program. British Columbia Ministry of Forests, For. Serv. Intern. Rep. PM-PB-15 (cited with approval of BC Ministry of Forests Protection Branch).
50. Hughes, P. R., 1973, *Dendroctonus:* Production of pheromones and related compounds in response to host monoterpenes, *Z. Angew. Entomol.* **73**:294–312.
51. Hynum, B. G., and Berryman, A. A., 1980, *Dendroctonus ponderosae* (Coleoptera: Scolytidae): Pre-aggregation landing and gallery initiation on lodgepole pine, *Can. Entomol.* **112**:185–191.
52. Klein, W. H., Bennett, D. D., and Young, R. W., 1980, Evaluation of panoramic reconnaisance, aerial photography for measuring annual mortality of lodgepole pine caused by the mountain pine beetle, USDA For. Serv. Rep. No. 80-2.
53. Lanier, G. N., and Wood, D. L., 1968, Controlled mating, karyology, morphology, and sex-ratio in the *Dendroctonus ponderosae* complex, *Ann. Entomol. Soc. Am.* **61**:517–526.
54. Larsson, S., Oren, R., Waring, R. H., and Barrett, J. W., 1983, Attack of mountain pine beetle is related to tree vigor of ponderosae pine, *For. Sci.* **29**:395–402.
55. Mahoney, R. L., 1978, Lodgepole pine/Mountain pine beetle risk classification methods and their application, in: *Theory and Practice of Mountain Pine Beetle Management in Lodgepole Pine Forests* (A. A. Berryman, G. D. Amman, R. W. Stark, and D. L. Kibbee, eds.), pp. 106–113, University of Idaho, Moscow, Idaho.
56. McCambridge, W. F., 1975, Scotch pine and mountain pine beetles. *Green Thumb* **32**:87.
57. McGregor, M. D., 1985, The conflict between people and the beetle, in: Insect and Disease Conditions in the United States (R. C. Loomis, S. Tucker, and T. H. Hofacker, eds.), USDA For. Serv. Gen. Rep. WO-46.
58. McGregor, M. D., and Cole, D. M., 1985, Practices and considerations for noncommercial forests, in: Integrating Management Strategies for the Mountain Pine Beetle with Multiple-Resource Management of Lodgepole Pine Forests (M. D. McGregor and D. M. Cole, eds.), pp. 56–59, USDA For. Serv. Gen. Tech. Rep. INT-174.
59. McMullen, L. H., Safranyik, L., and Linton, D. A., 1986, Suppression of mountain pine beetle infestations in lodgepole pine forests, Can. For. Serv. Inform. Rep. BC-X-276.
60. Miller, R. H., and Berryman, A. A., 1985, Energetics of conifer defense against bark beetles and associated fungi, in: The Role of the Host in the Population Dynamics of Forest Insects, Proceedings of the IUFRO Conference of Working Parties S2-07-05 and S2-07-06 (L. Safranyik, ed.), pp. 13–23, Can. For. Serv., Pac. For. Res. Cent., Victoria, B.C.
61. Miller, R. H., Berryman, A. A., and Ryan, C. A., 1976, Biotic elicitors of defense reactions in lodgepole pine, *Phytochemistry* **25**:611–612.
62. Mitchell, R. G., Waring, R. H., and Pitman, G. B., 1983, Thinning lodgepole pine increases tree vigor and resistance to mountain pine beetle, *For. Sci.* **29**:204–211.
63. Moeck, H. A., and Safranyik, L., 1984, Assessment of predator and parasitoid control of bark beetles, Can. For. Serv. Inform. Rep. BC-X-248.
64. Moeck, H. A., Wood, D. L., and Lindahl, K. Q., Jr. 1981, Host selection behavior of bark beetles

(Coleoptera: Scolytidae) attacking *Pinus ponderosae,* with special emphasis on the western pine beetle, *Dendroctonus brevicomis, J. Chem. Ecol.* **7**:49–83.

65. Nijolt, W. W., McMullen, L. H., and Safranyik, L., 1981, Pine oil protects living trees from attack by three bark beetle species, *Dendroctonus* spp. (Coleoptera: Scolytidae), *Can. Entomol.* **113**:337–340.
66. Pitman, G. B., 1971, *Trans*-verbenol and alpha-pinene. Their utility in manipulation of the mountain pine beetle, *J. Econ. Entomol.* **64**:426–430.
67. Pitman, G. B., Stock, M. W., and McKnight, R. C,, 1978, Pheromone application in mountain pine beetle/lodgepole pine management: Theory and practice, in: *Theory and Practice of Mountain Pine Beetle Management in Lodgepole Pine Forests,* Symposium Proceedings, (A. A. Berryman, G. D. Amman, R. W. Stark, and D. L. Kibbee, eds.), pp. 165–173, University of Idaho, Moscow, Idaho.
68. Pitman, G. B., Vite, J. P., Kinzer, G. W., and Fentiman, A. F., Jr., 1968, Bark beetle attractants: trans-Verbenol isolated from *Dendroctonus, Nature (Lond.)* **218**:168–169.
69. Raffa, K. F., 1981, The role of host resistance in the colonization behavior, ecology, and evolution of bark beetles, Ph.D. thesis, Washington State University, Pullman, Washington.
70. Raffa, K. F., and Berryman, A. A., 1980, Flight responses and host selection by bark beetles, in: *Dispersal of Forest Insects: Evaluation, theory, and Management Implications* (A. A. Berryman and L. Safranyik, eds.), pp. 213–233, Conference Office, Cooperative Extension Service, Washington State University, Pullman, Washington.
71. Raffa, K. F., and Berryman, A. A., 1982, Gustatory cues in the orientation of *Dendroctonus ponderosae* (Coleoptera: Scolytidae) to host trees, *Can. Entomol.* **114**:97–104.
72. Raffa, K. F., and Berryman, A. A., 1982, Physiological differences between lodgepole pines resistant and susceptible to the mountain pine beetle and associated microorganisms, *Environ. Entomol.* **11**:486–492.
73. Raffa, K. F., and Berryman, A. A., 1983, The role of host plant resistance in the colonization behavior and ecology of bark beetles (Coleoptera: Scolytidae), *Ecol. Monog.* **53**:27–49.
74. Raffa, K. F., and Berryman, A. A., 1983, Physiological aspects of lodgepole pine wound responses to a fungal symbiont of the mountain pine beetle, *Can. Entomol.* **115**:723–734.
75. Raffa, K. F., and Berryman, A. A., 1986, A mechanistic computer model of mountain pine beetle populations interacting with lodgepole pine stands and its implications for forest managers, *For. Sci.* **32**:789–805.
76. Raffa, K. F., and Berryman, A. A., 1987, Interacting selective pressures in conifer - bark beetle systems: A basis for reciprocal adaptations?, *Am. Natur.* **129**:234–262.
77. Reid, R. W., 1958, Internal changes in the female mountain pine beetle *Dendroctonus monticolae* Hopk., associated with egg laying and flight, *Can. Entomol.* **90**:464–468.
78. Reid, R. W., 1958, The behaviour of the mountain pine beetle, *Dendroctonus monticolae* Hopk. during mating, egg laying, and gallery construction, *Can. Entomol.* **90**:505–509.
79. Reid, R. W., 1962, Biology of the mountain pine beetle, *Dendroctonus monticolae* Hopkins in the East Kootenay region of British Columbia. I. Life cycle, brood development and flight periods, *Can. Entomol.* **94**:531–538.
80. Reid, R. W., 1962, Biology of the mountain pine beetle, *Dendroctonus monticolae* Hopkins, in the East Kootenay region of British Columbia. II. Behaviour in the host, fecundity, and internal changes in the female, *Can. Entomol.* **94**:605–613.
81. Reid, R. W., 1963, Biology of the mountain pine beetle, *Dendroctonus monticolae* Hopkins in the East Kootenay region of British Columbia. III. Interaction between the beetle and its host with emphasis on brood mortality and survival, *Can. Entomol.* **95**:225–238.
82. Reid, R. W., and Gates, H., 1970, Effect of temperature and resin on hatch of eggs of the mountain pine beetle (*Dendroctonus ponderosae*), *Can. Entomol.* **102**:617–622.
83. Reid, R. W., Whitney, H. S., and Watson, J. A., 1967, Reactions of lodgepole pine to attack by *Dendroctonus ponderosae* Hopkins and blue stain fungi, *Can. J. Bot.* **45**:115–116.
84. Renwick, J. A. A., and Vite, J. P., 1970, Systems of chemical communication in *Dendroctonus, Contrib. Boyce Thompson Inst.* **24**:283–292.
85. Rudinsky, J. A., Morgan, M. E., Libbey, L. M., and Putnam, T. B., 1974, Antiaggregative-rivalry pheromone of the mountain pine beetle and a new arrestant of the southern pine beetle, *Environ. Entomol.* **3**:90–98.
86. Safranyik, L., 1978, Effects of climate and weather on mountain pine beetle populations, in: *Theory and*

Practice of Mountain Pine Beetle Management in Lodgepole Pine Forests (A. A. Berryman, G. D. Amman, R. W. Stark, and D. L. Kibbee, eds.), pp. 77–86, University of Idaho, Moscow, Idaho.

87. Safranyik, L., Shrimpton, D. M., and Whitney, H. S., 1974, Management of lodgepole pine to reduce losses from the mountain pine beetle, *Environ. Can. For. Serv. For. Tech. Rep.* **1**:24 pp.
88. Safranyik, L., Shrimpton, D. M., and Whitney, H. S., 1975, An interpretation of the interaction between lodgepole pine, the mountain pine beetle and its associated blue stain fungi in western Canada, in: *Management of Lodgepole Pine Ecosystems* (D. M. Baumgartner, ed.), pp. 406–428, Washington State University Co-op. Ext. Ser., Pullman, Washington.
89. Safranyik, L., and Vithayasai, C., 1971, Some characteristics of the spatial arrangement of attacks by the mountain pine beetle, *Dendroctonus ponderosae* (Coleoptera: Scolytidae), on lodgepole pine, *Can. Entomol.* **103**:1607–1625.
90. Safranyik, L., and Whitney, H. S., 1985, Development and survival of axenically reared mountain pine beetle, *Dendroctonus ponderosae* (Coleoptera: Scolytidae) at constant temperatures, *Can. Entomol.* **117**:185–192.
91. Sartwell, C., and Stevens, R. E., 1975, Mountain pine beetle in ponderosa pine. Prospects for silvicultural control in second-growth stands, *J. For.* **73**:136–140.
92. Schmid, J. M., 1969, *Laphria gilva* (Diptera: Asilidae), a predator of *Dendroctonus ponderosae* (Coleo ptera: Scolytidae) in the Black Hills, *Can. Entomol.* **102**:969–977.
93. Schmid, J. M., 1970, *Enoclerus sphegeus* (Coleoptera: Cleridae) a predator of *Dendroctonus ponderosae* in the Black Hills of South Dakota, *Ann. Entomol. Soc. Am.* **62**:1237–1241.
94. Schmid, J. M., 1970, *Medetera aldrichii* (Diptera: Dolichopodidae) in the Black Hills. I. Emergence and behavior of adults, *Can. Entomol.* **102**:705–713.
95. Schmid, J. M., 1971, *Medetera aldrichii* (Diptera: Dolichopodidae) in the Black Hills. II. Biology and densities of the immature stages, *Can. Entomol.* **103**:848–853.
96. Schmidt, W. C., 1982, Alternative solutions to mountain pine beetle. The silviculture perspective, in: *Proceedings of the joint Canada/USA Workshop on Mountain Pine Beetle Related Problems in Western North America* (D. M. Shrimpton, ed.), pp. 33–40, Env. Can. BC-X-230.
97. Schoonhoven, L. M., 1982, Biological aspects of antifeedants, *Entomologia exper. applic.* **31**:57–69.
98. Shepherd, R. F., 1965, Distribution of attacks by *Dendroctonus ponderosae* Hopk. on *Pinus contorta* Dougl. var. latifolia Engelm, *Can. Entomol.* **97**: 207–215.
99. Shepherd, R. F., 1966, Factors influencing the orientation and rates of activity of *Dendroctonus ponderosae* (Coleoptera: Scolytidae), *Can. Entomol.* **98**:507–518.
100. Shrimpton, D. M., 1973, Extractives associated with the wound response of lodgepole pine attacked by the mountain pine beetle and associated microorganisms, *Can. J. Bot.* **51**:527–534.
101. Shrimpton, D. M., 1973, Age- and size-related response of lodgepole pine to inoculation with *Europhium clavigerum*, *Can. J. Bot.* **51**:115–1160.
102. Shrimpton, D. M., 1978, Resistance of lodgepole pine to mountain pine beetle infestation, in: *Theory and Practice of Mountain Pine Beetle Management in Lodgepole Pine Forests* (A. A. Berryman, G. D. Amman, R. W. Stark, and D. L. Kibbee, eds.), pp. 64–76, College of Forest Resources, University of Idaho, Moscow, Idaho.
103. Shrimpton, D. M., and Watson, J. A., 1971, Response of lodgepole pine seedlings to inoculation with *Europhium clavigerum* a blue stain fungus, *Can. J. Bot.* **49**:373–375.
104. Shrimpton, D. M., and Whitney, H. S., 1968, Inhibition of growth of blue stain fungi by wood extractives, *Can. J. Bot.* **46**:757–761.
105. Stock, M. W., and Amman, G. D., 1980, Genetic differentiation among mountain pine beetle populations from lodgepole pine and ponderosa pine in northeast Utah, *Ann. Entomol. Soc. Am.* **73**:472–478.
106. Stock, M. W., and Guenther, J. D., 1979, Isozyme variation among mountain pine beetle (*Dendroctonus ponderosae*) populations in the Pacific Northwest, *Environ. Entomol.* **8**:889–893.
107. Waring, R. H., and Pitman, G. B., 1980, A simple model of host resistance to bark beetles, Oregon State University, School of Forestry, Corvallis, Oregon, Res. Note 65.
108. Waring, R. H., and Pitman, G. B., 1983, Physiological stress in lodgepole pine as a precursor to mountain pine beetle attack, *Z. Angew. Entomol.* **96**:265–270.
109. Whitney, H. S., 1982, Relationships between bark beetles and symbiotic organisms, in: *Bark Beetles in North American Forests: A System for the Study of Evolutionary Biology* (J. B. Mitton and K. B. Sturgeon, eds.), pp. 183–211, University of Texas Press, Austin, Texas.

110. Whitney, H. S., and Farris, S. H., 1970, Maxillary mycangium in the mountain pine beetle, *Science* **167**:54–55.
111. Wood, D. L., 1972, Selection and colonization of ponderosa pine by bark beetles, in: *Royal Entomological Society Symposium No. 6: Insect/Plant Relationships* (H. F. van Emden, ed.), pp. 101–107, Blackwell Scientific, Oxford.
112. Wood, D. L., 1982, The role of pheromones, kairomones, and allomones in the host selection and colonization behavior of bark beetles, *Annu. Rev. Entomol.* **27**:411–446.
113. Wood, D. L., Stark, R. W., Waters, F., Bedard, W. D., and Cobb, F. W., Jr., 1985, Treatment tactics and strategies, in: *Integrated Pest Management in Pine–Bark Beetle Ecosystems* (W. F. Waters, R. W. Stark, and D. L. Wood, eds.), pp. 121–140, Wiley, New York.
114. Wood, S. L., 1982, The bark and ambrosia beetles of North and Central America (Coleoptera: Scolytidae) a taxonomic monograph, *Great Basin Nat.* Memoirs 6, Brigham Young University, Provo, Utah.

CHAPTER 25

THE SOUTHERN PINE BEETLE

Richard O. Flamm, Robert N. Coulson, and Thomas L. Payne

Richard O. Flamm and Robert N. Coulson • Department of Entomology, Texas A&M University, College Station, Texas 77843. **Thomas L. Payne** • Department of Entomology, Virginia Polytechnic Institute and State University, Blacksburg, Virginia 24061.

1. INTRODUCTION

The Southern pine beetle, *Dendroctonus frontalis* Zimmerman (Coleoptera: Scolytidae), is the most destructive insect pest of pine forests in the southeastern United States and in parts of Mexico and Central America (Fig. 1). This is a well-worn statement but is nonetheless richly deserved and quite accurate. The beetle epitomizes the definition of its genus, killer of trees. As such, *D. frontalis* can overcome even vigorous trees when its populations are large. In the United States, the southern pine beetle primarily colonizes loblolly (*Pinus taeda*) and shortleaf (*P. enchinata*) pines, both of which are important sources of pulp and timber. These pines are fast-growing species, inhabiting some of the

FIGURE 1. Known distribution of the southern pine beetle.

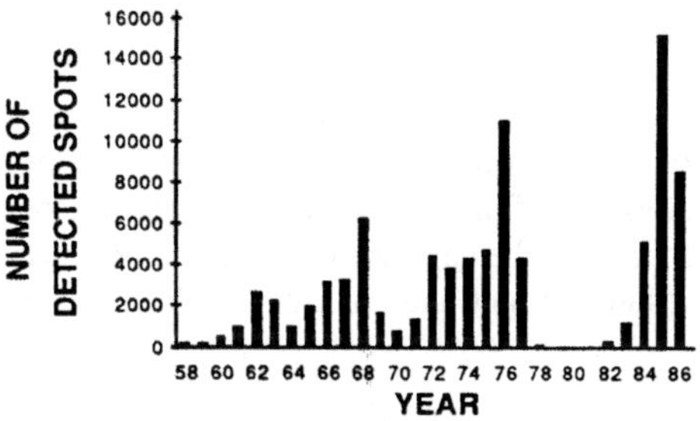

FIGURE 2. Number of *Dendroctonus frontalis* infestations (spots) detected in Texas, 1958–1986. (Texas Forest Service data.)

finest timber-growing land on the continent. In addition, most watersheds in the southern pine forest are dominated by these species, and in many recreational areas high-valued old growth pine prevail.

Adult *D. frontalis* kill trees by boring through the bark and feeding and ovipositing in the phloem. Tree death, a necessary event for successful development of the brood, results from the tree being girdled from the activity of the adult beetles and is hastened by the invasion of fungi that are carried by the beetle.

Somewhere within the beetle's range, epidemic populations may be found almost every year, but beetle activity fluctuates significantly in local areas and across the range of the insect (Fig. 2).

Long before formal records were kept, accounts suggest that *D. frontalis* plagued virgin southern pine forests over large areas throughout the 1700s and 1800s. Accounts from Moravian settlers dating back to 1750 describe the destruction of vast amounts of pine timber due to the mischief of what appears to have been bark beetles. Oldtimers in east Texas report that early in this century settlers used beetle infestations to clear land for pasture by hitting the trees with the back of an ax and then leaning infested sticks against them.

Early accounts of tree mortality caused by *D. frontalis* are fragmentary, but one can still determine its general impact. For example, timber killed from 1891 to 1929 had a value of at least \$50 million.[65] Additional records showed that *D. frontalis* was responsible for killing more than 200,000 cords and 500 million board-feet of timber from 1882 to 1960. Approximately 3 billion board-feet valued at \$225 million were destroyed between 1960 and 1978. In 1985, Texas suffered record destruction, losing more than \$49 million.

2. LIFE CYCLE AND BEHAVIOR

2.1. Host Selection

Host trees are selected for attack by female beetles, the first to attack being referred to as pioneers.[5] Males enter the picture only after the females have successfully attacked a host.

Some investigators have proposed that pioneer beetles find suitable hosts by responding to primary attractants produced by the tree.[33,55,58] Primary attraction has been demonstrated with ambrosia beetles[47] and several other scolytids,[48] but not with *D. frontalis*.

Specific host odors, however, do appear to function in *D. frontalis* host selection by arresting southern pine beetles after they have landed on a tree.[57,61]

Other studies suggest that host selection is determined by beetles landing randomly on vertical objects.[29] Once on the tree, the beetles bite the bark in response to chemical stimuli;[75] if the tree is suitable, attractive pheromones are released by the pioneers, setting in motion the aggregation phase. In the absence of the correct chemical stimuli, the beetle resumes flight and continues to land at random on both host and nonhost trees.

2.2. Aggregation and Attack

As soon as a few beetles have bored into a susceptible tree, secondary attraction begins. This phase of the beetle's life cycle is critical because it enables *D. frontalis* to accumulate on the tree rapidly enough and in sufficient numbers to overcome host defenses. Aggregation is dependent on beetle- and host-produced volatiles and their effects on flying beetles.[54] *D. frontalis* has a complex pheromone system that acts intraspecifically in aggregation[41,57] and inhibition.[56] It is also perceived by several insect associates, including the *Ips* bark beetles and many natural enemies.

After attacking a tree, pioneer females immediately begin to release frontalin,[41,61,62] considered the primary aggregation pheromone of the beetle.[43,57] Frontalin, along with host tree odors (particularly α-pinene), attracts large numbers of beetles (especially males) to the tree.[61] Arriving males release endo-brevicomin which is primarily attractive to females. In addition, *D. frontalis* releases other volatiles that operate as synergists, attractants or inhibitors. As the number of arriving beetles increases, higher concentrations of pheromones are released, many of which have an inhibitory effect at high concentrations. As a result, the tree loses its attractiveness, and the attack switches to a new tree.[28]

The duration of attack ranges from 8 days to 6 weeks, largely depending on seasonal conditions and the number of beetles in the area. Attacked trees can be identified by the characteristic pitch tubes found at the site of boring (Fig. 3) and by frass in bark crevices and suspended in spider webs. In most instances, the central section of the tree is attacked first and at higher density than the sections above and below. In some cases, the vertical arrangement of *D. frontalis* attacks on the tree may be influenced by *Ips* species. During outbreaks, Southern pine beetles almost always attack first[53,71] and colonize the mid to lower bole. *Ips* beetles may then fill up the vacant spaces in the upper bole. In non-outbreak areas, *Ips* species may arrive first, colonizing the mid-bole, while *D. frontalis* later attacks at the lower bole or crown, usually in low densities. These trees may serve as refuge trees, sustaining populations of *D. frontalis* when conditions are not conducive to population growth.

2.3. Mating and Oviposition

Dendroctonus frontalis is monogamous within individual egg galleries, but females may mate with other males following reemergence and attack of new host trees. Copulation takes place in the nuptial chamber, which is formed by the females in the inner

FIGURE 3. Characteristic pitch tubes on host tree mass attacked by *Dendroctonus frontalis*.

bark.[39,72] After mating, she digs an S-shaped egg gallery and cuts egg niches in the sides of the gallery wall (Fig. 4). Eggs are deposited in niches alternately, and at irregular intervals, on opposite sides of the gallery. The niches are then plugged with frass. Gallery density is highest at the center section of the infested bole, decreasing gradually toward the top and abruptly toward the bottom of the bole.[22] *D. frontalis* typically excavates 50–100 cm of gallery per 100 cm^2 of phloem surface area.[22]

Egg density is apparently regulated by a density-dependent negative feedback mechanism operating during gallery construction, so that the number of eggs laid per female decreases as an exponential function of the density of attacks.[16,25] The actual mechanism involves the reemergence of beetles from densely colonized trees. This behavior permits efficient use of the host tree and prevents crowding that would result in brood mortality. Furthermore, it enables the insect to respond quickly to exploit favorable host conditions.[24]

2.4. Reemergence

Once egg laying is completed in a particular host, adult beetles either die in the gallery or reemerge. Several important features of the behavior of reemerged adults have been identified from both field and laboratory studies. First, females are capable of establishing two or three cohorts in different trees.[10,73] Second, additional matings may not be necessary for production of viable eggs in later attacks.[79] Third, reemerged adults produce pheromones that attract field populations and also perceive and respond to pheromones and host odors.[11]

The process of reemergence adds another dimension to the resource utilization mechanism described earlier. Reemerged adults can lay all their eggs in one host or in each of several hosts depending on whether adult population density in a given area is low or high. This distribution of eggs through multiple reemergences in response to variable density of adult populations is clearly a survival-enhancing mechanism for the beetle. Reemergence and resource utilization are therefore complementary processes.

2.5. Brood Development

Eggs usually hatch in 3–27 days, depending on temperature.[26,46] Larvae initiate their galleries perpendicular to the egg gallery in the cambial layer of the host but, when nearly mature, bore into the outer bark.[30] By migrating to the outer bark, *D. frontalis* is probably responding to adverse conditions in the phloem or avoiding competition with other cambial feeders. Larval development is completed in 13–63 days, depending on temperature.[26,46,74]

Larvae pupate in the outer bark, transforming into callow adults in 3–36 days, again depending on temperature.[74] Callow adults remain in the bark until the cuticle hardens. Mature adults emerge through exit holes bored from the pupal chamber but, if conditions are not favorable, they will remain in the bark.[40] Development from egg to adult ranges from 26 to 140 days, depending on temperature,[4,73] with up to eight generations per year occurring in the Southern United States.

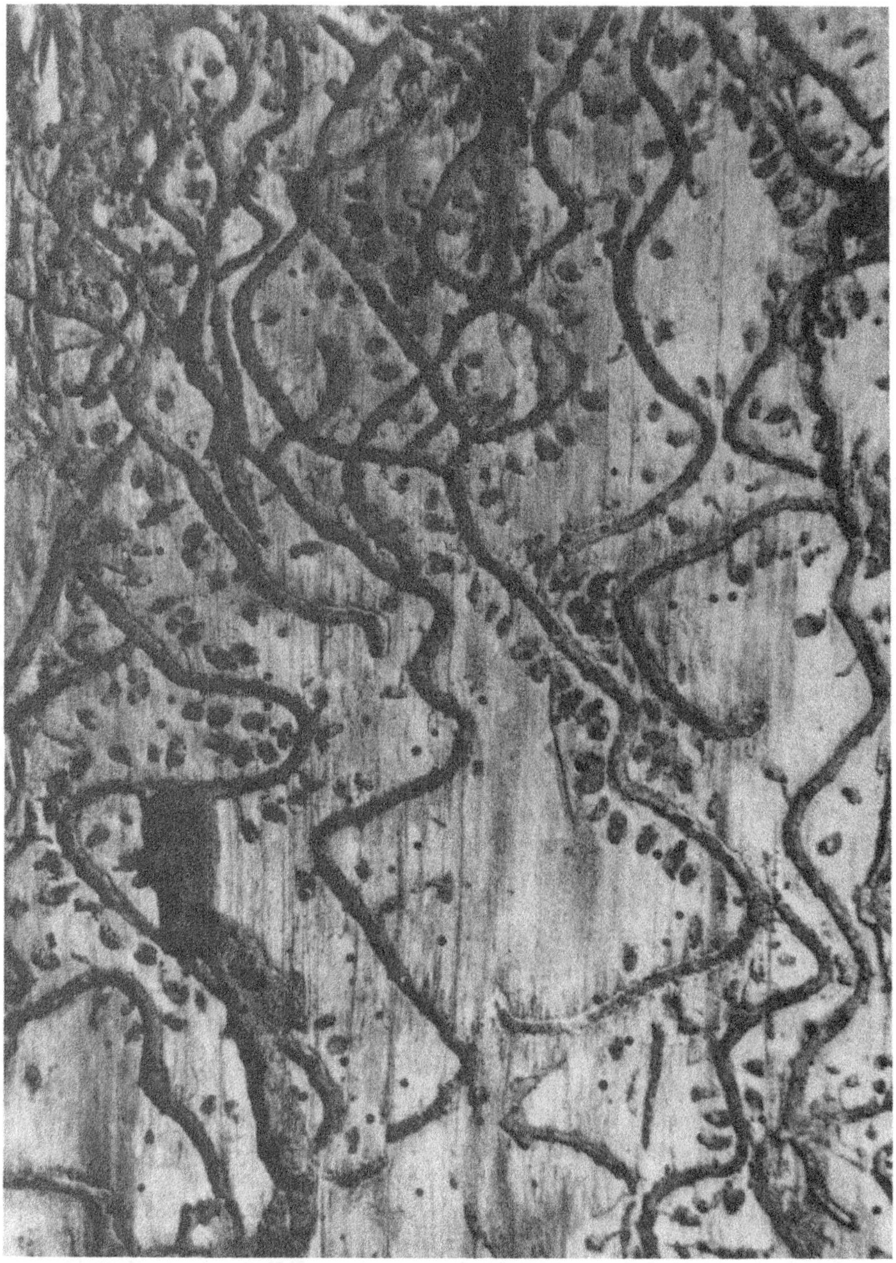

FIGURE 4. Bark removed from loblolly pine to expose the meandering parent galleries and the larval galleries of *Dendroctonus frontalis*. (Photograph courtesy of R. F. Billings.)

2.6. Emergence

The final process in the life cycle is emergence, which occurs in small daily increments over a period of 14–28 days during the warmer months.[13] This pattern probably has survival value for the species in that weather-related disasters would involve only a small part of the population. Beetles emerge in the morning and the afternoon,[40] corresponding to periods in which temperature inversions occur beneath the forest canopy.[22] Inversions greatly enhance the response to pheromones and tree odors, reducing mortality among dispersing adults by facilitating host identification and location.

Densities of emergent beetles vary widely, typically ranging from 2 to 42 beetles/dm^2 (= 100 cm^2) of bark surface,[13] but densities up to 126 beetles/dm^2 have been observed.

3. ECOLOGICAL RELATIONSHIPS

3.1. Relationships with Mycangial Fungi

The evolutionary success of bark beetles can be attributed, in part, to their symbiotic relationships with fungi. Beetle-introduced fungi are thought to be important in overcoming host resistance and in larval nutrition. Several bark beetle species, including *D. frontalis,* have specialized body structures (mycangia) in which some of these fungi are carried.[1,32] The two common mycangial fungi associated with *D. frontalis* are an unidentified basidiomycete and a *Sporothrix* sp. The former is thought to function in southern pine beetle nutrition, while the latter may be involved in tree death.[6] Another fungal associate, the blue stain fungus *Ceratocystis minor,* is not found in the mycangium but, instead, is carried externally by adult beetles and phoretic mites.[7] Even so, *C. minor* is carried by more than 90% of the attacking population and is considered a major tree-killing agent.[6]

3.2. Host Resistance and Susceptibility

Southern yellow pines, like most coniferous trees, have three components in their defensive response to bark beetles and fungi: (1) wound-cleansing performed by the resin duct system, (2) containment of infection performed by the hypersensitive reaction, and (3) wound-healing through the formation of wound periderm and callus tissue[3] (see Chapters 23, 24, and 26).

Resistance enforced by the resin duct system involves both physical and chemical properties.[19,38] The most important physical properties are total resin flow, flow rate, viscosity, and time to crystalization. Limonene is the most toxic chemical.[19] Because they are subject to many asynchronous beetle generations annually, southern pines may rely heavily on the resin duct system as a first line of defense.[31] For example, there is a strong relationship between physical properties of the duct resin and the success of *D. frontalis* attacks[36,37] in that more beetles are needed to colonize trees with high resin flow rates.

The hypersensitive reaction is initiated following invasion by *D. frontalis* or associated fungi.[3] This system is energy demanding and is initiated only at the sites of infection.[52,60] Infected areas become initially soaked with resin,[70] and the lesion so created is surrounded by callus cells originating from the cambium.[78] Eventually this resin-soaked area dies, is surrounded by newly formed periderm[67] and is sloughed off into the outer bark.

Both defense systems probably have the greatest impact on colonizing beetles when populations are low and attack rates slow.[52] The critical colonization rate and threshold attack density for overcoming host defenses are presumbaly functions of tree vigor, as with other aggressive bark beetles (Chapters 23 and 24). However, rapid colonization mediated by pheromones renders virtually any tree highly susceptible to infestation and death when beetle populations in the area are high enough.

3.3. Relationships with Other Members of the Bark Beetle Guild

The impact of *D. frontalis* is indisputable. However, it is a member of a guild of bark beetles, all of which are economically important (Fig. 5). This guild includes the black turpentine beetle, *D. terebrans* (Olivier); the four-spined engraver, *Ips avulsus* (Eichhoff); the eastern five-spined engraver, *I. grandicollis* (Eichhoff); and the six-spined engraver, *I. calligraphus* (Germar). As a group, these four insects are not noted for outbreaks as spectacular as those of *D. frontalis*. Rather, they colonize single or small groups of trees, generally in association with lightning strikes, logging, windthrow, fire, naval stores operations, and other natural or man-caused disturbances. They also attack trees colonized by *D. frontalis*.

Trees colonized by both *D. frontalis* and *Ips* are usually partitioned such that the lower bole is colonized by the southern pine beetle while the *Ips* species colonize the upper bole. *D. frontalis* may occupy the entire bole, and on rare occasions low densities will occur in the upper bole only. If *D. frontalis* is absent or in low densities, *Ips* can be found over the entire bole. Generally, olfactory interactions, arrival times, and attack densities dictate how the tree is partitioned among the species.

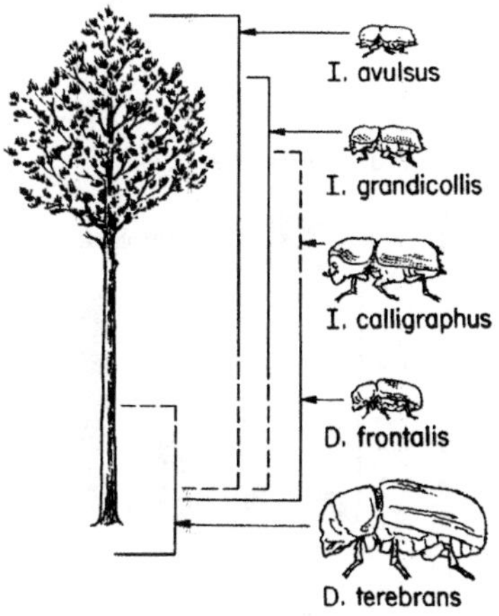

FIGURE 5. The bark beetle guild and the levels on the tree where they are usually found.

Interactions among the bark beetle species are considered primarily competitive. Competition is suggested because many more beetles arrive at a tree than attack it and egg-laying females must reemerge and lay their remaining eggs in subsequent host(s). However, because *Ips* beetles are relatively nonaggressive colonizers and rarely attain population sizes large enough for sustained infestation growth, they may actually benefit as guild members by exploiting the ability of *D. frontalis* to kill trees. By contrast, *D. frontalis*, being the most aggressive tree killer, usually has the capacity to kill trees on its own. Only when *D. frontalis* population densities are low would the *Ips* contribute significantly in overcoming tree resistance.

Two population parameters—emergence density and survivorship—illustrate why *D. frontalis* often reaches outbreak status and why the *Ips* species rarely do (Fig. 6). In trees traumatized[14] in September, *D. frontalis* showed high survivorship and emergence densities, while *Ips* emergence was never very high. In addition, emergence and survivorship in the September trees were significantly greater than the "typical" emergence in *D. frontalis* infestations (1977–1978 data). Relatively high emergence and survivorship seem to be a regular occurrence in the fall and spring for *D. frontalis*, compared to other times of the year.[75] No studies, however, have reported emergence and survivorship of the magnitude observed in trees colonized in September 1983. The large emergence at that time revealed an outbreak of *D. frontalis* that peaked in 1985 with record population

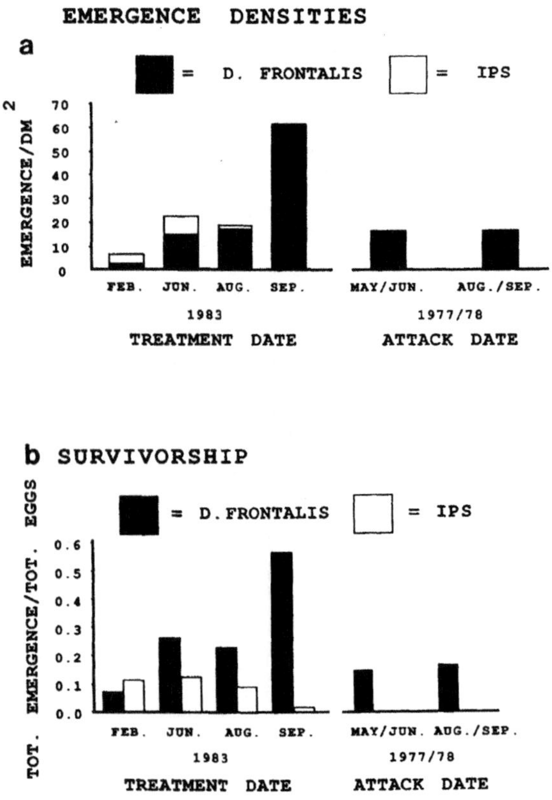

FIGURE 6. Emergence density (beetles/dm^2) (a) and survivorship (total emergence/total eggs) (b) from trees traumatized by means of a lightning strike simulation (1983 study) and naturally infested trees associated with *Dendroctonus frontalis* infestations (1977/1978 study). Data represents within-tree populations starting at ground level up to 11.0-m bole height.

levels in east Texas. *Ips* populations, in contrast, remained characteristically below epidemic proportions.

3.4. Parasites, Predators, and Pathogens

Dendrontonus frontalis is associated with approximately 35 parasitoids of which only a few are host specific.[20,72] Of the nonhost-specific parasitoids, host preference is determined by the insect from which they were reared. For example, when adult parasitoids emerging from *D. frontalis*-infested logs were given simultaneous choices between logs containing *D. frontalis* or Ips, they overwhelmingly chose the former.[42]

Adult parasitoids probably respond to insect- and host-produced odors to locate trees infested with advanced brood stages.[8] Once at the tree they may orient to sound[64] or heat.[63] Oviposition is generally through the bark onto third- or fourth-instar larvae, stages that have migrated to the outer bark. Thin bark sections of the tree typically contain the highest parasitoid densities. Most parasitoids apparently sting the host to immobilize it before depositing their eggs. But one of the most common, *Roptrocerus xylophagorum* Ratzeburg, enters the bark through entrance and air holes and oviposits in the egg galleries.

Parasitism by all species combined is usually low, averaging about 4% with a maximum of 25–30%. For this reason parasitoids are considered relatively unimportant in suppressing *D. frontalis* populations.

Insects and woodpeckers are the major predators of *D. frontalis*. The most abundant of the insect predators, the clerid beetle *Thanasimus dubius* (Fabricius), averages $0.65/dm^2$ of bark surface area.[50] *T. dubius* causes an average of 13% mortality, greater than any other insect predator.[50] Each insect consumes 2.2 prey/day, including insects besides *D. frontalis*.[74] It was estimated that 15 clerids/100 bark beetle adults would suppress *D. frontalis* populations.[74]

Woodpeckers, primarily the downy woodpecker, *Dendrocopus pubescens* (Linnaeus), destroy an average of 5% of the brood population. Most brood are killed indirectly by the removal of the outer bark which causes the inner bark to dry quickly and allows cold weather and disease to exert a greater effect on the brood.[50] Woodpeckers cause more brood mortality in the winter (36–63%) than in the summer (12–30%), even though the highest numbers of woodpeckers are active in *D. frontalis* infestations in the summer.

Dendroctonus frontalis is also attacked by a variety of pathogenic organisms that may reduce its egg production and survival or kill the beetle outright. Mortality averages 22%[49,68] with most occurring during cold temperatures and in the midportion of the infested bole. The major pathogens include the microsporidian, *Unikaryon minutum;* the fungi, *Paecilomyces* sp.; and the nematodes, *Cortylenchus* sp.

4. POPULATION PATTERNS AND THEIR CAUSES

4.1. Initiation of Infestations

Southern pine beetle infestations are thought to begin with the colonization of a single weakened tree. In an established infestation, the supply of beetles is usually

FIGURE 7. Procedure used to disturb trees. Detonator cord was wrapped around trees, secured at the base and at 12.5-m bole height, and ignited with a blasting cap. (Photograph courtesy of R. F. Billings.)

adequate for the successful colonization of even healthy trees. When beetle populations are low, however, colonization may depend on trees with diminished defense systems.[9] Disturbances, common in pine forest ecosystems, often create habitats that can be exploited by bark beetles.[34,35] Only recently, however, have specific disturbance regimes been associated with bark beetle population dynamics.[14,15,18]

Because of its consistent seasonal and annual patterns, lightning disturbance is thought to be central to the epidemiology of *D. frontalis*. In the southern pine forest, pines are the most frequent targets of cloud-to-ground lightning strikes because they grow taller and are more numerous than the hardwood species. Furthermore, in the geographical range of *D. frontalis*, 28–112 lightning strikes can be expected annually per square mile of forest. Not all pines struck by lightning are killed, but surviving trees are considered severely disturbed. These weakened trees may provide refuges for endemic beetle popula-

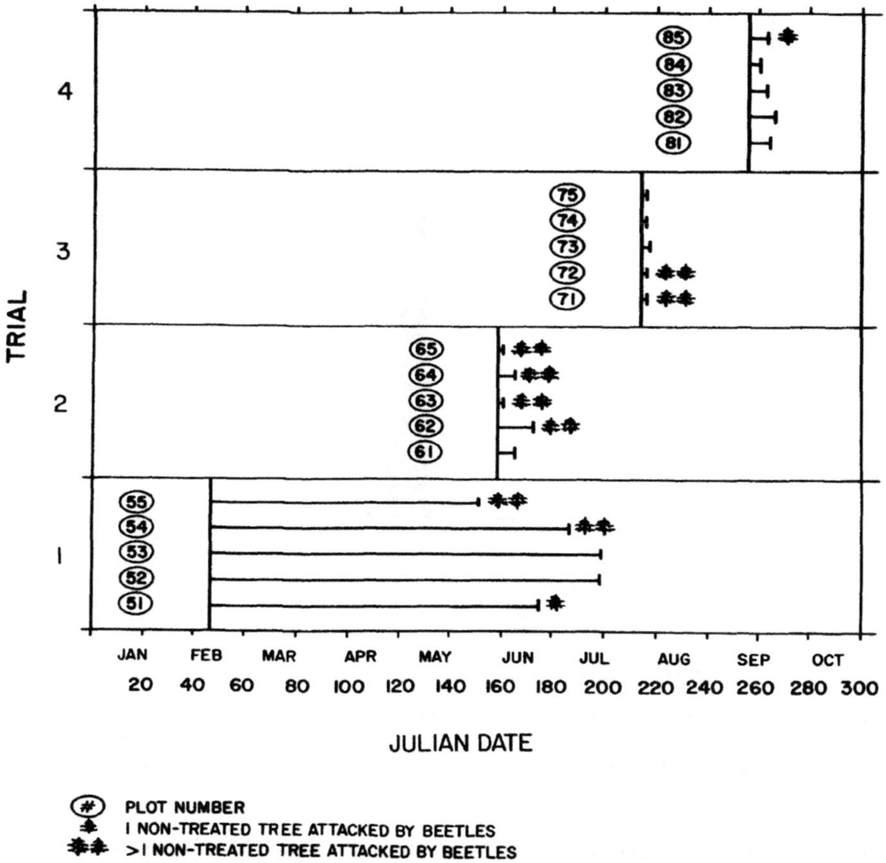

FIGURE 8. Response of the bark beetle guild to the 20 disturbed plots (numbers enclosed in circles) in four trials. The length of the horizontal lines is time between treatment of the trees and mass colonization by the bark beetle guild. (From Coulson *et al.*[14])

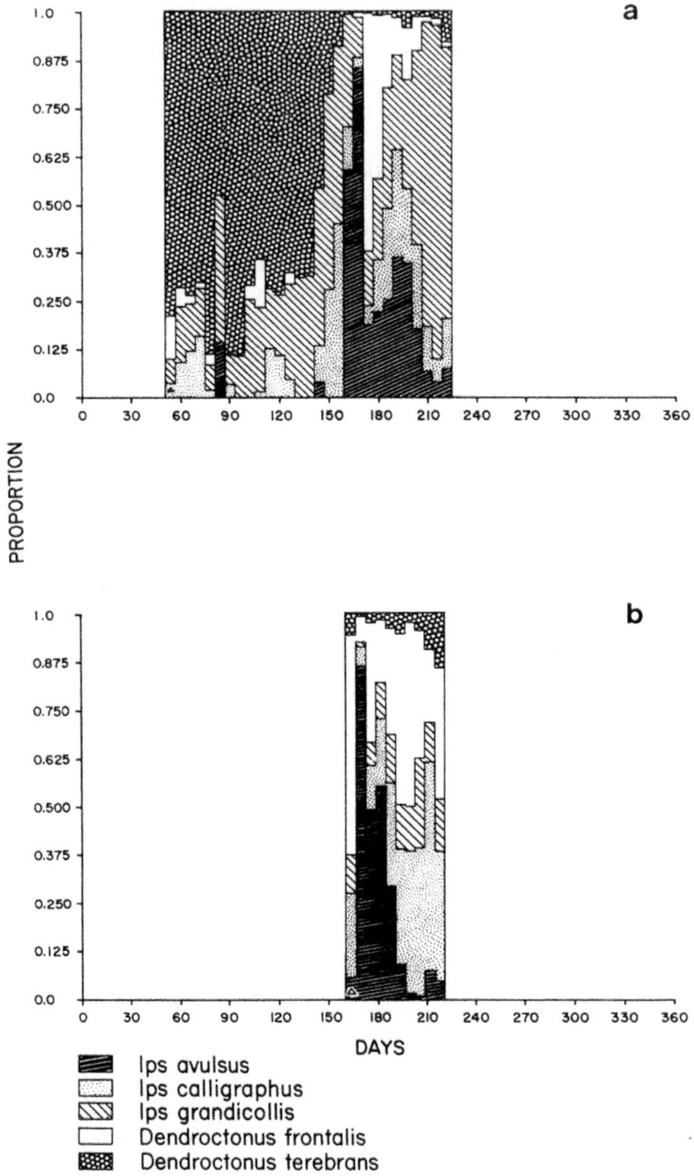

FIGURE 9. Relative proportions of *Ips avulsus, I. calligraphus, I. grandicollis, Dendroctonus terebrans,* and *D. frontalis* responding to the disturbed trees in plots of trial 1 (a), 2 (b), 3 (c), and 4 (d) from the initiation of treatment (Julian date) until mass colonization, plotted on 6 day interval. (From Coulson *et al.*[14])

tions[21] or may act as foci for beetle infestations,[44] depending on meteorological conditions, bark beetle population size, forest stand structure, and landscape structure. The potential for new infestations is related to meteorological conditions that may produce thunderstorms and associated lightning strikes on trees. In addition, meteorological conditions at a given time encourage or inhibit population growth. The size of the background

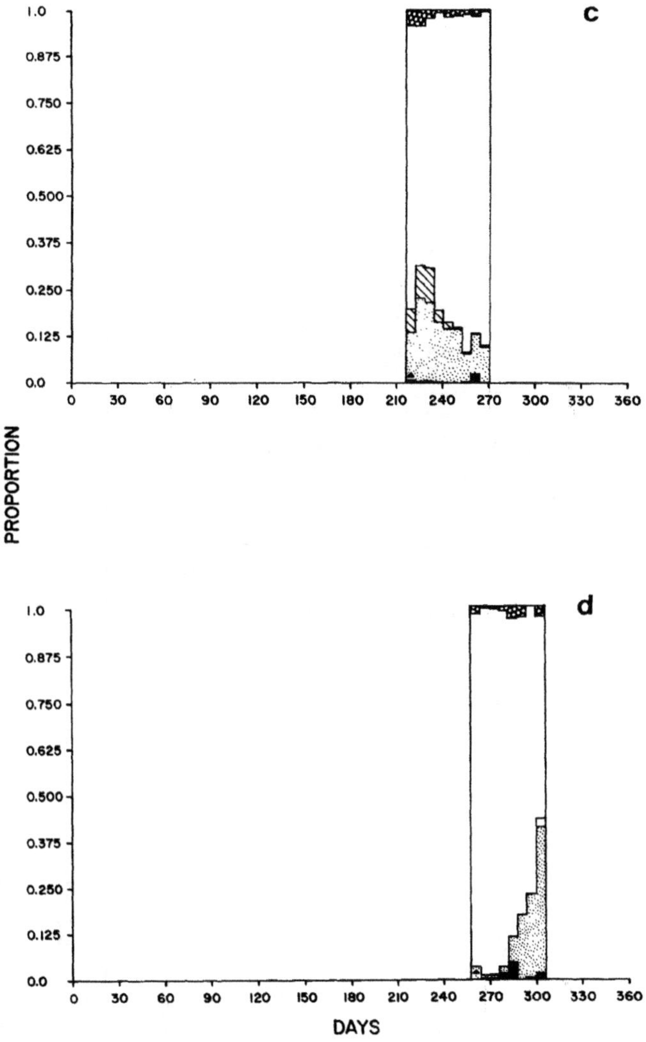

FIGURE 9. *(continued)*

population, including all species of the bark beetle guild, also influences the extent to which new lightning-struck trees are infested by beetles. Finally, variables of landscape and forest structure establish the potential for epizootic population growth.

The role of disturbance in the initiation of infestations has been demonstrated experimentally with simulated lightning strikes[45] (Fig. 7). These simulations demonstrated that trees disturbed during periods of beetle inactivity (during winter storms, for example) remain available for colonization later, when conditions for continuous dark beetle dispersal are favorable (Fig. 8). Disturbed trees also serve as centers for the initiation and subsequent growth of infestations. The simulations showed further that bark beetle guild

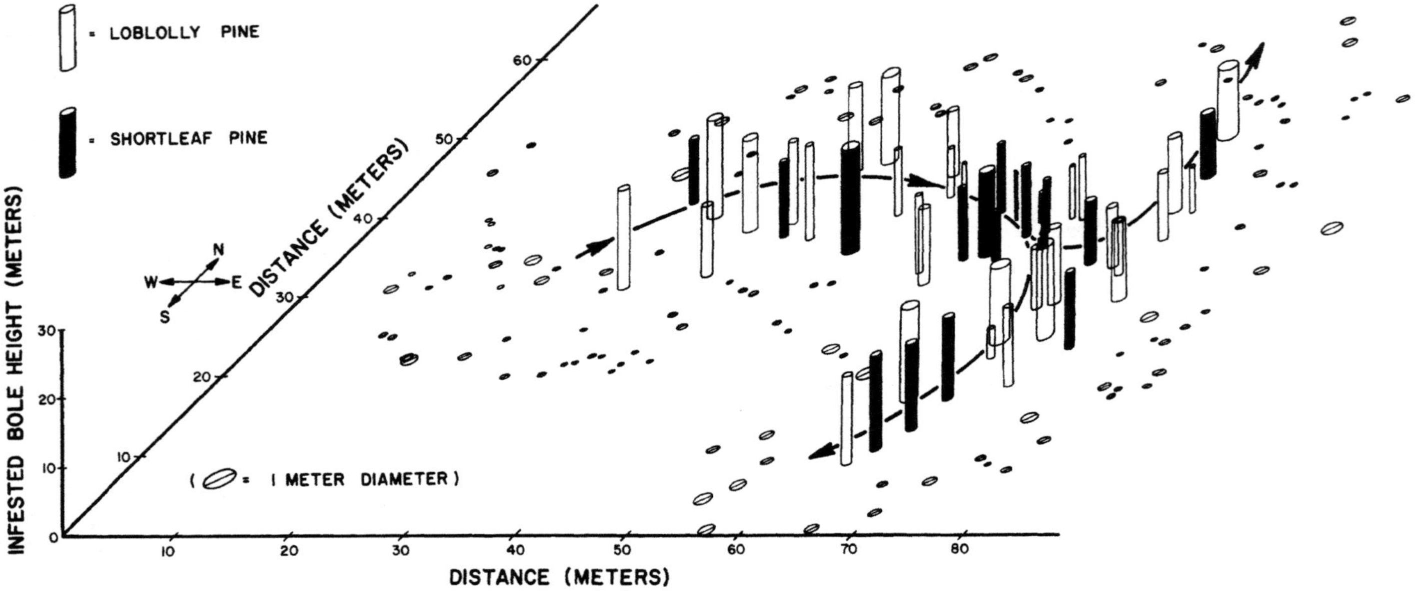

FIGURE 10. Spatial arrangement of attacked and unattacked trees in a *Dendroctonus frontalis* infestation. The cylinders represent attacked trees and are proportional to the actual size of the trees in the infestation. The ellipses represent peripheral unattacked trees. The two axes indicate the actual scale in meters.

members respond to disturbed trees in varying proportions during the year (Fig. 9). These seasonal population fluctuations may be characteristic of *D. frontalis* natural history.[24]

4.2. Growth of Infestations

In *D. frontalis* infestations, new trees are usually colonized along active fronts at points of highest tree density[66] (Fig. 10). Several factors are responsible for this pattern of development. First, re-emergence takes place in the more recently attacked trees, i.e., adjacent to the next trees that will come under attack. These trees form the active front(s). Second, the developmental rate of southern pine beetle populations in the field is rapid (30 to 50 days during the summer). Brood life stages therefore develop to the adult stage and enter the colonization process in a short period of time. Synchrony between emerging and re-emerging adults can be quickly achieved and maintained under favorable weather conditions, providing a continuous supply of adults for the purpose of colonization.[17] Third, the fact that new trees are being colonized means that pheromones provide a continuous focal point for attraction. Without this focal point, emergent adults, which are removed in space from the active front by about 10–20 ms, would probably disperse and suffer substantially higher mortality than actually occurs.[27] Fourth, high rates of population increase reveal that *D. frontalis* populations can reach epidemic proportions quickly.[24]

5. MANAGEMENT IMPLICATIONS

5.1. Prediction

Southern pine beetle activity is generally continuous across the southern United States, ranging from single infested trees to large infestations in various stages of growth or decline. Since forest managers cannot control all spots, they accept low levels of losses. Extensive, growing infestations, however, can have serious economic consequences. Therefore, research efforts in recent years have focused on improving predictive capabilities.

One method of predicting *D. frontalis* activity and resulting timber loss involves the use of mathematical models. The Tambeetle model,[12,15,23] a spot-growth model that accounts for the reproduction and mortality of beetles within an infestation, is organized around trees that are currently infested, vacated, and noninfested trees near active beetle sites. Tambeetle is based on a series of interconnected submodels that simulate brood emergence, oviposition, re-emergence, beetle allocation, pheromone emission and distribution, and tree drying. Another model (HOG) was developed in Arkansas to predict growth of infestations in that state.[69]

A second postulated approach to forecasting beetle activity is the development of a lightning strike detection system. This system would pinpoint locations of cloud-to-ground lightning strikes over a large forested area. The pattern of lightning strikes would be mapped and then superimposed on a habitat grid, pinpointing the stands at greatest risk of infestation growth. By using a lightning strike detection system and quantifying bark beetle epidemiology, forest managers may be able to plan suppression measures before infestation growth begins.

5.2. Population Management

The objective of *D. frontalis* population management is reduction of timber losses, not beetle mortality. Three approaches are available to forest managers: (1) prevention, (2) killing beetles, and (3) spot disruption.

5.2.1. Prevention

Infestations typically occur in slow-growing overstocked pine stands in which tree vigor and resistance to attack are low. Knowing this, forest managers can implement strategies that reduce competition and stress within stands, thus increasing vigor and resistance of individual trees.[51] Studies have shown, for example, that conditions favorable to infestation growth are readily alleviated by stand thinnings, especially those that remove the lower crown classes, which are the primary targets for *D. frontalis* attack. Treatments should be applied cautiously to avoid disturbances that would offset any of the treatment benefits.[2]

5.2.2. Killing Beetles

Forest managers have had only limited success with killing beetles directly, in part because the well-developed survival mechanisms of the beetle enable it to overcome high mortality rates.

Several chemicals, though effective to varying degrees, are not widely accepted among managers because of the expense or risk involved in their use. Chemicals are more appropriate against adult beetles infesting highly valued pines in recreational areas and in residential subdivisions. These areas, which typically contain old susceptible timber, are easily monitored and accessible so that control practices can be initiated before an infestation begins to grow.

Other attempts at direct control have involved wrapping explosive cables around infested trees, spraying the bole with fuel oil, and igniting brush piles around trees. The less said about these tactics the better. More reasonable control strategies include piling and burning of infested timber, destruction of trap trees, drowning, exposure to solar heat, and peeling and destruction of infested bark. None of these treatments, however, emphasize the ecology and behavior of *D. frontalis* or the preservation of its natural enemies. These tactics simply cause temporary perturbations in *D. frontalis* populations and have for the most part been rejected.

A growing number of forest managers agree that, because ecological considerations are neglected, killing beetles is usually ineffective. When beetles are killed directly, the conditions responsible for the problem in the first place are left largely unchanged. For the most part, control strategies that incorporate fundamental principles of forest development and bark beetle biology will show the best results.

5.2.3. Spot Disruption

5.2.3.1. Tree Removal. Present methods of tree removal include salvage and cut-and-leave. Salvage control consists of rapid removal of all trees containing *D. frontalis* brood or attacking adults from an infested area. A buffer strip of uninfested trees is also removed from the active head of the infestation.

Cut-and-leave is similar to salvage treatment, except that felled trees are not removed. Infested trees and a buffer strip are felled with the crowns pointing toward the center of the infestation. Generally, cut-and-leave is the procedure of choice when infestations are inaccessible or when the costs of the salvage operation would exceed the value of the timber that could be recovered.

Continuous spot growth requires a source of secondary attractants, adjacent host type, and a supply of adult beetles. Felling the most recently attacked trees appears to destroy the attractant source,[76] while the buffer strip eliminates potential hosts. With salvage control, beetles are either removed before emergence, or they disperse in the absence of attractants.[27] What becomes of the dispersing beetles in cut-and-leave operations is unknown. Because between-tree survival is low during hot weather,[59] salvage and cut-and-leave may be more effective in midsummer.

What concerns forest managers about these tactics (cut-and-leave primarily) is the possibility that dispersing beetles may initiate new spots by aggregating on trees in the vicinity, a hypothesis that is difficult to test. In the absence of rigorous evaluation, therefore, the value of cut-and-leave remains circumspect.

5.2.3.2. Behavioral Chemicals. In the context of population management, behavioral chemicals can be classified as attractants (aggregation pheromones) or interruptants (inhibitors). The response of both males and females to attractants ensures that beetles arrive over a relatively short period of time in sufficient numbers to overcome the resistance of the tree. Theoretically, spot growth could be disrupted by simultaneous application of attractants to previously infested trees, infested trees containing late-stage brood *D. frontalis*, and nonhost species. Indeed, when applied in small spots during endemic conditions, infestation growth did not occur, probably because the pheromone source was concentrated in a small area to begin with, and few beetles were drawn in from the surrounding area. By contrast, large spots under epidemic conditions were not controlled successfully, probably because they contained several areas of highly concentrated aggregation pheromone that attracted beetles from the high populations in the surrounding area.

When *D. frontalis* populations reach epidemic levels, interruptants (or a combination of interruptants, attractants, and cutting freshly attacked trees) are more effective. Interruptants are believed to halt mass aggregation on single host trees and cause flying beetles to land on adjacent host trees.[53,77] In theory, interruptants placed on a buffer zone of unattacked trees should disperse beetles coming from the active head away from the buffer strip, causing disruption of spot growth. Treatment with inhibitors has stopped spot growth in plantations with trees of small diameter, although not in stands of larger trees. Even in stands of sawlog-sized trees, however, the treatment has been effective when freshly attacked still-attractive trees are felled at the same time.

ACKNOWLEDGMENTS. We thank L. Gattis for editorial assistance and C. W. Agnew for the drawing of the southern pine beetle that serves as the chapter header. Figures 9 and 10 are reproduced with kind permission of the Entomological Society of America.

REFERENCES

1. Barras, S. J., and Perry, T., 1972, Fungal symbionts in the prothoracic mycangium of *Dendroctonus frontalis* (Coleoptera: Scolytidae), *Z. Angew. Entomol.* **71**:95–104.
2. Belanger, R. P., Coulson, R. N., Lewis, J., Payne, T. L., Stephen, F. M., Taylor, J. W., Wood, J. D., and Westbrook, R. F., 1979, Southern pine beetle research applications, and implementation activities for the southern forestry community, U.S.D.A. Forest. Service, Pineville, Louisiana.
3. Berryman, A. A., 1972, Resistance of conifers to invasion by bark beetle–fungus associations, *BioScience* **22**:589–602.
4. Billings, R. F., and Kibbe, C. A., 1978, Seasonal relationships between southern pine beetle brood development and loblolly pine foliage color in east Texas, *Southwest Entomol.* **3**:89–95.
5. Borden, J. H., 1974, Aggregation pheromones in the Scolytidae, in: *Pheromones* (M. C. Birch, ed.), pp. 135–160, American Elsevier, New York.
6. Bridges, J. R., 1985, Relationships of symbiotic fungi to southern pine beetle population trends, in: Proceedings, Integrated Pest Management Research Symposium (S. J. Branham and R. C. Thatcher, eds.), pp. 127–135, U.S.D.A. Forest Service, Gen. Tech. Rep. SO-56.
7. Bridges, J. R., and Moser, J. C., 1983, Role of two phoretic mites in transmission of blue-stain fungus, *Ceratocystis minor, Ecol. Entomol.* **8**:9–12.
8. Camors, F. B., and Payne, T. L., 1973, Sequence of arrival of entomophagous insects to trees infested with the southern pine beetle, *Environ. Entomol.* **2**:267–270.
9. Cates, R. G., and Alexander, H., 1982, Host resistance and susceptibility, in: *Bark Beetles in North American Conifers* (J. B. Mitton and K. B. Sturgeon, eds.), pp. 212–263, University of Texas Press, Austin, Texas.
10. Clark, E. W., and Osgood, E. A., Jr., 1964, Mass rearing the southern pine beetle and the coarse writing engraver, U.S.D.A. Forest. Service, Res. Note. SE-30.
11. Coulson, R. N., Fargo, W. S., Pulley, P. E., Foltz, J. L., Pope, D. N., Richerson, J. V., and Payne, T. L., 1978, Evaluation of the reemergence process of attacking adult *Dendroctonus frontalis* (Coleoptera: Scolytidae), *Can. Entomol.* **110**:475–456.
12. Coulson, R. N., Amman, G. D., Dahlsten, D. L., Demars, C. J., Jr., and Stephen, F. M., 1985, Forest–bark beetle interactions: Bark beetle population dynamics, in: *Integrated Pest Management in Pine–Bark Beetle Ecosystems* (W. F. Waters, R. W. Stark, and D. L. Woods, eds.), pp. 61–80, Wiley, New York.
13. Coulson, R. N., Fargo, W. S., Pulley, P. E., Pope, D. N., and Bunting, A. M., 1979, Spatial and temporal patterns of emergence for within-tree populations of *Dendroctonus frontalis* (Coleoptera: Scolytidae), *Can. Entomol.* **111**:273–287.
14. Coulson, R. N., Flamm, R. O., Pulley, P. E., Payne, T. L., Rykiel, E. J., and Wagner, T. L., 1986, Response of the southern pine bark beetle guild (Coleoptera: Scolytidae) to host disturbance, *Environ. Entomol.* **15**:850–858.
15. Coulson, R. N., Hennier, P. B., Flamm, R. O., Rykiel, E. J., Hu, L. C., and Payne, T. L., 1983, The role of lightning in the epidemiology of the southern pine beetle, *Z. Angew. Entomol.* **96**:182–193.
16. Coulson, R. N., Mayyasi, A. M., Foltz, J. L., and Hain, F. P., 1976, Resource utilization by the southern pine beetle, *Can. Entomol.* **108**:353–362.
17. Coulson, R. N., Pulley, P. E., Pope, D. N., Fargo, W. S., and Edson, L. J., 1980, Continuous population estimates for *Dendroctonus frontalis* occurring in infestations, *Res. Popul. Ecol.* **22**:117–135.
18. Coulson, R. N., Rykiel, E. J., Saunders, M. C., Payne, T. L., Flamm, R. O., Wagner, T. L., and Hennier, P. B., 1985, A conceptual model of the role of lightning in the epidemiology of the southern pine beetle, in: *The Role of the Host in the Population Dynamics of Forest Insects* (L. Safranyik and A. A. Berryman, eds.), pp. 136–146. Proceedings of the International Union of Forest Resource Organizations (Conference), Banff, Alberta, Canada.
19. Coyne, J. F., and Lott, L. H., 1976, Toxicity of substances in pine oleoresin to southern pine beetles, *J. Ga. Entomol. Soc.* **11**:297–301.
20. Dixon, J. C., and Osgood, E. A., 1961, *Southern pine beetle: review of present knowledge,* U.S. Dep. Agric. Forest Service. Res. Pap. SE-128.
21. Fares, Y., Sharpe, P. J. H., and Magnuson, C. E., 1980, Phermonone dispersion in forests, *J. Theor. Biol.*
22. Fargo, W. S., Coulson, R. N., Pulley, P. E., Pope, D. N., and Kelley, C. L., 1978, Spatial and temporal patterns of within-tree colonization by *Dendroctonus frontalis, Can. Entomol.* **110**:1213–1232.

23. Feldman, R. M., Curry, G. L., and Coulson, R. N., 1980, The use and structure of the TAMBEETLE spot dynamics model, in: *Modeling Southern Pine Beetle Populations.* (F. M. Stephen, J. L. Searcy, and G. D. Hertel, eds.), pp. 20–29, Proceedings U.S. Dep. Agric. Forest Service Symp. Tech. Bull. 1630.
24. Flamm, R. O., and Coulson, R. N., 1988. Population dynamics of the southern pine bark beetle guild in traumatized hosts, in: *Mechanisms of Woody Plant Resistance* (W. Mattson, ed.). Springer-Verlag, New York.
25. Flamm, R. O., Wagner, T. L., Cook, S. P., Pulley, P. E., Coulson, R. N., and McArdle, T. M., 1987, Host colonization by cohabiting *Dendroctonus frontalis, Ips avulsus,* and *I calligraphus* (Coleoptera: Scolytidae), *Environ. Entomol.,* **16:**390–399.
26. Gagne, J. A., 1980, The effects of temperature on population processes of the southern pine beetle, *Dendroctonus frontalis* Zimmerman, Ph.D. dissertation, Texas A&M University.
27. Gara, R. I., 1967, Studies on the attack behavior of the southern pine beetle. I. The spreading and collapse of outbreaks, *Contrib. Boyce Thompson Inst.* **23:**349–354.
28. Gara, R. I., and Coster, J. E., 1968, Studies on the attack behavior of the southern pine beetle. III. Sequence of tree infestations within stands, *Contrib. Boyce Thompson Inst.* **23:**55–66.
29. Gara, R. I., Vite, J. P., and Cramer, H. H., 1965, Manipulation of *Dendroctonus frontalis* by use of a population aggregating pheromone, *Contrib. Boyce Thompson Inst.* **23:**55–66.
30. Goldman, S. E., and Franklin, R. T., 1977, Development and feeding habits of southern pine beetle larvae, *Ann. Entomol. Soc. Am.* **70:**54–56.
31. Hain, F. P., Cook, S. P., Matson, P. A., and Wilson, K. G., 1985, Factors contributing to southern pine host resistance, in: Proceedings of the Integrated Pest Management Symposium (S. J. Branham and R. C. Thatcher, eds.), pp. 154–160, U.S.D.A. Forest Service, Gen. Tech. Rep. SO-56.
32. Happ, G. M., Happ, C. M., and Barras, S. J., 1971, Fine structure of the prothoracic mycangium, a chamber for the culture of symbiotic fungi, in the southern pine beetle, *Dendroctonus frontalis, Tissue and Cell* **3:**295–308.
33. Heikkenen, H. J., 1977, Southern pine beetle: a hypothesis regarding the primary attractant, *J. For.* **75:**412–413.
34. Hetrick, L. A., 1949, Some overlooked relationships of southern pine beetle, *J. Econ. Entomol.* **42:**466–469.
35. Hicks, R. R., Jr., 1980, Climate, site, and stand factors, in: The Southern Pine Beetle, (R. C. Thatcher, J. L. Searcy, J. E. Coster, and G. D. Hertel, eds.), pp. 55–70, U.S.D.A. Forest Service, Tech. Bull. 1631.
36. Hodges, J. D., Elam, W. W., and Watson, W. F., 1977, Physical properties of the oleoresin system of the four major southern pines, *Can. J. For. Res.* **7:**520–525.
37. Hodges, J. D., Elam, W. W., Watson, W. F., and Nebeker, T. E., 1979, Oleoresin characteristics and susceptibility of four southern pines to southern pine beetle (Coleoptera: Scolytidae) attacks, *Can. Entomol.* **111:**889–896.
38. Hodges, J. D., and Lorio, P. L., Jr., 1975, Moisture stress and composition of xylem oleoresin in loblolly pine, *For. Sci.* 283–290.
39. Hopkins, A. D., 1909, Practical information on the Scolytid beetles of North American forests. Bark beetles of the genus *Dendroctonus,* U.S.D.A. Entomol. Bull. 83, Part I.
40. Kinn, D. W., 1978, Diel emergence patterns of the southern pine beetle (*Dendroctonus frontalis* Zimm.), *J. Ga. Entomol. Soc.* **13:**80–85.
41. Kinzer, G. W., Fentiman, A. F., Page, T. F., Fultz, R. L., Vite, J. P., and Pitman, G. B., 1969, Bark beetle attractants: identification, synthesis, and field bioassay of a new compound isolated from *Dendroctonus, Nature (Lond.)* **221:**447–478.
42. Kudon, L. H., and Berisford, C. W., 1981, An olfactometer for bark beetle parasitoids, *J. Chem. Ecol.* **7:**359–366.
43. Linit, M. J., and Stephen, F. M., 1983, Parasite and predator component of within-tree southern pine beetle (Coleoptera: Scolytidae) mortality, *Can. Entomol.* **115:**679–688.
44. Lorio, P. L., and Bennett, W. H., 1974, Recurring southern pine beetle infestations near Oakdale, La., U.S.D.A., Forest Service. Res. Pap. SO-94.
45. Miller, M. C., 1983, Lightning strike simulation for studying southern pine bark and engraver beetle attacks, U.S.D.A., Forest Service. Pap. SO-296.
46. Mizell, R. F., III, and Nebeker, T. E., 1978, Estimating the development time of southern pine beetle *Dendroctonus frontalis* Zimmerman as a function of field temperatures, *Environ. Entomol.* **7:**592–595.

47. Moeck, H. A., 1970, Ethanol as the primary attractant for the ambrosia beetle, *Trypodendron lineatum* (Coleoptera: Scolytidae), *Can. Entomol.* **27:**11–12.
48. Moeck, H. A., Wood, D. L., and Lindahl, K. Q., Jr., 1981, Host selection behavior of bark beetles (Coleopera: Scolytidae) attacking *Pinus ponderosa* with special emphasis on the western pine beetle, *Dendroctonus brevicomis, J. Chem. Ecol.* **7:**49–83.
49. Moore, G. E., 1971, Mortality factors caused by pathogenic bacteria and fungi of the southern pine beetle in North Carolina, *J. Invert. Pathol.* **17:**28–37.
50. Moore, G. E., 1972, Southern pine beetle mortality in North Carolina caused by parasites and predators, *Environ. Entomol.* **1:**58–65.
51. Nebeker, T. E., and Hodges, J. D., 1983, Influence of forestry practices on host susceptibility to bark beetles, *Z. Angew. Entomol.* **96:**194–208.
52. Paine, T. D., 1984, Influence of the mycangial fungi of the western pine beetle on water conduction through ponderosa pine seedlings, *Can. J. Bot.* **62:**556–563.
53. Paine, T. D., Birch, M. C., and Svihra, P., 1981, Niche breadth and resource partitioning by four sympatric species of bark beetles (Coleoptera: Scolytidae), *Oecologia* **48:**1–6.
54. Payne, T. L., 1979, Pheromone and host odor perception in bark beetles, in: *Neurotoxicology of Insecticides and Pheromones* (T. Narahasi, ed.), pp. 25–27, Plenum, New York.
55. Payne, T. L., 1980, Life history and habits, in: *The Southern Pine Beetle,* (R. C. Thatcher, J. L. Searcy, J. E. Coster, and G. D. Hertel, eds.), pp. 7–28, U.S.D.A., Forest Service, Tech. Bull. 1631.
56. Payne, T. L., 1981, Disruption of southern pine beetle infestations with attractants and inhibitors in: *Management of Insect Pest with Semiochemicals* (E. R. Mitchell, ed.), pp. 365–383, Plenum, New York.
57. Payne, T. L., Coster, J. E., Richerson, J. V., Edson, L. J., and Hart, E. R., 1978, Field response of the southern pine beetle to behavioral chemicals, *Environ. Entomol.* **7:**575–582.
58. Person, H. L., 1931, Theory in explanation of the selection of certain trees by the western pine beetle, *J. For.* **17:**306–307.
59. Pope, D. N., Coulson, R. N., Fargo, W. S., Gagne, J. A., and Kelley, C. L., 1980, The allocation process and between-tree survival probabilities in *Dendroctonus frontalis* infestations, *Res. Popul. Ecol.* **22:**197–210.
60. Raffa, K. F., and Berryman, A. A., 1983, The role of host plant resistance in the colonization behavior and ecology of bark beetles (Coleoptera: Scolytidae), *Ecol. Monogr.* **53:**27–49.
61. Renwick, J. A. A., and Vite, J. P., 1969, Bark beetle attractants: mechanism of colonization by *Dendroctonus frontalis, Nature* **224:**1222–1223.
62. Renwick, J. A. A., and Vite, J. P., 1970, Systems of chemical communication in *Dendroctonus, Contrib. Boyce Thompson Inst.* **24:**283–292.
63. Richerson, J. V., and Borden, J. H., 1972, Host finding by heat perception in *Coeloides brunneri* (Hymenoptera: Braconidae), *Can. Entomol.* **104:**1877–1881.
64. Ryan, R. D., and Rudinsky, J. A., 1962, Biology and habits of the douglas-fir beetle parasite, *Coeloides brunneri* Viereck (Hymenoptera: Braconidae), in Western Oregon, *Can. Entomol.* **94:**748–763.
65. St. George, R. A., and Beal, J. A., 1929, The southern pine beetle: a serious enemy of pines in the South, U.S.D.A. Farm. Bull. 1586.
66. Schowalter, T. D., Pope, D. N., Coulson, R. N., and Fargo, W. S., 1981, Patterns of southern pine beetle infestation enlargement, *For. Sci.* **27:**837–849.
67. Shrimpton, D. M., 1978, Resistance of lodgepole pine to maintain pine beetle infestation, in: *Theory and Practice of Mountain Pine Beetle Management in Lodgepole Pine Forests* (A. A. Berryman, G. D. Amman, R. W. Stark, and D. L. Kibbee, eds.), pp. 64–76, University of Idaho Press, Moscow, Idaho.
68. Sikorowski, P. P., Pabst, G. S., and Tomson, O., 1979, *The impact of diseases on southern pine beetle in Mississippi,* Miss. Agric. and For. Exp. Stn. Tech. Bull. 99.
69. Stephen, F. M., and Loh, M. P., 1985, A *Dendroctonus frontalis* infestation growth model: organization, refinement, and utilization, *Proceedings of the Integrated Pest Management Symposium,* (S. J. Branham and R. C. Thatcher, eds.), pp. 186–194, U.S.D.A. Forest Service, Gen. Tech. Rep. SO-56.
70. Stephen, F. M., Paine, T. D., and Lih, M. P., 1983, Understanding bark beetle/host interactions: a means for improving decision strategies, *Z. angew. Entomol.* **96:**257–265.
71. Svihra, P., Paine, T. D., and Birch, M. C., 1980, Interspecific olfactory communication in southern pine beetles, *Naturwiss.* **67:**518–519.

72. Thatcher, R. C., 1960, *Bark beetles affecting southern pines: a review of current knowledge*, U.S.D.A. Forest Service. Occas. Pap. 180. Southwestern Forestry Experiment Station.
73. Thatcher, R. C., and Pickard, L. S., 1964, Seasonal variations in activity of the southern pine beetle in east Texas, *J. Econ. Entomol.* **57**:840–842.
74. Thatcher, R. C., and Pickard, L. S., 1967, Seasonal development of the southern pine beetle in east Texas, *J. Econ. Entomol.* **60**:656–658.
75. Thomas, H. H., Richmond, J. H., and Bradley, E. L., 1981, Bioassay of pine bark extracts as biting stimulants for the southern pine beetle, U.S.D.A. Forest Service Res. Note SE-302.
76. Vite, J. P., and Crozier, R. G., 1968, Studies on the attack behavior of the southern pine beetle. IV. Influence of host conditions on aggregation pattern, *Contrib. Boyce. Thompson Inst.* **24**:87–94.
77. Vite, J. P., and Franke, W., 1976, The aggregation pheromones of bark beetles: progress and problems, *Naturwis.* **63**:550–555.
78. Wong, B. L., and Berryman, A. A., 1977, Host resistance to the fir engraver beetle, 3. Lesion development and containment of infection by resistant *Abies grandis* inoculated with *Trichosporium symbioticum*, *Can. J. Bot.* **55**:2358–2365.
79. Yu, C. C., and Tsao, C. H., 1967, Gallery construction and sexual behavior in the southern pine beetle, *Dendroctonus frontalis* Zimm. (Coleoptera: Scolytidae), *J. Ga. Entomol. Soc.* **2**:95–98.

CHAPTER 26

THE FIR ENGRAVER BEETLE IN WESTERN STATES

Alan A. Berryman and George T. Ferrell

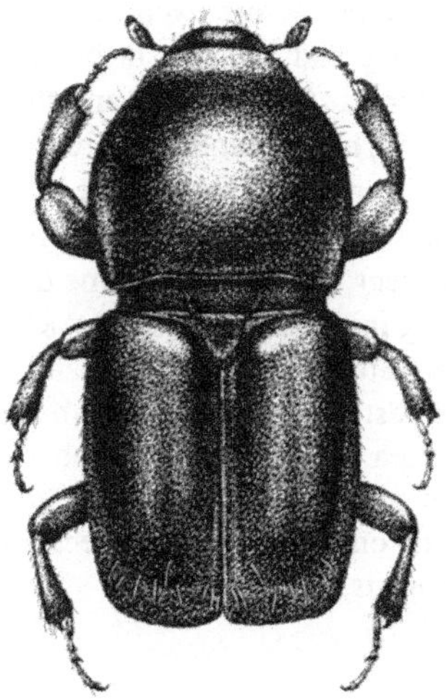

Alan A. Berryman • Department of Entomology, Washington State University, Pullman, Washington 99164-6432. **George T. Ferrell** • U.S. Forest Service, Pacific Southwest Forest and Range Experiment Station, Berkeley, California 94701.

1. INTRODUCTION

The fir engraver beetle, *Scolytus ventralis* LeConte (Coleoptera: Scolytidae), was first described from two specimens collected at Vancouver, Washington, but the species has been subsequently recorded throughout the western United States and British Columbia.[20,55] Its primary hosts are true firs, *Abies;* specifically white fir (*A. concolor*), California red fir (*A. magnifica*), and grand fir (*A. grandis*).[34] It has also been observed on occasion attacking subalpine fir, Douglas-fir, Engelmann spruce, mountain hemlock, and Western larch.[20,55]

Fir engraver adults and developing broods kill true firs by mining the cambium, phloem, and outer sapwood of the bole, thereby girdling the tree. The death of the tree is accelerated by infection with a brown-staining fungus, *Trichosporium symbioticum* Wright, which is carried in pits on the head of the beetle.[39,58] Some trees may survive light attack, in which case only portions of the bole are killed, while others may be attacked for several successive years before eventually succumbing to the invaders.

Heavy losses of white and grand fir have been caused by the fir engraver in the western United States. In California, for example, an estimated 450 million board-feet of timber are destroyed annually.[54] This amounts to more than one half the estimated annual growth of firs in this state.[55] Annual surveys in the state of Washington show the fir engraver infesting 10,000–20,000 acres. Defective overmature trees and low value stands at high elevations are most subject to attack, but some damage occurs in vigorous young stands. Resistant trees surviving initial unsuccessful attacks may develop defects and rots, rendering them useless for manufacture into lumber and wood products.[36,55]

White and grand firs are fast-growing trees that have assumed considerable importance in western North America as a source of timber and pulpwood. They are also important on many watersheds and recreational areas. Because of the increasing economic significance of true firs, the fir engraver beetle is receiving greater attention from both forest managers and entomologists.

2. BIOLOGY AND BEHAVIOR

2.1. Flight and Attack

Adult fir engravers are small black beetles (3–5 mm in length) with a prominent concavity of the posterior abdomen. Adult beetles emerge from infested trees and fly in search of new hosts from June to September, with flight peaking in mid-July. Daily flight patterns are determined to a large extent by temperature, with most activity between 1400 and 1700 hr, when ambient temperatures are normally above the 24°C threshold for flight.[23] Daily flight ceases when light intensity under the canopy drops below 100 foot candles, even though temperatures may be favorable. High winds within the stand also inhibit flight. Trapping studies indicate that most beetles fly at crown height but a slight concentration also occurs close to the ground.[2]

During the initial period of flight and dispersal, adult engravers crawl indiscriminantly on the boles of many tree species, but feeding scars are usually seen on the bark of twigs, branches, and stems of *Abies*. Attacks are apparently made at random on both resistant and susceptible firs.[14] Resistant reactions by certain trees, however, cause females to abandon their attacks.[9,14]. Successfully attacked trees can be recognized by the reddish brown or white boring dust expelled from attack holes and accumulating in crevices, cobwebs, or on the ground at the base of the tree.

Female engravers bore slightly upward into the bark and, on reaching the cambium, construct small nuptial chambers in which they await the males. These primary attacks are normally initiated in bark crevices, under lichens, or at the base of branches, but successive attacks are influenced by the presence of nearby galleries, so that the spacing between them is more or less uniform. This behavior results in decreased competition for food between the broods of adjacent galleries.[15]

Population aggregation on susceptible and to some extent on resistant trees usually follows initial attack.[14,23] Grand fir logs bearing a few *S. ventralis* attacks attract a larger proportion of the flying population, and aggregation occurs earlier and reaches higher densities, than on nearby unattacked logs.[2] Attraction persists for at least 2 weeks, and termination usually coincides with the decline in the number of beetles flying in the area.[23] Although aggregation pheromones have not yet been identified from *S. ventralis*, they have been found in other species of *Scolytus*.[44]

The fir engraver exploits weakened, dying, or recently killed fir trees, which form a temporary and scarce resource in the sense that food available at one point in time is not necessarily available in the future. Although the species is adapted to find these rare and widely scattered susceptible hosts, one would expect high mortality during the flight period, particularly when food is scarce. Based on this reasoning, and on the assumption that per-capita flight survival also depends on the number of beetles emerging from trees killed by the previous generation, a general model for bark beetle survival during flight has been proposed (Fig. 1). In this model, flight is directly related to the ratio of suitable hosts available for attack (logs or severely weakened trees) to those infested by the previous generation.

2.2. Mating and Oviposition

Male engravers are frequently observed with their heads pressed to the abdominal declivity of the female projecting from the entrance hole. The female is thus stimulated into mating, after which the male may enter the entrance hole and assist in removing debris. Although about one half the galleries observed contain no males, almost twice as many eggs are laid in those containing both sexes.[2]

After mating, the female usually constructs two horizontal egg galleries extending from the sides of the nuptial chamber (Fig. 2). Galleries cut across the grain of the wood, and about one half their depth is excavated in the outer sapwood. Eggs are deposited in niches cut in the lateral gallery walls; each niche is then plugged with frass and/or boring dust. The length of the egg galleries varies from less than 2 cm to 20 cm or so and is strongly influenced by the physiological state of the female and the conditions encountered in the host. Females parasitized by nematodes or phorid flies, for instance, bore

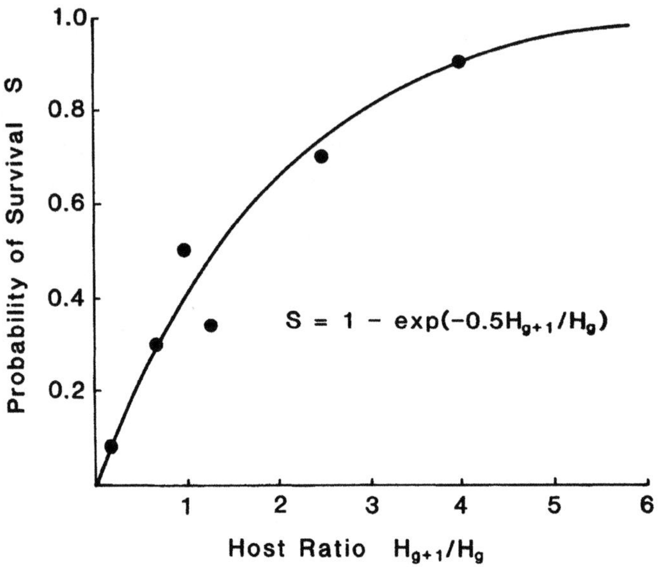

FIGURE 1. Relationship between the probability of a female fir engraver beetle surviving to mount a successful attack on a susceptible host tree and the ratio of susceptible hosts available in the current year to the number attacked in the previous year. (Redrawn from Berryman.[13])

short malformed galleries and lay few eggs, while resistant reactions by the host cause them to abandon their attacks.[2,14,27] Fecundity of individual females varies considerably, averaging 57 eggs, with a maximum of 300.[7,53,55]

2.3. Development and Life Cycle

Eggs hatch within 2 weeks, and the larvae feed by mining the phloem–cambium–sapwood region. Larvae sometimes mine into the outer bark, particularly when unfavorable conditions of resin or crowding are encountered at the cambium–sapwood interface. Larval mines generally radiate at right angles from the parent gallery and do not intersect unless high attack densities cause crowding (Fig. 2). The six larval periods last about 2 months under room conditions[51] with the sixth instar entering a quiescent nonfeeding prepupal stage before molting to the pupa. The pupal stage lasts about 3 days at room temperature.

Complete development from egg to adult usually takes 1 year in northern Idaho. Adult activity continues from mid-June to the end of September, and the eggs are deposited during this period. Eggs laid during the early part of the summer develop into fifth- or sixth-instar larvae before winter commences, but most larvae overwinter in the third or fourth instar. Some adults may emerge in September and October, depending on weather conditions, and create a partial second generation. It appears that about 40% of the larvae enter an obligatory diapause and overwinter in this state. The other 60% may emerge in

FIGURE 2. Bark removed from a grand fir stem to expose the horizontal parental egg galleries and the vertical larval galleries of *Scolytus ventralis*. (Photograph courtesy of the USDA Forest Service.)

late summer or overwinter in a facultative diapause, depending on weather conditions.[52] Thus, the fir engraver normally has one generation a year throughout most of its range, may have a partial second generation in warmer regions or years, and requires 2 years to complete a generation at high altitudes.[55]

3. ECOLOGICAL RELATIONSHIPS

3.1. Relationship with Phytopathogenic Fungi

A brown-staining fungus *Trichosporium symbioticum* is invariably associated with galleries of the fir engraver beetle.[58] This fungus is transported in mycangial pits on the head of both male and female beetles[39] (Fig. 3). *Trichosporium symbioticum* is always

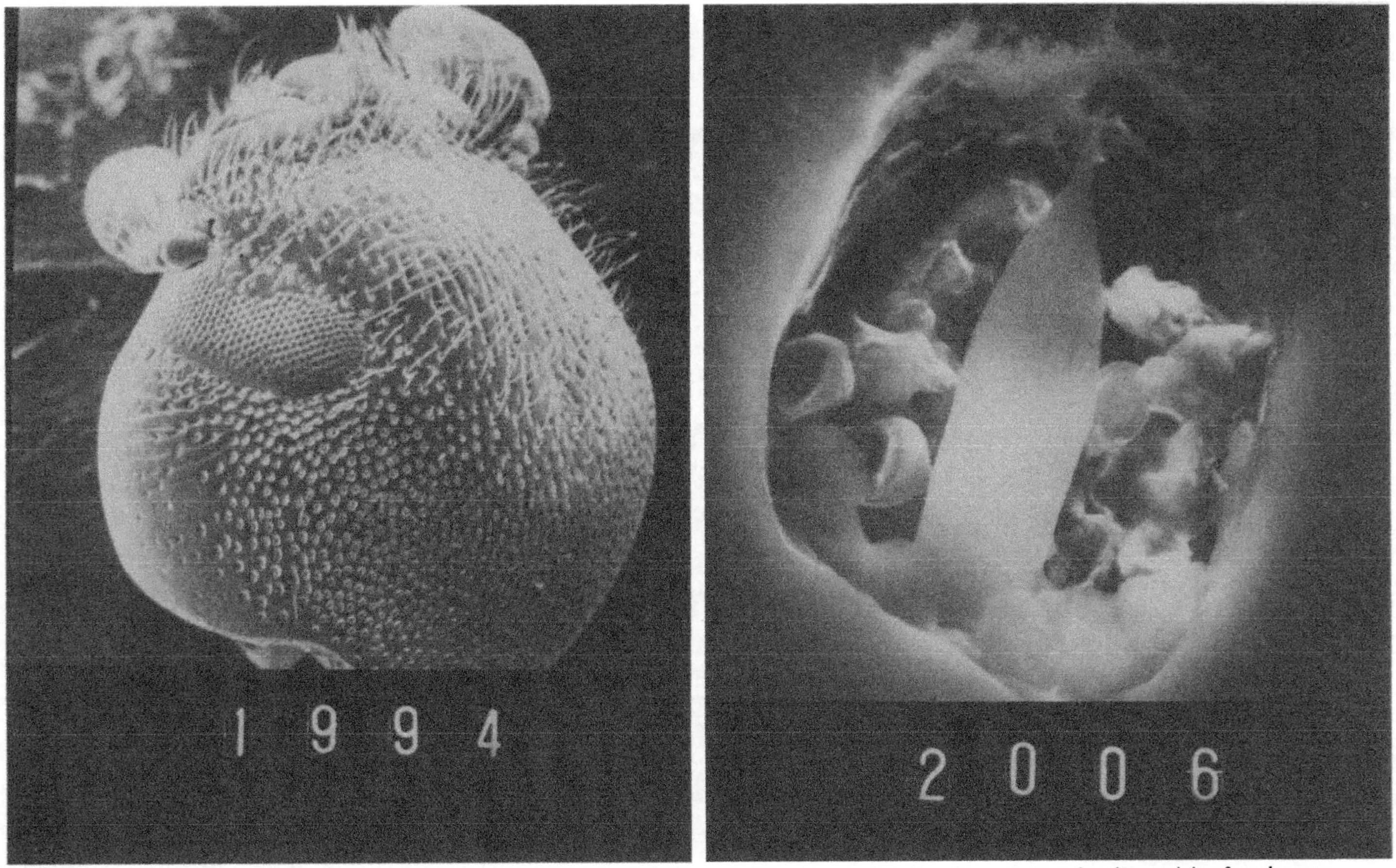

FIGURE 3. Scanning electron micrographs of the head of a female fir engraver showing specialized mycangial pits with a closeup of a pit containing fungal spores. (From Livingston and Berryman.[39])

observed growing in advance of developing larvae and is probably involved in the death and drying of phloem tissue prior to larval feeding. An unidentified yeast is also found associated with both larvae and adults.

Many scolytid bark beetles are associated with phytopathogenic fungi, e.g., the notorious Dutch elm disease carried by several elm bark beetles. *Trichosporium symbioticum* also has phytopathogenic properties, causing rapid embolism and disruption of water and nutrient transport in the outer xylem and phloem.[57] This fungus, together with the yeast, is probably essential for establishing conditions favorable for beetle development in living trees; i.e., it may assist in overcoming host resistance, decreasing the moisture content of the phloem tissue to a level suitable for beetle survival, inhibiting other fungi detrimental to beetle survival and development, or providing nutrients essential for larval growth, adult maturation, or egg production. For example, fir engraver pupal chambers are characteristically lined with a white mat of fungal hyphae, which include both *T. symbioticum* and the unidentified yeast. This lining is not found in pupal chambers after adult emergence, indicating that it has been devoured by pre-emergent beetles. The mycanigial pits probably become charged with fungal spores at the same time.

3.2. Relationship with the Host Tree

At the beginning of the flight period, fir engraver females attack living hosts in a more or less random pattern and apparently cannot distinguish between trees that will form a suitable substrate for brood development and those that will resist attack.[14] On resistant trees, however, attacking females are soon repelled by resin flow and aggregation then ceases. The degree of aggregation on an individual tree seems to vary with the time that females are able to remain in the tree and is reflected by the length of their egg galleries. In areas with large beetle populations, almost every tree receives some attacks and the resistance expressed by individual trees is reflected by the length of the beetle galleries as well as the density of their attacks;[14,27] i.e., shorter galleries and lower attack densities are found on the more resistant trees.

3.2.1. Host Resistance

Like most coniferous trees, true firs have two defenses against bark beetles and their associated fungi.[10] The first line of defense, the static or preformed resistance mechanism, consists of resin canals and pockets in the cortex of the bark (Fig. 4). Beetle galleries that contact these preformed structures almost always fail to produce larval galleries as the adults invariably abandon the attack.[29] However, because of the sparsity of resin containing structures, only a small proportion of the attacking beetles normally come into contact with them (about 20%). Thus, the preformed resin system does not seem to be an effective defense against an aggregated beetle attack.[29]

The second line of defense is a dynamic induced response, sometimes referred to as a hypersensitive reaction, that occurs after the beetle and fungus have invaded the tree[9,10] (Fig. 5). Resistant firs respond to a mechanical injury, similar to that caused by beetle penetration, or to inoculation of the phloem with *T. symbioticum,* by producing a lesion which becomes increasingly larger and darker with time. Although the lesions so induced

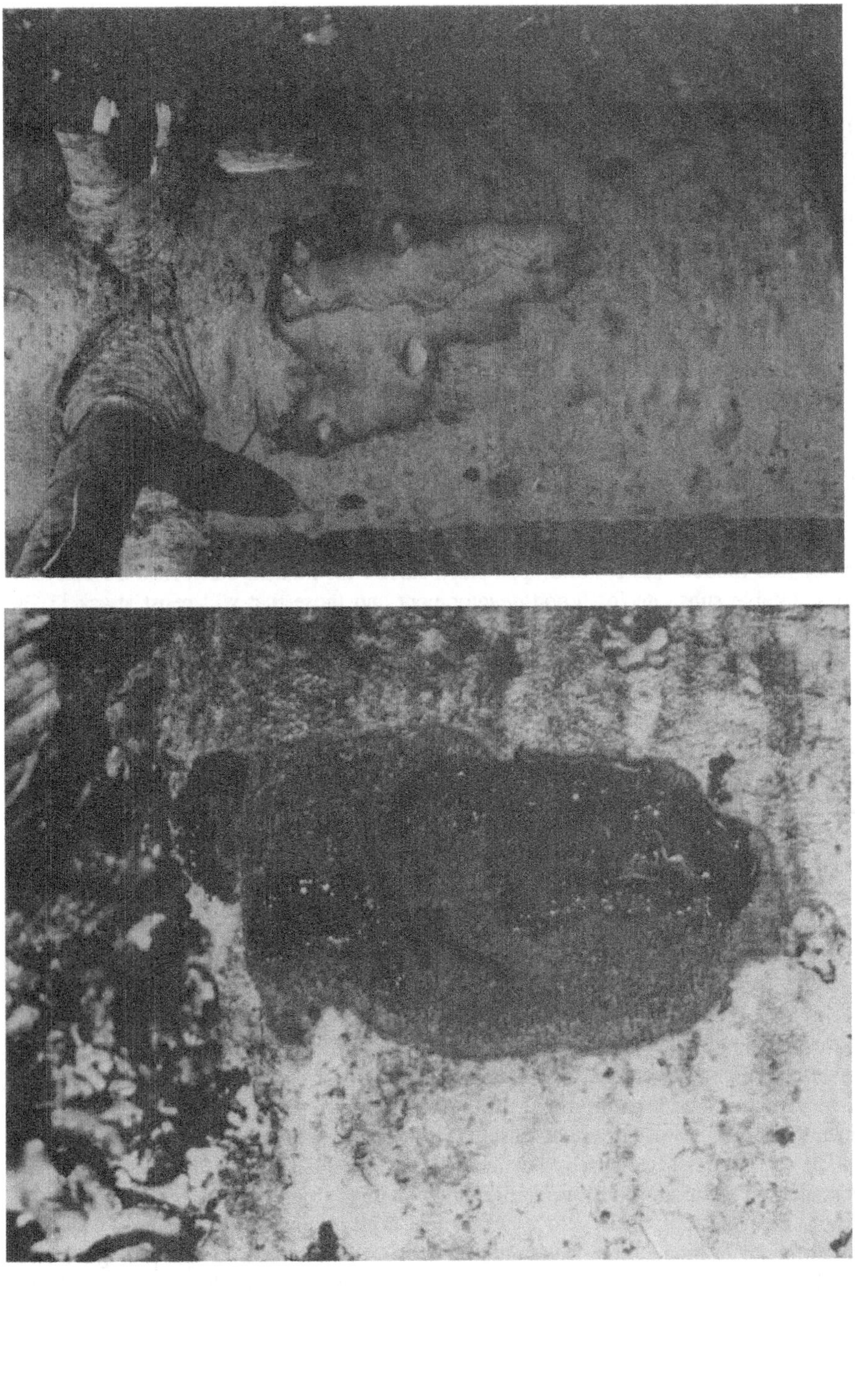

FIGURE 4. Outer bark shaved from a white fir stem showing cortical resin canals and pockets and a fir engraver attack hole (arrow). (From Ferrell.[29])

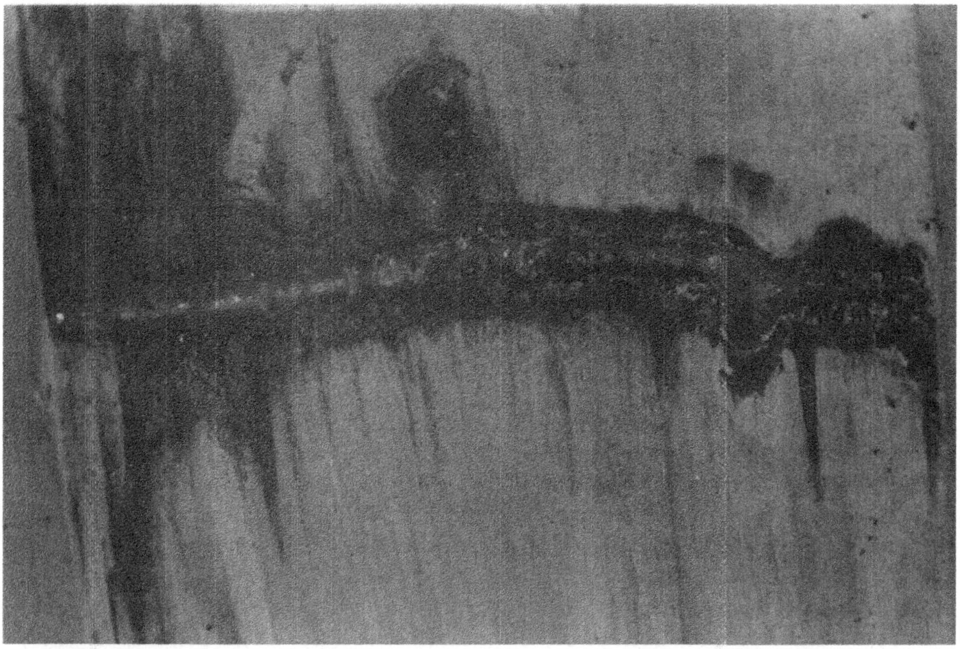

FIGURE 5. Unsuccessful *S. ventralis* attack surrounded by resin-impregnated tissue formed during the hypersensitive defense reaction.

are morphologically similar, the presence of the fungus results in a much more extensive reaction and the presence of different defensive chemicals.[47]

The induced reaction to beetle–fungus invasion involves the resinosis and death of phloem tissue and the formation of traumatic resin cavities in the outer sapwood at the edge of the lesion area. Phloem resinosis occurs in advance of fungal spread and is not caused directly by actual hyphal penetration.[57] In this way, the fungus is contained within the reaction area and is isolated from the healthy tissues of the host. After 28 days or so, the expanding lesion and the growth of the fungus levels off, with the lesion boundary some distance ahead of the fungus. At this time, traumatic resin ducts begin to form around the boundary of the lesion, and these secrete large quantities of resin into the wounded area.

Inhibition of fungal growth in the lesion is important for wound stabilization and healing. Phenolics and terpenes that accumulate in the lesion have been shown to inhibit fungal growth.[46,47,57] However, they do not drastically affect the viability or pathogenicity of the fungus. Thus, fungi are probably confined by the removal of nutrients, water, or oxygen, which are essential for growth (i.e., a scorched earth strategy) as well as by chemical inhibition.[46,57]

The effects of the hypersensitive reaction on the insect are twofold. First, the metabolites of active cellular degeneration appear to be produced under pressure and therefore to flow into the insect's tunnel.[9] The monoterpenes found in the reaction area are highly repellent to the beetle, so the attack is usually abandoned when this occurs.[14,19,27]

Second, the resin-soaked tissues of the reaction zone are completely unsuitable for survival of bark beetle eggs and larvae.[14] These may be killed by the toxic properties of the resin, by mechanical action of the hardened resin, by indigestibility of resin soaked tissue, or by lack of nutrients in the dead cells. Thus, even if resinosis occurs too slowly to repel the beetle, it can still kill the developing brood.

The resistance expressed by individual trees under natural forest conditions varies with the physiological condition of the tree. In vigorous trees, large concentrations of defensive chemicals accumulate rapidly, and beetles are repelled before mass attack can occur. In trees of intermediate resistance, resinosis occurs more slowly, galleries are longer, eggs may be deposited, and aggregated attack may occur. The beetle broods rarely survive, however, although the trees themselves sometimes die.[59] In susceptible trees, resin production occurs very slowly, if at all, and brood mortality due to resinosis rarely exceeds 50%.[14]

The capability of fir trees to resist bark beetle attack is affected by intrinsic properties of individual plants (e.g., age, genetics) and by external environmental conditions. For example, when trees are inoculated with *T. symbioticum,* smaller quantities of monoterpenes are produced by very young and very old trees, by suppressed trees, by diseased or heavily defoliated trees, and by firs that are under severe water stress.[27,45,59] The density of attacks has also been found to be higher on firs infested with true mistletoe and root-rot pathogens, and following periods of drought,[22,24,25,32] while the amount of tree mortality in a particular region has been associated with dense stands, drought periods, dry sites, defoliation, and root pathogens.[21,26,27,30,31,33,37,40,49,50,56,60] Stand disturbances due to logging, windthrow, lightning, and such also provide fir engravers with susceptible host material. Thus, the quantity of host material availble for *S. ventralis* attack is determined by intrinsic and extrinsic factors acting on individual trees and over large forest areas.

3.2.2. Food and Intraspecific Competition

Once fir engraver beetles have invaded and reproduced within a fir tree or log, the number of offspring produced is related to the condition of the phloem tissue on which the larvae feed, the density of beetle attacks, and interactions with other organisms.

The suitability of phloem tissue as food for developing larvae is affected by resin impregnation, moisture content, and possibly the concentration of sugars and other nutrients. Areas of phloem impregnated with resin as a result of induced defensive reactions are completely unsuitable for food. Eggs and larvae trapped within these areas usually die from intoxication or starvation.[14] Mortality of fir engraver eggs and larvae due to host resinosis and high phloem moisture is generally inversely related to the density of egg galleries. This is because increasing attack densities reduce the defensive capabilities of the tree, ensure a more uniform growth of fungi, and increase the rate of phloem drying.

By contrast, high attack densities result in crowding of the developing brood and greater mortality from competitive interactions. Intraspecific competition can result in extremely high larval mortality and is probably the most important cause of death among developing broods.[2,11,53] However, deaths from starvation and cannibalism in overcrowded regions are difficult to measure directly. This problem can be overcome by using models that stimulate the competition process[12,15] and that predict the unexplained mortality in *S. ventralis* life tables with reasonable accuracy.[11] These models incorporate the

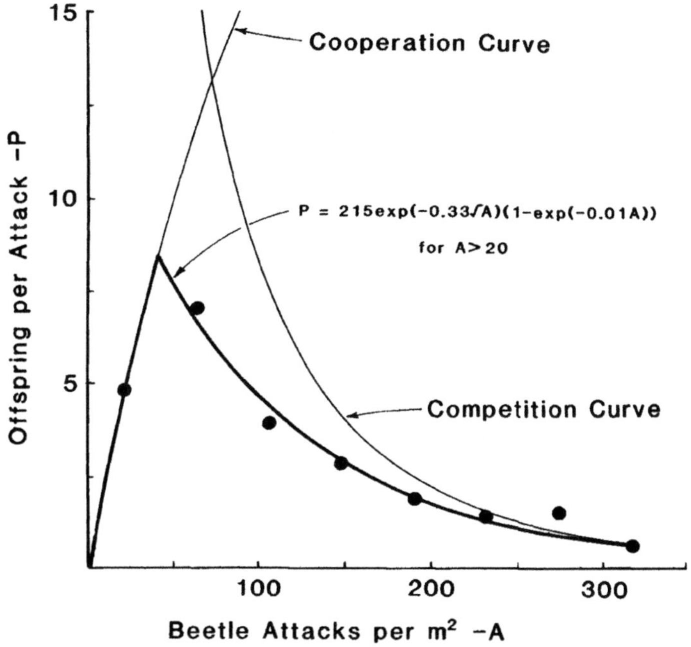

FIGURE 6. Relationship between *S. ventralis* productivity, expressed as offspring produced per attacking female, and the density of attacks per m^2 of bark. The composite function contains a cooperation effect, whereby beetles help each other in conditioning the host tissue, and a competition effect, whereby beetles struggle for a limited quantity of food (phloem). (Redrawn from Berryman.[13])

effects of cooperation between beetles in overcoming host resistance and conditioning the phloem, as well as the effects of intraspecific competition (Fig. 6).

3.3. Relationship with Other Organisms

Fir engraver beetles interact with many other organisms within the confines of infested trees (Table I).

3.3.1. Parasites

At least 10 species of insect parasitoids have been reared from logs infested by fir engraver beetles.[2] Of these, seven wasps (Hymenoptera: Braconidae: Pteromalidae: Torymidae: Ichneumonidae) attack the larval stages and one phorid fly (Diptera: Phoridae) attacks the adult beetle.

Parasitism is extremely variable both within and between trees, averaging between 0 and 800 attacked hosts per m^2, but densities over 1000 per m^2 are observed in individual samples.[8] Fir engraver larvae occupying portions of the tree with bark less than 3 mm thick are most heavily parasitized, and parasitism in thick-barked regions occurs mainly

TABLE I
Typical Life Table for *S. ventralis* Populations While within the Tree[a,b]

Age interval (*x*)	No. alive at start of *x* lx	No. dying during *x* *dx*	Factor causing *dx* *dxf*	*dx* as a % of lx 100*qx*
Parent females	128	56	Nematodes, resin, etc.	43.9
Potential eggs	7737	3400	As above	43.9
Egg laid	4337	463	Resin	10.67
		79	Mites	1.81
		4	Medetera	0.08
		15	Nematodes	0.34
		46	Unknown	1.06
		607	Combined	13.96
Larvae	3730	356	Resin	9.53
Instar I–II		1	Mites	0.03
		4	Nematodes	0.10
		2	Parasites	0.06
		12	Fungus	0.31
		34	Unknown	0.92
		409	Combined	10.95
Instar III–IV	3321	90	Resin	2.72
		3	Nematodes	0.08
		48	Parasites	1.43
		15	Fungus	0.44
		50	Unknown	1.51
		206	Combined	6.18
Instar V–VI	3115	5	Resin	0.18
		29	Parasites	0.92
		3	Fungus	0.09
		536	Predators	17.21
		1900[c]	"Disappearance"	60.96
		80	Unknown	2.58
		2553	Combined	81.94
Pupa and brood adult	562[d]	205	Nematodes, fungus, etc.	36.45
Emerged adults	357	?	Flight mortality	?
Females	134			
Generation				95.37

Trend index $= \frac{134}{128} = 1.05$; intrinsic rate of increase $= \log_e (1.05) = 0.05$

[a]Modified from Berryman.[11]
[b]Means of 14 trees; all counts converted to 1 m^2 of bark surface.
[c]Mortality occurred over all larval instars.
[d]Count estimated from last instar, pupae, and adults in sample.

beneath bark crevices. Measurements of parasitoid ovipositor lengths show that most species are incapable of reaching larvae more than 3 mm below the bark surface.[2]

There is some disagreement on the effect of larval parasitism on *S. ventralis* populations. Some authors claim that *Coeloides brunneri* Viereck is responsible for low engraver populations in certain years,[48] and parasitoids and predators together may destroy up to

84% of the broods in some places.[55] However, data from detailed life table studies only show 3–10% mortality from parasitism.[2,11,53] Moreover, there is no evidence from any of these studies that parasitoids exhibit density-dependent responses to their bark beetle prey.

Parasitic mites have been observed feeding on fir engraver eggs and larvae and being carried externally on the bodies of adult beetles (2). Egg parasitism varies considerably between and within trees, averaging 30% in individual trees, with patches up to 57%. Larval parasitism is generally low, but parasitized larvae often bear many gravid female mites (up to 35).

The most abundant parasite of *S. ventralis* is the nematode, *Sulphuretylenchus elongatus* (Massey) Nickle, which infects 45–93% of emerging beetles.[3] Occasional egg and larval mortality, usually less than 2%, has been attributed to this parasite.[2] The most important effect, however, is on the female beetle. Heavy nematode infections seriously impair flight, gallery construction, and oviposition potential.[3,4] Because of its abundance and severe impact on fecundity, *S. elongatus* may be responsible for suppressing local fir engraver populations. For example, this nematode was believed to have caused the decline of a beetle infestation in New Mexico.[41]

3.3.2. Predators

Of the predatory insects reared from *S. ventralis*-infested bolts, by far the most abundant are the black-bellied clerid, *Enoclerus lecontei* (Wolcott), and a dolichopodid fly, *Medetera aldrichii* Wheeler.

Clerid larvae average around 40 per m^2 in infested trees and, as each requires about eight bark beetle larvae to complete development,[6] we estimated that they cause around 9% mortality to fir engraver broods. Clerid adults also feed on dispersing and attacking bark beetle adults, but no estimate of mortality is available.

Larval populations of *Medetera aldrichii* average about 30 per m^2, but densities vary from 0 to 200. As each *M. aldrichii* consumes 5–10 bark beetle larvae,[5,25] we estimated that these predators caused an average brood mortality of 5%.

Predation by unknown species of woodpeckers is typically sporadic, yet intense. Woodpeckers work only only about 5% of the infested trees, but the birds often reduce broods by 80% or more in these trees.

3.3.3. Disease

Species of *Aspergillus* and *Penicillium* have been isolated from *S. ventralis* parent galleries and larval mines and from dead larvae.[2] Both fungi are known to be parasitic on insects, but their impact on *S. ventralis* populations is probably minimal.

3.3.4. Interspecific Competition

A number of beetles are commonly observed feeding in trees killed by fir engraver beetles. Of these, several species feed in the phloem region and compete with fir engraver broods for this limited food supply. High *S. ventralis* brood mortality has been observed under bark containing the galleries of *Pityophthorus pseudotsugae* Swaine and *Pityokteines elegans* Swaine and when more than 50% of the phloem is used by the other insects.[8,11,53]

In northern Idaho, *P. elegans* is the most common competitor, occurring in about 40% of the infested trees. When galleries of this scolytid are located near *S. ventralis* galleries, mortality occurs among the broods of both beetles, but fir engraver larvae suffer higher mortality and appear to be the weaker competitor. *Pityokteines* populations have been observed to increase during the decline of fir engraver outbreaks and may have hastened the collapse.[11]

4. POPULATION PATTERNS AND THEIR CAUSES

Based on detailed life-table studies over a 5-year period, the following hypothesis has been proposed for the numerical behavior of *S. ventralis* populations[11] (Fig. 7): The primary determinant of fir engraver abundance is the quantity of food, which is set by (1) The size of susceptible trees, determining the quantity of phloem in individual susceptibles, as conditioned by age, genetics and environmental variables governing tree growth. (2) The density of susceptibles per unit area of forest (per ha) as affected by tree characteristics (genetics, age) and environmental factors influencing tree vigor (e.g., disease, insects, precipitation).

Endemic populations of *S. ventralis* are maintained by the annual turnover of a few unhealthy trees whose vigor has been reduced by age, competition, disease, etc. In northern Idaho most of the grand fir trees killed by endemic engraver populations are heavily infected by root diseases,[33] while mistletoe appears to be a major factor predisposing white fir to attack by *S. ventralis* in California.[22]

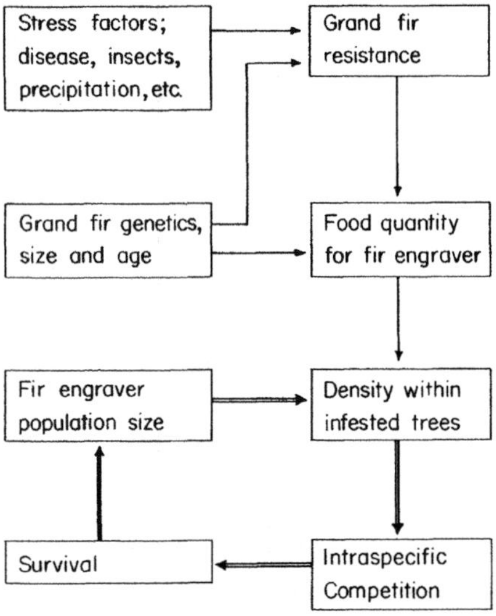

FIGURE 7. Conceptual model of the basic processes involved in regulating the abundance of *S. ventralis*. (From Berryman.[11])

Epidemics of the fir engraver erupt when food supplies increase substantially, i.e., when large numbers of trees are weakened by catastrophic events (e.g., defoliation, competition, drought, disease) or when stands become overmature and senescent.[22,56]

The fundamental mechanism regulating the size of *S. ventralis* populations is intraspecific competition operating as a negative feedback process (Fig. 7, double arrows). At endemic levels, competition for limited food acts as a "perfect" or instantaneous density-dependent mortality factor. When the food limitations are relaxed the population becomes more dispersed, competition is alleviated, and the population grows. As the outbreak progresses and the food supply is utilized, or the trees recover their resistance, intraspecific competition intensifies and the population declines to an endemic level at which a balance exists between the competitive use of the resource and a low-level replacement of susceptible hosts.

Other subsidiary processes may modify, to some degree, the behavior of this system. For instance, nematode parasitism and interspecific competition, operating as imperfect or delayed negative feedback processes, may accentuate fir engraver population fluctuations. These components are dependent to some extent on *S. ventralis* populations in previous generations and thus respond with a time-delay. In this way they tend to have the greatest impact during the latter phases of an epidemic and to be important determinants of the rate of population decline from the epidemic condition.

Independent studies and subsequent research provide evidence to support this explanation of fir engraver population behavior. For example, fir engraver attack scars have been observed to increase in years of low rainfall, particularly when 3 drought years follow each other and when logging activity occurs in an area[22,24,25,31] (Fig. 8).

In order to examine the hypothetical behavior of *S. ventralis* populations in more detail, two simulation models have been developed. The first is a high-resolution spatially defined model,[16] which contains all those density-dependent feedback processes thought to be important in regulating fir engraver population behavior, as well as stochastic density-independent effects due to weather, defoliation, root-rot, etc. (Fig. 9). The second is a very simple spatially-undefined model based on the flight and within-tree survival functions of Figs. 1 and 6. As there were no really significant differences in the qualitative behavior of these models, we will only consider the simple one here. The model was constructed from the following assumptions:

1. The number of trees available for attack (susceptibles) in any given year is a driving variable which is provided as an annual input. All trees are assumed to be of equal size.
2. Females emerging from one generation survive and attack trees according to the flight function (see Fig. 1).
3. Surviving females are allocated equally to available trees as long as the density of beetles is at or above the optimum attack density (i.e., the peak of the curve in Fig. 6 is about 45 beetles per m^2). Otherwise, the number of trees attacked is reduced until this density is attained (i.e., beetles optimize within-tree densities to achieve maximum survival if hosts are abundant).
4. Females lay a constant number of eggs.
5. Survival from eggs to emerging adults is determined by the composite cooperation-competition function (Fig. 6).
6. The sex ratio remains constant at 50 : 50.

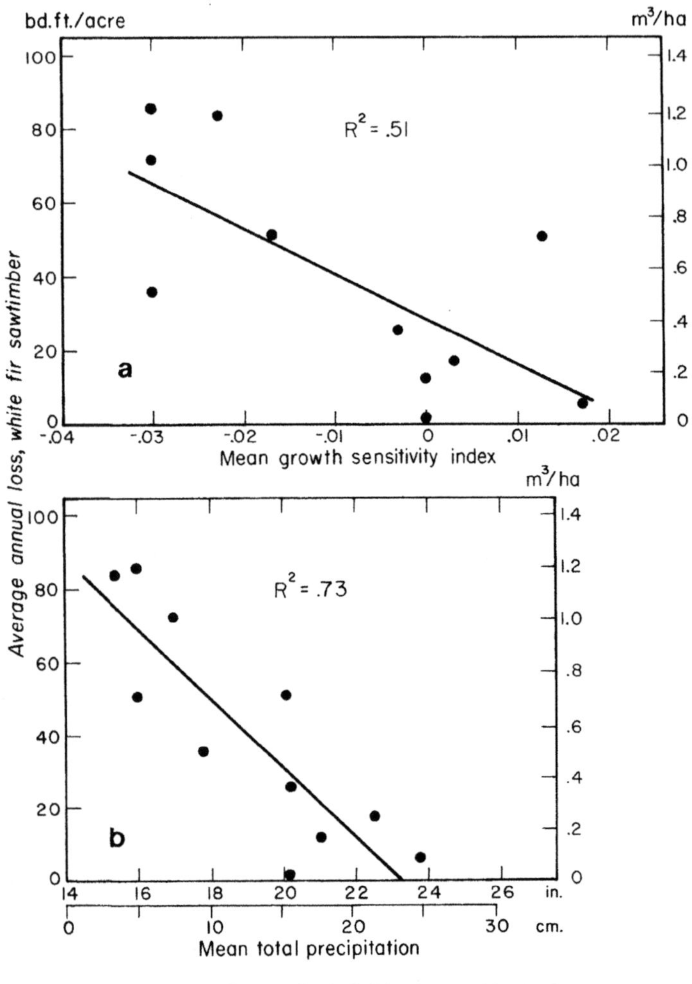

FIGURE 8. Relationship between white fir mortality in California caused by the fir engraver beetle and (a) a tree growth sensitivity index, which measures acceleration (>0) or deceleration (<0) of tree growth over the past 2 years, and (b) average annual precipitation. (From Ferrell and Hall.[31])

Some typical simulations on this model are shown in Figure 10. The salient features are that attack and emergence densities quickly attain stable levels when food availability is constant (upper trajectory). The system has a high degree of temporal stability as would be expected under the dominant influence of fast-acting negative feedback caused by intraspecific competition for food (Figs. 7 and 9). When the food supply increases, we observe a reduction in the density of attacks and a slight increase in emergence densities (middle trajectory). However, the productivity of the population (emergence per female) increases substantially and this causes the population to grow quickly to the limits set by its food

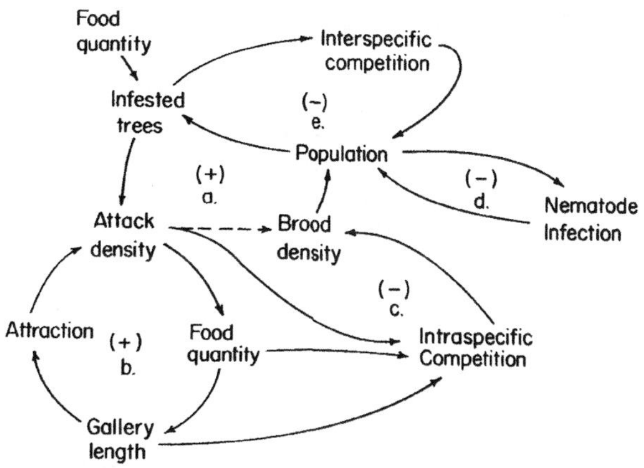

FIGURE 9. Feedback structure of a simulation model of *S. ventralis* populations interacting with their host plants, competing phloem-feeding beetles, and nematode parasites.

supply. When the food supply increases temporarily, as is more usual following drought, defoliation or similar stress, the attack densities fall at first, increase at the beginning of the food decline, and fall again as the food supply returns to its initial state (lower trajectory in Fig. 10). This dynamic behavior is qualitatively similar to natural fir engraver populations as they adjust to changes in the abundance of susceptible hosts.[11,17,60]

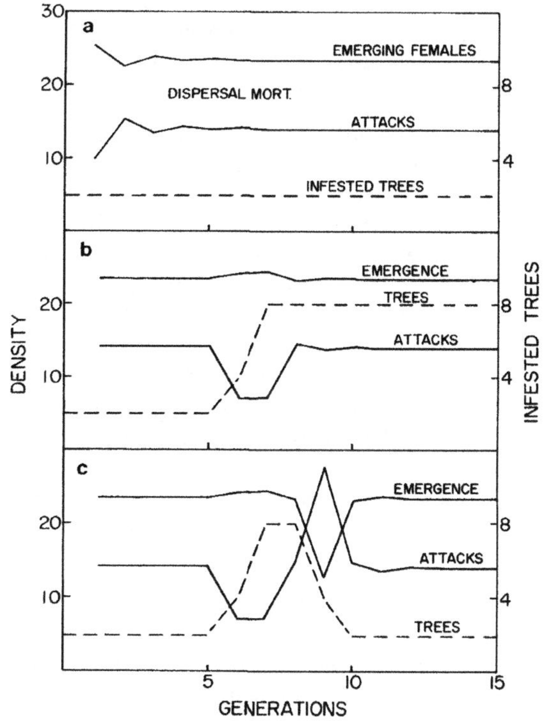

FIGURE 10. Simulated dynamics of *S. ventralis* attack and emergence under conditions of constant host breeding material (a), a single increase in breeding material (b), and a pulse increase and decrease in susceptible breeding substrate (c).

5. MANAGEMENT IMPLICATIONS

With a model for the population dynamics of the fir engraver, it is possible to explore its relevance to the management of this pest. Managers are interested in models for two purposes, first to predict pest population levels, and second to test alternative management treatments. In the following discussion we will examine the insight that the model provides into these activities.

5.1. Prediction

In the management of North American forests, low-level losses due to endemic *S. ventralis* populations are generally considered economically acceptable. However, occasional outbreaks may cause extensive timber losses which may have severe economic repercussions. The manager, therefore, is concerned with predicting the location and intensity of outbreaks.

It is apparent that fir engraver beetle populations are driven by the quantity of food available; i.e., changes in the number of available hosts determine whether the population will increase or decrease. Thus, a description of the quantity of food available per unit area will permit prediction of population behavior and, for this reason, methods have been developed for assessing the probability of tree and stand damage from engraver beetle attacks.[28,30,40,42,49,50] Aside from these risk-assessment methods, the population model leads to certain inferences concerning the behavior of bark beetle populations during incipient outbreaks, which may provide clues to future behavior. For example,

1. Attack densities will usually decrease, probably to the zone of maximum productivity (50–100 attacks per m^2) during the phase of population growth (Fig. 10, lower trajectory).
2. The number of hosts killed per unit area of forest will increase.
3. The mean distance between newly infested and abandoned trees, or clusters of trees, will decrease. This variable may be the most powerful in terms of predicting the intensity of outbreaks because of its influence on beetle mortality during flight. For instance, if beetles are moving into adjacent trees, then flight losses should be relatively light and total productivity approaches its theoretical maximum. However, if beetles have to move considerable distances between abandoned and new hosts, even though more hosts are infested, flight losses will be higher. Furthermore, close proximity of suitable hosts may indicate that much of the stand is available but cannot as yet be used because of the low beetle population and its tendency to aggregate.

Because of the inherent variability of forest ecosystems, the occurrence of only one of the above may not be a strong predictor of future events. However, their combined occurrence may have considerable predictive power, particularly when associated with other factors such as droughts or defoliator and disease epidemics.

5.2. Population Management

The model can also be used to evaluate management alternatives. There are two basic approaches to bark beetle control: (1) killing prereproductive beetles, which may be achieved by removing or spraying infested trees or trapping flying beetles with natural or synthetic pheromones; and (2) decreasing the food supply by removing susceptible trees, or possibly by increasing stand resistance with the application of fertilizers or precommercial thinnings. The objective of a bark beetle control is to reduce the number of trees killed by the beetles, not killing beetles per se.

5.2.1. Killing Beetles

Suppose that we have a beetle population occupying a constant food supply and we apply a treatment which kills 50% of the prereproductive adults each generation. The survivors may attack the same number of hosts, but at a lower density, resulting in higher productivity which compensates for the treatment effect (Table II). If we increase the killing efficiency to 80%, the population is reduced to such an extent that it appears that some trees are saved by the control effort. However, if we assume that these trees remain available (i.e., susceptible), they must be added to the food supply for the next beetle generation, thereby increasing the host ratio and beetle survival during flight. The continual addition of hosts to the available pool as the treatment is continued results in less and less flight loss and a slow return to the original rate of tree mortality (Table II). Thus, reducing beetle populations by 80% in permanently weakened stands will not affect the eventual rate of tree mortality.

TABLE II
Simulation of the Effect of Continuous Beetle Killing and Susceptible Host Harvest, Initiated in the Third Generation, on Fir Engraver Attack Density and Tree Mortality when 10 Susceptible Hosts Are Provided Each Generation

	50% Beetle kill		80% Beetle kill[a]		90% Beetle kill[a]		50% Tree harvest	
Generation	Attacks[b] (m^2)	Trees killed	Attacks[b] (m^2)	Trees killed	Attacks[b] (m^2)	Trees killed	Attacks[b] (m^2)	Trees killed
1	154	10	154	10	154	10	154	10
2	154	10	154	10	154	10	154	10
3	76	10	44	7	52	3	176	5
4	86	10	45	7	43	3	147	5
5	85	10	52	7	57	2	157	5
6	85	10	54	8	44	2	153	5
7	85	10	55	9	77	1	154	5
8	85	10	55	10	45	1	154	5
Trees saved		0		12		48		30

[a]The assumption is made that hosts available to one generation and not utilized are available to the next, thus increasing the host ratio and flight survival.
[b]It is assumed that beetles will concentrate on fewer hosts when the density of attacks per available host drops below 45 per m^2.

To ensure reduction of timber losses, the level of mortality inflicted on the beetle population must exceed the maximum productivity of the population. For the fir engraver model, the critical mortality level is 87%, and treatment with 90% effectiveness reduces tree losses significantly (Table II). However, it must be realized that, if the saved trees are permanently susceptible, a resurgence will occur if the treatment is discontinued.

Past control operations against other species of bark beetles generally support the results of simulations on the fir engraver model. For example, "all results from applied control indicate that the killing of beetles, no matter by what method, has only a limited effect in reducing tree mortality. The trend of epidemics is only temporarily altered by direct control"[43] "The recommendation that it is necessary to destroy only from 50 to 75 percent of the broods in order to bring a destructive outbreak under control is particularly dangerous. . . . At the present time all efforts at control of the Black Hills beetle are based on an attempt to treat as nearly 100 percent of the infestation as is physically possible."[18] Finally, an extensive study of tree survival in insecticide treated and untreated stands found that "[populations] declined in approximately the same number of years, and lodgepole pine survival in the two types of stands was comparable. However, suppression measures did slow the rate of tree mortality in two stands still under attack."[1]

To our knowledge, there have been no observations made on changes in attack density during a bark beetle control effort. However, it has been noted that, "after a severe freezing during which 90 percent of the overwintering brood were killed over extensive forest areas, western pine beetle attacks were less dense the following spring and brood production was much higher than usual."[43]

These observations of bark beetle population behavior support the model predictions and lead to the conclusion that killing beetles will generally fail to protect susceptible stands (e.g., weakened by old age, root diseases), although the rate of mortality may be reduced by effective and continued treatment. In temporarily weakened stands (e.g., those responding to temporary stresses such as drought, defoliation), killing beetles may be a useful strategy to hold the population in check until the trees recover from the stressed condition. In these cases simple population models may prove useful for identifying the minimal treatment effectiveness required to maintain, or reduce, timber losses.

5.2.2. Harvesting Trees

The harvest of host trees affects beetle populations by decreasing the food supply either directly, by removing susceptible (available) hosts, or indirectly, through effects on growing conditions and vigor of the remaining trees. For example, Table II shows that the food supply has been reduced by harvesting one half the available susceptible hosts. This increases flight losses and beetle attack densities, reduced productivity, and causes the beetle population to subside. Continued maintenance of this low level food supply results, after a period of damped oscillation, in a stable beetle population and a reduced level of tree mortality (Table II).

Unfortunately, the host harvesting approach, although being theoretically sound, is often impractical because of our general inability to recognize hosts that are available for attack. In cases in which susceptible hosts can be recognized and harvested, timber losses to bark beetles can be significantly reduced. For example, methods have been devised for rating fir trees in California to attack by fir engraver beetles.[28] Although this risk-rating system has yet to be verified, it is currently being used in timber harvesting experiments.

The indirect manipulation of food supplies, using treatments that increase the vigor of the residual stand, may be the best long-term approach to fir engraver control. However, these methods require intensive forest management practices, and this, as yet, is not possible on most North American forested lands dominated by the true firs.

ACKNOWLEDGMENTS. We thank John Madden and Kenneth Raffa for their helpful comments on an earlier draft of this chapter. The drawing heading the chapter was done by Celest Green and published with permission of the University of California Press. We also thank the Entomological Society of Canada for permission to republish Figs 3, 4, and 7.

REFERENCES

1. Amman, G. D., and Baker, B. H., 1972, Mountain pine beetle influence on lodgepole pine stand structure, *J. For.* **70:**207–209.
2. Ashraf, M., and Berryman, A. A., 1969, Biology of *Scolytus ventralis* (Coleoptera: Scolytidae) attacking *Abies grandis* in northern Idaho, *Melandera* **2:**1–23.
3. Ashraf, M., and Berryman, A. A., 1970, Biology of *Sulphuretylenchus elongatus* (Nematoda: Sphaerulariidae), and its effect on its host, *Scolytus ventralis* (Coleoptera: Scolytidae), *Can. Entomol.* **102:**197–213.
4. Ashraf, M., and Berryman, A. A., 1970, Histopathology of *Scolytus ventralis* (Coleoptera: Scolytidae) infected by *Sulphuretylenchus elongatus* (Nematoda: Sphaerulariidae), *Ann. Entomol. Soc. Am.* **63:**924–930.
5. Beaver, R. A., 1966, The biology and immature stages of two species of *Medetera* (Diptera: Dolichopodidae) associated with the bark beetle *Scolytus scolytus* (F.), *Proc. R. Entomol. Soc. Lond. (A)* **41:**145–154.
6. Berryman, A. A., 1967, Estimation of *Dendroctonus brevicomis* LeConte (Coleoptera: Scolytidae) mortality caused by insect predators, *Can. Entomol.* **99:**1009–1014.
7. Berryman, A. A., 1968, Estimation of oviposition by the fir engraver, *Scolytus ventralis* LeConte (Coleoptera: Scolytidae), *Ann. Entomol. Soc. Am.* **61:**227–228.
8. Berryman, A. A., 1968, Distribution of *Scolytus ventralis* LeConte attacks, emergence and parasites in grand fir, *Can. Entomol.* **100:**57–68.
9. Berryman, A. A., 1969, Responses of *Abies grandis* to attack by *Scolytus ventralis* (Coleoptera: Scolytidae), *Can. Entomol.* **101:**1033–1042.
10. Berryman, A. A., 1972, Resistance of conifers to invasion by bark beetle–fungus associations, *BioScience* **22:**598–602.
11. Berryman, A. A., 1973, Population dynamics of the fir engraver, *Scolytus ventralis* (Coleoptera: Scolytidae). I. Analysis of population behavior and survival from 1964 to 1971, *Can. Entomol.* **105:**1465–1488.
12. Berryman, A. A., 1974, Dynamics of bark beetle populations: Towards a general productivity model, *Environ. Entomol.* **3:**579–585.
13. Berryman, A. A., 1979, Dynamics of bark beetle populations: Analysis of dispersal and redistribution, *Bull. Soc. Entomol. Suisse* **52:**227–234.
14. Berryman, A. A., and Ashraf, M., 1970, Effects of *Abies grandis* resin on the attack behavior and brood survival of *Scolytus ventralis* (Coleoptera: Scolytidae), *Can. Entomol.* **102:** 1229–1236.
15. Berryman, A. A., and Pienaar, L. V., 1973, Simulation of intraspecific competition and survival of *Scolytus ventralis* broods (Coleoptera: Scolytidae), *Environ. Entomol.* **2:**447–459.
16. Berryman, A. A., and Pienaar, L. V., 1974, Simulation: A powerful method of investigating the dynamics and management of insect populations, *Environ. Entomol.* **3:**199–207.
17. Berryman, A. A., and Wright, L. C., 1978, Defoliation, tree condition and bark beetles, in: The Douglas-

fir Tussock Moth: A Synthesis (M. H. Brookes, R. W. Stark, and R. W. Campbell, eds.), pp. 81–86, USDA Tech. Bull. 1585.
18. Blackman, M. W., 1931, The Black-Hills beetle (*Dendroctonus ponderosae* Hopk.), N.Y. State Coll. For., Syracuse Univ. Bull. 4, Tech. Publ. 36.
19. Bordasch, R. P., and Berryman, A. A. 1977, Host resistance to the fir engraver beetle, *Scolytus ventralis* (Coleoptera: Scolytidae). 2. Repellency of *Abies grandis* resins and some monoterpenes, *Can. Entomol.* **109:**95–100.
20. Chamberlin, W. J., 1958, *The Scolytoidea of the Northwest,* Oregon State College, Corvallis, Oregon.
21. Cobb, F. W., Parmeter, J. R., Wood, D. L., and Stark, R. W., 1973, Root pathogen as agents predisposing ponderosa pine and white fir to bark beetles, in: Proceedings of the Fourth International Conference on *Fomes Annosus,* Sept. 17–21, 1973, Athens, Georgia, pp. 8–15.
22. Felix, L. S., Uhrenholdt, B., and Parmeter, J. R., 1971, Association of *Scolytus ventralis* (Coleoptera: Scolytidae) and *Phoradendron bolleanum* subspecies *pauciflorum* on *Abies concolor, Can. Entomol.* **103:**1697–1703.
23. Ferrell, G. T., 1971, Host selection by the fir engraver, *Scolytus ventralis* (Coleoptera: Scolytidae): Preliminary field studies, *Can. Entomol.* **103:**1717–1725.
24. Ferrell, G. T., 1973, Stand and tree characteristics influencing density of fir engraver beetle attack scars in white fir, U.S.D.A. For. Serv. Res. Paper PSW-97.
25. Ferrell, G. T., 1973, Weather, logging, and tree growth factors associated with fir engraver attack scars in white fir, U.S.D.A. For. Serv., Res. Paper PSW-92.
26. Ferrell, G. T., 1974, Moisture stress and fir engraver (Coleoptera: Scolytidae) attack in white fir infected by true mistletoe, *Can. Entomol.* **106:**315–318.
27. Ferrell, G. T., 1978, Moisture stress threshold of susceptibility to fir engraver beetles in pole-size white firs, *For. Sci.* **24:**85–92.
28. Ferrell, G. T., 1980, Risk-rating systems for mature red fir and white fir in northern California, U.S.D.A. For. Ser., General Tech. Rep. PSW-39.
29. Ferrell, G. T., 1983, Host resistance to the fir engraver, *Scolytus ventralis* (Coleoptera: Scolytidae): Frequencies of attacks contacting cortical resin blisters and canals of *Abies concolor, Can. Entomol.* **115:**1421–1428.
30. Ferrell, G. T., 1986, Use of indicator plants to assess susceptibility to the fir engraver in California red, and white firs, U·S·D·A· Forest Serv. Res. Note PSW-388.
31. Ferrell, G. T., and Hall, R. C., 1975, Weather and tree growth associated with white fir mortality caused by fir engraver and roundheaded fir borer, U.S.D.A. For. Ser., Res. Paper PSW-109.
32. Ferrell, G. T., and Smith, R. S., 1976, Indicators of *Fomes annosus* wood decay and bark beetle susceptibility in sapling white fir, *For. Sci.* **22:**365–369.
33. Hertert, H. D., Miller, D. L., and Partridge, A. D., 1975, Interaction of bark beetles (Coleoptera: Scolytidae) and root-rot pathogens in grand fir in northern Idaho, *Can. Entomol.* **107:**899–904.
34. Hopping, G. R., 1922, Coniferous hosts of the Ipidae of the Pacific Coast and Rocky Mountain Regions, *Can. Entomol.* **54:**128–134.
35. Hopping, G. R., 1947, Notes on the seasonal development of *Medetera aldrichii* Wheeler (Diptera: Dolichopodidae) as a predator of the Douglas-fir beetle, *Dendroctonus pseudotsugae* Hopkins, *Can. Entomol.* **79:**150–153.
36. Johnson, N. E., and Shea, K. R., 1963, White fir defects associated with attacks by the fir engraver, Weyerhauser Co., For. Res. Note 54.
37. Lane, B. B., and Goheen, D. J., 1979, Incidence of root disease in bark beetle-infected Eastern Oregon and Washington true firs, *Plant Dis. Rep.* **63:**262–266.
38. Zimmerman, C., Synopsis of the Scolytoidea of America north of Mexico, *Trans. Am. Entomol. Soc.* **2:**150–178.
39. Livingston, R. L., and Berryman, A. A., 1972, Fungus transport structures in the fir engraver, *Scolytus ventralis* (Coleoptera: Scolytidae), *Can. Entomol.* **104:**1793–1803.
40. Mahoney, R. L., Moore, J. A., and Schenk, J. A., 1979, Validation and refinement of a plant indicator model for grand fir mortality by the fir engraver, *J. Entomol. Soc. Br. Columbia* **76:**17–19.
41. Massey, C. L., 1969, The influence of nematode parasites and associates on bark beetles in the United States, *Bull. Entomol. Soc. Amr.* **12:**384–386.

42. Moore, J. A., Schenk, J. A., and Hatch, C., 1978, Validation and refinement of a hazard rating model for fir engraver-caused mortality in grand fir stands, *For. Sci.* **24**:309–312.
43. Miller, J. M., and Keen, F. P., 1960, Biology and control of the western pine beetle, U.S.D.A. Misc. Publ. 600.
44. Pearce, G. T., Gore, W. E., Silverstein, R. M., Peacock, J. W., Cuthbert, R. S., Lanier, G. N., and Simeon, J. B., 1975, Chemical attractants for the smaller European elm bark beetle *Scolytus multistriatus* (Coleoptera: Scolytidae), *J. Chem. Ecol.* **1**:115–124.
45. Raffa, K. F., and Berryman, A. A., 1982, Accumulation of monoterpenes and associated volatiles following inoculation of grand fir with a fungus transmitted by the fir engraver, *Scolytus ventralis* (Coleoptera: Scolytidae), *Can. Entomol.* **114**:797–810.
46. Raffa, K. F., Berryman, A. A., Simasko, J., Teal, W., and Wong, B. L., 1985, Effects of grand fir monoterpenes on the fir engraver, *Scolytus ventralis* (Coleoptera: Scolytidae) and its symbiotic fungus, *Environ. Entomol.* **14**:522–556.
47. Russell, C. E., and Berryman, A. A., 1976, Host resistance to the fir engraver beetle. I. Monoterpene composition of *Abies grandis* pitch blisters and fungus-infected wounds, *Can. J. Bot.* **54**:14–18.
48. Salman, K. A., 1933, Forest insects of the year 1932, *Calif. Dept. Agric. Monog. Bull.* **22**:131–137.
49. Schenk, J. A., Mahoney, R. L., Moore, J. A., and Adams, D., 1976, Understory plants as indicators of grand fir mortality due to the fir engraver, *J. Entomol. Soc. Br. Columbia* **73**:21–24.
50. Schenk, J. A., Moore, J. A., Adams, D., and Mahoney, R. L., 1977, A preliminary hazard rating of grand fir stands for mortality by the fir engraver, *For. Sci.* **23**:103–110.
51. Scott, B. A., Jr., and Berryman, A. A., 1971, Laboratory rearing techniques for *Scolytus ventralis* (Coleoptera: Scolytidae), Wash. Agric. Expt. Sta. Bull. 741.
52. Scott, B. A., Jr., and Berryman, A. A., 1972, Larval diapause in *Scolytus ventralis* (Coleoptera: Scolytidae), *J. Entomol. Soc. Br. Columbia* **69**:50–53.
53. Stark, R. W., and Borden, J. H., 1965, Observation of mortality factors of the fir engraver, *Scolytus ventralis* (Coleoptera: Scolytidae), *J. Econ. Entomol.* **58**:1162–1163.
54. Stevens, R. E., 1956, The fir engraver beetle, U.S.D.A. For. Pest. Leaflet 13.
55. Struble, G. R., 1957, The fir engraver, a serious enemy of western true firs, U.S.D.A. Prod. Res. Rep. 11.
56. Wickman, B E., 1963, Mortality and growth reduction of white fir following defoliation by the Douglas-fir tussock moth. U.S.D.A. For. Serv. Res. Paper PSW-7.
57. Wong, B. A., and Berryman, A. A., 1977, Host resistance to the fir engraver beetle. 3. Lesion development and containment of infection by resistant *Abies grandis* inoculated with *Trichosporium symbioticum*, *Can. J. Bot.* **55**:2358–2365.
58. Wright, E., 1935, *Trichosporium symbioticam* n. sp., a wood staining fungus associated with *Scolytus ventralis*, *J. Agric. Res.* **50**:525–538.
59. Wright, L. C., Berryman, A. A., and Gurusiddaiah, S., 1979, Host resistance to the fir engraver beetle, *Scolytus ventralis* (Coleoptera: Scolytidae). 4. Effects of defoliation on wound monoterpene and inner bark carbohydrate concentrations, *Can. Entomol.* **111**:1255–1262.
60. Wright, L. C., Berryman, A. A., and Wickman, B. E., 1984, Abundance of the fir engraver, *Scolytus ventralis*, and the Douglas-fir Beetle, *Dendroctonus pseudotsugae*, following tree defoliation by the Douglas-fir tussock moth, *Orgyia pseudotsugata*, *Can. Entomol.* **116**:293–305.

CHAPTER 27

THE STRIPED AMBROSIA BEETLE

John H. Borden

John H. Borden • Center for Pest Management, Department of Biological Sciences, Simon Fraser University, Burnaby, British Columbia, V5A 1S6 Canada.

1. INTRODUCTION

The striped ambrosia beetle, *Trypodendron lineatum* (Olivier) (Coleoptera: Scolytidae), is one of nature's recyclers.[10] Along with other ambrosia beetles, its ecological niche is to locate and infest dead and dying timber. It is a pioneer in the degradation process, exploiting this renewed habitat each spring, and leaving it within months to be further degraded by saprophytic fungi and other wood-infesting insects.

Trypodendron lineatum is holarctic in distribution throughout the coniferous forests of the north temperate zone.[102] It will infest almost any coniferous species but has also been recorded from four genera of deciduous trees: *Alnus, Betula,* and *Malus*[102] and *Acer.*[64]

Without the aid of symbiotic (ambrosial) fungi, *T. lineatum* could not exploit its habitat in the sapwood of fallen or broken trees. Early in the process of boring minute tunnels (pinholes) into the wood, the female beetles inoculate the wood in several places with the spores of these fungi,[38] primarily *Monilia ferruginea.*[5,43] The spores are carried and nurtured throughout the winter by the female beetles in specialized chambers called mycangia in the prothoracic cuticle.[42,96] The fungal hyphae penetrate the wood but also

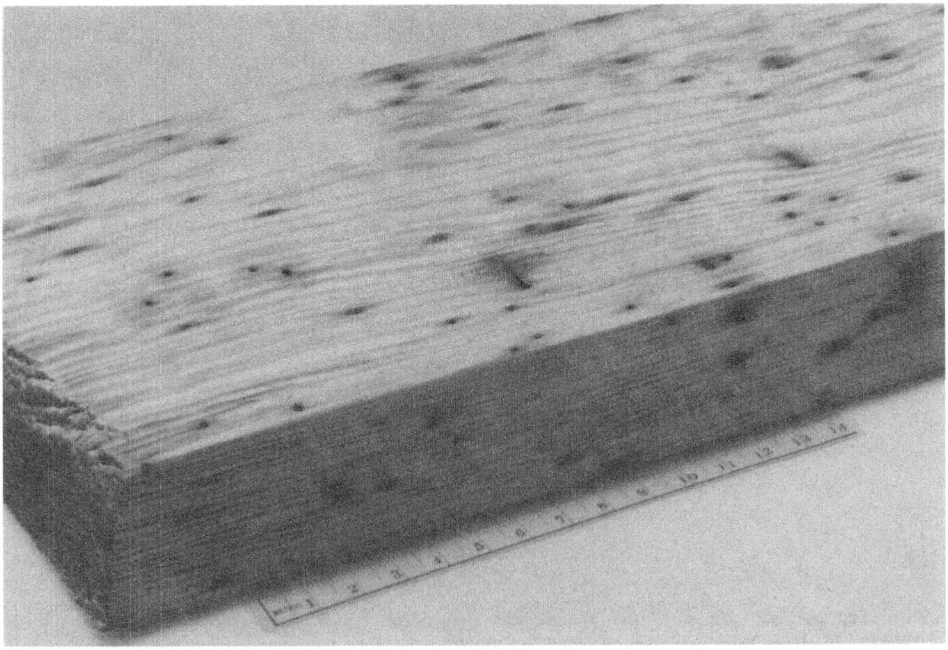

FIGURE 1. Outside surface of a board sawn from a Western hemlock log infested with *Trypodendron lineatum*. Note dark "pinholes" and discoloration extending into surrounding wood caused by fungal staining of the galleries bored by the female beetles.

grow into the tunnels. The mycelial growth provides a bountiful garden for the beetles, which consume this ambrosia instead of the wood they themselves cannot digest.

A side effect of infestation of the wood by the ambrosia fungi is a dark, almost black, fungus-produced stain that lines the beetle galleries and penetrates the surrounding wood (Fig. 1). These black-stained galleries indicate insect damage and may result in buyer resistance. Thus, lumber and plywood infested by *T. lineatum* is subject to severe degrade in quality and sells for a greatly reduced price.

On the British Columbia coast alone, the value loss from ambrosia beetle degrade to sawlogs (primarily due to *T. lineatum*) amounts to more than $63 million (Canadian) each year.[70] In addition, wood penetrated by ambrosia beetle pinholes will not be accepted by most timber-importing nations. Thus, further costs are incurred in initial sawmilling, remanufacturing and repackaging to eliminate damaged wood, altering forest harvesting practices to remove vulnerable inventory, and pest-management programs directed against the beetles.[12]

2. BIOLOGY AND LIFE HISTORY

2.1. Dispersal in the Spring

As snow recedes in the spring and sunlight penetrates to the forest floor, adult *T. lineatum* begin to stir in response to the increasing warmth of the litter and duff in which they have spent the winter. The beetles begin their dispersal flight (Fig. 2) when an ambient temperature threshold of 15.5°C is reached.[28,33] Mass flights (swarming) occur in clear, sunny weather, in winds of less than 5 km/hr,[92] when daily minimum temperatures rise above freezing and midday temperatures exceed 18°C.[50,56] Optimal temperatures for flight range between 19° and 26°C, and above 30°C flight is inhibited.[92]

Daily flight begins in late morning, peaking by mid-afternoon, and numbers of beetles in flight decline sharply in late afternoon.[50] Typically *T. lineatum* flies at a level just above the understory vegetation.[100] Within the forest, the beetles fly at 3.0–4.5 m above the ground but drop to 1.8–2.4 m in exposed areas,[23] a behavior that will facilitate contact with susceptible logs in forest openings, e.g., following winter blowdown.

At first, as the beetles disperse away from their overwintering site, their photopositive response predominates and they are unresponsive to host odor. They fly in a fairly constant upwind direction[23,56] and, in the laboratory, are capable of flying continuously for up to 8 hr.[19] However, after 6–24 hr of flight exercise on a tether, the beetles become indifferent to light or photonegative and readily bore into suitable hosts.[45,46] For male beetles, a threshold for this behavioral reversal was later established at 30 min of flight.[7] While the mechanism by which the internal feedback system functions is unknown, a plausible hypothesis is that the beetle can detect the buildup of metabolites generated during flight. During dispersal flight, the beetles use up approximately 25% of their original lipid deposits.[74]

The initial domination of photopositive responses over other behavior undoubtedly ensures that the beetles will disperse away from the overwintering site. Secondarily, it will promote outbreeding between populations as dispersing beetles from different sources intermix.

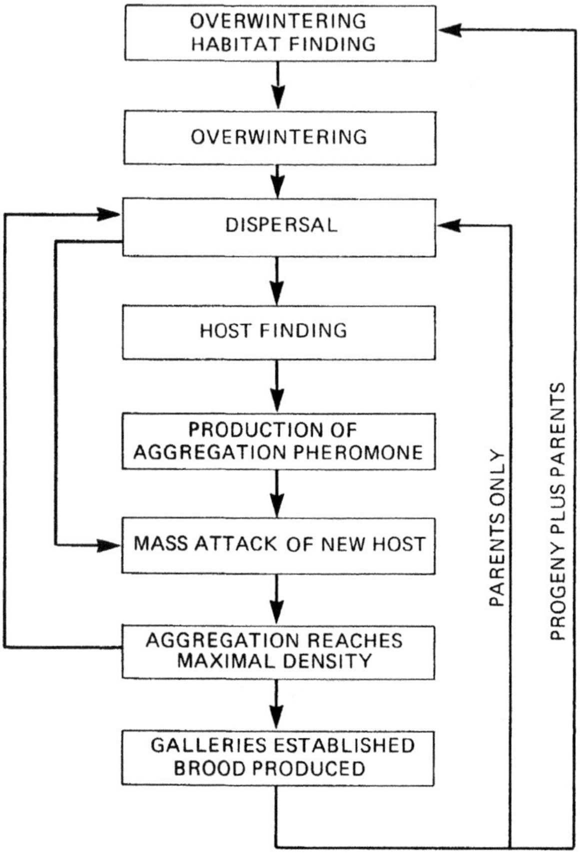

FIGURE 2. Flow diagram depicting major stages in life cycle of *Trypodendron lineatum* with emphasis on pheromone-mediated mass attack of new hosts by dispersing beetles (inside loop on left). Other loops depict role of antiaggregation pheromone in returning host-seeking beetles to the dispersal phase (outside loop on left); production by unsuccessful "parent" beetles (and possibly successful parents as well) of a mid-summer brood (inside loop on right); and migration of emergent brood beetles and re-emergent parent beetles (for the second year) to the overwintering habitat (outside loop on right).

2.2. Host Selection

Trypodendron lineatum is very specialized in the hosts it will infest. Recently-killed timber, such as blowdown arising from spring storms and redolent with the volatiles of fresh resin, is attacked by other species of ambrosia beetles, e.g. *Gnathotrichus sulcatus,* in the Pacific Northwest.[16,86] *T. lineatum* prefers aged timber, such as windfalls arising from autumn or winter storms, broken trees snapped by an overburden of snow or logs felled at least 3–5 months previously.[36,68,86] These hosts will have undergone anaerobic metabolism for some time,[47] building up a concentration of ethanol[72] that acts as a long range attractant to *T. lineatum,* either alone,[73] or in combination with low levels of the monoterpene α-pinene.[6,84] Ethanol is also hypothesized to be an arrestant and boring stimulant for ambrosia beetles.[69,71]

Although low levels of α-pinene may be attractive in combination with ethanol, attack by *T. lineatum* can be deterred for up to five days by the application of turpentine oil (a mixture of monoterpenes) to otherwise acceptable logs.[77] Therefore, the monoterpenes initially present in felled trees probably must dissipate to an acceptable level before these logs can become suitable hosts. In western North America, the first-emerging *T. lineatum* characteristically fly earlier in the spring than their major competitor, *G. sulcatus*.[33,65] Therefore, they can select the most suitable hosts, leaving those that are too old or too fresh to be attacked by other species.

Under normal circumstances, unless there has been a natural disaster, such as a catastrophic storm, acceptable hosts are rare and widely scattered in the forest. Like many other scolytid species, *T. lineatum* has evolved mechanisms that enable it to find and exploit these widely spaced, evanescent hosts.[2] The primary basis for this exploitation is a remarkably powerful aggregation pheromone, lineatin (3,3,7-trimethyl-2,9-dioxatricyclo[3.3.1.$0^{4,7}$]nonane).[14] Lineatin is the key component in a precise behavioral sequence of host selection (Fig. 2). It is released from the hindgut region of female beetles[13,97] when they bore into the xylem of a suitable host.[25,92,93] Lineatin is synergized by ethanol and α-pinene from the host.[3,15,60,99] Dispersing males are drawn by this aromatic bouquet to mate with the pheromone-producing females, while other females are attracted to the suitable host. In this way, populations are concentrated within the limited number of available hosts and can exploit them as fully as possible, ensuring that successful reproduction will occur.

Trypodendron lineatum infests a three-dimensional resource, the sapwood of its hosts. Thus it is not subject to extremely detrimental effects of overcrowding, as are bark beetles in the limited space of the nearly two-dimensional inner bark.[10] However, although adverse effects of overcrowding have not been investigated in *T. lineatum*, they probably exist.

There is evidence that *T. lineatum* can regulate its population density within a host. Although exceedingly high attack densities are occasionally reported,[86] the average attack densities are quite moderate, even on suitably-aged logs, e.g., 150/m^2 on Norway spruce, *Picea abies*,[32] and about 215/m^2 on Douglas-fir, *Pseudotsuga menziesii*.[36] At such densities, there is considerably more physical room for additional attacks. There is also convincing evidence that the male beetles produce a moderately strong antiaggregation pheromone that deters attraction to and attack on heavily populated hosts.[9,61,76,78] Therefore, even though there are limited hosts available in nature, it is evidently of adaptive advantage for the resident beetles to be able to regulate their attack density and to avoid the adverse effects of overpopulation. Host-seeking beetles encountering a fully populated host must then resume dispersal (Fig. 2).

2.3. Life inside the Host

Within 1 hr of landing on a suitable host the female beetle usually begins to bore into it. Before that time, she will usually have been joined by a male and have mated at least once.[50]

Males do not require a period of flight exercise before they mate,[40] and young, overwintered females in spring flights may already be fertilized.[18] These observations suggest that at least some females may mate before leaving the overwintering habitat.

On the new host, males mate with females outside of the galleries. Although the testes become inactive once the male is inside the host,[17] the male follows the female into the entrance tunnel and remains with her throughout the gallery construction and brood-

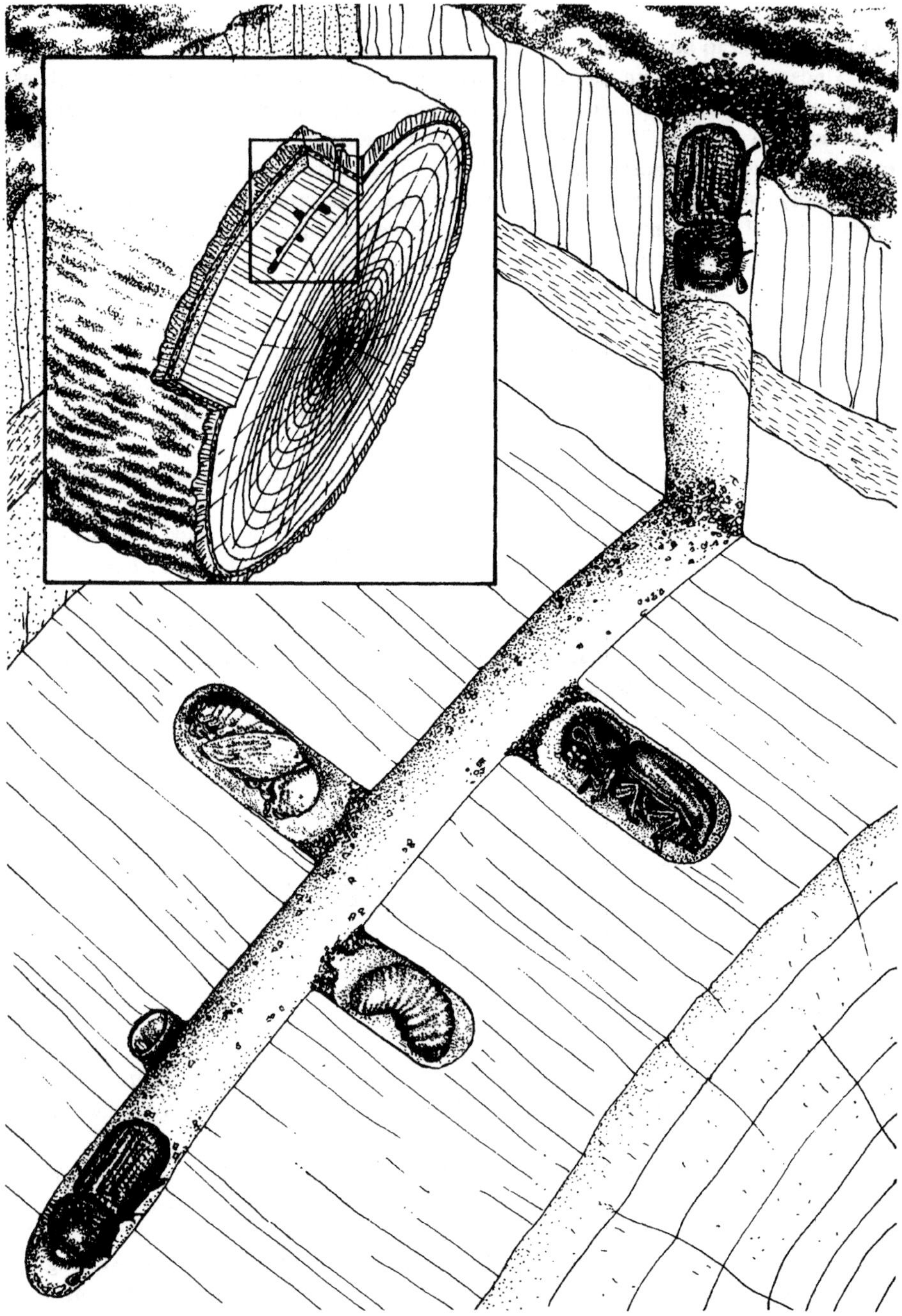

tending period (Fig. 3). Although little understood, this pair bond is apparently essential for optimum brood production. Females confined to logs with males produced an average of 17.0 eggs per gallery, while mated females confined without males produced only 9.5 eggs/gallery.[21] During the initial phases of gallery construction, the male removes boring dust that the female pushes into the entrance tunnel. Later, he removes boring dust and larval excreta from deep within the galleries. Thus the presence of the male may simply allow the female more time for egg laying and gallery extension.

The female beetles bore cylindrical entrance tunnels toward the centre of the tree. The tunnels are 1.75 mm in diameter, slightly greater than the diameter of the beetles.[50] Once they are well inside the sapwood, the females turn laterally, and bore one or more branch tunnels called egg galleries. The entire gallery system is in a plane perpendicular to the grain of the wood (Fig. 3).

Oviposition occurs within the first two weeks of gallery construction.[86] At 4–6-mm intervals within the egg gallery, the female lays eggs in niches chewed 0.6–0.7 mm deep into the walls of the galleries (Fig. 3). She then seals the eggs firmly in place with a protective packing of boring dust.[50]

The eggs hatch within 9–10 days and the white, legless larvae enlarge the egg niches into larval cradles, consuming ambrosia fungus and wood as they pass through two larval instars.[4,85] During this time, the parent beetles keep the egg galleries open by cropping the ambrosia fungus, and they remove larval feces ejected by the larvae through holes in the packing at the base of the cradles. This relatively advanced degree of parental care is apparently essential for brood survival. In the absence of parent beetles, the tunnels become overgrown with foreign microorganisms within 24 hr.[5] After the last larva has pupated, the parents may abandon the gallery, but by this time there are maturing callow adults to take over the task of cropping the fungus.[50]

Two or 3 weeks after hatching, the larvae reverse direction, plug the entrance to their cradles with frass and pupate. Approximately 1 week later adult eclosion occurs. The newly emerged callow adults rest until they are well sclerotized. They then chew their way out of the cradles, reverse direction, re-enter the cradle and consume all fungus, and the pupal and larval exuviae. After feeding for a further 3–4 weeks on the ambrosia fungus until their fat content reaches about 30% (males) or about 40% (females) of their dry weight,[74] the fully sclerotized brood adults leave the gallery via the main entrance hole. The male to female sex ratio of 6539 emerged beetles in British Columbia was 1 : 1,[11] a value in general agreement with that found by most researchers. In Europe[50] and British Columbia[101] the total period from initial invasion of the host to the first peak of brood emergence is 9–10 weeks.

For both sexes of *T. lineatum,* there is an almost obligatory association with the host once its suitability has been established. There is a profound change from a dispersive,

FIGURE 3. Schematic drawing of the gallery system of *Trypodendron lineatum* showing from the upper right: male beetle in the outer bark guarding the entrance hole, which is surrounded by frass (boring dust and feces) cleaned out from the gallery; callow (immature) female beetle in "cradle"; pupa in cradle; immature larva in cradle with hole through which larval feces are expelled into the main tunnel; niche with newly laid egg packed in place with boring dust; and parent female extending the tunnel through the sapwood. (In nature there would be many more progeny, which would be more synchronized in development). Inset shows gallery system in relation to entire log.

nonfeeding, sexually active insect to a feeding, reproductive insect that is unable to fly. In both sexes the flight muscles degenerate to a nonfunctional condition,[19,20] precluding further dispersal during the period of gallery construction and brood tending. This degeneration is reversible, and by the time the parent beetles finish feeding, they are once more fully capable of flight. Should the host log become unsuitable (e.g., through excessive drying), regeneration of the flight muscles occurs much earlier.

Many other internal changes occur. In females the fat body declines to half its original size within 5 days,[75] and the ovarioles enlarge and mature,[18,20] probably using metabolites from both the flight muscles and the fat body. In males there is a steady increase in fat content to triple the preattack level within 3 weeks[75] and deterioration of the testes.[17] The demonstration that application of juvenile hormone (JH III) to female beetles can stimulate ovariole maturation (41) suggests that some or all of these physiological events may be under hormonal control.

2.4. Emergence, Reattack, Migration to the Overwintering Habitat and Overwintering

Emergence and flight of parent and brood beetles occurs sporadically throughout the summer.[33,92] Emergence from caged logs occurs on days when the maximum ambient temperature is as low as 14.4°C, but 24°C is considered the critical ambient temperature to induce a mass emergence.[11]

An undetermined portion of the parent beetles are thought to seek out new hosts and initiate a second brood.[1,28,44,59] Although parent beetles do not abandon successful galleries until their brood has started to emerge, adults do re-emerge from unsuccessful galleries. Thus, midsummer attacks are apparently made by these unsuccessful "parents".[101]

Despite the warm mid-summer weather encountered by emergent brood beetles, they immediately fly to hibernation sites within the neighboring forest (Fig. 2). The beetles are photopositive at first and fly upward to a height of about 15 m before moving laterally. However, a photic reversal evidently occurs, and at the time they enter the forest most are apparently flying between 1–2 m above ground, and seek out hibernation sites where they encounter a sharp reduction in light intensity.[37] In dense young forests, these sites occur 6–18 m from the forest margin,[50] but in older forests, hibernating beetles are found up to 90 m within the stand.[65] The beetles overwinter 0–6 cm deep in the litter and duff within about 1 m of the base of trees and stumps, or in the outer layers of these obstacles, primarily within 1 m from the ground.[37,57,58]

While the beetles that reattack new hosts in mid-summer are capable of producing and responding to pheromone, the brood beetles (and possibly those parent beetles that overwinter for a second year) are not. Moreover, neither re-emerged parents nor brood beetles were inclined to mate.[40] This unresponsive condition has been a source of frustration to pest management specialists seeking to control *T. lineatum* by mass-trapping emergent beetles with pheromone baited traps.

A large percentage of the hibernating beetles may be overwintering for the second time. The proportion of such beetles was found to range from 6.9 to 52%.[20,30] In two successive years, more than 50% of beetles caught in the spring were old parents.[20]

In the overwintering habitat, the beetles enter a reproductive diapause[40,96] and are maintained metabolically throughout the winter by using up about 25% of their fat content.[74] Monthly tests for mating activity indicate that diapause is broken by February, 2–3 months before the beetles resume activity in the spring.[40] Thus, as spring approaches, the overwintering beetles lie ready and waiting for the first warm weather that will permit them to fly and seek out new hosts.

3. PATTERNS OF NUMERICAL BEHAVIOR

Probably the single most important determinant of the success of *T. lineatum* within the host habitat is the quality of the host itself. Critical factors that have been investigated include host species, age after felling, temperature, moisture content, and starch content.

Studies of clearcut logs disclosed that *T. lineatum* attacks occurred on 72% of Douglas-fir logs, 57% of Western hemlock, *Tsuga heterophylla,* 38% of Western red-cedar, *Thuja plicata,* and 25% of grand fir, *Abies grandis;* respective attack densities were 77.4, 35.4, 11.8, and 12.9 per m^2.[51] These data are in agreement with other findings that *T. lineatum* preferred the odor of Douglas-fir logs over that of other species.[24,101]

In both Europe[32] and North America,[36] *T. lineatum* in the spring show a striking preference for logs felled prior to January. However, for logs felled later than December short sections were more atractive than longer logs, suggesting that they suffered more rapid cell death and underwent more rapid anaerobic preconditioning than did longer logs.[36,55]

Brood production per gallery was approximately four times greater in Douglas-fir logs felled from October–December than in logs felled in February–April.[27] Similarly brood production was 2.2 and 1.4 times higher in logs from fast-growing trees than in logs from slow-growing trees of the same age, suggesting that *T. lineatum* will be a major threat to rapidly grown trees with deep sapwood in intensively managed forests. The moisture content of the sapwood was unrelated to host preference or brood production, but aged logs selected by the beetles characteristically had lower starch content in the sapwood and lower cell viability than unattractive, spring-felled logs.[27,31] These variables may be most important for establishment and survival of the ambrosia fungus.

Low sapwood moisture content will render logs unsuitable to for *T. lineatum*. Douglas-fir logs with a moisture content between 112 and 186% of oven-dry weight were heavily attacked, whereas a Western hemlock log with 94% moisture content was not.[54] Excessive moisture loss would also explain why felled Douglas-fir and Western hemlock trees with the limbs on were avoided by *T. lineatum*.[52,53] Moisture loss may also be associated with high temperature. When logs are exposed to the sun, the resultant local drying on the south side is considered responsible for approximately doubling the percentage of galleries failing to produce brood.[34] These factors would explain the absence of high attack densities on the upper part of logs.[35,50,66,86]

Although a few parasites and predators are associated with *T. lineatum* (82), apparently none has a regulatory impact on population levels. For example, the only natural enemy of *T. lineatum* found in British Columbia was a hymenopteran parasite, *Eurytoma polygraphi,* which parasitized only 2% of larval broods.[86] Similarly, of four species of coleopteran predators observed, none was recovered from *T. lineatum* galleries.[50] Only mites and one species of internal nematode were found associated with the beetles, in their galleries, but neither had any apparent impact.[50] In British Columbia, a 47.3% mortality rate of eggs and young larvae was recorded in Douglas-fir.[44] In another study, 28.9 and 44.5% mortality from eggs to the adult stage occurred in Douglas-fir and Western hemlock, respectively.[101] Although the galleries were covered one week after attack, natural enemies could have entered the galleries during that week. However, no entomophagous agents were noted.

The maximum recorded production of progeny per gallery in Europe[50] and North America[101] was 52 and 56, respectively. However, the estimated average production in Europe was 10–30 progeny per gallery, a range in agreement with the finding of 17.0 eggs per gallery in British Columbia.[21] Assuming that the two parent beetles emerge with their brood, the success rate of some *T. lineatum* infestations is not very high. Emergence from logs of Norway spruce, Sitka spruce, *Picea sitchensis,* and Japanese larch, *Larix kaempferi,* in Scotland averaged only 2.1 beetles per gallery.[8] From Douglas-fir logs in British Columbia, 4.3 beetles emerged per gallery.[11] Higher emergence rates averaging 8.5 per gallery and 13.3 per successful gallery were recorded from western hemlock and Pacific silver fir, *Abies amabilis.*[34] These rates probably approach the norm for this species.

Finally, although brood productivity in successful galleries is relatively low, overwintering mortality is negligible. For example, 95% of an experimental population of *T. lineatum* survived overwintering burial in a natural habitat.[22] By separating dead from live beetles in litter and duff in the spring, overwintering mortality was determined to be 5–7%.[29] However, this estimate may be somewhat low because predation in which the entire body was removed could not be detected.

4. HYPOTHESIS TO EXPLAIN OBSERVED POPULATION BEHAVIOR

The above observations led to the logical hypothesis that the abundance of *T. lineatum* is host dependent (Fig. 4). Such a hypothesis appears to provide a plausible explanation for the spatial-gradient behavior of *T. lineatum* populations.

Trypodendron lineatum has a demonstrated capacity to find virtually all the well-dispersed available hosts when the resource remains constant and to respond as well to an increase in the numbers of hosts. This capacity is predicated on a yearly surplus of beetles (Fig. 4). If the host resource were to increase from one year to another, most of the surplus overwintering beetles would be able to find and exploit it, resulting in an increase in the resident population. If the host resource were to remain constant or decline, the surplus beetles would be those that were unable to find a suitable host within their flight range, or were rejected by the presence of antiaggregation pheromone from an already fully occupied host. They would perish from a number of causes, including starvation, desiccation, and predation. Alternatively, they might survive to attack inferior hosts, only to produce minimal broods therein.

While this hypothesis has not been tested directly, there is circumstantial evidence for its support. My observations indicate that if widely scattered suitable hosts occur in a stable natural setting, there are usually sufficient *T. lineatum* to find and infest them with uncanny accuracy. Local populations do increase in response to increases in available host habitats. For example, about a fourfold increase has been observed in a population exposed to fresh logging slash, an increase of about 625,000 beetles per ha in 1 year.[34]

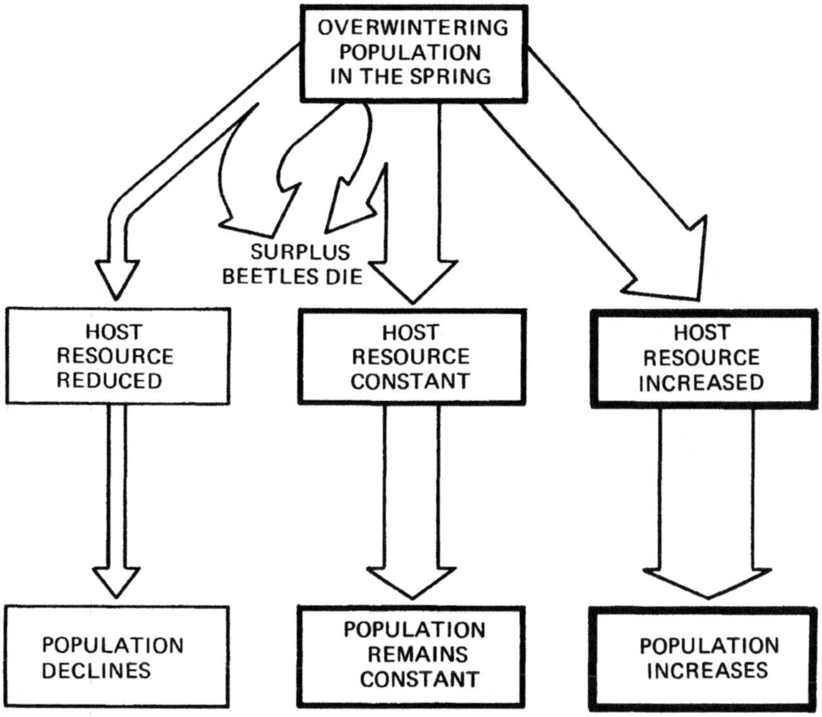

FIGURE 4. Conceptual model illustrating the hypothesis of host-dependent population regulation in *Trypodendron lineatum.*

Once established, high populations of *T. lineatum* may persist as long as the supply of hosts also remains high. Each year traditional logging methods on the west coast of North America simulate a natural disaster, leaving a residue of hosts equivalent to that following a typhoon or an earthquake. Hugh populations are maintained in logging slash and in logs felled and bucked (or in past times, cold-decked) in the forest.[88,89] Portions of these populations are transported in infested logs to timber-processing areas, e.g., dryland sorts and booming grounds,* where extremely large resident populations are also maintained by a ready supply of new wood each spring.[49,65,90,91]

There is little direct scientific evidence to support the hypothesis that a reduction in the host habitat resource will result in decline in *T. lineatum* populations. Nevertheless, there is such faith in this hypothesis that reduction of the host resource is almost universally accepted as an effective and realistic pest management strategy. If such a strategy is

*A dryland sort is a site where timber is sorted to species and grade prior to being transported to the mill. Dryland sorts have largely replaced "wet" sorts where timber was sorted in water. A "booming" ground is a sheltered portion of water in which sorted logs are assembled either as single logs or in bundles into an enclosed raft or boom that is then towed away by tugboat.

to be a success, the decline in host resource must be persistent and substantial. Otherwise, there will simply be an increase in attack density on the reduced host resource.[36]

5. IMPLICATIONS FOR PEST MANAGEMENT

A number of phenomena in the population biology of *T. lineatum* leave the species vulnerable to pest-management strategies and tactics[12] (Table I).

As population levels of *T. lineatum* are hypothesized to be host dependent, the most obvious pest-management strategy is to limit the amount of available habitat (Table I). Unfortunately, at the very time that host-seeking beetles are active in the spring, it is most

TABLE I
Pest-Management Strategies and Tactics That Exploit Observed or Hypothesized Behavior of *Trypodendron lineatum* Populations

	Pest management	
Observation or hypothesis	Strategy	Tactic
Population levels are dependent on the supply of hosts (logs, stumps, fallen trees) available for brood development	Regulate the amount of host resource	Reduce the amount of logging slash and culled logs left in the woods
		Remove felled and bucked timber from the woods before the beetles fly in the spring, especially that felled prior to January
		Process log inventory as rapidly as possible through timber-processing areas
	Protect the host from attack	Kill beetles on hosts with residual insecticide
		Deter beetles from boring into hosts by water misting
		Repel bettles from hosts with artificial repellent or antiaggregation pheromone
Aggregation of population on available hosts is odor mediated, particularly by aggregation pheromone	Manipulate host-seeking population by providing a false chemical message	Induce beetles to attack pheromone-baited trap logs which are then disposed of
		Intercept host-seeking beetles by mass-trapping them in pheromone-baited traps
Populations of overwintering beetles are concentrated in a restricted region of the forest	Modify the overwintering habitat around timber-producing areas	Cut back the forest margin to increase the hazard to beetles in flight to the overwintering habitat
		Manage overwintering habitat to create flight channels and to facilitate pest control with fire or insecticides.
	Increase the distance from an overwintering habitat in the woods to the supply of new hosts in the spring	Set up logging plans so that new cutblocks are as far away as possible from the previous year's cutblocks

likely that there will be a tempting array of suitably aged logs available to attack.[89] These will have accumulated as felled and bucked timber in logging settings and rights-of-way in the woods and as stored logs in dryland sorts and booming grounds awaiting transport to the mill.

It is now a standard recommendation that companies reduce as much as possible the amount of vulnerable timber inventory in the woods and processing areas in the spring.[80,98] The risk of degrade[70] and other losses such as export restrictions due to beetle attack may be greater than the risk of being unable to draw from a large inventory of stored logs to deliver on demand the desired species and grade of log to the mill or timber buyer.

In one combined dryland sort and booming ground in coastal British Columbia, ambrosia beetle attack (primarily by *T. lineatum*) at least doubled in susceptible logs awaiting departure in freshly made booms.[49] Therefore, if one cannot reduce the overall inventory, the next best recommendation is to process the inventory rapidly through areas where it is most vulnerable.

The second major strategy arising from the dependence of *T. lineatum* on fresh hosts in the spring is to protect the logs from attack (Table I). In the past, this was done by spraying logs with chemical pesticides such as DDT or lindane.[82] Applications could be made from the ground,[8,94,95] from rafts,[62] or from the air,[62,87] even when logs were stored in freshwater lakes. In British Columbia, increasing environmental awareness coupled with an unsubstantiated claim that lindane residues occurred in oysters at the mouth of the Cowichan River[39] signaled the end of aerial spraying of logs in the water. Throughout much of the temperate world, it is now difficult to gain approval for the use of chemical pesticides in the forest, and alternatives to such chemicals for the control of ambrosia beetles have been sought.

Alternative methods of protecting logs from attack by ambrosia beetles have not been widely adopted or have not proved operationally effective. A water-misting operation at one dryland sort was successful in protecting high value logs from attack, probably by altering the microenvironment on the log surface.[81,90] This operation was also very cost effective. However, because of the need for a large constant supply of water, substantial startup costs, maintenance requirements, inconvenience to workers, potential environmental hazards from leaching, and probably sheer inertia, water-misting has been used in few other dryland sorts.

Another proposed but not yet operationally proven method of protecting logs from attack involves the use of repellents such as turpentine, oleic acid, or pine oil.[77,83] The prospect of using an antiaggregation pheromone[9,61,76,78] must await identification of the naturally occurring semiochemical(s).

Probably the best successes in exploiting the host-dependent vulnerability of *T. lineatum* have been in the use of the aggregation pheromone, lineatin, to attract host-seeking beetles away from the product at risk (Table I). To date, this strategy has been applied only in timber processing areas, where the chance of success is greater than in the forest. In dryland sorts, two tactics are now common. One is to place piles of suitably aged, pheromone-baited, cull logs around periphery of the sorting area.[66] The other is to place pheromone-baited traps[63] around the periphery of the sort and also within the forest margin.[65] Dispersing beetles flying into the sort from the forest margin are then intercepted and killed in the traps. Failing that, they are induced to attack the cull-log piles, which when fully attacked (and prior to brood emergence) are removed from the sort and reduced to pulp chips, killing the beetles in the process.

Because the logs to be protected are continuously moving into and out of the dryland sort, it is impossible to monitor them to evaluate the effectiveness of a pheromone-based pest-management program. However, assessment of the effectiveness of the trapping program, as well as decisions on where to concentrate the majority of traps and trap logs can be done by sampling the litter and duff[26,79] to establish an estimated level of the overwintering population around a timber processing area.[65] A first prerequisite is to establish by sampling on transects into the forest, the area in which the majority of overwintering beetles lie. Within this area, the habitat is divided into three strata[37]: (1) 0–1 m high on the surface of stumps or tree boles; (2) in the litter and duff within a 0.45- or 0.90-m radius of the stump or tree (to give low or high estimates of the population, which is primarily concentrated at base of stumps and trees); and (3) in the litter and duff outside of this radius. All sampling is done on a per-unit area basis within 0.9 m of stumps or standing trees, and populations in the other strata are determined by extrapolation from known or estimated percentages of the population in the stratum sampled. After deducting the basal area of the stumps and trees, the total estimated high and low population levels are then calculated for the entire forested area around the timber processing area.

By this method (or some modification of it), a pest manager can estimate the target population to be trapped in an effort to protect the product. Thus, in three successive years in one dryland sort using successively greater and more efficient trapping intensities, 1–3%, 12–21%, and 44–77% of the estimated overwintering population of *T. lineatum* were trapped.[65] In the third year, 2,257,000 beetles were actually trapped.

A final pest-management strategy is based on the dependence of *T. lineatum* on a suitable overwintering site (Table I). Several tactics might make this habitat less favorable for the beetles or might decrease their chance of finding acceptable hosts in the spring. None has yet been attempted on a large-scale or rationally planned basis.

Around a dryland sort, it should be possible to cut back the forest to increase the distance that a beetle must fly to find an acceptable overwintering site and to return in the spring. In logging operations, establishing cutblocks far apart in successive years would have the same effect. The beetles would have to fly farther in the spring, decreasing the chance that they will all find a new source of vulnerable logs.

The forest margin around a dryland sort could also be modified, e.g., by spacing of trees that might allow prescribed burning or an insecticidal application to be made. Alternatively, channels could be cut through a dense forest margin, creating flyways in which dispersing beetles might be intercepted by strategically placed, pheromone-baited traps.[67]

To be truly effective, a pest-management program will probably demand the integration of various suitable management strategies and tactics (Table I). For example, the following four-point pest-management plan for a dryland sort on the British Columbia coast has been proposed[48]: (1) rapid inventory turnover to reduce the amount of vulnerable wood at peak flight times; (2) protection of the product, e.g., by water misting on decked logs, as well as booms; (3) pheromone-based mass trapping to suppress the beetle population; and (4) alteration of the overwintering habitat to the detriment of the beetles.

ACKNOWLEDGMENTS. I thank C. E. Fockler for assistance in library research and B. S. Lindgren, J. A. Mclean, W. W. Nijholt, and T. L. Shore for review of the manuscript.

REFERENCES

1. Annila, E., Bakke, A., Bejer-Petersen, B., and Lekander, B., 1972, Flight period and brood emergence in *Trypodendron lineatum* (Oliv.) (Col., Scolytidae) in the Nordic countries, *Commun. Inst. For. Fenn.* **76:**(4):7–28.
2. Atkins, M. D., 1966, Behavioural variation among scolytids in relation to their habitat, *Can. Entomol.* **98:**285–288.
3. Bakke, A., 1983, Dosage response of the ambrosia beetle *Trypodendron lineatum* (Olivier) (Coleoptera: Scolytidae) to semiochemicals, *Z. Angew. Entomol* **95:**158–161.
4. Balfour, R. M., 1962, Concerning the number of larval instars in *Trypodendron lineatum* (Oliv.) (Coleoptera: Scolytidae), *Entomol. Mon. Mag.* **98:**52–53.
5. Batra, L. R., 1966, Ambrosia fungi: Extent of specificity to ambrosia beetles, *Science* **153:**193–195.
6. Bauer, J., and Vité, J. P., 1975, Host selection by *Trypodendron lineatum, Naturwissenschaften* **62:**539.
7. Bennett, R. B., and Borden, J. H., 1971, Flight arrestment of tethered *Dendroctonus pseudotsugae* and *Trypodendron lineatum* (Coleoptera: Scolytidae) in response to olfactory stimuli, *Ann. Entomol. Soc. Am.* **64:**1273–1286.
8. Bletchly, J. D., and White, M. G., 1962, Significance and control of attack by the ambrosia beetle *Trypodendron lineatum* (Oliv.) (Col. Scolytidae) in Argyllshire Forests, *Forestry* **35:**139–163.
9. Borden, J. H., 1974, Pheromone mask produced by male *Trypodendron lineatum* (Coleoptera: Scolytidae), *Can. J. Zool.* **52:**533–536.
10. Borden, J. H., 1984, Semiochemical-mediated aggregation and dispersion in the Coleoptera, in: *Insect Communication* (T. Lewis, ed.), pp. 123–149, Academic, London.
11. Borden, J. H., and Fockler, C. E., 1973, Emergence and orientation behavior of brood *Trypodendron lineatum* (Coleoptera: Scolytidae), *J. Entomol. Soc. B.C.* **70:**34–38.
12. Borden, J. H., and McLean, J. A., 1981, Pheromone-based suppression of ambrosia beetles in industrial timber processing areas, in: *Management of Insect Pests with Semiochemicals* (E. R. Mitchell, ed.), pp. 133–154, Plenum, New York.
13. Borden, J. H., and Slater, C. E., 1969, Sex pheromone of *Trypodendron lineatum:* Production in the female hindgut–Malpighian tubule region, *Ann. Entomol. Soc. Am.* **62:**454–455.
14. Borden, J. H., Handley, J. R., Johnston, B. D., MacConnell, J. G., Silverstein, R. M., Slessor, K. N., Swigar, A. A., and Wong, D. T. W., 1979, Synthesis and field testing of 4,6,6-lineatin, the aggregation pheromone of *Trypodendron lineatum* (Coleoptera: Scolytidae), *J. Chem. Ecol.* **5:**681–689.
15. Borden, J. H., King, C. J., Lindgren, S., Chong, L., Gray, D. R., Oehlschlager, A. C., Slessor, K. N., and Pierce, H. D., Jr., 1982, Variation in response of *Trypodendron lineatum* from two continents to semiochemicals and trap form, *Environ. Entomol.* **11:**403–408.
16. Cade, S. C., Hrutfiord, B. F., and Gara, R. I., 1970, Identification of a primary attractant for *Gnathotrichus sulcatus* isolated from western hemlock logs, *J. Econ. Entomol.* **63:**1014–1015.
17. Chapman, J. A., 1954, Physiological and biological studies on the ambrosia beetle, *Trypodendron lineatum* (Oliv.) and the Douglas fir beetle, *Dendroctonus pseudotsugae* Hopk., Can. Dep. Agric., For. Biol. Div., For. Biol. Lab., Victoria, B.C. Interim Rep. 1954-2.
18. Chapman, J. A., 1955, Interpretation of adult history in the ambrosia beetle, *Trypodendron.* Can. Dep. Agric., For. Biol. Div., Bi-mon. Prog. Rep. **11**(6):3–4.
19. Chapman, J. A., 1956, Flight muscle changes during adult life in a scolytid beetle, *Nature (Lond.)* **177:**1183.
20. Chapman, J. A., 1958, Studies on the physiology of the ambrosia beetle *Trypodendron* in relation to its ecology, *Proc. 10th Int. Congr. Entomol.* **4:**375–380.
21. Chapman, J. A., 1959, Forced attacks by the ambrosia beetle, *Trypodendron,* Can. Dep. Agric., For. Biol Div., Bi-mon. Prog. Rep. **15**(5):3.
22. Chapman, J. A., 1960, Overwintering mortality in the ambrosia beetle *Trypodendron,* Can. Dept. Agric., For. Biol. Div. Bi-mon. Prog. Rep. **16**(4):3–4.
23. Chapman, J. A., 1962, Field studies on attack flight and log selection by the ambrosia beetle *Trypodendron lineatum* (Oliv.) (Coleoptera: Scolytidae), *Can. Entomol.* **94:**74–92.
24. Chapman, J. A., 1963, Field selection of different log odors by scolytid beetles, *Can. Entomol.* **95:**673–676.
25. Chapman, J. A., 1966, The effect of attack by the ambrosia beetle *Trypodendron lineatum* (Olivier) on log attractiveness, *Can. Entomol.* **98:**50–59.

26. Chapman, J. A., 1974, Ambrosia beetle. Guidelines to population estimates near dry land log-storage areas and damage hazard assessment, Can. For. Serv., Pac. For. Res. Cent. Inf. Rep. BC-X-103.
27. Chapman, J. A., and Dyer, E. D. A., 1969, Characteristics of Douglas-fir logs in relation to ambrosia beetle attack, *For. Sci.* **15**:95–101.
28. Chapman, J. A., and Kinghorn, J. M., 1958, Studies of flight and attack activity of the ambrosia beetle, *Trypodendron lineatum* (Oliv.) and other scolytids, *Can. Entomol.* **90**:362–372.
29. Chapman, J. A., and Neitsch, R., 1959, Autumn–winter mortality in the ambrosia beetle *Trypodendron*, Can. Dept. Agric., For. Biol. Div., Bi-mon. Prog. Rep. **15**(3):3–4.
30. Chapman, J. A., and Nijholt, W. W., 1965, Proportion of old and young adults in an overwintering population of the ambrosia beetle *Trypodendron lineatum* (Oliv.), Can. Dept. For., Entomol. and Pathol. Br., Bi-mon. Prog. Rep. **21**(1):3–4.
31. Chapman, J. A., Farris, S. H., and Kinghorn, J. M., 1963, Douglas-fir sapwood starch in relation to log attack by the ambrosia beetle, *Trypodendron, For. Sci.* **9**:430–439.
32. Christiansen, E., and Saether, T., 1968, Infestation density of *Trypodendron lineatum* (Olivier) (Coleoptera: Scolytidae) in relation to felling date of logs, *Norsk. Entomol. Tidssk.* **15**:28–30.
33. Daterman, G. E., Rudinsky, J. A., and Nagel, W. P., 1965, Flight patterns of bark and timber beetles associated with coniferous forests of Western Oregon, Oregon State University Agric. Exp. Sta. Tech. Bull. 87.
34. Dyer, E. D. A., 1963, Attack and brood production of ambrosia beetles in logging debris, *Can. Entomol.* **95**:624–631.
35. Dyer, E. D. A., 1963, Distribution of *Trypodendron* attacks around the circumference of logs, Can. Dept. For., For. Entomol. and Pathol. Br., Bi-mon. Prog. Rep. **19**(2):3–4.
36. Dyer, E. D. A., and Chapman, J. A., 1965, Flight and attack of the ambrosia beetle, *Trypodendron lineatum* (Oliv.) in relation to felling date of logs, *Can. Entomol.* **97**:42–57.
37. Dyer, E. D. A., and Kinghorn, J. M., 1961, Factors influencing the distribution of overwintering ambrosia beetles, *Trypodendron lineatum* (Oliv.), *Can. Entomol.* **93**:746–759.
38. Farris, S. H., and Chapman, J. A. 1957, A preliminary study of the deposition and early growth of fungus within the galleries of the ambrosia beetle *Trypodendron lineatum* (Oliv.), Can. Dept. Agric., For. Biol. Div., Bi-mon. Prog. Rep. **13**(6):3.
39. Finegan, R. D., 1967, Environmental pollution by benzene hexachloride used for control of ambrosia beetle, B.C. Fish and Wildl. Br. Rep. 41–06.
40. Fockler, C. E., and Borden, J. H., 1972, Sexual behavior and seasonal mating activity of *Trypodendron lineatum* (Coleoptera: Scolytidae), *Can. Entomol.* **104**:1841–1853.
41. Fockler, C. E., and Borden, J. H., 1973, Mating activity and ovariole development of *Trypodendron lineatum:* Effect of a juvenile hormone analogue, *Ann. Entomol. Soc. Am.* **66**:509–512.
42. Francke-Grossman, H., 1957, Über die Ambrosiazucht holzbrütender Ipiden im Hinblick auf das System, *Dtsch. Ges. Angew. Entomol.* **14**:139–144.
43. Funk, A., 1965, The symbiotic fungi of certain ambrosia beetles in British Columbia, *Can. J. Bot.* **43**:929–932.
44. Gibson, C., Kinghorn, J. M., and Chapman, J. A., 1958, Ambrosia beetle brood productivity, Can. Dept. Agric., For. Biol. Div., Bi-mon. Prog. Rep. **14**(5):2–3.
45. Graham, K., 1959, Release by flight exercise of a chemotropic response from photopositive domination in a scolytid beetle. *Nature (Lond.)* **184**:283–284.
46. Graham, K., 1962, Photic behavior in the ecology of the ambrosia beetle *Trypodendron lineatum, Proc. 11th Int. Congr. Entomol.* **2**:226.
47. Graham, K., 1968, Anaerobic induction of primary chemical attractancy for ambrosia beetles, *Can. J. Zool.* **46**:905–908.
48. Gray, D. R., 1985, Ambrosia beetles (Coleoptera: Scolytidae) in a Vancouver Island dryland sort: Their damage and proposed control, M.P.M. Professional Paper, Simon Fraser University, Burnaby, B.C.
49. Gray, D. R., and Borden, J. H., 1985, Ambrosia beetle attack on logs before and after processing through a dryland sorting area, *For. Chron.* **61**:299–302.
50. Hadorn, C., 1933, Recherches sur la morphologie, les stades évolutifs et l'hivernage du bostryche liseré (*Xyloterus lineatus* Oliv.), *Org. de la Soc. forest Suisse, Bern* (Suppl. 11).
51. Johnson, N. E., 1958, Ambrosia beetle infestation of coniferous logs on clearcuttings in northwestern Oregon, *J. For.* **56**:508–511.

52. Johnson, N. E., 1961, Ambrosia beetle attacks in young-growth western hemlock, Can. Dept. For., For. Entomol. and Pathol. Br., Bi-mon. Prog. Rep. **17**(5):3.
53. Johnson, N. E., 1964, Effects of different drying rates and two insecticides on beetle-attacks in felled Douglas-fir and western hemlock Weyerhauser For. Res. Note No. 58.
54. Kinghorn, J. M., 1956, Sapwood moisture in relation to *Trypodendron* attack, Can. Dept. Agric., For. Biol. Div., Bi-mon. Prog. Rep. **12**(5):3–4.
55. Kinghorn, J. M., 1957, An induced differential bark-beetle attack, Can. Dept. Agric., For. Biol. Div., Bi-mon. Prog. Rep. **13**(2):3–4.
56. Kinghorn, J. M., and Chapman, J. A., 1954, Observations on the flight of *Trypodendron* sp., Can. Dep. Agric., For. Biol. Div., Bi-mon. Prog. Rep. **10**(3):3.
57. Kinghorn, J. M., and Chapman, J. A., 1959, The overwintering of the ambrosia beetle *Trypodendron lineatum* (Oliv.), *For Sci.* **5**:81–92.
58. Kinghorn, J. M., and Dyer, E. D. A., 1960, Overwintering of ambrosia beetle, Can. Dept. Agric., For. Biol. Div., Bi-mon. Prog. Rep. **16**(1):4.
59. Klimetzek, D., 1984, Grundlagen einer Überwachung und Bekämpfung der Nutzholzborkenkäfer (*Trypodendron* spp.) mit Lock- und Ablenkstoffen, *Freiburger Waldschutz-Abhandlungen* **5**:1–227.
60. Klimetzek, D., Vité, J. P., and Mori, K., 1980, Zur Wirkung und Formulierung des Populationslockstoffes des Nutzholzborkenäfer *Trypodendron* (=*Xyloterus*) *lineatum*, *Z. Angew. Entomol.* **89**:57–63.
61. Klimetzek, D., Kiesel, K., Mohring, C., and Bakke, A., 1981, *Trypodendron lineatum:* Reduction of pheromone response of male beetles, *Naturwissenschaften* **68**:149–151.
62. Lejeune, R. R., and Richmond, H. A., 1975, Striped ambrosia beetle *Trypodendron lineatum* (Oliv.), in: *Aerial Control of Forest Insects in Canada* (M. L. Prebble, ed.), pp. 246–249, Canadian Department of Environment, Ottawa.
63. Lindgren, B. S., 1983, A multiple funnel trap for scolytid beetles (Coleoptera), *Can. Entomol.* **115**:299–302.
64. Lindgren, B. S., 1986, *Trypodendron lineatum* (Olivier) (Coleoptera: Scolytidae) breeding in bigleaf maple, *Acer macrophyllum*, *J. Entomol. Soc. B.C.* **83**:44.
65. Lindgren, B. S., and Borden, J. H., 1983, Survey and mass trapping of ambrosia beetles (Coleoptera: Scolytidae) in timber processing areas on Vancouver Island, *Can. J. For. Res.* **13**:481–493.
66. Lindgren, B. S., Borden, J. H., Gray, D. R., Lee, P. C., Palmer, D. A., and Chong, L., 1982, Evaluation of two trap log techniques for ambrosia beetles (Coleoptera: Scolytidae) in timber processing areas, *J. Econ. Entomol.* **75**:577–586.
67. Lindgren, B. S., Borden, J. H., Chong, L., Friskie, L. M., and Orr, D. B., 1983, Factors influencing the efficiency of pheromone-baited traps for three species of ambrosia beetles (Coleoptera: Scolytidae), *Can. Entomol.* **115**:303–313.
68. Mathers, W. G., 1935, Time of felling in relation to injury from ambrosia beetles or pinworms, *B. C. Lumberman* **19**(8):14.
69. MacConnell, J. G., Borden, J. H., Silverstein, R. M., and Stokkink, E., 1977, Isolation and tentative identification of lineatin, a pheromone from the frass of *Trypodendron lineatum* (Coleoptera: Scolytidae), *J. Chem. Ecol.* **3**:549–561.
70. McLean, J. A., 1985, Ambrosia beetles: A multimillion dollar degrade problem of sawlogs in coastal British Columbia, *For. Chron.* **61**:295–298.
71. McLean, J. A., and Borden, J. H., 1977, Attack by *Gnathotrichus sulcatus* (Coleoptera: Scolytidae) on stumps and felled trees baited with sulcatol and ethanol, *Can. Entomol.* **109**:675–686.
72. Moeck, H. A., 1970, Ethanol as the primary attractant for the ambrosia beetle *Trypodendron lineatum* (Coleoptera: Scolytidae), *Can. Entomol.* **102**:985–995.
73. Moeck, H. A., 1971, Field test of ethanol as a scolytid attractant, Canada Dept. Fish. and For., Bi-mon. Res. Notes **27**:11–12.
74. Nijholt, W. W., 1967, Moisture and fat content during the adult life of the ambrosia beetle *Trypodendron lineatum* (Oliv.), *J. Entomol. Soc. B.C.* **64**:51–55.
75. Nijholt, W. W., 1969, Fat content of the ambrosia beetle, *Trypodendron lineatum* (Oliv.) during attack and brood production, *J. Entomol. Soc. B.C.* **66**:29–31.
76. Nijholt, W. W., 1970, The effect of mating and the presence of the male ambrosia beetle, *Trypodendron lineatum*, on "secondary" attraction, *Can. Entomol.* **102**:894–897.

77. Nijholt, W. W., 1973, Ambrosia beetle attacks delayed by turpentine oil, *Can. Dept. Fish. and For., Bi-mon. Res. Notes* **29**:36.
78. Nijholt, W. W., 1973, The effect of male *Trypodendron lineatum* (Coleoptera: Scolytidae) on the response of field populations to secondary attraction, *Can. Entomol.* **105**:583–590.
79. Nijholt, W. W., 1976, A simplified procedure for sampling overwintering populations of the ambrosia beetle *Trypodendron lineatum* (Oliv.), Can. For. Serv., Pac. For. Res. Cen. BC-X-103 (addendum).
80. Nijholt, W. W., 1978, Ambrosia beetle. A menace to the forest industry, Can. For. Serv., Pac. For. Res. Cent. Rep. BC-P-25.
81. Nijholt, W. W., 1978, Evaluation of operational watermisting for log protection from ambrosia beetle damage, Can. For. Serv., Pac. For. Res. Can. Rep. BC-P-22.
82. Nijholt, W. W., 1979, The striped ambrosia beetle *Trypodendron lineatum* (Olivier). An annotated bibliography, Environ. Can., For. Serv. Rep. BC-X-121.
83. Nijholt, W. W., 1980, Pine oil and oleic acid delay and reduce attacks on logs by ambrosia beetles (Coleoptera: Scolytidae), *Can. Entomol.* **112**:199–204.
84. Nijholt, W. W., and Schonherr, J., 1976, Chemical response behavior of scolytids in West Germany and Western Canada, Can. For. Serv., Bi-mon. Res. Notes **32**:31–32.
85. Novak, V., 1963, Studie vyvojorych stupnu drevokaza carkovaneho *Trypodendron lineatum* Ol. (Coleoptera: Scolytidae), *Zool. Listy* **12**:135–138.
86. Prebble, M. L., and Graham, K., 1957, Studies of attack by ambrosia beetles in softwood logs on Vancouver Island, British Columbia, *For. Sci.* **3**:90–112.
87. Richmond, H. A., 1961, Helicopters protect log booms in B. C., *Can. Lumberman* **81**(12)40–42.
88. Richmond, H. A., 1968, The ambrosia beetle on the British Columbia coast, B.C. Loggers Div., Council of For. Ind. of B.C. Mimeo Rep.
89. Richmond, H. A., 1969, Appetite for wood, *B.C. Lumberman* **53**(8):34–36.
90. Richmond, H. A., and Nijholt, W. W., 1972, Water misting for log protection from ambrosia beetles in B.C., Can. For. Serv., Pac. For. Res. Cent. Inform. Rep. B.C. P-4-72.
91. Richmond, H. A., and Radcliffe, D. N., 1961, Ambrosia beetle attack on sawlogs in water storage, *B.C. Lumberman* **45**(10):28–32.
92. Rudinsky, J. A., and Daterman, G. E., 1964, Field studies on flight patterns and olfactory responses of ambrosia beetles in Douglas fir forests of western Oregon, *Can. Entomol.* **96**:1339–1352.
93. Rudinsky, J. A., and Daterman, G. E., 1964, Response of the ambrosia beetle *Trypodendron lineatum* (Oliv.) to a female-produced pheromone, *Z. Angew. Entomol.* **54**:300–303.
94. Schindler, U., 1969, Experience from last year's control operations against *Trypodendron lineatum, Allg. Forst.* **24**:205–206.
95. Schindler, U., 1970, Further experience in the control of *Trypodendron lineatum, Holz. Zentralbl.* **96**:505–506.
96. Schneider, I., and Rudinsky, J. A., 1969, Anatomical and biological changes in internal organs of adult *Trypodendron lineatum, Gnathotrichus retusus* and *G. sulcatus* (Coleoptera: Scolytidae), *Ann. Entomol. Soc. Am.* **62**:995–1003.
97. Schneider, I., and Rudinsky, J. A., 1969, The site of pheromone production in *Trypodendron lineatum* (Coleoptera: Scolytidae): bioassay and histological studies of the hindgut, *Can. Entomol.* **101**:1181–1186.
98. Shore, T. L., 1985, Ambrosia beetles, Forest Pest Leaflet No. 72, Pacific Forestry Centre, Canadian Forestry Service, Victoria, B.C.
99. Shore, T. L., and McLean, J. A., 1983, A further evaluation of the interactions between the pheromones and two host kairomones of the ambrosia beetles *Trypodendron lineatum* and *Gnathotrichus sulcatus* (Coleoptera: Scolytidae), *Can. Entomol.* **115**:1–5.
100. Shore, T. L., and McLean, J. A., 1984, The effect of height of pheromone-baited traps on catches of the ambrosia beetle, *Trypodendron lineatum, J. Entomol. Soc. B.C.* **81**:17–18.
101. Shore, T. L., McLean, J. A., and Zanuncio, J. C., 1987, Reproduction and survival of the ambrosia beetle *Trypodendron lineatum* (Coleoptera: Scolytidae) in Douglas-fir and western hemlock logs, *Can. Entomol.* **119**:131–139.
102. Wood, S. L., 1957, Ambrosia beetles of the tribe Xyloterini (Coleoptera: Scolytidae) in North America, *Can. Entomol.* **89**:337–354.

TAXONOMIC INDEX

SUBJECT INDEX

GPSR Compliance

The European Union's (EU) General Product Safety Regulation (GPSR) is a set of rules that requires consumer products to be safe and our obligations to ensure this.

If you have any concerns about our products, you can contact us on ProductSafety@springernature.com

In case Publisher is established outside the EU, the EU authorized representative is:

Springer Nature Customer Service Center GmbH
Europaplatz 3
69115 Heidelberg, Germany

Batch number: 10755248

Printed by Printforce, the Netherlands